Sperm Competition and Sexual Selection

Sperm Competition and Sexual Selection

Edited by

T. R. Birkhead
Department of Animal and Plant Sciences
The University, Sheffield S10 2TN, UK

and

A. P. Møller
Laboratoire d'Ecologie, CNRS URA 258
Université Pierre et Marie Curie
F-75252 Paris Cedex 05, France

Academic Press
SAN DIEGO LONDON BOSTON
NEW YORK SYDNEY TOKYO TORONTO

This book is printed on acid-free paper

Academic Press
525 B Street, Suite 1900, San Diego, California 92101-4495, USA
http://www.apnet.com

Academic Press
24–28 Oval Road, London NW1 7DX, UK
http://www.hbuk.co.uk/ap/

ISBN 0-12-100543-7

A catalogue record for this book is available from the British Library

Typeset by Paston Press Ltd, Loddon, Norfolk
Printed in Great Britain by Cambridge University Press, Cambridge

98 99 00 01 02 03 CUP 9 8 7 6 5 4 3 2 1

Contents

List of Contributors

Bruno Baur, Department of Integrative Biology, Section of Conservation Biology (NLU), University of Basel, St Johanns-Vorstadt 10, CH-4056 Basel, Switzerland. e-mail: baur@ubaclu.inibas.ch

Tim R. Birkhead, Department of Animal and Plant Sciences, The University, Sheffield S10 2TN, UK. e-mail: T.R.Birkhead@sheffield.ac.uk

William G. Breed, Department of Anatomical Sciences, University of Adelaide, Adelaide, SA 5005, Australia. e-mail: wbreed@medicine.adelaide.edu.au

Lynda F. Delph, Department of Biology, Indiana University, Bloomington, IN 47405, USA. e-mail: ldelph@ucs.indiana.edu

William G. Eberhard, Smithsonian Tropical Research Institute, and Biología, Universidad de Costa Rica, Ciudad Universitaria, Costa Rica. e-mail:

Mark A. Elgar, Department of Zoology, University of Melbourne, Parkville, Victoria 3052, Australia. e-mail: elgar@ariel.ucs.unimelb.edu.au

Montserrat Gomendio, Departmento de Ecologia Evolutiva, Museo Nacional de Ciencias Naturales (CSIC), Jose Gutierrez Abascal 2, E-28006 Madrid, Spain. e-mail: MCNMG41@cc.csic.es

Tim R. Halliday, Biology Department, Open University, Milton Keynes MK7 6AA, UK. e-mail: T.R.Halliday@open.ac.uk

Alexander H. Harcourt, Department of Anthropology, University of California, Davis, CA 95616, USA. e-mail: AHHARCOURT@UCDAVIS.EDU

Kayri Havens, Chicago Botanic Garden, 1000 Lake Cook Road, Glencoe, IL 60022, USA.

Don R. Levitan, Department of Biological Science, Florida State University, Tallahassee, FL 32306-2043, USA. e-mail: levitan@bio.fsu.edu

Thomas Madsen, The University of Sydney, School of Biological Sciences, Zoology Building AO8, NSW 2006, Australia.

Nico K. Michiels, Max-Planck Institut für Verhaltensphysiologie, PO Box 1564, D–82305, Starnberg, Germany. e-mail: michiels@mpi-seewiesen.mpg.de

Anders Pape Møller, Laboratoire d'Ecologie, CNRS URA 258, Université Pierre et Marie Curie, Bât. A. 7ème étage, 7 quai St Bernard, Case 237, F–75252 Paris Cedex 05, France. e-mail: apmoller@biobase.dk

Mats Olsson, The University of Sydney, School of Biological Sciences, Zoology Building AO8, NSW 2006, Australia. e-mail: m.olsson@liszt.zool.gu.se

Geoff A. Parker, Population Biology Research Group, School of Biological Sciences, University of Liverpool, Liverpool L69 3BX, UK. e-mail: gap@liverpool.ac.uk

Chris W. Petersen, College of the Atlantic, 105 Eden Street, Bar Harbour, Maine 04609, USA. e-mail: CHRISP@ecology.coa.edu

Andy Purvis, Department of Biology, Imperial College, Silwood Park, Ascot, Berks SL5 7PY, UK.

Eduardo R. S. Roldán, Instituto de Bioquímica (CSIC-UCM), Ciudad Universitaria, E–28040 Madrid, Spain.

Glenn Shimmin, Department of Anatomy, Monash University, Clayton, Victoria 3168, Australia.

Leigh W. Simmons, Department of Zoology, The University of Western Australia, Nedlands, WA 6009, Australia. e-mail: lsimmons@liverpool.ac.uk

Mike T. Siva-Jothy, Department of Animal and Plant Sciences, The University, Sheffield S10 2TN, UK. e-mail: zo1mts@sunc.sheffield.ac.uk

David A. Taggart, Department of Anatomy, Monash University, Clayton, Victoria 3168, Australia. e-mail: David.Taggart@med.monash.edu.au

Peter D. Temple-Smith, Conservation and Research Unit, Zoological Parks and Gardens Board of Victoria, PO Box 74, Parkville, Victoria 3052, Australia.

Robert R. Warner, Department of Ecology, Evolution and Marine Biology, University of California, Santa Barbara, CA 93106, USA. e-mail: warner@lifesci.ucsb.edu

Jonathan Wright, School of Biological Sciences, University of Wales, Bangor, Gwynedd LL57 2UW, UK. e-mail: jw10015@hermes.cambridge.ac.uk

Preface

Scientific disciplines are like the fossil record. Some taxa burst on to the scene only to disappear rapidly, others like the ammonites, evolve to a certain state and then remain unchanged for aeons, while others develop and undergo extensive adaptive radiation. The field of sperm competition has been around for 27 years, a relatively short time as scientific disciplines go, but during this time it has expanded enormously, and is now radiating into other disciplines. The factors that determine whether or not particular areas of science 'take off' are not well understood and when Geoff Parker published his ground-breaking paper on insect sperm competition in 1970, he could have had little idea of its consequences.

Within 14 years sperm competition studies, particularly on insects, were widespread, elegantly demonstrated by Bob Smith's *Sperm Competition and the Evolution of Animal Mating Systems* (1984), this volume's predecessor. Since then studies of sperm competition have continued apace, and have been undertaken in an increasing diversity of taxa. The new knowledge and understanding, and recognition of what we still need to know, in the mere 13 years since Smith's volume, are remarkable. Our aim in this volume is to document those changes and suggest avenues for future research. The present volume is divided into two parts, a general section and a taxonomic section. Before describing briefly the book's contents, it is probably worth saying something about its title. The inclusion of 'sexual selection' within the title was simply to remind people that sperm competition is a central part of Darwin's theory of sexual selection. Sexual selection does not stop at copulation, and the fact that females in virtually every animal group copulate with several males means that sperm competition is a central and ubiquitous part of sexual selection.

The general section comprises four chapters, each dealing with a fundamental issue. It is particularly fitting that Geoff Parker should write the opening chapter to both this, and Smith's volume. His sustained contribution to this field has been remarkable. In this opening chapter he presents an overview of his work during the past 13 years which provides us with a theoretical framework for the study of sperm competition. In the second chapter Anders Møller discusses the relationship between sperm competition and sexual selection. In chapter three Bill Eberhard deals with a rapidly expanding and controversial area, the role of females in sperm competition. In chapter four Jon Wright discusses another controversial issue: the relationship between paternal care and sperm competition.

The taxonomic part of the book comprises 13 chapters. Compared with Smith's volume, the taxonomic coverage here is more diverse, reflecting the expansion of the field. However, it has not been possible to be fully comprehensive in our taxonomic coverage. There are certain animal groups, such as Crustacea, where sperm competition is known to occur, but where we felt (possibly mistakenly) there was insufficient information to merit a full separate chapter. We start with plants. Lynda Delph and Kayri Havens (chapter 5) have reviewed pollen competition in flowering plants and in doing so have highlighted the similarities and differences with sperm competition in animals showing that there is much that animal biologists can learn from botanists. Don Levitan (chapter 6) then discusses animal external fertilizers – marine invertebrates, a group that have numerous parallels with flowering plants. In chapter 7 Nico Michiels considers the special sperm competition problems faced by simultaneous hermaphrodites. This is followed by Bruno Baur's review (chapter 8) of sperm competition and its potential in the molluscs. Not only is the potential for sperm competition in this group considerable, but, as Baur points out, with their multiple sperm storage structures and other bizarre features, the potential of molluscs as study organisms is also tremendous. Mark Elgar's discussion of spiders and other arachnids in chapter 9 shows the extent to which our understanding of sperm competition in this taxon has developed since the two chapters in Smith's volume. The insects caused us some problems. (Bob Smith devoted four chapters – some 200 pages and 30% of the entire volume – to this group.) After much deliberation we decided to go for a single, large chapter which dealt with more general issues in insect sperm competition: Leigh Simmons and Mike Siva-Jothy have provided a broad and comprehensive review in chapter 10. Most of the vertebrate taxa we present here were also covered in Smith's volume. However, in almost every case our knowledge of sperm competition has increased substantially; this is true for fishes (chapter 11 by Chris Petersen and Bob Warner), amphibia (chapter 12 by Tim Halliday), reptiles (chapter 13 by Mats Olsson and Thomas Madsen), birds (chapter 14 by Tim Birkhead) and mammals (chapter 16 by Montserrat Gomendio, Sandy Harcourt and Eduardo Roldán). The chapter (15) on monotremes and marsupials by Dave Taggart, Bill Breed, Peter Temple-Smith, Andy Purvis and Glenn Shimmin is new: we included it because a substantial amount of information exists on monotreme and marsupial reproductive anatomy and biology, but relatively little on their behaviour in the field. Finally, in chapter 17 we provide an overview by discussing the relationship between sperm competition and sexual selection in terms of both function and mechanism and how these translate into fitness.

T. R. Birkhead
A. P. Møller

Foreword

This is a book dedicated to the notion that sexual selection is not so much the competition among males to mate with females as it is the contest among ejaculates to fertilize eggs. It recognizes that much of biologically meaningful sexual intrigue occurs after the mating is done.

The study of sperm competition is a young but precocious science. The phenomenon has proven to be nearly ubiquitous and is widely acknowledged as a conceptual engine that has driven much empirical work in Behavioral Ecology over the past decade and a half. Accordingly, there seems little need for me to use this space to explain the importance of sperm competition or to tout this volume which is already anticipated with great enthusiasm. Nor would it serve for me to underscore a few of the book's all too numerous strengths. Most readers will by now be aware of many issues and controversies in sperm competition, and researchers in the field will bring their own agendas and sophisticated queries to this new resource.

Having thus liberated myself of obligation to review or promote, I propose an alternative undertaking. I will briefly attend the historical development of our ideas about fertilization, inheritance, gametes, sexual selection and other issues in an attempt to discern how we finally came to the scientific recognition of sperm competition, why it may have taken us so long to get there, and were we may be headed in the future.

At a minimum, three basic ideas were needed to 'discover' sperm competition and begin to divine some of its implications. First we should have observed that males are reproductively selfish and compete for mates. Second, we needed to know that semen convey paternal characteristics to offspring. Finally, we would have to have recognized that females routinely mate more than one male within a reproductive cycle. With these three observations in hand, sperm competition can be deduced by applying the simplest logic: if two or more males compete to inseminate a female and the competition ends in a draw (*i.e.* the contested female mates more than one of her suitors), then the competitor's ejaculates will renew the contest by competing to fertilize the female's eggs. So, when in the history of biology could we have assembled the understanding needed to work this logic? Male competitiveness for access and control of sex is highly conspicuous and recognition of the pattern in animals dates to antiquity.

Likewise inheritance of paternal traits via the semen has long been considered probable. Aristotle (384–322 BC) noted the existence of the two competing models for reproduction which have alternated in popularity over most of the last two millennia. He correctly rejected his teacher Plato's **preformation** ideas in favor of **epigenesis**, but Aristotle held the distinctly chauvinistic notion that 'male sexual product' designed offspring by directing development of unorganized 'female matter'. Galen (130–200 AD) took a more egalitarian view. In his *On the Semen*, he disputed Aristotle's androcentric scheme and proposed a surprisingly accurate gender-equal role in heredity. Galen posited that embryos developed from semen mixed with the female sexual product and [nourished by] the mother's blood. Antony van Leeuwenhoek (1632–1723) together with John Ham, a medical student, first observed spermatozoa in 1677. Leeuwenhoek described the appearance, motility, survival, and distinguishing characteristics of sperm from an array of different species. But these wonderfully accurate observations were somewhat diminished by Leeuwenhoek's staunch sperm preformationist view that deluded him into believing he could microscopically distinguish 'boy' and 'girl' sperm.

Forgiving their deficiencies, all of the foregoing constructs imply a male hereditary contribution to offspring through semen. Thus, the second of the concepts needed to recognize sperm competition endured without challenge for almost 2000 years. Then came a bizarre, historical twist in the form of an influential cadre who denied any heredity through the semen thereby demolished this long-established part of the conceptual framework.

Albrecht von Haller (1708–1777), Charles Bonnet (1720–1793) and Lazzaro Spallanzaini (1729–1799) among others, were consecrate ova preformationists who held that eggs contained the *homunculi* or *animalunculi*, and that semen was required only to trigger growth of tiny preexisting egg-bound individuals. During the second half of the eighteenth century and well into the nineteenth century, Spallanzaini was held to be the authority on fertilization and his view went largely undisputed. Among Spallanzaini's influential contributions were the first scientific investigations into artificial insemination and his disproof of the strange but then prevalent notion that the factor triggering egg growth was some incorporeal agent or '*aura seminalis*', emanating from the semen. However, Spallanzaini erroneously concluded that the liquid portion of the semen was the developmental initiator. He specifically denied any role for spermatozoa, which he was convinced were parasites passed from generation to generation by intercourse. When Charles Darwin (1809–1882) read Spallanzaini, he must have been profoundly confused! Perhaps it was this confusion that pressed Darwin to his own fuzzy 'gemmule' theory of inheritance, which despite its own vagaries at least restored a heritable male contribution to reproduction.

Ideas about fertilization and heredity remained extremely amorphous through the eighteenth and most of the nineteenth centuries. In fact, the union of nuclei of sperm and egg was not observed until the end of the nineteenth century and Mendel's laws would wait the century's turn to be rediscovered. The lack of solid concepts about reproduction and heredity could have obstructed Darwin's view of sperm competition, but why did so many later evolutionary biologists overlook the issue for nearly 70 years into the twentieth century.

Sexual selection was the process offered by Charles Darwin to explain the evolution of extravagant, apparently maladaptive traits found mostly in males (Darwin, 1871). Darwin recognized two kinds of sexual selection. The first, intrasexual selection or the competition among males for the privilege of mating with females, is of course one of the insights in the minimal array we required to deduce sperm competition. It would seem Darwin's observations and musings on intersexual selection or the preferences exercised by females in their selection of mates might have led him to another requisite insight: multiple mating by females. But did it?

In the preface to this volume's predecessor, I quoted Darwin and quipped that the elucidator of sexual selection may have delayed recognition of sperm competition because he believed females to be monogamous. Uncomfortable with this view, Tim Birkhead delved deeply into Darwin's writings and discovered accounts of several examples of polyandry, a few cases of extra pair copulation, and at least one instance of mixed paternity that had not escaped our founder's notice. Tim also reminds us that some of Darwin's beloved barnacles brought him tantalizingly close to our topic. Females of some barnacle species are equipped with pockets that house from two to fourteen diminutive males. Darwin described these 'little husbands' as 'mere bags of spermatozoa' and observed that all of the tiny males within a particular female's shell had potential to fertilize her eggs. Clearly, Darwin teetered on the brink of adding the corollary of sperm competition to his sexual selection theorem—but he didn't do it, and we are left to ponder why.

In his essay, *Darwin and Sex*, Birkhead proposes that Darwin's personal sensitivities and aversion to censure in the restrictive Victorian environment may have deterred him from pursuing sperm competition and other explicit sexual topics. This conclusion is perfectly consistent with what we know of the character of the man and I've no doubt that social inhibition caused Darwin to skirt some delicate subjects. But I believe a powerful human bias is the more important explanation for Darwin's failure to grasp, and his descendent disciples' ninety-nine year deferred recognition of, sperm competition. This bias is the cross-cultural conviction (still largely intact) that males are 'naturally' polygynous, females 'naturally' monogamous. The bias is reinforced by what we are able to observe of vertebrate (including human) reproductive behavior: Males are overt, exhibitory, and eager; females are covert, clandestine, and reticent. Case-in-point: On the five year voyage, Darwin spent tens of thousands of hours observing hundreds of the world's animal species (most of whose females must multiply mate) and yet his life's journals apparently contain fewer than a half-dozen accounts of polyandry mostly involving domesticated species. Consider also that the best-loved and most ardently observed vertebrates, birds, were generally regarded as the animal paragon of nuptial fidelity until the middle of the last decade when our innocence was disillusioned by the probing curiosity of this volume's editors. We now know (though we might still wish to deny it) that extra pair copulation and mixed paternity occur in virtually all avian species. I concede that Darwin knew of a few cases of polyandry, but I believe he probably thought these instances were exceptional and thus relatively unimportant.

Optimality thinking may also have helped perpetuate the female monogamy myth. Beginning early in this century, it was determined that the typical ejaculate delivered enough sperm to fertilize multiples of the typical female's lifetime production of eggs. So it seemed a female would have no shortage of sperm and thus no reason to mate multiply. Therefore, we concluded her 'optimal tactic' should be to choose one ideal mate from among available suitors, then get on with her reproduction. To mate with more than one male would waste her precious time and energy! Ironically, prior to the discovery of sperm competition, it seems we had no clear grasp of the pervasiveness of polyandry and not even an inkling of its possible fitness benefits to females.

Finally, it has been my observation that we humans often behave as if we have perfect understanding of the evolutionary forces that have shaped us when in fact we often don't understand them at all. It seems possible that natural selection has erected psychological and social barriers against our cognition of selected facts to prevent rationality from interfering with our all-important reproductive impulses. For example, I can imagine that society might be protected and the individual fitness of both men and women served by a psychology that generally denied the occurrence of polyandry.

Eons before they were aware of their gametes, our ancestors almost certainly experienced powerful emotions evolved in the context of sperm competition. Romantic attractions coupled with the mate choice ambivalence felt and acted upon by human females must have created both the reproductive opportunities and uncertainties that selected the human male compulsion to copulate and his masculine sexual jealousies. These passions with their profound impacts on the lives of individuals persisted compounding through the ages to shape our cultures—and yet their ultimate cause apparently never yielded to human introspection, for alas, sperm competition, was first discovered not in humans, but in insects!

In 1970 Geoff Parker outlined the theory with the publication of his review entitled: *Sperm competition and its evolutionary consequences in the insects*. Why was Parker the discoverer? And, why were insects the revealing organisms? Let's begin with the second question. Two characteristics of female insects must have tweaked Parker's interest: first their ability to store sperm for long periods in special organs, the spermathecae; second their relatively easily observed patterns of multiple mating. His initial insight was that sperm from different males could reside in the female tract until days, weeks or even years later, when some few would fertilize the female's eggs. Now we come to Parker's brilliance (which only Geoff himself would deny) and probably more important, his passion for 'games'. Insects provided him the initial insight; then Parker the game theorist began to play with the evolutionary implications of the potential for sperm to compete. He quickly recognized the paradox of opposing forces sperm competition would place on males. On the one hand, he reasoned, selection should favor adaptations that would enable males to pre-empt stored sperm from earlier rivals. But, it should also favor mechanisms enabling males to prevent remating by their mates and/or to disable or impair future ejaculates, thereby preventing or minimizing competition from future mates' sperm.

Parker's validation of his theory is a rather involved entomological subplot I call *A Story of Three Flies*. The three are, the fruit fly (*Drosophila melanogaster*), the screwworm fly (*Cochlionmyia hominovorax*), and the yellow dung fly (*Scatophaga stercoraria*).

Beginning with the last, the yellow dung fly is a common barnyard species with which Geoff Parker amused himself while pursuing his Ph.D. degree at University of Bristol in the 1960's. As Parker's focal species, *S. stercoraria* was the first animal to be studied specifically to investigate male patterns possibly evolved in the context of sperm competition. Very briefly, Parker determined that the last male to mate with a given female in a set of competitive matings always fertilized about 80% of the eggs she laid until another mating occurred. Parker wondered why males didn't extend the time spent *in copula* in order to transfer more sperm and thereby obtain 100% of the fertilizations. The answer was that yellow dung fly males were playing the sperm competition game to maximize their lifetime fertilization rates and not the number of egg gains from each mating. Geoff's dung flies behaved as his model predicted and this result provided empirical evidence that sperm competition had indeed acted as a selective force. As it turns out, dozens of excellent sperm competition experiments and other appropriate observations on insects had been accumulating in the literature for years. This body of data was the product of geneticists and applied entomologists who knew little and cared less about the role of sperm competition in evolution.

It has been said God created the fruit fly for Thomas Hunt Morgan (1866–1945) and if this is so, the founder of the gene theory and his followers certainly justified the divine gift. *Drosophila melanogaster* probably originated in Southeast Asia and was introduced to America in the 1870's. Morgan began culturing the affable dipteran at Columbia University in 1907 and by 1925 he had identified about 100 of its genes. The genes provided heritable markers that served to keep track of sperm derby winners and losers. Over the years, scores of drosophila 'gene jockeys' conducted their matings and incidentally revealed a plethora of patterns and processes that took on whole new meanings in Parker's (1970) sperm competition review.

The third tale in the two-winged trilogy is expansive. The story initially features one fly and one human, but climaxes with a cast of hundreds (both human and insectan). In the 1930's, E. F. Knipling, an ambitious economic entomologist with the United States Department of Agriculture, began to explore the theoretical underpinnings of what came to be known as the sterile insect technique (SIT) for managing pest populations. This idea would eventually set into motion an extraordinary research effort on insect reproductive behavior and sperm competition, motivated by purely pragmatic objectives. Knipling developed simple models based on a female monogamy assumption. His calculations predicted that pest populations could be dramatically suppressed within a few generations if a large number of sterilized laboratory-reared males could be released into the field coincident with the seasonal emergence of each new adult cohort. Knipling eventually selected *Cochlionmyia hominivorax*, a devastating pest of range cattle, upon which to test the SIT population management scheme. Whether by incredible luck or brilliant design, this species was an extremely fortuitous choice because among all of the candidate pests, only screwworm fly females are monogamous.

Beginning in the late 1950's screwworms were factory reared in enormous numbers and mass sterilized by irradiation from a cobalt 60 source. Sterilized pupae were distributed by airdrop over the target areas. In a little over a decade, Knipling succeeded in eradicating the screwworm from Florida and reduced to negligible its impact in a 2000 by 500-mile wide corridor along the Mexican borders of Texas, Arizona, and California. He became a hero to ranchers and the darling of the budding environmental movement that heartily applauded pest control *sans* insecticides.

Small wonder that applied entomologists, working on everything from migratory locusts to mosquitoes eagerly jumped aboard the SIT bandwagon. Unfortunately, none of the early copycat SIT programs was able to duplicate the screwworm success. In fact, most were dismal failures. Why? The answer lies in female mating behavior. Unlike the monogamous screwworm females, the females of virtually all other pests species were polygynous. When females have multiple mates, the SIT equation is dramatically altered such that the number of sterile males required to depress a population becomes astronomical.

When the cause of their disappointment became apparent, applied entomologists began to take great interest in the frequency and timing of their pests' multiple mating, the virility of males both wild and laboratory reared, and the patterns of fertilization by stored sperm from different males (*i.e.* sperm competition!). There followed a spate of studies on mating behavior and sperm utilization patterns facilitated by the sterilization technology developed for SIT. Sterility became the marker that produced sperm competition data for a host of insect pests.

It is rare to find a body of data in the literature more than marginally useful for testing a new theory. Typically, the needed data are accumulated over a period of years, but the 'three fly story' illustrates a remarkable exception. Parker presented his theory, and was then able to buttress it with not only his own *Scatophaga* studies, but also perfectly appropriate data of great depth from drosophila genetics and of rich diversity from applied entomology. Parker's theory provided instant adaptive explanations for such hitherto mysterious phenomena as mate guarding, prolonged copulation, mating plugs, takeovers, traumatic insemination, insemination reaction, extravagant sperm production, sperm displacement, polymorphic sperm and others. But despite Parker's splendid enunciation of the theory and demonstration of its explanatory power, his review seems not to have received the attention it deserved for almost a decade. However, a handful of people working independently on insect reproductive behavior took special notice of the review.

I was one, and Jonathan K. Waage was another. In 1979, each of us, then unknown to the other, published a paper in *Science*. Jon was first, reporting that male damselflies had evolved intromittent organs that functioned not only to inseminate, but also to remove previously stored competitive sperm from the female reproductive tract. My article followed with a demonstration that paternal caring male water bugs were able to avoid cuckoldry by insisting on repeated copulation alternating with oviposition. Our papers seemed to capture imaginations. Jim Lloyd at University of Florida was moved to comment that the damselfly male genitalia functioned as a '... veritable Swiss army knife of sexual gadgetry', and referring to male water bugs he suggested they behave '... as errant macho Californians, demanding proof of fatherhood before paying out paternity benefits'. Together, the papers excited a great deal of interest in the subject, and brought belated attention to their foundation in the Parker review.

In June of 1980, Geoff Parker was featured speaker at a symposium on sperm competition sponsored by the Society for the Study of Evolution and the American Society of Naturalists held in Tucson. At the conclusion of the symposium participants discussed the possibility of publishing a book on sperm competition. The decision to do so was delayed for a year while possible contributors were queried and a publisher was sought. The book finally appeared in 1984 and its publication gave the study of sperm competition another big boost. Although the volume featured insects, it expanded the application of sperm competition to vertebrates including everything from fishes to humans. Excepting Don Dewsbury's pioneering work on rodents, few original data were to be found in the vertebrate chapters because at the time, almost none existed. By design the book was replete with conjecture. I pressed authors to put forth their most imaginative ideas as a stimulus to the field and most writers accommodated, selflessly braving the risk of being wrong.

The book promoted a general interest in sperm competition, stimulated entomological work, and most importantly expanded the study to vertebrates. Allozymes provided the initial tools needed to investigate paternity in vertebrates, which could not be irradiated and often lacked useful morphological markers.

The 90's revolution in molecular genetics probably saved the field of sperm competition from stagnation. DNA fingerprinting technologies, especially those involving PCR, have permitted investigators to explore crucial but hitherto unanswerable questions about what happens in nature. For example, RAPD PCR has been used in staged competition experiments involving more than two contestants and PCR of microsatellite DNA has permitted the detection of multiple paternity among field-collected clutches of eggs, putative siblings, and offspring from females of unknown mating history. This brings us to the current state of our knowledge and a test of my prophetic powers. Where will the study of sperm competition go from here?

Over the past decade we have learned much about patterns of fertilization following multiple matings and this work will continue at an accelerated rate with the simplification and improved accessibility of molecular genetics techniques.

To date we understand only a very few of the mechanisms underlying patterns of fertilization and most of what we know involves behavior of whole organisms. Many patterns themselves remain inexplicable, and events that occur within females of most species are truly enigmatic. Thus, I believe a principal future focus will involve the study of mechanisms underlying the patterns. This is not to suggest that behavioral ecologists interested in sperm competition have developed a fascination with mechanisms for their own sake or that we have acceded to the reductionists' program. We have simply followed our subject from the level of individuals in ecological space, to events that begin when ejaculation confines populations of sperm to the much smaller but labyrinthine, dynamic biosystem of the female reproductive tract.

William Eberhard, a contributor to this volume, is also author of a treatise entitled *Sexual Selection and Animal Genitalia* published in 1985 just following the edited volume on sperm competition. In my opinion Bill Eberhard's genitalia book, and a more recent one, *Female Control: Sexual Selection by Cryptic Female Choice* places him on the same stage with equal billing to Geoff Parker. For sperm competition and cryptic choice (first proposed by Randy Thornhill) are simply male and female sides of the same coin. Eberhard's first book offered a possible solution to the conundrum: Why are male intromittent organs so extravagantly variable among species and why do they apparently evolve so rapidly? His answer was that these structures are tools of cryptic courtship forged in the fires of intersexual selection. If this is correct, females must have an active role in sperm utilization. And if females can favor the sperm of preferred mates, they will have wrenched at least some post-copulatory control away from males. The extent to which cryptic choice has pre-empted or modulated contests among sperm is an area of future study that has the potential to amaze.

Finally, there are many ideas about why females multiply mate but few data to demonstrate enhanced fitness for this nearly universal feminine behavioral predilection. Only in insects has it been shown that females of some species benefit from male ejaculatory nutrient and defensive compound transfers. Much additional work on the problem of female benefits will be crucial to our understanding patterns of sperm utilization as well as female tolerances for sperm competition. I predict (with some foreknowledge of research in progress) that new adaptive explanations for polyandry will be proposed and both old and new will be tested. This work could yield exciting new insights.

I conclude with the observation that not even our reckless speculation of nearly fifteen years ago seems to have adequately anticipated the extent and elaboration of the subject we still call sperm competition. But readers of this volume who contributed to the first book on sperm competition will recognize that if nothing else we have accomplished one of our objectives, which was to stimulate the field. Speaking on behalf of those contributors, the editorial 'we' is unabashedly delighted to note that the excellent current contribution seems to confirm the value of our earlier efforts.

June 1998

Robert L. Smith
Department of Entomology
University of Arizona
Tucson
USA

Acknowledgements

Our main thanks go to the contributors of this volume: their efforts have been remarkable. In addition to producing their chapters they have responded swiftly and willingly to our questions and have provided careful and constructive comments on the chapters of others. We are also indebted to all those other individuals who commented on individual chapters: R. Babcock, J. D. Bishop, J. Dickinson, M. G. Gage, P. Gowaty, B. Kempenaers, H. D. M. Moore, S. Pitnick, J. Shykoff, B. Siverno, Paul Watson and D. F. Westneat. T.R.B. is grateful to Jayne Pellat, Bobbie Fletcher and especially to Jayne Young for a great deal of help in the production of this volume.

Part One

General Themes

1 Sperm Competition and the Evolution of Ejaculates: Towards a Theory Base

G. A. Parker

Population Biology Research Group, School of Biological Sciences, Nicholson Building, University of Liverpool, Liverpool L69 3BX, UK

I. INTRODUCTION

My initial review on sperm competition (Parker 1970a), which focused on the insects, attempted to demonstrate that ejaculates commonly overlap in the female tract and to indicate how paternity could be estimated. Its main aim was to outline the wide variety of adaptations which can arise through sperm competition as a selective force in evolution. These adaptations occur at many biological levels: they may be behavioural (e.g. mate-guarding), physiological (e.g. male accessory gland fluids inducing unreceptivity after mating), anatomical (e.g. copulatory

Sperm Competition and Sexual Selection
ISBN 0-12-100543-7

plugs), or all three (e.g. internal fertilization). Sperm competition is now widely recognized as a major and pervasive force in evolution; its study has matured into a discrete discipline within behavioural ecology and currently attracts considerable interest both empirically and theoretically (Smith 1984; Birkhead and Møller 1992; Baker and Bellis 1995; Birkhead and Parker 1997).

I originally defined sperm competition as 'competition within a single female between the sperm from two or more males for the fertilization of the ova'. I now favour the more general definition: 'competition between the sperm from two or more males for the fertilization of a given set of ova', so that external fertilization is not excluded. Sperm competition is thus equivalent to inter-ejaculate competition; we can, however, recognize a second component – intra-ejaculate competition – which refers to competition between the sperm within one ejaculate. Whether this can have importance in adaptation depends on whether a sperm's characteristics can be determined by the expression of genes in the haploid sperm genotype rather than by the genes of the diploid parent which creates it (see Parker and Begon 1993). While it is clear from studies of sex-linked drive and other phenomena (see Hecht *et al.* 1986) that haploid genes can be expressed, it is more usual to expect that the characteristics of a sperm are determined by the diploid genotype of the male that produces it (Beatty 1972).

Sperm competition studies are in a stage of rapid progress and interest throughout a wide diversity of groups. The theoretical basis for such studies has also expanded since 1990 but no comprehensive synthesis yet exists. In this chapter, I consider the problem of how a male should allocate sperm among different ejaculates and summarize a model framework for the analysis of this problem; my aim is to consolidate a prospective theory base for empirical advances. It is a view mainly from the male perspective. Because sperm competition can involve sexual conflict in which the interests of male and female differ, it is clear that the mating or ejaculatory strategy which is best for a male need not be best for the female (Parker 1984a; Stockley 1997). Earlier work on sperm competition tended to focus on male interests, but emphasis has shifted towards the views that female interest will either have some effect (Knowlton and Greenwell 1984; Parker 1984a; Birkhead 1995, 1996) or may even override male interests (Eberhard 1985, 1996; Baker and Bellis 1995). My own view is that resolution of mating conflict (i.e. whether mating takes place) will depend on the circumstances, and that either sex can exert a strong or even overriding influence (see Parker 1979, 1984a; Clutton-Brock and Parker 1995). Whether the female can have a strong influence on the evolution of ejaculate characteristics depends on how much control she can exercise on an ejaculate within her reproductive tract. Resolution could, in principle, lie at any point in the continuum between 'male win' and 'female win' extremes, although intermediate solutions are likely (see Section VIII). However, in many instances, there will be no conflict between the male strategy considered here (e.g. how much of his reproductive effort should a male spend on sperm) and female interests, so that the present models may serve as fair approximations of the selective forces shaping ejaculate characteristics.

II. THE LOGIC OF SPERM COMPETITION GAMES

Recently, a new theoretical basis for the analysis of the evolution of ejaculate characteristics in terms of sperm size and number has been developed (Parker 1990a,b, 1993; Parker and Begon 1993; Ball and Parker 1996; Parker *et al.* 1996; Ball and Parker 1997). It derives from an earlier analysis of ejaculate expenditure (Parker 1982; see also Parker 1984b). I have termed these models 'sperm competition games' to stress that they are evolutionary games between rival males, whose ejaculates compete for the fertilization of the ova. Pay-offs to competing males depend on the ejaculation strategies played by other males in the population; their analysis requires an evolutionarily stable strategy (ESS; see Maynard Smith 1982) approach and hence forms a part of evolutionary game theory. Sperm competition games ask such questions as: 'how much energy should be spent on each ejaculate?' Or 'what are the ESS ejaculate characteristics such as sperm size and number?' Or 'when is it best to mate singly, and when is it best to mate multiply with the same female'?

The games have the following structure:

(1) The competing males have specified information and states;
(2) Males have a specified range of ejaculate strategies;
(3) Their ejaculates enter into competition;
(4) The sperm competition mechanism determines which sperm from each ejaculate enter the 'fertilization set' – i.e. the set of sperm used randomly for fertilization;
(5) The female is assumed to exert no preference over the strategy being considered. This last assumption can be altered: some games analyse the effect on ejaculate characteristics of a female preference for the sperm of a particular type of male (see Section VIII).

The following account gives a brief outline of the mathematical logic of sperm competition games. Most games examine ejaculation strategies that can vary continuously in evolutionary time, and use the technique of 'differentiating the fitness function' to deduce the ESS. The 'fitness function' is a plot of the fitness, W, of mutants in the ESS population against deviations from the ESS in some strategic aspect of ejaculate expenditure, x; x is best envisaged initially as sperm number. In a population at the ESS (called x^*), all mutants playing $x \neq x^*$ must do worse than x^*, so the plot of $W(x,x^*)$ rises to a peak where the 'mutant' plays the ESS (see Fig. 1.1a). So, we can solve for the ESS by setting

$$\left[\frac{W(x,x^*)}{x}\right]_{x=x^*} = 0; \qquad (1a)$$

subject to

$$\left[\frac{\partial^2 W(x,x^*)}{\partial x^2}\right]_{x=x^*} < 0; \qquad (1b)$$

to ensure that x^* is a maximum (see Maynard Smith 1982).

To determine the ESS for some aspect of sperm expenditure, we must assume that there is some constraint on sperm production. We assume that there is a fixed budget for total reproductive expenditure, so that there is a trade-off between its two components: ejaculate expenditure and mating expenditure (see Fig. 1.1b). Expenditure on gaining matings determines the number of matings a male achieves, and his ejaculate expenditure determines the expected value of each mating: increasing ejaculate expenditure increases the gain from each mating, but reduces number of matings that a male can achieve. There is some evidence for such a trade-off: for example, Warner *et al.* (1995) showed that in the blue head Wrasse, *Thalassoma bifasciatum*, larger males divert energy from gamete production into mate-guarding activities. They gain a higher number of matings, but lower fertilization success per mating.

A mutant male's fitness is simply the number of matings achieved, $n(x,x^*)$, times expected value of each mating, $v(x,x^*)$:

$$W(x,x^*) = n(x,x^*) \,.\, v(x,x^*) \qquad (2)$$

(see Fig. 1.1b) and from equation (1a) this occurs when (Parker 1993):

$$-\left[\frac{n(x,x^*)}{n'(x,x^*)}\right]_{x=x^*} = \left[\frac{v(x,x^*)}{v'(x,x^*)}\right]_{x=x^*}; \qquad (3)$$

where the primes denote the differential coefficient of n or v with respect to x. Note that both sides of equation (3) are positive, since n' is negative (the number of matings decreases as the ejaculate expenditure increases) and v' is positive (gains from a given mating increase with ejaculate expenditure).

The exact forms of n and v depend on the model. By way of example, consider first n (the number of matings achieved). The logic most commonly used to deduce $n(x,x^*)$, the number of matings achieved by a mutant playing ejaculate expenditure x in a population at the ESS x^* is as follows. Each male has a fixed total energy budget of R units, and the average cost of obtaining each mating (finding a female, etc.) is C units. The cost of each unit mass of ejaculate is D, so the cost of an ejaculate containing x units of sperm is Dx (note that D includes the cost of the seminal fluid; this will often exceed the cost of the sperm). Note that R, C, and D are all expressed in the same energy units. Let $\langle x\rangle$ and $\langle x^*\rangle$ be the average ejaculate expenditures by a mutant and by an ESS player. Note

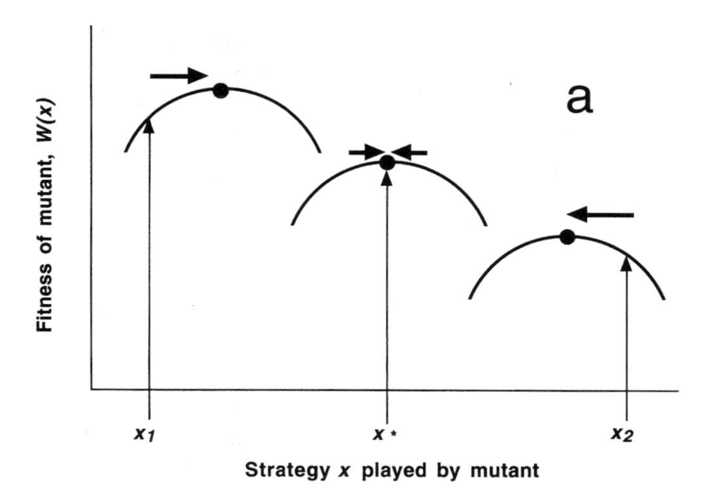

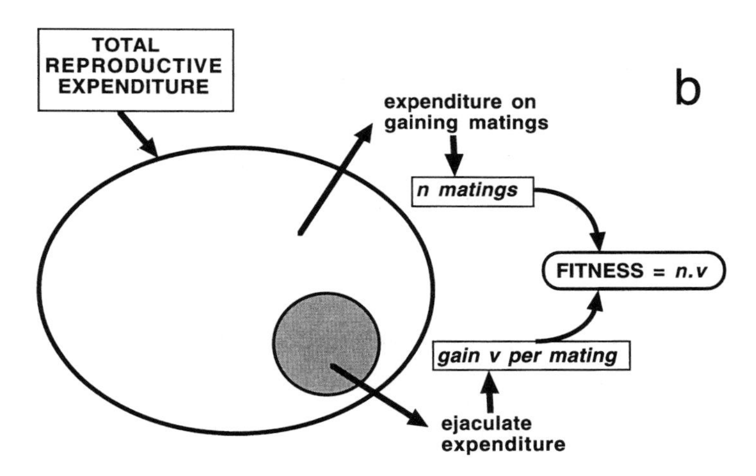

Fig. 1.1. **a.** *A hypothetical example of the fitness of a mutant playing a given value of a continuously increasing strategy* x *in three populations,* x_1, x^*, *and* x_2. *In the population playing* x_1, *a range of values of* x *that are higher than* x_1 *can spread (the optimal mutant for invasion is given by the black dot). Similarly, in the population playing* x_2, *a range of values of* x *that are higher than* x_2 *can spread. In a population at the ESS,* x^*, *no mutant strategy can spread, since all* $x \neq x^*$ *have lower fitness. The ESS can be solved by differentiation the fitness function* $W(x,x^*)$ *with respect to* x *and evaluating at* $x = x^*$, *following equation 1. (Modified from Parker and Maynard Smith 1990).* **b.** *In sperm competition games, the fixed total reproductive expenditure is divided up between expenditure on gaining matings (e.g. mate-searching, fighting for access to females, etc.) and expenditure on the ejaculate. There is therefore a trade-off between the two expenditures. Effort spent on the ejaculate increases the gain per mating,* v, *but reduces the expenditure on matings, which reduces the number of matings achieved,* n. *Fitness is the product of matings and gain per mating,* n.v.

that $\langle x \rangle$ and $\langle x^* \rangle$ may differ from x and x^* if the latter relate to just one mating context, such as what to do when mating first, or what to do when mated with a previously mated female. Then $\langle x \rangle$ and $\langle x^* \rangle$ are the average across all mating types for individuals playing x and x^* in a particular context. The ESS number of matings per male is thus $R/(C + D\langle x^* \rangle)$, whereas a mutant playing $x \neq x^*$ achieves $R/(C + D\langle x \rangle)$ matings. Hence, the number of matings gained by the mutant male, relative to the population average, is

$$n(x,x^*) = (C + D\langle x^* \rangle)/(C + D\langle x \rangle). \tag{4}$$

The function $v(x,x^*)$, the value of each mating of a mutant playing ejaculate expenditure x in a population at the ESS x^*, depends on the mechanism of sperm competition. It is often convenient to model the range of exposure to sperm competition between two extremes:

(1) *Risk models*: in a population, sperm competition between two ejaculates occurs with low probability q; on $(1 - q)$ occasions, just one male fertilizes the eggs. Such a pattern is frequent in internally fertilizing species, but note that in some instances there may be competing ejaculates from many different males within a female tract. Suppose that sperm competition obeys the 'raffle' principle (Parker 1982), i.e. a male's fertilization success is proportional to his relative contribution to the sperm pool competing for a mean of ε eggs. As an example, under the risk model, if males play the same ejaculate strategy at all matings:

$$v(x,x^*) = \varepsilon(1 - q)\frac{x}{x} + 2q\left(\frac{\varepsilon x}{x + x^*}\right). \tag{5a}$$

Note that the 2 on the right hand side arises because for each of the $(1 - q)$ matings with a female that will mate only once, a male encounters q females which will mate again in the future and q females which will have already mated. The mean number of matings per female is thus $(1 + q)$. I recently discovered a confusion in the calculations involving 'risk' models in sperm competition games (see below), stemming from the first approach (Parker 1982) and continuing into recent analyses (Parker 1984b; Parker 1993; Parker and Begon 1993; Ball and Parker 1996). This is discussed more fully elsewhere (Parker *et al.* 1997). It relates to an ambiguity over what is meant by the risk of sperm competition. All analyses prior to Parker *et al.* (1997) relate to risk defined strictly as the probability, p, that a given male, when ejaculating, will face sperm competition from another ejaculate. This is not equivalent to q, the probability of occurrence of sperm competition in the population (i.e. the probability of double mating for a given female), although in the prior analyses, p was incorrectly regarded as synonymous with q: in fact $p = 2q/(1 + q)$. ESS solutions in p and q differ only in their exact form, and not in their qualitative predictions. However, for biological purposes, solutions are most usefully expressed in terms of q, the probability of double mating. These versions are presented here; they relate to the earlier solutions if we substitute $q = p/(2 - p)$.

(2) *Intensity models*: sperm competition typically involves N competing ejaculates; e.g. in fish, which are group spawners, males compete in groups of N and simultaneously shed sperm at the moment the female releases her eggs. Although intensity models are not restricted in application to externally fertilizing species (they are relevant to internal fertilizing species which commonly mate multiply), they have been applied mainly to such groups. As an example, under the intensity model, if males play the same ejaculate strategy at all matings and sperm competition obeys the raffle principle, when there are, on average, N competing ejaculates, of which one is a mutant playing x;

$$v(x,x^*) = \frac{\varepsilon x}{x + (N - 1)x^*}; \tag{5b}$$

cf. Parker and Begon (1993).

Note that when $q = 1$, $N = 2$, two males compete in both the risk and the intensity models: they generate continuously increasing levels of sperm competition. This dual approach has the advantage of mathematical simplicity and tractability; it is most inaccurate where the two ranges meet. Ideally, one would use probability distribution to describe the distribution of the number of competing ejaculates within a species, and some sperm competition game models do just this. The drawback is that the probability distribution of competing ejaculates must be either measured directly, or deduced from first principles.

III. SPERM COMPETITION AND EJACULATE EXPENDITURE ACROSS SPECIES: A FAIR RAFFLE WITH NO INFORMATION

How should increases in sperm competition affect sperm expenditure, viewed *across species*? The answer is intuitively obvious: sperm expenditure should increase with the magnitude of sperm competition. However, as we shall see in Section IV.C.2, this conclusion cannot be generalized to predict what should happen in relation to changes in the intensity of sperm competition *within a species*.

Assume that (i) a fair raffle applies, so that $v(x,x^*)$ follows equation 5;

and (ii) males of a given species have no information about sperm competition – their ejaculation strategy is therefore tuned by natural selection to the population level of N or p.

To enable a 'dimensionless' comparison across species (see Charnov 1993), it is convenient to express the ESS ejaculate effort as a proportion of total effort expended per mating (i.e. cost of finding and/or guarding a female plus cost of ejaculation). Calling the proportion of effort expended on the ejaculate

$$E = \left(\frac{Dx}{C + D\langle x\rangle}\right), \tag{6}$$

and applying the logic of sperm competition games, the ESS for the risk model is

$$E^* = \frac{q}{2}, \tag{7}$$

(see Parker *et al.* 1997; previously reported as $p/(4 - 2p)$ by Parker 1984b, 1993). Thus, the proportion of mating effort which should be expended on the ejaculate is roughly one half the proportion of females that engage in 'double mating'.

The ESS for the intensity model is

$$E^* = \left(\frac{N-1}{N}\right), \tag{8}$$

(Parker *et al.* 1996). Thus, group-spawning species should allocate virtually all of their reproductive resources to sperm.

The relationship between increasing sperm competition and relative ejaculate expenditure for the two models is shown in Fig. 1.2: when there is a low risk of sperm competition, virtually all reproductive resources should be spent on mate acquisition (Fig. 1.2a), when there are many competing ejaculates, virtually all resources should be spent on sperm production (Fig. 1.2b). As expected, the two sets of equations give the same solutions if $q = 1$ and $N = 2$, since in both these cases there is always sperm competition between two males. Equations 7 and 8 therefore generate a continuous range from the lowest level of sperm competition ($q \rightarrow 0$) to the highest intensity ($N \rightarrow \infty$): Fig. 1.2a and b represent a type of continuation of each other (but note their different scales).

The proportion of reproductive effort expended on the ejaculate rises to 1.0 as the number of competing ejaculates, N, becomes very large (see Fig. 1.2), therefore expenditure on gaining matings declines to zero at the maximum intensity of sperm competition. It may not 'pay' group-spawners to spend much on mating effort, thus the reason why they show, for example, low mobility or do not fight much compared with internal fertilizers, may be that it does not pay them to do so rather than that they cannot do so (Parker *et al.* 1996).

The first evidence for a positive relationship between sperm production and sperm competition came from studies of fish and primates.

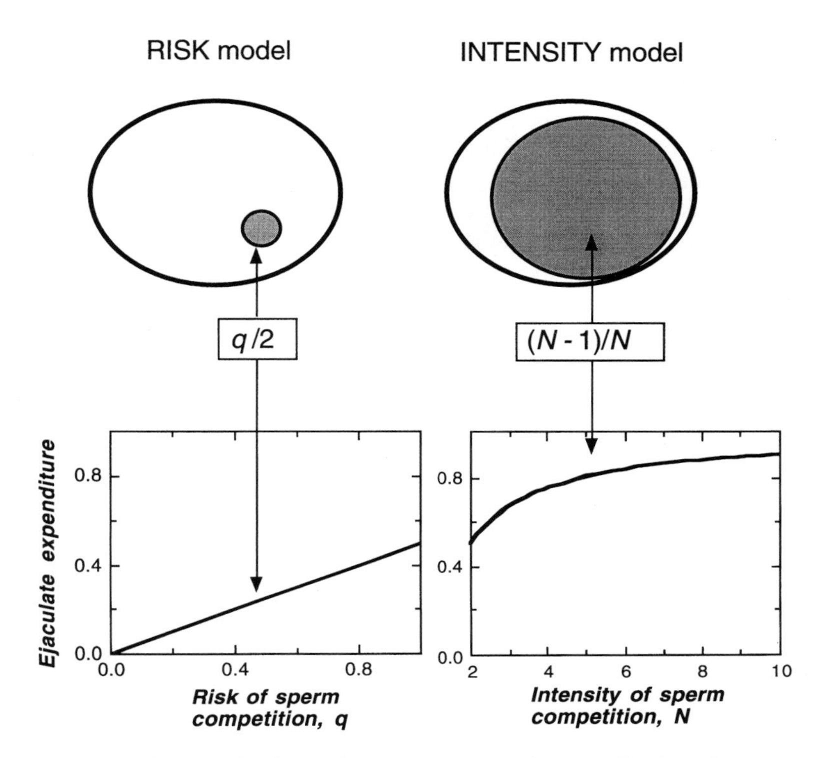

Fig. 1.2. *Predicted relationship across species between* E, *the relative expenditure per mating on an ejaculate and the population risk (probability of double mating,* q*) or intensity (number of competing ejaculates,* N*) of sperm competition for a 'fair raffle' with no information (see equation 6). Note that the scales on the* q *(probability of double mating) or* N *(mean number of competing ejaculates) axes are not equivalent in the two graphs. When there is a low risk of sperm competition (*q *is low), very little expenditure on the ejaculate is predicted; when there is high intensity of sperm competition (*N *is high), most reproductive effort will be allocated to the ejaculate.*

Warner and Robertson (1978) and Robertson and Warner (1978) developed and produced good evidence for an earlier hypothesis of Robertson and Choat (1974) that the difference in relative testis size in initial and terminal phase males of certain fish species (labrids and scarids) relates to sperm competition: initial-phase males typically show much greater relative testis size and, typically, have group spawning or multi-male spawning, whereas terminal-phase males normally spawn singly with a female. Short (1979; see also Short 1977 for relation between sperm production and sperm demand), Clutton-Brock and Harvey (1977) and Harcourt *et al.* (1981), pointed out that relative testis size increases across species with sperm competition in primates. There is now considerable evidence that sperm competition increases relative testis size in many groups.

Note that the predictions in Fig. 1.2 are for expenditure on an ejaculate, expressed as a proportion of the entire effort spent on a mating. Unfortunately, there are very few direct measures of ejaculate size that can be related to the effort spent on gaining a mating, which is needed for quantitative testing of the models. Although it is not immediately obvious that relative ejaculate expenditure (E) is equivalent to relative testis size or gonado-somatic index (GSI = testis mass as a proportion of body mass), we could view a male's soma as equivalent to his effort spent on obtaining matings, and his gonad as his effort spent on ejaculates. A full examination of the relationship between GSI and E, as defined in equation 6, is necessary, but as a first approximation it appears reasonable to claim that they will respond to increased sperm competition in a roughly equivalent manner. Møller (1989) provides some support for this view from mammal ejaculate data: species which have relatively large testes for their body size have high daily sperm production rates, large sperm reserves, and relatively high numbers of sperm per ejaculate. The relative sperm production rate per gram of testis appears to be independent of relative testis size, suggesting that high sperm counts must be paid for by increased testis mass (Møller 1989). Further, in mammals, high sperm competition leads to high sperm counts per ejaculate (Møller 1991).

IV. HOW SHOULD INFORMATION OR PHENOTYPIC STATE INFLUENCE A MALE'S SPERM ALLOCATION?

Against this background across species (Fig. 1.2), we can ask more detailed questions about individual behaviour within a species. I now examine some examples of how males may tune their sperm allocation in relation to their own phenotypic state or to cues correlating with the risk or intensity of sperm competition. Many such games could be devised; the following is a summary of those currently analysed.

A. *Perfect information in loaded raffles*

First, consider a sperm competition game similar to the one devised for Fig. 1.2, but with two important differences: (i) when there is sperm competition between two males, the ensuing raffle is not 'fair' – sperm from the first male to mate (male 1) or the second male to mate (male 2) may be favoured, and (ii) males 1 and 2 'know' their roles (first to mate, second to mate) perfectly. Thus, we can define one role as *favoured* and the other role as *disfavoured*. Specifically, suppose that sperm in role 2 are disfavoured by a factor r ($0 < r < 1$) relative to sperm in role 1, so that proportionate gains to each male under sperm competition are:

$$\left(\frac{s_1}{s_1 + rs_2}\right) \text{ for male 1;}$$

$$\left(\frac{rs_2}{s_1 + rs_2}\right) \text{ for male 2;}$$

(the conclusions are, of course, exactly reversed if we make the opposite assumption, where $r > 1$, and role 2 is favoured).

There are two extreme possibilities to consider about roles (although in reality, we would expect a continuum). Roles might be occupied entirely randomly, so that each male in the population is equally likely to occur as male 1 or male 2. Alternatively, roles may be occupied entirely non-randomly, so that a given male always occurs in role 1 or in role 2. The assumptions about random or nonrandom roles critically influence the predictions of the model (for mathematical details of the case where there are always just two males in competition, see Parker 1990a). The ESS for 'random roles' is that each male should spend equally on the ejaculate, despite having perfect information as to which role is being played at any given time. This effect arises through a type of mathematical balance: if roles are random, the costs or benefits of deviations in sperm expenditure become exactly equal in each role, despite the loading in the raffle. In contrast, the ESS for nonrandom roles is for the male in the favoured role to spend less on the ejaculate than the male in the disfavoured role. The greater the unfairness or loading in the raffle, the greater the disparity in ejaculate expenditures. Effectively, if a male is always disfavoured, he compensates by producing more sperm per ejaculate at the expense of mating opportunities; if he is equally likely to find himself in either role, he plays the same sperm allocation strategy despite his role (Fig. 1.3).

Some empirical studies appear to support these conclusions. A detailed survey of the biological evidence relating to the predictions of these models for insects is given in Chapter 10. Circumstantial evidence is also found in mammals. The random roles model may apply for 13-lined ground squirrels, *Spermophilus tridecemlineatus* (Schwagmeyer and Parker 1994). These rodents have a very short reproductive season. Typically, a female in oestrus emerges from her burrow early in the morning to copulate. A male usually arrives quickly, followed later by another and possibly a third (the most common number is two). The males show a form of 'queuing' for mating; the first male pursues the female and performs many mounts and apparent copulations before quitting; the second male then begins his sequence of mounting and copulating (Schwagmeyer and Parker 1987). The biology fits quite well with the assumptions of the random roles model (Parker 1990a). Males sometimes overlap in time and, therefore, often have information about roles 1 and 2. The commonest number of mates per female in nature is two. The raffle is loaded:

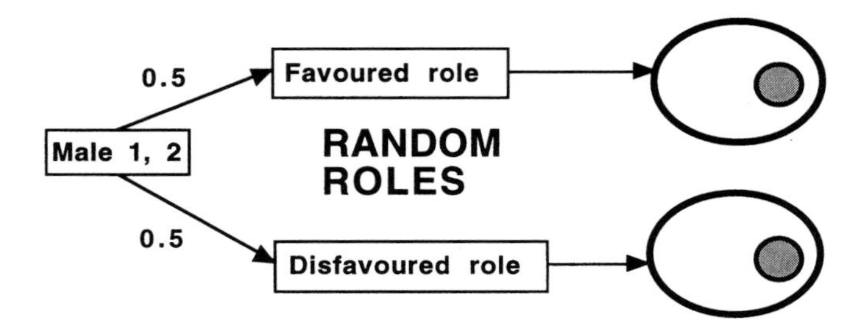

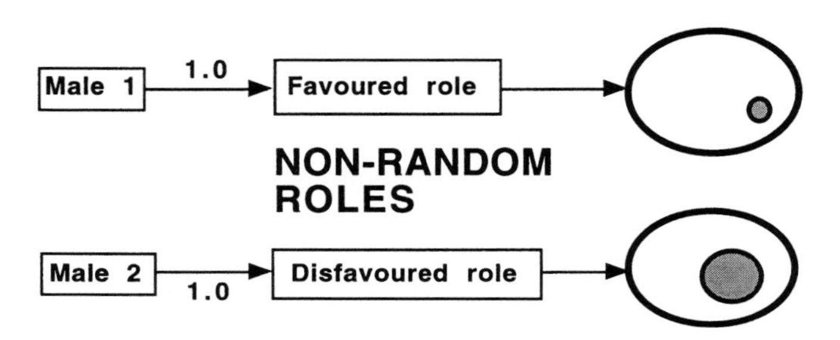

Fig. 1.3. *The loaded raffle model, in which each sperm of male 2 count as r (0 < r < 1) in the raffle relative to each sperm of male 1. In the random roles version of the model, each male has an equal chance (0.5) of being in role 1 or role 2; roles are drawn randomly. The ESS is for equal expenditure in each role, despite the loading. In the nonrandom roles version, each male in the population always occupies the same role, i.e. with probability 1.0. The ESS is now for males in the favoured role (1) to spend less on the ejaculate relative to males in the disfavoured role (2); this conclusion is reversed if role 2 is the favoured role.*

there is a first-male advantage (because copulation is induced, see Chapter 16) so that the value of mating second gradually declines with the delay after the first mating (Schwagmeyer and Foltz 1990). Further, it also appears that any given male is equally likely to be first or second (Schwagmeyer and Parker 1994). P. L. Schwagmeyer did not want to kill and dissect females to count sperm, so she recorded behaviour which might correlate with the total sperm transferred by a given male: the total time in copula, the number of mating bouts, the mean length of longest copulation (the most important measure correlating with paternity in laboratory studies), etc. No difference was detectable between first and second males in any of these measures, suggesting that the total ejaculate expenditure is equal, fitting the prediction for random roles in a loaded raffle with perfect information about roles.

Stockley and Purvis (1993) give further circumstantial evidence for the present predictions from mammal breeding systems. They argue that where there is continuous breeding, the operational sex ratio (OSR, Emlen and Oring 1977) is highly male-biased so that there is typically mate guarding, with optimal timing of mating by larger males who will occupy the favoured role (fitting the non-random roles assumption). They predicted and found typically relatively greater sperm expenditure by smaller males, as measured by GSI. For seasonal breeders, the OSR will be less male-biased, offering reduced opportunity for mate guarding, so that roles were predicted to be more random. More equal relative sperm expenditures were found in such species.

B. Asymmetric levels of information in fair raffles

The last games described above showed that loading a fertilization raffle may or may not lead to unequal sperm expenditures, depending on roles. What happens if there is a fair raffle, but one male has more information about the probability of sperm competition than his opponent? This asymmetry of information can also lead to unequal sperm expenditure.

To show this, we again use the risk model: males perform extra-pair copulations (EPCs) or sneak matings with a low probability, p; so that the probability that a paired or guarding male does not face sperm competition is $(1 - p)$ (for further details see Parker 1990b). Sperm competition obeys the raffle principle and the raffle is a fair one so that all sperm count equally (equation 5a). But when mating occurs we have the following asymmetry of information: one male (labelled 1) has no information about sperm competition, so that his strategy is shaped by the mean risk (he only 'knows' p). The other (labelled 2) has information that when he mates there will always be sperm competition, with probability 1.0. There are again two versions of the model, which are biologically analogous to the random and nonrandom roles models of the previous section. The EPC model is a form of random roles model in which all the reproductive males and females are envisaged as being paired, but perform EPCs (Birkhead 1988) with a probability of p – such an assumption might fit socially monogamous birds (Birkhead and Møller 1992). The 'sneak-guarder' model is a nonrandom roles model in which males are either sneaks or guarders, but not both simultaneously. Essentially, there are two male mating strategies related to phenotype: guarders (typically larger, older males) attempt to guard females against sneaks (typically smaller, younger males), which attempt to steal matings opportunistically. Such dual strategies are common in many groups (Dunbar 1982; Taborsky 1994). For both models we assume that each male 'knows' which role (EPC male or paired male; sneak or guarder) he occupies when mating.

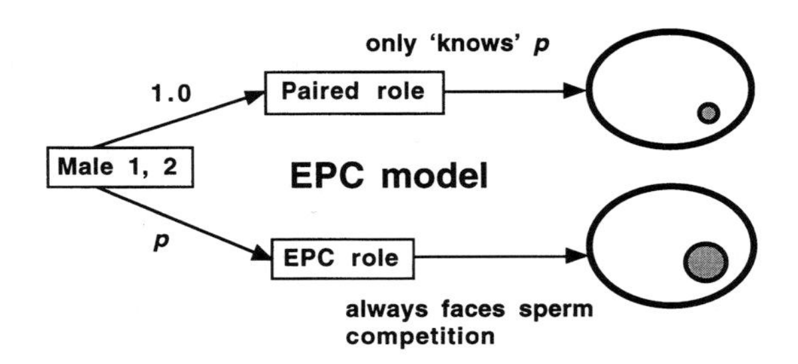

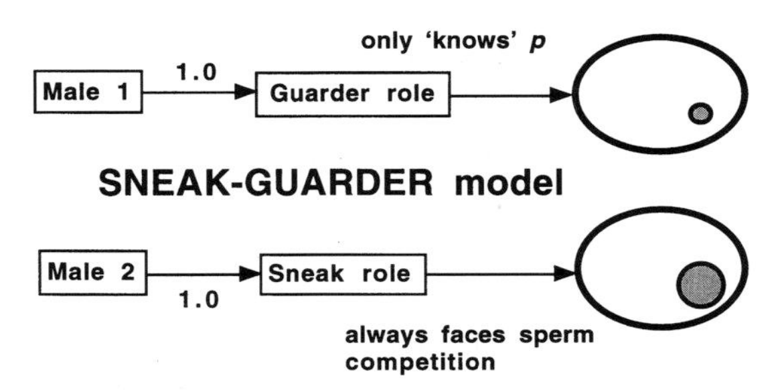

Fig. 1.4. *The model with asymmetric information and asymmetric sperm competition risk. In the EPC version, each male is paired (with probability 1.0) and has the opportunity to perform an EPC with probability* p. *In the paired role (1) he only knows the probability* p *that he will face sperm competition; in the EPC role he has perfect information that he will face sperm competition (probability 1.0). The ESS is to expend less on the ejaculate in the paired role than in the EPC role. In the sneak–guarder version, each male always occupies either a sneak or a guarder role (probability 1.0 of being in either role for a given male). The information rules are the same: guarders only 'know'* p *and sneaks always face sperm competition with a probability of 1.0. The ESS is also the same (qualitatively): males in the guarder role spend less on each ejaculate than sneaks.*

The ESSs for the two cases are qualitatively and quantitatively very similar, despite the different assumptions (Fig. 1.4): the male with perfect information about sperm competition (the EPC or the sneak male) should spend more than the male with imperfect information (the paired male or the guarder). At the ESS, the difference between the expenditures under the two models at a given level of sperm competition risk is small (Parker 1990b). The disparity in expenditures increases as sperm competition risk decreases. If the relative cost of sperm is the same in each role, then higher sperm expenditure in the EPC or sneak role generates the further prediction that EPCs and sneak matings should yield higher paternity under those matings where sperm competition occurs. This conclusion depends on the cost of sperm being equal for both males; for example, if paired males or guarders can produce sperm more cheaply, they may achieve higher paternity. Again, the average expenditure on ejaculates always increases with the average sperm competition risk across species.

Some evidence exists for the predictions in Fig. 1.4. It is very typical for male fish acting as sneaks to have a higher relative testis mass (GSI) than males of the same species playing guarding strategies, and sometimes – despite their smaller size – the absolute testis mass for sneaks is greater (see Chapter 11). Gage *et al.* (1995) showed that the number of sperm in artificially induced ejaculates in male salmon, *Salmo salar*, is relatively greater for parr (sneaks) than for anadromous males (guarders). The paternity success of EPC matings in socially monogamous birds is variable but the EPC success is sometimes much greater than would be expected from the relative number of matings, especially at low risk (Birkhead and Møller 1995). Recent work on zebra finch, *Taeniopygia guttata*, appears to support the prediction of greater sperm numbers when EPCs are most likely (see Chapter 14).

It is sometimes difficult to determine whether differences in sperm allocation arise from the present effects (unequal information and unequal risk) or from those of the previous model (unequal loadings in the sperm raffle, and nonrandom roles). For example, as Stockley and Purvis (1993) point out, an alternative explanation of their results could relate to the sneak-guarder effect, rather than to disfavoured roles in a loaded raffle – although it seems entirely plausible that both effects operate simultaneously. In salmon, because the guarding male may typically gain the best position for spawning, he is probably in the favoured role since each sperm shed more distantly by a sneaking parr will count for less in the raffle. However, sometimes there will be no parr present and, even if there are, the guarding male may have poor information about the risk of sperm competition because of the very small size of parr. Again, both effects could plausibly operate simultaneously.

C. Assessment of sperm competition risk or intensity

All of the sperm competition games outlined show that sperm allocation should increase across species as the sperm competition risk increases. Does this mean that if an individual male estimates the level of sperm competition to be higher than average, he should allocate more sperm? Following intuition, the risk model suggests that if a male assesses a higher than expected sperm competition risk at a mating, he should increase his ejaculate expenditure. Contrary to intuition, the intensity model suggests that if a male can estimate the number of competing ejaculates he should *reduce* his expenditure as the number of competing ejaculates *increases* above two. These predictions appear at first sight contradictory, although they are not, as we shall see after first dealing separately with the two models.

I. Sperm allocation in relation to risks

Parker *et al.* (1997) have examined the risk model (double mating occurs in a species with low probability q) under conditions where males have imperfect information about the risk of double mating. There are three possible assessments for a male to make, corresponding to the possible sperm competition conditions that he may face on encountering a female. He can assess that there is 'no risk' (for any given female encountered there is a probability $(1-q)$ that she will mate only once), or that there is a 'future risk' (with probability q a female encountered will not yet have mated but will mate again in the future), or that there is a 'past risk', i.e. that the female has already mated (which will have occurred with probability q). This sperm competition game asks what a male should do when he assesses that one of these particular contingencies is more likely than the baseline probability of risk = q. The analysis is rather complex (for details see Parker *et al.* 1997). But in summary, the ESS at any given q is for a male to increase his ejaculate expenditure when he can assess that he faces future or past risk of sperm competition. The better his estimate, the more ejaculate he expends. The result conforms to intuitive expectations – the more likely a given mating will face sperm competition, the more the male should ejaculate.

To demonstrate, we look at some special cases, chosen for their biological realism. The first compares perfect information with no information (Fig. 1.5a). If males have no information on past or future risks on meeting a female, the relation between ESS ejaculate allocation and the average risk q for the species follows equation 7 (Fig. 1.5a, dotted line). In contrast, if a male has perfect information and can discriminate between females which will mate only once and those which will mate twice, he should expend an arbitrary minimum amount of sperm on meeting a virgin female that will not remate (the 'no competition case'; Fig. 1.5a, the continuous line on x-axis):

$$E^*_{nr} \to 0, \tag{9a}$$

and should expend an ESS effort of

$$E^*_r = \frac{(1+q)}{4} \tag{9b}$$

(Parker *et al.* 1997) on meeting a female which has already mated or which will mate again in the future (the 'competition case'; Fig. 1.5a, continuously increasing line).

Interestingly, the average ejaculate expenditure in such a population is exactly the same as in a population in which males have no information – it follows the dotted line in Fig. 1.5a (equation 7). So relative size of testis in the two populations might be much the same whether males have perfect information or no information about females. However, the loading in the fertilization raffle does affect the average expenditure, although the 'direction' of the loading (whether the first male or the

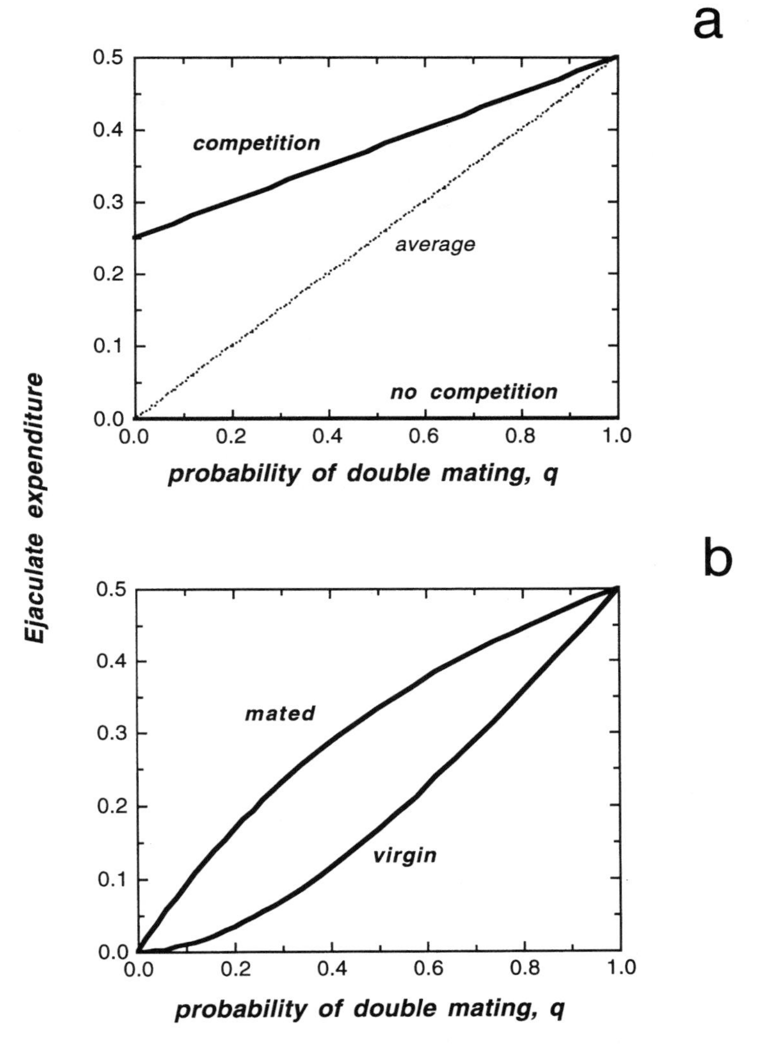

Fig. 1.5. a. *Predicted relationship across species between* E, *the relative expenditure per mating on an ejaculate and the population risk* (q) *of sperm competition when males have perfect information. A male encountering a female that will mate only once should allocate close to zero to the ejaculate, whereas a male encountering a female mate twice (i.e. where his ejaculate will face sperm competition) should allocate a relative expenditure as in equation 9b (upper bold line). The average ejaculate expenditure is given by the faint dotted line; it is equal to the case of no information (equation 7).* **b.** *Predicted relationship across species between the* q *of sperm competition and* E *when males have partial information: they can distinguish between virgin females and mated females, but cannot tell which virgin females will mate again in the future. The ESS allocation to mated females (upper line) exceeds that to virgins (lower line) at all values of* q *except zero and one. (See also Parker* et al. *1997).*

second male is favoured) does not (Parker *et al.* 1997). In both cases, the more unfair the raffle, the less the average expenditure at any given probability of double mating.

These limits (perfect information/zero information) can now be compared with a case of incomplete information. Suppose that a male can differentiate between mated and virgin females, but cannot assess which virgin females will remate. He can assess past risk perfectly, but has no information about future risk. Parker *et al.* (1997) obtained the following results. When he meets a virgin, a male should expend

$$E_{\mathrm{v}}^{*} = \frac{q^2}{(1+q)}, \tag{10a}$$

and when he meets a mated female, he should expend

$$E_{\mathrm{m}}^{*} = \frac{q}{(1+q)}, \tag{10b}$$

see Fig. 1.5b. Thus, for all cases where the risk of sperm competition is intermediate between the two possible extremes ($0 < q < 1$), males spend more on ejaculates with mated females, as intuition would suggest. Of course, if q approaches 1, all females will mate twice, and so if a male meets a virgin, he should expend the same as if he meets a mated female. If q approaches 0, the risk that an initially virgin female will remate after her first mating is infinitesimal, so that when a male meets a virgin, it pays to make almost zero expenditure. Hence if a male meets a mated female, he need ejaculate only a very small amount of sperm to ensure high fertilization success – this is why both effort with mated females and effort with virgins converge to zero as risk, q, approaches zero. It explains the contrast with the effort spent on mated females when there is perfect information (Fig. 1.5a, continuous line). Here, males should expend a large ejaculate effort ($E \to 0.25$) when they mate with mated females in conditions when double matings are extremely rare ($q \to 0$). This occurs because males have perfect information at the time of a female's first mating whether she will remate, so for females which will remate the male must gear his expenditure accordingly. However, if, at the time of a female's first mating, males cannot predict whether she will remate (Fig. 1.5b), he must gear his strategy to the average level of future risk which is infinitesimal as q approaches zero.

An alternative case of partial information is when a male cannot estimate past risk (so he treats mated females and virgin females that will not remate with a second male in the future equally), but can identify virgin females which will mate again in the future. Such a possibility seems at first less plausible, but could be relevant to circumstances where a male has 'privileged information' about a likely mating by a second male (if both males know about each other, the situation becomes one of perfect information). The male has two strategies, as before, but now these are s_{f}^{*} (sperm strategy with females that are likely to mate again in the future) and s_{nf}^{*} (sperm strategy with females that are not likely to mate again in the future). For a fair raffle, we find that this future risk assessment model gives parallel results to the past risks assessment model: E^*_{nf} is equal to E^*_v in equation 10a, and E^*_f is equal to E^*_m in equation 10b. It makes no difference which of the two risk categories of females cannot be recognized – the outcome is the same. If the raffle is loaded, we get different outcomes for the two cases: then E^*_{nf} becomes equal to E^*_v (and E^*_f equal to E^*_m) with the direction of the competitive loading reversed. To explain this, suppose that in the past risks model each of the second male's sperm are loaded by competitive weight, *r*, then the two equations for the future risk model equal the corresponding two for past risk if we replace *r* with $1/r$ (see Parker *et al.* 1997).

Some evidence that males can perform assessments of risk is available in a few internally fertilizing species. For example, Baker and Bellis (1989) found, from analysis of condom samples, that the number of sperm ejaculated by men depends on the proportion of time that they had been separated from their partner in the interval between copulations. (The data were corrected to exclude the effect that sperm numbers increase with time since last mating.) Sperm number increased with proportion of time apart, which the authors viewed as an adaptive response in the 'past risk' sense. Some experiments have shown that males transfer more sperm in matings with a second male present than in the absence of a second male (rats, Bellis *et al.* 1990; the beetle *Tenebrio molitor*, Gage and Baker 1991; medflies, *Ceratitis capitata*, Gage 1991; a bushcricket, *Requena verticalis*, Simmons *et al.* 1993). This can be interpreted as either assessment of past or future risk, or a combination of both. No such effect could be detected in the zebra finch (Birkhead and Fletcher 1995). In the Indian meal moth, *Plodia*, there is no effect of the presence of a second male, but not only can males detect whether females have been mated previously, remarkably, they appear to assess how much rival sperm females contain – the number of eupyrene (but not apyrene) sperm transferred by a second male increases with the ejaculate size of the first male to mate (Cook and Gage 1995). In contrast, male dungflies cannot differentiate between mated and virgin females (Parker *et al.* 1993), even though it would give them advantage to do so. Wedell (1992) found that bigger spermatophores containing more sperm are transferred by males mating with virgin females in the orthopteran *Decticus verrucivorus*. The situation here is further complicated by the fact that a large part of the spermatophore constitutes a nuptial gift. If future sperm competition is likely, greater ejaculate investment may be favourable because of the higher reproductive value of virgins (Wedell 1992).

2. Sperm allocation in relation to intensity

We obtain very different results if sperm competition is intense so that there are often several competing ejaculates. Parker *et al.* (1996)

analysed models in which there are typically several males competing for the same set of eggs, and hence have the possibility to monitor the intensity of sperm competition at a given mating. Although their models were framed in terms of group-spawning fish, the conclusions would apply to an internal fertilizing species in which males assembled around a female and each one mated with her.

Fertilization occurs when each of N competing males ejaculates simultaneously at the moment of spawning, and all eggs are fertilized instantaneously. Parker *et al.* (1996) examined three cases. In the first, males cannot estimate the number of competitors and sperm allocation is shaped only by the mean level of sperm competition. This gives the result in equation 8, where N is the mean number of males present at spawning; the ESS ejaculate expenditure increases with sperm competition intensity across species after Fig. 1.2. In the second case, males have imperfect information and can assess only whether the number of competitors is greater or less than average number, N. Suppose that on average, N_H males are present in high competition, and N_L males in low competition. Half the spawns occur in each state, and s_H^* and s_L^* are the ESS sperm allocations for the two states. The ratio of sperm in the two states is

$$\frac{s_H^*}{s_L^*} = \left(\frac{N_H - 1}{N_L - 1}\right)\left(\frac{N_L}{N_H}\right)^2. \tag{11}$$

If N_H and N_L are quite large, this ratio approximates to N_L/N_H: counterintuitively, more sperm should be released when there are fewer than average competitors, and less sperm when there are more than average competitors. Although it can apply for a wider range, this conclusion was found always to hold if the difference between each condition (N_L, N_H) and the mean (N) is less than the mean, and if the mean number of competitors exceeds two.

In the third model of Parker *et al.* (1996), males have perfect information about the number of competitors present at each spawning: they assess local sperm competition intensity. To make progress, some assumption must be made about the probability distribution of males at spawnings. Parker *et al.* (1996) assumed that this probability is not under male control, and that the distribution of males follows a Poisson process. They obtained the following ESS solution for the relative proportion of effort to be spent on the ejaculate when there are N_i males present at a spawning of type i:

$$E_i^* = \left(\frac{(N_i - 1}{N_i^2}\right)N, \tag{12a}$$

where N is the average number of males present across all spawnings in a given species. A male assesses N_i (the actual number of males present at a particular spawning, i, and expends E_i^*. Figure 1.6 shows the ESS ejaculate expenditure in equation 12a plotted against N_i, for three

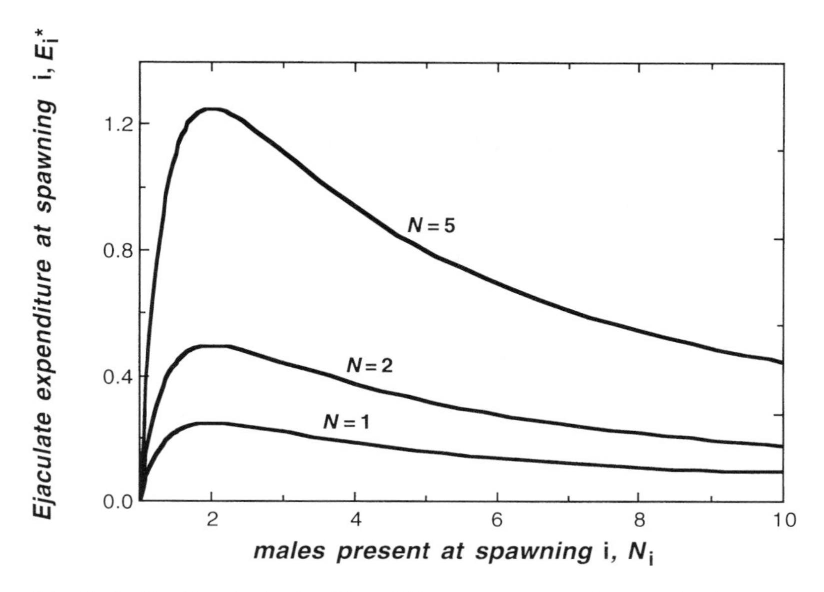

Fig. 1.6. *Predicted relationship within a group-spawning species between expenditure on the ejaculate (relative to the average total reproductive effort per spawning), E, and the number, N_i, of males competing at spawning of type i (from equation 12). The curves are for three different populations or species with different mean number of males present at a spawning, N = 1, 2 and 5. (See also Parker* et al. *1996.)*

species differing only in average number of males present, N. The result is similar to the previous model; counterintuitively, males should decrease their expenditure as the number of competing males increases, for all $N_i > 2$. Expenditure has a maximum at $N_i = 2$, but if $N_i = 1$, there is no sperm competition, and so an arbitrary minimum ejaculate should be released, as in the risk model with perfect information. Optimal ejaculate characteristics in the absence of sperm competition were analysed by Shapiro and Giraldeau (1996) and Ball and Parker (1996, 1997).

Following earlier models, ejaculate expenditure increases with the average intensity of sperm competition, N (see Fig. 1.6). The average ejaculate expenditure in this model increases to 1.0 as the mean number of males, N, becomes very large, conforming to equation 8. In fact, Parker *et al.* (1996) found that, provided the distribution of competitors at spawnings is the same, the average ejaculate expenditure in a species where sperm allocation is adjusted perfectly to the number of competitors is identical to one where males cannot count and hence allocate sperm equally at all spawnings. (This parallels the conclusion in the risk models that average expenditure with perfect information equals average expenditure with no information.)

Why does this difference between average expenditure and local expenditure arise? It always pays males to expend more ejaculate effort as the general intensity of sperm competition increases. But if, in a given spawning, there are many competitors, it pays to spend less and to conserve resources for more profitable (less competitive) spawnings. Put in terms of a human analogy, the value of buying extra tickets in a raffle for a fixed prize becomes greater the fewer the competitors.

The conclusion that males should reduce ejaculate expenditure as local competition increases above $N_i = 2$ appears at first to be in sharp contrast to the risk model predictions, all of which reveal increasing expenditure with increasing risk of sperm competition. However, the risk models deal with the situation where there is either just one mating (no sperm competition, $N_i = 1$) or two matings (i.e. $N_i = 2$). If there is sometimes no competition, and sometimes competition from one other competitor, in terms of the raffle analogy it pays to buy a single ticket (the arbitrary minimum) if there is no competitor and many tickets if there is just one competitor.

Very little evidence exists for the above predictions, but Thomaz *et al.* (1997) have reported that the relative success of each salmon parr (in spawnings where several parr and one anadromous male compete) declines with the number of parr, although their summed success increases slightly. It is clear that parr do relatively worse as their numbers increase, in a way that cannot be explained simply by their increase in number. With just one parr, the total parr success is about 0.25; with 12 parr it is only about 0.40, so either the average number of sperm ejaculated by each parr reduces as their number increases, or the competitive weight of each parr sperm reduces. If parr are more constrained in their costs of sperm production than anadromous males (they have higher expenditure on the testis; see Gage *et al.* 1995), then the first interpretation would be compatible with the predictions in Fig. 1.6.

Since sperm expenditure in relation to sperm competition intensity within a species (equation 11; Fig. 1.2) shows the opposite relationship to the mean response across a species (equation 7: Fig. 1.6), great care must be taken when extrapolating between within-species models and across-species models in behavioural ecology (Parker *et al.* 1996).

V. SPERM ALLOCATION UNDER OTHER MECHANISMS OF SPERM COMPETITION

The previous models have been based on a form of raffle. I shall outline approaches to optimal sperm allocation under other mechanisms of sperm competition in this section: models have not yet been developed to a comparable stage for other mechanisms and so this serves merely to indicate how the logic of sperm competition games can be modified to fit other mechanisms. As we shall see, the loaded raffle can serve as an approximation for several mechanisms.

A. Passive sperm loss

The term 'passive sperm loss' was proposed by Lessells and Birkhead (1990) to describe a sperm competition mechanism in birds in which sperm are lost passively at a random rate from the sperm stores. A similar mechanism (random sperm death) had been used previously by Parker (1984a) in models of multiple vs. single ejaculation (see Section VII) and subsequently in models of sperm size and number (Parker 1993). The passive sperm-loss model accounts well for the paternity pattern obtained both in zebra finch and domestic fowl, *Gallus gallus*, sperm competition (Colegrave *et al.* 1995; Birkhead *et al.* 1995).

If sperm die or are otherwise passively lost after being transferred to the female, the relative numbers of sperm present in the fertilization set (the set of sperm used at the moment of fertilization; Parker *et al.* 1990) from competing ejaculates will depend on the timing of copulations and fertilization events. Imagine two passive loss processes: (i) during movement of sperm from the site of insemination to the fertilization set (which we shall assume to be the sperm store), involving random loss at rate k_a; and (ii) during storage in the sperm store, at rate k_b. Suppose that there is competition between the ejaculates of two males, mating at a time delay, d, apart. Let the average time to reach the sperm store = t_a, and let fertilization occur at time T after the first copulation ($T > t_a$). Call $t_b = T - t_a$, the time spent by the sperm of male 1 in storage before fertilization. If s_1 sperm are ejaculated by the first male (1) and s_2 by the second, the numbers of sperm present in the store at time T from each will be

$$s_1(T) = s_1 e^{-k_a t_a - k_b t_b}, = 0 \text{ if } T < t_a, \text{ from male 1;}$$

$$s_2(T) = s_2 e^{-k_a t_a - k_b (t_b - d)}, = 0 \text{ if } T < t_a + d, \text{ from male 2;}$$

and hence the relative gains for the two males, assuming that both have sperm present in the sperm store at the time of fertilization ($T > t_a + d$), will be

$$\frac{s_1}{s_1 + s_2 e^{k_b d}} \text{ to male 1,}$$

$$\frac{s_2 e^{k_b d}}{s_1 + s_2 e^{k_b d}} \text{ to male 2.}$$

The differential effect of survival is dependent only on the delay, d, between the two matings. For each surviving sperm from male 1, there will be $e^{k_b d}$ surviving sperm from male 2 at the time of fertilization, assuming both have ejaculated equal sperm numbers. The passive loss mechanism is simply a special case of the loaded raffle, with the loading factor $r = e^{k_b d}$ (compare the equations above with those in Section IV.A), so that role 2 (mating second) is the favoured role ($e^{k_b d} > 1$, if $d > 0$). We can use the loaded raffle model to make predictions about ESS sperm expenditures, simply by setting $r = e^{k_b d}$.

Thus, if two males mate with the same female and 'know' their roles, 1 or 2, and the time delay between matings, *d*, then if their roles are determined randomly, they should ejaculate equal amounts of sperm even though there will be less sperm from male 1 surviving at the time of fertilization (Parker 1990a); paternity prospects should be in the ratio $1:e^{k_b d}$ in favour of the second male. But if their roles are not random, so that male 1 tends usually to be in role 1 and male 2 usually in role 2, more sperm should be ejaculated by male 1 to compensate for the discounting owing to sperm loss; however, the second male should achieve greater paternity (Parker 1990b).

B. Sperm displacement

In insects, evidence exists in some species for sperm displacement as a mechanism of sperm competition (see Chapter 10). By sperm displacement, it is envisaged that there is some volumetric exchange of previously stored sperm by new sperm. The various possible mechanisms for sperm displacement (or replacement, *sensu* Waage 1979; in which there is active removal of previous sperm and replacement by self's sperm) remain to be fully categorized (but see Parker *et al.* 1990). Many insects have a fixed volume sperm store, and effort expended on an ejaculate by a mating male displaces previously stored sperm either directly (e.g. *Locusta locusta*, where new sperm are introduced into the spermatheca; Parker and Smith 1975) or indirectly (e.g. *Scatophaga stercoraria*, where new sperm are introduced into the bursa; Simmons *et al.* in prep.). In contrast, some insects have elastic sperm stores and sperm are simply added numerically and mix randomly with each successive mating (e.g. *Gryllus bimaculatus*, Simmons 1987), where the outcome is one of a fair raffle. Many systems may be intermediate between these extremes. Sakaluk and Eggert (1996) have investigated such a case in which there is only partial displacement, so that some expansion of the sperm stores occurs.

To obtain a feel for the properties of sperm displacement systems, I summarize below some aspects of sperm allocation under displacement at two extremes of mixing.

1. Constant displacement with zero mixing

Suppose that each new sperm introduced displaces *b* previously stored sperm (*b* is expected to be less than 1 unless displacement is direct) and that there is zero mixing of new sperm with old sperm during displacement. It is convenient to express displacement as bs/s_{max}, where *s* is the number of new sperm introduced during copulation and s_{max} is the maximum number of sperm in the sperm stores; bs/s_{max} is the total displacement relative to the maximum capacity of the stores. Suppose that the stores are full to capacity at the time of mating by the last male to mate before fertilization occurs. With zero mixing, the expected gains (or P_2, the proportion offspring fathered by the last male; Boorman and Parker 1976) rise linearly at a rate b/s_{max}, since

$$P_2 = \frac{bs}{s_{max}}, \text{ or } 1.0 \text{ if } bs \geq s_{max},$$

(see Fig. 1.7). If it pays to mate at all, the ESS is to fill stores completely, so that $P_2 = 1.0$; i.e. to displace all of the previous sperm by transferring

$$s^* = \frac{s_{max}}{b}, \tag{12b}$$

and for 'one to one' displacement, $b = 1$, and $s^* = s_{max}$. Note that the ESS in equation 12b is not dependent on the risk of sperm competition, p. This special case is, admittedly, an unrealistic one (to my knowledge, displacement with zero mixing has not been reported any species); it is featured to demonstrate that increased risk of sperm competition need not lead to increased ejaculate expenditure if gains are linear. In this hypothetical example, we would expect that males would always ejaculate an amount of sperm that just filled the stores.

For comparison, note that in the raffle models, gains are nonlinear: the gains of male 2 transferring s_2 sperm to a female that already contains s_{max} sperm are

$$P_2 = \frac{s_2}{s_m + s_2}.$$

As s_2 increases, P_2 increases with decreasing gradient, attaining an asymptote of 1.0 only when s_2 is huge relative to s_m (see the broken curve in Fig. 1.7). A further difference is that under the raffle, there is no upper limit on the maximum amount of sperm transfer. An upper limit is present in the linear gains model owing to the discontinuity in gradient when $P_2 = 1.0$.

2. Constant displacement with instant random mixing

Suppose now that each new sperm introduced again displaces c previously stored sperm and that there is instant mixing of new sperm with old sperm during displacement. P_2 gains are now nonlinear, as in the raffle model, and the ESS is for partial (not maximal) displacement. Further, the ejaculate effort will depend on p, as in the raffle. Parker and Simmons (1991) have shown that the last male to mate transferring s_2 sperm will achieve exponentially diminishing returns so that he gains

$$P_2 = \frac{1 - \exp\left(-\frac{bs_2}{s_{max}}\right)}{1 - \exp\left(-\frac{b(s_1+s_2)}{s_{max}}\right)}, \tag{13a}$$

where s_1 is the sperm from the first male(s) to mate (see also Chapter

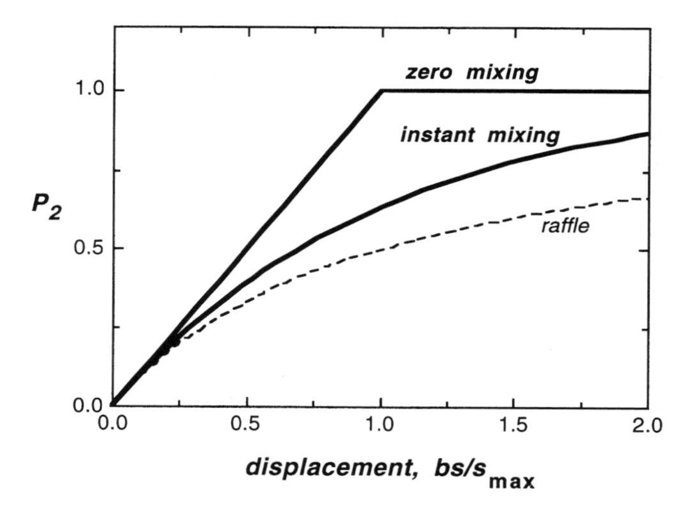

Fig. 1.7. *Expected probability of paternity of the last male to mate (P_2), in relation to total displacement expressed as a proportion of the total volume of the sperm stores, bs/s_{max}. In the case of zero mixing, only previously stored sperm is displaced and gains rise linearly up to the maximum P_2 of 1.0. For random mixing during displacement, P_2 rises with exponentially diminishing returns towards its asymptotic value of 1.0. In the case of a raffle in which the second of two males plays against s_{max} sperm from male 1, P_2 follows the faint broken lowest curve and also reaches an asymptotic value of 1.0.*

10). Note that if the total sperm transferred, $b(s_1 + s_2)$, becomes large relative to s_{max}, the maximum that can be stored, the sperm stores are 'full' and the denominator in the above equation approaches 1.0, giving

$$P_2 \approx -1 - \exp\left(-\frac{bs_2}{s_{max}}\right), \tag{13b}$$

i.e. the P_2 gains become independent of the previous sperm transfer, s_1 (see Parker *et al.* 1990).

A much-analysed example of this system is the yellow dung fly, *Scatophaga stercoraria*. Males arrive quickly at cattle droppings and outnumber gravid females with which they copulate and then guard during oviposition until separation. The present model of displacement with immediate random mixing of sperm appears to fit the P_2 pattern: if one male is allowed to copulate normally, and a second male's copulation terminated at a given time to manipulate the amount of sperm he transfers (sperm is transferred at a constant rate during copulation; Simmons *et al.* in prep.), P_2 gains to the last male show exponentially diminishing returns, fitting equation 13b. We can therefore describe the paternity gains (and sperm transferred) simply in terms of the sperm displacement rate, $c = bs/s_{max}$, and the copula duration, t, since sperm are transferred at constant rate. Although P_2 can be made experimentally to depend on the number of sperm transferred by a previous male (by early termination of the first copulation, Simmons *et al.* 1996), with one full first copulation over 80% of asymptotic sperm density has been achieved, so that the best strategy for the last male becomes virtually independent of the behaviour of previous males. Further, in nature, virtually all (98%; Parker *et al.* 1993) of females arriving at a dropping already contain sperm from previous matings. A variant of the approximate equation 13b has therefore been used to calculate the optimal male strategy in dungflies (e.g. Parker and Stuart 1976; Parker *et al.* 1993; Parker and Simmons 1994). The analyses have used the marginal value theorem (MVT; Charnov 1976; but see also Parker and Stuart 1976), which expresses all expenditures in terms of their time costs (see Fig. 1.8a). This differs from the sperm competition game approach (Section II) in that with MVT, expenditures are expressed in terms of time rather than energy: the product of number of matings and the value of each mating is maximized for a long period of mate-searching. Thus, the animal has a fixed total time rather than fixed total energy budget. There are two time expenditures, which are directly analogous to the two energetic expenditures in Section II. For dungflies these are μ, time spent mate-searching and guarding the female (analogous to C), and t, time spent copulating (analogous to Dx). With MVT, the optimal copula duration is given by the tangent method (Fig. 1.8a). It gives a reasonably close fit to observed copula duration, depending on the refinements that are added (e.g. Parker 1992; Parker *et al.* 1993; Charnov and Parker 1995).

In its simplest form, the optimal copula duration, t^*, is solved numerically by iteration of

$$e^{ct^*} = 1 + c(\mu + t^*), \tag{14a}$$

(Parker and Stuart 1976). However, a good approximation which can be solved explicitly is

$$t^* \approx -\frac{1}{c}\left(1 + \frac{1}{c(\mu + 1)}\right) \ln c(\mu + 1) \tag{14b}$$

(Stephens and Dunbar 1993).

A second difference between MVT and sperm competition game approaches is that MVT is not frequency-dependent, since the best strategy for the last male to mate does not depend (at least as an approximation) on the strategy of other males (see equation 13b). Although the effect of gains in future batches on the optimal copula duration was considered in the earliest analysis (Parker 1970b), its effect is trivial, especially since the probability of survival of females between batches of eggs may not be high.

Copula duration in dungflies appears to be geared to a male's size (Parker and Simmons 1994), which varies greatly owing to variations in larval competition. Smaller males copulate for longer periods than larger males. Large males have a different form for fertilization gains with time spent copulating from that of small males, because (unsurprisingly) they

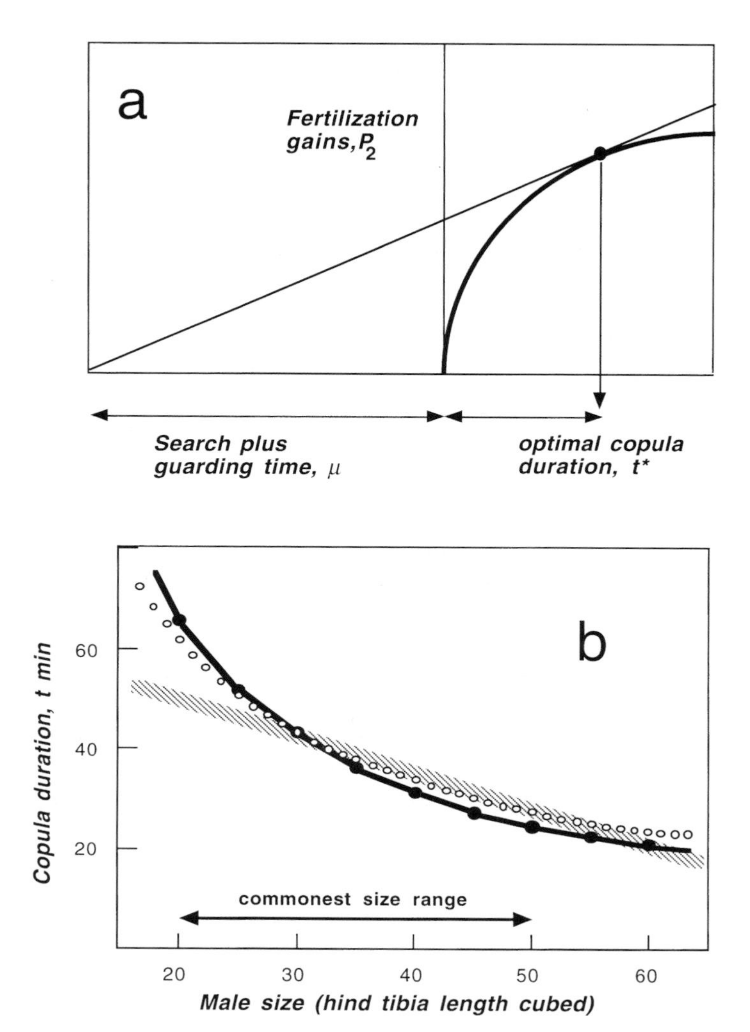

Fig. 1.8 a. *Marginal value solution for optimal copula duration in dungflies. The optimal copula duration, t*, is given as the value of t at which a tangent drawn from a distance, μ, to the curve relating fertilization gains (P_2) with copulation time, t.* **b.** *Copula duration in relation to male size (measured as hind tibia length cubed). The cross-hatched line is the observed relationship, and the curve with open circles is the prediction from the exact calculation (Parker and Simmons 1994). The curve with filled circles is the approximate prediction based on constant, ct, for all males. Modified from Charnov and Parker (1995).*

displace sperm at a faster rate. Larger males also have a much greater chance of taking over a female from another male during oviposition than a smaller male, and this reduces their time, μ. This size-related effect on μ could be quantified from a knowledge of the distributions of ovipositing pairs and searching males at droppings. Both the take-over effect and the displacement effect tend to reduce optimal copula duration for larger males, and the two together give a good fit across the commonest size range (Fig. 1.8b), although very small males are predicted to copulate longer than is observed (Parker and Simmons 1994). There is thus evidence that ejaculate expenditure is fine-tuned to a male's phenotype in a rather remarkable way. Dungfly copula duration has been examined in some detail because it is arguably the best evidence available to date that ejaculate expenditure is shaped by evolution in accordance with sperm competition theory.

The work on size-dependent copula duration in dungflies has led to a very general insight affecting optimal foraging in patches under MVT (Charnov and Parker 1995): if the gain rate (equivalent to displacement rate, c) is directly proportional to forager quality (equivalent to male size), and the travel time, μ, between patches is inversely proportional to forager quality, then for systems where gains follow exponentially diminishing returns, the product ct^* should be constant across all forager qualities (male sizes). Using just this approximation, we get virtually the same prediction for dungflies (Fig. 1.8b, filled circles) as the exact calculation (Fig. 1.8b, open circles). Further, the prediction is that all individuals, whatever their forager quality, should achieve approximately the same total gains from a resource: the effect of increasing the gain rate and reducing the travel time exactly counterbalance. This applies for dungflies; there was no detectable difference in P_2 across the male size range for males allowed to complete mating (Simmons and Parker 1992; Charnov and Parker 1995). It should also apply for many analogous foraging situations in which 'good' phenotypes have quicker rates of uptake from a resource, and shorter travel times between resources.

VI. SPERM SIZE

Not all sperm are minute. For example, certain species of *Drosophila* produce giant sperm which, although of fairly typical width, can be well over 20 times the length of the male that produces them (Pitnick *et al.* 1995a,b). However, such cases are rather uncommon and tiny sperm are the general rule. Nevertheless, sperm do vary considerably in size (mainly length) between species. There has recently been interest in sperm size and sperm competition, and there is some evidence that sperm size may either increase (mammals, Gomendio and Roldan 1991; butterflies, Gage 1994; birds, Briskie *et al.* 1997) or decrease (fishes, Stockley *et al.* 1997) across species with the risk or intensity of sperm competition.

A. *Sperm size and sperm competitive ability*

Parker (1993) investigated various risk model sperm competition games obeying the protocol outlined in Section II, in which a unit of ejaculate, x, is the product of the number of sperm, s, and mass, m of each one (i.e. $x = sm$). The ejaculate expenditure can now be increased by increasing either sperm size or sperm number unilaterally at the expense of expenditure on gaining matings. The original analysis (Parker 1993) is analysed in terms of risk defined from the mating male's perspective (i.e. as p, see Section II), but the conclusions below are framed in terms of q, the probability of double mating.

Parker (1993) asked the question, if we assume that larger sperm outcompete smaller sperm, does it follow that sperm size should change with the risk of sperm competition across species? If the main effect of altering a sperm's size (most plausibly sperm length) is to alter its competitive ability, it appears that the answer to this question is no.

Suppose that there is a raffle with no information, following Section III. The E^*, is exactly as in equations 6 and 7. Let m be the mass (i.e. size) of an individual sperm; at the ESS this is m^*. We solve for the ESS sperm number, s^*, by considering a mutant deviating by having $s \neq s^*$, and allowing x to be sm^* in equations 1–4 and 5a in Section II, and evaluating all differentiations at $s = s^*$. This gives the result that ESS ejaculate expenditure,

$$E^* = \left(\frac{Ds^*m^*}{C + D\langle s^*m^*\rangle}\right) = \frac{q}{2},$$

exactly as in equation 7. A mutant male producing sperm that has size m has a competitive ability that gives it a relative loading of $r(m)$ in the fertilization raffle. When sperm competition occurs, the mutant will play against an ESS player with sperm of size m^*. The mutant's gains in the fertilization raffle are thus:

$$\left(\frac{r(m)s^*}{r(m)s^* + r(m^*)s^*}\right).$$

We solve for the ESS sperm size m^* by allowing x to be s^*m^* in equations 1, 2, 3, 4 and 5a in Section II, and evaluating all differentiations at $m = m^*$. This gives

$$E^* = \left(\frac{Ds^*m^*}{C + D\langle s^*m^*\rangle}\right) = \left(\frac{qr'(m^*)m^*}{2r(m^*)}\right),$$

and combining the two versions of E^* above, we get

$$r'(m^*) = \left(\frac{r(m^*)}{m^*}\right), \tag{15}$$

where $r'(m^*)$ is the gradient of the relation between relative competitive ability, r, and size m. (cf. corresponding analysis in Parker 1993; where

15 is correctly reported). Thus, provided that there is some finite risk, equation 15 shows that the ESS value for sperm size, m^*, is not dependent on the risk of sperm competition. (The same conclusion (equation 15) can be derived from the intensity model).

If sperm size m^* is independent of the risk, q, of sperm competition, then obviously sperm number must increase in order that E^* rises linearly as $q/2$. Sperm numbers increase with risk, but sperm size remains constant.

Equation 15 is the mathematical version of the familiar MVT solution (see Fig. 1.9).

Suppose that competitive ability, r, is an increasing function of sperm size, m. An intermediate optimum between being maximum or minimum

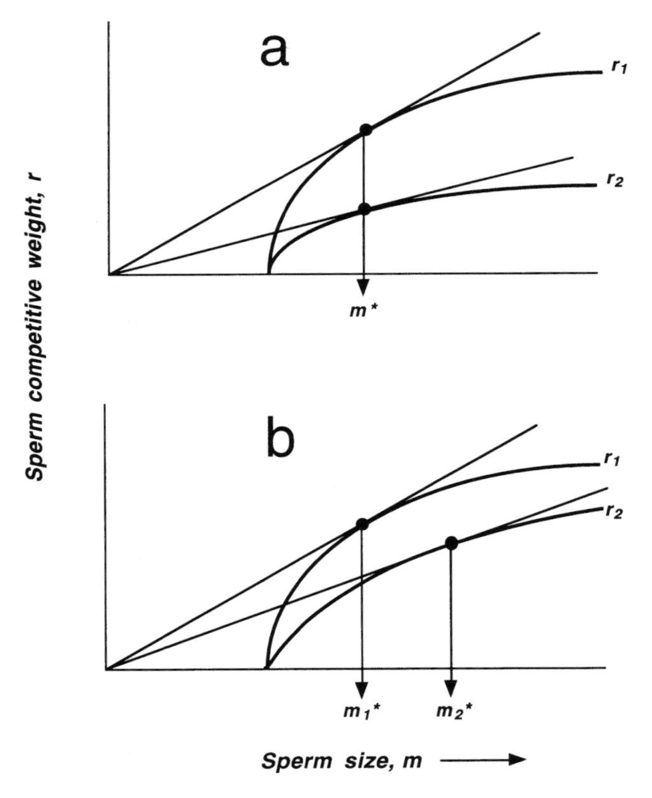

Fig. 1.9. *Marginal value solutions for optimal sperm size,* m^*, *when the competitive weight,* r, *of a sperm is an increasing function of sperm mass,* m. *Each graph shows two different functions for* $r(m)$, r_1 *and* r_2, *which correspond to two different average sperm density conditions (for example,* r_1 *may be a population with a higher level of sperm competition than* r_2*). In a, the asymptotic value of* $r(m)$ *changes with sperm density, since* r_1 *and* r_2 *are related by a constant factor. Curves* r_1 *and* r_2 *both have the same optimal value for* m^*. *In b, the rate of approach to the asymptote of* $r(m)$ *changes with sperm density. Curve* r_1, *which rises more steeply, has a lower optimal value for* m^* *than curve* r_2. *Modified from Parker (1993).*

size can occur only if $r(m)$ has a form of the type shown in Fig. 1.9, such that a tangent to $r(m)$ can be drawn from the origin (see also the MVT diagram of optimal copula duration, Fig. 1.8).

What could generate a dependency between sperm size and the risk or intensity of sperm competition? One possibility is that the relative competitive ability of sperm increases at a different rate with sperm size as the number of competing sperm increases (Parker 1993). Figure 1.9 shows two ways in which relative competitive ability of a sperm may be altered by the total number of sperm present. Suppose that increasing sperm numbers alters relative competitive ability, $r(m)$, by a constant factor, so that in two sperm density conditions, 1 and 2, the ratio r_1/r_2 is constant at all m, and the asymptotic value of $r(m)$ is different in r_1 and r_2 (Fig. 1.9a). There is then no effect of sperm competition on sperm size because the optimal size will be the same in both conditions. However, if increasing sperm numbers changes the rate at which $r(m)$ rises to its asymptote, the optimal size, m^*, will differ in the two populations (Fig. 1.9b). The slower the rise in $r(m)$ to its asymptote, the greater the sperm size expected. It is not immediately obvious why this effect should account for the increase in sperm size with sperm competition in the three internally fertilizing groups (mammals, Gomendio and Roldan 1991; butterflies, Gage 1994; birds, Briskie *et al.* 1997) and yet decrease in externally fertilizing fishes (Stockley *et al.* 1997).

B. Sperm size and sperm survival

Increased size of a sperm may conceivably either increase its survival if size indicates some form of energy reserve, or it may decrease its survival if size increases swimming speed by increasing the tail length plus its associated mitochondria in the mid-piece. The available evidence (mammals, Gomendio and Roldan 1991, 1993a,b; Roldan *et al.* 1992; fishes, Stockley *et al.* 1997) suggests that increasing sperm size (in this case length) decreases sperm survival, probably because of the increased energy demands of the longer tail.

Two rather different models have been constructed to examine the interaction between sperm size and sperm survival, depending on the mode of fertilization (Parker 1993; Ball and Parker 1996). For most internal fertilizers, sperm from different ejaculates arrive in the female tract at different times so that some sperm may die before any competition for fertilization can occur (e.g. passive sperm loss). Hence, differential sperm survival can affect the relative numbers of sperm from each ejaculate surviving to the instant of fertilization (Fig. 1.10a). It is assumed that all (or a fixed proportion) of the eggs are fertilized. In contrast, for external fertilizers in which a female spawns synchronously with N males present, all sperm enter competition in a continuous fertilization process. Differential sperm survival affects the relative numbers of

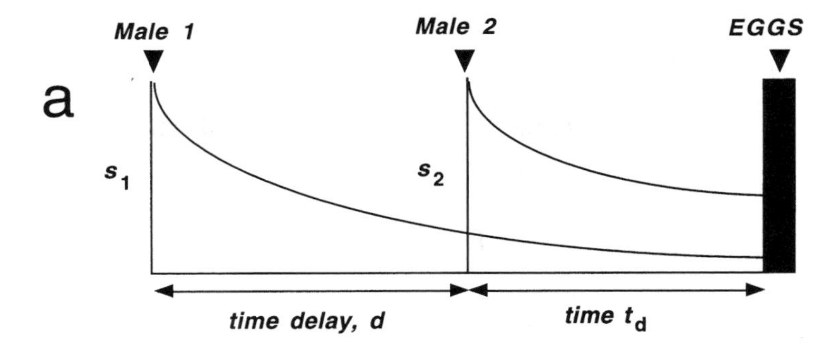

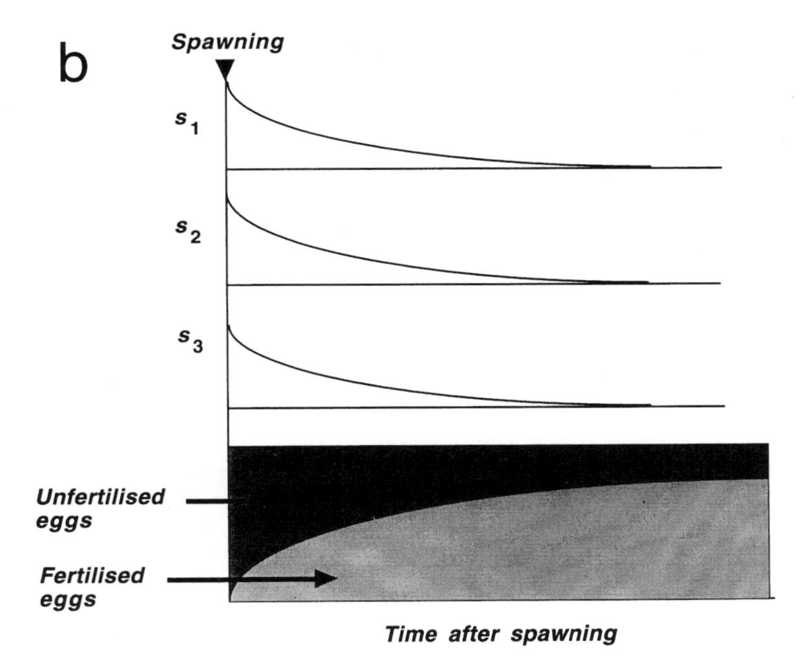

Fig. 1.10. *Comparison of instant and continuous fertilization models.* **a.** *Instantaneous fertilization model for an internally fertilizing species. Male 1 mates first and his sperm number,* s_1*, begins to decline in number in the female tract. After time delay,* d*, male 2 mates and so when fertilization occurs time* t_d *later, he has more sperm (assuming the same number were ejaculated) because of passive sperm loss. The process of fertilization is instantaneous, at timed +* t_d *after the first mating.* **b.** *Continuous fertilization model for an externally fertilizing group-spawner species. The males (here* N = *3) all ejaculate at the moment of release of the eggs. With time after spawning, sperm numbers* (s_1, s_2, s_3) *decline, and the cumulative proportion of eggs fertilized increases, as a continuous process.*

sperm from each ejaculate that are present during fertilization, which is analysed as a continuous process (Fig. 1.10b). The proportion of eggs that are fertilized increases from the start of the fertilization process, but not all the eggs need be fertilized by the time that all the sperm are dead (or lost from the system).

1. Instant fertilization (the 'internal fertilizer' model)

The assumptions of this model (Parker 1993) follow the usual 'fair raffle' protocol of Sections II and III, and roles (first male to mate, second male to mate) are random. Fertilization is instantaneous, but ejaculation occurs some time beforehand so that some sperm are lost (e.g. by passive loss; Section V.A). Sperm size determines both sperm longevity (both positive and negative relations with size were examined), and sperm competitiveness (which is assumed to be positively related to size, as in Fig. 1.9).

The ESS results are intuitively obvious: (i) where size increases sperm survival, then size should increase with the mean time between mating and fertilization, and (ii) where size decreases survival, then size should decrease with the mean time between mating and fertilization. The random roles conclusion applies – if a male is equally likely to occur in either role (first or second), he should expend the same on the ejaculate whatever role he is in. If t = the average time between ejaculation and fertilization, for two males mating with a time delay of d between the two ejaculations, and fertilization occurring a time t_d after the second mating, $t = d/2 + t_d$ (see Fig. 1.10a). With passive sperm loss (see Section V.A.) we would expect the proportion of the ejaculate surviving to time t to be $l(m,t) = e^{-k(m)t}$, where $k(m)$ is the random loss rate in the female tract as a function of sperm size. Then at the ESS, equation 15 becomes

$$m^* = \frac{r(m^*)}{(r'(m^*) - r(m^*)k'(m^*)t)}, \qquad (16)$$

where the primes denote the gradients of $r(m)$ and $k(m)$ (equation 16 is correctly reported by Parker 1993). Again, there is no dependency of sperm size on risk of sperm competition. So, as before, sperm numbers must increase with risk, but sperm size must remain constant unless some special effect is advocated.

One plausible special effect that could cause a relation between sperm size and sperm competition risk concerns the time that a female is receptive and available for mating (e.g. the oestrus period in mammals). All other things being equal, one might predict that the probability, q, of double mating of a given female will increase with the duration of receptivity. If the average time between mating and fertilization also correlates positively with the duration of receptivity, then a relation between risk and sperm size can be generated. The exact relation depends on the forms of $r(m)$, and $k(m)$. With simple explicit forms, Parker (1993) was able to generate numerical solutions which showed an increase in sperm size with increased sperm competition risk if $k'(m)$ is positive (sperm size increases sperm survival), and a decrease in sperm size with increased risk if $k'(m)$ is negative (sperm size decreases sperm survival). However, in both cases, sperm numbers were found to increase with sperm competition risk.

2. Continuous fertilization (the 'external fertilizer' model)

Ball and Parker (1996) constructed a model of continuous fertilization for group-spawning, externally fertilizing species to investigate the effect of sperm survivorship on sperm size.

Assumptions are as before, except that fertilization is continuous, and sperm die during the process of fertilization (Fig. 1.10b). Again, sperm size affects both sperm longevity (positively or negatively), and sperm competitiveness is viewed as equivalent to sperm speed, which increases with sperm length (with diminishing returns in Fig. 1.9).

The model is mathematically more complex than previous models because of the continuous-time dimension. For tractability, Ball and Parker (1996) used an all or nothing (step) function to model sperm longevity in relation to sperm size: they assumed that each sperm has a fixed longevity which depends on its size, m, so that the proportion of sperm that survive to time t after spawning is

$$l\,(m,t) = 1, \text{ for times } t < \tau(m);$$

$$l\,(m,t) = 0, \text{ for times } t > \tau(m).$$

They obtained the following general results for ESS sperm size at the two extremes of sperm competition. The optimum sperm size at zero sperm competition maximizes the total distance travelled by the entire ejaculate in its lifetime, and is

$$m^* = \left[\frac{r(m^*)}{r'(m^*)} + \frac{\tau(m^*)}{\tau'(m^*)}\right]^{-1} \tag{17}$$

cf. equations 15 and 16. This can be interpreted as follows: if there is no sperm competition, sperm size should represent the best compromise between speed and longevity to optimize the number of collisions with the ova during the entire lifetime of the ejaculate. It is not necessary to 'out-compete' other ejaculates in a scramble for fertilization. In contrast, when sperm competition is at a maximum, the ESS sperm size simply maximizes the product of sperm speed and sperm number for the ejaculate, which is given by the marginal value theorem equation (15) for instant fertilization. This solution is independent of sperm longevity: when sperm competition is maximal, the sperm density tends to be vast so that all the eggs are fertilized in the first instant of time. Sperm survival does not matter – the best thing to do is to maximize the product of speed and number to get the most collisions in the first time instant.

Although not immediately obvious, the following conclusions can be deduced by comparing equations 15 and 17:

(1) If longevity *decreases* with sperm size ($\tau'(m)$ is negative), the noncompetitive optimum is less than that for maximum competition, so that sperm size *increases* with sperm competition intensity;
(2) If longevity *increases* with sperm size ($\tau'(m)$ is positive), the noncompetitive optimum is greater than that for maximum competition, so that sperm size *decreases* with sperm competition intensity.

Using exponential diminishing returns to describe the way that speed (i.e. competitive ability), r, increases with sperm size, m, and simple linear forms to describe how longevity, τ, decreases or increases with size, Ball and Parker (1996) obtained the relationship between sperm size and sperm competition intensity shown in Fig. 1.11a, confirming the above conclusions. They also found the following:

(1) As in all other models, ESS ejaculate expenditure always increases with sperm competition intensity across species;
(2) At the ESS, not all eggs are fertilized by the time that all sperm are dead: from the male point of view this represents 'adaptive infertility'. The ESS degree of infertility decreases with sperm competition intensity;
(3) ESS sperm numbers typically increase with sperm competition intensity, and always so if sperm competition is high enough, although decreases are possible over a range of low sperm competition intensity if sperm longevity decreases with sperm size and infertility is high enough.

A more detailed approach to sperm size with continuous fertilization (Ball in prep.) divides a sperm's mass into two strategically variable components (i) tail length, which increases speed, r, and reduces survivorship, τ, and (ii) energy reserves, which increase τ. This gives rather different results from the above model in which a sperm's mass is treated as a single strategic variable. Although ESS tail length increases with sperm competition intensity, the energy reserves component decreases; however, the total mass of a sperm decreases with sperm competition intensity, rather than increases, as in Ball and Parker's (1996) model in which energy reserves were undefined. Ball (1998) gets a similar result to Ball and Parker if, in his new model, the energy reserves are fixed as a constant.

D. *Sperm size strategies within species*

What of sperm size in relation to intraspecific behaviour? While good evidence exists that a given male can alter sperm numbers in response to

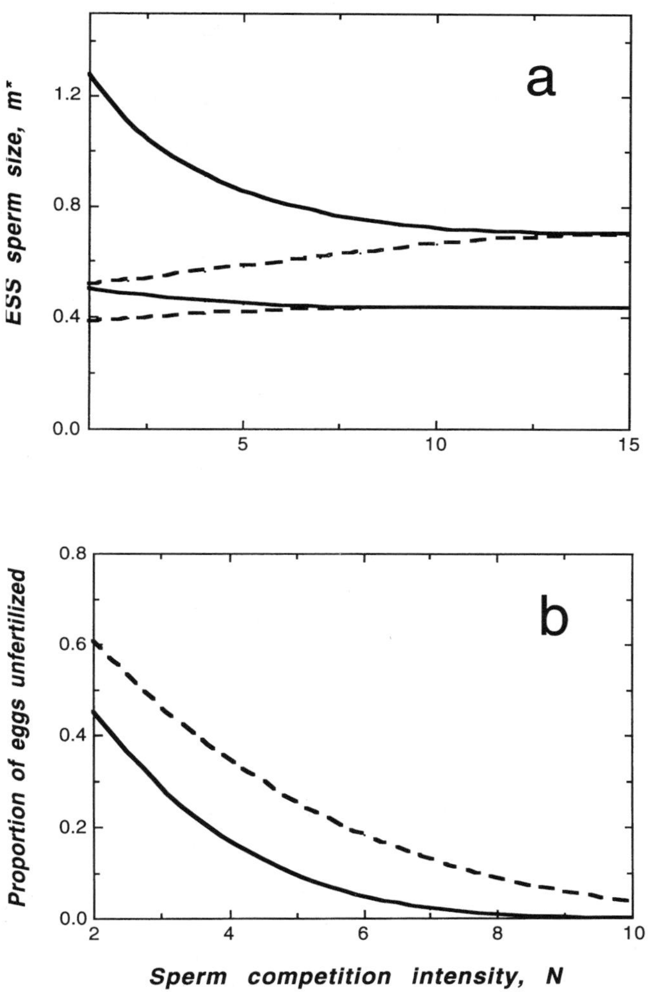

Fig. 1.11. *Some results of numerical calculations for the continuous fertilization model. For description of the model and the parameter values used, see Ball and Parker (1996).* **a.** *Relation between ESS sperm size,* m*, *and sperm competition intensity. In the upper pair of curves, the competitive weight function* r(m) *rises less quickly to its asymptotic value than in the lower two curves. In each pair of curves, the upper (unbroken) curve is for the case where sperm longevity, τ, decreases with size, and the lower (broken) curve is for the case where sperm longevity increases with size. When the number of competing males,* N, *is small, the ESS sperm size differs in each pair of curves because of the different effects of sperm size on longevity, τ, as predicted by equation 17, but as* N *gets large the sperm lifetime, τ, has no effect on* m*, *as predicted by equation 15.* **b.** *Relation between the proportion of eggs remaining unfertilized and sperm competition intensity in the continuous fertilization model. In the upper (broken) curve, the competitive weight function* r(m) *rises less quickly to its asymptotic value than in the lower (unbroken) curve. Modified from Ball and Parker (1996).*

local mating conditions, facultative changes in the form of individual sperm appear unlikely unless future conditions can be anticipated. If a given male meets unpredicted local conditions, time would be needed to produce the relevant sperm type for each condition, unless selective ejaculation of given sperm types can be achieved. Sperm characteristics are more likely to represent the weighted average of the distribution of local conditions. Ball and Parker (1997) used this assumption to extend the instant fertilization group-spawner model of Parker *et al.* (1996) to the case of continuous fertilization. Because sperm size is assumed to be fixed at an average optimum, the aim of the model is to predict how males should change sperm numbers in relation to local spawning competition (see Section IV.C.2, equation 12 and Fig. 1.6). Ball and Parker (1997) found that under certain circumstances, ESS sperm numbers could decline with the number of competing males, N_i, from $N_i = 1$, rather than from $N_i = 2$ with instantaneous fertilization (see Fig. 1.6). Such situations occur only when the level of infertility is high. Sperm size, m^*, was found to increase with the average of N_i (as expected from the across-species predictions of Ball and Parker 1996). The sum of the sperm ejaculated by all males at a spawning increases with the number of competitors, N_i, and so the fertility of the spawning increases with N_i. This suggests that selection may act on females to promote male–male competition.

While it may be difficult for a given male to alter sperm size in relation to unpredictable mating conditions, when given males in a population employ different mating strategies, it would appear possible for them to have different sperm size optima, 'tuned' to the conditions they face. However, the relevant theoretical analyses have not yet been undertaken and what little evidence exists does not yet support differences between alternative mating strategies and sperm size (Gage *et al.* 1995).

E. The biological data

Biological evidence in support of the above predictions about sperm size and sperm competition is equivocal. Sperm size has mainly been measured as length, and comparisons across species have shown sperm longevity to decline with length, probably because of the increased energy demands of the longer tail (see above). Birds (Briskie *et al.* 1997), mammals (Gomendio and Roldan 1991, 1993a,b; Roldan *et al.* 1992) and butterflies (Gage 1994) show increasing sperm size with increased sperm competition. Depending on the model, with internal fertilization, we expect sperm size to increase with risk (i) if risk increases with the delay between matings, or (ii) if sperm competitiveness rises less steeply to its asymptotic value as sperm density increases. To date, there is no real basis for claiming either of these effects. Other predictions, such as increases with sperm competition across species in sperm number and ejaculate expenditure (as measured by GSI) are found as expected (present volume).

The evidence from comparative studies on fish shows that both GSI and sperm numbers increase with sperm competition intensity (Stockley *et al.* 1997), as predicted by the continuous external fertilization model. Across fish species, sperm longevity reduces with sperm length (Stockley *et al.* 1997), so we would predict from the model of Ball and Parker (1996) that sperm length would increase. In fact, contrary to the prediction, it decreases with sperm competition intensity (Stockley *et al.* 1997). The new model of Ball (in prep., see above) predicts a decrease in sperm mass (tail plus energy reserves) with sperm competition intensity. However, it still predicts an increase in tail length: energy reserves decline faster than tail length increases. The observed decrease of sperm tail length with sperm competition intensity across fish species therefore remains rather mysterious.

F. *Other issues: zygote provisioning, giant sperm, and sperm polymorphism*

1. Sperm size and zygote provisioning

Should sperm ever contribute to the provisioning of the zygote? Should a portion of each sperm be allocated to zygote provisioning? There are good theoretical reasons why no part of sperm resources should be related to this function (Parker 1982). The argument relies on an analysis of the condition required to favour increased sperm size as a result of selection to assist with zygote provisioning. Assume that males produce sperm of mass m_{min}, at which size no resources in the sperm are allocated towards provisioning the zygote (it is the size selected by sperm competition pressures, or other fertilization constraints). Females produce gametes of size m_{opt}, the size which maximizes female reproductive success given that there is no contribution of resources to the zygote from the sperm. Let b be the benefit to the zygote (in terms of viability and future reproduction) of receiving m units of provisioning. If the sperm contributes no reserves at all, m_{opt} is found by the marginal value theorem (cf. Figs 1.8 and 1.9) in which the gradient of the benefit function at the optimum is

$$b'(m_{\text{opt}}) = \left(\frac{b(m_{\text{opt}})}{m_{\text{opt}}}\right), \tag{18}$$

see Parker (1982).

We next seek a condition such that the m_{min} strategy will be stable. Consider first the risks model, in which the population probability of sperm competition between two competing ejaculates is q, and for simplicity assume that males have no information (see Section III). Suppose that ejaculate expenditure is held constant at its ESS value, and that there is a size-number trade-off, so that sperm number is inversely related to size. A mutant male that increases sperm size to m by reducing sperm number therefore achieves the same number of matings, but produces a relative number of sperm of m^{-1} instead of $m_{\min}^{-1}$. Thus, a male deviating by increasing sperm size to $m > m_{\min}$ supplies $(m - m_{\min})$ resources towards zygote provisioning and increases the fitness of his zygotes from $b(m_{\text{opt}})$ to $b(m_{\text{opt}} + m - m_{\min})$, but reduces his chances of fertilization under sperm competition by having less sperm. His fitness is the number of zygotes times their viability:

$$W(m, m_{\min}) = \left[(1-q) + 2q\left(\frac{m^{-1}}{m^{-1} + m_{\min}^{-1}}\right)\right] b(m_{\text{opt}} + m - m_{\min})$$

see the similar equation in Parker (1982) which is expressed in terms of p (see Section II). In order that the $m_{\min}$ strategy will be locally stable against any such mutant increasing sperm investment away from $m_{\min}$, we require that the differential coefficient of $W(m,m_{\min})$ with respect to m is negative around $m = m_{\min}$. This gives the condition that

$$b'(m_{\text{opt}}) = \left(\frac{qb(m_{\text{opt}})}{2m_{\min}}\right), \tag{19}$$

and combining equations 18 and 19 gives the result

$$q > 2\frac{m_{\min}}{m_{\text{opt}}}, \tag{20}$$

which differs by a factor of two from the approximate equivalent given in Parker (1982) and elsewhere in terms of p. Thus, provided that the probability of sperm competition is greater than twice the ratio sperm size/ovum size, no contribution should be made by the sperm to the provisioning of the zygote. This is indeed a very easy condition to satisfy, but if it is violated, there is a danger of losing anisogamy itself (Parker 1982).

Performing the equivalent calculation for the intensity model in which N males compete for the set of ova, we get the condition that to prevent zygote provisioning by sperm,

$$\frac{N-1}{N} > \frac{m_{\min}}{m_{\text{opt}}}; \tag{21}$$

(see Parker 1982). If N is very large, then we only need that a sperm is smaller than an ovum to satisfy equation 21; if N is only 2, it is satisfied provided that a sperm does not exceed half the size of an ovum.

At any high level of difference between the sizes of the male and female gametes, both the risk or intensity approaches to sperm competition suggest that sperm should not increase in size to assist with zygote provisioning under the raffle model of fertilization. Indeed, it has been argued that in conditions 20 and 21 ovum investment, m_{opt}, should be replaced by the total energetic investment by the female in each offspring (Parker 1984b), which makes condition 20 several orders of magnitude more robust in mammals.

Note, however, that the above arguments do not preclude the possibility that sperm become larger for competitive reasons, and then after fertilization, some of the competitive investment in sperm size is made use of for provisioning. The point is that this would represent a secondary opportunistic use of a feature evolved in response to other selective pressures, rather in the way that a female mammal may eat a placenta after parturition.

Zero provisioning could be maintained for reasons other than sperm competition. A rather similar analysis generates robust conditions against zygote provisioning by sperm when conception probability increases with sperm numbers (Parker 1982). An entirely unrelated argument also suggests that cytoplasmic reserves in sperm should be reduced to a minimum: Hurst (1990) has suggested that a strong division into two gametic sizes reduces the possibility of transfer of cellular parasites during fertilization. From this view, sperm become smaller to the point where no sperm cytoplasm is included in the zygote so that the diversity of cellular parasites is restricted to those already present in the ovum; this selective force would also act against any zygote provisioning role for sperm.

2. Giant sperm

Some species have giant sperm; although the ratio of ovum/sperm mass is usually still high, the gigantism stems from the extraordinary sperm length (see Chapter 10, for a review of the insects). The function of the remarkable tail remains obscure. Pitnick and Markow (1994) have found that giant sperm are expensive to produce and that the increase in the tail length appears to be a trade-off against sperm numbers. The costs of the tail are considerable both in terms of reduced sperm production and delayed maturity (Pitnick *et al.* 1995a). Theoretical arguments (see above) indicate that zygote provisioning is unlikely to be the initial selective force favouring increased sperm size, although this could be a secondary adaptation after size has increased for other reasons relating to fertilization (e.g. for increased competitiveness). Bressac *et al.* (1995) suggest such a function from the observation that the sperm tail enters the ovum. However, Pitnick *et al.* (1995b) demonstrate that in many species only a small fraction of the tail enters the egg, and further, Karr and Pitnick (1996) show from phylogenetic analysis that the increase in tail length has evolved independently of the component that enters the egg, arguing against zygote provisioning.

Possibly the remarkable tail acts as a coiled purchase against the internal wall of the spermatheca or its duct, or even around the ovum, to propel more effectively the male pronucleus forward towards the egg (outcompeting shorter rivals), or to propel it within the egg cytoplasm towards the female pronucleus (outcompeting shorter rivals in conditions of polyspermy, rather similarly to the growth of the pollen tube down the stigma in plants).

3. Sperm polymorphism

Diversity of sperm within an ejaculate is relatively common in some animal groups, and pollen polymorphism is also not uncommon. For example, many insects show polymorphic sperm, occasionally reaching up to five different morphs, and Lepidoptera almost universally show dimorphic sperm with smaller, nonfertilizing 'apyrene' sperm in numerical predominance with their fertilizing 'eupyrene' counterparts (see Silberglied *et al.* 1984). A detailed review of sperm polymorphism and the theories proposed for its evolution in insects is given elsewhere in Chapter 10).

VII. EVOLUTION OF COPULATORY PATTERNS: MULTIPLE VERSUS SINGLE EJACULATION

In some species, copulatory patterns are complex involving several ejaculations by the same male for one set of eggs (e.g. some mammals, Dewsbury 1972; most birds, Birkhead and Møller 1992; a few insects, e.g. Smith 1979). Why is sperm sometimes delivered repetitively, rather than – as is more typical – in one large single dose? Below, we consider, mainly for mammals, two different analyses of the possible function of multiple ejaculation. One concerns sperm competition and the other relates to 'topping up', i.e. maintaining sperm numbers at a level high enough to ensure fertilization throughout a period when conception is unpredictable.

A. Sperm competition

Multiple ejaculation has been seen as a way of increasing a male's sperm dose in a female in a competitive mating situation (Lanier *et al.* 1979). This hypothesis may be most plausible if there are constraints on facultative adjustments in ejaculate volume which make it difficult to alter sperm dose at a single mating. The behavioural method of multiple mating may then be an easier and less costly option than varying ejaculate volume (Parker 1984a).

Parker (1984a) considered the effects of sperm competition in a game between the two strategies (single (S) vs. multiple (M) ejaculation). Specifically, S plays a single mating at start of oestrus, and M plays n matings spaced at equal intervals through oestrus, each delivering $1/n$ the dose of sperm of S (see Fig. 1.12a). For simplicity, conception was assumed to be unpredictable through oestrus: a sperm alive at any given time during this period has an equal chance of entering a fair raffle for fertilization. Sperm were assumed to show passive loss in the female tract from the

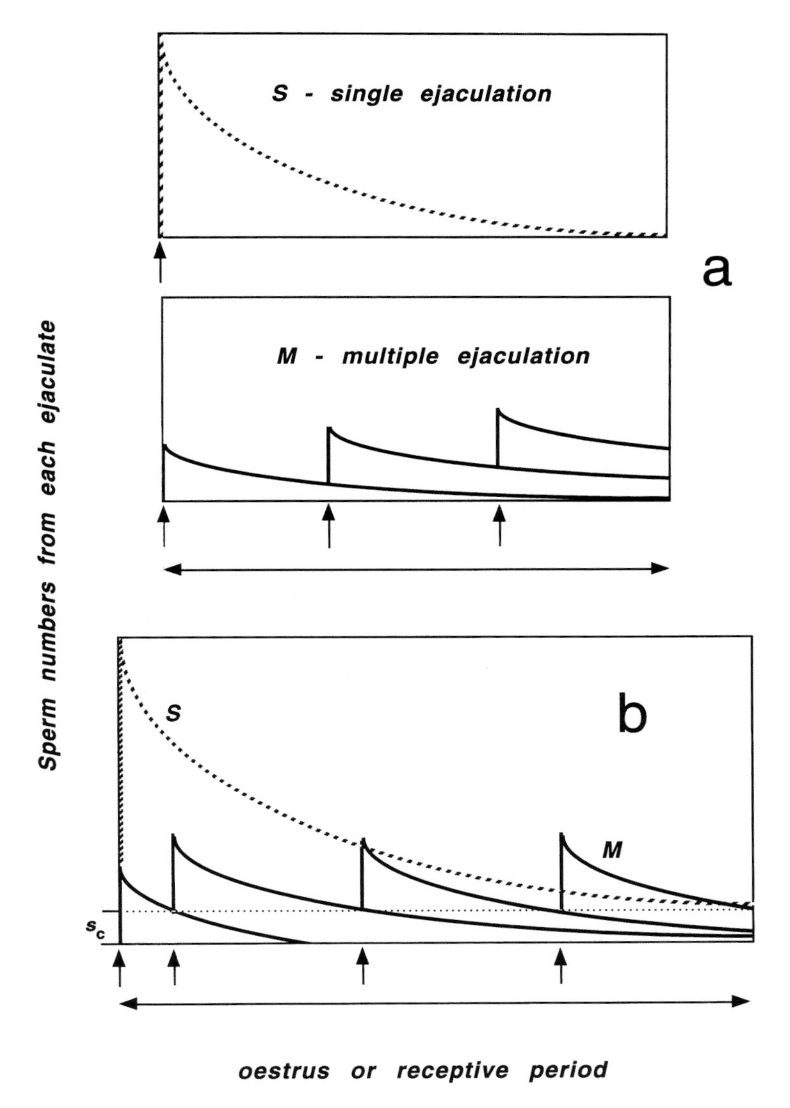

Fig. 1.12. **a.** *The single vs. multiple ejaculation strategy sperm competition game. Two males mate with a female during oestrus, or some receptive period in which fertilization can occur. One male plays S (a single ejaculation at the start of oestrus, and the other plays M (many ejaculations spread through the oestrus period at equal intervals). Both males expend the same total amount of ejaculate, and both ejaculates suffer the same rate of sperm death in the female tract.* **b.** *Similar alternative strategies in a noncompetitive situation (only one male mates with a given female). S is the same as in a, but M is slightly different, as explained in the text, and mates each time the sperm dose in the female drops to the critical level, s_c. This means that the first interval between ejaculations differs from the rest (see text).*

time they are ejaculated. A calculation of the pay-offs of S vs. M in a population fixed at either S or M revealed S to be the ESS with the biggest zone of attraction; it wins unless there is an extremely high sperm death during oestrus. Depending on the number of ejaculations, n, performed by the M player, unless 90–95% of sperm ejaculated at the start of oestrus are dead by end of oestrus, S wins because it has a huge predominance of sperm for most of the period. If the sperm loss is greater than this, M is the ESS because although S wins when fertilization occurs early in the period, M wins later when most of the S sperm have died. The model suggests that under sperm competition, S is usually favourable unless there is extreme sperm mortality. This offers some exciting possibilities. High levels of sperm competition risk might cause a switch from multiple ejaculation to single ejaculation. Males whose sperm suffer high death rates in the female tract may do best to play M. Candidate examples of when it would pay to play M may those where females use EPCs to get 'better' sperm by selecting against their partner's sperm, or spermicidal immune responses by females.

However, it appears likely that the length of life of the sperm will be shaped by the duration of oestrus, and there is some evidence for this (Parker 1984a; Gomendio and Roldan 1993a). If selection has tuned the sperm life in such a way that only 5–10% survive by the end of oestrus (but enough to permit fertilization by a single ejaculate), then the M strategy may be favoured under sperm competition (Parker 1984a). What is required is a more complete optimality analysis of coevolution of ejaculation strategies, sperm lifetimes and oestrus durations under varying degrees of sperm competition.

B. *Topping up*

The sperm competition model above makes the assumption typical of sperm competition games, that fertilization occurs if there is any surviving sperm. In the present model (G. A. Parker, unpublished results) there is no sperm competition: either an M or S player mates with a female without competition and conception occurs only if sperm numbers stay above a critical threshold, s_c (see Fig. 1.12b). An optimal M would always copulate every time that the sperm number from the previous ejaculate drops to s_c, and the best number of ejaculations depends on the costs of mating. Because there are no sperm at the first ejaculation, but s_c at each subsequent ejaculation, the sperm number and the first intermating interval are different from subsequent ones (Fig. 1.12b). The result is that in this no-sperm-competition model, an optimal multiple mating strategy (M) is always better than a single mating strategy (S) because M conserves sperm by the topping up process. S loses because some of the single large dose of sperm transferred could have been 'safe' inside the male rather than subject to loss inside the female.

So, in the absence of sperm competition, multiple ejaculation will always be better than single ejaculation unless there are significant costs of multiple mating, e.g. in terms of (i) time spent around the same female and (ii) energy spent on several matings. Much more modelling could yet be done on the theory of timing and number of copulations within a given female cycle.

The topping up hypothesis has been discussed in relation to human and other data by Baker and Bellis (1993, 1995).

VIII. FEMALE INFLUENCES ON SPERM ALLOCATION

Female interests, the potential for female control of paternity, and other possible female influences have formed a major focus of interest in sperm competition studies (see Section I). A problem is that detecting female influences is technically difficult (see Chapter 17), and although there is some evidence for sperm preference by females (e.g. Eberhard 1996), much of it is flawed, and anything that hints of female control tends to be highlighted enthusiastically. Some effects can often be explained simply by purely male differences (e.g. bigger males have a higher sperm transfer rate, so if both are given equal copulation durations, bigger males achieve greater paternity; see Chapter 10 for examples). Nevertheless, plausible mechanisms have been proposed which could, in principle, allow females to influence sperm competition and paternity. These include ejection or flowback of nonpreferred ejaculates as in certain birds (e.g. Davies 1983) and mammals (e.g. Ginsberg and Huck 1989; Baker and Bellis 1995), and various immunological mechanisms both direct (e.g. Birkhead *et al.* 1993) and indirect (e.g. Folstad and Skarstein 1997).

There is now ample evidence, particularly from heterospermic inseminations (females inseminated with ejaculates from different males) in domestic mammals, that genetic variation among males is correlated with differences in paternity prospects. An excellent review is given by Dziuk (1996), who concludes that paternity rarely follows relative sperm numbers inseminated (raffle principle). It is not known whether this reflects genetic differences among males directly, or indirectly through female effects.

A prospective model of sperm preference by females (G. A. Parker, unpublished results) allows (for tractability) two types of male, A and B: when in competition, females favour the sperm of A males against those of B. The analysis follows the loaded raffle in that when A and B ejaculates compete, each sperm of A counts as 1.0 relative to r ($0 < r < 1$) for each sperm of B. The situation therefore resembles the nonrandom roles model of the loaded raffle (Section IV.A), but differs in that males must now optimize across three conditions:

(1) No sperm competition (either A or B alone);
(2) Competition with same type (A, A or B, B);
(3) Competition with opposite type (A, B).

The major difference is category (3). Preliminary results follow the non-random roles model in the loaded raffle – males in the disfavoured role (B) spend more on sperm. However, to determine the ESS, one would ideally make this a coevolutionary game in which the female could increase the loading, at greater cost to herself, while males increase sperm numbers and resistance to female preferences. A random roles version of the same model would involve the case where a given male has a set of females that favour his sperm, and a set that disfavour his sperm, as might occur if females exercise some sperm filtering on a basis of genetic compatibility (e.g. related to inbreeding). Such games should be modelled as 'arms races' between males and females, although it is difficult to do this other than in a very general way which does not readily allow empirical testing.

IX. FUTURE

After over a quarter of a century of interest in sperm competition and its evolutionary consequences, we are still advancing rapidly in our understanding of mechanisms and adaptations. This volume – like its predecessor (Smith 1984) – highlights the ubiquity and all-pervasive nature of sperm competition as a force in evolution. However, much remains to be done and there are many topics which remain largely unexplored, both empirically and theoretically.

One of the most exciting areas is the adaptive value of sperm structure, which is hugely variable (Sivinski 1984). Associated with this is the origin and evolution of sperm polymorphism, and the function of giant sperm. To date, sperm competition games have focused on sexually reproducing diploids with separate sexes but there is scope for consideration of different systems; the effects of kin selection on sperm expenditure remain unexplored (one might expect a reduction in competitive ejaculate expenditure when competing males are related). Recently, Charnov (1996) examined sperm competition and sperm allocation in simultaneous hermaphrodites, using a variety of assumptions about sperm displacement. Results all have the form that the ESS displacement can be expressed in terms of total sperm input divided by the total sperm store capacity (rather as in the case of the dungfly).

It is undoubtedly an exciting and productive era in the study of sperm competition. My own view is that progress is best achieved by an integrated mix of empirical work with theoretical development.

DEDICATION AND ACKNOWLEDGEMENTS

I wish to dedicate this paper to 'the sperm competition group' in the (then) Department of Environmental and Evolutionary Biology, University of Liverpool (1990–96): Leigh Simmons, Nina Wedell, Paula Stockley, Matt Gage, and Penny Cook. I owe to them more than they realize for making those years so stimulating and such fun, and for their work and discussions which helped to shape my thinking on sperm competition theory.

I also owe a large debt of gratitude to Mike Ball, whose stimulating collaboration in recent years has greatly expanded the scope of the analyses of sperm competition games, and to the BBSRC for the Senior Research Fellowship (1990–95) which enabled me to do most of the work summarized here. Some of the earlier models (including my unpublished 'topping up' model of Section VII.B) were presented at the EGI Conference 1989 (University of Oxford), and many were presented in my Tinbergen Lecture at the ASAB Conference 1995 (Zoological Society of London, Regents Park) and elsewhere; I am most grateful for the stimulus provided by these occasions. Tim Birkhead and P. L. Schwagmeyer helped greatly to improve the first draft of the manuscript, and I am also indebted to A. P. Møller, D. Mock and P. Stockley for their comments.

REFERENCES

Baker RR & Bellis MA (1989) Number of sperm in human ejaculates varies in accordance with sperm competition theory. *Anim. Behav.* **37:** 867–869.

Baker RR & Bellis MA (1993) Human sperm competition: ejaculate adjustment by males and the function of masturbation. *Anim. Behav.* **46:** 861–885.

Baker RR & Bellis MA (1995) *Human Sperm Competition: Copulation, Masturbation and Infidelity*. Chapman and Hall, London.

Ball MA (1998) Sperm competition games: incorporation of tail length into the continuous fertilization model. (in prep.).

Ball MA & Parker GA (1996) Sperm competition games: external fertilization and 'adaptive' infertility. *J. Theor. Biol.* **180:** 141–150.

Ball MA & Parker GA (1997) Sperm competition games: inter- and intra-species results of a continuous external fertilization model. *J. Theor. Biol.* **186:** 459–466.

Beatty RA (1972) The genetics of size and shape of spermatozoan organelles. In *The Genetics of the Spermatozoan*. RA Beatty & S Gluecksohn-Waelsch (eds), pp. 97–115. Edinburgh University Press, Edinburgh.

Bellis MA, Barker RR & Gage MJG (1990) Variation in rat ejaculates is consistent with the kamikaze sperm hypothesis. *J. Mammal.* **71:** 479–480.

Birkhead TR (1988) Behavioral aspects of sperm competition in birds. *Adv. Study Behav.* **18:** 35–72.
Birkhead TR (1995) Sperm competition: evolutionary causes and consequences. In *Seventh International Symposium on Spermatology: Plenary Papers. Reproduction, Fertility Development* **7:** 755–775.
Birkhead TR (1996) Sperm competition: evolution and mechanisms. *Curr. Top. Develop. Biol.* **33:** 103–158.
Birkhead TR & Fletcher F (1995) Depletion determines sperm numbers in male zebra finches. *Anim. Behav.* **48:** 451–456.
Birkhead TR & Møller AP (1992) *Sperm Competition in Birds: Evolutionary Causes and Consequences*. Academic Press, London.
Birkhead TR & Møller AP (1995) Extra-pair copulation and extra-pair paternity in birds. *Anim. Behav.* **49:** 843–848.
Birkhead TR & Parker GA (1997) Sperm competition and mating systems. In *Behavioural Ecology: an Evolutionary Approach,* 4th edn. JR. Krebs & NB Davies (eds), pp. 121–145. Blackwell, Oxford.
Birkhead TR, Møller AP & Sutherland WJ (1993) Why do females make it so difficult for males to fertilize their eggs? *J. Theor. Biol.* **161:** 51–60.
Birkhead TR, Wishart GJ & Biggins JD (1995) Sperm precedence in the domestic fowl. *Proc. Roy. Soc. Lond. B* **261:** 285–292.
Boorman E & Parker GA (1976) Sperm (ejaculate) competition in *Drosophila melanogaster*, and the reproductive value of females to males in relation to female age and mating status. *Ecol. Entomol.* **1:** 145–155.
Bressac C, Fleury A & Lachaise D (1995) Another way of being anisogamous in *Drosophila* subgenus species: giant sperm, one-to-one gamete ratio, and high zygote provisioning. *Proc. Natl Acad. Sci. USA* **91:** 10399–10402.
Briskie JV, Montgomerie E & Birkhead TR (1997) The evolution of sperm size in birds. *Evolution* **51:** 937–945.
Charnov EL (1976) Optimal foraging: the marginal value theorem. *Theor. Popul. Biol.* **9:** 129–136.
Charnov EL (1993) *Life History Invariants: Some Explorations of Symmetry in Evolutionary Ecology*. Oxford University Press, Oxford.
Charnov EL (1996) Sperm competition and sex allocation in simultaneous hermaphrodites. *Evol. Ecol.* **10:** 457–462.
Charnov EL & Parker GA (1995) Dimensionless invariants from foraging theory's marginal value theorem. *Proc. Natl Acad. Sci. USA* **92:** 1446–1450.
Clutton-Brock TH & Harvey P (1977) Primate ecology and social organization. *J. Zool.* **183:** 1–39.
Clutton-Brock TH & Parker GA (1995) Sexual coercion in animal societies. *Anim. Behav.* **49:** 1345–1365.
Colegrave N, Birkhead TR & Lessells CM (1995) Sperm precedence in zebra finches does not require special mechanisms of sperm competition. *Proc. Roy. Soc. Lond. B* **259:** 223–228.
Cook PA & Gage MJG (1995) Effects of risks of sperm competition on the numbers of eupyrene and apyrene sperm ejaculated by the male moth *Plodia interpunctella* (Lepidoptera: Pyralidae). *Behav. Ecol. Sociobiol.* **36:** 261–268.
Davies NB (1983) Polyandry, cloaca pecking and sperm competition in dunnocks. *Nature* **302:** 334–336.
Dewsbury DA (1972) Patterns of copulatory behavior in male mammals. *Quart. Rev. Biol.* **47:** 1–33.
Dunbar RIM (1982) Intraspecific variations in mating strategy. In *Perspectives in*

Ethology, Vol. 5. PPG Bateson & PH Klopfer (eds), pp. 382–431. Plenum Press, New York.

Dziuk PJ (1996) Factors that influence the proportion offspring sired by a male following heterospermic insemination. *Anim. Reprod. Sci.* **43:** 65–88.

Eberhard WG (1985) *Sexual Selection and Animal Genitalia.* Harvard University Press, Cambridge, MA.

Eberhard WG (1996) *Female Control: Sexual Selection by Cryptic Female Choice.* Princeton University Press, Princeton.

Emlen ST & Oring LW (1977) Ecology, sexual selection and the evolution of mating systems. *Science* **197:** 215–223.

Folstad I & Skarstein F (1997) Is male germ line control creating avenues for female choice? *Behav. Ecol.* **8:** 109–112.

Gage MJG (1991) Risk of sperm competition directly affects ejaculate size in the Mediterranean fruit fly. *Anim. Behav.* **42:** 1036–1037.

Gage MJG (1994) Associations between body size, mating pattern, testis size and sperm lengths across butterflies. *Proc. Roy. Soc. Lond. B* **258:** 25–30.

Gage MJG & Baker RR (1991) Ejaculate size varies with socio-sexual situation in an insect. *Ecol. Entomol.* **16:** 331–337.

Gage MJG, Stockley P & Parker GA (1995) Effects of alternative male mating strategies on characteristics of sperm production in the Atlantic salmon (*Salmo salar*): theoretical and empirical investigations. *Phil. Trans. Roy. Soc. Lond. B* 391–399.

Ginsberg JR & Huck UW (1989) Sperm competition in mammals. *Trends Ecol. Evol.* **4:** 74–79.

Gomendio M & Roldan ERS (1991) Sperm competition influences sperm size in mammals. *Proc. Roy. Soc. Lond. B* **247:** 89–95.

Gomendio M & Roldan ERS (1993a) Coevolution between male ejaculates and female reproductive biology in eutherian mammals. *Proc. Roy. Soc. Lond. B* **252:** 7–12.

Gomendio M & Roldan ERS (1993b) Mechanisms of sperm competition: linking physiology and behavioural ecology. *Trends Ecol. Evol.* **8:** 95–100.

Harcourt AH, Harvey PH, Larson SG & Short RV (1981) Testis weight, body weight and breeding system in primates. *Nature* **293:** 55–57.

Hecht NB, Bower PA, Waters SH, Yelick PC & Distel RJ (1986) Evidence for haploid expression of mouse testicular genes. *Exp. Cell Res.* **164:** 183–190.

Hurst L (1990) Parasite diversity and the evolution of diploidy, multicellularity and anisogamy. *J. Theor. Biol.* **144:** 429–443.

Karr TL & Pitnick S (1996) The ins and outs of fertilization. *Nature* **379:** 405–406.

Knowlton N & Greenwell SR (1984) Male sperm competition avoidance mechanisms: the influence of female interests. In *Sperm Competition and the Evolution of Animal Mating Systems.* RL Smith (ed.), pp. 62–85. Academic Press, London.

Lanier DL, Estep DQ & Dewsbury DA (1979) Role of prolonged copulatory behaviour in facilitating reproductive success in a competitive mating situation in laboratory rats. *J. Comp. Physiol. Psychol.* **93:** 781–792.

Lessells CM & Birkhead TR (1990) Mechanisms of sperm competition in birds: mathematical models. *Behav. Ecol. Sociobiol.* **27:** 325–337.

Maynard Smith J (1982) *Evolution and the Theory of Games.* Cambridge University Press, Cambridge.

Møller AP (1989) Ejaculate quality, testes size and sperm production in mammals. *Func. Ecol.* **3:** 91–96.

Møller AP (1991) Sperm competition and sperm counts: an evaluation of the empirical evidence. In *Comparative Spermatology 20 Years Later.* B Baccetti (ed.), pp. 775–777. Raven Press, New York.

Parker GA (1970a) Sperm competition and its evolutionary consequences in the insects. *Biol. Rev.* **45:** 525–567.

Parker GA (1970b) Sperm competition and its evolutionary effect on copula duration in the fly, *Scatophaga stercoraria. J. Insect Physiol.* **16:** 1301–1328.

Parker GA (1979) Sexual selection and sexual conflict. In *Sexual Selection and Reproductive Competition in Insects.* MS Blum & NA Blum (eds), pp. 123–166. Academic Press, New York.

Parker GA (1982) Why are there so many tiny sperm? Sperm competition and the maintenance of two sexes. *J. Theor. Biol.* **96:** 281–294.

Parker GA (1984a) Sperm competition and the evolution of animal mating strategies. In *Sperm Competition and the Evolution of Animal Mating Systems.* RL Smith (ed.), pp. 1–60. Academic Press, London.

Parker GA (1984b) The producer/scrounger model and its relevance to sexuality. In *Producers and Scroungers: Strategies of Exploitation and Parasitism.* CJ Barnard (ed.), pp. 127–153. Croom Helm, London.

Parker GA (1990a) Sperm competition games: raffles and roles. *Proc. Roy. Soc. Lond. B* **242:** 120–126.

Parker GA (1990b) Sperm competition games: sneaks and extra-pair copulations. *Proc. Roy. Soc. Lond. B* **242:** 127–133.

Parker GA (1992) Marginal value theorem with exploitation time costs: diet, sperm reserves, and optimal copula duration in dung flies. *Am. Natur.* **139:** 1237–1256.

Parker GA (1993) Sperm competition games: sperm size and number under adult control. *Proc. Roy. Soc. Lond. B* **253:** 245–254.

Parker GA & Begon ME (1993) Sperm competition games: sperm size and number under gametic control. *Proc. Roy. Soc. Lond. B* **253:** 255–262.

Parker GA & Maynard Smith J (1990) Optionality theory in evolutionary biology. *Nature* **348:** 27–33.

Parker GA & Simmons LW (1991) A model of constant random sperm displacement during mating: evidence from *Scatophaga. Proc. Roy. Soc. Lond. B* **246:** 107–115.

Parker GA & Simmons LW (1994) The evolution of phenotypic optima and copula duration in dungflies. *Nature* **370:** 53–56.

Parker GA & Smith JL (1975) Sperm competition and the evolution of the precopulatory passive phase behaviour in *Locusta migratoria migratorioides. J. Entomol. (A)* **49:** 155–171.

Parker GA & Stuart RA (1976) Animal behaviour as a strategy optimizer: evolution of resource assessment strategies and optimal emigration thresholds. *Am. Natur.* **110:** 1055–1076.

Parker GA, Simmons LW & Kirk H (1990) Analysing sperm competition data: simple models for predicting mechanisms. *Behav. Ecol. Sociobiol.* **27:** 55–65.

Parker GA, Simmons LW & Ward PI (1993) Optimal copula duration in dungflies: effects of frequency dependence and female mating status. *Behav. Ecol. Sociobiol.* **32:** 157–166.

Parker GA, Ball MA, Stockley P & Gage MJG (1996) Sperm competition games: assessment of sperm competition intensity by group spawners. *Proc. Roy. Soc. Lond. B* **263:** 1291–1297.

Parker GA, Ball MA, Stockley P & Gage MJG (1997) Sperm competition games: a prospective analysis of risk assessment. *Proc. R. Soc. Lond. B* **264:** 1793–1802.

Pitnick S & Markow TA (1994) Male gametic strategies: sperm size, testes size,

and the allocation of ejaculate among successive mates by the sperm-limited fly *Drosophila pachea* and its relatives. *Am. Natur.* **143:** 785–819.
Pitnick S, Markow TA & Spicer GS (1995a) Delayed maturity is a cost of producing large sperm in *Drosophila. Proc. Natl Acad. Sci. USA* **92:** 10614–10618.
Pitnick S, Spicer GS & Markow TA (1995b) How long is a giant sperm? *Nature* **375:** 109.
Robertson DR & Choat JH (1974) Protogynous hermaphroditism and social systems in labrid fishes. *Proc 2nd Int. Symp. Coral Reefs* **1:** 217–225.
Robertson DR & Warner RR (1978) Sexual patterns in the labrid fishes of the western Caribbean. Part II. The parrotfishes (Scaridae). *Smithsonian Contrib. Zool.* **255:** 1–26.
Roldan ERS, Gomendio M & Vitullo AD (1992) The evolution of eutherian spermatozoa and underlying selective forces: female selection and sperm competition. *Biol. Rev.* **67:** 551–593.
Sakaluk SK & Eggert A-K (1996) Female control of sperm transfer and intraspecific variation in sperm precedence: antecedents to the evolution of a courtship food gift. *Evolution* **50:** 694–703.
Schwagmeyer PL & Foltz DW (1990) Factors affecting the outcome of sperm competition in 13-lined ground squirrels. *Anim. Behav.* **39:** 156–162.
Schwagmeyer PL & Parker GA (1987) Queuing for mates in 13-lined ground squirrels. *Anim. Behav.* **35:** 1015–1025.
Schwagmeyer PL & Parker GA (1994) Mate quitting rules for male 13-lined ground squirrels. *Behav. Ecol.* **5:** 142–150.
Shapiro DY & Giraldeau LA (1996) Mating tactics in external fertilizers when sperm is limited. *Behav. Ecol.* **7:** 19–23.
Short RV (1977) Sexual selection and the descent of man. In *Proceedings of the Canberra Symposium on Reproduction and Evolution*, pp. 3–19. Australian Academy of Sciences, Canberra.
Short RV (1979) Sexual selection and its component parts, somatic and genital selection, as illustrated by man and the great apes. *Adv. Study Behav.* **9:** 131–158.
Silberglied RE, Shepherd JG & Dickinson JL (1984) Eunuchs: the role of apyrene sperm in Lepidoptera? *Am. Natur.* **123:** 255–265.
Simmons LW (1987) Sperm competition as a mechanism of female choice in the field cricket *Gryllus bimaculatus* (DeGeer). *Behav. Ecol. Sociobiol.* **21:** 197–202.
Simmons LW & Parker GA (1992) Individual variation in sperm competition success of yellow dung flies, *Scatophaga stercoraria. Evolution* **46:** 366–375.
Simmons LW, Craig M, Llorens T, Schinzig M & Hosken D (1993) Bushcricket spermatophores vary in accord with sperm competition and parental investment theory. *Proc. Roy. Soc. Lond. B* **251:** 183–186.
Simmons LW, Stockley P, Jackson RL & Parker GA (1996) Sperm competition and sperm selection: no evidence for a female influence over paternity in the yellow dung fly, *Scatophaga stercoraria. Behav. Ecol. Sociobiol.* **38:** 199–206.
Simmons LW, Stockley P & Parker GA (1998) The mechanism of sperm displacement in the yellow dung fly *Scatophaga stercoraria.* (in prep.).
Sivinski J (1984) Sperm in competition. In *Sperm Competition and the Evolution of Animal Mating Systems.* RL Smith (ed.), pp. 86–115. Academic Press, London.
Smith RL (1979) Repeated copulation and sperm precedence: paternity assurance for a male brooding water bug. *Science* **205:** 1029–1031.
Smith RL (1984) (ed.) *Sperm Competition and the Evolution of Animal Mating Systems.* Academic Press, London.

Stephens DW & Dunbar SR (1993) Dimensional analysis in behavioural ecology. *Behav. Ecol.* **4:** 172–183.

Stockley P (1997) Sexual conflict resulting from adaptations to sperm competition. *Trends Ecol. Evol.* **12:** 127–166.

Stockley P & Purvis A (1993) Sperm competition in mammals: a comparative study of male roles and relative investment in sperm production. *Func. Ecol.* 7560–7570.

Stockley P, Gage MJG, Parker GA & Møller AP (1997) Sperm competition in fish: the evolution of testis size and ejaculate characteristics. *Am. Natur.* **149:** 933–954.

Taborsky M (1994) Sneakers, satellites, and helpers: parasitic and co-operative behaviour in fish reproduction. *Adv. Study Behav.* **23:** 1–100.

Thomaz D, Beall E & Burke T (1997) Alternative reproductive tactics in Atlantic salmon: factors affecting mature parr success. *Proc. Roy. Soc. Lond. B* **264:** 219–226.

Waage J (1979) Dual function of the damselfly penis: sperm removal and transfer. *Science* **203:** 916–918.

Warner RR, Shapiro DY, Marconato A & Petersen CW (1995) Sexual conflict: males with the highest mating success convey the lowest fertilization benefits to females. *Proc. Roy. Soc. Lond. B* **262:** 135–139.

Warner RR & Robertson DR (1978) Sexual patterns in the labroid fishes of the Western Caribbean, I: the Wrasses (Labridae). *Smithsonian Contrib. Zool.* **254:** 1–27.

Wedell N (1992) Protandry and mate assessment in the wartbiter *Decticus verrucivorous* (Orthoptera: Tettigoniidae). *Behav. Ecol. Sociobiol.* **31:** 301–308.

2 Sperm Competition and Sexual Selection

Anders Pape Møller

Laboratoire d'Ecologie, CNRS URA 258, Université Pierre et Marie Curie, Bât. A, 7ème étage, 7 quai St Bernard, Case 237, F–75252 Paris Cedex 5, France

I. SEXUAL SELECTION AND SPERM COMPETITION

Sexual selection was described by Charles Darwin (1871) as arising from 'the advantages that certain individuals have over others of the same sex and species in exclusive relation to reproduction'. Sexual selection gives rise to costly secondary sexual characters and is therefore opposed by natural selection. Extravagant characters such as the horns of horned beetles and antelopes and the fantastic plumes of many male birds were supposed to be the product of sexual selection. Two processes were suggested to give rise to sexual selection, namely, competition among individuals of the chosen sex for access to individuals of the 'choosy' sex, usually male–male competition, and choice of attractive partners by individuals of the choosy sex, usually female choice.

Sexual selection was originally assumed to arise from variation in number of females mated, and a recent review of sexual selection by Andersson (1994) follows this tradition. Darwin (1871) assumed that females basically were sexually monogamous, and sperm competition therefore did not play a role in nineteenth century sexual selection. Interestingly, this point of view is still perpetuated in the current literature,

Sperm Competition and Sexual Selection
ISBN 0-12-100543-7

with sperm competition being considered of little importance for sexual selection (Andersson 1994; p. xvi). In my opinion, sexual selection is a continuous process that takes place at a number of different stages in reproduction, from the actual process of mate acquisition through copulation, subsequent fertilization, differential abortion and infanticide, and differential investment in offspring (Møller 1994). After all, variation in male reproductive success due to variation in the number of social mates does not differ in any respect from variation in male reproductive success resulting from extra-pair paternity, or because one male kills the offspring of another male and indirectly alters the distribution of reproductive success among males. Similarly, male mating success may have one distribution when females are allocated to males, but this distribution may be altered at the time of copulation owing to lack of mate fidelity. This variance in male success may be further skewed at the time of fertilization if females favour the sperm of certain males over others, if the sperm of particular males are competitively superior, and if subsequent changes in variance may take place as a consequence of differential abortion, infanticide, and differential investment in offspring dependent on the phenotype of the mate (Møller 1994; Fig. 2.1).

Sexual selection can be considered to consist of a number of components that affect total fitness, with two major routes: mating success and fecundity per mate. Both routes and their component parts may be related to the expression of a male secondary sexual character (or a female secondary sexual character in cases of mutual sexual selection or

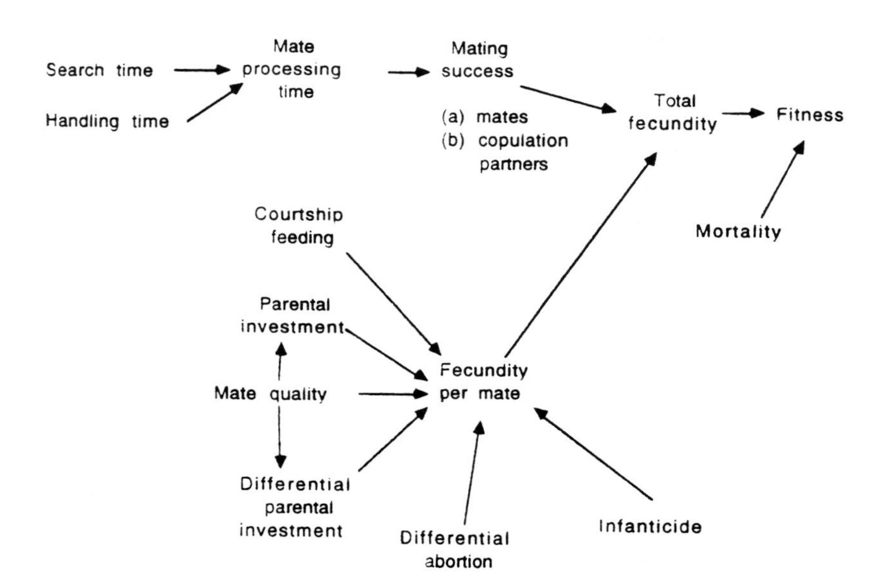

Fig. 2.1. *Sexual selection as a continuous process from mate acquisition to the production of independent offspring. Adapted from Møller (1994).*

sex-role reversed species). The relative importance of each of the components in terms of sexual selection will depend on the intensity of selection, which is simply the magnitude of the relationship between each of these fitness components and the expression of the secondary sexual character. Male mating success will, at the proximate level, depend on the time it takes for a male to process a female, which obviously will depend on male investment in each reproductive event and female preference for a particular male phenotype (Fig. 2.1). The mate processing time can be separated into two different components: the mate searching time and the mate handling time (Fig. 2.1). These apply to both social mates and sexual mates. Total fecundity will be a product of mating success (social and sexual mates) and fecundity per mate, and total fecundity will, together with survival, comprise total fitness (Fig. 2.1).

The second major route to fitness through sexual selection is via effects on fecundity per mate (Fig. 2.1). This will in turn depend on the relationship between the expression of male secondary sexual characters and factors such as the intensity of courtship feeding, male resources such as territories, male parental investment, differential parental investment, differential abortion, and infanticide (Fig. 2.1). If females invest differentially in the offspring of particular males, because such males have certain attractive features, this will affect the distribution of reproductive success among males and the intensity of sexual selection on males. Female mate choice due to differential investment may be less obvious than female choice of social mates, but differential female choice and other types of cryptic female choice (Thornhill 1983) are not necessarily less important as components of sexual selection than other types of sexual selection (Eberhard 1996).

This view of sexual selection as a continuous process that starts with the mate selection or acquisition process and finishes with the production of dependent offspring contrasts with the current view of other students of sexual selection (Andersson 1994). However, I argue that each of the steps of selection presented in Fig. 2.1 potentially gives rise to sexual selection, and that the entire process of sexual selection can be understood only by analysis of each of these steps. If sexual selection takes place at the different stages, we should expect sexual selection to be a reinforcing process that continuously gives rise to an increase in variance in male mating success. The relative importance of the different components of sexual selection will depend on the opportunities for selection. The social mating system may severely constrain the opportunities for sexual selection at the time of choice of social mates (Møller 1992). For example, most females of socially monogamous species may be severely constrained in their mate choice if the sex ratio is equal and females have a continuously decreasing number of males as their mating options. Females of lek breeding species, with fewer constraints on mate choice, will have more opportunity to choose among available males and may have less incentive to adjust their original mate choice through sperm competition. The relative importance of sperm competition as a component of sexual selection will thus depend on the importance of this single component compared with the entire process of sexual selection. We need information on the relative magnitude of the sperm competition component of sexual selection before any general statements can be made about the relative importance of sperm competition in the context of sexual selection.

Theoretical models of sexual selection are traditionally classified with respect to the benefits accruing to individuals of the choosy sex (Andersson 1994; Møller 1994). Models of direct fitness benefits envisage that the expression of a male secondary sexual character reliably reflects the ability of males to provide direct fitness benefits (Heywood 1989; Hoelzer 1989; Grafen 1990; Price *et al.* 1993). Such models and their predictions have not traditionally been considered a major challenge to theoreticians because it is intuitively easy to understand how an exaggerated female mate preference may evolve in response to any direct fitness benefits acquired by choosy females. If females in the current generation obtain direct benefits from their choice of the most extravagantly ornamented males, this process will be relatively more important than any gains obtained in subsequent generations. Interestingly, a number of species demonstrate negative relationships between the magnitude of direct fitness benefits and the expression of male secondary sexual characters, implying that there is a cost in terms of male parental care associated with mate choice (see review by Møller 1994). Differential investment by females can in this situation be evolutionarily stable only if females benefit indirectly from their mate choice. Two different indirect benefits are traditionally considered. If the male trait and the female preference both have a genetic basis, both characters may become exaggerated as a consequence of coevolution between the male trait and the female mate preference (Fisher 1930). In other words, females may benefit in terms of the sexual attractiveness of their 'sons', as envisaged by Fisher. This also applies where females pay a cost for their mate choice (Pomiankowski *et al.* 1991). The other indirect benefit is the genetic benefit signalled by the male trait either in terms of general viability (Heywood 1989; Iwasa *et al.* 1991) or specific parasite resistance (Hamilton and Zuk 1982). Both sons and daughters of choosy females that mate with an attractive male under this scenario will acquire genetically based greater viability. These different types of fitness benefits from sexual selection also apply to sexual selection arising from sperm competition, as discussed in the following paragraphs.

Sexual selection may reduce reproduction and survival if females spend limited time and energy on their mate choice. Individuals mated with conspecifics with the most extravagant secondary sexual characters are presumed to acquire direct or indirect fitness benefits. In a sperm competition context, direct benefits include a higher probability of fertilization, a reduced risk of sexual harassment or infanticide, or the acquisition of male parental care from males other than the social mate (see Table 2.1). There is some evidence for the existence of these direct fitness

Table 2.1. Direct and indirect fitness benefits accruing to females from their choice of multiple copulation partners. Adapted from Birkhead and Møller (1992).

Benefit	Reference
Direct fitness benefits	
Higher probability of fertilization	Walker (1980), Sheldon (1994)
Lower probability of sexual harassment	Parker (1974)
Lower probability of infanticide	Hrdy (1977)
Acquisition of parental care from additional males	Emlen (1978), Burke *et al.* (1989)
Prospecting for future mates	Colwell and Oring (1989)
Ejaculate nutrients	Thornhill (1976)
Food and courtship food	Wolf (1975)
Indirect fitness benefits	
Genetically more viable offspring	Møller (1988a), Smith (1988)
Genetically more resistant offspring	Hamilton (1990)
Genetically more diverse offspring	Williams (1975)
Genetically more attractive sons	Birkhead and Møller (1992)

benefits of copulations with multiple mates. The females of many insects show increased fertility as a direct consequence of multiple copulations (see review by Ridley 1988). Males of a number of different animal species ranging from insects to mammals are consistent in their ability to fertilize eggs in a competitive situation (Lewis and Austad 1990; Dziuk 1996), and females may benefit from copulations with several males simply because they increase the probability of copulating with a male of superior fertilizing ability.

Studies of multiple paternity and viability in reptiles have provided direct evidence for an advantage of copulation with multiple males. Swedish populations of two species of reptiles, the European adder *Vipera berus* and the viviparous sand lizard *Lacerta agilis*, have demonstrated an association between offspring viability and multiple paternity because copulations with multiple males reduce the frequency of still-born offspring (Madsen *et al.* 1992; Olsson *et al.* 1994a,b, 1996). This advantage of female copulations with multiple males apparently arose as a consequence of female choice of sperm from particularly viable males (Olsson *et al.* 1996). The two Swedish populations of reptiles are characterized by being small and isolated and therefore potentially genetically depauperate. For example, an Italian study of the European adder did not find any evidence of an association between multiple paternity and offspring viability (Capula and Luiselli 1994); this Italian population was not subject to the same effects of population isolation and small effective population size as the Swedish population.

Females of a range of insects benefit from copulations with multiple males because of the direct benefits acquired in terms of nutrients provided by males for developing eggs (Boggs and Gilbert 1979; Thornhill 1983; Gwynne 1984; Simmons 1990; Wiklund *et al.* 1993). This occurs in taxa as diverse as scorpionflies, bush crickets and butterflies.

In contrast, females that do not engage in copulations with multiple males could also benefit in a number of different ways. They would spend a reduced amount of time and energy on copulations, have reduced risks of predation or other causes of death during copulation, and have reduced risks of contracting a sexually transmitted disease. Because copulations appear to be of very low cost but of potentially large benefit to males, females may engage in copulations with multiple males in order to reduce the cost of avoidance of persistent males (Parker 1974). This hypothesis is based on the argument that a female can reduce this cost by complying with the interests of males. Females may directly benefit from association with a guarding or copulating male by avoiding harassment and even death in water striders, odonates and butterflies (Rubenstein 1984; Wilcox 1984; Miller 1987; Svärd and Wiklund 1988; Tsubaki *et al.* 1994). Females of some species may also forage at a greater rate owing to lack of harassment if guarded by a male, but at a lower rate in other species as a result of interference between foraging of the guarded female and the guarding male (Rubenstein 1984; Wilcox 1984; Davies 1985). In some bird species females frequently die during extra-pair copulations (Huxley 1912; McKinney *et al.* 1983), suggesting that female engagement in extra-pair copulations in order to avoid the costs of persistent males is not a feasible option in all cases. The hypothesis that females engage in copulations with additional males in order to avoid infanticide has gained support in primates (Hrdy 1977). In contrast, Møller (1988c) found that in the barn swallow *Hirundo rustica* the most likely perpetrators of infanticide had a lower probability of successful engagement in extra-pair copulations than other males. The final direct benefit of engagement in extra-pair copulations concerns the possibility of a female re-mating with the extra-pair male during future reproductive events. There is some evidence from the spotted sandpiper *Tringa macularia* and the oystercatcher *Haematopus ostralegus* that females use extra-pair copulations for assessment of potential mates (Colwell and Oring 1989; Heg *et al.* 1993). However, the actual fitness consequences of such a mate assessment strategy have not been determined for any long-lived species. Such a mate assessment strategy would not be feasible in small species with a high mortality rate. This chapter suggests that there is some evidence of direct fitness benefits to females engaging in copulations with multiple males, particularly in insects.

Copulations with multiple males may resemble copulations at leks in the sense that females do not appear to obtain much other than ejaculates, and indirect fitness benefits thus appear to remain as the most obvious benefit from such activity. A number of different kinds of indirect fitness benefits may accrue to females engaging in copulations with multiple males (Table 2.1). These range from the production of genetically more viable offspring, including offspring with greater resistance to parasites, genetically more diverse offspring, and genetically more attractive 'sons'. The hypothesis about genetic diversity of offspring seems unlikely because meiosis alone gives rise to a diverse array of genetic make-up, and the fusion of male and female gametes can potentially provide a huge diversity offspring (Williams 1975). There are no empirical tests of this hypothesis, although the tree swallow *Tachycineta bicolor* and the aquatic warbler *Acrocephalus paludicola* may be good candidates simply because of the large number of sires – up to five different males – of single broods. It is difficult to explain such a distribution of paternity from the point of view of good genes, and direct fitness benefits of extra-pair paternity also appear unlikely. As we will see in a later section, there is some evidence for the hypothesis that extra-pair offspring are indeed more viable than other offspring, in accordance with the viability hypothesis. To date, there are no tests of the Fisherian male attractiveness hypothesis.

In this review of sperm competition and sexual selection I view sperm competition as an integral part of sexual selection theory. In the next section I review sperm competition and the many anatomical and physiological adaptations to win at sperm competition as mechanisms to increase success in sexual selection. In the subsequent section I review how sperm competition and paternity are related to characters that traditionally have been considered secondary sexual characters. The consistency of females in their choice (or avoidance) of sires and the benefits that females obtain through the choice of particular sires are described in detail. In Section IV I review sexually transmitted diseases as a cost of sperm competition and sexual selection and determine to what extent they have played a role in the evolution of sperm competition.

II. SPERM COMPETITION AND SEXUAL SELECTION ON GENITALIA

The notion that sexual selection may affect the evolution of animal genitalia, accessory organs, and their associated physiology was initially proposed by Short (1979), who discriminated between somatic and genital sexual selection. Males that were able to produce large ejaculates would be at a selective advantage in sperm competition, as would males that by means of a longer penis were able to deposit sperm closest to the site of fertilization. These ideas received support from comparative studies of relative testes size in primates (Harcourt *et al.* 1981) and positive relationships between relative testes size and the intensity of sperm competition have subsequently been found in a range of other taxa (reviewed by Møller and Briskie 1995). Evolutionarily stable strategy modelling of sperm competition has suggested that it may sometimes be advantageous for small males to invest more in sperm production to compensate for their suboptimal timing of copulations (Parker 1990a,b). Empirical tests of these models using mammals with continuous or seasonal breeding and Atlantic salmon *Salmo salar*, respectively, revealed greater investment in sperm production by small males in continuously breeding mammals and in salmon (Stockley and Purvis 1993; Gage *et al.* 1995).

Ejaculate features and nuptial gifts associated with sperm transfer (either nuptial gifts donated to the female or accessory substances associated with the spermatophore) have also been subject to selection pressures arising from sperm competition. For example, the number of sperm, sperm motility and sperm production rates have increased as a consequence of the selection pressures arising from female copulations with multiple males (Gage 1994; Møller 1988d,e, 1989; Svärd and Wiklund 1989; Wedell 1993, 1994). The size of the nuptial gift associated with copulations in several species of insects protects the sperm during sperm transfer, since females eat any remaining sperm after the nuptial gift has been eaten (Wedell and Arak 1989; Wedell 1991, 1993, 1994; Simmons 1995). A comparative study demonstrated that an evolutionary increase in the mass of the ampulla (which contains the sperm) and the number of sperm, respectively, was associated with an increase in the mass of the spermatophylax (which is the nuptial gift; Vahed and Gilbert 1996).

Genitalia themselves may be a target of sexual selection, as emphasized by Short (1979) and Eberhard (1985). If males with a particular phenotype of genitalia have an advantage in terms of fertilization of a female, such males will leave more offspring, and provided a heritable basis of the genitalia – sons will be equally beneficially equipped. This scenario has received some support from comparative studies of a range of invertebrates (Eberhard 1985), although these analyses have never strictly controlled for similarities due to common ancestry, but there is virtually no evidence based on intraspecific studies. Such studies are badly needed to allow us to discriminate between the different explanations for the evolution of extravagant genitalia.

The extreme divergence of animal genitalia (genitalia are frequently used by systematists as diagnostic characters for the identification of species) has led some scientists to suggest that genital sexual selection mainly could be interpreted in terms of a Fisherian process of exaggeration of arbitrary traits of attractiveness (Eberhard 1985, 1993). However, it is easy to imagine direct fitness benefits from this selection process if males with a particular phenotype have a greater propensity for fertilization. The process could also be associated with a 'good genes' process of sexual selection if only males of high phenotypic and genetic quality were able to express a particular genitalia phenotype. Eberhard (1993) suggested that good genes explanations for the evolution of genitalia are unlikely because their costs are presumably small (the handicap models of sexual selection require a differential cost for signals to be reliable), or because they are not useful indicators of parasite resistance. However, the latter explanation has not been tested, and the considerable amount of genital manipulation or inspection in a wide range of species may render a revealing handicap function a possible alternative, for example in connection with sexually transmitted diseases. Animal genitalia often show little intraspecific phenotypic variation (Eberhard 1985), while true secondary sexual characters generally demonstrate high levels of variation (Møller 1994). However, Cordero and Miller (1992) found considerable variation in the length of the horn on the penis of a damselfly and this variation in penile morphology was directly related to male body size. A small amount of phenotypic variance makes it difficult for males to provide perceivably different stimuli to females during copulation, and for females to perceive differences among males. Obviously, condition-dependent expressions of phenotypes with their large phenotypic variances are difficult to reconcile with such a scenario.

There is considerable intraspecific variation in ejaculate features of males, and some of this variation has important consequences for fertilization success (Dziuk 1996). It has been suggested that male external phenotypes reliably reflect ejaculate features (Sheldon 1994), although this hypothesis has rarely been investigated in any detail. A study of the zebra finch *Taeniopygia guttata*, in which females prefer males with the highest song rate and brightest beak colour, found no evidence of covariation between male phenotype and ejaculate features (Birkhead and Fletcher 1995).

If sexual selection acts on males via their ability to fertilize eggs in competition for females, the evolutionary scene may be set for a coevolutionary 'arms race' between the female and the participating male parties owing to conflicts of reproductive interest (Parker 1984). The male and the female reproductive systems thus may be predicted to coevolve if females are able to select males of better phenotypic or genetic quality by making the passage of sperm to the site of fertilization more difficult (Birkhead *et al.* 1993; Keller and Reeve 1995). Although there is evidence for female choice of sperm provided by a particular male in a sperm competition situation in an ascidian and a reptile (Bishop 1996; Bishop *et al.* 1996; Olsson *et al.* 1996), but not in an insect (Simmons *et al.* 1996), there is considerable comparative evidence for coevolution of female reproductive biology and ejaculate characteristics in fish and mammals (Stockley *et al.* 1996b; Gomendio and Roldan 1993). Furthermore, experimentally induced arrest of the evolution of female reproductive defences in *Drosophila* fruitflies gave rise to a dramatic negative effect of ejaculate features of evolving males on female fitness (Rice 1996). Male adaptations that increase fertilization success, when females mate with multiple males, may often lead to a conflict with female interests because of the resulting reductions in female fitness (Stockley 1997). Female adaptations may reduce such male-induced costs although complete elimination may not be feasible (Stockley 1997). Cryptic female choice based on male seminal products may be involved as a mechanism in sexual selection (Cordero 1995; Eberhard and Cordero 1995). Evidence from a range of invertebrates and birds suggest that interactions between the female and the ejaculates of different males competing for the fertilization of the eggs of a single female may determine the outcome of sexual selection, but also generally reduce female fitness (Birkhead and Møller 1992; Cordero 1995; Eberhard and Cordero 1995). For example, female birds mated to multiple males often experience reduced duration of sperm storage compared with singly mated females (Birkhead and Møller 1992). This area of research provides a novel way to address the physiological mechanisms of sexual selection and sperm competition, but also raises questions about coevolution of male and female sexual strategies.

III. SEXUAL SELECTION DURING DIFFERENT STAGES OF THE REPRODUCTIVE CYCLE

As indicated in the introduction, postcopulatory sexual selection may be predicted to reinforce pre-mating sexual selection if females benefit continuously during the different stages of sexual selection. Females may be unable to achieve an optimal mate choice from the outset because of a number of constraints, and the entire sexual selection process can thus be viewed as involving continuous adjustment of female mate choice (Møller 1992). Constraint on female mate choice will depend on the relative reproductive rates of the sexes. If males have a high relative reproductive rate, females may not be particularly restricted in their mate choice because males of preferred phenotypes will be able to copulate with most females that prefer them as mates. Male relative reproductive rates may often be low and thereby prevent many females access to the most preferred individuals. If there is even a small amount of variance in the quality of males, females will be constrained in their mate choice, since not all females with a similar mate preference will have access to a male of the preferred phenotype. This will be the case particularly in pair-bonded species, but also in species without a pair bond, as discussed below. Such constraints will play an important role in the evolution of sperm competition since the competition can be viewed as the outcome of female readjustment of their mate choice at the time of copulation (and fertilization). The clearest example of this phenomenon occurs among socially monogamous animals, such as most birds, in which only the first female to choose a mate will be able to obtain the highest quality social partner (Fig. 2.2). Once the first female is mated, the remaining females are unable to acquire the most attractive male and the discrepancy between the quality of the male that a female obtains and the preferred male increases continuously as a larger number of

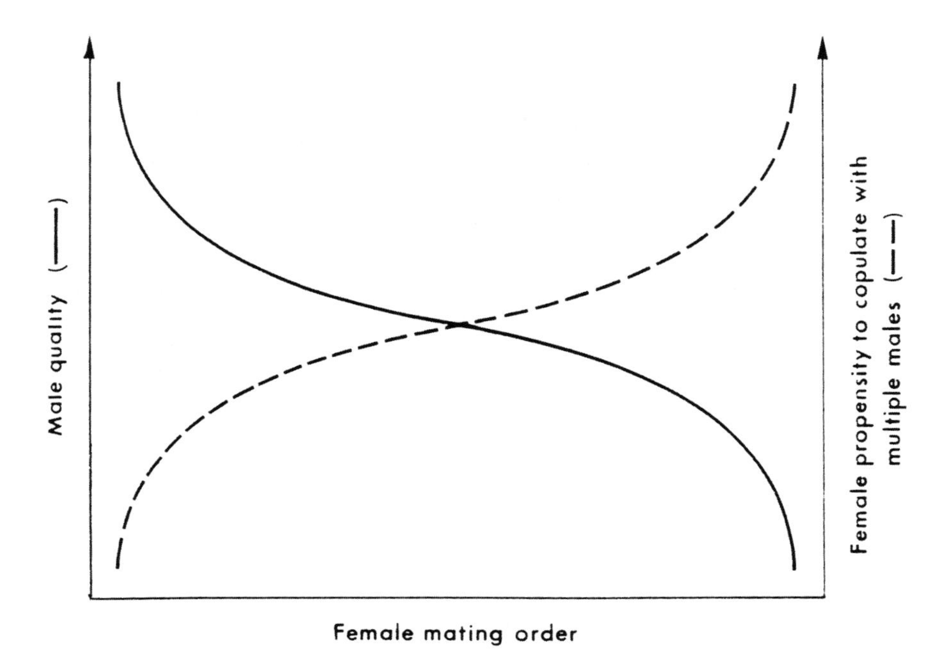

Fig. 2.2. *The propensity of females to engage in copulations with multiple males in relation to the discrepancy between the phenotype of the preferred mate and that of the mate acquired. Adapted from Møller (1992).*

females become mated. Eventually, only a single female will be unmated, and she has to choose among the least preferred males available. Females can adjust their poor choice in terms of a sire for their offspring by raising offspring with their social mate, but engaging in copulations with a more attractive male. A similar scenario applies to other social mating systems, since females even in a lek-breeding situation may be constrained in their choice of sire by the current availability of males. For example, single males account for up to only three-quarters of the total number of copulations in lekking species. This maximum does not result from differences in female mate preference, because females that have copulated with a less preferred male are more likely to engage in a second copulation with another male than females that have copulated with a preferred male. The extent of this constraint on mate choice based on differences in the quality of males will be reflected in the frequency of copulations with multiple males and ultimately paternity. Obviously, males may attempt to control the activities of females to ensure their certainty of paternity, but females have a number of mechanisms available for adjustment of their mate choice at different stages of the reproductive cycle. If females are relatively unconstrained in their mate choice, as in some lek-breeding mating systems, or if males do not differ markedly in genetic quality, there will be little discrepancy between the genetic quality of the male acquired and the genetic quality of males available; females will not therefore be inclined to adjust their mate choice through copulations with other males. Female adjustment of precopulation mate choice at various stages of the reproductive cycle is related to cryptic female mate choice as one mechanism allowing female control of fertilization and ultimately the outcome of sexual selection (Thornhill 1983; Møller 1994; Eberhard 1996).

In the following subsections various aspects of sperm competition and sexual selection will be reviewed: (A) developmental selection and sperm competition; (B) repeatability of paternity and sperm competitive ability; (C) fitness costs to females of sperm competition; (D) sperm competition, paternity and sexual selection; and (E) paternity and the intensity of sexual selection. The first subsection deals with variation in the genetic and phenotypic quality of gametes and how this variation may affect sperm competition and sexual selection. Sexual selection will only arise from sperm competition if individuals are consistent in their mating behaviour. Evidence for such consistency, measured as the repeatability of behaviour and paternity, is reviewed in the second subsection. In the third subsection I briefly review the costs to females which are associated with sperm competition – costs that have to be balanced by fitness benefits in order to render sperm competition an evolutionarily stable strategy for females. In subsection D I review the evidence for sperm competition and paternity being associated with sexual selection. Finally, in subsection E I present evidence for the assumption that sperm competition indeed results in an increase in the variance in male reproductive success and hence gives rise to sexual selection.

A. *Developmental selection*

Gamete phenotypes are usually assumed to be expressed under the control of the parental genome, although gamete control of gamete phenotypes should not be completely ruled out (Parker 1993; Parker and Begon 1993). Individuals differ considerably in their ability to control developmental processes, and the types of cells that differ most among species, such as gametes, often have the least degree of developmental stability, as measured by their phenotypic variance. Abnormal gametes often have inferior fertilizing abilities and may represent genotypes with poor abilities to control developmental processes under current environmental conditions. Developmental selection is a term from the botanical literature referring to the situation when plants (or other organisms) preferentially rear outcrossed offspring in favour of self-fertilized offspring of poorer quality (Buchholz 1922). However, developmental selection may be applied to a greater range of phenomena related to parental selection offspring of variable genetic or phenotypic quality. A recent review of developmental selection against gametes and embryos with developmentally unstable phenotypes has suggested that such selection may be a widespread phenomenon in a wide range of plant and animal species (Møller 1997a).

Because developmental instability has a statistically significant additive genetic component, and offspring therefore resemble their parents with respect to the extent of asymmetry of their phenotype (Møller and Thornhill 1997), individuals that choose partners with respect to symmetry may benefit from such mate choice in terms of production of a greater number of offspring with a symmetric, regular phenotype. A pollination experiment in rosebay willow-herb *Epilobium angustifolium* provides evidence for this mechanism (Møller 1996). The proportion of embryos aborted is frequently very high in outcrossing plants, and may exceed three-quarters in rosebay willow-herb; abortion occurs as a result of developmental errors during the first cell divisions of the embryo. Abortion rate of embryos in different flowers has been directly related to the degree of floral asymmetry in both pollen donors and recipients (Møller 1996). Since floral asymmetry was a heritable character, as estimated from a common garden experiment in which plants were grown under similar environmental conditions (Møller 1996), this result implies that the lack of developmental control was expressed during both development of embryos and development of flowers. A second example is a study of paternity in the common shrew *Sorex araneus* (Stockley *et al.* 1996a). Males with highly asymmetry scent glands, and hence a high degree of developmental instability, had fewer surviving offspring than male shrews with symmetric glands, although they initially produced a similar number (Stockley *et al.* 1996a). Unfortunately, the direct cause of this differential survival is unknown. The very high rates of abortion and infanticide in many plants and animals may result from developmental selection against developmentally unstable phenotypes, and since sexual selection often acts against asymmetry, this type of developmental selection may be considered a component of sexual selection (Møller 1997a). Sperm competition can be considered as a mechanism that allows females to have their eggs fertilized by sperm with superior developmental stability. Similarly, embryos formed by gametes with regular phenotypes may have a greater probability of success than embryos formed by fusion of developmentally unstable gametes.

B. Repeatability of paternity

If sperm competition and the resultant paternity gives rise to sexual selection, we would expect that females mated to a particular male would rate their mating situation similarly on different occasions. In other words, females that were mated to an attractive male, and for that reason remained sexually faithful, would also remain faithful during later reproductive events if they remained paired to the same male. Similarly, a female mated to an unattractive male should repeatedly engage in copulations with other males during subsequent reproductive events. These predictions assume that the attractiveness of a mate and the neighbours does not change between reproductive events. Such changes are unlikely for different reproductive events in the same breeding season, but more likely when considering different breeding seasons. A measure of the consistency of sperm competition is the repeatability of propensity for multiple mating, sperm storage, sperm precedence, sperm and ejaculate features, and ultimately paternity which expresses the variation in paternity among females relative to the variation within females (Falconer 1989). The repeatability of a character is also an upper limit to the heritability of the character (Falconer 1989). There is some evidence for statistically significant consistency in propensity for multiple mating, duration of sperm storage, sperm precedence, and sperm and ejaculate features in various taxa (see Chapters 10, 14 and 16). The repeatability measure of consistency of paternity has not previously been calculated, and the following calculations are based on data available in the literature and by personal communication (Table 2.2). There was a statistically significant repeatability for three out of five species (Table 2.2). Estimates based on very small sample sizes will have a low power of the statistical test, implying that the null hypothesis cannot readily be accepted. Indirect fitness benefits of sperm competition and female choice of sires require a significant amount of variation in paternity among broods or females, which is not necessarily the case for direct fitness benefits of female choice. The current estimates of repeatability of paternity suggest that females may benefit indirectly from extra-pair paternity in some species, while the estimate for the tree swallow is inconsistent with indirect fitness benefits of extra-pair paternity.

Table 2.2. Repeatability (R) of percentage offspring fathered by the resident male in his own nest in different broods by a female mated to the same male, or by different females mated to the same male.

Species	R	*F*	df	*P*	Reference
Agelaius phoeniceus (red-winged blackbird)		Significant			P. Weatherhead, pers. comm.
Dendroica petechia (yellow warbler)	0.29	1.74	27.30	0.071	Yezerinac *et al.* (1995)
Hirundo rustica (barn swallow)	0.72	6.06	15.16	0.0004	Møller and Tegelström (1997)
Parus caeruleus (blue tit)	0.51	3.18	27.31	0.001	B. Kempenaers, pers. comm.
Tachycineta bicolor (tree swallow)	–0.25	0.60	6.8	0.73	Dunn *et al.* (1994) Table 1

C. Fitness costs to females of mixed and extra-pair paternity

If females pay a considerable cost for engaging in copulations with multiple males and raising the resulting offspring, this cost could only be offset if females acquired a balancing benefit in terms of indirect fitness benefits. A range of potential costs of extra-pair paternity have been identified (Birkhead and Møller 1992). I now discuss direct costs of multiple mating, mate desertion, male parental care and nest predation as costs of such paternity.

Females may incur fitness costs from copulating with multiple males, and these range from time and energy expenditure on copulations and risks of predation, parasitism and sexually transmitted disease, to physiological interactions between females and the ejaculates of different males within the female reproductive tract (reviewed by Birkhead and Møller 1992, Eberhard 1996 and Chapter 3 in this volume). The magnitude of these fitness costs have not been precisely quantified.

A number of theoretical models have suggested that females might suffer mate desertion as a consequence of extra-pair paternity (Trivers 1972; Maynard Smith 1977). However, despite a large number of studies there is almost no data suggesting that females incur such a cost in any particular intraspecific study. A comparative study of divorce and extra-pair paternity in birds revealed a positive association between these two variables (Cezilly and Nager 1995). It remains unclear whether the positive association between extra-pair paternity and divorce across species was due to mate desertion being a consequence of sperm competition, or whether a third factor such as sexual selection gave rise to high 'divorce' rates and high frequencies of extra-pair paternity.

Reductions in male parental care as a consequence of copulations with multiple males by the female have been suggested by some theoretical models (Trivers 1972; Ridley 1978; Houston and Davies 1985; Winkler 1987; Whittingham *et al.* 1992; Xia 1992; Westneat and Sherman 1993; Houston 1995), while other models with different assumptions have not predicted such a cost (Maynard Smith 1978; Wittenberger 1979; Grafen 1980; Werren *et al.* 1980; Houston and Davies 1985; Whittingham *et al.* 1992; Houston 1995). If certainty of paternity is the same in all breeding attempts, then paternity should have no effect on the optimal paternal behaviour (Maynard Smith 1978; Grafen 1980). However, certainty of paternity may, under certain conditions affect the optimal male parental effort. This is the case when the probability of future reproduction is high, and when fitness gains from activities other than parental effort are large (Houston 1995). The relationship between paternity and male parental effort may, under other circumstances, be virtually flat or have a shallow slope, and the probability of finding a negative relationship between paternity and male parental effort in empirical studies with their traditionally small sample sizes is negligible (Houston 1995).

Butterfly and bush cricket males adjust the amount of paternal investment in relation to their confidence of paternity (Simmons *et al.* 1993; Cook and Wedell 1996). Early studies of the water bug *Abedus herberti* also suggested a direct relationship between paternal investment and the evolution of courtship patterns and the frequency of multiple copulations (Smith 1979a,b).

A comparative study of male parental care and extra-pair paternity in birds revealed a negative relationship during provisioning of nestlings, but not at other stages of the reproductive cycle such as nest building and incubation (Møller and Birkhead 1993). Males provide a larger share of the food for nestlings in bird species with a low frequency of extra-pair paternity, while males provide little or no food for the offspring in species with a high frequency of extra-pair paternity. A number of studies have investigated whether male parental care is related to certainty of paternity within species, although the majority of purely observational studies cannot be used to evaluate the hypothesis because of the potentially large number of confounding variables (Kempenaers and Sheldon 1997). Studies of the reed bunting *Emberiza schoeniclus* and barn swallow based on comparisons of male parental care for different broods that varied in the proportion offspring fathered by the male, have found clear negative relationships (Dixon *et al.* 1994; Møller and Tegelström 1997), and similar results have been reported for field crickets *Gryllus bimaculatus* (Simmons *et al.* 1993). Such studies are very sensitive owing to the efficient control of a range of potentially confounding variables that may affect the level of male parental care. A number of other studies have not found a relationship between paternity and male parental care (see Chapter 4). However, it is not obvious whether this is due to no relationship being present, the slope between male parental effort and paternity being shallow, or the effect of uncontrolled confounding variables. In conclusion, in some species there is clear evidence for a cost of extra-pair paternity in terms of reduced male parental care.

Females engaged in extra-pair copulations may also pay a cost of extra-pair paternity in terms of increased risks of nest predation. Two mechanisms may be involved. First, the level and intensity of begging calls by the offspring may increase owing to the reduction in relatedness among the offspring and the increase in selfish offspring behaviour (Briskie *et al.* 1994); nest predators may be differentially attracted to nests by the loud calls of unrelated offspring. Second, the risk of nest predation may be elevated in broods with extra-pair paternity if males are less eager to defend offspring where there is a low certainty of paternity (Møller 1991a; Weatherhead *et al.* 1994). A study of the barn swallow revealed that males were less eager to defend their offspring against a model predator when their mates had engaged in extra-pair copulations compared with pairs in which females had not had similar opportunities (Møller 1991a). A second study, of red-winged blackbirds *Agelaius phoeniceus*, demonstrated that males behaved less aggressively towards a potential predator if their nest contained extra-pair offspring (Weatherhead *et al.* 1994). Risks of nest predation were increased by cuckoldry since nests without extra-pair offspring were more likely to survive than nests with some extra-pair offspring, which again were more likely to survive than nests only containing extra-pair offspring (Weatherhead *et al.* 1994). If this cost of extra-pair paternity generally applied to birds, we should expect that nest predation rates would be directly related to the proportion extra-pair offspring present.

In conclusion, there is evidence for copulations with multiple copulations and extra-pair paternity being costly to females in terms of reduced male parental care and increased risks of nest predation. Females can balance such costs of sperm competition only by direct or, more likely, indirect fitness benefits.

D. Copulations with multiple males, paternity and sexual selection

If sperm competition gives rise to sexual selection, we should expect female copulations with multiple males to be related to the expression of male secondary sexual characters or sexual behaviour. The relationship between sperm competition, male display and secondary sexual characters, and sexual selection will be briefly reviewed in the following paragraphs. I will start by reviewing (1) the relative reproductive rates of the two sexes and the opportunity for sexual selection via sperm competition, (2) the relationship between sexual size dimorphism and sperm competition, and, finally (3) the relationship between other aspects of secondary sexual characters, sexual display and sperm competition.

The direction of mating competition, and the control of sexual selection, has been hypothesized to depend on the relative reproductive rates of males and females (Clutton-Brock and Vincent 1991). In species with parental care, the sex with the lower potential reproductive rate should be the main competitor for mates, because mate searching and parental care are mutually exclusive activities (Clutton-Brock and Parker 1992). Recent modelling of the effects of sperm competition on potential reproductive rates, and thus the direction of sexual selection, has provided some general predictions (Simmons and Parker 1996). Sperm competition will generally increase the time males spend not engaged in mate-searching activity because of the time expenditure on sperm replenishment arising from the greater demand for sperm under sperm competition. This should reduce the potential reproductive rate of males, although the effect may not be great because sperm replenishment often accounts for only a small fraction of male time budgets. In species with male parental care, sperm competition may be associated with a decrease in male care, which will generally increase male potential reproductive rate. Hence, sperm competition is predicted to increase the intensity of sexual selection through female choice in this situation. A reduction in male parental care has been found to be associated with the risk of sperm competition in birds (Møller and Birkhead 1993) and the intensity of sexual selection appears to be greater in these species, as determined from their greater degree of sexual dichromatism (Møller and Birkhead 1994). Comparative evidence from birds has also demonstrated that sperm competition favours an increase in sperm expenditure, as demonstrated by the positive relationship between relative testes size and extra-pair paternity (Møller and Briskie 1995). Studies of other taxa should be made to test the generality of these conclusions.

Sexual size dimorphism is common in animals, with the majority of species having larger females than males (Darwin 1871). Large male body size has been investigated with respect to precopulatory male–male competition (Darwin 1871; Andersson 1994), but it has only recently been appreciated that large male size may also be advantageous in sperm competition. Male size covaries with ejaculate size and increased fertilization success in many insects and fish (Rutowski *et al.* 1983; Svärd and Wiklund 1989; Parker 1992; Wedell 1993). A comparative study of the importance of male size for ejaculate and nuptial gift production revealed a positive effect only on ejaculate size (Wedell 1997). This finding suggests that selection for increased ejaculate volume may select simultaneously for increased male body size.

Female copulations with multiple males may sometimes reduce female fitness, and precopulatory struggling may be attempts by females to avoid the costs of multiple copulations (Thornhill and Alcock 1983; Arnqvist 1992). An alternative explanation is that precopulatory struggling functions as a means of female assessment of the male's fitness by testing his endurance and stamina (Cox and LeBoeuf 1977; Arnqvist 1992). A number of independent evolutionarily modified appendages that facilitate coercive copulations for males have been described in a range of different insects such as water striders, scorpionflies, crickets and dung flies (Thornhill 1980; Arnqvist 1989; Sakaluk *et al.* 1995; Allen and Simmons 1996).

Experimental evidence from pheromone trials used for assessment of female mate preferences in the flour beetle *Tribolium castaneum* (Lewis and Austad 1994) demonstrated that male fertilization success (in particular male sperm precedence) is positively correlated with attractiveness. Spermatophylax size (which represents the size of the nuptial gift) is a direct predictor of paternity in bush crickets (Wedell and Arak 1989; Wedell 1991, 1993, 1994; Simmons 1995). Previous studies of scorpionflies have demonstrated a similar function for the size of male nuptial gifts in terms of prey size (Thornhill 1976).

An observational study of the black-capped chickadee *Parus atricapillus* over a period of several years revealed that some females made excursions out of the territory of their mate in order to copulate with an additional male (Smith 1988). The males visited were generally dominant and Smith argued that females may obtain indirect fitness benefits by copulating with such males. A subsequent study of paternity and social dominance of males revealed a direct relationship between the number of extra-pair offspring sired by males and their rank (Otter *et al.* 1994).

An experimental test of the same hypothesis was provided for the barn swallow by means of tail-length manipulation (male tail-length is a secondary sexual character currently subject to a directional female mate preference; Møller 1988a). Female barn swallows were more likely to engage in extra-pair copulations with males that had their tails elongated than with tail-shortened males. Interestingly, it was the females mated to males with shortened tails that were more likely to engage in extra-pair copulations, while females mated to those with elongated tails were less likely to engage in extra-pair copulations. A subsequent study of extra-pair copulations and parasite loads in the barn swallow revealed that males that were successful in engaging in extra-pair copulations had fewer ectoparasites than unsuccessful males (Møller 1991b). Females involved in extra-pair copulations thus may have a small risk of contracting ectoparasites from their copulation activities, even though they prefer to copulate with males that frequently engage in extra-pair copulations.

Not all copulations result in sperm transfer and hence in fertilization, and it is therefore essential to determine whether extra-pair paternity is directly related to secondary sexual characters. For example, it is possible that males simply exchange paternity in female copulations with multiple males, and that such copulations do not result in an increase in the variance in male reproductive success. The most extensive data set based on an experimental approach is from the barn swallow: extra-pair paternity was determined in another tail manipulation experiment in which tail feathers were either shortened, elongated, manipulated, or left as untreated controls (Saino *et al.* 1997). Paternity was determined by analysis of microsatellite polymorphisms. The proportion offspring sired by the resident male was strongly positively related to experimental tail length, increasing from slightly more than 40% among broods of tail-shortened males to 60% among broods of the two control groups and to almost 90% among broods of tail-elongated males (Saino *et al.* 1997; Fig. 2.3a). The number of extra-pair offspring sired in other nests showed a similar pattern (Saino *et al.* 1997). Hence, reproductive success increased with experimental tail length (Fig. 2.3b). There was also an effect of original tail length because males with originally long tails sired a larger number offspring in their own nest and had a larger number of extra-pair offspring in other nests than short-tailed males. A previous Canadian study of extra-pair paternity in a small sample of 11 pairs of barn swallows revealed a positive association between original tail length and paternity, while experimental treatment (in this case either tail shortening or tail elongation) did not affect extra-pair paternity (Smith *et al.* 1991). However, experimental treatment was confounded by original phenotype in this study since males that received the elongation treatment had a tail length that was almost significantly shorter than that of males with shortened tails.

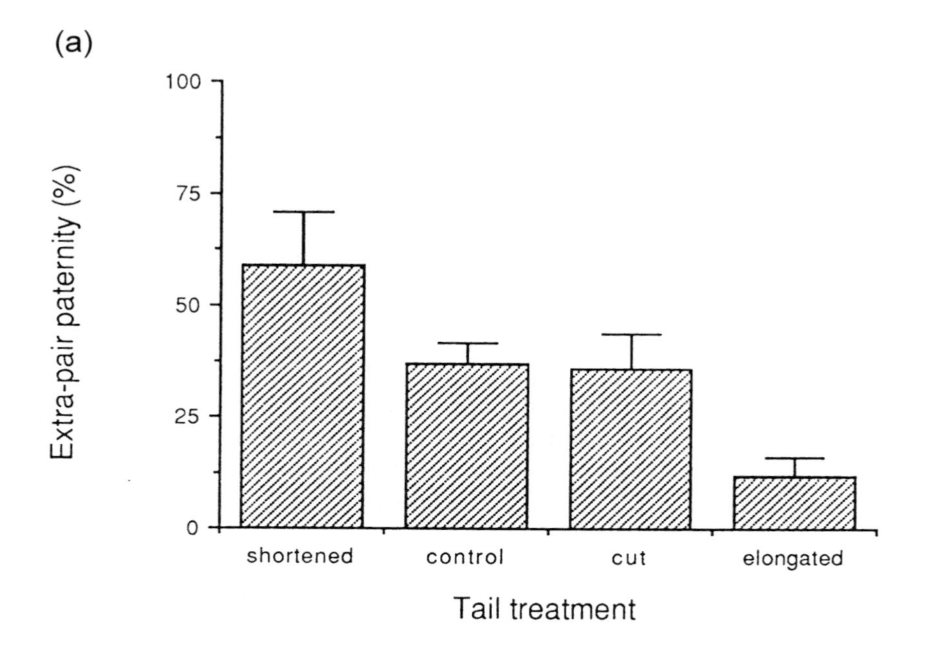

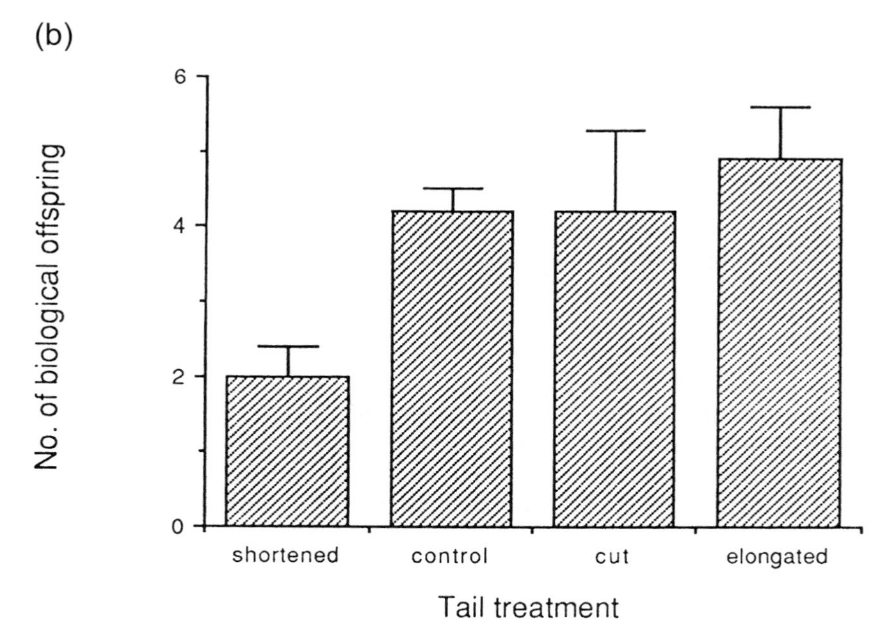

Fig. 2.3. *Percentage of extra-pair offspring in the nest of the resident male (a), and the number of biological offspring ((b) number of offspring sired in own nest and other nests) in relation to tail length manipulation of male barn swallows* Hirundo rustica. *Adapted from Saino* et al. *(1997).*

A study of extra-pair copulations in house sparrows *Passer domesticus* investigated female engagement in female-solicited and in apparently forced extra-pair copulations (Møller 1992). Male house sparrows participating in extra-pair copulations had very large black throat badges (a secondary sexual character subject to a directional female mate preference; Møller 1988b), and the female was more likely to be involved in an extra-pair copulation particularly when the male involved had a badge larger than that of her social mate. A similar result was obtained from the zebra finch in which male attractiveness depends on the colour of the bill of males (Burley and Price 1991; Houtman 1992; Burley *et al.* 1994). Individual males with attractive red bills had greater success in extra-pair copulations than males with a less attractive colour. Experimental manipulation of the colour of rings in captive zebra finches resulted in females mated to males wearing the preferred red-coloured rings having no or very few extra-pair offspring, while females mated to males with less preferred green-coloured rings more often had extra-pair young in their broods (Burley *et al.* 1996).

A relationship between male coloration and extra-pair paternity was documented in the sexually dichromatic yellowhammer *Emberiza citrinella* (Sundberg and Dixon 1996). Males are, on average, more brightly coloured than females, and males successful in acquiring extra-pair paternity were more brightly yellow than cuckolded males. In this species male coloration is closely associated with age, and yellow colour thus appears to reliably reflect male survival prospects. Males with extra-pair offspring were therefore also on average older than cuckolded males.

Three other studies of extra-pair paternity and the expression of sexual characters in birds did not reveal any relationship between plumage characters and paternity (Hill *et al.* 1994; Rätti *et al.* 1995; Weatherhead and Boag 1995). The studies by Hill *et al.* (1994) and Weatherhead and Boag (1995) did not consider the possibility that male attractiveness relative to neighbours may have influenced extra-pair paternity. The experimental study of extra-pair paternity in the pied flycatcher *Ficedula hypoleuca*, in which males have, on average, darker coloration than females, did not reveal a difference in the probability of broods containing extra-pair offspring related to male colour, although females had easy access to two males differing considerably in coloration (Rätti *et al.* 1995).

Bird song has evolved in the context of sexual selection with song having a dual function of mate attraction and deterrence of competitor males (Catchpole and Slater 1995). Repertoire size has been found to be attractive to females, with females giving more intense copulation solicitation displays to playback of songs with larger repertoires (Catchpole and Slater 1995). Song features and paternity was investigated in the great reed warbler *Acrocephalus arundinaceus* by Hasselquist *et al.* (1996). Previous studies had determined that females indeed gave more intense copulation displays to a larger repertoire (Catchpole and Slater 1995). Although extra-pair paternity was relatively uncommon in the great reed warbler, with only 3.1% of all offspring being sired by extra-pair males, it was possible to determine the association between extra-pair paternity and song features of males. A total of seven broods with extra-pair paternity were identified, and the extra-pair male was identified by means of multilocus DNA-fingerprinting (Hasselquist *et al.* 1996). The extra-pair male had a larger repertoire in all seven cases, differing significantly from the null expectation of equally many extra-pair males with smaller and larger repertoires.

A second study of song characteristics and extra-pair paternity, in the barn swallow, revealed an interaction between the expression of a secondary sexual character (tail length) and sexual behaviour (song rate) (Møller *et al.* 1997). The proportion offspring sired by the resident male was positively related to both tail length and male song rate, although song rate was of greater importance for males with long tails, compared with males with short tails. Hence, an unattractive male benefited less from a high song rate than an attractive male. Song rate varies according to male condition while male tail length is fixed each year. Hence, if females use both secondary sexual characters and sexual behaviour as cues in their choice of extra-pair males, they should place less emphasis on behavioural characters that are subject to short-term temporal changes in male condition. The differential importance of song rate in relation to the size of the secondary sexual character prevents unattractive males from cheating by singing at a level that exceeds their quality (Møller *et al.* 1997).

Indirect evidence of sperm competition giving rise to sexual selection has been found in a number of studies. In the blue tit *Parus caeruleus* some females make excursions out of the territories of their mates to copulate with an extra-pair male (Kempenaers *et al.* 1992). The rate of such excursions was inversely proportional to the rate of visits to a focal territory by other females, suggesting that some males that had faithful mates received many visits by neighbouring females, while males with unfaithful females received few or no visits by neighbouring females. Female blue tits appeared to obtain an indirect fitness benefit from their extra-pair copulations because extra-pair offspring were much more likely to recruit to the local breeding population than within-pair offspring. This suggests that extra-pair offspring had greater viability. Unfortunately, it was unknown which male character females used for selecting a sire for their extra-pair offspring. Data on extra-pair copulations and extra-pair paternity in the shag *Phalacrocorax aristotelis* also suggest that females choose high quality males as extra-pair sires (Graves *et al.* 1993).

Several non-avian examples of copulations with multiple males relating to a character subject to a female mate preference have been described. Studies of crickets have demonstrated that females take more sperm from preferred males, allowing spermatophores to remain attached for a longer time and multiply mating with preferred males, thereby biasing paternity through the outcome of sperm competition (Simmons

1987; Sakaluk 1984; see further examples Chapter 10). Men with faces that were rated attractive by women have a high degree of body symmetry, reflecting an ability to undergo stable development of the phenotype during ontogeny (Thornhill and Gangestad 1993; Gangestad *et al.* 1994). A subsequent study of self-reported sexual behaviour in relation to body asymmetry revealed a strongly positive relationship between the age of first sexual intercourse and the number of lifetime sexual partners, respectively, and the degree of body symmetry (Thornhill and Gangestad 1994). A subsequent observational study of symmetry and sexual behaviour in humans in another population demonstrated a direct relationship between body symmetry and the frequency of extra-pair copulations (Baker 1997). The probability of fertilization in humans is supposed to be related to the frequency of high sperm retention orgasms, and women had such orgasms particularly more frequently when copulating with men with a high degree of body symmetry (Thornhill *et al.* 1995).

If sperm competition is currently playing an important role in sexual selection, we should also expect extra-pair paternity to be associated with secondary sexual characters across species. There is some empirical evidence for this prediction. First, a comparative study of extra-pair paternity and plumage brightness in birds revealed a strong positive association between extra-pair paternity and sexual dichromatism (Møller and Birkhead 1994). Extra-pair paternity was virtually absent in sexually monochromatic bird species, but reached quite high levels in species with considerable sexual dichromatism. Interestingly, this relationship was independent of confounding variables such as the frequency of polygyny which traditionally has been assumed to result in intense sexual selection (Darwin 1871). This result suggested that variation in reproductive success due to sperm competition rather than variation in reproductive success due to polygyny was the most important component of sexual selection.

A second comparative study investigated the relationship between immune defence, secondary sexual characters and extra-pair paternity in birds (Møller 1997b). The relative size of the spleen, which plays an important role in the production of antibodies, was used as a measure of immune defence. Bird species with a high frequency of extra-pair paternity had both a large degree of sexual dichromatism and a relatively large spleen (Møller 1997b). If sexual dichromatism reliably signals immunocompetence, then we should not expect a relationship between sexual dichromatism and the relative size of the spleen after having accounted for the relationship with extra-pair paternity. This was exactly the result found, suggesting that interspecific differences in sexual dichromatism have evolved to signal immunocompetence, and that this signalling is directly involved in the evolution of copulations with multiple males for acquisition of genetic benefits associated with parasite resistance (Møller 1997b).

In conclusion, there is evidence for an association between female copulations with multiple males, paternity and secondary sexual characters, implying that sperm competition plays an important role in sexual selection. Females appear to obtain indirect fitness benefits in terms of viability and parasite resistance from their choice of sires.

E. Paternity and measures of the intensity of sexual selection

Although variation in paternity is related to sexual display or secondary sexual characters, this does not imply that variation in paternity results in an increase in the variance in male reproductive success. A number of indices of sexual selection have been proposed, but I will concentrate on the standardized variance in male success which is defined as the variance divided by the squared mean value (Wade and Arnold 1980). Studies of paternity have found positive correlations between the paternity in a male's own nest and the number of chicks fathered on other males' territories (Gibbs *et al.* 1990; Weatherhead and Boag 1995; Webster *et al.* 1995). Estimates of the standardized variance in male success was higher for realized as compared to apparent male reproductive success in all eight studies available (Table 2.3). Hence, there is evidence that variance in paternity results in an increase in the variance in male reproductive success and thus increases the opportunity for sexual

Table 2.3. The standardized variance in apparent and realized male reproductive success (after correcting for paternity) in different bird species.

Species	Standardized variance in			
	Apparent success (V_A)	Realized success (V_R)	V_R/V_A	Reference
Agelaius phoeniceus (red-winged blackbird)	0.25	0.39	1.56	Gibbs *et al.* (1990)
	1.02	1.205	1.18	Westneat and Sherman (1993)
Delichon urbica (house martin)	0.06	0.31	5.17	Whittingham and Lifjeld (1995)
Dendroica petechia (yellow warbler)	0.035	0.532	15.20	Yezerinac *et al.* (1995)
Hirundo rustica (barn swallow)	0.482	0.494	1.02	A. P. Møller and N. Saino unpublished data
Parus atricapillus (black-capped chickadee)	0.043	0.351	8.16	K. Otter pers. comm.
Progne subis (purple martin)	0.049	0.328	6.69	R. Wagner pers. comm.
Wilsonia citrina (hooded warbler)	0.18	0.46	2.56	Stutchbury *et al.* (1997)

selection. Webster *et al.* (1995) provided a method for partitioning of variance and covariance components of sexual selection into components due to social and sexual mates, respectively. This method allows for sophisticated tests of how components of sexual selection such as the number of social and sexual mates, and their interactions, translate into reproductive success.

IV. SPERM COMPETITION, SEXUALLY TRANSMITTED DISEASES, AND SEXUAL SELECTION

Sperm competition and multiple mating results in enhanced probabilities of horizontal transfer of parasites among individual hosts. Parasite virulence is defined as the amount of damage inflicted by parasites on their hosts, and a number of hypotheses for the evolution of virulence have been proposed. Two hypotheses with empirical, experimental and comparative support posit that the evolution of virulence is directly related to the relative frequency of horizontal transfer of parasites among hosts, and that the evolution of virulence is related to the frequency of multiple parasite infections, respectively (Anderson and May 1982; Ewald 1983; reviewed by Bull 1994). Horizontal transmission of parasites as opposed to vertical transfer can take place as efficiently among social mates in polygynous or polyandrous mating systems as among sexual mates during copulations. Similarly, contact between social and sexual mates also increases the probability of multiple parasite infections. As sexual selection, as mediated by increased variance in male mating success, becomes more important the virulence of parasites transmitted during copulation (sexually transmitted diseases) is expected to increase (Hamilton 1990). It is important to distinguish between sexually transmitted diseases and parasites that are fortuitously transmitted during copulation. The latter will be transmitted during any kind of body contact, even among individuals without a sexual relationship, and the frequency of their transmission will therefore not be very dependent on the frequency of copulations with more than a single partner. The evolution of mating systems and strict sexual monogamy may thus depend on interactions with sexually transmitted diseases. Reductions in the variance in male mating success will result in reduced parasite virulence and give rise to a new increase in the variance in mating success (Hamilton 1990; Sheldon 1993).

The empirical evidence for the effects of sexually transmitted diseases is remarkably poor, with the exception of humans (Anderson and May 1992; Lockhart *et al.* 1996). There is some evidence that human sexual behaviour has changed as a result of a sexually transmitted disease becoming common, but the extent to which this is due to information campaigns or to the biological impact of the sexually transmitted disease on sexual behaviour remains unknown (Anderson and May 1992). A possible exception was reported by Ewald (1994) who suggested that differences in transmission of human immunodeficiency virus (HIV) have had a direct impact on the evolution of virulence of this microparasite; however, further confirmation of this relationship is needed. A number of sexually transmitted diseases have been described in animals, but many of these appear not to be truly sexually transmitted (Sheldon 1993; Lockhart *et al.* 1996). For example, *Mycoplasma cloacale*, which is sometimes an important cause of disease in geese *Anser anser* (Stipkovits *et al.* 1986), cannot be classified as a true sexually transmitted disease because the dynamics of the spread of the disease and the high rates of mortality reported cannot be accounted for by the relatively rare cases of extra-pair copulations in this species, even in captivity. A recent study of the tree swallow demonstrated a striking similarity in the cloacal bacterial fauna of pair members, which deviated significantly from random expectation (Lombardo *et al.* 1996). Some of these bacteria subsequently had significant negative effects on the growth of offspring.

Sexually transmitted diseases in the form of smuts fungi are common in flowering plants of the family Caryophyllaceae (Roy 1994). Sexual reproduction in many plants differs from sexual reproduction in animals by the involvement of a third party, the pollinator, for transfer of pollen among flowers. The dynamics of the pathogen and the host thus become more complicated than the dynamics of sexually transmitted diseases in animals. There is clear evidence of parasite manipulation of host plants, since infected plants flower earlier and for a longer time than uninfected plants. The spread of the disease is promoted by the size and the number of flowers produced and the timing of flowering, and the plant thus faces the dilemma of trading reproductive success against the risk of becoming infected with the fungus (Klatz and Schmid 1995). The presence of the venereal fungus thus can be considered to select against floral attractiveness. There is genetically based resistance to the fungus and this resistance is genetically correlated with flowering time (Klatz and Schmid 1995). Resistant plants flower later than susceptible plants, and such a delay in flowering may be the cost that plants pay for being resistant. The relative importance of plant venereal diseases for pollination frequency (multiple mating) remains unknown, although venereal diseases caused by smuts fungi in plants are unlikely to be a general factor determining the extent of pollen competition owing to their restricted taxonomic distribution.

In conclusion, although sexually transmitted diseases may have important implications for the frequency of copulations with multiple mates, there is very little theoretical work and even less empirical animal work available to test theoretical assumptions and predictions. Pollinator-transmitted smuts fungi act as sexually transmitted diseases and provide a well-studied model system for investigations of the importance of parasites for the evolution of multiple mating.

V. CONCLUSIONS

The main aims of this chapter have been twofold; to place sperm competition within the framework of sexual selection, and to review evidence for sexual selection being an outcome of sperm competition. Sperm competition may result in sexual selection at a number of different levels. These range from the anatomical and physiological adaptations that enhance fertilization by individuals of superior phenotypic or genetic quality, or superior attractiveness in the context of sperm competition, to female choice of multiple copulation partners based on the secondary sexual characteristics of males. There is virtually no intraspecific evidence for sperm competition at the anatomical or physiological level resulting in sexual selection, and the scope for novel insights thus is tremendous. For example, there are no rigorous empirical studies of sexual selection arising from variation in male genitalia and female reproductive physiology. The magnitude of the sexual selection components thus remains unknown, as does the potential types of fitness benefits acquired by females. Given the interspecific patterns of anatomical and physiological adaptations to sperm competition it is likely that these mechanisms will also prove to be important at the intraspecific level. A particularly interesting question is whether anatomical and physiological adaptations will be found to be positively correlated with the expression of secondary sexual characters among individuals adopting a particular mating strategy. This is the case for testes size in some bird species (Møller 1994), but apparently also applies to other anatomical and physiological properties such as sperm velocity, as demonstrated for the barn swallow (T. R. Birkhead *et al.* unpublished data). This identifies a very important model system that allows integration of the study of sexual selection mediated through sperm competition with the study of precopulation sexual selection. For alternative mating strategies, males that adopt sneak behaviour generally have the least elaborate secondary sexual characters, but invest heavily in sperm production, while males of the other mating strategy have elaborate secondary sexual characters and invest relatively little in sperm production (Parker 1990a,b).

An increasing number of empirical studies have revealed that females pay considerable costs for engaging in copulations with multiple males. These costs range from reduced paternal care to increased risks of predation on offspring. Females must acquire considerable indirect fitness benefits from their copulations with multiple males in order to balance the evolutionary accounts. Empirical work has suggested that there is sometimes consistency in paternity among different broods by the same female or among different females of the same male. This suggests that some males are considered better mates, or alternatively are better able to use efficient paternity guards than others. The latter explanation appears less likely because neither frequent within-pair copulations nor mate guarding appear to be efficient paternity guards when comparisons are made among pairs of a single species.

Several studies have now shown that females apparently use the expression of secondary sexual characters of males when choosing a copulation partner, and such copulations transfer directly into paternity. The intensity of sexual selection also appears to increase as a result of sperm competition. The fitness benefits that females acquire as a result of copulations with multiple males generally remain obscure with a few notable exceptions (Kempenaers *et al.* 1992; Hasselquist *et al.* 1996). Sperm competition in animals in general has great promise as a model system for studies of indirect fitness benefits because confounding environmental effects and direct fitness benefits can be excluded more easily as alternative explanations.

Finally, the role of sexually transmitted diseases in sperm competition and sexual selection remains to be elucidated. There is virtually no empirical data available from free-living animals while an accumulating amount of evidence suggests that smuts fungi transmitted by pollinating insects may play a very important role in the evolution of some plant mating systems.

ACKNOWLEDGEMENTS

I am grateful for critical comments provided by T. R. Birkhead, B. Sheldon and L. W. Simmons. P. Dunn, B. Kempenaers, K. Otter, R. J. Robertson, B. J. Stutchbury, R. Wagner, and P. J. Weatherhead kindly provided unpublished information. My research was supported by grants from the Swedish and Danish Natural Science Research Councils.

REFERENCES

Allen GR & Simmons LW (1996) Coercive mating, fluctuating asymmetry and male mating success in the dung fly *Sepsis cynipsea*. *Anim. Behav.* **52:** 737–741.

Anderson RM & May RM (1982) Coevolution of hosts and parasites. *Parasitology* **85:** 411–426.

Anderson RM & May RM (1992) *Infectious Diseases of Humans: Dynamics and Control.* Oxford University Press, Oxford.

Andersson M (1994) *Sexual Selection.* Princeton University Press, Princeton.

Arnqvist G (1989) Sexual selection in a water strider: the function, mechanism of selection and heritability of a male grasping apparatus. *Oikos*, **56:** 344–350.

Arnqvist G (1992) Spatial variation in selective regimes: sexual selection in the water strider, *Gerris odontogaster*. *Evolution* **46:** 914–929.

Baker RR (1997) Copulation, masturbation and infidelity: State-of-the-art. In *Proceedings of International Society of Human Ethology*. Plenum, New York (in press).

Birkhead TR & Fletcher F (1995) Male phenotype and ejaculate quality in the zebra finch *Taeniopygia guttata*. *Proc. Roy. Soc. Lond. B* **262:** 329–334.

Birkhead TR & Møller AP (1992) *Sperm Competition in Birds: Evolutionary Causes and Consequences*. Academic Press, London.

Birkhead TR, Sutherland WJ & Møller AP (1993) Why do females make it so difficult for males to fertilize their eggs? *J. Theor. Biol.* **161:** 51–60.

Bishop JDD (1996) Female control of paternity in the internally fertilizing compound ascidian *Diplisoma listerianum*. I. Autoradiographic investigation of sperm movements in the female reproductive tract. *Proc. Roy. Soc. Lond. B* **263:** 369–376.

Bishop JDD, Jones CS & Noble LR (1996) Female control of paternity in the internally fertilizing compound ascidian *Diplisoma listerianum*. II. Investigation of male mating success using RAPD markers. *Proc. Roy. Soc. Lond. B* **263:** 401–407.

Boggs CL & Gilbert LE (1979) Male contribution to egg production in butterflies: evidence for transfer of nutrients at mating. *Science* **206:** 83–84.

Briskie JV, Naugler CT & Leech SM (1994) Begging intensity of nestling birds varies with sibling relatedness. *Proc. Roy. Soc. Lond. B* **258:** 73–78.

Buchholz JT (1922) Developmental selection in vascular plants. *Bot. Gaz.* **73:** 249–286.

Bull JJ (1994) Virulence. *Evolution* **48:** 1423–1437.

Burke T, Davies NB, Bruford MW & Hatchwell BJ (1989) Parental care and mating behaviour of polyandrous dunnocks *Prunella modularis* related to paternity by DNA fingerprinting. *Nature* **338:** 249–251.

Burley NT & Price DK (1991) Extra-pair copulation and attractiveness in zebra finches. *Acta XX Congr. Int. Ornithol.* 1367–1372.

Burley NT, Enstrom DA & Chitwood L (1994) Extra-pair relations in zebra finches: differential male success results from female tactics. *Anim. Behav.* **48:** 1031–1041.

Burley NT, Parker PG & Lundy K (1996) Sexual selection and extrapair fertilization in a socially monogamous passerine, the zebra finch (*Taeniopygia guttata*). *Behav. Ecol.* **7:** 218–226.

Capula M & Luiselli L (1994) Can female adders multiply? *Nature* **369:** 528.

Catchpole CK & Slater PJB (1995) *Bird Song: Biological Themes and Variations*. Cambridge University Press, Cambridge.

Cezilly F & Nager RG (1995) Comparative evidence for a positive association between divorce and extra-pair paternity in birds. *Proc. Roy. Soc. Lond. B* **262:** 7–12.

Clutton-Brock TH & Parker GA (1992) Potential reproductive rates and the operation of sexual selection. *Quart. Rev. Biol.* **67:** 437–456.

Clutton-Brock TH & Vincent ACJ (1991) Sexual selection and the potential reproductive rates of males and females. *Nature* **351:** 58–60.

Colwell MA & Oring LW (1989) Extra-pair mating in the spotted sandpiper: A female mate acquisition tactic. *Anim. Behav.* **38:** 675–684.

Cook PA & Wedell N (1996) Ejaculate dynamics in butterflies: a strategy for maximizing fertilization success? *Proc. Roy. Soc. Lond. B* **263:** 1047–1051.

Cordero C (1995) Ejaculate substances that affect female reproductive physiology and behavior – honest or arbitrary traits. *J. Theor. Biol.* **174:** 453–461.

Cordero C & Miller PL (1992) Sperm transfer, displacement and precedence in *Ischnura graellsii* (Odonata: Coenagrionidae). *Behav. Ecol. Sociobiol.* **30:** 261–267.

Cox CR & LeBoeuf B (1977) Female incitation of male competition: a mechanism in sexual selection. *Am. Nat.* **111:** 317–335.

Darwin C (1871) *The Descent of Man, and Selection in Relation to Sex*. John Murray, London.

Davies NB (1985) Cooperation and conflict among dunnocks, *Prunella modularis*, in a variable mating system. *Anim. Behav.* **33:** 628–648.

Dixon A, Ross D, O'Malley SLC & Burke T (1994) Paternal investment inversely related to degree of extra-pair paternity in the reed bunting. *Nature* **371:** 698–700.

Dunn PO, Whittingham LA, Lifjeld JT, Robertson RJ & Boag PT (1994) Effects of breeding density, synchrony, and experience on extra-pair paternity in tree swallows. *Behav. Ecol.* **5:** 123–129.

Dziuk PJ (1996) Factors that influence the proportion offspring sired by a male following heterospermic insemination. *Anim. Reprod. Sci.* **43:** 65–88.

Eberhard WG (1985) *Sexual Selection and Animal Genitalia*. Harvard University Press, Cambridge, MA.

Eberhard WG (1993) Evaluating models of sexual selection: Genitalia as a test case. *Am. Nat.* **142:** 564–571.

Eberhard WG (1996) *Female Control: Sexual Selection by Cryptic Female Choice*. Princeton University Press, Princeton.

Eberhard WG & Cordero C (1995) Sexual selection by cryptic female choice on male seminal products: a new bridge between sexual selection and reproductive physiology. *Trends Ecol. Evol.* **10:** 493–496.

Emlen ST (1978) Cooperative breeding. In *Behavioural Ecology: An Evolutionary Approach*. JR Krebs & NB Davies (eds), pp. 245–281. Blackwell, Oxford.

Ewald PW (1983) Host-parasite relations, vectors, and the evolution of disease severity. *Annu. Rev. Ecol. System.* **14:** 465–485.

Ewald PW (1994) *Evolution of Infectious Disease*. Oxford University Press, Oxford.

Falconer DS (1989) *An Introduction to Quantitative Genetics*, 3rd edn. Longman, New York.

Fisher RA (1930) *The Genetical Theory of Natural Selection*. Clarendon Press, Oxford.

Gage MJG (1994) Association between body size, mating pattern, testis size and sperm lengths across butterflies. *Proc. Roy. Soc. Lond. B* **258:** 247–254.

Gage MJG, Stockley P & Parker GA (1995) Effects of alternative male mating strategies on characteristics of sperm production in the Atlantic salmon (*Salmo salar*): theoretical and empirical investigations. *Phil.l Trans. Roy. Soc. Lond. B* **350:** 391–399.

Gangestad SW, Thornhill R & Yeo RA (1994) Facial attractiveness, developmental stability, and fluctuating asymmetry. *Ethol. Sociobiol.* **15:** 73–85.

Gibbs HL, Weatherhead PJ, Boag PT, White BN, Tabak LM & Hoysak DJ (1990) Realized reproductive success of polygynous red-winged blackbirds revealed by DNA markers. *Science* **250:** 1394–1397.

Gomendio M & Roldan ERS (1993) Coevolution between male ejaculates and female reproductive biology in eutherian mammals. *Proc. Roy. Soc. Lond. B* **252:** 7–12.

Grafen A (1980) Opportunity cost, benefit and degree of relatedness. *Anim. Behav.* **28:** 967–968.

Grafen A (1990) Sexual selection unhandicapped by the Fisher process. *J. Theor. Biol.* **144:** 473–516.

Graves J, Ortega-Ruano J & Slater PJB (1993) Extra-pair copulations paternity in shags: do females choose better males? *Proc. Roy. Soc. Lond. B* **253:** 3–7.

Gwynne DT (1984) Courtship feeding increases female reproductive success in bushcrickets. *Nature* **307:** 361–363.

Hamilton WD (1990) Mate choice near and far. *Am. Zool.* **30:** 341–352.

Hamilton WD & Zuk M (1982) Heritable true fitness bright birds: A role for parasites? *Science* **218:** 384–387.

Harcourt AH, Harvey PH, Larson SG & Short RV (1981) Testes weight, body weight and breeding system in primates. *Nature* **293:** 55–57.

Hasselquist D, Bensch S & von Schantz T (1996) Correlation between male song repertoire, extra-pair fertilizations and offspring survival in the great reed warbler. *Nature* **381:** 229–232.

Heg D, Ens BJ, Burke T, Jenkins L & Kruijt JP (1993) Why does the typically monogamous oystercatcher *Haematopus ostralegus* engage in extra-pair copulations? *Behaviour* **126:** 247–289.

Heywood JS (1989) Sexual selection by the handicap mechanism. *Evolution* **43:** 1387–1397.

Hill GE, Montgomerie R, Roeder R & Boag P (1994) Sexual selection and cuckoldry in a monogamous songbird: Implications for sexual selection theory. *Behav. Ecol. Sociobiol.* **35:** 193–199.

Hoelzer GA (1989) The good parent process of sexual selection. *Anim. Behav.* **38:** 1067–1078.

Houston AI (1995) Parental effort and paternity. *Anim. Behav.* **50:** 1635–1644.

Houston AI & Davies NB (1985) The evolution of cooperation and life history in the dunnock. In *Behavioural Ecology: Ecological Consequences of Adaptive Behaviour.* RM Sibly & RH Smith (eds), pp. 471–488. Blackwell, Oxford.

Houtman AE (1992) Female zebra finches choose extra-pair copulations with genetically attractive males. *Proc. Roy. Soc. Lond. B* **249:** 3–6.

Hrdy SB (1977) *The Langurs of Abu.* Harvard University Press, Cambridge, MA.

Huxley JS (1912) A 'disharmony' in the reproductive habits of the wild duck (*Anas boschas* L.). *Biologisch. Zentralbl.* **32:** 621–623.

Iwasa Y, Pomiankowski A & Nee S (1991) The evolution of costly mate preferences. II. The 'handicap' principle. *Evolution* **45:** 1431–1442.

Keller L & Reeve HK (1995) Why do females mate with multiple males? The sexually selected sperm hypothesis. *Adv. Study Behav.* **24:** 291–315.

Kempenaers B & Sheldon BC (1997) Studying paternity and paternal care: pitfalls and problems. *Anim. Behav.* **53:** 423–427.

Kempenaers B, Verheyen GR, Broeck, M. v. d, Burke T, Broeckhoven CV & Dhondt AA (1992) Extra-pair paternity results from female preference for high-quality males in the blue tit. *Nature* **357:** 494–496.

Klatz O & Schmid B (1995) Plant venereal disease: A model for integrating genetics, ecology and epidemiology. *Trends Ecol. Evol.* **10:** 221–222.

Lewis SM & Austad SN (1990) Sources of intraspecific variation in sperm precedence in red flour beetles. *Am. Nat.* **135:** 351–359.

Lewis SM & Austad SN (1994) Sexual selection in flour beetles: the relationship between sperm precedence and male olfactory attractiveness. *Behav. Ecol. Sociobiol.* **5:** 219–224.

Lockhart AB, Thrall PH & Antonovics J (1996) Sexually transmitted disease in animals: ecological and evolutionary implications. *Biol. Rev.* **71:** 415–471.

Lombardo MP, Thorpe PA, Cichewicz R, Henshaw M, Millard C, Steen C & Zeller TK (1996) Communities of cloacal bacteria in tree swallow families. *Condor* **98:** 167–172.

Madsen T, Shine R, Loman J & Håkansson T (1992) Why do female adders copulate so frequently? *Nature* **355:** 440–441.

Maynard Smith J (1977) Parental investment: A prospective analysis. *Anim. Behav.* **25:** 1–9.

Maynard Smith J (1978) Optimization theory in evolution. *Annu. Rev. Ecol. System.* **9:** 31–56.

McKinney F, Derrickson SR & Mineau P (1983) Forced copulation in waterfowl. *Behaviour* **86:** 250–294.

Miller PL (1987) Sperm competition in *Ischnura elegans* (Vander Linden) (Zygoptera: Coenagrionidae). *Odontologica* **16:** 201–208.

Møller AP (1988a) Female choice selects for male sexual tail ornaments in the monogamous swallow. *Nature* **322:** 640–642.

Møller AP (1988b) Badge size in the house sparrow *Passer domesticus*: effects of intra- and intersexual selection. *Behav. Ecol. Sociobiol.* **22:** 373–378.

Møller AP (1988c) Infanticidal and anti-infanticidal strategies in the swallow *Hirundo rustica. Behav. Ecol. Sociobiol.* **22:** 365–371.

Møller AP (1988d) Ejaculate quality, testes size and sperm competition in primates. *J. Hum. Evol.* **17:** 479–488.

Møller AP (1988e) Testes size, ejaculate quality and sperm competition in birds. *Biol. J. Linn. Soc.* **33:** 273–283.

Møller AP (1989) Ejaculate quality, testes size and sperm competition in mammals. *Funct. Ecol.* **3:** 91–96.

Møller AP (1991a) Defence offspring by male swallows, *Hirundo rustica*, in relation to participation in extra-pair copulations by their mates. *Anim. Behav.* **42:** 261–267.

Møller AP (1991b) Parasites, sexual ornaments and mate choice in the barn swallow *Hirundo rustica*. In *Ecology, Behavior, and Evolution of Bird-Parasite Interactions.* JE Loye & M Zuk (eds), pp. 328–343. Oxford University Press, Oxford.

Møller AP (1992) Frequency of female copulations with multiple males and sexual selection. *Am. Nat.* **139:** 1089–1101.

Møller AP (1994) *Sexual Selection and the Barn Swallow.* Oxford University Press, Oxford.

Møller AP (1996) Floral asymmetry, embryo abortion, and developmental selection in plants. *Proc. Roy. Soc. Lond. B* **263:** 53–56.

Møller AP (1997a) Developmental selection against developmentally unstable offspring and sexual selection. *J. Theor. Biol.* **185:** 415–422.

Møller AP (1997b) Immune defence, extra-pair paternity and sexual selection in birds. *Proc. Roy. Soc. Lond. B* **264:** 561–566.

Møller AP & Birkhead TR (1993) Certainty of paternity covaries with paternal care in birds. *Behav. Ecol. Sociobiol.* **33:** 361–368.

Møller AP & Birkhead TR (1994) The evolution of plumage brightness in birds is related to extra-pair paternity. *Evolution* **48:** 1089–1100.

Møller AP & Briskie JV (1995) Extra-pair paternity, sperm competition and the evolution of testes size in birds. *Behav. Ecol. Sociobiol.* **36:** 357–365.

Møller AP & Tegelström H (1997) Extra-pair paternity and tail ornamentation in the barn swallow *Hirundo rustica. Behav. Ecol. Sociobiol.* (in press).

Møller AP & Thornhill R (1997) A meta-analysis of the heritability of developmental stability. *J. Evol. Biol.* **10:** 1–16.

Møller AP, Saino N, Taramino G, Galeotti P & Ferrari S (1997) Paternity and multiple signalling: Effects of a secondary sexual character and song on paternity in the barn swallow. *Am. Nat.* (in press).

Olsson M, Gullberg A, Tegelström H, Madsen T & Shine R (1994a) Can female adders multiply? *Nature* **369:** 528.

Olsson M, Madsen T, Shine R, Gullberg A & Tegelström H (1994b) Rewards of promiscuity. *Nature* **372:** 230.

Olsson M, Shine M, Madsen T, Gullberg A & Tegelström H (1996) Sperm selection by females. *Nature* **383:** 585.

Otter K, Ratcliffe L & Boag PT (1994) Extra-pair paternity in the black-capped chickadee. *Condor* **96:** 218–222.

Parker GA (1974) Courtship persistence and female guarding as male time investment strategies. *Behaviour* **48:** 157–184.

Parker GA (1984) Sperm competition and the evolution of animal mating strategies. In *Sperm Competition and the Evolution of Animal Mating Systems.* RL Smith (ed.), pp. 1–60. Academic Press, Orlando.

Parker GA (1990a) Sperm competition games: raffles and roles. *Proc. Roy. Soc. Lond. B* **242:** 120–126.

Parker GA (1990b) Sperm competition games: guards and extra pair copulations. *Proc. Roy. Soc. Lond. B* **242:** 127–133.

Parker GA (1992) The evolution of sexual size dimorphism in fish. *J. Fish Biol.* **41:** 1–20.

Parker GA (1993) Sperm competition games: sperm size and sperm number under adult control. *Proc. Roy. Soc. Lond. B* **263:** 245–254.

Parker GA & Begon ME (1993) Sperm competition games: sperm size and number under gametic control. *Proc. Roy. Soc. Lond. B* **253:** 255–262.

Pomiankowski A, Iwasa Y & Nee S (1991) The evolution of costly mate preferences. I. Fisher and biased mutation. *Evolution* **45:** 1422–1430.

Price T, Schluter D & Heckman NE (1993) Sexual selection when the female directly benefits. *Biol. J. Linn. Soc.* **48:** 187–211.

Rätti O, Hovi M, Lundberg A, Tegelström H & Alatalo RV (1995) Extra-pair paternity and male characteristics in the pied flycatcher. *Behav. Ecol. Sociobiol.* **37:** 419–425.

Rice WR (1996) Sexually antagonistic male adaptation triggered by experimental arrest of female evolution. *Nature* **381:** 232–234.

Ridley M (1978) Paternal care. *Anim. Behav.* **26:** 904–932.

Ridley M (1988) Mating frequency and fecundity in insects. *Biol. Rev.* **63:** 509–549.

Roy BA (1994) The use and abuse of pollinators by fungi. *Trends Ecol. Evol.* **9:** 335–339.

Rubenstein DI (1984) Resource acquisition and alternative mating strategies in water striders. *Am. Zool.* **24:** 345–353.

Rutowski RL, Newton M & Schaefer J (1983) Interspecific variation in the size of the nutrient investment made by male butterflies during copulation. *Evolution* **37:** 708–713.

Saino N, Primmer C, Ellegren H & Møller AP (1997) An experimental study of paternity and tail ornamentation in the barn swallow (*Hirundo rustica*). *Evolution* **51:** 562–570.

Sakaluk S (1984) Male crickets feed females to ensure complete sperm transfer. *Science* **223:** 609–610.

Sakaluk SK, Bangert PJ, Eggert, A-K, Gack C & Swanson LV (1995) The gin trap as a device facilitating coercive mating in sagebrush crickets. *Proc. Roy. Soc. Lond. B* **261:** 65–71.

Sheldon BC (1993) Sexually transmitted disease in birds: occurrence and evolutionary significance. *Phil. Trans. Roy. Soc. Lond. B* **339:** 491–497.

Sheldon BC (1994) Male phenotype, fertility, and the pursuit of extra-pair copulations by female birds. *Proc. Roy. Soc. Lond. B* **257:** 25–30.

Short RV (1979) Sexual selection and its component parts, somatic and genital selection, as illustrated by man and the great apes. *Adv. Study Behav.* **9:** 131–158.

Simmons LW (1987) Sperm competition as a mechanism of female choice in the field cricket *Gryllus bimaculatus. Behav. Ecol. Sociobiol.* **21:** 197–202.

Simmons LW (1990) Nuptial feeding in tettigonids: male costs and the rate of fecundity increase. *Behav. Ecol. Sociobiol.* **27:** 43–47.

Simmons LW (1995) Male bushcrickets tailor spermatophores in relation to their remating interval. *Funct. Ecol.* **9:** 881–886.

Simmons LW & Parker GA (1996) Parental investment the control of sexual selection: can sperm competition affect the direction of sexual competition? *Proc. Roy. Soc. Lond. B* **263:** 515–519.

Simmons LW, Craig M, Llorens T, Schinzig M & Hosken D (1993) Bushcricket spermatophores vary in accord with sperm competition and parental investment theory. *Proc. Roy. Soc. Lond. B* **251:** 183–186.

Simmons LW, Stockley P, Jackson RL & Parker GA (1996) Sperm competition or sperm selection: no evidence for female influence over paternity in yellow dung flies *Scatophaga stercoraria. Behav. Ecol. Sociobiol.* **38:** 199–206.

Smith HG, Montgomerie R, Poldmaa T, White BN & Boag PT (1991) DNA fingerprinting reveals relation between tail ornaments and cuckoldry in barn swallows, *Hirundo rustica. Behav. Ecol.* **2:** 90–98.

Smith RL (1979a) Paternity assurance and altered roles in the mating behaviour of a giant water bug, *Abedus herberti* (Heteroptera, Belostomatidae). *Anim. Behav.* **27:** 716–725.

Smith RL (1979b) Repeated copulation and sperm precedence: paternity assurance for a male brooding water bug. *Science* **205:** 1029–1031.

Smith SM (1988) Extra-pair copulations in black-capped chickadees: The role of the female. *Behaviour* **107:** 15–23.

Stipkovits L, Varga Z, Czifra G & Dobos-Kovacs M (1986) Occurrence of mycoplasmas in geese infected with inflammation of the cloaca and phallus. *Avian Pathol.* **15:** 289–299.

Stockley P (1997) Sexual conflict resulting from adaptations to sperm competition. *Trends Ecol. Evol.* **12:** 154–159.

Stockley P & Purvis A (1993) Sperm competition in mammals: a comparative study of male roles and relative investment in sperm production. *Funct. Ecol.* **7:** 560–570.

Stockley P, Searle JB, Macdonald DW & Jones CS (1996a) Correlates of reproductive success within alternative reproductive tactics of the common shrew. *Behav. Ecol.* **7:** 334–340.

Stockley P, Gage MJG, Parker GA & Møller AP (1996b) Female reproductive biology and the coevolution of ejaculate characteristics in fish. *Proc. Roy. Soc. Lond. B* **263:** 451–458.

Stutchbury BJM, Piper WH, Neudorf DL, Tarof SA, Rhymer JM, Fuller G & Fleischer G (1997) Correlates of extra-pair fertilization success in hooded warblers. *Behav. Ecol. Sociobiol.* **40:** 119–126.

Sundberg J & Dixon A (1996) Old, colourful male yellowhammers, *Emberiza citrinella*, benefit from extra-pair copulations. *Anim. Behav.* **52:** 113–122.
Svärd L & Wiklund C (1988) Prolonged mating in the monarch butterfly *Danaus plexippus* and nightfall as a cue for sperm transfer. *Oikos* **52:** 351–354.
Svärd L & Wiklund C (1989) Mass and production rate of ejaculates in relation to monadry/polyandry in butterflies. *Behav. Ecol. Sociobiol.* **24:** 395–402.
Thornhill R (1976) Sexual selection and nuptial feeding behavior in *Bittacus apicalis* (Insecta: Mecoptera). *Am. Nat.* **110:** 529–548.
Thornhill R (1980) Rape in *Panorpa* scorpionflies and a general rape hypothesis. *Anim. Behav.* **28:** 52–59.
Thornhill R (1983) Cryptic female choice and its implications in the scorpionfly *Harpobittacus nigriceps. Am. Nat.* **122:** 765–788.
Thornhill R & Alcock J (1983) *The Evolution of Insect Mating Systems.* Harvard University Press, Cambridge, MA.
Thornhill R & Gangestad SW (1993) Human facial beauty: Averageness, symmetry, and parasite resistance. *Hum. Nat.* **4:** 237–269.
Thornhill R & Gangestad SW (1994) Human fluctuating asymmetry and sexual behavior. *Psychol. Sci.* **5:** 297–302.
Thornhill R, Gangestad SW & Comer R (1995) Human female orgasm and mate fluctuating asymmetry. *Anim. Behav.* **50:** 1601–1615.
Trivers RL (1972) Parental investment and sexual selection. In *Sexual Selection and the Descent of Man, 1871–1971.* B Campbell (ed.), pp. 136–179. Aldine-Atherton, Chicago.
Tsubaki Y, Siva-Jothy MT & Ono T (1994) Re-copulation and post-copulatory mate guarding increase immediate female reproductive output in the dragonfly *Nannophya pygmaea* Rambur. *Behav. Ecol. Sociobiol.* **35:** 219–225.
Vahed K & Gilbert FS (1996) Differences across taxa in nuptial gift size correlate with differences in sperm number and ejaculate volume in bush crickets (Orthoptera: Tettigoniidae). *Proc. Roy. Soc. Lond. B* **263:** 1255–1263.
Wade MJ & Arnold SJ (1980) The intensity of sexual selection in relation to male sexual behavior, female choice and sperm precedence. *Anim. Behav.* **28:** 446–461.
Walker WF (1980) Sperm utilization strategies in nonsocial insects. *Am. Nat.* **115:** 780–799.
Weatherhead PJ & Boag PT (1995) Pair and extra-pair mating success relative to male quality in red-winged blackbirds. *Behav. Ecol. Sociobiol.* **37:** 81–91.
Weatherhead PJ, Montgomerie R, Gibbs HL & Boag PT (1994) The cost of extra-pair fertilizations to female red-winged blackbirds. *Proc. Roy. Soc. Lond. B* **258:** 315–320.
Webster MS, Pruett-Jones S, Westneat DF & Arnold SJ (1995) Measuring the effects of pairing success, extra-pair copulations and mate quality on the opportunity for sexual selection. *Evolution* **49:** 1147–1157.
Wedell N (1991) Sperm competition selects for nuptial feeding in a bushcricket. *Evolution* **45:** 1975–1978.
Wedell N (1993) Spermatophore size in bushcrickets: comparative evidence for nuptial gifts as a sperm competition device. *Evolution* **47:** 1203–1212.
Wedell N (1994) Variation in nuptial gift quality in bushcrickets. *Behav. Ecol.* **5:** 418–425.
Wedell N (1997) Ejaculate size in bushcrickets: the importance of being large. *J. Evol. Biol.* **10:** 315–325.
Wedell N & Arak A (1989) The wartbiter spermatophore and its effect on female

reproductive output (Orthoptera: Tettigonidae, *Decticus verrucivorus*). *Behav. Ecol. Sociobiol.* **24:** 117–125.

Werren JH, Gross MR & Shine R (1980) Paternity and the evolution of male parental care. *J. Theor. Biol.* **82:** 619–631.

Westneat DF (1993) Polygyny and extrapair fertilizations in eastern red-winged blackbirds (*Agelaius phoeniceus*). *Behav. Ecol.* **4:** 49–60.

Westneat DF & Sherman PW (1993) Parentage and the evolution of parental behaviour. *Behav. Ecol.* **4:** 66–77.

Whittingham LA & Lifjeld JT (1995) Extra-pair fertilizations increase the opportunity for sexual selection in the monogamous house martin *Delichon urbica*. *J. Avian Biol.* **26:** 283–288.

Whittingham LA, Taylor PD & Robertson RJ (1992) Confidence of paternity and male parental care. *Am. Nat.* **139:** 1115–1125.

Wiklund C, Kaitala A, Lindfors V & Abenius J (1993) Polyandry and its effect on female reproduction in the green-veined white butterfly (*Pieris napi* L.). *Behav. Ecol. Sociobiol.* **33:** 25–33.

Wilcox RS (1984) Male copulatory guarding enhances female foraging in a water strider. *Behav. Ecol. Sociobiol.* **15:** 171–174.

Williams GC (1975) *Sex and Evolution*. Princeton University Press, Princeton.

Winkler DW (1987) A general model for parental care. *Am. Nat.* **130:** 526–543.

Wittenberger JF (1979) The evolution of mating systems in birds and mammals. In *Handbook of Behavioral Neurobiology*, Vol. 3. P Marler & J Vandenbergh (eds), pp. 271–349. Plenum, New York.

Wolf LL (1975) 'Prostitution' behavior in a tropical hummingbird. *Condor* **77:** 140–144.

Xia X (1992) Uncertainty of paternity can select against paternal care. *Am. Nat.* **139:** 1126–1129.

Yezerinac SM, Weatherhead PJ & Boag PT (1995) Extra-pair paternity and the opportunity for sexual selection in a socially monogamous bird (*Dendroica petechia*). *Behav. Ecol. Sociobiol.* **37:** 179–188.

3 Female Roles in Sperm Competition

William G. Eberhard

Smithsonian Tropical Research Institute and Biología, Universidad de Costa Rica, Ciudad Universitaria, Costa Rica

I. INTRODUCTION

Just over 25 years ago, Parker called attention to the fact that reproductive competition between males may continue after copulation and insemination, and described this competition as 'sperm competition'. As currently defined (see Chapter 1), sperm competition is 'competition between the sperm from two or more males for the fertilization of a given set of ova.' The phrase sperm competition evokes the image of armies of tiny, one-tailed soldiers racing up female ducts, or struggling in hand-to-hand combat to gain access to large, passive treasures. It emphasizes, as

Sperm Competition and Sexual Selection
ISBN 0-12-100543-7

Table 3.1. Labels for sexual selection involving male reproductive competition before and after copulation begins. The phrase 'sperm competition' has often been used in an inclusive sense, to designate all male competition that occurs after copulation has begun. As is clear from the analogies in the table, this terminology can result in an underestimation of female roles in determining paternity.

	Intrasexual interactions	**Intersexual interactions**
Prior to initiation of copulation	Male–male battles	'Classic' female choice
After initiation of copulation	Sperm competition *(sensu stricto)*	Cryptic female choice

has been traditional in biology, the active male role in male–female interactions. The theme of this chapter is that images of this sort can be misleading, because they can lead to underestimates of the likelihood that females influence how males compete, and which males win.

At least to a first approximation, sperm competition is the post-intromission or postcopulatory equivalent of precopulatory male competition (Table 3.1). Darwin (1871) distinguished between precopulatory male competition involving direct physical conflicts between males (resulting in sexual selection by male–male battles) and male competition involving attempts to induce females to allow them to mate (resulting in sexual selection by female choice). A similar division can be made after copulation has begun (the classic dividing line in Darwinian discussions of sexual selection) or after insemination has occurred (the dividing line suggested by Parker's definition). In the interests of making the discussion as general as possible, I will use the earlier event, the initiation of copulation, as the dividing line between 'precopulatory' and 'postcopulatory' interactions. 'Postcopulatory male competition' will refer to male competition that occurs after copulation has begun. 'Sperm competition' will be restricted to 'sperm competition *sensu stricto*' as in Table 3.1, involving both direct interactions among sperm from different males (e.g. dilution or raffle competition), and direct interactions between males and sperm from other males (e.g. physical displacement) (see Chapter 10 for further discussion, and a more restricted definition of sperm competition). I will not discuss competition among sperm in the same ejaculate.

Some postcopulatory male competition involves direct battles between the participants themselves, such as dilution of the sperm or physical removal (Waage 1979, 1984). A large part of this book concerns these types of interactions. Other types of male competition can occur with respect to differences between males in their abilities to elicit appropriate female responses. Since female-imposed paternity biases of this sort would be missed using classic Darwinian measures of male reproductive success (e.g. numbers of matings), Thornhill and Alcock (1983) described them with the phrase 'cryptic female choice'. Female responses that select among competing males and their sperm can involve overt behaviour (e.g. oviposition, remating). They can also be much more subtle, such as changes in hormone titres, or transport or killing of sperm within her body.

'Sperm competition' is often used to refer to all postcopulatory male competition. Unfortunately, it is something of a misnomer, as the phrase emphasizes direct male–male interactions; I will argue that most postcopulatory male competition probably does not involve sperm competition *sensu stricto*, but rather cryptic female choice – postcopulatory biases imposed by females on male reproductive success. This chapter is written as a complement to my book on this subject (Eberhard 1996). The reader is referred to the book for more complete discussion and documentation of many of the points that follow. I have attempted to relate ideas on cryptic female choice to discussions in other chapters of this book.

II. NATURAL VERSUS SEXUAL SELECTION AND CRITERIA FOR RECOGNIZING CRYPTIC FEMALE CHOICE

Female reproductive traits can evolve due to selection of two basic types. Many selective pressures have nothing to do with competition among males, but rather with the female's ability to survive and produce a maximum number of healthy offspring. For instance, the low pH in the vagina of many mammals apparently reduces the danger of microbial infections of the female's reproductive tract. Such female traits resulting from natural selection can also incidentally result in selection on males (e.g. favour the ability of sperm to survive at a low pH, or ability to produce seminal products that can raise the pH). Female reproductive traits may also sometimes evolve because of their screening effects on males. For example, the advantage of having offspring sired by males whose semen has particularly good abilities to resist a low pH might favour a further reduction in vaginal pH as a mate-screening device.

Either of these two types of benefit could be responsible for the establishment and maintenance of a female trait that affects the outcome of postcopulatory competition among males. A focus on determining which type of selection has acted on females will lead to asking whether or not the female gains in terms of her own fitness from having biased the competition. For obvious practical reasons, direct tests to determine female benefits have seldom been performed, and the selective advantages of many female reproductive traits that have the potential to screen possible sires have yet to be established.

In contrast, my focus here, and elsewhere (Eberhard 1996) is on the factors that have influenced the evolution of male traits, as has been traditional in studies of sexual selection (see Chapter 1 of Andersson 1994). From a male point of view, it makes no difference whether a female trait that biases his reproductive chances exists because of selection in another evolutionary context or not. Selection on the male will favour characteristics that enable him to adjust to, or compensate for, such a female trait, whatever its advantage to the female may be. This focus leads to different criteria for the occurrence of cryptic female choice (see Chapter 3 of Eberhard 1996). In general terms, if a female's traits, whether they be behavioural, morphological, or physiological, have the effect of consistently favouring some conspecific males that have copulated with her and that possess a particular trait that in other mates is less fully developed, it is reasonable to conclude that the female traits bias postcopulatory male competition in a way that favours possession of that male trait, thus producing cryptic female choice. In an attempt to explain why males and their ejaculates have the traits that they have, I include in discussions of mechanisms of cryptic female choice possible 'incidental' screening by females due to responses that may have evolved in other contexts, as well as 'purposive' screening which evolved in the mate-choice context (see also Section III).

III. PRECOPULATORY VERSUS POSTCOPULATORY FEMALE INFLUENCE ON MALE COMPETITION

Female influence over events in the classic Darwinian arena of precopulatory sexual selection can influence whether or not a female's first mate will ever be exposed to postcopulatory competition with other males (see Chapter 2). Either of the two types of precopulatory sexual selection, male–male battles, or female choice, could affect a male's chances of being a female's second or subsequent mate. The possible role of the female in determining the outcome of male–male battles would appear to be small. Prior to copulation, males are relatively free to interact with each other, independent of the female. For example, if two male dungflies battle for control of a pat of cow dung soon after it is deposited on the ground, a female that arrives later to oviposit in the dung would appear to have no opportunity to influence the outcome of that battle. Unless she goes elsewhere to copulate, a female will be more likely to be courted and mated by a winner of such a battle (this, of course, is why the males fight in the first place). Recently, Wiley and Poston (1996) noted that even at this stage females usually have important but previously neglected effects on the outcomes of male competition.

Darwin emphasized that female choice involves active selective behaviour by females: '. . . females, which no longer remain passive, but select the more agreeable partners' (Darwin 1871, p. 916). Wiley and Poston (1996) call attention to the fact that a female can also systematically bias the likelihood of some males rather than others mating with her even in the absence of either active choosing behaviour, or of any ability to discriminate among males (see also Ahnesjö *et al.* 1992). For example, synchrony and location of mating can result in strong biases in paternity in the Montezuma oropendula *Psaurocolius montezuma*. Females nest more or less synchronously in dense colonies, and often mate on or near their nests. The result is that a single dominant male can effectively exclude other males from copulations with the 10–30 females at such a site. By choosing this particular location and timing of copulation, the female effectively creates the opportunity for biases in paternity that favour dominant males, even without overtly choosing one male over another. Other female traits that can result in such 'indirect' precopulatory mate choice (Wiley and Poston 1996) include advertising receptivity to mating, advertising that copulation is occurring, indiscriminate flight from males, and aggregating with other receptive females. These are similar to what I have termed 'female-imposed rules of the game' in the context of postcopulatory competition between males (Eberhard 1996). Wiley and Poston (1996, p. 1378) speculate that indirect mate choice is ubiquitous: 'It seems likely that competition for mates by one sex always depends on conditions set by indirect mate choice by the other sex'. Thus, some female traits which have not generally been considered important in precopulatory sexual selection by female choice can, nevertheless, have important consequences.

If females often appear likely to affect male–male competition prior to copulation, consider how much more likely they are to affect it after intromission and ejaculation have occurred. While precopulatory male competition often occurs with relatively little direct female influence, postcopulatory competition is generally played out within the female's own body. Even small changes in female reproductive morphology and physiology can have huge consequences for fertilization success. A more wrinkled lining of the oviduct that will stretch to accommodate a larger egg might favour sperm that lurk along the folds and avoid the macrophages that often patrol the female tract to combat infections; a more elongate spermatheca in an insect could result in a greater tendency for last male sperm precedence (Walker 1980; but see Ridley 1989), which could in turn result in increased paternity biases favouring males that are able to defend oviposition sites. The extreme asymmetry between the tiny, delicate sperm and the hulking, complex female with her large array of morphological, behavioural and physiological capabilities does not mean that males and their sperm are necessarily completely powerless, but it certainly loads the deck in favour of the possibility of female influence. In an analogy with human sporting events, the female's body constitutes the field on which males compete, and her behaviour and physiology set some of the rules by which they must abide (see Chapter 16 on the mammalian female tract as an arena). Small changes in female morphology, behaviour or physiology can tilt the playing field and change the rules, thus biasing paternity.

IV. FEMALE PREADAPTATIONS TO BIAS FERTILIZATION SUCCESS OF SPERM AND THUS EXERCISE SEXUAL SELECTION

There are several reasons to think that females have important effects on postcopulatory male competition. First, there is the enormous disparity between sperm cells and adult females in size and resources to influence outcomes of male competition. In many respects, sperm are relatively helpless once they have been introduced into the female's body. They are, of course, usually mobile but evidence regarding their movements within the females of the two best-studied animal groups, mammals and insects, is consistent: a major portion of the movement of sperm within the female is brought about by the female (see Chapter 16; Eberhard 1996; see also Chapter 8 on molluscs; Birkhead and Møller 1993 on birds; and Eberhard 1985 on other groups). Simple, relatively insignificant movements of the female, such as a few 'downstream' peristaltic twitches of her reproductive tract, or beating in the wrong direction of the cilia lining the reproductive tract, can result in the army of little 'warriors' being unceremoniously dumped from her body, or being diverted to other internal sites where they will be digested (Chapters 8 and 16; LaMunyon unpublished data; see Eberhard 1996 for other examples). The image of valiant swimmers racing toward the waiting egg is thus misleading; perhaps it would be more appropriate to think of a school of tiny fish swirled in the ebb and flow of waves breaking on a rocky shore.

Second, because sperm are almost never deposited directly onto the female's eggs in species with internal fertilization (Eberhard 1985), insemination of a female is seldom precisely equivalent to fertilization of her eggs. If a given male is to sire the maximum number of a female's offspring, it is not enough that his sperm move or be moved to fertilization sites (or, as is often the case, first to a storage site and later to a fertilization site) as just discussed. Many other female processes that occur during and after copulation must also be triggered. The list is long: the female must ovulate, she must make large rather than small eggs, she must oviposit, and she must nourish his sperm; she must refrain from digesting his sperm, from aborting his zygotes, and from mating with additional males. In animals with maternal care, she must also commit resources to caring for the offspring. In mammals, she must prepare the lining of her uterus for implantation of the embryos, and then respond positively to the hormonal blandishments of the embryos to provide them with resources (Haig 1993). If females respond to copulation with anything less than complete responses (and incomplete responses are not uncommon – see Eberhard 1996), selection can act on males to favour those who have superior abilities to elicit these types of female response.

This multiplicity of necessary female responses has an important consequence in relation to the possibility that males will obtain complete control over fertilization of a female's eggs. Because many female responses are sequential, reproduction requires that all of them occur (e.g. sperm transport, ovulation, implantation, avoidance of abortion). Take, for example, a male that is able reliably to induce maximum sperm transport responses in his mates, or one that is able to directly place his sperm in storage or fertilization sites in the female. Such a male may, nevertheless, fail completely to reproduce unless he is also able to induce ovulation. In fact, biases in female control of ovulation could even conceivably act against those males best able to introduce their sperm directly into storage or fertilization sites, causing reductions in such male abilities. Thus, even if a female loses the ability to influence a given process, the multiplicity of postcopulatory processes that affect reproduction makes it unlikely that if she mates with several males she will lose all ultimate control of the paternity of her offspring. A male must have all the female processes under control if he is to be certain of fertilizing all of her eggs; the female need only have one to exercise a veto (see Section IX for problems with the concept of 'control').

Finally, and possibly most importantly, there is a logical, a priori reason to expect that females will tend not to be indiscriminate in exercising their influence over the fate of the sperm from different males. The expectation is that females will evolve to accentuate the biasing effects that they have on the fertilization chances of sperm from different males, and that this will often result in sexual selection by cryptic female choice that will tend to produce relatively rapid evolutionary divergence. This last point is especially important because one of the clear trends in the male and female traits that appear to be closely associated with possible biasing of sperm use is that they very often differ among closely related species, and thus have diverged relatively rapidly. This divergence is extensively documented in male and female genital morphology, and is apparently also widespread in male copulatory courtship behaviour (Eberhard 1985, 1996). It may also occur in physiological interactions involving male seminal products, although the data here are much more limited (Eberhard 1996).

The reason to expect cryptic female choice is related to ideas associated with male 'sensory traps' that are used to influence female choice (West-Eberhard 1979, 1983; Ryan 1990; Christy 1995). The basic concept of sensory traps is that males can take advantage of female traits that have evolved under natural selection in other contexts to gain sexual advantage in competition with other males. Natural selection on females will often result in the evolution of female abilities to sense certain stimuli, and to respond to these stimuli in certain ways. In the context of post-copulatory male competition, natural selection on females will undoubtedly favour those that perform some reproductive processes only after copulation has occurred. For instance, a female that oviposits prior to rather than following copulation will not usually be favoured. Sexual selection can act on males to use this triggering mechanism in the female as a sensory trap – to convey the message 'copulation has occurred' more loudly and clearly and thus more reliably trigger her reproductive responses to this message.

Courting males can only make use of sensory traps if two conditions are fulfilled: (1) they are able to produce the stimuli to which the female responds, or at least imitate them well enough to elicit female responses; and (2) the female response that evolved under previous selection is appropriate to increase the male's chances of fertilizing her eggs. Both of these conditions are especially likely to obtain when a female uses stimuli associated with copulation to trigger the postcopulatory reproductive processes.

It seems inevitable that stimuli from the male himself or from his semen will provide especially reliable signals to the female that copulation has occurred. Natural selection is thus expected often to favour the use by a female of stimuli from males and their ejaculates to trigger several of her crucial reproductive processes. The result, with respect to the first condition, is that males will be especially preadapted to mimic and accentuate the stimuli that convey the 'copulation has occurred' message; in essence, they will be mimicking or exaggerating themselves.

Take, for example, one of the original illustrations of a sensory trap, the *Xylocopa* carpenter bee males that release floral scents from thoracic glands at their mating stations (West-Eberhard 1984; Minkley *et al.* 1991). A male bee will only be able to evolve to use a female's pre-existing responses to sweet floral fragrances to heighten his sexual allure if he previously happened to produce, in some other selective context, substances that have such odours or can acquire such odours with little modification. Utilization of this particular sensory trap may evolve only occasionally, when this precondition in males is met (thoracic glands of this sort have not been noted in other bees – W. T. Wcislo, pers. comm.). In contrast, if a female triggers a reproductive process such as ovulation by the stimuli resulting from the friction of the male's genitalia against hers, minor modifications of many male traits that are already present could increase the friction stimuli (e.g. addition of wrinkles or spines to the surface of his genitalia, thicker or longer genitalia, addition of a swelling at the tip, repeated thrusting movements instead of simple insertion) (see Eberhard 1996 for evidence that in a variety of animals friction is an important triggering stimulus).

With respect to the second condition, that existing female responses be appropriate to increase a male's reproduction, natural selection of female responses to copulation will again predispose them to be utilized by males as sensory traps. This is because not only are the female responses to copulation reproductive responses that can affect a male's reproduction, the responses also have the correct polarity: all female responses likely to evolve under natural selection will potentially *increase* the male's chances of reproduction – *increased* likelihood of ovulation, oviposition, sperm transport, and sperm maintenance and *decreased* likelihood of further mating with other males and intensity of resistance to invading microorganisms in the reproductive tract. Thus, positive pay-offs to the male in terms of more offspring are likely already to be 'built' into the female by natural selection. What will happen in a species in which males begin to utilize such sensory traps? Unless male signals become so powerful that they trigger radical responses that are disadvantageous to the female under natural selection (see Section VIII), there will be selection on females to discriminate among males, and bias fertilization of their eggs in favour of some males over others. If some aspect of the male signal is correlated with the male's heritable viability (an 'indicator mechanism' – see Andersson 1994), the female may evolve to respond more reliably to this aspect of the signal and thus obtain more vigorous male and female offspring. In other cases she may obtain sons that are superior signallers by favouring males that are especially capable of using the sensory trap. One simple, easily evolved mechanism for biasing male reproductive success is for the female to change the response threshold. Take, for example, a species in which males are evolving to increase friction by using multiple thrusts and females that have offspring sired by males that have made multiple thrusts are favoured. If a female's chances of responding are increased from 50% with single thrusts to 70% with multiple thrusts, then if a female arises that has only a 10% chance of responding to only a single thrust, she could be favoured over the others because she would be more likely to have sons sired by males that made multiple thrusts. Female response criteria are expected to change rather than remain static.

In addition, the types of stimuli that males could use in such sensory traps associated with copulation are much less limited than in the usual precopulatory sensory traps, especially when stimuli are not indicator mechanisms. This is because, in contrast with other contexts, natural selection on female responses will not tend to work against sexual selection favouring elaboration of male signals in the trap. For example, unless the female *Xylocopa* bee begins to dissociate her responses to floral odours in a sexual context from those in the context of foraging for food, extensive modification of male pheromonal signals will be unlikely. Sexual selection on males to add further components to their odours will not favour scents that are not flower-like. The more crucial the original, naturally selected context in which the female response evolved (e.g. predator escape cues in *Uca* crabs; Christy 1995), the more strict the mimicry restraint on male cues is likely to be.

In contrast, a male that informs his mate during copulation that copulation has occurred is not telling a lie, but is providing her with a potentially useful cue, and natural selection on females will not be likely to favour disregarding his signal or failing to respond to it. Males can thus add 'frills' or additional manipulations to their signals without having natural selection of females act against their incorporation. Male courtship signals during copulation are thus much freer to wander over the range of possible stimuli. This can help explain the rapid divergence and the apparently arbitrary nature of the strange antics often associated with courtship during copulation and thought to have evolved under cryptic female choice. For instance, males of different insect species lick, tap, rub, push, kick, stroke, shake, squeeze, feed, sing to, and vibrate the female during copulation (Eberhard 1994).

Further consideration of the properties of nervous systems is important in this context. Animal nervous systems tend to be relatively highly interconnected, rather than strictly compartmentalized. In addition, the activity of a given set of neurones is often influenced not only by nerve impulses from other neurones, but also by neuromodulation that alters cellular or synaptic properties and thus reshapes the output of neural circuits, sometimes effectively rewiring or reprogramming them (Katz and Frost 1996). This flexibility and interconnectedness can have important consequences for the evolution of sensory traps. It is not unlikely that positive responses in the neurones of the female that influence her reproductive responses to copulation (for example ovulation) will be affected both positively and negatively by some other stimuli in addition to those, for example, resulting from friction in her genitalia. A male that is able to add any extra stimuli that have the property of increasing female responses (a bite on her antenna, or a whiff of a strong smell) can gain an advantage over other males that do not have such extra tricks. Once males that are able to utilize such a trick appear in a population, selection will act on females to screen their mates with respect to this characteristic in addition to the ability to thrust. As described above, a change in the female's response threshold is one likely mechanism, although others are also possible (she might, for example, turn her head away from the male and make it more difficult for him to bite her or to smell his pheromone, or add hair to the site that is bitten to ensure that only strong bites succeed in stimulating her, etc.). Each readjustment of the female's sense organs and nervous system stemming from such a round of selection will alter the array of interconnected portions of her nervous system that are brought into play in this context, thus altering the set of female sensory and effector responses on which males can act (Eberhard 1993). Provided that the male's messages are not in themselves damaging to the female (see Section VII), they are relatively free to roam over this changing landscape provided by the female's nervous system.

V. THE RELATION BETWEEN CRYPTIC FEMALE CHOICE AND SPERM COMPETITION *SENSU STRICTO*

Direct postcopulatory male competition and cryptic female choice are not mutually exclusive alternatives. Just as precopulatory female choice and male–male battles can occur sequentially, or even simultaneously (as when a female fur seal calls out when a satellite male makes sexual advances, and thus attracts territorial males and precipitates fights – see Wiley and Poston 1996), so female behavioural, morphological, and physiological traits brought into play in postcopulatory contexts can, for example, facilitate raffle competition among sperm from multiple males (e.g. with multiple, rapid copulations, low rates of termination of copulation before ejaculation, large sperm receiving and storage organs, benign receiving environments where sperm can survive, and no sperm dumping). Some female traits that affect the outcome of sperm competition *sensu stricto* may have evolved in other contexts (e.g. a large-volume birth canal for large offspring that also acts as a sperm receiving structure), and thus bias this type of male competition only incidentally. Other female traits may be explicable only as means of influencing the amounts of sperm transferred and their subsequent chances in raffle competition (e.g. tortuous female spermatheca ducts).

In addition, there are many cryptic female choice mechanisms whose effects can be completely independent of sperm competition *sensu stricto*, such as changes in rates of oviposition, ovulation, preparation for implantation of the embryo (mammals), abortion, amounts of reserves in eggs, or provisioning offspring. Some cryptic female choice mechanisms may sometimes influence sperm competition, and sometimes not, depending on how and when they are implemented (sperm transport, sperm nourishment, sperm utilization efficiency, and sperm dumping – see Eberhard 1996).

VI. WHAT CAN A FEMALE GAIN FROM POLYANDRY?

There are numerous possible advantages that a female might derive from mating with more than one male, including opportunities for sperm competition *sensu stricto* and cryptic female choice (see, for example, lists in Walker 1980; Keller and Reeve 1995; and Chapters 2, 14, and 16). Testing the many different possibilities empirically is a demanding task, especially because some important possibilities involve subtle effects (Keller and Reeve 1995), and several are not exclusive. Discrimination of the relative importance of competing hypotheses will require precise, technically difficult quantification. An example of the crucial role of such precise quantification is the debate over whether large spermatophores in some butterflies are mechanisms by which males feed potential offspring, or mechanisms by which they inhibit future matings by the female (see Chapter 10). Although some preliminary discriminations can be made, it is not yet possible to judge which possible advantages of multiple mating tend to predominate.

One important point is, however, clear. The discovery that at least occasional polyandry is widespread in many socially monandrous species, and that the females of many of these species actively seek out additional mates rather than be forced to mate (see Chapter 14 on birds; Chapter 16 on mammals), means that whatever the relevant advantages of polyandry prove to be, they are probably substantial. Mating can be risky, and other things being equal, it is best avoided by females (Daly 1978; see Chapter 16). If polyandry is so advantageous, then one can confidently predict that it will also be found to occur in many additional animal groups whose mating systems have not yet been studied. If so, this means that postcopulatory male competition will probably prove to be even more widespread than is presently appreciated.

VII. A WORST-CASE ILLUSTRATION OF FEMALE-IMPOSED BIASES

One way to illustrate the likelihood of selective female influence on postcopulatory male competition is to examine a worst-case example, in which males appear initially to have seized complete control of paternity and female influence appears to be impossible. Damselflies and dragonflies are a good example, because the secondary male genitalia of many species can pull out the sperm of previous males from inside the female, or to pack it deep into the recesses of the female's reproductive tract where it is less likely to have access to her eggs (Waage 1979, 1984, 1986; see Chapter 10). How could a female odonate influence paternity when the male is so well equipped to manipulate the sperm inside her?

There are, in fact, several ways in which a female could still affect which males sire her eggs. Perhaps the most direct mechanism involves altering her pattern of oviposition. The number of eggs laid immediately following a copulation and before mating with another male could vary according to characteristics of the male. In *Erythemis simplicicollis*, for example, females refused to lay any eggs at all after some copulations (27% of copulations with satellite males, 7% of copulations with territorial males) (McVey 1988). A male able to clear out sperm from a female but unable to induce her to oviposit immediately afterward may not sire her offspring. A second, more subtle effect of female regulation of oviposition is the following. In some (possibly many) odonates, the male does not completely clear out all the previous sperm in the female, but leaves some which are distant from the fertilization site. Over the course of a day or so, the sperm in the female slowly mix. For example, in *E. simplicicollis* the last males' sperm fertilize essentially all eggs laid immediately after copulation, but only 65% of those laid 24 h later. Thus, if the female mates with a male, but then refrains from ovipositing until the next day (20% of the eggs in *E. simplicicollis* are laid when the female returns to the pond without remating), she is simultaneously favouring the reproduction of males who mated with her prior to her last copulation, and acting against the reproductive interests of the males present at the pond on the day when she finally oviposits.

Other female traits can also be important. The morphology of the reproductive tract of female odonates influences the male's ability to remove previous sperm. For example, in the damselfly *Mnais pruinosa* the spermathecal duct is so narrow that the male's genitalia cannot enter, and they can thus only remove sperm from her bursa (Siva-Jothy and Tsubaki 1989). The enlarged distal area of the spermatheca of *Argia* is also probably inaccessible to the male (Waage 1986), and the sperm in the spermathecae of *Calopteryx splendens xanthostomo* are apparently untouched during copulation (Siva-Jothy and Hooper 1995) (see summary in Chapter 10). In addition, it is possible that females facultatively alter their morphology by muscle contractions. The arrangement of muscles in the walls of the bursa and the spermatheca in some odonates shows that a female can reduce the volume of her spermatheca and bring the bursa closer to the external genital opening (Siva-Jothy 1987). Both of these movements would probably facilitate more complete sperm removal if the female performs them during copulation, and variations in female movements of this type could explain the substantial intraspecific variation observed in the degree to which even those sperm in the most accessible portions of the female are cleared out in some odonates (e.g. the coefficient of variation of the volume of sperm remaining after long copulations in *M. pruinosa* was 71%; Siva-Jothy and Tsubaki 1989).

An additional possibility is associated with the recently recognized fact that at least some female odonates can also apparently eject sperm themselves (E. Gonzalez in Eberhard 1996). Sperm ejection behaviour in one species occurs more frequently following matings with one of two male morphs than after those with the other. Another recently recognized capacity of female odonates is to alter the proportions of sperm stored in the spermathecae (often from several previous males) and those in the bursa (predominantly from her last mate) that are used to fertilize her eggs (Siva-Jothy and Hooper 1996).

Even the properties of the sperm themselves, which determine how successful a subsequent male's sperm removal attempts are likely to be, may represent adjustment to female-imposed conditions. The tangled mass of odonate sperm, which is not typical of many other animals, facilitates both the entanglement and removal with the spines and bristles of the male's genitalia, and probably also the relatively slow sperm mixing which are essential determinants of successful male strategies. If any female-imposed conditions have resulted in the evolution and maintenance of these sperm characteristics, then they will have indirectly influenced the reproductive payoffs of these types of male morphology and behaviour. It would appear difficult, in fact, to explain these sperm traits solely on the basis of male-determined advantages, since a male could presumably reduce the chances that his sperm would be removed by subsequent males by making them smaller or less disposed to tangle with each other. Thus, females may have indirectly affected postcopulatory male competition by making sperm removal and packing more advantageous.

In conclusion, even in a group of animals in which males have relatively complete control of one aspect of paternity determination as a result of sperm competition *sensu stricto*, closer examination shows that variation in a number of other female-controlled processes and characters can impose significant biases in paternity probabilities (see also Fincke 1997). One could equally well choose a different male technique, such as using large ejaculates to flood out sperm from other males in the female's reproductive tract, as occurs in mammals (see Chapter 16), and produce another list of female-imposed conditions that affect the effectiveness of the technique itself (e.g. volume of female sites where sperm are received and stored, fine structure of the sites that could allow sperm to find hiding places and avoid being diluted, physiological conditions in receiving and storage sites that affect sperm survival, transport toward fertilization sites, sperm dumping, termination of copulation before sperm transfer is complete), and that could affect paternity regardless of the flooding behaviour (e.g. biases in remating, ovulation, or triggering of the hormonal cycle that prepares the uterus for implantation). The general conclusion is that even in 'worst case' situations cryptic female choice is capable of having substantial effects on the final outcome of male competition that also involves strong sperm competition *sensu stricto*.

VIII. MALE–FEMALE CONFLICT

One thread left dangling in the discussion above involves the possibility that the male will be able to manipulate the female so completely that he will cause her to respond in ways that are costly to her in terms of her survival or reproduction. It is possible, for example, that a male could induce oviposition powerfully, and thereby benefit because the female would be less likely to meet further males before her eggs were laid, and that the female's reproductive potential would be reduced by this induction because she would not have time to search out the very best oviposition sites available. Recognition of possible conflicts of interest has provided many exciting and productive insights into many different evolutionary processes, so it seems reasonable to use a 'conflicts of interests' approach in interpreting male–female interactions in sexual reproduction. This line of thinking is encouraged by the fact that there is a clear conflict of reproductive interest between males and females: a male will be best served if he fertilizes all of the eggs of every female he meets; and a female will often be best served by not having all her eggs fertilized by the first male she encounters. Many authors refer to an 'arms race' between males and females in postcopulatory interactions (Parker 1984; Alexander *et al.* 1997; Wiley 1997); Parker (Chapter 1) refers to 'male win' and 'female win' situations. This approach can be rewarding, especially in interpreting studies in fields such as the study of female reproductive morphology and physiology, where authors have traditionally tended to view reproduction as cooperation between males and females, and in untangling complex issues in the evolution of human sexual behaviour (Hardy 1997). The arms race metaphor also draws attention to the unending nature of evolution under sexual selection.

However, this metaphor can be misleading when applied to sexual interactions between males and females. In an arms race, the interests of each party are best served by outright winning over those of the other. Consider the biological 'nuclear arms race' in which the nuclear genome of a eukaryote interacts with a selfish cytoplasmic genome that causes deviant sex ratios. The nuclear genome will be best able to survive and reproduce if it can completely repress the traits promoting the damaging selfish interests of the cytoplasmic genome; the cytoplasmic genes will be best served if they can completely repress these effects of the nuclear genes. Sexual interactions, in contrast, are asymmetrical in that complete wins are only favourable on the male side of the ledger. If a female successfully prevents all males from manipulating her, she loses. At worst, she may sometimes end up not having sperm to fertilize her eggs. If males are utilizing sensory traps, a female that avoids male manipulation by altering her responses to his stimuli may reduce her fitness in other contexts. At the very best, a female will tend to lose because the males who sire her offspring will be only average in their abilities as manipulators, and so her sons will thus tend to be only mediocre manipulators.

A female's reproductive interests are best served by a more subtle tactic – that of selective resistance to male manipulations, on one hand defending against overly damaging manipulative effects, and on the other selectively accepting some males, thus achieving a paternity bias in favour of those males most capable of manipulation. The optimum degree of female resistance will depend on the traits of males in the same population, and can result in situations that superficially resemble an arms race. If, for example, it is advantageous for a female insect to be fertilized by males with relatively long intromittent organs in a population in which the lengths of male genitalia vary from 0.7 to 1.0 mm, it may pay the female to have a duct about 0.9 mm long rather than one that is either 0.5 or 1.5 mm. Although escalation by the female is easy, her 'resistance' should be carefully adjusted to male capabilities. For example, if males with 1.2 mm genitalia arise, the best female duct length might increase to 1.1 mm (see Keller and Reeve 1995 for a similar discussion).

In other words, selective cooperation (cooperation with males that have certain traits, and rejection of others) probably often describes more accurately the most advantageous female tactic. More precisely, the benefit a female will derive from increasing her resistance to male manipulations will be determined by the balance between the reduction in the losses that she would otherwise suffer as a result of being manipulated, and the intrinsic cost of the resistance mechanism plus the resulting change in the number and quality of her offspring. Females are probably more likely to retain influence over disputed processes because of the relatively lower investments needed to affect paternity (Stockley 1997), the variety of potential control mechanisms at hand evolutionarily (see above), and their preadaptation of already having traits that have evolved to trigger and modulate their own reproductive processes (Eberhard 1996). Apparent 'balances of power' or even male 'wins' may often result from restraint by the female (see also next section).

The importance of these considerations regarding the asymmetry of selection on sexual interactions is shown by a widely cited recent description of the postcopulatory manipulative and apparently damaging properties of male seminal products in *Drosophila melanogaster*. Chapman *et al.* (1995) found that some male seminal products cause increased mortality rates in females. They interpret their findings in terms of 'the cost of mating' in 'an evolutionary conflict between the sexes'. A similar interpretation of these results is made in Chapters 10 and 17, where the authors conclude that 'males are ahead in this particular arms race.' The claim that the effects of the male substances impose a 'cost' on the female depends, however, on the balance between two factors – the reproductive losses suffered by females from their reduced longevity, and the possible gains from their additional effect of inducing more rapid oviposition. No data were given on this crucial point by Chapman *et al.* (1995) and data from a subsequent study (Chapman and Partridge 1996) argue against the original interpretation. Females did not suffer fitness losses due to male manipulations under normal conditions, but only at unnaturally high levels of nutrition. Additional important points regarding the possible costs of mating also remain to be clarified. For example, more eggs may be gained by early oviposition than are lost by reduced longevity, especially under conditions in which females are likely to be killed by predators or by natural causes before laying all their eggs. It is thus still possible that female thresholds and responses to male seminal stimuli have evolved to be entirely in accord with female interests, and that the male products are simply adaptive triggering devices.

The important point is that one should take into account the fact that both the female's responses to male products and the thresholds at which these responses are triggered are evolved characters, and can be expected to change in relation to male traits (as indeed appears to occur in the sexual physiology of *Drosophila*; see Rice 1996). Thresholds may even be facultatively adjusted in adaptive ways by individual females (see Simmons and Gwynne 1991 on a katydid). The female should not necessarily be expected to suffer a cost, even when she responds to invasive male signals such as seminal products that act on her nervous system. She is likely to be a 'moving target' rather than stationary in these evolutionary interactions, and she is far from powerless to respond to male advances.

An enlightening example of how female restraint and selectivity can occur in an apparently conflict-laden interaction comes from studies of the transfer of male seminal products in another set of related flies, the housefly *Musca domestica* and two species of blowflies in the genus *Lucilia*. In these cases, female morphology and physiology is involved, making it easier to visualize male–female conflicts of interest. Male seminal products also have powerful effects on female oviposition and remating in these flies (Riemann and Thorson 1969; Smith *et al.* 1988, 1989; Barton-Browne *et al.* 1990), and the details of how male effects manifest themselves suggest an arms race. In the housefly the male products move rapidly from the female's reproductive tract into her body cavity and to her brain where they apparently exercise their effects (Leopold *et al.* 1971a,b). This invasion is accomplished when the male products apparently digest an opening in the wall of a sac in the female reproductive tract where they accumulate during copulation (Leopold *et al.* 1971a). The male blowfly *Lucilia sericata* uses another technique to the same end: rows of sharp teeth on his genitalia abrade the lining of the female's reproductive tract at the points where the accessory gland products are deposited (Fig. 3.1) (Lewis and Pollock 1975).

These might appear to be exotic cases of male control to the detriment of helpless females, but data from a third species, *L. cuprina* bring the relevant questions into sharper focus. Males of *L. cuprina* also have abrasive structures on their genitalia in the area where the accessory gland products emerge during copulation, but the lining of the female's reproductive tract is thickened to form pads where the products are deposited (Fig. 3.1) (Merrett 1989). The male's genitalia also tear the lining of the wall in this species, but his accessory gland products accumulate within the thick pads, and are retained there for up to several days rather than moving directly into her body cavity.

This species illustrates two points. The first is that it is probably often quite simple for females to 'win' in such conflicts with males. A relatively small modification by the female can counteract and nullify even apparently powerful male manipulations. One is led to reconsider the first two

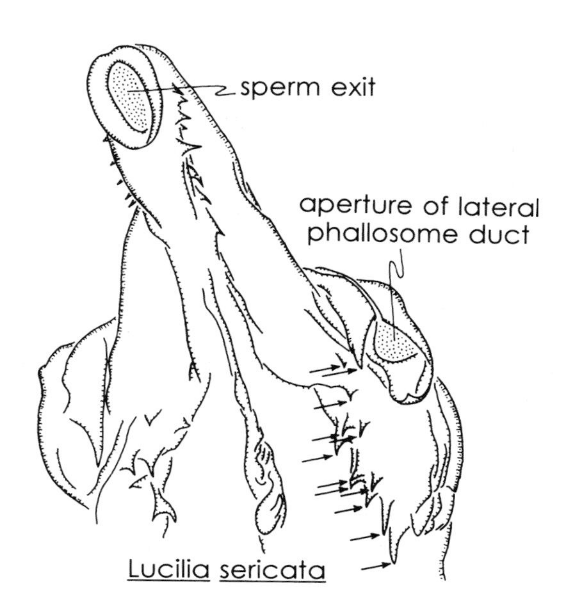

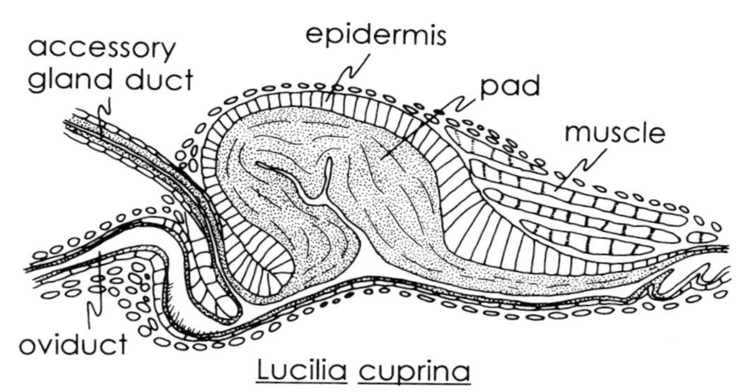

Fig. 3.1. *The apparent ease of female control contrasted with its paradoxically incomplete action in* Lucilia *blowflies illustrates problems with an indiscriminate application of the 'arms race' metaphor for the evolution of male and female reproductive traits. The male genitalia of these flies have many sharp structures (arrows) that apparently abraid the lining of the female reproductive tract near the sites (the apertures of the lateral phallosome ducts) where male accessory gland substances emerge (drawing of* Lucilla sericata, *above). This is thought to give the male products (which inhibit female remating) improved access to the female's body cavity, where they act. The female reproductive tract of* L. cuprina *(below) has a thickened lining (pad) where the male products are sequestered, constituting an apparent 'win' for the female. However, females of this species continue to respond to male products transferred during normal copulations. The pad may be better understood as a filtering device to favour certain males over others, rather than as a device to provide comprehensive resistance to male manipulations. (Above after Lewis and Pollock 1975; below after Merrett 1989.)*

species: perhaps the male's powerful effects on females only occur on sufferance by the female. The thin-walled sac in *Musca* that is breached by male seminal products makes more sense, for example, as a filter to favour those males whose seminal products have a digestive capacity, than as a 'defence' against these seminal traits.

The second, more important point is related to the first. The apparent female win in the third species, *L. cuprina*, turns out not to be a 'win' at all. Both remating and oviposition behaviour in females of this species are influenced by normal transfer of male accessory gland products (Smith *et al.* 1989, 1990). Why does the apparent female defence in this species fall short of eliminating the male effect? One arms race interpretation would be that perhaps the female defence has been extended as far as it can and that there is still enough power in the small amounts of male products that leak out of the pads to exert manipulative effects, despite the female's resistance. Perhaps problems with the passage of eggs down the duct place limits on the thickening and stiffness of the pad, or perhaps this is a very recent advance by the males to which the females have not yet completely adjusted. The alternative, selective cooperation interpretation is simpler. Perhaps these female flies are resisting just enough to screen males on the basis of their accessory gland products and their delivery techniques, favouring those males able to overcome certain modest barriers in a population in which not all males are able to do so completely.

Although this case illustrates the possibility of female restraint and selectivity, these arguments are not completely satisfactory. While it appears probable that female *L. cuprina* flies could increase their resistance to the degree of male manipulation if the manipulations were actually damaging their reproductive interests, there is still some doubt. Ideas of this type are difficult to test. How is one to decide whether, for example, a given female response threshold to a male seminal product is the best defence the female has been able to evolve, or whether this threshold is a screening mechanism designed to allow only some males to pass, and which could become more restrictive if this were favoured by natural selection? This same problem occurs in many other types of male–female interactions. If a female flees from a courting male, is she really attempting to escape, or is she screening males on the basis of their pursuit abilities? I do not have any easy answers; the point that I am making is that situations that appear to be male–female conflicts often have an alternative explanation – that of selective female cooperation.

This distinction between adversative and selectively cooperative interactions is more easily made in the evolution of genitalia. In this morphological context it is easier to distinguish comprehensive resistance from selective cooperation because a female genital structure that functions to resist male genital coupling would (at least sometimes) be relatively simple to recognize. In addition, establishing functional significance in external female genitalia is sometimes easier because some structures have no function other than interaction with male genitalia. Natural selection for complete resistance would be expected to result in facultative female genital mechanisms for categorical lack of cooperation. One would expect structures such as erectable spines or inflatable sacs that would be capable of fending off all male coupling attempts. Conversely, selective cooperation would result in structures such as grooves or slots or pits, which facilitate coupling (e.g. increase a male's ability to hold onto the female), but only by males that have the appropriate genital morphology that fits with the female's cooperative structure.

When these predictions are compared with the multitude of female genital structures that have been studied, the data are unequivocal. Pure resistance structures are almost completely unknown. I know of one possible example, in a water strider (Arnqvist and Rowe 1995), and the possibility that the female spines are also used for screening males in this species has not been tested. Selective cooperation structures that lack other likely functions, in contrast, are widespread (see Eberhard 1997 for examples).

In genitalia, then, selective female cooperation rather than all-out female resistance predominates, yet males evolve highly aggressive structures and tactics (Eberhard 1985). The impression of conflict derives from focus on the male side; the female side, while also coevolutionary, is selectively cooperative rather than adversative. If the evolution of genital morphology is representative of other phases of male–female sexual interactions, then the selective cooperation metaphor is generally more appropriate than that of the arms race. The less extensive data available on copulatory courtship also indicate selective cooperation, in this case because the male behaviour patterns are almost entirely 'non-forceful' in the sense that they do not physically oblige the female to cooperate (Eberhard 1996). There are no clear patterns in the few careful studies of physiological manipulations by males (Rice 1996; Chapman and Partridge 1996 on *Drosophila*; see discussion of muscoid flies above).

IX. WHO IS WINNING – A NON-QUESTION?

One is accustomed in discussions of reproductive conflicts of interest and the resulting evolutionary races to suppose that either one of the two parties may be 'winning' at any particular moment (see, in the context of postcopulatory male competition Parker 1984; Holland and Rice 1997; Chapters 10 and 17). The very question of winners and losers may often be misleading, however, in cases in which females benefit from choosing among males by obtaining superior male offspring (above) or through having more vigorous offspring of both sexes (see Cordero 1995 on male seminal products; see Chapter 2 and Chapter 10 on genital evolution – but see next section). At least some members of both sexes can win simultaneously when selective female cooperation occurs: the female can 'win' (by having higher quality offspring) by ensuring that only certain types of males will 'win' (by outcompeting rival males in their ability to adapt to or overcome certain female-imposed conditions).

Similarly, the question of which sex is 'in control' of postcopulatory events in any particular species can also be misleading since the word 'control' has more than one meaning. In one sense it implies that one or the other sex completely overrides the effects of the other (complete or outright control); but another meaning is that implied in 'partial' as opposed to 'complete' control. Both the male and the female can influence the 'control' of a given process and, as noted by Parker in Chapter 1, this may be common.

A second problem with 'control' occurs at the level of analysis. Certain traits of the male can 'control' the outcome of the male's competition with rival males to overcome female barriers, while at the same time the female 'controls' which male traits will be useful in this competition ('the rules of the game'), and thus, indirectly, which males will sire offspring (see Wiley and Poston 1996 for this same argument in the context of precopulatory sexual selection).

The problem of levels of control can be illustrated with an example from this book. Simmons and Siva-Jothy (Chapter 10) propose, as an alternative explanation to female influence on paternity in the beetle *Chelymorpha alternans*, that males with longer intromittent organs 'may simply be more efficient at delivering their ejaculate to the site of storage'. However, females of this species have a very long, highly convoluted spermatheca duct, so this is not an alternative explanation, but is simply a statement of possible male strategies once the female has set the playing field (in this case the tortuous spermatheca duct). The question of which sex 'controls' processes that result in differences in paternity can have two answers: differences in the lengths of male organs can explain paternity biases and, at a deeper level, the reason why the lengths of male organs affect paternity is explained by the details of the design of the female reproductive tract. There will be an explanation based on female traits lurking behind many explanations based on male traits.

X. ARE THESE ARGUMENTS APPLICABLE IN NATURE?

One reasonable response to the theoretical points discussed above is to question whether or not they are biologically relevant. There are many attractive theories in biology that have not proven to be of general interest, not because they were not logical, but because they applied to few, if any, living organisms (reproductive character displacement comes to mind). I have summarized data elsewhere (Eberhard 1996) indicating that female effects on postcopulatory male competition are likely to prove to be common in nature. I compiled and discussed diverse types of evidence from over 100 species regarding 20 different types of mechanisms that could result in cryptic female choice. The evidence ranged from direct experimental tests for selective female effects, to observations of processes that are capable of producing biases but that have not been demonstrated to do so (a possible mechanism (number 21), selective digestion of sperm within the female, is mentioned by Baur in Chapter 8). In addition I have discussed several major patterns in the evolution of reproductive traits, and argued that they are easily intelligible and expected under the hypothesis of cryptic female choice. These traits included behaviour (especially the widespread occurrence of courtship behaviour during copulation), morphology (in particular, the widespread occurrence of tortuous female reproductive tracts that prevent or screen male access to sperm storage and fertilization sites, and the even more widespread trend for male genitalia to diverge rapidly), and physiology (in particular the widespread existence of male seminal products that have manipulative effects on female reproductive processes). I will not attempt to review these data here, other than to note that accounts of several additional possible cases of cryptic female choice have been published recently (Côte and Hunte 1989; de Lope and Møller 1993; Adkins-Regan 1995; Burley *et al.* 1996; Bishop *et al.* 1996; Weigensburg and Fairbairn 1996). There is also further documentation (of the likelihood of female discrimination among males, and female influence on paternity, respectively) for two cases that were discussed there (Sakaluk and Eggert 1996; LaMunyon unpublished data), and evidence that copulatory courtship is widespread in a further large sample of spiders (Huber unpublished data).

Several authors in this volume present challenges to the idea of cryptic female choice. Parker (Chapter 1) gives the impression that there is little convincing evidence that females influence sperm allocation. He states that 'much of it is flawed' without citing specific examples, thus robbing me of the chance to try to reply only two chapters later. Parker stresses that there are alternative interpretations based on male differences such as ejaculate sizes rather than on female roles in some species, but fails to take into account the 'rules of the game' consideration that even the opportunity for males to compete in sperm raffles can be influenced by numerous female traits (see Section VII, this chapter).

Møller (Chapter 2), in reference to genital evolution, expresses doubts about the Fisherian type of selective advantages to females that were discussed above. He argues that additional factors that might be associated with genital morphology have not been taken into account, mentioning specifically the possibility of a link between a male's genital morphology and his resistance to parasites. Exercises of this sort can, of course, be continued *ad infinitum* – we also still lack tests for possible associations between genital form and male renal function, intelligence, and running speed. Such objections are compelling to the degree that other considerations suggest that the factors that were omitted previously are likely to be important. Møller claims that there is a 'considerable amount of genital manipulation or inspection in a wide range of species', but does not give even one reference. I know of no direct evidence, or any a priori reason suggesting an association between the kinds of differences in genital morphology commonly seen between closely related species (a tuft of bristles at one site rather than another on a clasping organ, an intromittent sclerite that is pointed instead of rounded, recurved spines rather than sandpaper-like scales on the tip of a penis, etc.) and differences in parasite resistance (or renal function, intelligence, or running speed for that matter).

Simmons and Siva-Jothy (Chapter 10) also mention the possibility of a link between complex male genitalia and male quality. The only supporting reason they provide, however, hinges on the possibility that moulting in hemimetabolous insects is more likely to fail when the male has complex genitalia (they cite one species). No explanation is offered for why different genital forms should evolve in different species. Their hypothesis also leaves rapid divergent evolution of genitalia unexplained in quite a few other groups: the rest of the insects (most of which are holometabolous), snakes, lizards, bony fish, sharks and rays, opilionids, spiders, mites, mammals, nematodes, polychaete and oligochaete worms, turbellarian flatworms, and several others (see Eberhard 1985).

Birkhead and Møller summarize the status of ideas regarding those possible mechanisms of cryptic female choice that involve internal manipulation of sperm (Chapter 17) as 'controversial'. Their conclusion that there are very few studies that directly demonstrate these kinds of choice is true; there are simply very few studies of these kinds of phenomena, or, for that matter, of other possible cryptic female choice mechanisms such as different oviposition rates associated with differences between males. There are many critical tests yet to be performed, as well as some still unexplained facts (e.g. the extremely elaborate and diverse sperm morphology in apparently monogamous termites, which is paradoxical from the perspective of cryptic female choice, as well as from that of sperm competition *sensu stricto*; Eberhard 1996). But the hypothesis of cryptic female choice has survived several general tests that could have resulted in its being discarded or severely modified (Eberhard 1985, 1996). There are also some impressively large trends in nature, such as the widespread occurrence of copulatory courtship, and the rapid divergent evolution of male genitalia, that seem difficult to explain otherwise. The field is still too young for confident final decisions.

ACKNOWLEDGEMENTS

I thank John Christy, Bernhard Huber, Bill Wcislo, and Mary Jane West-Eberhard for detailed, useful criticisms at short notice.

REFERENCES

Adkins-Regan E (1995) Predictors of fertilization in the Japanese quail, *Coturnix japonica*. *Anim. Behav.* **50:** 1404–1415.

Ahnesjö I, Vincent A, Alatalo R, Halliday T & Sutherland WJ (1992) The role of females in influencing mating patterns. *Behav. Ecol.* **4:** 187–189.

Alexander RD, Marshall D & Cooley J (1997) Evolutionary perspectives on insect mating. In *Social Competition and Cooperation in Insects and Arachnids. I. Evolution of Mating Systems* J Choe & B Crespi (eds), pp. 4–31. Cambridge University Press, Cambridge.

Andersson M (1994) *Sexual Selection*. Princeton University Press, Princeton.

Arnqvist G & Rowe L (1995) Sexual conflict and arms races between the sexes: a morphological adaptation for control of mating in a female insect. *Proc. R. Soc. Lond. B* **261:** 123–127.

Bishop JDD, Jones CS & Noble LR (1996) Female control of paternity in the internally fertilizing compound ascidian *Diplosoma listerianum*. II. Investigation of male mating success using RAPD markers. *Proc. Roy. Soc. Lond. B* **263:** 401–407.

Burley NT, Parker PG & Lundy K (1996) Sexual selection and extrapair fertilization in a socially monogamous passerine, the zebra finch (*Taeniopygia guttata*). *Behav. Ecol.* **7:** 218–226.

Chapman T, Liddle LF, Kalb JM, Wolfner MF & Partridge L (1995) Cost of mating in *Drosophila melanogaster* females is mediated by male accessory gland products. *Nature* **373:** 241–244.

Chapman T & Partridge L (1996) Female fitness in *Drosophila melanogaster*: an interaction between the effect of nutrition and of encounter rate with males. *Proc. Roy. Soc. Lond. B* **263:** 755–759.

Christy JH (1995) Mimicry, mate choice, and the sensory trap hypothesis. *Am. Nat.* **146:** 171–181.

Cordero C (1995) Ejaculate substances that affect female insect reproductive physiology and behavior: Honest or arbitrary traits? *J. Theor. Biol.* **174:** 453–461.

Côte IM & Hunte W (1989) Male and female mate choice in the redlip blenny: why big is better. *Anim. Behav.* **38:** 78–88.

Daly M (1978) The cost of mating. *Am. Nat.* **112:** 771–774.

de Lope F & Møller AP (1993) Female reproductive effort depends on the degree of ornamentation of their mates. *Evolution* **47:** 1152–1160.

Eberhard WG (1985) *Sexual Selection and Animal Genitalia*. Harvard University Press, Cambridge, MA.

Eberhard WG (1991) Copulatory courtship and cryptic female choice in insects. *Biol. Rev.* **66:** 1–31.

Eberhard WG (1993) Evaluating models of sexual selection: genitalia as a test case. *Am. Nat.* **142:** 564–571.

Eberhard WG (1994) Evidence for widespread courtship during copulation in 131 species of insects and spiders. *Evolution* **48:** 711–733.

Eberhard WG (1996) *Female Control: Sexual Selection by Cryptic Female Choice*. Princeton University Press, Princeton.

Eberhard WG (1997) Sexual selection by cryptic female choice in insects and arachnids. In *Social Competition and Cooperation in Insects and Arachnids. I. Evolution of Mating Systems*. J Choe & B Crespi (eds), pp. 32–57. Cambridge University Press, Cambridge, UK.

Fincke OM (1997) Conflict resolution in the Odonata: implications for understanding female mating patterns and female choice. *Biol. J. Linn. Soc. Lond.* **60:** 201–220.

Haig D (1993) Genetic conflicts in human pregnancy. *Quart. Rev. Biol.* **68:** 495–531.

Holland B & Rice WR (1997) Cryptic sexual selection – more control issues. *Evolution* **51:** 321–324.

Katz PS & Frost WN (1996) Intrinsic neuromodulation: altering neuronal circuits from within. *Trends Neurosci.* **19:** 54–61.

Keller L & Reeve HK (1995) Why do females mate with multiple males? The sexually selected sperm hypothesis. *Adv. Study Behav.* **24:** 291–315.

Leopold RA, Terranova AC & Swilley EM (1971a) Mating refusal in *Musca domestica*: Effects of repeated mating and decerebration upon frequency and duration of copulation. *J. Exp. Zool.* **176:** 353–360.

Leopold RA, Terranova AC, Thorson BJ & Degrugillier ME (1971b) The biosynthesis of the male housefly accessory secretion and its fate in the mated female. *J. Insect Physiol.* **17:** 987–1003.

Lewis CT & Pollock JN (1975) Engagement of the phallosome in blowflies. *J. Entomol. (A)* **49:** 137–147.

McVey ME (1988) The opportunity for sexual selection in a territorial dragonfly, *Erythemis simplicicollis*. In *Reproductive Success – Studies of Individual Variation in Contrasting Breeding Systems.* TH Clutton-Brock (ed.), pp. 44–58. University of Chicago Press, Chicago.

Merrett DJ (1989) The morphology of the phallosome and accessory gland material transfer during copulation in the blowfly, *Lucilia cuprina* (Insecta, Diptera). *Zoomorphology* **108:** 359–366.

Minkley RL, Buchmann S & Wcislo WT (1991) Bioassay evidence for a sex attractant pheromone in the large carpenter bee, *Xylocopa varipuncta* (Anthophoridae: Hymenoptera). *J. Zool.* **224:** 285–291.

Parker GA (1984) Sperm competiton and the evolution of animal mating strategies. In *Sperm Competition* and *the Evolution of Animal Mating Systems.* RL Smith (ed.), pp. 1–60. Academic Press, New York.

Peretti AV. Cortejo copulatorio en escorpiones (Arachnida, Scorpiones) (in prep.).

Rice WR (1996) Sexually antagonistic male adaptation triggered by experimental arrest of female evolution. *Nature* **381:** 232–234.

Ridley M (1989) The incidence of sperm displacement in insects: four conjectures, one corroboration. *Biol. J. Linn. Soc.* **38:** 349–367.

Riemann JG & Thorson BJ (1969) Effect of male accessory material on oviposition and mating by female house flies. *Ann. Entomol. Soc. Amer.* **62:** 828–834.

Ryan MJ (1990) Sexual selection, sensory systems and sensory exploitation. *Oxford Surv. Evol. Biol.* **7:** 157–195.

Sakaluk SK & Eggert A-K (1996) Female control of sperm transfer and intraspecific variation in sperm precedence: antecedents to the evolution of a courtship food gift. *Evolution* **50:** 694–703.

Simmons LW & Gwynne DT (1991) The refractory period of female katydids (Orthoptera: Tettigoniidae): sexual conflict over the remating interval? *Behav. Ecol.* **2:** 276–282.

Siva-Jothy MT (1987) The structure and function of the female sperm-storage organs in libellulid dragonflies. *J. Insect Physiol.* **33:** 559–567.

Siva-Jothy MT & Hooper RE (1995) The disposition and genetic diversity of stored sperm in females of the damselfly *Calopteryx splendens xanthostoma* (Charpentier). *Proc. Roy. Soc. Lond. B* **259:** 313–318.

Siva-Jothy MT & Hooper RE (1996) Differential use of stored sperm during oviposition in the damselfly *Calopteryx splendens xanthostoma* (Charpentier). *Behav. Ecol. Sociobiol.* **39:** 389–393.

Siva-Jothy M & Tsubaki Y (1989) Variation in copulation duration in *Mnais pruinosa pruinosa* Selys (Odonata: Calopterygidae) 1. Alternative mate-securing tactics and sperm precedence. *Behav. Ecol. Sociobiol.* **24:** 39–45.

Smith PH, Barton-Browne L & van Gerwen ACM (1988) Sperm storage and utilisation and egg fertility in the sheep flowfly, *Lucilia cuprina. J. Insect Physiol.* **34:** 125–129.

Smith PH, Barton-Browne L & van Gerwen ACM (1989) Causes and correlates of loss and recovery of sexual receptivity in *Lucilia cuprina* females after their first mating. *J. Insect Behav.* **2:** 325–337.

Smith PH, Gillott C, Barton-Browne L & van Gerwen ACM (1990) The mating-induced refractoriness of *Lucilia cuprina* females: manipulating the male contribution. *Physiol. Entomol.* **15:** 469–481.

Stockley P (1997) Sexual conflict resulting from adaptations to sperm competition. *Trends Ecol. Evol.* **12:** 154–159.

Thornhill R & Alcock J (1983) *The Evolution of Insect Mating Systems.* Harvard University Press, Cambridge, MA.

Waage J (1979) Dual function of the damselfly penis: sperm removal and transfer. *Science* 203: 916–918.

Waage J (1984) Sperm competition and the evolution of odonate mating systems. In *Sperm Competition and the Evolution of Animal Mating Systems.* RL Smith (ed.), pp. 251–290. Academic Press, New York.

Waage JK (1986) Evidence for widespread sperm displacement ability among Zygoptera (Odonata) and the means for predicting its presence. *Biol. J. Linn. Soc.* **28:** 285–300.

Walker W (1980) Sperm utilization strategies in nonsocial insects. *Am. Nat.* **115:** 780–799.

Weigensberg I & Fairbairn DJ (1996) The sexual arms race and phenotypic correlates of mating success in the water strider, *Aquarius remigis* (Hemiptera: Gerridae) *J. Insect Behav.* **9:** 307–319.

West-Eberhard MJ (1979) Sexual selection, social competition, and evolution. *Proc. Am. Phil. Soc.* **123:** 222–234.

West-Eberhard MJ (1983) Sexual selection, social competition, and speciation. *Quart. Rev. Biol.* **58:** 155–183.

West-Eberhard MJ (1984) Sexual selection, competitive communication and species-specific signals in insects. In *Insect Communication.* T. Lewis (ed.), pp. 283–324. Academic Press, New York.

Wiley RH (1997) Coevolution of the sexes. *Science* **275:** 1075–1076.

Wiley RH & Poston J (1996) Indirect mate choice, competition for mates, and coevolution of the sexes. *Evolution* **50:** 1371–1381.

4 Paternity and Paternal Care

Jonathan Wright

School of Biological Sciences, University of Wales, Bangor, Gwynedd, LL57 2UW, UK

I. INTRODUCTION

A. Chapter aims

Sperm competition has important consequences for the evolution of paternal care, the most obvious of which is that males should be selected not to care for young that they are unlikely to have fathered. Many authors have assumed a simple causal relationship between a male's probability of paternity and the level of paternal care he should provide. Unfortunately, most tests carried out to date on avian systems have failed to demonstrate changes in the levels of male care in response to experimentally induced reductions in apparent paternity (Whittingham *et al.* 1993; Wright and Cotton 1994). This lack of an obvious effect in birds has raised doubts about the reliability of theoretical models and the experimental techniques being used. The aim of this chapter is to show that our original approach to this problem was over simplistic, both theoretically and empirically. Adaptive adjustments in the level of care by males in response to reduced paternity must be considered in the context of the mating system as a whole, including patterns of extra-pair matings. This is because life history decisions regarding paternity and paternal care are closely linked to other components of the mating system, such as the extent of male investment in extra-pair matings with additional females. Using this

Sperm Competition and Sexual Selection
ISBN 0-12-100543-7

whole mating system approach, it may be possible to explain the failure of a number of recent empirical studies to demonstrate a reduction in male care following loss of paternity. An additional aim of this chapter is to clarify some of the theoretical issues surrounding paternity and parental care, especially regarding mathematical models and the use of their predictions in experimental tests by field biologists.

B. Paternity

Sperm competition is competition for paternity, and results in males achieving differential success in fertilizing eggs and passing their genes on to future generations. Although actual paternity is 'all-or-nothing' – a male either is or is not the father of a particular juvenile – the distinction is rarely so absolute from the male's point of view. As a result of females mating with more than one male, genetic relatedness between any one male and the offspring produced is essentially reduced to a probability. This is due to the mixing of sperm, either during external fertilization (Gross and Charnov 1980), or during internal fertilization inside the reproductive tract of the female (Smith 1984). Paternity is therefore often dealt with as a proportion relative to unity, the latter representing actual fatherhood (e.g. 'P'; Houston and Davies 1985). A male that has 50% probability of paternity in a brood of four young (i.e. $P = 0.5$) has not necessarily fathered two chicks, he simply has a 50:50 chance of siring any one of the four young, which, on average, provides the evolutionary equivalent of two offspring. Therefore, throughout this chapter I use 'paternity' in the sense of a continuously varying probability of paternity, rather than the strict definition of paternity as absolute fatherhood.

This distinction between actual fatherhood and a probability of paternity becomes important if males cannot recognize their own young, either between or within broods. If a male adjusting his level of care could use 'discriminate cues' (Westneat and Sherman 1993) to recognize and discriminate between his own young and the offspring of other males, then a loss of paternity is simply equivalent to a reduction in number of young. Such kin recognition is, however, unlikely because we might expect selection to act against offspring (as well as the mothers of offspring) that signal the identity of their father, because they would be more easily discriminated against whenever they themselves were fathered by a male other than the one providing care (Beecher 1988). Despite some early analysis of the problem by Beecher (1991), such reasoning has only recently been verified using appropriate mathematical models (Pagel 1997). In addition, for the types of avian system in which paternity is thought to be important in regulating male care, the data suggest that males do not discriminate between different chicks in their brood (Burke *et al.* 1989; Westneat *et al.* 1995; but see Kempenaers and Sheldon 1996). Most empirical evidence suggests that males tend to rely upon 'indiscriminate cues' to their paternity (Westneat and Sherman 1993), such as polyandrous male dunnocks (*Prunella modularis*) which appear to use their proportion of exclusive mating access with the female to estimate their probability of paternity in the brood as a whole (Burke *et al.* 1989; Davies *et al.* 1992). Hence, much of the fieldwork on paternity and paternal care has been focused upon male 'certainty' or 'confidence' of paternity in its 'perceptual sense' (Schwagmeyer and Mock 1993).

C. Male parental care

Trivers (1972) defined parental investment as 'any investment by the parent in an individual offspring that increases the offspring's chance of surviving (and hence its reproductive success) at the cost of the parent's ability to invest in other offspring'. Evolutionary arguments involving the relationship between paternity and paternal care are based on exactly this life history trade-off for males between current and future reproduction. Therefore, whenever considering adaptive reductions in male care in response to reduced probability of paternity, we need to think in terms of the potential fitness pay-offs arising from an increased ability to invest in alternative and future reproductive events (for full discussion see Section II.A, below). Extensive male parental care occurs in relatively few taxonomic groups, but is especially widespread among fish and birds (Clutton-Brock 1991). Discussions of paternity and paternal care have therefore tended to concentrate on these vertebrate classes, and are limited to those in which females mate with multiple males (Westneat and Sherman 1993). In addition, although not normally considered in the context of paternal care, infanticide by males can be thought of as negative paternal investment (Gowaty 1996a). Even in taxa where male investment in care is low (e.g. mammals; Woodroffe and Vincent 1994), infanticide may represent an adaptive response to low paternity by more quickly providing males with greater shares of paternity in future breeding attempts (Trivers 1972; Hrdy 1979).

Many of the mating systems in which paternity may have an influence on levels of paternal care involve both male and female care of the young (e.g. socially monogamous birds; Lack 1968; Birkhead and Møller 1992). Cooperation and conflict between carers adds additional complexity to strategies of adaptive adjustment in levels of care by males. A common prediction from both ESS (Chase 1980; Houston and Davies 1985) and static optimization (Winkler 1987; Kacelnik and Cuthill 1990) models of biparental care is that if one parent changes its work rate, the other should compensate for the change, but such compensation should be incomplete. Hence, a reduction in male care following reduced paternity may well prompt females to increase the amount of care that they provide, but not to the extent that the total level of care is unchanged. Incomplete female compensation should therefore result in the total amount of care being reduced, to the detriment of the young. Such incomplete compensation has been clearly demonstrated within avian biparental care systems (Wright and Cuthill 1989; Markman *et al.* 1995). However, as with avian male-removal studies (see Markman *et al.* 1996; and references therein), the data from these biparental care studies sometimes show apparently complete compensation on the part of the female (Wright and Cuthill 1990a,b; Ketterson and van Nolan 1992; Saino and Møller 1995; Lozano and Lemon 1996). These mixed results may be due to variation in the costs of female care (e.g. variation in foraging conditions), or insufficient sensitivity in the measures used to record the consequences of the withdrawal of male care, either for the female, her provisioning effort, or for the young concerned. Either way, changes in levels of male parental care as a result of lost paternity will have knock-on effects on the caring effort of other individuals involved in raising the young. The presence and abilities of additional carers in both biparental and cooperative systems are clearly important in determining the extent to which males will adjust their care in response to paternity (Hatchwell and Davies 1990). These behavioural adjustments between carers also have important implications for the interpretation of data demonstrating decreases in paternal care. This is because the causal relationship could be working in reverse, with males simply showing a compensatory decrease in care in response to an increase in female care. Hence, measures of paternal care as a proportion of the total care provided are inadequate on their own and can be misleading when assessing effects of paternity: absolute levels of care by both males and females should be presented (Wright 1992).

D. Paternity and parental care

Trivers (1972) first noted the problem of loss of paternity for males in species with internal fertilization and extensive male parental investment. Confidence of paternity should be greater in species where males can see the eggs and where females cannot store sperm from previous matings with other males (Bulmer 1979; Perrone and Zaret 1979). In taxa such as fish, where there is both internal and external fertilization, we do indeed see extensive paternal care restricted to species with external fertilization and therefore, presumably, greater confidence of paternity (Gross and Shine 1981). However, 'sneaking' male strategies can still make it possible for males to lose paternity, even in externally fertilizing species (Baylis 1981; Gross 1984). In addition, male care has been suggested to be more common in species with external fertilization because the latter allows females to desert first, thereby leaving males in the 'cruel bind' of having to care or risk losing everything (Trivers 1972; Dawkins and Carlisle 1976). The link between paternity and paternal care in fish species becomes further complicated by the fact that species with external fertilization tend to have larger clutches laid by a greater number of females and, with relatively low costs of male care, a large healthy brood can then be used to court additional females at the nest site (for a discussion, see Clutton-Brock 1991).

Despite the difficulties of demonstrating the original effect suggested by Trivers (1972), variation in paternity was subsequently cited as an explanation of widespread differences in levels of paternal care both within and between species (e.g. Alexander 1974, 1979; Ridley 1978; Alexander and Borgia 1979; Bulmer 1979; Perrone and Zaret 1979). More recently, certainty of paternity has been used to explain individual behavioural decisions concerning paternal care within species, especially in birds (Møller 1988a). Thus, the same adaptive argument concerning paternity and paternal care has been used to explain everything from minute-to-minute differences in individual male behaviour to large-scale taxonomic differences in the extent of male care. However, when discussing adaptive responses to reduced paternity in terms of paternal effort, an important distinction must be made between 'facultative' behavioural responses occurring over 'ecological' time and 'nonfacultative' responses occurring over 'evolutionary' time (Westneat and Sherman 1993). We almost always think of individual males making adaptive facultative decisions regarding their level of care in response to specific behavioural evidence that they have been cuckolded during that particular breeding episode. However, nonfacultative ('hard-wired') adjustments in parental effort might instead be predetermined according to the breeding situation, regardless of the specific behavioural events experienced by an individual which might relate to any estimate of paternity. This is because individuals may do better by ignoring unreliable local cues and instead following a fixed, nonfacultative strategy of paternal investment, depending upon more reliable estimates of paternity as determined by the species, the particular population, or even the age or dominance rank of the male concerned. Levels of male care could therefore be the result of differences in paternity, yet we would be unable to detect such an effect through behavioural manipulations of certainty of paternity. It is important to note that the processes which lead to facultative rather than nonfacultative adjustments in levels of paternal care may be different and, as I shall argue below, these processes may depend critically upon the pattern and predictability of extra-pair paternity within a particular mating system.

II. THEORETICAL TREATMENTS

A. Theoretical framework

Westneat and Sherman (1993) provide the most useful scheme for considering parentage and parental care. They describe a hierarchical organization of the possible paternal care responses to changes in paternity at both the evolutionary and behavioural levels. The key to their approach lies in dividing the reproductive effort within one breeding attempt into 'parental effort' (PE, effort invested in parental behaviour to gain greater fitness from the current brood), 'mating effort' (ME, effort invested in acquiring fertilizations and so higher levels of paternity), and 'somatic effort' (SE, effort invested to increase an individual's chances of surviving to future breeding attempts, where it will face the same reproductive investment decisions again). These three alternatives can be traded off against each other, such that increased investment in one leads to fewer resources being available for investment in the other two (i.e. an extension of 'parental investment' as defined by Trivers (1972); see Section IC, above). For example, higher levels of mate guarding (a component of ME) might have costs owing to lost foraging time, which may then reduce a male's ability to provide care within the same breeding attempt (a component of PE) and/or reduce his body condition and future reproductive potential (a component of SE). Increased investment in each form of reproductive effort yields greater fitness benefits, but at the expense of other forms of effort. The curves described by each fitness benefit function can be thought of as constraints upon, or preconditions to, the resultant optimum behavioural strategy employed by a particular species, population, age class or type of individual male. For any one male, there is a range of possible reproductive options, expressed in terms of various combinations of investment in PE, ME and SE. The life-history decision for males is therefore to optimize the proportions of investment in PE, ME and SE in successive breeding attempts such that they maximize the overall fitness returns over their lifetime.

An important property of this theoretical framework is that it encompasses aspects of both within-pair and extra-pair mating systems in paternity and paternal investment decisions. Therefore, mixed reproductive strategies by males can be considered as a trade-off between investment in ME and investment in PE. For example, evidence from field studies of birds suggests that, compared with females, males will tend to reduce chick feeding PE more quickly when the latter provides low marginal fitness returns; this could be because males gain differentially from additional investment in ME, either from extra-pair activity or from some other social interaction with neighbours (Carey 1990; Wright and Cuthill 1990a,b; Wright and Cotton 1994). It should be noted that the fitness benefits from ME, in the form of increased paternity, can be divided into paternity of the pair young which receive that male's care, and paternity of extra-pair young which do not. Mate guarding and within-pair copulation effort (e.g. number or size of ejaculates) are examples of ME to gain paternity of pair young, whereas effort spent seeking extra-pair copulations (EPCs) is ME to gain paternity of extra-pair young which will not receive care from the male in question.

Reduced paternity of within-pair young in the current breeding attempt will lower the fitness returns for males per unit of investment in

PE. Hence, the optimum level of investment in PE will be reduced when paternity is lost within the brood for which the male provides care, because greater overall fitness will be obtained by males switching their investment into ME and/or SE. Interestingly, this prediction is dependent upon a male's probability of paternity in one breeding attempt changing somewhat independently of all present or future fitness pay-offs. For example, if loss of paternity in the present results from a male's inability to guard his mate from the attentions of other males, then it is possible that any loss of paternity will covary with the same male's ability to get extra-pair matings during the same breeding attempt. So, a male that loses paternity in this way would not necessarily gain from switching his investment from PE to ME, because the extra investment in ME will yield no extra fitness benefits per unit of reproductive effort. In the same way, conditions in future breeding attempts, and the per-unit fitness pay-offs from investment in SE, may also be related to present paternity. If paternity is lost in the present as a result of inability to guard a mate, this may be reflected in similarly low levels of paternity from poor mate guarding in the future. If this is the case, there would be no advantage in switching investment from PE to SE because the fitness returns from present investment in SE are equally devalued by losses in paternity that will occur in the future. Therefore if conditions in future reproductive episodes are, on average, the same as in the present (e.g. individual levels of paternity), there will be nothing to be gained by males reducing their levels of care in any one breeding attempt (Maynard Smith 1978; Grafen 1980; Westneat and Sherman 1993).

The key to the Westneat and Sherman (1993) theoretical framework is that the magnitude of a reduction in male care (PE) as a result of lost paternity is dependent not only upon the costs of paternal care in terms of reduced SE, but also upon the fitness benefits available from extra-pair ME. Importantly, PE and paternity are linked not just via the PE benefit function, but also through the trade-off between PE and ME. Greater investment in ME may increase paternity in pair and/or extra-pair broods, but increased ME will also be at the expense of a male's ability to invest in PE.

B. *Paternity and paternal care models*

Paternity appears as a parameter in a number of early parental care models, although little explicit information is provided on the precise form of its relationship with male care (e.g. Werren *et al.* 1980; Houston and Davies 1985; Winkler 1987). The lack of a precise prediction led early workers to suggest a linear relationship between paternity and parental care (Møller 1988a). However, more recent models have been designed specifically to tackle the question of how paternity will influence levels of paternal care (Whittingham *et al.* 1992; Westneat and

Sherman 1993; Houston 1995). Figure 4.1 provides a graphical representation of the most reasonable shape for the causal relationship between paternity and paternal care. From the arguments above (see Section II.A), it is clear that such a relationship is a product of models which assume that in the present breeding episode paternity varies independently of other parameters in alternative and future reproductive episodes. As paternity increases the fitness benefits of male care increase proportionally as a result of improving the chances of survival and future reproduction of more related young. In general, therefore, the optimum amount of parental effort will increase with paternity. The function is discontinuous because a certain 'threshold' of paternity exists below which care is never worthwhile (Houston 1995). Above this threshold the optimum level of care increases with paternity, but at a decelerating rate because higher levels of paternal care carry accelerating costs to males and/or diminishing returns in terms of the increase in fitness of young per unit of paternal investment.

The exact shape of the paternity-paternal care relationship in Fig. 4.1 is important when testing for an effect of paternity upon male care. For example, the lack of any obvious effect in many field studies on birds might simply be explained by the systems under investigation existing in the relatively flat part of the curve near the asymptote (Whittingham *et al.* 1993). Specifically, we need to know both the minimum level of paternity necessary for male care to occur, and the level of paternity at which we might expect no further increase in levels of male care. In all models, the shapes of the cost–benefit curves have a critical effect upon the resultant shape of the paternity–parental care function (Westneat and Sherman 1993; Houston 1995). For particular sets of cost–benefit

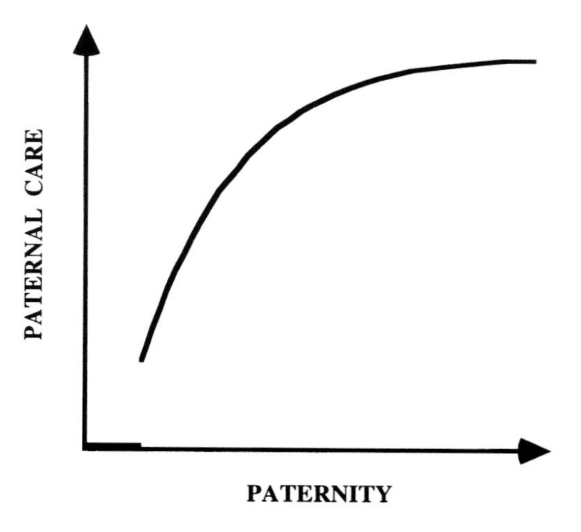

Fig. 4.1. *The basic prediction for an effect of paternity on the optimum level of male parental care.*

functions, it has been shown that only very low levels of paternity will have substantial negative effects upon male parental care (Whittingham *et al.* 1992; Westneat and Sherman 1993; Houston 1995). Unfortunately, it is notoriously difficult in empirical studies to actually quantify the form of functions describing the fitness costs and benefits of parental care (Clutton-Brock 1991). However, by examining the results of certain experimental studies, it may still be possible to confirm the general applicability of models which assume particular forms in these cost–benefit functions. For example, Wright and Cuthill (1990a) performed a comprehensive set of experimental manipulations of partner contribution and brood size in European starlings, *Sturnus vulgaris*. The resulting patterns of parental care are compatible with predictions from both ESS (Chase 1980; Houston and Davies 1985) and static optimization (Winkler 1987; Kacelnik and Cuthill 1990) models of biparental care. It therefore seems reasonable to use the same types of function to explore optimum levels of male care in response to variation in paternity. In contrast, recent modelling efforts have involved cost–benefit curves of unproven shape, designed to produce 'discontinuous' or step-like 'threshold' response curves with extensive and almost flat asymptotes, so that they are able to match the empirical studies which fail to find an effect of certainty of paternity upon paternal care (Whittingham *et al.* 1992; but see Houston 1995).

It should also be noted that Fig. 4.1 represents an oversimplification, because most models do not consider adjustments in female parental care when calculating optimum levels of paternal care (Whittingham *et al.* 1992; Westneat and Sherman 1993). As mentioned above (see Section I.B), any adjustments in the level of male care as a result of changes in paternity should result in associated compensatory changes in the levels of female care (Wright and Cuthill 1989). Adjustments in optimum levels of male care in response to changes in paternity in a biparental system should therefore be modelled in terms of an ESS which also incorporate optimal adjustments in female care. A recent model by Houston (1995) has done just that, reassuringly confirming for biparental care systems the hypothetical shape of the paternity-paternal care effect shown in Fig. 4.1.

III. EXTRA-PAIR MATING AND 'FIXED' PATERNITY

A. *Differential male quality*

In real populations, males will differ in phenotypic quality. This may consist of variation in physical characters such as body size or physiological condition, or in sexually selected traits used in female choice, or in terms of individual success in behavioural activities such as foraging or the ability to acquire a breeding site. When differences in male quality have a genetic basis, variation between males is expected to lead to adaptive female choice for males containing these 'good genes' (see Andersson 1994), resulting in genetic polygyny which may occur irrespective of the observed social mating system (Petrie and Lipsitch 1994). It is becoming increasingly clear that females cannot be regarded as passive participants in the competition between males to gain matings and guard their females from other males. Sperm competition, and therefore patterns of paternity, are often the result of active extra-pair mate choice by females (Kempenaers *et al.* 1992; Sheldon 1994a; Gowaty 1995). Empirical studies have shown that females are often in control of copulations (Eens and Pinxten 1995; Kempenaers *et al.* 1995), and that individual females show consistent levels of extra-pair activity over and above the ability of their mate to guard them (Lifjeld and Robertson 1992; Lifjeld *et al.* 1993). Evidence of female choice for good genes comes from avian data demonstrating that females preferentially copulate with more 'attractive' extra-pair males of higher quality than their own partners (Møller 1988c; Smith 1988; Norris 1990; Houtman 1992; Kempenaers *et al.* 1992; Graves *et al.* 1993; Otter *et al.* 1994; Hasselquist *et al.* 1995). It should be noted that such patterns in extra-pair activity are not in themselves evidence for female choice for good genes, because we would predict similar patterns if males of better quality were more attractive to females simply because of their greater fertility (Sheldon 1994b). Either way, the argument presented here is that better quality males are nearly always expected to obtain more matings per unit time spent in extra-pair ME. Females paired with better quality males will also gain little from mating with other males in the population. Hence, better quality males have been shown to experience reduced loss of paternity and to guard mates less as a result (Gowaty and Bridges 1991; Gowaty *et al.* 1989; Smith *et al.* 1991; Kempenaers *et al.* 1992; Møller 1994; Johnsen and Lifjeld 1995; Kempenaers *et al.* 1995). However, we know that females do not have it all their own way in extra-pair mate choice, because poor quality males still obtain within-pair copulations and paired males must be gaining some fitness return from all their investment in mate guarding behaviour. In fact, the dynamics of mate guarding and levels of extra-pair paternity may be best understood not only in terms of relative male phenotypic quality in the population, but also in terms of the relative phenotypic quality differences between the pair male and the pair female. For example, in mismatched pairs of birds where one sex is larger than the other, the larger sex may be able to gain control over the female's extra-pair mating activity, with the level of extra-pair paternity reflecting the evolutionary interests of the larger and physically more powerful sex (Wagner *et al.* 1996a). Therefore extra-pair mating systems, and possibly all aspects of all mating systems, are best understood in terms of intersexual conflict, with males attempting to control female mate choice and females evolving mechanisms to resist such male control (Gowaty 1996b). The patterns we see in extra-pair matings are probably a compromise between male and female evolutionary interests, with relative male quality playing a major role via female choice and pair male ability to constrain female choice. Field studies on birds tend to support the notion that, to some extent, females are successful in exercising extra-pair mate choice, and that patterns of paternity are to a large extent predictable on the basis of individual male quality.

B. Predicting no relationship between paternity and paternal care

From this knowledge of the effects of individual male quality, we might predict that the paternity a male achieves in his present brood will be positively related to his ability to gain paternity from additional pairings, from extra-pair activity, or from future breeding opportunities. This covariation with male quality may cause each male to experience (on average) a fixed level of paternity. This is an example of Westneat and Sherman's (1993) category of 'unrestricted' patterns of paternity, where the average levels of paternity are expected to be the same for all broods during the lifetime of the male. Ignoring for the moment any effects of male age (see Section III.C, below), good quality males might be expected to obtain high paternity throughout their lives, and poor quality males might be expected to consistently lose paternity to the same degree each time they try to breed. The differences in the levels of paternity experienced by different males will remain roughly the same from one breeding episode to the next on the basis of the relative differences in their quality. This argument also applies to any fitness pay-off per unit of investment in ME, in terms of time and energy spent on additional mating opportunities concurrent with the present brood, either from extra-pair activities or additional mates. If individual levels of within-pair paternity were, on average, the same in successive breeding episodes, and the benefits per unit investment in ME do not alter, there would be no adaptive advantage accruing to any individual that varied his level of paternal care between breeding attempts (Maynard Smith 1978; Grafen 1980). It is clear that, 'If the same factors that led to reduced parentage affect the alternatives to providing care, then no changes in PE are expected ...' (Westneat and Sherman 1993, p. 75). The suggestion here is that in many natural systems, such as birds, differential male quality is the primary factor leading to variation in paternity. Each male will experience a 'fixed' level of paternity commensurate with his quality during his lifetime and so perhaps we should not be so surprised when we do not find evidence for behavioural adjustment by males in their level of care in response to variation in certainty of paternity (see Section V.B, below).

C. The effect of male age

The relationship between paternity and levels of paternal care is essentially one of life history trade-offs, involving present vs. future reproductive investment. Therefore, any effect of male age on the fitness pay-offs of investment in PE, ME or SE must have important consequences for adaptive strategies of paternity and paternal care. Age can have a positive effect on individual male phenotype, owing to growth or the acquisition of skills through breeding experience. In addition, a correlation between male age and relative genetic quality could arise owing to the selective death of low-quality males in the population at younger ages. Females may prefer older males as partners because they can defend more resources or are better able to care for young (i.e. benefits per unit investment in PE will increase for older males). However, only when older males are actually of superior genetic quality will they be of interest to females seeking EPCs for good genes (i.e. benefits per unit investment in ME will increase for older males). In birds, older males tend to gain a greater share of paternity both in their own broods (e.g. indigo buntings *Passerina cyanea*; Westneat 1990), and the broods of other younger males (e.g. purple martins *Progne subis*; Morton *et al.* 1990). Such patterns may result from older males being better able to deter other males (e.g. yellowhammer *Emberiza citrinella*; Sundberg 1992), or from females of older males being less likely to seek EPCs, thereby requiring lower levels of mate guarding by their mates (e.g. Eastern Bluebirds, *Sialia sialis*; Gowaty *et al.* 1989; Gowaty and Bridges 1991). Male age does not always appear as a factor explaining variation in male phenotypic quality and patterns of paternity (Dunn *et al.* 1994a,b; Hansen and Price 1995; Yezerinac *et al.* 1996), but without following males throughout their lifetime it will always be difficult to distinguish between the effect of male age and the potentially confounding effect of male quality.

Male age is an important component of any theoretical treatment of paternity and paternal care, and it has the potential to create a range of effects. Westneat and Sherman (1993) categorize male age effects as 'restricted' patterns of paternity, where a male's paternity differs systematically between groups of a male's offspring. The result is likely to be a nonfacultative age-dependent strategy for male investment in PE, based upon predictable patterns of paternity during the lifetime of each individual. Male age effects will be expected to occur at an evolutionary level, rather than as facultative (i.e. behavioural) adjustments in paternal care on an ecological time-scale, because they are likely to be the most reliable way to ensure a known level of effort. Greater short-term behavioural assessment by males, and the potentially costly withdrawal of care that may result from a miscalculation, would be unnecessary and is likely to be avoided when a nonfacultative age-dependent mechanism could be used instead. If this is the case, then males might pay no attention to any direct experience of loss of paternity, but instead rely upon their age as a predictable cue. Younger males with low average paternity may hold back their investment in PE, and instead have a set strategy to put relatively more resources into SE (i.e. survival and future reproduction). Older males may obtain a greater share of paternity because of greater competitive ability or active female choice for proven genotypes. Hence, older males may put more investment into PE because of the greater per unit fitness benefits from helping to raise more related young. There could therefore be a positive effect of paternity on paternal care as a result of fixed age-dependent male strategies. Alternatively, if there is a strong trade-off between investment in PE and extra-pair ME within the same breeding attempt, then older males may gain more from their attractiveness to females and invest in a consistently greater level of ME. For young males, investment in PE would be the only source of fitness benefits in the present breeding attempt, and so younger males with lower paternity might actually be predicted to have consistently higher investment in PE than older males. For example, in purple martins, young males obtain fewer EPCs and experience consistently low paternity, and yet appear to provide care at rates similar to those of older males (Wagner *et al.* 1996b). The exact nature of the relationship between male age, paternity and paternal care in any particular system will therefore depend upon the precise shape of age-dependent fitness functions and the trade-offs between PE, ME and SE.

IV. HOW PATERNITY AFFECTS PATERNAL CARE

A. Interspecific comparisons

In a recent comparison of molecular and behavioural data across bird species, Møller and Birkhead (1993) reported a positive effect of average within-brood paternity upon the proportion of costly parental care that is provided by males. Despite problems regarding these analyses (Dale 1995; Møller and Birkhead 1995), and the fact that such a comparative approach cannot strictly be used to deduce the existence of evolutionary trade-offs (see Lessells 1991), it is worth considering the theoretical basis of such a result. The best way to think about the effect of paternity on paternal care in comparisons between species is as a covariance between ME, PE and paternity. For example, in species where females may more easily resist mate guarding attempts by their males (by using features of the environment which allow females to conceal their movements) there might be a larger proportion of extra-pair fertilizations. In such species, the fitness benefits to males per unit of investment in extra-pair ME will increase. If males optimize life history trade-offs, it will pay them to put proportionally more effort into ME at the expense of PE (assuming that conditions between years, and therefore investment in SE, remain constant). Therefore, in species where males exhibit higher ME, we might expect to see lower average levels of paternity and lower PE. Note that this is not necessarily as a result of the direct causal relationship in Fig. 4.1 between paternity and paternal care, as previously claimed. For the paternity–paternal care prediction in Fig. 4.1 to explain the Møller and Birkhead (1993) result across species, it would have to be as a consequence of covariation between paternity, investment in PE and investment in ME. Although hard to imagine in real mating systems, if male ME were to be held constant across species, then low levels of paternity within a species as a whole would not necessarily select for low levels of PE. There would simply be no point in males reducing their PE if there was, in turn, no benefit to be obtained from increased ME. There would also be no point in males putting less into PE, and producing fewer viable young per breeding attempt, so that they themselves might live for longer (i.e. increased SE). This is because levels of paternity in all future breeding attempts would be the same as in the present, providing no advantage to reduced PE and increased investment in SE instead. Any and all future reproduction would be devalued by the same species-wide reduction in paternity and living longer in order to invest more in future young would produce just the same low fitness return per unit of PE invested as it would from PE in current young. However, any variation in levels of extra-pair paternity across populations or species is necessarily accompanied by differences in the extra-pair mating system, and these will alter the fitness pay-offs to males from investment in extra-pair ME. It is this variation in ME investment at the expense of PE, as well as the direct paternity–paternal care effect, which produces the link between paternity and paternal care shown in studies such as Møller and Birkhead (1993). Therefore, interspecific comparisons should look at not only paternity and paternal care levels, but also at the indirect ecological and demographic factors which give rise to variation in male investment in ME, and within the context of inter-sexual conflict over the extent of female extra-pair activity. Only by examining extra-pair mating systems and the differential benefits per unit investment in extra-pair ME, can we really understand the variation in levels of paternity and paternal care across species.

B. *Intraspecific comparisons*

The vast majority of studies which have investigated the relationship between paternity and paternal care have done so within one species or breeding population (Whittingham *et al.* 1993). Individual males that are cuckolded are expected to estimate their loss of paternity and respond by facultatively reducing their levels of care. Such patterns of paternity and parental care can now be identified with relative ease. Observations of copulation behaviour and DNA profiling are commonly used to assess variation in paternity within breeding populations, and these can be easily compared with behavioural measures of male care (e.g. the provisioning of nestlings by birds; Burke *et al.* 1989). One problem is that a positive correlation obtained between measures of paternity and male care (Westneat 1988; Lubjuhn *et al.* 1993) could be due to covariance with male phenotypic quality. As discussed above (see Sections III.A and III.C), older and/or better quality males might be expected to gain more matings, to mate guard more effectively, and to provide greater levels of care than younger and/or poorer quality males. By comparing between different males we are not looking at the paternity–paternal care relationship in Fig. 4.1, but rather at factors that covary with the relative phenotypic differences between males. It is probable that each male within the population will be under different life history constraints, and that the shapes of their individual fitness benefit curves for investment in PE, ME and SE will therefore differ. Each male is expected to follow its own optimum solution to the trade-off between these three forms of reproductive effort, and this is what we see in graphs of variation in paternity and paternal care within a species or population. The appropriate comparison would be to look at the adaptive trade-off decision that occurs within the range of options open to any one individual male (Lessells 1991). In fact, this has recently been achieved by carrying out within-male comparisons, where paternity and paternal care have been recorded for the same male between different breeding attempts (Dixon *et al.* 1994; Weatherhead and Boag 1995). Therefore, studies looking for correlations between paternity measures and male parental effort may find them (Westneat 1988; Weatherhead *et al.* 1994), or they may not (Westneat 1995; Whittingham and Lifjeld 1995); either way, such data do not tell us anything about the cause and the adaptive nature of adjustments in paternal care in response to loss of paternity. Intraspecific tests for an effect of paternity on paternal care therefore require experimental manipulations, although these possess their own methodological problems (see Section V.C, below).

C. The importance of uncertainty

Males are assumed to use their recent experience of events to estimate their probability of paternity, and this appears to be justified in birds by the fact that males will perform retaliatory copulations following suspected extra-pair activity by their female (Barash 1977; Birkhead and Møller 1992; Hatchwell and Davies 1992). It is worth considering, however, that a male's estimate of his paternity from copulation rates or relative mating access also needs to be accurate enough to reveal differences in his paternity between different breeding opportunities, if an adaptive adjustment of male care is to be made (Davies *et al.* 1992). Some models claim to deal with 'assessment of parentage' rather than actual parentage itself (Whittingham *et al.* 1992; Westneat and Sherman 1993), and therefore may seem more applicable for intraspecific comparisons in the field. However, since these assume perfect assessment of parentage by males, they do not appear to be very different from other theoretical treatments in this respect. Theoretical studies of paternity and parental care appear to have ignored the problems associated with male assessment of threats to their paternity, such as how individual males should deal with inaccuracy and variation in their certainty of paternity estimates. One interesting recent suggestion proposes that if males do use certainty of paternity estimates in order to adjust their levels of care, females should be selected to make their extra-pair mating even less amenable to accurate assessment (Westneat and Sargent 1996). Therefore, failure of males to show a parental response to behavioural cues of loss of paternity (experimental or otherwise) could be explained by a history of intersexual conflict in which females have been able to successfully hide any evidence of extra-pair paternity useful to their partner.

Behavioural assessments of paternity and facultative adjustment of male care will evolve only as a response to uncertainty or unpredictability in the level of paternity that a male experiences. Nonfacultative changes in levels of paternal care are expected to suffice for any predetermined or systematic variation in paternity (e.g. male phenotypic quality or age; see Sections III.B and III.C, above). Interestingly, mathematical models have shown that random variation in paternity within a population can also produce reductions in male care at the evolutionary level (Xia 1992), and such an effect may be expected if the level of paternity obtained is not amenable to direct behavioural assessment by males prior to the period of paternal care. Therefore it appears that the nature of paternity – whether predetermined in any one breeding attempt or instead unpredictable, and whether males are able to detect loss of paternity in the latter case – will determine whether facultative or nonfacultative mechanisms of adjustment in male care will evolve. Although correlations of paternity with male care are common (see Section IV.B, above), little evidence exists on whether or not individual male paternity is consistent between different breeding attempts (see Chapter 2). The few field studies carried out on birds have shown both correlations of individual male paternity between seasons (Weatherhead and Boag 1995), and a lack of such correlations (Dunn *et al.* 1994b; Yezerinac *et al.* 1996).

The probability of individual loss of paternity will be influenced by the proximity and accessibility of additional breeding conspecifics. For example, in systems such as colonially nesting birds, which breed at very high densities, we might expect females to have greater access to extra-pair males (Birkhead *et al.* 1987), and thus patterns of extra-pair activity to be predetermined because of the high correlation with relative male phenotypic quality (e.g. Møller 1987, 1988c). The degree of freedom that females have in mating access to males will determine the scope of female choice for good genes, and hence the strength of the correlation between relative male quality and paternity. If there is spatial and/or temporal separation of breeding conspecifics, then paternity becomes more uncertain. It is this uncertainty in levels of individual paternity within each breeding attempt which selects for facultative vs. nonfacultative adjustments in paternal care.

V. RE-EXAMINING THE DATA

A. *Paternity and infanticide*

The most obvious evidence that paternity influences paternal care is that males rarely care indiscriminately for young within the breeding population. Associations with particular nest sites or particular females mean that males direct their care towards young that they have had at least some chance of fathering (Trivers 1972). Likewise, when males take over nest sites or females from other males, they are likely to have very little or no paternity in any of the young present and so might be expected to withhold paternal care. In such cases, the killing of dependent young by males is thought to have evolved to prevent investment in the young of other males and, more importantly, to bring females into reproductive condition rapidly so that they themselves obtain high levels of paternity in the next breeding attempt (Trivers 1972; Hrdy 1979). One of the best studied examples of this is in lions (*Panthera leo*) where males taking over a pride will kill or evict all the young of previous males, thereby bringing all females in the group into oestrus as quickly as possible and more or less synchronously (Packer and Pusey 1983). One might expect infanticide to play a part in mating systems where the duration of male tenure is short and take-overs are common, investment in young is extensive and time-consuming, and especially if female parental care delays subsequent fertile periods. The prevalence of monogamy and relatively short breeding seasons in birds means that infanticide is rare (Mock 1984; Rohwer 1986), although male removal experiments have confirmed its existence in some species (e.g. tree swallow *Tachycineta bicolor*; Robertson and Stutchbury 1988; Robertson 1990). Unmated male barn swallows (*Hirundo rustica*) have been shown to be infanticidal, but there is no evidence that these same males ever gained EPCs as a result (Møller 1988b). In mammalian mating systems, and especially in rodents, the threat of infanticide by males with low certainty of paternity plays a major role in structuring the genetic as well as the social mating system. Male white-footed mice (*Peromyscus leucopus*) are infanticidal on the basis of certainty of paternity, only killing young when female home ranges do not overlap with their own, or when they have not copulated with a female within the appropriate time period

(Wolff and Cicirello 1989). To avoid wasted investment in young that may then be killed, female rodents have evolved mechanisms whereby they fail to implant fertilized eggs or spontaneously abort in the presence of unfamiliar males (i.e. the Bruce effect, Bruce 1959). Alternatively, female rodents may avoid infanticide by conceiving only in the presence of a single male (e.g. Djungarian hamster *Phodopus sungorus campbelli*; Wynne-Edwards and Lisk 1984).

Female mating strategies have been intrepreted as attempts to avert the potential threat of infanticide, either by copulating with several different males within a group in order to confuse paternity (Hrdy 1979), or alternatively by mating preferentially with the male that can best protect their young (Hrdy 1981). In the latter case, the threat of withdrawal of any male care or protection could then prevent females from seeking matings outside the group. Though largely applied to work on cooperative groups, this line of thinking could be extended to other mating systems, such as monogamous birds, and in some cases this might explain the low levels of extra-pair mating by females (Westneat *et al.* 1990; Gowaty 1996b). However, in order to hypothesize the existence of such evolutionary games regarding intersexual conflict over paternal care, we first need evidence that certainty of paternity does actually cause behavioural responses in paternal care.

B. Experimental tests in birds

Although applicable to a range of taxonomic groups, it is evident that this chapter is heavily biased toward avian mating systems, mainly because of the nature of bird mating and parental care systems, but also due to the paucity of useful data on other taxa. A large amount of relevant work exists on birds, especially in the area of experimental studies of certainty of paternity and parental care. Such experiments are of particular interest here, if only because a number of studies have failed to find an effect of certainty of paternity upon paternal care in birds (Whittingham *et al.* 1993), which has prompted recent reassessments of both theory (see Section II.B, above) and methodology (see Sections V.C and V.D below). Hopefully, the theoretical framework described in this chapter will provide an alternative and more biologically relevant explanation for the inconsistent evidence for facultative responses to experimental manipulations of certainty of paternity in birds.

One of the most striking results is that, compared with studies on monogamous species, it is those conducted upon cooperatively breeding species which provide the most consistent evidence for facultative adjustments of parental effort on the basis of paternity (Whittingham *et al.* 1993). In polyandrous dunnocks, temporary male removals create asymmetries in mating access between alpha and beta males, which lead to adjustments in male care that correspond to actual levels of paternity

(Burke *et al.* 1989; Davies *et al.* 1992). As in infanticidal species, cooperatively breeding dominant male acorn woodpeckers (*Melanerpes formicivorus*) will destroy clutches laid during periods when they were experimentally removed from the group (Koenig 1990). A potential explanation for these results is that in cooperatively breeding species an individual male's level of paternity is particularly uncertain, as compared with monogamous pairings where consistent female choice for certain types of extra-pair partners will provide each male with a predetermined expectation of paternity (see Section III.A, above). As described above (see Section V.A), females in cooperative groups of birds may use copulations in order to persuade males to care for young. Females have been shown to maximize total male care, and therefore personal reproductive output, by equalizing paternity between males within the group (Hatchwell and Davies 1990; Davies *et al.* 1996). Thus, perhaps the paternity–paternal care effect seen in dunnocks occurs only because females within polyandrous groups make males compete for matings, such that the only method by which individual males can assess paternity is by using some measure of relative mating access (Davies *et al.* 1992), rather than paternity being predetermined by male quality, dominance rank or age. Hence, in this case it is worthwhile for males to respond to certainty of paternity manipulations and adjust their levels of care accordingly. Alternatively, it is possible that in cooperatively breeding species males can better afford to withdraw their care, due to the presence of greater numbers of individuals attending the nest which can respond with compensatory increases in their own levels of care (Hatchwell and Davies 1990; Mulder *et al.* 1994; but see Jamieson *et al.* 1994). While the latter point is undoubtedly true, it is perhaps insufficient as an explanation of why facultative adjustments in male care appear more prevalent in cooperatively breeding species.

In noncooperative mating systems, patterns of extra-pair mating are often based upon factors such as male age and/or relative phenotypic quality (see Sections III.A and III.C, above). These may produce such predetermined and predictable levels of individual paternity that males have evolved only nonfacultative responses in their levels of care (see Sections III.B and IV.C, above). Since experimental manipulations of certainty of paternity test for facultative behavioural responses in male care, perhaps we should not be surprised by the equivocal results from monogamous avian mating systems (Whittingham *et al.* 1993; Wright and Cotton 1994). In contrast to polyandrous groups in the same population, paternity in monogamous dunnocks has been shown to be almost always completely assured and, provided that these males gain some access to their female during egg laying, they fail to show the same behavioural responses to manipulations of paternity certainty (Davies *et al.* 1992). Perhaps when paired monogamously, male dunnocks fall back on a nonfacultative fixed level of care which has evolved to match their predetermined and consistently high levels of paternity. In eastern bluebirds, variation in extra-pair mating correlates well with male quality (Gowaty and Bridges 1991), and certainty of paternity experiments have been reported to have little effect on paternal care (Meek 1991, cited in Whittingham *et al.* 1993). Only when paternity is inconsistent or unpredictable between broods, such as in reed buntings (*Emberriza schoeniclus*), do we see clear male care responses to paternity in monogamous species (Dixon *et al.* 1994). A further example concerns polygynous European starlings, in which paternity is unpredictable on the basis of female dominance when the two broods are synchronous (Smith and von Schantz 1993), and an experimental paternity–paternal care effect has been shown with males allocating their care between two different broods according to behavioural cues (Smith *et al.* 1996).

Thus there appears to be a clear relationship between experimental evidence for an effect of certainty of paternity upon paternal care and the overall predictability of individual male paternity. Unfortunately, the experimental data on certainty of paternity and paternal care in birds are not always easy to interpret. In a classic study on barn swallows, the temporary removal of males during female fertile periods was associated with a clear reduction in the proportion of chick feeding and nest guarding performed by males (Møller 1988a, 1991). However, this reduction in male care was confounded by a reduction in brood size within the key manipulation group (Wright 1992). Therefore, despite some excellent information on the extra-pair mating system, including a positive effect of male secondary sexual characters on access to matings (Møller 1994), the experimental data on levels of paternal care in swallows are hard to interpret. Conversely, in a system such as that of the European starling where patterns of biparental care and pairing are well understood (Wright and Cuthill 1989, 1990a,b, 1992), an experimentally demonstrated effect of certainty of paternity upon paternal care provides an incomplete picture because we lack sufficient understanding of the extra-pair mating system – e.g. there was a conspicuous lack of mate guarding by males in the population under study (Wright and Cotton 1994). In one of the most intensively studied species, the tree swallow, we still have difficulty understanding why certainty of paternity manipulations appear to have no effect upon levels of paternal care (Whittingham *et al.* 1993). In this study population, individual male paternity appears to be unpredictable on the basis of age or morphological variables (Dunn *et al.* 1994a,b), and males are known to respond to low paternity with infanticide when taking over the broods of other males (Robertson and Stutchbury 1988; Robertson 1990). Of all monogamous avian systems, this is the one in which we might expect certainty of paternity to have an effect upon paternal care, yet the use of experimental manipulations has revealed little or no effect (Whittingham *et al.* 1993). However, despite substantial investigation of extra-pair activity in this species, it is still not known exactly which cues males use in assessing paternity, and therefore we cannot determine whether the experimental techniques used in manipulating certainty of paternity are appropriate (see Section V.C, below). We also do not have information regarding where the EPCs take place outside the colony let alone which males gain additional paternity from all this extra-pair activity (Dunn *et al.* 1994a). This suggests that the effective size of the tree swallow breeding population is much larger than the nestbox colony in which the observations and experiments were carried out. In the above cases interpretation of the experimental results concerning manipulations of certainty of paternity is difficult unless we obtain more information about both the mating and extra-pair mating systems.

C. Shortcomings of experimental tests

A further difficulty in assessing the results of experimental tests for an effect of certainty of paternity on paternal care arises from the wide range of experimental techniques employed. Actual and perceived paternity have been experimentally manipulated in a number of different ways, including permanent male removals leading to take-overs by incoming males (Robertson and Stutchbury 1988; Smith *et al.* 1996), temporary male removals (Møller 1988a; Davies *et al.* 1992), temporary male removals in which captive males had visual access to their females (Whittingham *et al.* 1993), temporary female removals (Sheldon *et al.* 1997), and temporary female removals in which males had visual access to their females in cages with decoy males (Wright and Cotton 1994). For the purpose of comparison and in order to understand male parental care strategies, we need to know whether males themselves actually perceived a loss of paternity as a result of any or all of these types of manipulation. For example, the success of more recent experimental manipulations might simply be due to the fact that these improve upon earlier male removal techniques and attempt to mimic real extra-pair activity (Wright and Cotton 1994; Sheldon *et al.* 1997). It is difficult to draw conclusions regarding the utility of particular techniques, given that some may be suited to certain species but not others; however, the variety of techniques does make comparison of experimental results across species more difficult.

More fundamentally, there is a serious problem when assessing the results of any certainty of paternity experiment, in that we do not know what happens inside the heads of the animals being tested. We have no way of knowing whether experimental manipulations have successfully reduced a male's certainty of paternity, other than the fact that we may see a significant effect on the level of male care. Experimental results which fail to find such an effect are therefore sometimes dismissed on the grounds that the method used to manipulate male certainty of paternity may not have done its job, and failures to produce an effect go unreported because these results become 'uninterpretable' (Schwagmeyer and Mock 1993). Thus, the hypothesis that a reduction in perceived paternity (through some measure of mating access) will result in facultative adjustment of male care becomes irrefutable as far as these types of experiments are concerned. This is compounded by the fact that any experimental study which fails to show a positive effect of certainty of paternity on paternal care is more likely to go unpublished. The result is a decidedly poor scientific methodology, which hinders the rigorous testing of theoretical predictions about certainty of paternity and the existence of facultative responses in levels of paternal care.

D. Future perspectives

This chapter serves to make the point that paternity and paternal care are more intricately linked than they might first appear, and that their effective investigation requires a whole mating systems approach. We need to be aware of not only levels of paternity and parental care, but also of the functioning of extra-pair mating systems, including the effectiveness of mate guarding and paternity assessment by males of different quality and the scope for free choice of extra-pair mates by females. Previous studies have been hampered by the fact that researchers have studied the mating and parental care aspects of natural systems separately, either because of the area of interest of the researchers concerned, or because of the limitations of data collection in any one aspect of the system. Only with a detailed understanding of patterns of extra-pair mating can we make correct predictions regarding the effect of paternity on paternal care, and especially the existence of facultative vs. nonfacultative responses to reduced probability of paternity. Such understanding is vital in order to explain the contrasting results from experimental manipulations of male certainty of paternity.

In the future, it is clear that data and experiments on a wider range of taxa are required. Avian studies have dominated this field and have been a very useful first step, but other taxa such as fish may prove more amenable to studies involving the experimental manipulation of whole mating systems. Such experiments should concentrate upon manipulations of the behavioural and ecological battleground between the sexes (Gowaty 1996b), because it is the balance of power in intersexual conflicts which drives the all-important patterns of extra-pair mating that we see in paternity analyses. There is also a fascinating role for sexual selection in such studies, because both pair and extra-pair mate choice must utilize signals of quality between and within the sexes. Female perception of male quality, based on sexual signals, can affect the quality of female with which a male is able to pair and raise young, his level of paternity and extra-pair share in the paternity of other broods, as well as the willingness of females to care for his young (Burley 1986). All the above factors, of which the effect of paternity is but one, should influence levels of paternal care, but thus far the research areas of sexual selection and parental care have not been effectively combined. Paternity and paternal care is a complex subject, but it is one which we are getting to grips with, both theoretically and empirically, and it promises much by way of future study.

ACKNOWLEDGEMENTS

This chapter benefited from discussions with many people, most notably, Pete Brotherton, Innes Cuthill, Nick Davies, Patricia Gowaty, Alasdair Houston, Rufus Johnstone, Kate Lessells, Sue McRae, Anders Møller, Marion Petrie, Ben Sheldon, Kyle Summers, David Westneat, Rosie Woodroffe and Yoram Yom-Tov. This work was supported by a Fellowship from the Natural Environmental Research Council, UK.

REFERENCES

Alexander RD (1974) The evolution of social behavior. *Annu. Rev. Ecol. System* **5:** 325–383.

Alexander RD (1979) Darwinism and human affairs. University of Washington Press, Seattle.

Alexander RD & Borgia G (1979) On the origin and basis of the male-female phenomenon. In *Sexual Selection and Reproductive Competition In Insects.* MS Blum & NA Blum (eds). Academic Press, New York.

Andersson M (1994) *Sexual Selection.* Princeton University Press, New Jersey.

Barash DP (1977) Sociobiology of rape in mallards (*Anas platyrhynchos*): responses of the mated male. *Science* **197:** 788–789.

Baylis JR (1981) The evolution of parental care in fishes, with reference to Darwin's rule of male sexual selection. *Environ. Biol. Fish* **6:** 223–251.

Beecher MD (1988) Kin recognition in birds. *Behavior. Genet.* **18:** 465–482.

Beecher MD (1991) Successes and failures of parent-offspring recognition in animals. In *Kin Recognition.* PG Hepper (ed.), pp. 94–124. Cambridge University Press, Cambridge.

Birkhead TR & Møller AP (1992) *Sperm Competition in Birds. Evolutionary Causes and Consequences.* Academic Press, London.

Birkhead TR, Atkin L & Møller AP (1987) Copulation behaviour in birds. *Behaviour* **101:** 101–138.

Bruce H (1959) An exteroceptive block to pregnancy in the mouse. *Nature* **184:** 105.

Bulmer LS (1979) Male parental care in bony fishes. *Quart. Rev. Biol.* **54:** 149–161.

Burke T, Davies NB, Bruford MW & Hatchwell BJ (1989) Parental care and mating behaviour of polyandrous dunnocks *Prunella modularis* related to paternity by DNA fingerprinting. *Nature* **338:** 249–251.

Burley N (1986) Sexual selection for aesthetic traits in species with biparental care. *Am. Nat.* **127:** 415–445.

Carey M (1990) Effects of brood sizes and nestling age on parental care by male field sparrows (*Spizella pusilla*). *Auk* **107:** 580–586.

Chase I (1980) Cooperative and non-cooperative behavior in animals. *Am. Nat.* **115:** 827–857.

Clutton-Brock TH (1991) *The Evolution of Parental Care.* Princeton University Press, Princeton.

Dale J (1995) Problems with pair-wise comparisons: does certainty of paternity covary with paternal care? *Anim. Behav.* **49:** 519–521.

Davies NB, Hatchwell BJ, Burke T & Robson T (1992) Paternity and parental effort in dunnocks *Prunella modularis:* how good are male chick-feeding rules? *Anim. Behav.* **43:** 729–745.

Davies NB, Hartley IR, Hatchwell BJ & Langmore NE (1996) Female control of copulations to maximise male help: a comparison of polyandrous alpine accentors, *Prunella collaris* and dunnocks *P. modularis. Anim. Behav.* **51:** 27–47.

Dawkins R & Carlisle TR (1976) Parental investment, mate desertion and a fallacy. *Nature* **262:** 131–133.

Dixon A, Ross D, O'Malley SLC & Burke T (1994) Paternal investment inversely related to degree of extra-pair paternity in the reed bunting. *Nature* **371:** 698–700.

Dunn PO, Robertson RJ, Michaud-Freeman D & Boag PT (1994a) Extra-pair paternity in tree swallows: why do females mate with more than one male? *Behav. Ecol. Sociobiol.* **35:** 273–281.

Dunn PO, Whittingham LA, Lifjeld JT, Robertson RJ & Boag PT (1994b) Effects of breeding density, synchrony and experience on extra-pair paternity in tree swallows. *Behav. Ecol.* **5:** 123–129.

Eens M & Pinxten R (1995) Inter-sexual conflicts over copulations in the European starling: evidence for the female mate guarding hypothesis. *Behav. Ecol. Sociobiol.* **36:** 71–81.

Gowaty PA (1995) *Darwinian Feminism: Intersections and Boundaries.* Chapman Hall, New York.

Gowaty PA (1996a) Field studies of parental care in birds: new data focus questions on variation among females. In *Advances in the Study of Behavior.* CT Snowdon & JS Rosenblatt (eds), pp. 476–531. Academic Press, New York.

Gowaty PA (1996b) Battles of the sexes and origins of monogamy. In *Partnerships in Birds.* JL Black (ed.), pp. 21–52. Oxford University Press, Oxford.

Gowaty PA & Bridges WC (1991) Behavioral, demographic, and environmental correlates of extra-pair fertilizations in eastern bluebirds, *Sialia sialis. Behav. Ecol.* **2:** 339–350.

Gowaty PA, Plissner JH & Williams TG (1989) Behavioural correlates of uncertain parentage: mate guarding and nest guarding by eastern bluebirds, *Sialia sialis. Anim. Behav.* **38:** 272–284.

Grafen A (1980) Opportunity cost, benefit and the degree of relatedness. *Anim. Behav.* **28:** 967–968.

Graves J, Ortega-Ruano J & Slater PJR (1993) Extra-pair copulations and paternity in shags: do females choose better males? *Proc. Roy. Soc. Lond. B* **253:** 3–7.

Gross MR (1984) Sunfish, salmon and the evolution of alternative reproductive tactics in fishes. In *Fish Reproduction: Strategies and Tactics.* RJ Wootton & G Potts (eds), pp. 55–75. Academic Press, New York.

Gross MR & Charnov EL (1980) Alternative male life histories in bluegill sunfish. *Proc. Natl Acad. Sci. USA* **77:** 6937–6940.

Gross MR & Shine R (1981) Parental care and mode of fertilization in ectothermic vertebrates. *Evolution* **35:** 775–793.

Hansen TF & Price DK (1995) Good genes and old-age – do old mates provide superior genes? *J. Evol. Biol.* **8:** 759–778.

Hasselquist D, Bensch S & von Schantz T (1995) Low frequency of extra-pair paternity in the polygynous great reed warbler, *Acrocephalus arundinaceus*. *Behav. Ecol.* **6:** 27–38.

Hatchwell BJ & Davies NB (1990) Provisioning of nestlings by dunnocks, *Prunella modularis*, in pairs and trios: compensation reactions by males and females. *Behav. Ecol. Sociobiol.* **27:** 199–210.

Hatchwell BJ & Davies NB (1992) Provisioning of nestlings by dunnocks, an experimental study of mating competition in monogamous and polyandrous dunnocks *Prunella modularis*: I. Mate guarding and copulations. *Anim. Behav.* **43:** 595–609.

Houston AI (1995) Parental effort and paternity. *Anim. Behav.* **50:** 1635–1644.

Houston AI & Davies NB (1985) The evolution of cooperation and life history in the dunnock, *Prunella modularis*. In *Behavioural Ecology: the Ecological Consequences of Adaptive Behaviour*. R Sibly & R Smith (eds), pp. 471–487. Blackwell Scientific Publications, Oxford.

Houtman AM (1992) Female zebra finches choose extra-pair copulations with genetically attractive males. *Proc. Roy. Soc. Lond. B* **249:** 3–6.

Hrdy SB (1979) Infanticide among animals: a review, classification, and examination of the implications for the reproductive strategies of females. *Ethol. Sociobiol.* **1:** 13–40.

Hrdy SB (1981) *The Woman that Never Evolved*. Harvard University Press, Cambridge, MA.

Jamieson IG, Quinn JS, Rose PA & White BN (1994) Shared paternity among non-relatives is a result of an egalitarian mating system in a communally breeding bird, the pukeko. *Proc. Roy. Soc. Lond. B* **257:** 271–277.

Johnsen A & Lifjeld JT (1995) Unattractive males guard their mates more closely: an experiment with Bluethroats (Aves, Turdidae: *Luscinia s. svecica*). *Ethology* **101:** 200–212.

Kacelnik A & Cuthill IC (1990) Central place foraging in starlings: II food allocation to chicks. *J. Anim. Ecol.* **59:** 655–674.

Kempenaers B & Sheldon B (1996) Why don't male birds discriminate between their own and extra-pair offspring? *Anim. Behav.* **51:** 1165–1173.

Kempenaers B, Verheyen GR, van de Broeck M, Burke T, van Broeckhoven C & Dhondt AA (1992) Extra-pair paternity results from female preference for high quality males in the blue tit. *Nature* **357:** 494–496.

Kempenaers B, Verheyen GR & Dhondt AA (1995) Mate guarding and copulation behaviour in monogamous and polygynous blue tits: do males follow a best-of-a-bad-job strategy? *Behav. Ecol. Sociobiol.* **36:** 33–42.

Ketterson E & van Nolan, Jr (1992) Hormones and life histories: an integrative approach. *Am. Nat.* **140:** S33-S62.

Koenig WD (1990) Opportunity of parentage and nest destruction in polygynandrous acorn woodpeckers, *Melanerpes formicivorus*. *Behav. Ecol.* **1:** 55–61.

Lack D (1968) *Ecological Adaptations for Breeding in Birds*. Chapman and Hall, London.

Lessells CM (1991) The evolution of life histories. In *Behavioural Ecology: An*

Evolutionary Approach. JR Krebs & NB Davies (eds), pp. 32–68. Blackwell Scientific Publications, Oxford.

Lifjeld JT & Robertson RJ (1992) Female control of extra-pair fertilizations in tree swallows. *Behav. Ecol. Sociobiol.* **31:** 89–96.

Lifjeld JT, Dunn PO, Robertson RJ & Boag PT (1993) Extra-pair paternity in monogamous tree swallows. *Anim. Behav.* **45:** 213–229.

Lozano GA & Lemon RE (1996) Male plumage, paternal care and reproductive success in yellow warblers, *Dendroica petechia. Anim. Behav.* **51:** 265–272.

Lubjuhn T, Curio E, Muth SC, Brün J & Epplen TJ (1993) Influence of extra-pair paternity on parental care in great tits (*Parus major*). In *DNA Fingerprinting: State of the Science.* SDJ Pena, R Chakraborty, JT Epplen & AJ Jefferys (eds), pp. 379–385. Birkhäuser Verlag, Bern.

Markman S, Yom-Tov Y & Wright J (1995) Male parental care in the Orange-Tufted Sunbird: behavioural adjustment in provisioning and nest guarding effort. *Anim. Behav.* **50:** 655–669.

Markman S, Yom-Tov Y & Wright J (1996) The effect of male removal on female parental care in the orange-tufted sunbird. *Anim. Behav.* **52:** 437–444.

Maynard Smith J (1978) *The Evolution of Sex.* Cambridge University Press, Cambridge.

Meek SB (1991) Parental investment and the maintenance of monogamy in eastern bluebirds. PhD thesis, Queen's University, Ontario.

Mock DW (1984) Infanticide, siblicide and avian nestling mortality. In *Infanticide: Comparative and Evolutionary Perspectives.* G Hausfater & SB Hrdy (eds), pp. 3–30. Aldine, New York.

Møller AP (1987) Advantages and disadvantages of coloniality in the swallow, *Hirundo rustica. Anim. Behav.* **35:** 819–832.

Møller AP (1988a) Paternity and parental care in the swallow, *Hirundo rustica. Anim. Behav.* **36:** 996–1005.

Møller AP (1988b) Infanticide and anti-infanticidal strategies in the swallow *Hirundo rustica. Behav. Ecol. Sociobiol.* **22:** 365–371.

Møller AP (1988c) Female choice selects for male sexual tail ornaments in the monogamous swallow. *Nature* **332:** 640–642.

Møller AP (1991) Defence offspring of male swallows, *Hirundo rustica*, in relation to participation in extra-pair copulations by their mates. *Anim. Behav.* **42:** 261–267.

Møller AP (1994) *Sexual Selection and the Barn Swallow.* Oxford University Press, Oxford.

Møller AP & Birkhead TR (1993) Certainty of paternity covaries with parental care. *Behav. Ecol. Sociobiol.* **33:** 261–268.

Møller AP & Birkhead TR (1995) Certainty of paternity and paternal care in birds: a reply to Dale. *Anim. Behav.* **49:** 522–523.

Morton ES, Forman L & Braun M (1990) Extra-pair fertilizations and the evolution of colonial breeding in purple martins. *Auk* **107:** 275–283.

Mulder RA, Dunn PO, Cockburn A, Lazenby-Cohen KA & Howell MJ (1994) Helpers liberate female fairy-wrens from constraints on extra-pair mate choice. *Proc. Roy. Soc. Lond. B* **255:** 223–229.

Norris KJ (1990) Female choice and the evolution of the conspicuous plumage coloration of monogamous male great tits. *Behav. Ecol. Sociobiol.* **26:** 129–138.

Otter K, Ratcliffe L & Boag PT (1994) Extra-pair paternity in the black-capped chickadee. *Condor* **96:** 218–222.

Packer C & Pusey AE (1983) Male takeovers and female reproductive parameters:

a simulation of oestrus synchrony in lions (*Panthera leo*). *Anim. Behav.* **31:** 334–340.

Pagel M (1997) Desperately concealing father. A theory of parent–infant resemblance. *Anim. Behav.* **53:** 973–981.

Perrone M & Zaret TM (1979) Parental care patterns of fishes. *Am. Nat.* **113:** 351–361.

Petrie M and Lipsitch M (1994) Avian polygyny is most likely in populations with high variability in heritable male fitness. *Proc. Roy. Soc. Lond. B* **256:** 275–280.

Ridley M (1978) Paternal care. *Anim. Behav.* **26:** 904–932.

Robertson RJ (1990) Tactics and counter-tactics of sexually selected infanticide in tree swallows. In *Population Biology of Passerine Birds: an Integrated Approach. Proc. NATO Advanced Workshop, Corsica, 1989.* J Blondel, A Gosler, JD Lebreton & R McCleery (eds), pp. 381–390. Springer, Berlin.

Robertson RJ & Stutchbury BJ (1988) Experimental evidence for sexually selected infanticide in Tree Swallows. *Anim. Behav.* **36:** 749–753.

Rohwer S (1986) Selection for adoption vs. infanticide by replacement mates in birds. *Curr. Ornithol.* **3:** 353–395.

Saino N & Møller AP (1995) Testosterone-induced depression of male parental behavior in the barn swallow: female compensation and effects on seasonal fitness. *Behav. Ecol. Sociobiol.* **36:** 151–157.

Schwagmeyer PL & Mock DW (1993) Shaken confidence of paternity. *Anim. Behav.* **46:** 1020–1022.

Sheldon BC (1994a) Sperm competition in the chaffinch: the role of the female. *Anim. Behav.* **47:** 163–173.

Sheldon BC (1994b) Male phenotype, fertility, and the pursuit of extra-pair copulations by female birds. *Proc. Roy. Soc. Lond. B* **257:** 25–30.

Sheldon BC, Räsänen K & Dias PC (1997) Certainty of paternity and parental care in collared flycatchers: an experiment. *Behav. Ecol.* **8:** 421–428.

Smith HG, Mongomerie R, Põldman T, White BN & Boag PT (1991) DNA fingerprinting reveals relation between tail ornaments and cuckoldry in barn swallows, *Hirundo rustica. Behav. Ecol.* **2:** 90–98.

Smith HG & von Schantz T (1993) Extra-pair paternity in the European starling: the effect of polygyny. *Condor* **95:** 1006–1015.

Smith HG, Wennerberg L & von Schantz T (1996) Sperm competition in the European starling (*Sturnus vulgaris*) – an experimental-study of mate switching. *Proc. Roy. Soc. Lond. B* **263:** 797–801.

Smith RL (1984) *Sperm Competition and the Evolution of Animal Mating Systems.* Academic Press, Orlando.

Smith SM (1988) Extra-pair copulations in black-capped chickadees: the role of the female. *Behaviour* **107:** 15–23.

Sundberg J (1992) Absence of mate guarding in the Yellowhammer *Emberiza citrinella*: a detention experiment. *J. Avian Biol.* **25:** 135–141.

Trivers RL (1972) Parental investment and sexual selection. In *Sexual Selection and the Descent of Man, 1871–1971.* B Campbell (ed.), pp. 136–179. Aldine Press, Chicago.

Wagner RH, Schug MD & Morton ES (1996a) Condition-dependent control of paternity by female purple martins: implications for coloniality. *Behav. Ecol. Sociobiol.* **38:** 379–389.

Wagner RH, Schug MD & Morton ES (1996b) Confidence of paternity, actual paternity and parental effort by purple martins. *Anim. Behav.* **52:** 123–132.

Weatherhead PJ & Boag PT (1995) Pair and extra-pair mating success relative to male quality in red-winged blackbirds. *Behav. Ecol. Sociobiol.* **37:** 81–91.

Weatherhead PJ, Montgomerie R, Gibbs HL & Boag PT (1994) The cost of extra-pair fertilizations to female red-winged blackbirds. *Proc. Roy. Soc. Lond. B* **258:** 315–320.

Werren JH, Gross MR & Shine R (1980) Paternity and the evolution of male parental care. *J. Theor. Biol.* **82:** 619–631.

Westneat DF (1988) Male parental care and extrapair copulations in the indigo bunting. *Auk* **105:** 149–160.

Westneat DF (1990) Genetic parentage in the indigo bunting: a study using DNA fingerprinting. *Behav. Ecol. Sociobiol.* **27:** 67–76.

Westneat DF (1995) Paternity and paternal behaviour in the red-winged blackbird, *Agelais pheoniceus. Anim. Behav.* **49:** 21–35.

Westneat DF & Sargent RC (1996) Sex and parenting: the effects of sexual conflict and parentage on parental strategies. *Trends Ecol. Evol.* **11:** 87–91.

Westneat DF, Sherman PW & Morton ML (1990) The ecology and evolution of extra-pair copulations in birds. In *Current Ornithology*, Vol. 7. DM Power (ed.), pp. 331–369. Plenum, New York.

Westneat DF & Sherman P (1993) Parentage and the evolution of parental behaviour. *Behav. Ecol.* **4:** 66–77.

Westneat DF, Clark AB & Rambo KC (1995) Within-brood patterns of paternity and paternal behaviour in red-winged blackbirds. *Behav. Ecol. Sociobiol.* **37:** 349–356.

Whittingham LA, Taylor PD & Robertson RJ (1992) Confidence of paternity and male parental care. *Am. Nat.* **139:** 1115–1125.

Whittingham LA, Dunn PO & Robertson RJ (1993) Confidence of paternity and male parental care: an experimental study in tree swallows. *Anim. Behav.* **46:** 139–147.

Whittingham LA & Lifjeld JT (1995) High paternal investment in unrelated young: extra-pair paternity and male parental care in house martins. *Behav. Ecol. Sociobiol.* **37:** 103–108.

Winkler DW (1987) A general model for parental care. *Am. Nat.* **130:** 526–543.

Wolff JO & Cicirello DM (1989) Field evidence for sexual selection and resource competition infanticide in white-footed mice. *Anim. Behav.* **38:** 637–642.

Woodroffe R & Vincent A (1994) Mother's little helpers: patterns of male in mammals. *Trends Ecol. Evol.* **9:** 294–297.

Wright J (1992) Certainty of paternity and parental care. *Anim. Behav.* **44:** 380–381.

Wright J & Cotton PA (1994) Experimentally induced sex differences in parental care: an effect of certainty of paternity? *Anim. Behav.* **47:** 1311–1322.

Wright J & Cuthill I (1989) Manipulation of sex differences in parental care. *Behav. Ecol. Sociobiol.* **25:** 171–181.

Wright J & Cuthill I (1990a) Manipulation of sex differences in parental care: the effect of brood size. *Anim. Behav.* **40:** 426–471.

Wright J & Cuthill I (1990b) Biparental care: short term manipulation of partner contribution and brood size in the starling *Sturnus vulgaris. Behav. Ecol.* **1:** 116–124.

Wright J & Cuthill I (1992) Monogamy in the European starling. *Behaviour* **120:** 262–285.

Wynne-Edwards KE & Lisk RD (1984) Djungarian hamsters fails to conceive in the presence of multiple males. *Anim. Behav.* **32:** 626–628.

Xia X (1992) Uncertainty of paternity can select against paternal care. *Am. Nat.* **139:** 1126–1129.

Yezerinac SM, Weatherhead PJ & Boag PT (1996) Cuckoldry and lack of parentage-dependent paternal care in yellow warblers: a cost–benefit approach. *Anim. Behav.* **52:** 821–832.

Part Two

Taxonomic Treatments

5 Pollen Competition in Flowering Plants

Lynda F. Delph[1] *and Kayri Havens*[2]

[1] Department of Biology, Indiana University, Bloomington, IN 47405, USA and [2] Chicago Botanic Garden, 1000 Lake Cook Road, Glencoe, IL 60022, USA

'While the behaviour of pollen grains depends to a considerable extent on the genes which they carry, this is fortunately not in general the case with spermatozoa . . .' J. B. S. Haldane (1932)

I. INTRODUCTION

Unlike animals with internal fertilization or those that broadcast sperm and eggs in direct response to potential mates in close proximity, flowering plants that cross-fertilize are unable to control directly the number of matings they achieve. Rather, they are at the mercy of pollinating agents, be they abiotic or biotic, to deliver male gametes from plant to plant. Flowering plants may influence the likelihood of fertilization by altering such characters as flower number, pollen packaging, and stigma (female) receptivity, but these characters act only indirectly to determine the number of matings.

Sperm Competition and Sexual Selection
ISBN 0-12-100543-7

Nevertheless, pollen competition in plants is analogous to sperm competition in animals, in that it refers to the process whereby male gametes compete for access to eggs. Pollen competition is a post-pollination phenomenon, just as sperm competition is a post-copulation phenomenon. It occurs in flowering plants whenever more pollen grains are deposited on stigmas than there are ovules to fertilize. The process of fertilization in plants is most closely related to that which occurs in animals that have internal fertilization, but there are some important differences that have influenced how pollen competition has been studied.

One important difference between flowering plants and animals is that, in plants, sperm do not swim towards the egg cell via the open female reproductive tract. Instead, they make their way through the cellular matrix of the style via a pollen tube that grows through the style and which enters a small opening in the ovule (Fig. 5.1). Given that pollen tubes have to grow to deliver the sperm nucleus to the egg (from 1 mm to 50 cm, depending on the species), studies of pollen competition have focused primarily on variation in the rate of pollen tube growth,

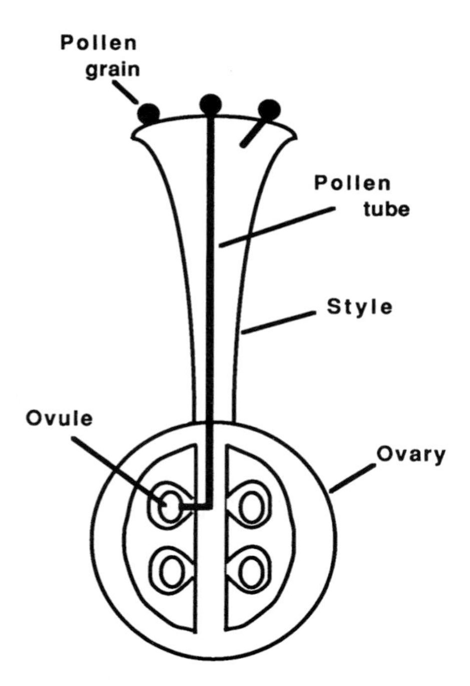

Fig. 5.1. *Diagram of fertilization in flowering plants. When pollen lands on a compatible, receptive stigma, it germinates and forms a pollen tube which carries the sperm to the egg. The tube penetrates the tissues of the stigma, style and ovary in its growth towards an ovule. Upon reaching the ovule, the tube passes through the micropyle and nucellus (megasporangium), and into the embryo sac (megagametophyte) where it bursts, releasing two sperm, one of which fuses with the egg nucleus.*

control of the rate of growth, and the consequences of variation in this trait.

Another difference between studies of pollen and sperm competition is the emphasis on the competition between the pollen from a single donor (one-donor competition) in plants vs. competition between the sperm from several different males (multiple-donor competition) in animals. This difference in emphasis is related to the greater degree of gene expression by pollen vs. sperm, in that the pollen grains from a single donor express their genetic differences more so than do the spermatozoa from a single donor. In plants, a large portion of the pollen genome is transcribed prior to fertilization and embryo formation, as opposed to transcription of relatively few genes in the sperm of animals (Erickson 1990). For example, in *Tradescantia paludosa*, ≈20 000 genes are expressed in the pollen compared with 30 000 in the diploid 'adult' – the sporophyte (Willing and Mascarenhas 1984). A similar study found 24 000 unique mRNA sequences in the pollen of *Zea mays*, and 31 000 mRNA sequences in the sporophyte (Willing *et al.* 1988). In addition, cytoplasmic channels between developing pollen grains are usually broken during meiosis so that the gene products of haploid transcription are restricted to individual pollen grains (Ottaviano and Mulcahy 1989; Mascarenhas 1992). Finally, there is substantial overlap between the genes expressed in the gametophytic (pollen) and sporophytic generation. Estimates of the extent of overlap range from ≈50% to 80% (Tanksley *et al.* 1981; Ottaviano and Mulcahy 1989; and references therein).

The high level of gene expression by pollen may be partly due to the fact that the haploid phase in plants is more complex than that of animals. In animals, the haploid phase is represented by a single cell, the gamete, while in plants meiosis produces spores that divide mitotically to produce a multicellular haploid (gametophytic) 'generation' (Fig. 5.2). In bryophytes (liverworts, hornworts, and mosses) the gametophyte is the dominant, free-living, nutritionally independent generation, while in flowering plants it has become highly reduced, consisting of three nuclei

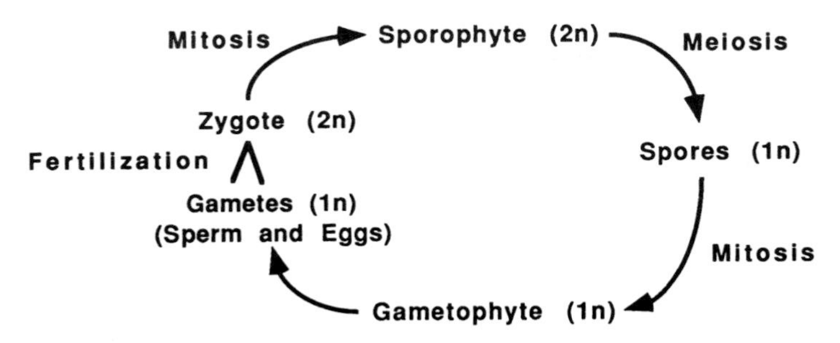

Fig. 5.2. *A generalized plant life cycle demonstrating the alternation of generations. All plants alternate between diploid sporophyte (spore-producing) and haploid gametophyte (gamete-producing) stages in their life cycle.*

in the case of the male gametophyte (pollen grain) and eight nuclei in the case of the female gametophyte (embryo sac).

Gamete competition in plants is not restricted to the angiosperms (flowering plants), although most of the work has been done with this group. Gymnosperms, like angiosperms, make pollen that must grow a tube through maternal sporophyte and female gametophyte tissue to achieve fertilization. However, in gymnosperms, pollen tube growth can take a year, or longer, so relatively few studies have addressed pollen competition in the cone-bearing plants (but see, for example, Schoen and Cheliak 1987; Apsit *et al.* 1989; Nakamura and Wheeler 1992). In the pteridophytes (ferns and their allies), many species are homosporous, (they produce only one type of spore that germinates into a bisexual gametophyte, rather than producing megaspores and microspores that give rise to separate female and male gametophytes). The bisexual gametophytes produce gametes through mitotic divisions; thus, all sperm and eggs are genetically identical. Self-fertilization can occur in homosporous ferns and results in a sporophyte that is homozygous at all loci, although in many species outcrossing is promoted because the female organs (archegonia) mature before the male organs (antheridia) (see Cousens 1988 for a review).

Instead of focusing on differences among the male gametes of a single individual, sperm competition is generally defined as competition within a single female between the sperm of two or more males for the fertilization of her ova (Parker 1970; Birkhead and Møller 1992). In essence, sperm competition is usually viewed as an extension of precopulatory inter-male competition; the number of competing phenotypes is equal to the number of inseminating males, because sperm characteristics are usually dictated by the male's diploid genotype (Sivinski 1984; Braun *et al.* 1989). Competition between sperm in a single ejaculate requires that the quality of the gamete must be partially determined by the genes that are transcribed in the haploid condition, and, in order to be biologically significant, there must be some correlation between pre- and post-zygotic genetic factors (Hartl 1970; Mulcahy 1975). While such haploid effects are well-known in pollen, they are rare in sperm; the sperm genome is, for the most part, unable to influence its phenotype (Haig and Bergstrom 1995 and references therein). In many animals, sperm development takes place in a syncytium and the spermatids remain cytoplasmically connected until the final stages of differentiation. Although some genes are transcribed during the haploid stage, the gene products appear to be shared among developing sperm, thus eliminating differences between the individual gametes (Willison *et al.* 1988; Braun *et al.* 1989).

A notable exception to diploid control of sperm phenotype is the case of 'meiotic drive', where there are differences in the competitive abilities of sperm from a single ejaculate. Examples of meiotic drive include the segregation distortion (SD) system of *Drosophila melanogaster* or the *t*-haplotype system of mice. In both cases half of the sperm – those not containing the driving allele – fail to function. This loss of functional sperm can have a negative effect on the male's fitness; selection is therefore expected to favour genes that suppress competition within an ejaculate and this may explain why haploid effects are rare in sperm (Haig and Bergstrom 1995).

Now that we have outlined why approaches to studying sperm and pollen competition differ, we present an historical perspective of why researchers first became interested in studying pollen competition. We then review experimental approaches to pollen competition and some methodological concerns. We end by discussing the evolutionary significance of pollen competition and the direction in which future work is headed.

II. HISTORICAL PERSPECTIVE

An understanding that competition among pollen grains could potentially lead to consequences that were interesting from an evolutionary perspective came about because of work in the 1920s with *Silene* spp., *Oenothera* spp., and *Zea mays* (Correns 1917, 1928; Heribert-Nilsson 1923; Brink 1927). The term 'certation' was coined to describe the phenomenon whereby certain types of pollen lose in competition with other types. For example, Correns worked with dioecious species of *Silene* in which sex is determined by sex chromosomes, with males as the heterogametic sex. He observed female-biased sex ratios and believed that this bias came about because pollen carrying X chromosomes either germinated earlier or grew faster than pollen carrying Y chromosomes. Hence, under conditions of pollen competition, the X-carrying pollen would sire more seeds than Y-carrying pollen and result in an over-abundance of female offspring. However, his results were quite likely marred by the fact that the one male he used for his crosses was an interspecific hybrid (see Carroll and Mulcahy 1993). Nevertheless, work of this type caused other researchers to consider the fact that pollen tube growth rates may vary.

In his book *The Causes of Evolution*, Haldane (1932) recognized that pollen competition was likely, because of 'serious overcrowding' of pollen grains on stigmas. He differentiated between two types of pollen competition: competition among individuals, which would select for increased pollen production, and competition between pollen grains from the same individual, which is analogous to competition among sperm from the same ejaculate. This intra-individual selection would come about if there were variation among grains in the genes that are expressed during pollen tube growth. He also pointed out that a pollen grain has a 'physiology of its own' and that pollen tube growth rate depends more on the genes it carries than does that of a spermatozoon.

The perspective that pollen tube growth rate is influenced by the genes each individual pollen grain carries has had a major impact on the types of studies conducted. It was an essential component of an extremely influential paper by Mulcahy (1979), in which he argued that pollen competition was of fundamental importance in the evolutionary success of angiosperms based on three propositions: (1) insect pollination delivers more pollen grains simultaneously than does abiotic pollination; (2) pollen tube growth rates vary and this reflects genetic differences among grains; and (3) many genes are expressed and exposed to selection in the haploid pollen (the gametophytes), and some of these are also expressed in the adult plants (sporophytes). In essence, he stated that because vast numbers of haploid individuals compete, selection intensity is high and, furthermore, the efficacy of this selection is enhanced by overlapping gene expression, as this could result in a positive correlation between pollen vigour and sporophytic vigour.

III. NECESSARY REQUIREMENTS FOR POLLEN COMPETITION

By definition, for competition to occur, there must be more pollen grains deposited than there are ovules to fertilize, otherwise every grain would have access to an ovule and the competitive ability of any given pollen grain would not influence its fertilization success. Studies on whether excess pollen is typically deposited on stigmas in natural populations have found evidence to support excess deposition for the majority of species observed (Casper 1983; Mulcahy *et al.* 1983; Snow 1986; Levin 1990). However, measuring the total number of pollen grains deposited by multiple insect visits will not necessarily be indicative of the degree of competition among grains because there is an advantage to getting there first. When sequential pollination occurs, pollen deposited subsequent to the first deposition of pollen (analogous to second-male matings in animals) may be ‘outcompeted’ by the pollen arriving first and achieve disproportionately fewer sirings, even if the delay is relatively short in time and the later-delivered pollen is of higher quality (Epperson and Clegg 1987 and references therein). In general, rapid pollen tube growth rates, combined with low variance in growth rate and short styles, should reduce the effectiveness of second-pollinations and weaken selection based on these growth rates (Mulcahy 1983; Thomson 1989).

The first-male advantage seen in plants is contrary to the pattern observed in a number of animal systems. For example, in most insects, second-sperm precedence is the most commonly observed pattern (see Chapter 10). Last-male sperm precedence is also common in birds (Birkhead and Møller 1992; see also Chapter 14).

The effectiveness of second-pollination events can be enhanced in plants by delaying stigmatic receptivity (Galen *et al.* 1986) or by requir-

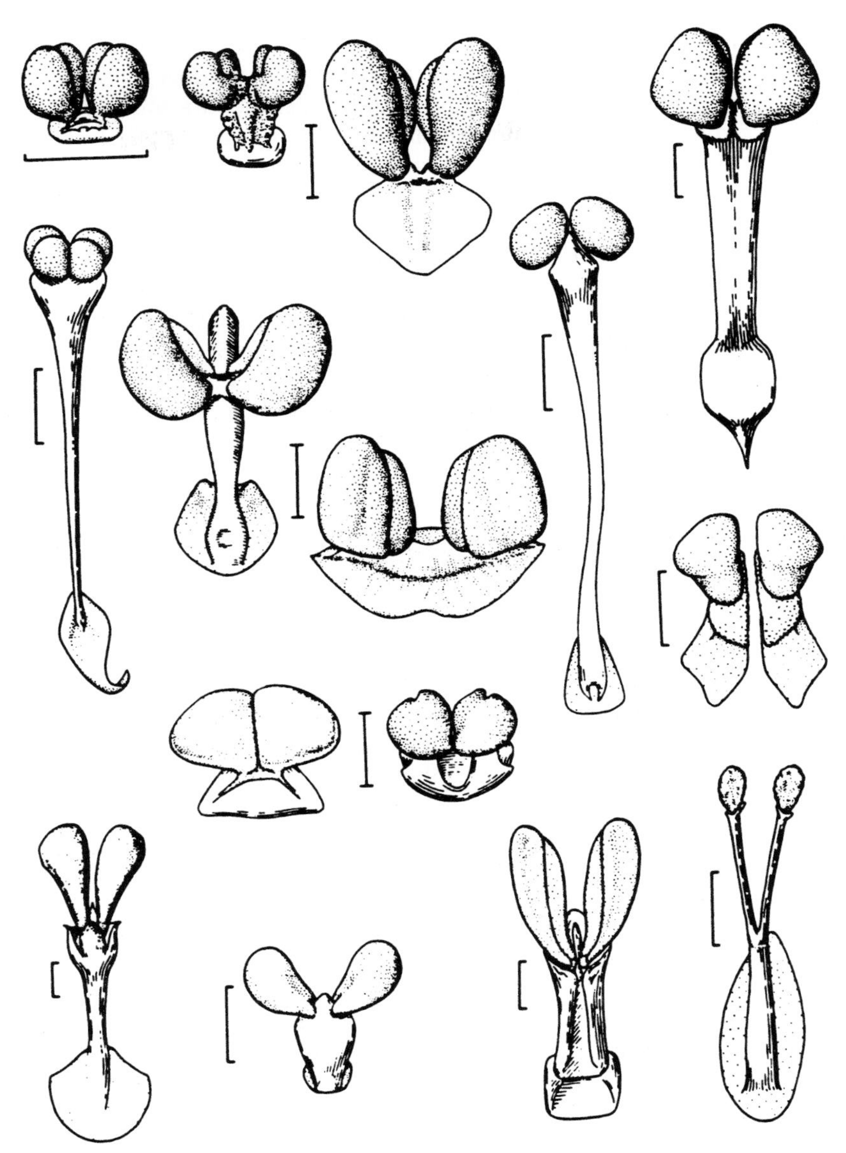

Fig. 5.3. *Pollinia are membrane-bound sacks containing all of the pollen grains from an anther locule. The pollinia shown here are from members of the Vandoideae tribe of the orchids (scale 1 mm). Reprinted with permission of Harvard University Press, from* The Orchids: Natural History and Classification *by Robert L. Dressler, illustration by Arlee Montalvo, Copyright © 1981 The Smithsonian Institution.*

ing that a threshold number of pollen grains be deposited before any pollen grains germinate. This threshold effect appears to have two possible underlying causes, the pollen-population effect and stigmatic inhibition. With the pollen-population effect, the percentage of pollen grains germinating is influenced by the number of pollen grains present (Brewbaker and Majumder 1961; Cruzan 1986). With stigmatic inhibition, pollen grains germinate only after an inhibitor of germination has been inactivated. This inactivation has been shown to be dependent on the number of pollen grains rising above a critical threshold that alters the pH of the stigma and destroys the pH-sensitive inhibitor (Ganeshaiah *et al.* 1986; Ganeshaiah and Shaanker 1988). In addition, Ganeshaiah *et al.* (1986) hypothesized that selection to overcome this inhibition may have been involved in the evolution of traits that bring large numbers of pollen grains to stigmas simultaneously, such as sticky threads (viscin) that hold pollen grains together.

In addition to large numbers of pollen grains arriving simultaneously because of viscin threads, pollen is also sometimes packaged into polyads and pollinia. These groupings of pollen are not unlike multisperm bundles (Sivinski 1984) and spermatophores (see Fig. 5.3). Polyads occur in 15% of flowering plant families (Kenrick and Knox 1982), and pollinia occur in the plant families Orchidaceae (orchids) and Asclepiadaceae (milkweeds), in which large numbers of ovules exist within each individual flower. The pollinia of orchids contain from 500 to 5000 pollen nuclei and, typically, the number of ovules per flower and the number of nuclei per pollinium are roughly equal (Richards 1986). Hence, although successful pollinations are relatively rare in orchids (Zimmerman and Mitchell Aide 1989; Ackerman and Montalvo 1990; Calvo 1993), potentially only one pollinator visit is required to achieve almost full fertilization.

IV. THE EVOLUTIONARY SIGNIFICANCE OF POLLEN COMPETITION

A. One-donor pollinations

An intriguing correlation has been found in several pollen competition studies involving pollinations with pollen from one donor: the greater the degree of pollen competition, the greater the vigour of the resulting offspring, including such characteristics as increased germination, more rapid growth of seedlings, enhanced competitive ability, and enhanced reproductive output (Mulcahy and Mulcahy 1975; McKenna and Mulcahy 1983; McKenna 1986; Richardson and Stephenson 1992; Quesada *et al.* 1993; Palmer and Zimmerman 1994). Not only does competition appear to enhance the vigour of individual seeds within fruits, it may also enhance offspring vigour by affecting which fruits are aborted by the maternal plant and which fruits are matured. Lee (1984) hypothesized that if there is a positive correlation between pollen vigour and offspring vigour, then plants could increase their fitness by selectively aborting those fruits with no or low levels of pollen competition and selectively maturing those with high levels of pollen competition. Results consistent with this hypothesis were found for *Lindera benzoin*: the fewer the pollen tubes, the more likely a fruit was to be aborted (Niesenbaum and Casper 1994).

Several studies have used the approach of comparing seed, fruit, and offspring characteristics between flowers that received high pollen vs. low pollen loads from a single donor. Presumably, with high pollen loads, only the fastest growing, and potentially genetically superior, pollen achieves fertilization, resulting in more vigorous progeny. High pollen loads have been correlated with high rates of seed germination and rapid seedling growth in *Oenothera organensis* (Fingerett 1979), with increased seed weight and seedling height and weight in *Turnera ulmifolia* (McKenna 1986), and with seedling growth in *Campanula americana* (Richardson and Stephenson 1992). Other studies have similarly found a correlation between high pollen loads from a single pollen donor and some aspects of offspring vigour (Ottaviano *et al.* 1983; Stephenson *et al.* 1986; Davis *et al.* 1987; Winsor *et al.* 1987; Ramstetter and Mulcahy 1988; Bertin 1990; Palmer and Zimmerman 1994), although this is not universally the case (Snow 1990).

While pollen competition frequently has been invoked as an explanation for this correlation, other factors may also play a role (Charlesworth *et al.* 1987). Pollen quantity can affect the growth rate of fruit and its likelihood to mature because pollen tubes produce hormones, primarily auxins and gibberellins, that stimulate fruit development (Nitsch 1952; Lund 1956; Vasil 1974; Lee and Bazzaz 1982). Pollen quantity can also indirectly affect fruit set because it is often positively correlated with seed number and developing seeds produce hormones which can increase the sink strength (demand for resources) of the fruit, in essence allowing it to garner more resources (Stephenson 1981). A second confounding factor with single-donor experiments involving the relationship between pollen competition and progeny vigour was pointed out by Charlesworth (1988). She noted that results of experiments comparing the vigour of seedlings from high and low pollen loads (e.g. Winsor *et al.* 1987) may be confounded by an effect of seed size. For example, high pollen loads produce fruits with many smaller seeds, relative to low pollen loads, which produce fruits with a few large seeds. When seeds of the same weight from the two treatments are matched, the comparison is between the relatively larger (presumably better) seeds from the high pollen treatment and the relatively smaller seeds from the low pollen treatment, and this may bias the results in favour of the high pollen treatment. Similarly, not comparing seeds of equal size could bias the results in favour of the low pollen treatment. While Charlesworth's explanation alone cannot account for the differences in progeny vigour seen in the Winsor *et al.* (1987) study (Stephenson *et al.* 1988), comparisons between fruits with equal numbers of seeds (e.g. Quesada *et al.* 1993, 1996) can completely eliminate this methodological concern altogether.

Experimental pollinations that vary competition by altering the placement of pollen on the stigmatic surface rather than increasing pollen loads also avoid some of these confounding effects (Mulcahy 1979). Several species of plants have elongate stigmatic surfaces that allow pollen to be placed distal to the ovary at the 'tip' of the style, or proximal to the ovary at the 'base' of the style (Fig. 5.4). Tip pollinations presumably result in more intense competition because pollen tubes must grow further, thus increasing the separation between fast- and slow-growing pollen tubes. This approach has been taken in studies of *Dianthus chinensis*, which found that tip as opposed to base pollinations resulted in offspring that germinated earlier, weighed more, and were more successful competitors when grown with perennial rye grass (*Lolium perenne*) (Mulcahy and Mulcahy 1975; McKenna and Mulcahy 1983). In addition, seeds of *Silene latifolia* resulting from tip pollinations emerged earlier than those from base pollinations (Purrington 1993). Similarly, distylous species (those with two floral morphs, one with long styles and one with short styles) have been used to alter the intensity of pollen competition by changing the distance travelled by pollen tubes. McKenna (1986) found that long-styled morphs of *Anchusa officinalis* produced

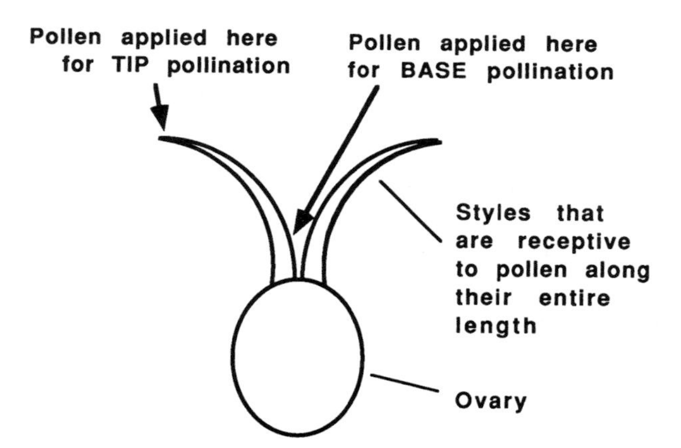

Fig. 5.4. *Some one-donor pollen competition experiments involve 'TIP' vs. 'BASE' pollinations. As shown here, pollen placed at the tip of the style must grow further than pollen placed at the base of the style. The greater distance travelled by pollen placed at the tip accentuates differences in pollen tube growth rates and, presumably, allows only relatively fast-growing pollen from tip pollinations to sire seeds, whereas both fast-growing and slow-growing pollen from base pollinations may sire seeds.*

heavier seeds with a higher germination percentage than did short-styled morphs.

B. Multiple donor pollinations

While flowers may receive more pollen from a single donor than needed to fertilize all of their ovules, they may also receive pollen loads consisting of pollen from more than one donor, allowing for pollen competition between pollen grains from different donors. Several studies, using a genetic marker to assess paternity of seeds, have examined the relative postpollination success of multiple pollen donors, usually using a genetic marker to assess paternity of seeds (Marshall and Ellstrand 1986; Bertin 1990; Marshall 1991; Quesada *et al.* 1991; Snow and Spira 1991a; Rigney *et al.* 1993; Havens and Delph 1996).

Some of the best evidence that pollen competitive ability may affect male reproductive success in natural populations comes from studies of *Hibiscus moscheutos* and involve pollinations with pairs of competing donors. This species receives excess pollen in natural populations (Spira *et al.* 1992) and exhibits among-individual variation in pollen tube growth rates (Snow and Spira 1991a). Individuals with relatively fast growing pollen tubes consistently sire a greater proportion of seeds on recipient plants than those with less vigorous pollen, and are therefore referred to as 'super-males' by the authors (Snow and Spira 1991a,b, 1996). If the variation seen among the pollen donors is heritable (see below), sexual selection is occurring via pollen competition.

In other work involving paired-donor pollinations with *Erythronium grandiflorum*, the growth of pollen from different donors was not consistent, but depended on the source of other pollen also growing in the style (Cruzan 1990). For example, this lily has a three-lobed stigma leading to one style and when pollen from the same donor is placed on two of the lobes it grows faster down the style than when the lobes contain pollen from different donors. This could occur either because of direct pollen interactions or because of mediation by the style.

Of the numerous studies attempting to document competition between pollen grains for access to ovules, only a few differentiated between fertilization success and success in siring seeds actually matured by the maternal parent. The importance of this differentiation has been noted theoretically (Walsh and Charlesworth 1992) and shown empirically in *Drosophila* (Kaufmann and Demerec 1942) and two species of flowering plants (Rigney 1995; Havens and Delph 1996). If the maternal parent aborts zygotes differentially, depending on paternity, then fertilization success may not be accurately measured by determining the paternity of seeds that survive to maturity. Hence, in order to measure fertilization success directly, the effect of seed abortion must be removed or measured. Rigney (1995) was able to remove and determine the genotype of aborting seeds through windows cut in the ovaries of *Erythronium grandiflorum*. She found that seed abortion is not random with respect to paternal genotype, and success in fertilization is not correlated with success in siring seeds brought to maturity in that species. Havens and Delph (1996) used a genetically engineered marker gene (β-glucuronidase or GUS) to determine the paternity of developing seeds, prior to seed abortion, and mature seeds in *O. organensis*. They found that fertilization success and success in siring seeds that were matured were not correlated (Fig. 5.5), and also showed that plants with faster pollen tube growth rates *in vitro* fertilized more ovules *in vivo*.

Most flowering plant species are hermaphroditic ($\approx$90%; Yampolsky and Yampolsky 1922); hence, combinations of pollen on stigmas can include pollen from the same plant (self pollen) and pollen from one or more donors (outcross pollen). Some studies have found that differences in male reproductive success are unrelated to the genetic distance between pollen and seed parents (Hessing 1986; Schlichting and Devlin

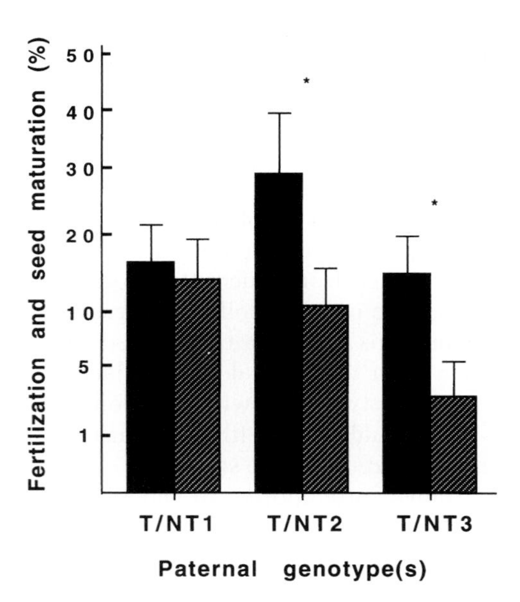

Fig. 5.5. *The mean (+ SE) percentage of ovules fertilized (solid bars, n = $\approx$1400) and seeds sired that were brought to maturity (hatched bars, n = $\approx$4900) by a transformed donor (T) of* Oenothera organensis *in competition with three nontransformed donors (NT1–3). Significant differences between stages within each cross are indicated by asterisks ($P < 0.05$), and indicate that the number of mature seeds sired by the transformed donor is not necessarily a reliable indication of the number of seeds fertilized by the donor ($r = 0.19$, $P = 0.72$). Redrawn from Havens and Delph (1996).*

1992; Johnston 1993; Snow and Spira 1993), although most of these studies were unable to uncouple success in fertilization from success in seed siring. More typically, pollen competition studies have discovered that differences in competitive ability are linked to genetic relatedness. For example, some studies have found that self pollen is less likely to fertilize ovules and/or sire seeds than outcross pollen (Bowman 1987; Cruzan 1989; Weller and Ornduff 1989; Aizen *et al.* 1990; Ramirez and Brito 1990; Rigney *et al.* 1993; Jones 1994). When outcross pollen is found to have a competitive advantage, it is often due to incompatibility that is detected only when compatible (outcross) pollen is present (Bateman 1956; Casper *et al.* 1988; Cruzan and Barrett 1993).

In a few cases, self pollen has been more successful than outcross pollen in competitive studies. Johnson and Mulcahy (1978) found that self pollen was preferred in highly inbred lines of maize, perhaps owing to specialization for a particular stylar environment over several generations. Baker and Shore (1995) found that self pollen was more successful in siring seeds than pollen from other populations in *Turnera ulmifolia*. They suggested that pollen competition has led to partial reproductive isolation between populations and ultimately may be a driving force behind speciation in this taxon. Wide crossing effects, such as those in *T. ulmifolia*, are not uncommon in plants. In fact, many studies showing differential paternity success have used different varieties (often highly inbred strains) of cultivated plants (Pfahler 1967; Mulcahy 1971; Sari-Gorla *et al.* 1975; Ottaviano *et al.* 1983; Gawel and Robaker 1986; Elgersma *et al.* 1989; Quesada *et al.* 1991), or plants from different populations (Bookman 1984). The extent to which these results represent what happens within natural populations of wild plants is not known, but these studies probably overestimate the amount of genetic variability for traits conferring paternity success within noncultivated plants.

V. MULTIPLE PATERNITY

When stigmatic pollen loads derive from multiple donors, multiple paternity can result. Mixed-donor pollen loads are possible if pollen carryover occurs (pollen carryover refers to pollen from a particular flower being deposited on more than just the next flower visited because not all of the pollen is deposited in one visit – some of it is carried over to subsequent flowers). High rates of pollen carryover are common (see reviews by Robertson 1992 and Morris *et al.* 1994) and are likely to enhance the opportunity for female choice (Galen and Rotenberry 1988).

Multiply sired fruits have been documented in a number of species (Brown *et al.* 1986; Ellstrand and Marshall 1986; Broyles and Wyatt

1990; Dudash and Ritland 1991; Muona *et al.* 1991; D. Campbell, unpublished data; L. Delph, unpublished data) and this phenomenon is likely to be widespread in plants with multiple-seeded fruit. Ibarra-Perez *et al.* (1996) pointed out that they knew of 'no study in which within-fruit multiple paternity was sought and not found'. Even species in which pollen is packaged in pollinia or polyads (units with several pollen grains), such as *Asclepias exaltata* and *Acacia melanoxylon*, have been found to have some multiple paternity within fruits, although the percentages are relatively low (Broyles and Wyatt 1990; Muona *et al.* 1991).

Some studies have been able to show that multiple paternity is not merely the result of random paternal success following mixed-donor pollen load deposition, but that both seed abortion and fruit abortion by the maternal plant can alter the percentage of multiply-sired fruits matured by a plant. In other words, the maternal plant appears to be involved in the postfertilization sorting of mates. For example, in *Raphanus sativus* (wild radish) maternal plants preferentially abort stylar ovules (those closest to the style), thereby selecting against those pollen donors that preferentially fertilize stylar ovules (Marshall and Ellstrand 1988). Instead, they favour maturation of ovules towards the basal end of the fruit, especially when stressed. In addition, multiply-sired fruits of wild radish are selectively matured over singly-sired fruits (Marshall and Ellstrand 1986; Marshall 1988, 1991). Additional evidence of maternal influence on mate choice comes from work with *O. organensis*: when pollen from two donors was applied to stigmas, nonrandom seed abortion occurred and resulted in a more equal paternity of the seed crop than would be predicted from fertilization success (Havens and Delph 1996). In this species, the pattern of seed abortion was not related to ovule position in the ovary.

Multiple paternity may enhance the fitness of the maternal parent by increasing the number of offspring, the quality of offspring, or both. In some crosses in *Costus allenii* (Schemske and Pautler 1984) and in *R. sativus* (Marshall and Ellstrand 1986, 1988), mixed pollen loads increased the fitness of the female parent, measured as total seed weight. However, several other studies have not found a fitness benefit associated with mixed pollen loads (Lee and Bazzaz 1982; Bawa and Webb 1984; Sork and Schemske 1992). Karron and Marshall (1990, 1993) have tested two hypotheses pertaining to the fitness advantages of multiply-sired progeny in *R. sativus*. The 'elbow room' hypothesis predicts that resource partitioning will increase with genetic diversity; therefore, half sibs in competition with each other will be more fit than full sibs. The 'lottery' hypothesis predicts that with genetically diverse offspring, a maternal plant will be more likely to produce successful phenotypes for each microsite in a patchy environment. Their studies did not provide empirical support for either hypothesis and they suggest further work is needed to determine why *R. sativus* preferentially matures multiply-sired fruit.

VI. HERITABLE VARIATION IN POLLEN TUBE GROWTH RATES

Pollen competition will not result in a response to gametophytic selection unless genetic variation exists for traits that confer success in competitive interactions. As Snow (1986) points out, 'Without establishing this fact, studies of pollen tube competition merely show that pollen mortality is high prior to fertilization'. To date, heritable variation for pollen tube growth rates has been documented only in maize (*Zea mays*) and zucchini (*Cucurbita pepo*). Ottaviano *et al.* (1983) found that pollen from a line of maize with high gametophytic selection sired more seeds than pollen from a line with weak selection when each was mixed with pollen from an unrelated line. In a more recent study employing a pollen mixture technique, pollen tube growth rate in maize was estimated by determining the proportion and location of ovules fertilized by one of two lines of pollen applied to the stigma, and the heritability of pollen tube growth rate was estimated to be 71%. However, widely different genetic lines were used which could, at least in part, explain the high heritability estimates obtained (Sari-Gorla *et al.* 1992). In a hybrid between cultivated zucchini and a wild gourd (*Cucurbita texana*), Quesada *et al.* (1996) found that selection for rapid pollen tube growth rate improved seedling growth rate and flower and fruit production of the offspring. Their study, designed to eliminate maternal effects and nonrandom seed abortion, suggests a heritable component in pollen performance and a genetic correlation between pollen and offspring vigour. Although the results of the maize and *Cucurbita* studies are intriguing, the genetic differentiation between cultivars and genetic variation in hybrids may not be typical of what is found in natural populations of noncultivated species. Studies attempting to document heritability of pollen tube growth rate within populations of noncultivated species have thus far been unsuccessful. Snow and Mazer (1988) found no effect of previous pollen competition on the proportion of seeds sired in *Raphanus raphanistrum* and thus, no evidence for heritable variation in pollen competitive ability. Havens (1994) found the clonal repeatability (an upper limit of heritability) of pollen tube growth rates in *O. organensis* to be less than 10%.

VII. MAINTAINING VARIATION IN POLLEN TUBE GROWTH RATES

Given that male reproductive success has been shown to be affected by pollen tube growth rates in several studies (see above), additive genetic variation in pollen competitive ability should have been largely eroded by gametophytic selection (i.e. selection for pollen traits) (Snow and Mazer

1988; Walsh and Charlesworth 1992). What then accounts for the variation seen in pollen competitive ability? Additive genetic variation might not be entirely responsible. For example, pollen–style interactions involving nonadditive genetic variation have been shown to occur: the pistil promotes or inhibits the growth rate of pollen from various donors, not because of differences in the quality of the pollen, but because of genetic complementarity effects (Gawel and Robacker 1986; Waser *et al.* 1987; Marshall 1988). In addition, recent studies clearly indicate that some of the variation in pollen tube growth rates is not genetically based, but is environmentally induced. Pollen tube growth rates have been found to be highly plastic, responding phenotypically to many types of environmental heterogeneity during pollen development, such as soil temperature and nutrient levels, and leaf loss from herbivory (Young and Stanton 1990; Lau and Stephenson 1993, 1994; Quesada *et al.* 1995; Mutikainen and Delph 1996; Delph *et al.* 1997). However, some of the variation may be heritable, since there are ways in which additive genetic variation in pollen competitive ability could be maintained.

Gene flow from populations without pollen competition could maintain additive genetic variation in pollen competitive ability (Snow and Spira 1996), but this has not been investigated empirically. Mutation is an additional factor that could maintain such variation, especially given the large number of genes expressed during pollen tube growth and the fact that gametophyte populations are large (Mulcahy 1979; Walsh and Charlesworth 1992). Mulcahy (1979) suggested that deleterious mutations in the haploid pollen would be exposed to selection and purged more efficiently than in the diploid sporophyte. In support of this premise, the insertion of a marker gene, which could be considered analogous to mutation, decreased pollen tube growth rates and fertilization ability in *O. organensis*, but had no apparent effect on sporophyte vigour, as measured by seed mass, seedling emergence, and dry weight after 8 weeks (Havens and Delph 1996).

Mutations in gametes may increase with age in plants, because plants do not have a differentiated germ line (Fig. 5.2). Instead, they may produce gametophytes from differentiating meristems throughout their life. These meristems are themselves the product of varying numbers of mitotic divisions, depending on where on the plant they are produced and the age of the plant. Mutations caused by copying errors may accumulate over time, resulting in older meristems having a higher genetic load than younger meristems (Klekowski 1988). Support for this hypothesis is limited, but includes a positive relationship between tree age and the proportion of nonviable pollen produced in a conifer (Aizen and Rovere 1995). While this study controlled for differences that might be caused by the resource status of the trees, the authors point out that they did not show directly that the pollen abortion had a genetic cause, and encourage further examination of this hypothesis.

In addition to gene flow and mutation, negative genetic correlations may maintain additive genetic variation. Such correlations would result in alleles for rapid pollen tube growth rate (which should enhance fitness) to be associated with traits that reduce fitness. Walsh and Charlesworth (1992) suggest that correlations of this type would lower fitness at 'some other life stage', and that this would be opposite to what is typically found (experiments generally show that pollen and progeny vigour are positively correlated). However, the antagonistic pleiotropic effects could also exist within a stage as well. As indicated by Delph *et al.* (1997), resource allocation during pollen development can greatly affect pollen tube growth rates; hence, if there is a trade-off between allocation to pollen vs. ovules, a plant that invests more in its pollen would invest less in its ovules, and be a better pollen than seed parent. Recent work on where resources from the plant are allocated within developing flowers supports the possibility of such trade-offs: hierarchic nutritional correlations exist between flower organs of a lily flower and the anther is the director controlling where resources go (Clément *et al.* 1996). In addition, phenotypic correlations showing this type of trade-off have been found for *Hibiscus moscheutos* (Fundyga 1996), but it is not known if these are caused by underlying negative genetic correlations.

Trade-offs between traits of pollen have also been found that are likely to have important consequences for pollen competition. In *Viola diversifolia* performance trade-offs are associated with a polymorphism in aperture number: pollen grains with four apertures germinate more quickly than those with three apertures, but they also have slower-growing pollen tubes and lower longevity (Dajoz *et al.* 1993). The trade-off between rapid germination and longevity acts to maintain a mixed evolutionarily stable strategy, such that all individuals produce both types of pollen (Till-Bottraud *et al.* 1994). However, the proportion of pollen types produced varies by maternal family and could lead to differential fitness, depending on whether pollination occurs soon after flower opening (favouring rapid germination) or is delayed (favouring long-lived pollen) (Dajoz *et al.* 1993).

Finally, given that pollen competitive ability can be influenced by the environment as well as genotype, genotype–environment interactions may exist and these have been shown theoretically to maintain genetic variation within populations (Gillespie and Turelli 1989). These interactions would be likely if the pollen performance of some genotypes is better buffered against environmental perturbation than others (Delph *et al.* 1997).

VIII. FUTURE DIRECTIONS

Despite of the number of pollen competition studies performed so far, the magnitude of the effect of pollen performance on male reproductive success in natural populations of plants is still unclear. These effects, if small compared with other factors, including prefertilization factors such as the timing of pollen production or amount of pollen produced by a plant, may essentially be swamped out and therefore difficult to investigate. More multiple-donor studies using genetic markers should help to address this problem.

Of utmost importance in terms of the evolutionary significance of pollen competition is the extent to which pollen performance is heritable. With only two studies having investigated heritability (or clonal repeatability) with naturally existing species (i.e. noncultivated and nonhybridized), it is difficult to make generalizations about the extent to which the variation seen is attributable to additive genetic variation and therefore available for selection to act on. If further studies measuring either broad- or narrow-sense heritability reveal little genetically based variation, then two additional lines of investigation should be fruitful. First, alternative hypotheses concerning the positive correlation between pollen and progeny vigour would need to be investigated. One possible hypothesis worth considering is that differences in the timing of fertilization of ovules sired by fast- vs. slow-growing pollen tubes lead to a 'relatively early start in postfertilization growth' by those seeds sired by fast-growing pollen (Mulcahy 1974). Second, genotype by environment interactions should be investigated as they may be widespread and be major contributors to the observed variation in pollen performance (Delph *et al.* 1997). Understanding the developmental basis of these interactions would then greatly aid our understanding of why pollen performance is variable.

ACKNOWLEDGEMENTS

Helpful comments on a draft of the manuscript were made by the editors and S. Carroll, S. Davis, D. Dudle, J. Gehring, E. Ketterson, D. Marr, S. Raouf, J. Shykoff, and A. Snow. Our work on pollen competition and the writing of this chapter were supported in part by grants from the National Science Foundation (BSR-9010556 and DEB-9319002) to L. Delph and the Indiana Academy of Sciences to K. Havens.

REFERENCES

Ackerman JD & Montalvo AM (1990) Short- and long-term limitations to fruit production in a tropical orchid. *Ecology* **71:** 263–272.

Aizen MA & Rovere AE (1995) Does pollen viability decrease with aging? A cross-population examination in *Austrocedrus chilensis* (Cupressaceae). *Int. J. Plant Sci.* **156:** 227–231.

Aizen MA, Searcy KB & Mulcahy DL (1990) Among- and within-flower comparisons of pollen tube growth following self- and cross-pollinations in *Dianthus chinensis* (Caryophyllaceae). *Am. J. Bot.* **77:** 671–676.

Apsit VJ, Nakamura RR & Wheeler NC (1989) Differential male reproductive success in Douglas fir. *Theor. Appl. Genet.* **77:** 681–684.

Baker AM & Shore JS (1995) Pollen competition in *Turnera ulmifolia* (Turneraceae). *Am. J. Bot.* **82:** 717–725.

Bateman AJ (1956) Cryptic self-incompatibility in the wall flower: *Cheiranthus cheiri* L. *Heredity* **10:** 257–261.

Bawa KS & Webb CJ (1984) Flower, fruit and seed abortion in tropical forest trees: implications for the evolution of paternal and maternal reproductive strategies. *Am. J. Bot.* **71:** 736–751.

Bertin RI (1990) Paternal success following mixed pollinations of *Campsis radicans*. *Am. Midl. Nat.* **124:** 153–163.

Birkhead TR & Møller AP (1992) *Sperm Competition in Birds*. Academic Press, London.

Bookman SS (1984) Evidence for selective fruit production in *Asclepias*. *Evolution* **38:** 72–86.

Bowman RN (1987) Cryptic self-incompatibility and the breeding system of *Clarkia unguiculata* (Onagraceae). *Am. J. Bot.* **74:** 471–476.

Braun RE, Behringer RR, Peschon JJ, Brinster RL & Palmiter RD (1989) Genetically haploid spermatids are phenotypically diploid. *Nature* **337:** 373–376.

Brewbaker JL & Majumder SK (1961) Cultural studies of the pollen population effect and the self-incompatibility inhibition. *Am. J. Bot.* **48:** 457–464.

Brink RA (1927) The sugary gene in maize as a modifier of the waxy ratio. *Genetics* **12:** 461–491.

Brown AHD, Grant JD & Pullen R (1986) Outcrossing and paternity in *Glycine argyrea* by paired fruit analysis. *Biol. J. Linn. Soc.* **29:** 282–294.

Broyles SB & Wyatt R (1990) Paternity analysis in a natural population of *Asclepias exaltata*: multiple paternity, functional gender, and the 'pollen-donation hypothesis'. *Evolution* **44:** 1454–1468.

Calvo RN (1993) Evolutionary demography of orchids: intensity and frequency of pollination and the cost of fruiting. *Ecology* **74:** 1033–1042.

Carroll SB & Mulcahy DL (1993) Progeny sex ratios in dioecious *Silene latifolia* (Caryophyllaceae). *Am. J. Bot.* **80:** 551–556.

Casper BB (1983) The efficiency of pollen transfer and rates of embryo initiation in Cryptantha (Boraginaceae). *Oecologia* **59:** 262–268.

Casper BB, Sayigh LS & Lee SS (1988) Demonstration of cryptic incompatibility in distylous *Amsinckia douglasiana*. *Evolution* **42:** 248–253.

Charlesworth D (1988) Evidence for pollen competition in plants and its relationship to progeny fitness: a comment. *Am. Nat.* **132:** 298–302.

Charlesworth D, Schemske DW & Sork VL (1987) The evolution of plant reproductive characters: sexual vs. natural selection. In *The Evolution of Sex and its Consequences*. S Stearns (ed.), pp. 317–335. Birkhauser-Verlag, Basel.

Clément C, Burrus M & Audran JC (1996) Floral organ growth and carbohydrate content during pollen development in *Lilium*. *Am. J. Bot.* **83:** 459–469.

Correns C (1917) Ein Fall experimenteller Verschiebung des Geschlechtsverhaltnisses. *Sitz. Koenigli. Preuss. Akade. Wiss.* 685–717.

Correns C (1928) Bestimmung, Vererbung und Verteilung des Geschlechtes bei den hoheren Pflanzen. *Handbuch der Vererbungswissenschaft* **2:** 1–38.

Cousens MI (1988) Reproductive strategies of the pteridophytes. In *Plant Repro-*

ductive Ecology: Patterns and Strategies. J Lovett Doust & L Lovett Doust (eds), pp. 307–328. Oxford University Press, New York.

Cruzan MB (1986) Pollen tube distributions in *Nicotiana glauca*: evidence for density dependent growth. *Am. J. Bot.* **73:** 902–907.

Cruzan MB (1989) Pollen tube attrition in *Erythronium grandiflorum*. *Am. J. Bot.* **76:** 562–570.

Cruzan MB (1990) Pollen–pollen and pollen–style interactions during pollen tube growth in *Erythronium grandiflorum* (Liliaceae). *Am. J. Bot.* **77:** 116–122.

Cruzan MB & Barrett SCH (1993) Contribution of cryptic incompatibility to the mating system of *Eichhornia paniculata* (Pontederiaceae). *Evolution* **47:** 925–934.

Dajoz I, Till-Bottraud I & Gouyon P-H (1993) Pollen aperture polymorphism and gametophyte performance in *Viola diversifolia*. *Evolution* **47:** 1080–1093.

Davis LE, Stephenson AG & Winsor JA (1987) Pollen competition improves performance and reproductive output of the common zucchini squash under field conditions. *J. Am. Soc. Hortic. Sci.* **112:** 711–716.

Delph LF, Johannsson MH & Stephenson AG (1997) How environmental factors affect pollen performance: ecological and evolutionary perspectives. *Ecology* **78:** 1632–1639.

Dressler RL (1981) *The Orchids: Natural History and Classification*. Harvard University Press, Cambridge, MA, USA.

Dudash MR & Ritland K (1991) Multiple paternity and self-fertilization in relation to floral age in *Mimulus guttatus* (Scrophulariaceae). *Am. J. Bot.* **78:** 1746–1753.

Elgersma A, Stephenson AG & den Nijs APM (1989) Effects of genotype and temperature on pollen tube growth in perennial ryegrass (*Lolium perenne* L.). *Sex. Plant Reprod.* **2:** 225–230.

Ellstrand NC & Marshall DL (1986) Patterns of multiple paternity in populations of *Raphanus sativus*. *Evolution* **40:** 837–842.

Epperson BK & Clegg MT (1987) First-pollination primacy and pollen selection in the morning glory, *Ipomoea purpurea*. *Heredity* **58:** 5–14.

Erickson RP (1990) Post-meiotic gene expression. *Trends Genet.* **6:** 264–269.

Fingerett ER (1979) Pollen competition in a species of evening primrose, *Oenothera organensis*, Munz. Master's Thesis. Washington State University, Pullman, WA, USA.

Fundyga RE (1996) A comparison of biomass, nitrogen and phosphorus as measures of floral resource allocation in the marsh hibiscus, *Hibiscus moscheutos*. Master's Thesis. Indiana University, Bloomington, IN, USA.

Galen C & Rotenberry JT (1988) Variance in pollen carryover in animal-pollinated plants: implications for mate choice. *J. Theor. Biol.* **135:** 419–429.

Galen C, Shykoff JA & Plowright RC (1986) Consequences of stigma receptivity schedules for sexual selection in flowering plants. *Am. Nat.* **127:** 462–476.

Ganeshaiah KN & Shaanker RU (1988) Regulation of seed number and female incitation of mate competition by a pH-dependent proteinaceous inhibitor of pollen grain germination in *Leucaena leucocephala*. *Oecologia* **75:** 110–113.

Ganeshaiah KN, Shaanker RU & Shivashankar G (1986) Stigmatic inhibition of pollen grain germination – its implication for frequency distribution of seed number in pods of *Leucaena leucocephala* (Lam) de Witt. *Oecologia* **70:** 568–572.

Gawel NJ & Robaker CD (1986) Effect of pollen–style interaction on the pollen tube growth of *Gossypium hirsutum*. *Theor. Appl. Genet.* **72:** 84–87.

Gillespie JH & Turelli M (1989) Genotype–environment interactions and the maintenance of polygenic variation. *Genetics* **121:** 129–138.

Haig D & Bergstrom CT (1995) Multiple mating, sperm competition and meiotic drive. *J. Evol. Biol.* **8:** 265–282.

Haldane JBS (1932) *The Causes of Evolution.* Harper, London.

Hartl DL (1970) Population consequences of non-mendelian segregation among multiple alleles. *Evolution* **24:** 415–423.

Havens K (1994) Clonal repeatability of *in vitro* pollen tube growth rates in *Oenothera organensis* (Onagraceae). *Am. J. Bot.* **81:** 161–165.

Havens K & Delph LF (1996) Differential seed maturation uncouples fertilization and siring success in *Oenothera organensis* (Onagraceae). *Heredity* **76:** 623–632.

Heribert-Nilsson N (1923) Zertationsversuche mit durchtrennung des griffes bei *Oenothera lamarckiana. Hereditas* **4:** 177–190.

Hessing MB (1986) Pollen growth following self- and cross-pollination in *Geranium caespitosum* James. In *Biotechnology and Ecology of Pollen.* DL Mulcahy, GB Mulcahy & E Ottaviano (eds), pp. 467–472. Springer-Verlag, New York.

Ibarra-Perez FJ, Ellstrand NC & Waines JG (1996) Multiple paternity in common bean (*Phaseolus vulgaris* L., Fabaceae). *Am. J. Bot.* **83:** 749–758.

Johnson CM & Mulcahy DL (1978) Male gametophyte in maize: II. Pollen vigor in inbred plants. *Theor. Appl. Genet.* **51:** 211–215.

Johnston MO (1993) Tests of two hypotheses concerning pollen competition in a self-compatible, long-styled species (*Lobelia cardinalis*: Lobeliaceae). *Am. J. Bot.* **80:** 1400–1406.

Jones KN (1994) Nonrandom mating in *Clarkia gracilis* (Onagraceae): a case of cryptic self-incompatibility. *Am. J. Bot.* **81:** 195–198.

Kaufmann BP & Demerec M (1942) Utilization of sperm by the female *Drosophila melanogaster*. *Am. Nat.* **76:** 445–469.

Karron JD & Marshall DL (1990) Fitness consequences of multiple paternity in wild radish, *Raphanus sativus. Evolution* **44:** 260–268.

Karron JD & Marshall DL (1993) Effects of environmental variation on fitness of singly and multiply sired progenies of *Raphanus sativus* (Brassicaceae). *Am. J. Bot.* **80:** 1407–1412.

Kenrick J & Knox RB (1982) Function of the polyad in reproduction of *Acacia. Ann. Bot.* **50:** 721–727.

Klekowski EJ (1988) *Mutation, Developmental Selection, and Plant Evolution.* Columbia University Press, New York.

Lau TC & Stephenson AG (1993) Effects of soil nitrogen on pollen production, pollen grain size, and pollen performance in *Cucurbita pepo* (Cucurbitaceae). *Am. J. Bot.* **80:** 763–768.

Lau TC & Stephenson AG (1994) Effects of soil phosphorus on pollen production, pollen size, pollen phosphorus content, and the ability to sire seeds in *Cucurbita pepo* (Cucurbitaceae). *Sex. Plant Reprod.* **7:** 215–220.

Lee TD (1984) Patterns of fruit maturation: a gametophytic competition hypothesis. *Am. Nat.* **123:** 427–432.

Lee TD & Bazzaz F (1982) Regulation of fruit maturation in an annual legume, *Cassia fasiculata. Ecology* **63:** 1374–1388.

Levin DA (1990) Sizes of natural microgametophyte populations in pistils of *Phlox drummondii. Am. J. Bot.* **77:** 356–363.

Lund HA (1956) Growth hormones in the styles and ovaries of tobacco responsible for fruit development. *Am. J. Bot.* **43:** 562–568.

Marshall DL (1988) Postpollination effects on seed paternity: mechanisms in addition to microgametophytic competition operate in wild radish. *Evolution* **42:** 1256–1266.

Marshall DL (1991) Nonrandom mating in wild radish: variation in pollen donor success and effects of multiple paternity among one- to six-donor pollinations. *Am. J. Bot.* **78:** 1404–1418.

Marshall DL & Ellstrand NC (1986) Sexual selection in *Raphanus sativus*: experimental data on nonrandom fertilization, maternal choice, and consequences of multiple paternity. *Am. Nat.* **127:** 446–461.

Marshall DL & Ellstrand NC (1988) Effective mate choice in wild radish: evidence for selective seed abortion and its mechanism. *Am. Nat.* **131:** 739–756.

Mascarenhas JP (1992) Pollen gene expression: molecular evidence. In *Sexual Reproduction in Flowering Plants*. SD Russell and C Dumas (eds), pp. 3–16. Academic Press, New York.

McKenna M (1986) Heterostyly and microgametophytic selection: the effect of pollen competition on sporophytic vigor in two distylous species. In *Biotechnology and Ecology of Pollen*. DL Mulcahy, GB Mulcahy & E Ottaviano (eds), pp. 443–448. Springer-Verlag, New York.

McKenna M & Mulcahy DL (1983) Ecological aspects of gametophytic competition in *Dianthus chinensis*. In *Pollen: Biology* and *Implications for Plant Breeding*. DL Mulcahy & E Ottaviano (eds), pp. 419–424. Elsevier, Amsterdam.

Morris WF, Price MV, Waser NM, Thomson JD, Thomson B & Stratton DA (1994) Systematic increase in pollen carryover and its consequences for geitonogamy in plant populations. *Oikos* **71:** 431–440.

Mulcahy DL (1971) A correlation between gametophytic and sporophytic characteristics in *Zea mays* L. *Science* **171:** 1155–1156.

Mulcahy DL (1974) Correlation between speed of pollen tube growth and seedling height in *Zea mays* L. *Nature* **249:** 491–493.

Mulcahy DL (1975) The biological significance of gamete competition. In *Gamete Competition in Plants and Animals*. DL Mulcahy (ed.), pp. 1–4. North-Holland Publishing Co., Amsterdam.

Mulcahy DL (1979) The rise of the angiosperms: a genecological factor. *Science* **206:** 20–23.

Mulcahy DL (1983) Models of pollen tube competition in *Geranium maculatum*. In *Pollination Biology*. L Real (ed.), pp. 152–161. Academic Press, Orlando, FL.

Mulcahy DL & Mulcahy GB (1975) The influence of gametophytic competition on sporophytic quality in *Dianthus chinensis*. *Theor. Appl. Genet.* **46:** 277–280.

Mulcahy DL, Curtis P & Snow A (1983) Pollen competition in natural populations. In *Handbook of Experimental Pollination Biology*. CE Jones & RJ Little (eds), pp. 330–337. Van Nostrand-Reinhold, New York.

Muona O, Moran GF & Bell JC (1991) Hierarchical patterns of correlated mating in *Acacia melanoxylon*. *Genetics* **127:** 619–626.

Mutikainen P & Delph LF (1996) Effects of herbivory on male reproductive success in plants. *Oikos* **75:** 353–358.

Nakamura RR & Wheeler NC (1992) Pollen competition and paternal success in Douglas fir. *Evolution* **46:** 846–851.

Niesenbaum RA & Casper BB (1994) Pollen tube numbers and selective fruit maturation in *Lindera benzoin*. *Am. Nat.* **144:** 184–191.

Nitsch JP (1952) Plant hormones and the development of fruits. *Quart. Rev. Biol.* **27:** 33–57.

Ottaviano E & Mulcahy DL (1989) Genetics of angiosperm pollen. *Adv. Genet.* **26:** 1–65.

Ottaviano E, Sari-Gorla M & Arenari I (1983) Male gametophytic competitive ability in maize: selection and implications with regard to the breeding system. In *Pollen: Biology and Implications for Plant Breeding*. DL Mulcahy & E Ottaviano (eds), pp. 367–373. Elsevier, New York.

Palmer TM & Zimmerman M (1994) Pollen competition and sporophyte fitness in *Brassica campestris*: does intense pollen competition result in individuals with better pollen? *Oikos* **69:** 80–86.

Parker GA (1970) Sperm competition and its evolutionary consequences in the insects. *Biol. Rev.* **45:** 525–567.

Pfahler PL (1967) Fertilization ability of maize pollen grains. II. Pollen genotype, female sporophyte, and pollen storage interactions. *Genetics* **57:** 513–521.

Purrington CB (1993) Parental effects on progeny sex ratio, emergence, and flowering in *Silene latifolia* (Caryophyllaceae). *J. Ecol.* **81:** 807–811.

Quesada M, Schlichting CD, Winsor JA & Stephenson AG (1991) Effects of genotype on pollen performance in *Cucurbita pepo*. *Sex. Plant Reprod.* **4:** 208–214.

Quesada M, Winsor JA & Stephenson AG (1993) Effects of pollen competition on progeny performance in a heterozygous cucurbit. *Am. Nat.* **142:** 694–706.

Quesada M, Bollman K & Stephenson AG (1995) Leaf damage decreases pollen production and hinders pollen performance in *Cucurbita texana*. *Ecology* **76:** 437–443.

Quesada M, Winsor JA & Stephenson AG (1996) Effects of pollen selection on progeny vigor in a cucurbit. *Theor. Appl. Genet.* **92:** 885–890.

Ramirez N & Brito Y (1990) Reproductive biology of a tropical palm swamp community in the Venezuelan Llanos. *Am. J. Bot.* **77:** 1260–1271.

Ramstetter J & Mulcahy DL (1988) Consequences of pollen competition for *Aureolaria flava* seedlings. *Bull. Ecol. Soc. Am. Suppl.* **69:** 269–270.

Richards AJ (1986) *Plant Breeding Systems*. Allen and Unwin, London.

Richardson TE & Stephenson AG (1992) Effects of parentage and size of the pollen load on progeny performance in *Campanula americana*. *Evolution* **46:** 1731–1739.

Rigney LP (1995) Postfertilization causes of differential success of pollen donors in *Erythronium grandiflorum* (Liliaceae): nonrandom ovule abortion. *Am. J. Bot.* **82:** 578–584.

Rigney LP, Thomson JD, Cruzan MB & Brunet J (1993) Differential success of pollen donors in a self-compatible lily. *Evolution* **47:** 915–924.

Robertson AW (1992) The relationship between floral display size, pollen carry-over and geitonogamy in *Myosotis colensoi* (Kirk) Macbride (Boraginaceae). *Biol. J. Linn. Soc.* **46:** 333–349.

Sari-Gorla M, Ottaviano E & Faini D (1975) Genetic variability of gametophytic growth rate in maize. *Theor. Appl. Genet.* **46:** 289–294.

Sari-Gorla M, Pe ME, Mulcahy DL & Ottaviano E (1992) Genetic dissection of pollen competitive ability in maize. *Heredity* **69:** 423–430.

Schemske DW & Pautler LP (1984) The effects of pollen composition on fitness components in a neotropical herb. *Oecologia* **62:** 31–36.

Schlichting CD & Devlin B (1992) Pollen and ovule sources affect seed production of *Lobelia cardinalis* (Lobeliaceae). *Am. J. Bot.* **79:** 891–898.

Schoen DJ & Cheliak WM (1987) Genetics of the polycross. 2. Male fertility variation in Norway spruce, *Picea abies* (L.) Karst. *Theor. Appl. Genet.* **74:** 554–559.

Sivinski J (1984) Sperm in competition. In *Sperm Competition and the Evolution of Animal Mating Systems.* RL Smith (ed.), pp. 86–116. Academic Press, Orlando, FL.

Snow AA (1986) Evidence for and against pollen competition in natural populations. In *Biotechnology and Ecology of Pollen.* DL Mulcahy, GB Mulcahy & E Ottaviano (eds), pp. 405–410. Springer-Verlag, New York.

Snow AA (1990) Effects of pollen load size on sporophyte competitive ability in two *Epilobium* species. *Am. Midl. Nat.* **125:** 348–355.

Snow AA & Mazer SJ (1988) Gametophytic selection in *Raphanus raphanistrum*: a test for heritable variation in pollen competitive ability. *Evolution* **42:** 1065–1075.

Snow AA & Spira TP (1991a) Pollen vigor and the potential for sexual selection in plants. *Nature* **352:** 796–797.

Snow AA & Spira TP (1991b) Differential pollen-tube growth rates and nonrandom fertilization in *Hibiscus moscheutos* (Malvaceae). *Am. J. Bot.* **78:** 1419–1426.

Snow AA & Spira TP (1993) Individual variation in the vigor of self pollen and selfed progeny in *Hibiscus moscheutos* (Malvaceae). *Am. J. Bot.* **80:** 160–164.

Snow AA & Spira TP (1996) Pollen-tube competition and male fitness in *Hibiscus moscheutos. Evolution* **50:** 1866–1870.

Sork VL & Schemske DW (1992) Fitness consequences of mixed donor pollen loads in the annual legume *Chamaecrista fasiculata. Am. J. Bot.* **79:** 508–515.

Spira TP, Snow AA, Whigham DF & Leak JL (1992) Flower visitation, pollen deposition, and pollen-tube competition in *Hibiscus moscheutos* (Malvaceae). *Am. J. Bot.* **79:** 428–433.

Stephenson AG (1981) Flower and fruit abortion: proximate causes and ultimate functions. *Annu. Rev. Ecol. Syst.* **12:** 253–279.

Stephenson AG, Winsor JA & Davis LE (1986) Effects of pollen load size on fruit maturation and sporophyte quality in zucchini. In *Biotechnology and Ecology of Pollen.* DL Mulcahy, GB Mulcahy & E Ottaviano (eds), pp. 429–434. Springer-Verlag, New York.

Stephenson AG, Winsor JA, Schlichting CD & Davis LE (1988) Pollen competition, nonrandom fertilization, and progeny fitness: a reply to Charlesworth. *Am. Nat.* **132:** 303–308.

Tanksley SD, Zamir D & Rick CM (1981) Evidence for extensive overlap of sporophytic and gametophytic gene expression in *Lycopersicon esculentum. Science* **213:** 453–455.

Thomson JD (1989) Germination schedules of pollen grains: implications for pollen selection. *Evolution* **43:** 220–223.

Till-Bottraud I, Venable DL, Dajoz I & Gouyon P-H (1994) Selection on pollen morphology: a game theory model. *Am. Nat.* **144:** 395–411.

Vasil IK (1974) The histology and physiology of pollen germination and pollen tube growth on the stigma and in the style. In *Fertilization in Higher Plants.* HF Linskens (ed.), pp. 105–118. North Holland Publishing Co., Amsterdam.

Walsh NE & Charlesworth D (1992) Evolutionary interpretations of differences in pollen tube growth rates. *Quart. Rev. Biol.* **67:** 19–37.

Waser NM, Price MV, Montalvo AM & Gray RN (1987) Female mate choice in a perennial herbaceous wildflower, *Delphinium nelsonii. Evol. Trends Plants* **1:** 29–33.

Weller SG & Ornduff R (1989) Incompatibility in *Amsinckia grandiflora* (Boraginaceae): distribution of callose plugs and pollen tubes following inter- and intramorph crosses. *Am. J. Bot.* **76:** 277–282.

Willing RP & Mascarenhas JP (1984) Analysis of the complexity and diversity of mRNAs from pollen and shoots of *Tradescantia. Plant Physiol.* **75:** 865–868.

Willing RP, Bashe D & Mascarenhas JP (1988) An analysis of the quantity and diversity of messenger RNAs from pollen and shoots of *Zea mays. Theor. Appl. Genet.* **75:** 751–753.

Willison K, Marsh M & Lyon MF (1988) Biochemical evidence for sharing of gene products in spermatogenesis. *Genet. Res.* **52:** 63.

Winsor JA, Davis LE & Stephenson AG (1987) The relationship between pollen load and fruit maturation and the effect of pollen load on offspring vigor in *Cucurbita pepo. Am. Nat.* **129:** 643–656.

Yamplosky C & Yamplosky H (1922) Distribution of sex forms in the phanerogamic flora. *Bibl. Gene.* **3:** 1–62.

Young HJ & Stanton ML (1990) Influence of environmental quality on pollen competitive ability in wild radish. *Science* **248:** 1631–1633.

Zimmerman JK & Mitchell Aide T (1989) Patterns of fruit production in a Neotropical orchid: pollinator vs. resource limitation. *Am. J. Bot.* **76:** 67–73.

6 Sperm Limitation, Gamete Competition, and Sexual Selection in External Fertilizers

D. R. Levitan

Department of Biological Science, Florida State University, Tallahassee, FL 32306–2043, USA

I. INTRODUCTION

External fertilization is a common and widespread reproductive strategy in aquatic environments (Giese and Kanatani 1987) and is generally thought to be ancestral to internal modes of reproduction (Jägersten 1972; Parker 1984; Wray 1995; but see Rouse and Fitzhugh 1994). Therefore, estimates of male and female fertilization success in external fertilizers may provide not only information on sperm competition for the majority of animal phyla but also insight into the evolution of sexual dimorphism and internal fertilization.

Despite the need to understand the patterns and consequences of variation in both male and female fertilization success, little is known

Sperm Competition and Sexual Selection
ISBN 0-12-100543-7

about the fate of gametes released in aquatic environments. Historically, discussions about reproductive success in external fertilizers were based on speculation or laboratory studies (reviewed by Levitan 1995a). It has only been in the last decade that some of the practical obstacles associated with 'chasing' gametes in an aquatic medium have been overcome. Estimates of gamete concentration and fertilization have been made, but there is still no direct information on sperm competition and multiple paternity.

In contrast to most other organisms, the available evidence on external fertilizers suggests that sperm is limiting. Evidence from field experiments (Table 6.1) and natural observations of spawning (Table 6.2) demonstrate that the proportion of a female's eggs that are fertilized is often much less than 100%, and a majority of the variation in female fertilization success can be explained by male abundance, proximity, or synchrony. This somewhat different view of sexual selection has implications for the generality of Bateman's principle (Bateman 1948) and the evolution of sexual dimorphism in this presumptive ancestral reproductive strategy.

In this chapter, I offer the possibility that in externally fertilizing organisms sexual selection is intense but approximately symmetrical across sexes. This is a result of (1) sperm limitation, which results in (2), increased variation in the proportion of a female's eggs that are fertilized, and hence increased variation in female reproductive success relative to taxa with internal fertilization, and in turn results in (3), selection for enhanced fertilization success not only for males but also for females. As a consequence, sexual dimorphism in both primary and secondary sexual characteristics is reduced or absent. This hypothesis leads to the notion that anisogamy and copulation evolved because of sperm limitation rather than sperm competition, an adaptation that in this scenario benefits both males and females. In order to build these arguments, I must first define the relevant terms, review what is known about fertilization in externally fertilizing organisms and patterns of sexual dimorphism, and then attempt to place this evidence in a theoretical framework. This is not a completed project, and my goal is to stimulate interest in sexual selection on external fertilizers.

II. DEFINITIONS AND THE SCOPE OF THIS REVIEW

Free spawning is defined as the release of sperm into the environment, whereas broadcast spawning is defined as the release of both eggs and sperm into the environment. Males can free spawn, pseudocopulate (release sperm directly on females or transfer a spermatophore), or copulate (release sperm within a female's reproductive tract). Females can broadcast spawn, brood eggs on an external surface, or brood eggs

Table 6.1. Experimental evidence of variation in female fertilization success in free-spawning invertebrates. Mean and range of the percentage of eggs fertilized and a summary of the major factors influencing variation in fertilization.

Taxa	% Fertilization		Effect	Reference
	Mean	**Range**		
Cnidarians				
Hydrozoans				
Hydractinia echinata	41	0–91	Female fertilization decreased with male distance	Yund (1990)
Bryozoans				
Celloporella hyalina	100	100[a]	Selfing inversely related to number of conspecific male zooids	Yund and McCartney (1994)
Echinoderms				
Asteroids				
Acanthaster planci	32	0–90	Female fertilization decreased with male distance	Babcock *et al.* (1992)
Asterias forbesi	52	2–99	Female fertilization decreased with depth	Present study
Echinoids				
Clypeaster rosaceus	30	2–72	Female fertilization decreased with male distance	Levitan and Young (1995)
Diadema antillarum	23	0–99	Female fertilization increased with male density and decreased with male distance, male size not significant	Levitan (1991)
Strongylocentrotus droebachiensis	30	1–95	Female fertilization increased with male abundance, decreased with male distance and flow	Pennington (1985)
Strongylocentrotus franciscanus	18	0–82	Female fertilization increased with abundance and aggregation, decreased with flow	Levitan *et al.* (1992)
Chordates				
Ascidians				
Botryllus schlosseri	41	25–60[b]	Female fertilization increased with male density, male success decreased with male competition	Yund and McCartney (1994)

[a] Selfing hermaphrodite.
[b] Range of means across treatments.

Table 6.2. Natural observations of variation in female fertilization success in free-spawning invertebrates.

Taxa	% Fertilization Mean	% Fertilization Range	Comments	Reference
Cnidaria				
Gorgonians				
Briareum asbestinum	4	< 0.01–6.5	Variation related to density	Brazeau and Lasker (1992)
Plexaura kuna	*c.* 20	0–100	Few male clones	Lasker *et al.* (1996)
Pseudoplexaura porosa	51	0–80[a]	Common	Lasker *et al.* (1996)
Scleractinians				
Montipora digitata	*c.* 30	0–75	Variation related to spawning synchrony	Oliver and Babcock (1992)
Arthropoda				
Merostomata				
Limulus polyphemus	74	0.6–100	No effect of satellite males	Loveland and Botton, in review
Echinodermata				
Asteroids				
Acanthaster planci	44	23–83	Variation related to spawning synchrony	Babcock and Mundy (1992)
Holothuroids				
Cucumaria frondosa	*c.* 70	45–82[b]	High density and synchrony	Hamel and Mercier (1966)
Cucumaria miniata	92	1–100	High density and synchrony	Sewell and Levitan (1992)
Actinopyga lecanora	73	67–78	Several individuals spawning	Babcock *et al.* (1992)
Bohadshia argus	57	0–96	Variation related to distance from males	Babcock *et al.* (1992)
Holothuria coluber	33	9–83	Variation related to synchrony	Babcock *et al.* (1992)

[a] Range of means across days.
[b] Range of means across multiple samples.

internally (after internal fertilization). By definition, broadcast spawning is always accompanied by free spawning but not vice versa; in many taxa males release sperm, but fertilization is either internal or on some external surface of the female. Giese and Kanatani (1987) appear to define free and broadcast spawning as above, but they sometimes use these terms interchangeably. Because holding and releasing eggs are alternative reproductive strategies with different consequences, these terms are strictly defined in this chapter. This review concentrates on broadcast spawning (leading to external fertilization) but also mentions studies in which males free spawn but females brood eggs (leading to internal fertilization). Including the latter group addresses the influence on sexual selection of release of sperm into an aquatic environment.

Sperm competition is defined as direct competition among males for a limited number of unfertilized eggs. For example, a case in which sperm from one male cannot fertilize an egg because it has already been fertilized by sperm from another male constitutes sperm competition. A case in which sperm from one male collide with more virgin eggs than that from another male is not sperm competition. The reason for drawing this distinction is that interesting differences in sexual selection arise under conditions of intense sperm competition (in which sperm are abundant) and sperm limitation (in which sperm are rare).

Arnold (1994a, p. S9) defined sexual selection as 'selection that arises from differences in mating success,' where mating success is defined as the 'number of mates that bear or sire progeny over some standardized time interval.' However, this definition of mating success does not address the issue of multiple paternity in a clutch and does not include selection that arises from differences in fertilization efficiency. For example, if five males and five females spawn synchronously, multiple paternity is likely. In addition, the selection that can influence the proportion of eggs fertilized by any one male should be included as sexual selection. Sexual selection is defined here as selection that arises from intrasexual differences in the proportion of an individual's gametes that fuse to become zygotes. This definition allows for multiple paternity and includes differences arising from mating success or fertilization efficiency. By this definition, sexual selection becomes unimportant to that sex when fecundity in that sex is no longer limited by fertilization. This definition is different from those that include mate fecundity (Arnold 1994a; Møller 1994) because it is independent of the number offspring produced – which can be influenced by natural selection – and depends simply on per-gamete success relative to others of the same sex.

I focus on dioecious marine invertebrates because hermaphroditic species (see Chapter 7) and fish (see Chapter 11) are covered elsewhere in this volume, but I refer to these other groups for comparative purposes. Detailed taxon-by-taxon coverage of the reproductive biology of free-spawning invertebrates is beyond the scope of this review but is available in the excellent set of volumes entitled *Reproduction in Marine Invertebrates*, edited by Giese, Pearse and Pearse (1975–91).

III. FACTORS INFLUENCING FERTILIZATION SUCCESS

The fertilization ecologies of internally and externally fertilizing species differ in a number of ways. Because in externally fertilizing species sperm are not deposited within the female, the probability of sperm–egg encounters can be highly variable and can depend on a variety of factors ranging from attributes of the environment to those of the population, individual, and gamete (discussed in Section III.A, B, C and D, respectively).

A. Environment

In broadcast spawners, the probability of fertilization, from either a male or a female perspective, is primarily a function of gamete concentration. In an aquatic environment, gametes can diffuse and become diluted quickly. For example, along high-wave-intensity shores, gametes diffuse so quickly that males spaced greater than 10 cm upcurrent from a female have reduced fertilization success (predicted by Denny and Shibata 1989; own unpublished data).

Under less severe conditions, such as protected coastlines, subtidal environments buffered from sea-surface conditions, or calmer weather, fertilization becomes more likely but is still constrained by the flow conditions. In the Bahamas, even under moderate-flow conditions (10 cm s^{-1}), dye particles and sperm were diluted by five orders of magnitude in only 20 s (Levitan and Young 1995). In the San Blas Islands, off the coast of Panama, under low-flow conditions (0.4–2.9 cm s^{-1}) dye particles were diluted by two orders of magnitude within 2 min of release (Lasker and Stewart 1992). Similarly, on the Great Barrier Reef, crown-of-thorns *Acanthaster planci* sea star sperm was diluted by two orders of magnitude within 1 m of release under low (2 cm s^{-1}) flow conditions (Benzie *et al.* 1994).

Several field experiments have documented a decline in the percentage of fertilized eggs with increasing water flow (Pennington 1985; Levitan *et al.* 1992; Petersen *et al.* 1992). Increased water velocity increases the rate of mixing, making gamete plumes larger and more diffuse, reducing the probability of gamete collision.

Even if gametes are in sufficient concentration, fertilization may be inhibited by water movement. Shear forces on gametes released into turbulence have been predicted to cause eggs to spin at up to 100 revolutions per second (Denny *et al.* 1992). Laboratory experiments demonstrate that the shear forces experienced by gametes in highly exposed environments can disrupt sperm–egg interactions, resulting in decreased fertilization success (Epel 1991; Mead and Denny 1995).

The depth of water in which the gametes are mixed can also influence levels of fertilization (Fig. 6.1). As the volume of water decreases, the concentration of gametes increases. Some marine invertebrates move into shallow water to spawn (Giese and Kanatani 1987; Pearse 1979), wait until low tide (McEuen 1988; Sewell and Levitan 1992), or circumvent the dilution problem by releasing buoyant gamete bundles (Oliver and Willis 1987).

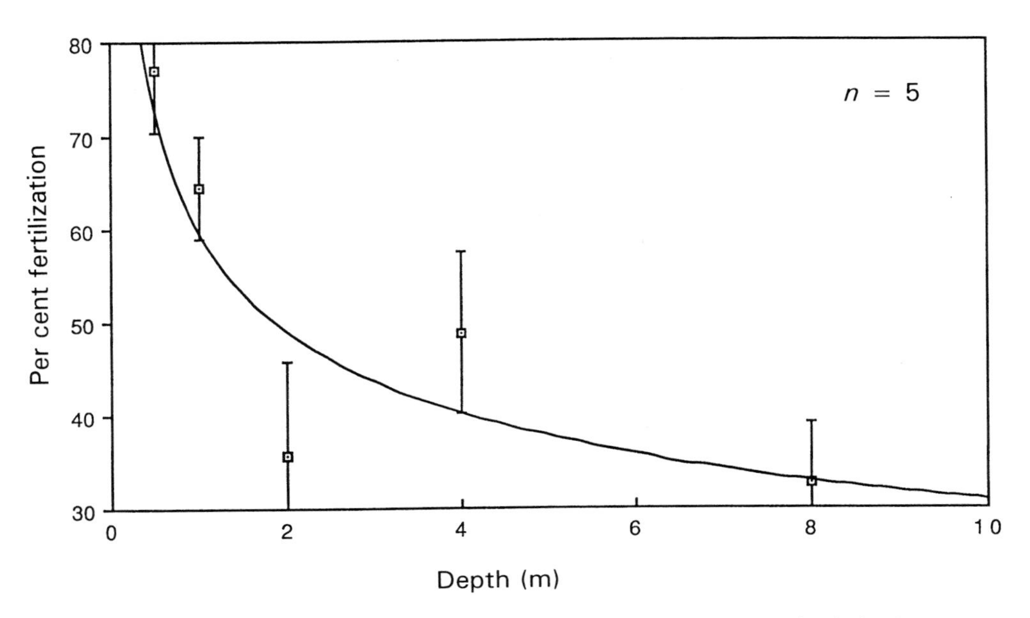

Fig. 6.1. *Female fertilization success decreases with increasing water depth in the sea star* Asterias vulgaris. *Eggs and sperm were collected in the laboratory; eggs were placed in Nitex containers, and sperm were placed in syringes. Gametes were brought to the field (sandy beach habitat, Nahant, MA, USA), and at each depth three Nitex bags were placed at the points of a triangle of 0.5 m side and suspended 15 cm off the bottom. Sperm were released into the centre of the triangle and 30 min later the containers were brought to the laboratory and examined for evidence of fertilization (mean and standard deviation reported). One replicate was conducted per day for 5 days during July 1989. Data collected by D. Levitan and S. Rumrill.*

B. Population

Gamete concentration and the probability of sperm–egg encounters are controlled not only by the rate of diffusion but also by the rate and timing of gamete release. At the population level, factors that can influence the local concentration of gametes are the distance between individuals and the abundance of individuals. There are also group behaviours such as aggregation and spawning synchrony that determine the distribution and abundance of animals that spawn simultaneously.

Most experimental studies of female fertilization success as a function of distance from a spawning male indicate that fertilization decreases to about 20% at 1 m and about 1% at 10 m (Pennington 1985; Yund 1990; Levitan 1991, 1995a; Yund and McCartney 1994; Levitan and Young 1995). Data from natural spawns also indicate that females spawning several metres from a male have very low levels of fertilization (Table 6.2). One exception to this finding is an experimental study with the large crown-of-thorns sea star in which fertilization of 6% was noted

100 m downstream of a single spawning male (Babcock *et al.* 1994). This exception may be a result of the high sperm output of this large-bodied species (Babcock *et al.* 1994) or the ability of eggs of this species to be fertilized at low sperm concentrations (Benzie and Dixon 1994).

In addition to the distance between males and females, the abundance of spawning individuals also influences the levels of fertilization. Experimental studies indicate that fertilization success increases as either the degree of aggregation or the number of spawning individuals increases. In the temperate sea urchin *Strongylocentrotus droebachiensis*, when the number of males increased from one to three, fertilization increased from around 5% to 60% 1 m from the spawning males (Pennington 1985). In the congeneric *Strongylocentrotus franciscanus*, fertilization increased by 12% as the number of males increased from two to eight in a 64-m^2 area (Levitan *et al.* 1992). In the tropical sea urchin *Diadema antillarum*, fertilization increased from 7% to 45% as the number of males increased from one to four and the experimental area decreased from 1 m^2 to 0.25 m^2 (Levitan 1991). In the brooding ascidian *Botryllus schlosseri*, an increase of up to 25% of eggs fertilized was noted when the number of free-spawning males increased from one to three colonies per experiment (Yund 1995).

Patterns of synchrony and aggregation are still poorly understood for the majority of free spawners (but see review by Giese and Kanatani 1987). In some instances synchrony appears to be high – sponges (Reiswig 1970), cnidarians (Harrison *et al.* 1984; Shlesinger and Loya 1985; Babcock *et al.* 1986; Minchin 1992; Lasker *et al.* 1996), nemerteans (Wilson 1900), polychaetes (Hornell 1894; Hargitt 1910; Schroeder and Hermans 1975; Caspers 1984; Babcock *et al.* 1992), molluscs (Gutsell 1930; Battle 1932; Wilborg 1946; Coe 1947; Babcock *et al.* 1992; Minchin 1992; Stekoll and Shirley 1993), enteropneusts (Hadfield 1975), echinoderms (Minchin 1987, 1992; McEuen 1988; Babcock *et al.* 1992; Sewell and Levitan 1992; Hamel and Mercier 1996) – whereas other observations indicate sporadic and unpredictable patterns of spawning – sponges (Reiswig 1970; Babcock *et al.* 1992), cnidarians (Shlesinger and Loya 1985; Babcock *et al.* 1992), polychaetes (Babcock *et al.* 1992), molluscs (Babcock *et al.* 1992), echinoderms (Levitan 1988; Pearse *et al.* 1988; Babcock *et al.* 1992). In the sea urchin *D. antillarum*, there was little evidence of spawning synchrony or aggregation in a systematic survey of aggregation and reproductive readiness (indicated by release of mature gametes on stimulation with KCl) in over 100 observations of natural spawning (Levitan 1988). In the north-east Pacific, individuals from several species were seen to spawn in isolation (Pearse *et al.* 1988). On the Great Barrier Reef (Babcock *et al.* 1992) a systematic survey of spawning times and number of animals seen spawning revealed a large number of isolated individuals spawning alone (Fig. 6.2A). The degree of conspecific synchrony increased during observations of multispecies spawning events (Fig. 6.2B).

Patterns of aggregation are also mixed. There are observations of

A

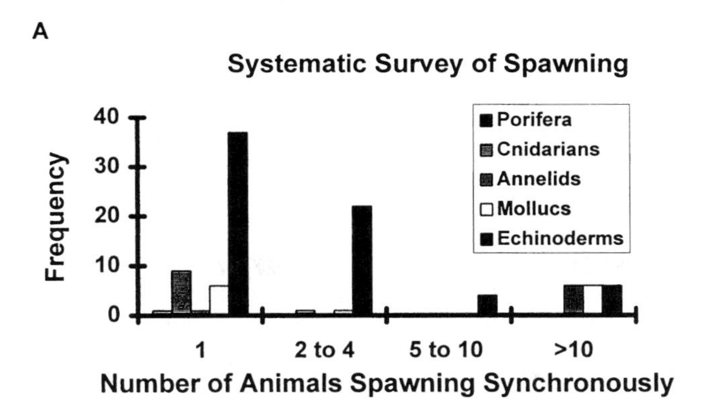

B

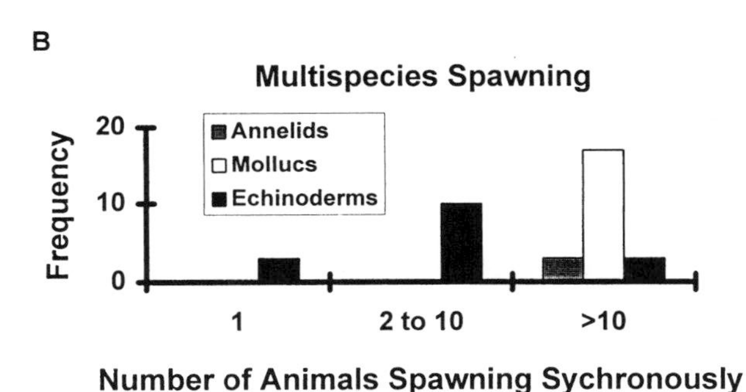

Fig. 6.2. *Frequency distribution of the number of synchronously spawning conspecifics on the Great Barrier Reef. Data from Babcock* et al. *(1992).* **A.** *Data collected during a systematic survey.* **B.** *Data collected during multispecies spawning events.*

large-group spawning (Schroeder and Hermans 1975; Pennington 1985; Giese and Kanatani 1987; Pearse and Cameron 1991; Minchin 1992), small-group spawning (Randall *et al.* 1964; Pennington 1985; Minchin 1987; Stekoll and Shirley 1993; Tominga *et al.* 1994; Unger and Lott 1994), pair spawning (Uchida and Yamada 1968; Schroeder and Hermans 1975; Run *et al.* 1988; Smiley *et al.* 1991; Tyler *et al.* 1992; Young *et al.* 1992), and isolated spawning (Randall *et al.* 1964; Pennington 1985; Minchin 1987; Levitan 1988; Pearse *et al.* 1988; Smith 1989; Pearse and Cameron 1991; Babcock *et al.* 1992).

Simulation models of aggregation in broadcast-spawning organisms predict that fertilization success will increase in a nonlinear fashion with increased aggregation (Levitan and Young 1995). If aggregation incurs costs (e.g. the energetic cost of finding mates, or food limitation at high density) in addition to fertilization benefits, and the benefits of aggregation decrease with population size (because abundance and distribution influences fertilization success; Levitan *et al.* 1992), aggregative behaviour may not always be advantageous, even if fertilization is not maximized (Levitan and Young 1995).

One of Thorson's rules (1950) is that males often spawn first during broadcast spawning events. Sperm, along with many other environmental cues (phytoplankton, temperature, salinity, light) can trigger spawning in both males and females (reviewed by Giese and Kanatani 1987). Thorson's rule has support across many taxa: 37 species across seven phyla cited by Thorson (1950) and more recently cnidarians (Cambell 1974), priapulids (van der Land 1975), sipunculids (Rice 1975), enteropneusts (Hadfield 1975), molluscs (Pearse 1979), and echinoderms (McEuen 1988; Hendler 1991). It is not absolute because several taxa have been noted in which females spawn first – sipunculids (Rice 1975), echinoderms (Hendler 1991; Chia and Walker 1991) – and numerous studies (as cited in the preceding paragraphs) make no note of sexual differences in spawning times. Differences in spawning times and in gamete attributes may be the only instances of sexual differences commonly found among broadcast spawners. A delay in female spawning may represent active choice or simply reflect constraints on egg release (e.g. maturation; Giese and Kanatani 1987) not present in sperm. The former alternative may represent evidence for female choice in external fertilizers; females may wait until the ambient sperm concentration is acceptable before spawning. If so, research is needed into changes in the threshold of choice with spawning season, female age, gamete age, decreasing opportunity, or increasing risk of adult mortality. If the last alternative is an important constraint, then the question becomes 'why do males spawn early?' rather than 'why do females delay?' The lag time may allow males to accumulate sperm to a critical concentration before females release eggs. For example, if water movement is slight, males that spawn first may have a higher relative sperm concentration than males that delay. Obviously, males that spawn too early, like females that delay too long, risk reproductive failure. Additional studies, particularly in situations where females spawn first, may shed light on the selective pressure on sexual differences in the timing of spawning.

Multispecies spawning events are often witnessed (Babcock *et al.* 1986, 1992; Minchin 1987, 1992; Pearse *et al.* 1988; Sewell and Levitan 1992). The best-known example is on the Great Barrier Reef off Australia, where over 100 species of corals spawn in a single annual mass-spawning event (Harrison *et al.* 1984; Babcock *et al.* 1986). Most species spawn during the week following the full moon within a 2- to 3-day period. Spawning is restricted to a 4-h period starting at sunset. Although there is temporal separation among some species, at any one time numerous species spawn simultaneously (Babcock *et al.* 1986). Interestingly, in the Red Sea, species spawn at various times during the year, and interspecific overlap is small (Shlesinger and Loya 1985). Multispecies spawning may result from selection for satiation of egg/embryo predators or simply from multiple species' using the same cues for optimal spawning times for enhanced fertilization or offspring survival

(Babcock *et al.* 1986). At least on the Great Barrier Reef, it appears that the costs of species-specific spawning times outweigh those of hybridization or egg or embryo death caused by fertilization with heterospecific sperm (see Willis *et al.* in press, for heterospecific fertilization of mass-spawning reef corals).

C. Individual behaviour

Individuals also can influence the distribution of gametes in the environment through changes in the rate and timing of gamete release, spawning behaviour, and the amount of energy invested in reproduction. Sexual dimorphism in reproductive allocation is discussed in Section V.

The rate at which gametes are extruded from the gonopore influences the distribution of gametes in the environment. A high spawning rate increases the gamete concentration (Denny and Shibata 1989) but decreases the spatial extent of the gamete cloud. A high spawning rate should be correlated with high spawning synchrony, close male–female pairing, and intense sperm competition.

The relative concentration of sperm may be of key importance in sperm competition. Studies of reproductive success on sessile free-spawning brooding ascidians and bryozoans indicate that males that are closer to females, and hence have the opportunity to place more concentrated sperm at the female than more distant males, can outcompete more distant males for fertilizations (Yund and McCartney 1994).

Reef fish tend to spawn within seconds during tightly paired and highly synchronized spawning rushes (Robertson and Hoffman 1977; Johannes 1978; Thresher 1984) that often involve multiple males (Warner *et al.* 1975). Swarming polychaetes also spawn quickly in tight synchrony (Schroeder and Hermans 1975). Observations of benthic invertebrates indicate that, although some may spawn within seconds (echinoderms, Hendler 1991; Holland 1991), at least in some cases, individuals can spawn for intervals ranging from nearly an hour to days. Examples include sponges (Reiswig 1970; Fell 1974), enteropneusts (Hadfield 1975), molluscs (Pearse 1979), brachiopods (Chuang 1959), and echinoderms (McEuen 1988; Hendler 1991; Hamel and Mercier 1996). Measurement of spawning rate and the correlation of spawning rate with aspects of sperm limitation and competition, although neglected, might be fruitful areas of research.

Typical spawning behaviour for many broadcast-spawning fishes and invertebrates is to rise into the water column before spawning. The distance reef fish rise from the substratum into the water column is positively correlated with body size (Thresher 1984). Echinoderms typically climb onto structures, rise up on arms, or rear their entire bodies upward during gamete expulsion (McEuen 1988; Hendler 1991; Babcock *et al.* 1992; Minchin 1992). Anemones have been noted to extend their columns upward during spawning (Minchin 1992), abalone climb kelp stipes (Stekoll and Shirley 1993), and chitons can assume a vertical orientation (Pearse 1979). This behaviour is thought to increase post-zygotic survivorship by getting the gametes away from benthic filter feeders and egg predators (Robertson and Hoffman 1977; Johannes 1978), but it could also keep negatively buoyant eggs from getting trapped in the sediment before the larvae can swim. These factors are independent of fertilization. Alternatively, these behaviours could serve to facilitate gamete mixing.

I have seen two individuals of the sea urchin *D. antillarum* spawn at a leisurely rate until they bumped into each other, at which time they rapidly spawned bursts of gametes as they tried to hop on top of each other; they were the only individuals seen to spawn at that time, but they were both male. This anecdote, like many others, typifies the dilemma associated with external fertilization in many benthic invertebrates; natural selection has programmed these organisms in ways that seem adaptive – increased aggregation and spawning intensity when detecting a spawning conspecific – but nonetheless their limited mobility, perceptual ability, and opportunity promote gamete wastage and sperm limitation.

D. Gamete

The numerous factors described above determine the concentration of gametes, the rate of gamete dispersion, and the degree of mixing of gametes from different individuals. This information, although critical, is not sufficient to predict levels of fertilization. Variation in gamete performance has been noted both among species (Branham 1972; Levitan 1993) and within species (Hultin and Hagström 1956; Levitan *et al.* 1991; Benzie and Dixon 1994; Levitan 1996a) in the laboratory. Variation in gamete performance has also been documented within (Levitan 1996a) and among species (Levitan 1995b) in the field.

Variation in gamete performance is linked to variation in gamete attributes. Models that predict the degree of fertilization generally include not only gamete concentration but also a variety of gamete attributes (Rothschild and Swann 1951; Vogel *et al.* 1982; Denny and Shibata 1989), which include egg size, the receptiveness of the egg surface to sperm, sperm velocity, and sperm longevity. Egg traits that have been demonstrated to vary across taxa are buoyancy, size, the proportion of sperm–egg collisions that result in fertilizations, the size and presence of jelly coats or other structures that can capture sperm, and the presence of sperm chemoattractants. Variation has been documented in sperm traits such as velocity, longevity, behaviour, and buoyancy (Levitan 1995a).

The size of the egg target can influence the number of sperm–egg collisions. Models generally consider the target size of the egg to be cross-sectional area (Rothschild and Swann 1951; Vogel *et al.* 1982; Denny and

Shibata 1989; Levitan 1993, 1996a,b; Podolsky and Strathmann 1996). Empirical data from the laboratory indicate that, among sea urchins in the genus *Strongylocentrotus*, increased egg cross-sectional area is correlated with a decrease in the concentration of sperm needed to fertilize it (Levitan 1993). This correlation is also evident within *S. franciscanus*, where 45% of the variation among females in the amount of sperm needed to fertilize 50% of the eggs could be explained by mean egg size (Levitan 1996a). Egg size also appears to be important within a single clutch of eggs. In the sea urchins *S. franciscanus* and *Strongylocentrotus purpuratus*, larger eggs were preferentially fertilized when sperm were limiting (Levitan 1996a).

Varying egg shape could provide a mechanism to increase egg target size and sperm–egg collisions without increasing egg volume (Podolsky and Strathmann 1996), but there could be developmental constraints on egg shape. Experiments in which echinoderm eggs are artificially deformed result in abnormal cleavage patterns (Rappaport and Rappaport 1994). Fish eggs, which come in a variety of shapes, have a small restricted area for sperm attachment, the micropyle (Amanze and Iyengar 1990). Because, in some taxa, the sperm attachment site determines planes of embryonic symmetry (Schroeder and Hermans 1975), it would be interesting to determine whether deviations from a symmetrical egg shape are correlated with restrictions in the surface available for fertilization.

Other attributes, such as jelly coats, accessory cells, or the effective range of chemoattractants, may also contribute to effective egg target size (Levitan 1995a; Podolsky and Strathmann 1996). These mechanisms have, to varying degrees, been shown to influence either levels of fertilization (jelly coats: Rothschild and Swann 1951; Podolsky 1995; follicle cells: T. Bolton and J. Havenhand, personal communication) or sperm behaviour (chemotaxis, Miller 1985) in the laboratory, but the efficiency of these potentially energetically economical solutions may be reduced under field conditions (e.g. jelly coats degrade to approximately 25% of original size in less than 1 min in moving water; own unpublished data).

Another mechanism for increasing egg target size, without increasing the level of energy investment, is to inflate eggs with water (Levitan 1993); however, developmental constraints may limit dilution (Podolsky and Strathmann 1996) and if all species dilute egg material to the same extent, it would be difficult to establish any patterns across taxa.

The ratio of predicted sperm-egg collisions to successful fertilizations varies among and within species (Vogel *et al.* 1982; Levitan 1993, 1996a), perhaps as a result of differences in properties of the egg surface (e.g. the number or distribution of sperm receptor sites) or of variation in sperm quality (e.g. inability to fertilize a particular egg or any egg).

Variation in sperm attributes may influence fertilization dynamics. Sperm morphology is correlated with function: 'In the Metazoa the primitive sperm is a small cell with a short rounded-conical head, a small and short middle piece containing a few (often four) mitochondria, and a tail consisting of a flagellum about 50 mm long' (Franzén 1987; page 34). Across taxa, 'primitive' sperm are associated with free-spawning strategies and more 'modified' sperm with pseudocopulation and copulation (reviewed by Franzén 1987). Although primitive sperm are generally similar across taxa, sperm head size does vary and is positively correlated with egg size in echinoderms (Eckelbarger *et al.* 1989). The issue of whether primitive sperm are ancestral or simply reflect convergent evolution for swimming in the sea requires attention (Rouse and Fitzhugh 1994).

Greater sperm swimming velocity should increase the rate of fertilization (Vogel *et al.* 1982; Levitan *et al.* 1991) and could be important under conditions of sperm competition (Levitan 1995a). Similarly, greater sperm longevity should increase the probability of fertilization at greater distances from males and could be important when spawning synchrony is low (Levitan 1995a). Sperm velocity varies among (Gray 1955; Levitan 1993) and within species (Levitan *et al.* 1991; Levitan 1993) of sea urchins. Sperm longevity also varies among species of sea urchins (Levitan 1993) and across taxa (Levitan 1995a; Levitan and Petersen 1995). Although no direct tests have addressed the effects of sperm longevity or velocity on fertilization success, some interesting correlations exist.

Among three congeneric species of sea urchins, *S. purpuratus*, *S. franciscanus*, and *S. droebachiensis*, there is a fivefold range in egg size, an inverse relationship between egg size and sperm velocity, and an inverse relationship between sperm velocity and sperm longevity (Levitan 1993). The inverse relationship between sperm longevity and velocity may represent a trade-off in per-spermatozoon energy allocation (Levitan 1993; Bolton and Havenhand 1996). Small eggs and fast sperm would be expected in situations of high sperm concentration and sperm competition (decreased selective pressure for attracting sperm and increased selective pressure for fast, competitive sperm). Larger eggs and longer-lived sperm would be expected in situations of low sperm concentration and sperm limitation (increased selective pressure for attracting sperm and increased selective pressure for retaining sperm viability at greater times or distances from the point of spawning). Recent empirical studies appear to confirm this prediction; *S. purpuratus* (smallest egg and fastest but shortest-lived sperm) lives in tight aggregations and is least sperm-limited, and *S. droebachiensis* (largest egg and slowest but longest-lived sperm) is more dispersed and is most sperm-limited (own unpublished data). *Strongylocentrotus franciscanus* has intermediate values of gamete attributes, aggregation, and fertilization success.

Although data are few, other groups appear to follow this pattern. The tropical reef fish *Thalassoma bifasciatum* has high levels of fertilization, often mates in groups or in pairs with streaker males (Petersen *et al.* 1992), and seems to have very short-lived sperm (*c.* 15 s). At the other extreme, the sperm of deep-sea echinoids, whose densities may be low and spawning cues rare, have been observed to swim for several days (Eckelbarger *et al.* 1989), as have sperm from cold-water Antarctic echinoderms (J. Pearse, personal communication). It is unclear whether extreme sperm longevities in the colder environments are a function of selection for longevity *per se* or simply the physiological outcome of lower temperatures.

Sperm behaviour may also influence patterns of fertilization. Fertilization-kinetics models assume that sperm move randomly (Rothschild and Swann 1951; Vogel *et al.* 1982), or that sperm swimming is negligible under natural conditions (Denny and Shibata 1989), but sperm chemotaxis (Miller 1985), variation in activity related to sperm concentration (Chia and Bickell 1983) and egg products (Epel 1978), and sperm–sperm interactions (Rothschild and Swann 1951), all of which have been observed in the laboratory, may play important roles in the sea.

Spawned materials other than the gametes themselves can influence the viscosity and dispersability of the gametes. Eggs and sperm can be released in a cloud, in stringy masses, or in clumps (McEuen 1988; Thomas 1994a,b). Eggs of different species of polychaetes disperse at different rates (Thomas 1994a). Among sea urchins, intersex and interspecific variation in dispersal rates of gametes have been observed (Thomas 1994b). Sea urchin species that live in more exposed habitats spawn more viscous materials than do shallow-water species (Thomas 1994b). Varying the susceptibility of gametes to dispersion can influence gamete concentration and sperm longevity.

The subtle influences on fertilization that gamete traits show in the laboratory reveal the potential for selection on those traits to influence reproductive success. However, determining the intensity of these selective pressures requires field tests, because environmental, population, or individual-level factors may overwhelm gamete-level factors under natural conditions.

Experiments using the sea urchin *S. franciscanus* addressed this issue. Gametes were released in a uniform manner, through a syringe, into an environment that varied substantially from day to day (flow varied from 0 to 85 cm s^{-1}) with location and sea conditions typically present off the coast of western Canada. Despite the variation in environmental conditions, gamete performance in the laboratory explained over 50% of the daily variation in field levels of fertilization (Levitan 1996a). This result shows that the influence of gamete traits on fertilization is unlikely to be swamped by other factors and that natural selection on gamete traits for enhanced fertilization success is likely to be important.

IV. BROADCAST SPAWNING AND BATEMAN'S PRINCIPLE

Bateman's principle suggests that, because sperm are more numerous than eggs, sperm will compete for fertilizations and males will have highly variable reproductive success (Bateman 1948). Some males, because of circumstance or quality, will garner a high proportion of fertilizations while others will have reduced fertilization success or complete reproductive failure because most, or all, females are mated, or eggs fertilized, before they get an opportunity. By the same token, a female will, on average, have a much higher percentage of her gametes fertilized and reduced variance in fertilization success. Females will not be limited by fertilization success, and their reproductive success will be determined by the availability of resources influencing egg production or postzygotic success.

The mechanism driving Bateman's predictions is the relative concentration of eggs and sperm. Sperm are much more abundant than eggs at the point of production in almost all taxa (but see Pitnick *et al.* 1995), and when sperm are deposited within females or directly on eggs, this disparity of numbers results in sperm competition and high female fertilization success. When gametes are released into the environment, however, the distribution and abundance of conspecifics, interacting with a turbulent aquatic environment, yield a highly unpredictable distribution of egg and sperm concentrations. Although, on average, sperm will be more concentrated than eggs, sperm can be limiting, and female fertilization success is no longer assured and can be highly variable.

Gamete interactions in an aquatic environment can be described as a two-dimensional continuum, with the ratio of sperm to eggs on one axis and the absolute concentration of gametes on the other (Fig. 6.3). The sperm–egg ratio determines the degree of sperm competition or egg competition. The gamete concentration axis determines the gradient of gamete limitation and the likelihood that competitive interactions will occur at all.

From the female perspective, the difference between sperm limitation and egg competition is that, in the former, removing eggs from other females would not increase the chances of fertilization for the remaining females' eggs; sperm rarely encounter any egg. In egg competition, removing eggs from other females would increase the chances of fertilization for the remaining females' eggs; sperm numbers are significantly depleted by the number adhering to eggs. From the male perspective, egg limitation and sperm competition differ in analogous ways.

This spectrum of gamete conditions influenced by environment, population, individual and gamete considerations can vary at a number of levels among and within species, from one spawning event to another over the lifetime of an individual, and among gametes in a single clutch. At the extreme, eggs released simultaneously from a single female may drift apart, such that some eggs experience intense sperm competition while others – because of the chaotic nature of water turbulence – miss a patch of sperm entirely. Thus, in a single spawning bout, both blocks to polyspermy and sperm chemotaxis could provide useful adaptations increasing reproductive success (Levitan 1995a).

Because sperm are more abundant than eggs, sperm competition

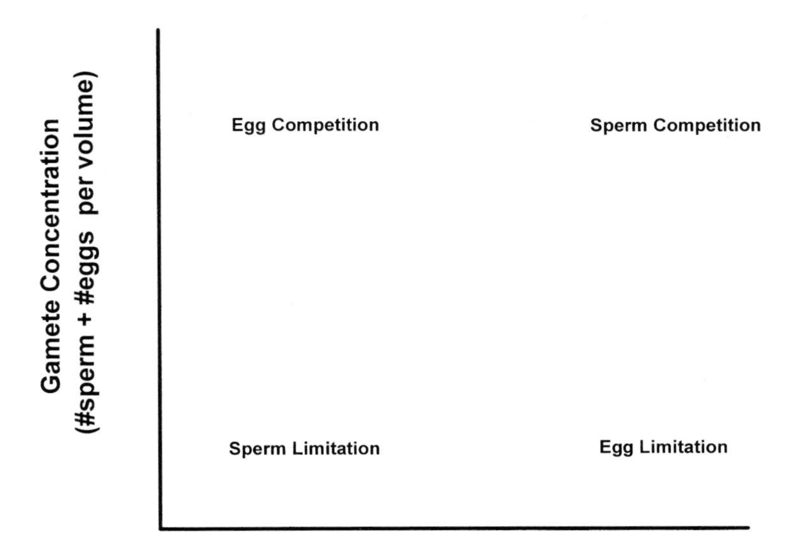

Fig. 6.3. *Gradient of relative and absolute gamete concentrations. There is a need for data from which to create a contour plot of the frequency distribution of absolute and relative gamete concentration during natural spawning events. These data will provide information on the intensity and symmetry of sexual selection in externally fertilizing species.*

should be more common than egg competition, but aspects of spawning behaviour and the nature of spawned materials may influence the relative dispersal of eggs and sperm. For example, if large numbers of eggs tend to remain clumped on or near the spawning female (McEuen 1988; Smith 1989; Thomas 1994a,b) and wisps of sperm occasionally pass by, then egg competition (certainly within a female and potentially among females) could occur. Once gametes have dispersed away from the spawning adults and gamete concentrations become exponentially reduced, then gamete competition will diminish and gamete interactions become rare.

Whether the present lack of field evidence for egg competition in broadcast spawners is due to our investigative biases or the general lack of information on gamete competition in broadcast spawners, or simply reflects the reality that eggs do not compete has yet to be determined. Laboratory studies on sea urchins have documented that, in a vial, larger eggs are preferentially fertilized over smaller eggs when sperm are limiting, so egg competition may exert a strong selective influence on egg size (Levitan 1996a). An important step in assessing gamete competition in broadcast spawners will be construction of a frequency distribution of gamete distributions on the gradients illustrated in Fig. 6.3.

There is very little information on how variability in fertilization success in males compares with that in females. Because each zygote is the product of one sperm and one egg, the average number of zygotes produced will be identical across sexes. What is less obvious is the distribution of fertilizations among individual males and females.

The key to determining the distribution of fertilization among sexes is the use of genetic markers to estimate parentage. In broadcast spawners the problem of identifying parentage is much more difficult because, in many cases, neither the male nor the female parent is known. In such cases, determining parentage by exclusion is not a very efficient mechanism because a marker must be absent in all the potential parents of one sex before it can be used to exclude parents of the other. The use of rare markers for inclusion or statistical clustering of highly polymorphic markers is preferable (Levitan and Grosberg 1993). Although such parentage assignments have been made in broadcast spawners in controlled experiments in the laboratory (Levitan and Grosberg 1993), the only field test has involved egg-laying horseshoe crabs (Brockmann *et al.* 1994; as discussed in Section V).

In situations or taxa where maternity can be observed with confidence, the problem of assigning paternity is more tractable. Genetic markers have been successfully used to determine paternity in free-spawning brooding ascidians (Grosberg 1991; Yund and McCartney 1994; Yund 1995) and bryozoans (Yund and McCartney 1994). Only one of these studies (Grosberg 1991) used a natural population and, in that case, the dispersal of a rare allele from a focal sperm source was investigated rather than comparisons of paternity among males. In Grosberg's study, the density of animals was high enough that variation in female fertilization success caused by sperm limitation was assumed to be negligible (R. K. Grosberg, personal communication). In the other ascidian studies (Yund and McCartney 1994; Yund 1995), variation in both male and female fertilization was noted, and the distribution was attributed to both body size (sperm production) and positional effects (male–female distance). In these experimental studies, however, the positions of individuals were assigned and isolated from natural populations, so the overall levels of natural variation have yet to be determined. In the bryozoan study (Yund and McCartney 1994), variation in fertilization success was confounded by the animals' ability to self-fertilize.

V. SEXUAL SELECTION AND SEXUAL DIMORPHISM

The notion that sperm and egg competition should be viewed as a continuum has been expressed in another way by Arnold (1994b). He suggests that the intensity of sexual selection can be expressed by the steepness of the slope between mating success and fecundity and that

Bateman's experiments highlight one pair of slopes (linear for males and single-mate saturation for females) out of a range of possibilities in which sexual selection can be intense for either males or females. One appealing aspect of this approach is that it provides a mechanism for comparing the shape of the sexual-selection relationship between sexes and matching these differences with patterns of sexual dimorphism. Sexual selection can be intense, but if the curves are similar across sexes then sexual dimorphism in, for example, energy allocation or body size may not evolve. Arnold (1994b) predicted a curve of diminishing returns in functionally male plants for which mate distance and pollination success are related. Because mate distance seems to be critically important in broadcast spawners (see Section III.B above), a similar curve for diminishing returns seems appropriate for both male and female spawners and has been observed in experimental studies (Levitan 1991). If the curves are similar over the range in mating success typically experienced in nature, then a reduction in sexual dimorphism is predicted. This would be the case if sperm limitation is common.

Plots of reproductive success as a function of investment in gametes and as a function of mating success provide a tool for understanding the relationship between gamete interactions, investment, and sexual dimorphism (Fig. 6.4). Equal investment in gametes across sexes suggests that the return on reproductive investment is similar across sexes, because of either ubiquitous sperm limitation or intense sperm competition (Levitan and Petersen 1995). In both cases a linear return on investment is predicted for both sexes; if twice as many eggs (or sperm) are produced, then twice as many zygotes should be produced relative to other individuals (Fig. 6.4a,b). If sperm competition or limitation is less important, then diminishing returns on investment are predicted for males but not for females because sperm typically greatly outnumber eggs (Charnov 1982; Petersen 1991).

The distinction between sperm competition and sperm limitation can be illuminated by the relationship between mating success and reproductive success (Fig. 6.4c,d). Sperm limitation results in increased offspring production with increased mating success for both males and females equally. For example, if a spawning male encounters double the number of spawning females, he doubles the total number of zygotes produced (e.g. 10% of eggs fertilized from twice as many eggs). Similarly, if a female encounters double the number of spawning males, she doubles the chances that sperm will fertilize her eggs. Males and females gain equally from increases in mating success, provided that sperm are limiting and competition is absent. Sperm competition results in sexual asymmetries in fecundity gains with increasing mating success. When all of a female's eggs are fertilized, then female fecundity is at a maximum, but a male's fecundity will continue to increase with mating success until all his sperm form zygotes (an unlikely event).

Sperm competition should result in sexual dimorphism in nongonadal traits (e.g. body size), as male–male competition and female choice

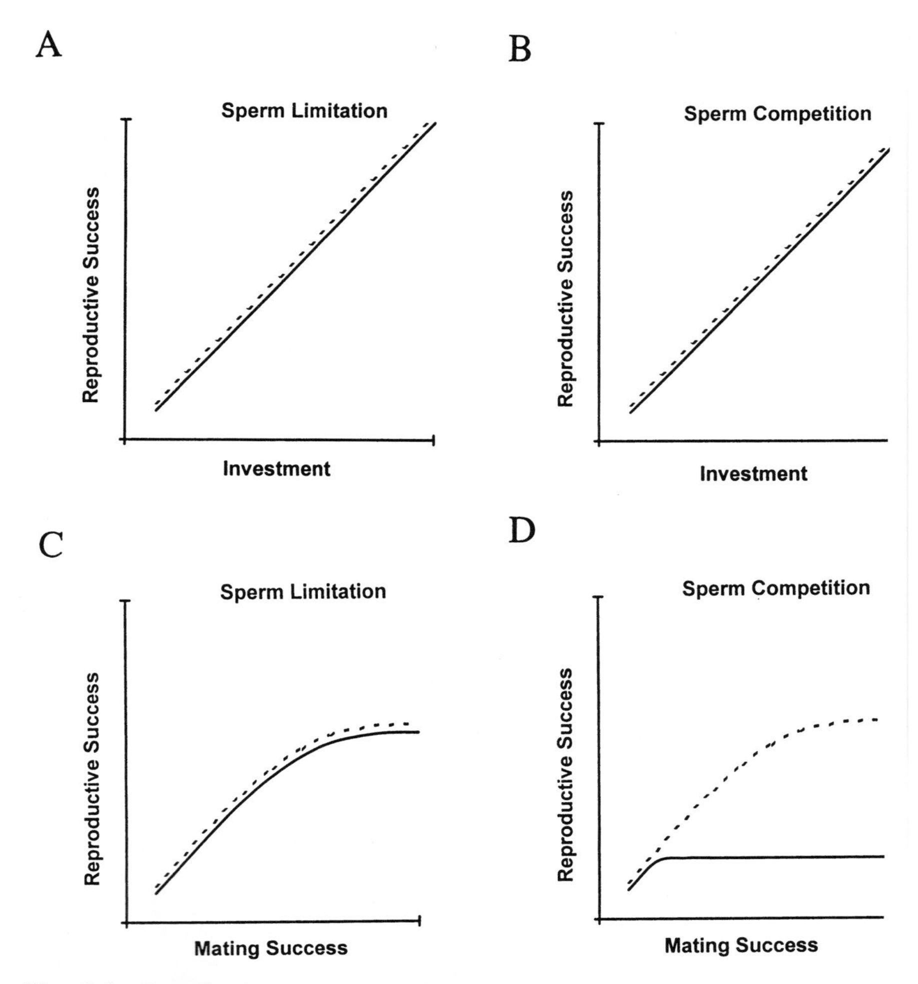

Fig. 6.4. *Reproductive success as a function of gamete investment (A and B) and mating success (C and D). Females are represented by the solid line and males by the dotted line. Under conditions of either sperm limitation or sperm competition, a linear relationship between gamete investment and reproductive success is predicted. In contrast, when reproductive success is a function of mating success ('the number of mates that bear or sire progeny over some standardized time interval', Arnold 1994a, p. S9) the response curves are predicted to be identical when sperm are limiting (C) but to differ when sperm compete in the same water mass for virgin eggs (D).*

become increasingly important. In contrast, sperm limitation should result in reduced sexual dimorphism, as sexual selection is equally intense across sexes.

However, because the intensity of sexual selection decreases to zero when reproductive success is no longer limited by fertilization, and

Fig. 6.5. *Reproductive success as a function of potential mating success. Females are represented by the solid line and males by the dotted line. If the actual range of mating success is below the threshold at which the male and female curves diverge, then sperm is limiting, the sexes experience equally intense sexual selection, and sexual dimorphism should be reduced. If the actual range in mating success is above the threshold at which male and female curves diverge, then sperm is not limiting, sperm compete, sexual selection is more intense for males than for females, and sexual dimorphism should be enhanced. If the range in mating success spans the threshold, intermediate levels of sperm competition and sexual dimorphism are predicted.*

because males produce many more gametes than females do, it is unlikely that sexual selection will be equally intense over the entire range in potential mating success. In other words, the level of mating success needed to ensure that 100% of a female's eggs are fertilized should be much lower than the level of mating success needed to ensure that 100% of a male's sperm fertilize eggs. The important criterion is how the relationship between potential mating success and reproductive success relates to the actual range in mating success realized in nature. If the actual range in mating success is below the threshold at which the intensity of male and female sexual selection diverge, then sexual dimorphism should be lower than in situations in which the range in mating success is above this threshold (Fig. 6.5).

These graphical models (Fig. 6.4a,b) predict that, if either sperm competition or sperm limitation is the dominant selective regime in broadcast spawners, then the two sexes' investments in gametes should be identical. This prediction is consistent with available data. Sexual dimorphism in gonadal indices appears to be absent in broadcast spawners. Gonadal indices should be used with caution if there is an allometric relationship between body size and gonad size. However, because sexual dimorphism in body size is rare or absent in broadcast spawners (Table 6.3, p. 197), comparisons of gonadal indices between sexes within a species should be less problematic. Among the echinoderms, which generally broadcast spawn, no difference in gonadal indices were noted in any class (Table 6.4, p. 200). Because eggs and sperm have different functions, it is not surprising that the composition of gonads differs between the sexes in the asteroids and echinoids examined (Table 6.5, p. 202). Males have higher ash and protein components, females have higher lipid levels, and carbohydrate levels seem equivalent across sexes. Among the free-spawning mollusc classes, gonadal indices appear to be similar in species tested in the Polyplacophora (Pearse 1979) and Pelecypoda (Sastry 1979; Newell and Bayne 1980; Choi *et al.* 1993). In one case, males had higher gonadal indices in one out of two populations of mussels (Newell *et al.* 1982). These data, which indicate equal investment in gametes in the two sexes, cannot be used to differentiate between sperm limitation and sperm competition but can be used as evidence that noncompetitive pair spawning, with high levels of female fertilization, is uncommon in these taxa.

The graphical models (Fig. 6.4c,d) also predict that, if sperm limitation is the dominant selective regime in broadcast spawners, then reductions in sexual dimorphism in nongonadal traits such as body size should be evident. Alternately, if sperm competition is common in broadcast spawners, then sexual dimorphism in nongonadal traits should be evident as male competition and female choice become divergent selective forces. The evidence matches the hypothesis that sperm limitation is the common selective environment in broadcast spawners. A correlation between external fertilization and reduced sexual dimorphism has often been stated without presenting data or a citation (e.g. Parker 1984; Strathmann 1990). A glance at Table 6.3 may indicate why – the pattern is striking. When both eggs and sperm are released, sexual dimorphism of any kind is overwhelmingly absent. In groups where males free spawn but females brood eggs, sexual dimorphism is also reduced and restricted to anatomical modifications associated with brooding. Size dimorphisms are common when fertilization is internal.

Perhaps most convincing are taxa that generally have one mode of reproduction with exceptions. In the Echiura, most free spawn, and no sexual dimorphism has been noted, but in the family Bonellidae, size dimorphism is extreme (males are dwarf) and fertilization is internal. In the polychaetes, although free spawning is common, species that copulate can have dwarf males that have been described as 'no more than a swimming penis' (Schroeder and Hermans 1975). Echinoderms typically free spawn, but the protandric sea cucumber *Leptosynapta clarki* has internal fertilization and small males (Sewell 1994), and concentricycloid males tend to be smaller and to have anatomical modifications for pseudocopulation (Rowe *et al.* 1991). Gastropod molluscs generally have internal fertilization, but the archaeogastropods tend to free spawn and to have reduced sexual dimorphism (Webber 1977).

Potential exceptions to this correlation between mode of reproduction and sexual dimorphism include a few species of pelecypods with slight shell differences, a few species of brittle stars (echinoderms), the males of which

Table 6.3. Patterns of gamete release and nongametic sexual dimorphism in invertebrates. In some cases, differences in the anatomy or colour of the reproductive structures were noted. These instances are reported as 'no external dimorphism'. The bottom of the table reports a summary of these patterns.

Taxa	Male	Female	Sexual dimorphism	Reference
Porifera	Free spawn	Broadcast/brood	None	Fell (1974)
Cnidaria	Free spawn	Broadcast/brood	None	Cambell (1974)
Ctenophora	Free spawn	Most broadcast/ some brood	All hermaphrodites	Pianka (1974)
Platyhelminthes				
Turbellaria	Copulate	Brood	Dwarf males in dioecious species	Henley (1974)
Gnathostomulida	Copulate	Brood	All hermaphrodites	Sterrer (1974)
Nemertinea	Free spawn/ pseudo-copulate	Broadcast/brood	Males smaller in some species that pseudo-copulate	Riser (1974)
Nematoda	Copulate	Brood	Yes, in dioecious species	Hope (1974)
Rotifera	Copulate	Brood	Males smaller in dioecious species	Thane (1974)
Gastrotricha	Copulate	Brood	All hermaphrodites	Hummon (1974)
Kinorhyncha	Copulation (assumed)	Brood	Some morphological differences	Higgins (1974)
Entoprocta	Free spawn	Brood	No consistent evidence for dimorphism	Mariscal (1975)
Tardigrada	Copulate/ pseudo-copulate	Brood	Males often smaller	Pollock (1975)
Priapulida	Free spawn	Broadcast	No external dimorphism	van der Land (1975)
Sipuncula	Free spawn	Broadcast	None	Rice (1975)
Pogonophora	Spermato-phore	Brood	Males can be smaller	Southward (1975)
Echiura	Free spawn/ copulate	Broadcast/brood	In family Bonellidae fertilization is internal with dwarf males; in other groups fertilization is external, with no size dimorphism	Gould-Somero (1975)
Annelida				
Polychaeta	Free spawn/ copulate	Broadcast/brood	Species that copulate often have dwarf males; otherwise no dimorphism	Schroeder and Hermans (1975)
Clitellata	Copulate	Brood	All hermaphrodites	Lasserre (1975)
Phoronida	Free spawn	Broadcast/brood	?	Zimmer (1991)
Bryozoa	Free spawn	Broadcast/brood	All hermaphrodites	Reed (1991)
Brachiopoda				
Articulata	Free spawn	Broadcast/brood	None (except brood chambers) majority no dimorphism	Long and Strickler (1991)

Continued

Table 6.3. Continued

Taxa	Male	Female	Sexual dimorphism	Reference
Mollusca				
Aplacophora	Free spawn	Assumed broadcast	None in dioecious species	Hadfield (1979)
Monoplacophora	Free spawn	Broadcast	None	Gonor (1979)
Polyplacophora	Free spawn	Broadcast	No external dimorphism	Pearse (1979)
Pelecypoda	Free spawn	Broadcast/brood	Rare shell differences, vast majority no external dimorphism	Sastry (1979)
Gastropoda	Most copulate/few free spawn	Most brood/few broadcast	Archaeogastropods tend to free spawn and to have little dimorphism; the other groups tend to copulate and to have size dimorphisms	Beeman (1977), Berry (1977), Webber (1977)
Scaphopoda	Free spawn	Broadcast	No external dimorphism	McFadien-Carter (1979)
Cephalopoda	Copulate	Brood	Size dimorphisms noted	Haven (1977), Arnold and Williams-Arnold (1977), Wells and Wells (1977), Charniaux-Cotton *et al.* (1992)
Arthopoda				
Crustacea	Copulate	Brood	Size and appendage dimorphisms	Giese and Kanatani (1987, 1992)
Insecta	Copulate, spermatophore	Brood	Primary and secondary sexual characters commonly dimorphic	Giese and Kanatani (1987), Gillot *et al.* (1992)
Merostomata	Free spawn (close pairing)	Broadcast	Males smaller and appendage dimorphism	Giese and Kanatani (1987)
Pycnogonida	Free spawn (close pairing)	Brood	?	Giese and Kanatani (1987)
Echinodermata				
Asteroida	Free spawn (pairing rarely)	Broadcast/brood	None (except brood pouches)	Chia and Walker (1991)
Ophiuroida	Free spawn (pairing noted)	Broadcast/brood	Some males smaller in rare pairing species	Hendler (1991)
Echinoidea	Free spawn (pairing rarely)	Most broadcast/some brood	None (except brood pouches)	Pearse and Cameron (1991)

Table 6.3. Continued

Taxa	Male	Female	Sexual dimorphism	Reference
Holothuroidea	Free spawn (pairing rarely)	Broadcast/brood	None (except brood pouches)	Smiley *et al.* (1991)
Crinoidea	Free spawn	Broadcast/brood	None (except brood pouches)	Holland (1991)
Concentricycloidea	Pseudo-copulate (assumed)	Brood (assumed)	Males often smaller	Rowe *et al.* (1991)
Chaetognatha	Pseudo-copulate	Brood	All hermaphrodites	Reeve and Cosper (1975)
Hemichordata				
Enteropneusta	Free spawn	Broadcast	None	Hadfield (1975)
Pterobrancha	Free spawn	?	None	Hadfield (1975)
Chordata				
Tunicata	Free spawn	Broadcast/brood	All hermaphrodites	Berrill (1975)
Acrania	Free spawn	Broadcast	None	Wickstead (1975)

Summary of sexual dimorphism and reproductive mode

	Sexual dimorphism		
Reproductive mode	Absent	Rare	Common
Free spawning and brood	2	1	5[a]
Free spawning and broadcast	20	2	1
Pseudocopulation and brood	0	0	4
Copulation and brood	0	0	11

[a] Only dimorphism is the presence of brood chambers.

are often smaller, and horseshoe crabs (arthropods), the males of which are generally smaller. Size dimorphism in brittle stars appears to be associated with male–female pairing. Observations of pairing appear to be associated with dwarf males, although there are few data (Hendler 1991). When males pair and attach to females, the likelihood of sperm limitation decreases and sexual dimorphism would be more likely (i.e. the sexes will differ in the costs and benefits of increased gamete production).

Horseshoe crabs characteristically spawn while clasped in pairs. This behaviour results in relatively high female fertilization success (R. Loveland and M. Botton, unpublished data; Table 6.1) and increases the possibility of male–male competition. Competition and multiple paternity among male horseshoe crabs are common (Brockmann *et al.* 1994). Male–female pairs climb up the shore and spawn in the presence of satellite males that, on average, sire 40% of progeny (Brockmann *et al.* 1994), an average higher than documented for fish competition. The high rate of satellite-male success is correlated with the position of the satellite male. Satellite males in the optimal position (anterior margin of

Table 6.4. Gonadal indices by sex in echinoderms. Publications prior to 1987 compiled by Lawrence and Lane (1982) and Lawrence (1987). Gonadal index is the gonadal wet weight divided by the body wet weight. Because authors often multiply this value by various constants, comparisons of indices across taxa may not be appropriate.

Taxa	Male	Female	Reference
Asteroidea			
Acanthaster planci	12	17	Conand (1985)
Asterias amurensis	14	17	Kim (1968)
Asterias rubens	17	24	Kowalski (1955); von Bismark (1959), Jangoux and Vloebergh (1973)
Asterias vulgaris	5	10	Lowe (1978)
Astropecten latespinosus	35	35	Nojima (1979)
Astrotole scabra	19	19	Town (1980)
Echinaster echinophorus	22	12	Ferguson (1974)
Echinaster sp.	30	12	Ferguson (1975)
Echinaster type I	11.8	17.5	Scheibling and Lawrence (1982)
Echinaster type II	25	27.7	Scheibling and Lawrence (1982)
Leptasterias hexactis	19	5	Menge (1975)
Leptasterias pusilla	270	300	Smith (1971)
Luidia clathrata	No difference	(6)	Lawrence (1973)
Oreaster hedemanni	3.8	3.4	Rao (1965)
Oreaster reticulatus	8	16	Scheibling (1979)
Patiriella gunnii	No difference	(13.2)	Byrne (1992)
Patiriella calcar	No difference	(12.5)	Byrne (1992)
Patiriella exigua	4.1	8.6	Byrne (1992)
Patiriella pseudoexigua	9.5	12.5	Chen and Chen (1992)
Pisaster giganteus	5	3	Farmanfarmaian *et al.* (1958)
Pisaster brevispinus	8	6	Farmanfarmaian *et al.* (1958)
Pisaster ochraceus	12	17	Farmanfarmaian *et al.* (1958)
Solaster stimpsoni	28	18	Engstrom (1974)
Mean	25.6	26.6	
Standard error	11.3	12.5	
Echinoidea			
Diadema setosum	30	40	Kobayashi and Nakamura (1967)
Echinarachius parma	11	15	Cocanour and Allen (1967)
Echinocardium cordatum	0.06	0.11	Moore (1936)
Echinometra lucunter	22	40	McPherson (1969)
Echinus esculentus	2	1.5	Moore (1934)
Eucidaris tribuloides	10.5	9.5	McPherson (1968a,b)

Table 6.4. Continued

Taxa	Male	Female	Reference
Evechinus chloroticus	11.8	2	Dix (1970)
Heliocidaris erythrogramma	9.9	9.1	Lawrence and Byrne (1994)
Heliocidaris tuberculata	13.2	13.0	O'Conner *et al.* (1976), Lawrence and Byrne (1994)
Hygrosoma petersii	32.3	31.8	Ahlfield (1977)
Lytechinus variegatus	1.6	1.8	Moore *et al.* (1963), Moore and McPherson (1965)
Meoma ventricosa	0.53	0.58	Chesher (1969)
Mespilia globulus	0.40	0.61	Kobayashi (1967)
Moira atropos	0.4	0.6	Moore and Lopez (1966)
Paracentrotus lividis	8.9	9.1	Byrne (1990)
Strongylocentrotus franciscanus	8	8	Bennett and Giese (1955)
Strongylocentrotus nudus	30	30	Fuji (1960a)
Strongylocentrotus intermedius	27.6	29.2	Fuji (1960a,b)
Strongylocentrotus purpuratus	20.8	19.8	Bennett and Giese (1955), Giese *et al.* (1958)
Tripneustes gratilla	14.5	15.5	O'Conner *et al.* (1976)
Tripneustes ventricosus	1.2	1.2	Moore *et al.* (1963)
Mean	12.2	13.4	
Standard error	2.4	2.9	
Holothuroidea			
Actinopyga echinites	6.1	7.8	Conand (1982)
Aslia lefevrei	17.5	18.0	Costelloe (1985)
Cucumaria lubrica	52	35	Engstrom (1974)
Holothuria mexicana	10	15	Engstrom (1980)
Holothuria scabra	0.1	0.35	Krishnan (1967)
Holothuria floridana	9	12	Engstrom (1980)
Microthele fuscogilva	0.8	2.4	Conand (1981)
Microthele nobilis	2.9	5	Conand (1981)
Stichopus japonicus	12	18	Choe (1962)
Thelenota ananas	1.1	1.6	Conand (1981)
Mean	11.2	11.5	
Standard error	4.9	3.4	
Ophiuroidea			
Amphioplus abditus	0.5	0.5	Hendler (1973)
Amphioplus sepultus	0.3	0.2	Hendler (1973)
Bathypectinura heros	10.3	9.0	Ahlfield (1977)
Ophiomusium lymani	6.4	6.7	Ahlfield (1977)
Ophiomusium spinigerum	13.9	11.0	Ahlfield (1977)
Mean	6.3	5.5	
Standard error	2.7	2.2	

Table 6.5. Proximate composition (per cent dry weight) of gonads in sea stars and sea urchins.

Species	Ash		Carbo-hydrate		Soluble Lipid		Soluble Protein		Insoluble Protein		Reference
	M	F	M	F	M	F	M	F	M	F	
Asteroids											
Anthenoides piercei	8.9	4.0	1.0	1.2	20.9	25.9	28.0	25.4	41.2	43.5	McClintock *et al.* (1995)
Tosia parva	18.9	5.3	0.5	0.5	17.6	54.7	19.3	19.2	43.7	20.3	McClintock *et al.* (1995)
Echinoids											
Heliocidaris erythrogramma	9	4	11	8	26	50	37	29	11	9	Lawrence and Byrne (1994)
Heliocidaris tuberculata	9	8	11	12	22	27	42	45	2	9	Lawrence and Byrne (1994)
Mean	11.5	5.3	5.9	5.4	21.6	39.4	31.6	29.6	24.5	20.4	

satellite male underneath anterior margin of clasping male) average 49% of sired progeny, compared with only 7% for males in other positions (Brockmann *et al.* 1994). The greater success of satellite horseshoe crabs than of fish may be a result of morphological or behavioural constraints that prevent the clasping male from sequestering the female or her eggs during their protracted spawning bouts.

Although sexual dimorphism may evolve from reproductive constraints other than sexual selection, the combined empirical evidence of sperm limitation in general (Tables 6.1 and 6.3), the evidence that both male and female fertilization success increases with the number of spawning animals (mating success) (Levitan 1991; Levitan *et al.* 1992; Yund and McCartney 1994; Yund 1995), and the reduced sexual dimorphism in both gonadal (Table 6.3) and nongonadal (Table 6.4) traits suggest that sexual selection is intense, but similar across sexes, in external fertilizers and that the mechanism driving sexual selection is mutual fertilization limitation rather than sperm competition.

VI. SPERM LIMITATION, SELECTION ON EGG SIZE, AND THE EVOLUTION OF ANISOGAMY

Parker *et al.* (1972) developed a model for the evolution of anisogamy based on sperm competition. They assumed that, in free-spawning organisms, 'sperm competition is rampant because all ejaculates must compete in the same external medium for fusions with ova' (Parker 1984, p. 6).

Shortly after Parker's (1984) review was published, Pennington (1985) published the first *in situ* experiment and measurement of external fertilization, demonstrating the potential for sperm limitation. Since then, numerous studies have been conducted, and all have demonstrated some degree of sperm limitation (Tables 6.1 and 6.2; review by Levitan 1995a). If ubiquitous sperm limitation in external fertilizers results in reduced sexual dimorphism, and external fertilization is an ancestral trait, how did the original morphological sexual dimorphism, anisogamy, evolve from an isogamous ancestor?

Incorporating the emerging information on external fertilization into the Parker *et al.* model requires two considerations (Levitan 1996a). First, sperm–sperm interactions (sperm competition for fertilization) are likely to be low, so although sperm may still compete in the Darwinian sense (males with more fertilizations will have greater fitness), racing, battling, or preventing fertilizations by other sperm and the notion of female choice may have reduced importance compared with sperm limitation.

Second, because sperm are limiting, selection for enhanced fertilization can act on females as well as on males (Levitan 1996a). This departure from Bateman's principle and the evidence that egg size influences the proportion of eggs fertilized (Levitan 1993, 1996a) suggest that selection for fewer, larger eggs is a function not only of postzygotic success, as argued by Parker *et al.* (1972), but also of enhanced fertilization success (Levitan 1993, 1996a,b).

The notion that fertilization rate can influence selection on gamete size is not a recent one. Kalmus (1932) and Scudo (1967) presented models for how gamete encounter rate can result in selection for anisogamy using group-selection arguments. Schuster and Sigmund (1982) developed a model indicating that collisions between gametes become more likely as size asymmetries increase.

Because females produce eggs, the effect of selection on egg size must be viewed from the maternal perspective. Optimal egg size is the one that maximizes maternal fitness, by balancing the number and fitness offspring (Vance 1973; Smith and Fretwell 1974). When sperm are limiting, fertilization success is an important component of egg fitness (Levitan 1993, 1996a,b).

A model for optimal egg size that incorporates both pre- and post-zygotic factors has been constructed for free-spawning echinoids (Levitan 1996a,b). In echinoids, size at metamorphosis tends to be similar across taxa. In the three *Strongylocentrotus* species mentioned earlier (*S. purpuratus*, *S. franciscanus*, and *S. droebachiensis*), although there is a fivefold difference in egg volume (Emlet *et al.* 1987), size at settlement for all three species is 0.20 mm (Emlet *et al.* 1987; Sinervo and McEdward 1988). This pattern suggests that selection for variation in egg size is likely to occur either pre- or post-zygotically but before settlement, during the larval, planktonic phase.

In the model, egg number is estimated to be the inverse of egg volume, and the total amount of egg material is assumed to be constant. The proportion of eggs fertilized (ϕ_∞) is estimated by a fertilization-kinetics model (Vogel *et al.* 1982) which incorporates the sperm–egg collision rate (b_O, $mm^3\ s^{-1}$; the product of egg cross-sectional area and sperm velocity), egg (E_O, eggs ml^{-1}) and sperm (S_O, sperm ml^{-1}) concentration, the sperm–egg contact time (t) and a fertilization constant (b; $mm^3\ s^{-1}$), which, when divided by the collision rate, provides the proportion of sperm–egg collisions that result in fertilization.

$$\phi_\infty = 1 - \exp(-\ bS_O/b_OE_O(1 - e^{-b_OE_Ot})). \tag{1}$$

The number of settling individuals (N_s) is calculated from the number of fertilized eggs (N_e), the instantaneous mortality rate (m), and the development time in the plankton (dt), where development time is a function of egg size (Vance 1973).

$$N_s = N_e e^{(-dtm)}. \tag{2}$$

The average empirical estimate of the daily planktonic mortality rate for the *Strongylocentrotus* species is 0.1615 day^{-1} (range = 0.06–0.27; Rumrill 1990). The relationship between development time and egg size is calculated to be: time (days) = 18.987 (egg volume$[mm^3])^{-0.1156}$ (Levitan 1996b).

Predictions of the egg size that maximizes parental fitness vary as a function of ambient sperm concentration and planktonic mortality (Fig. 6.6). This model predicts how sperm limitation results in selection for larger eggs than would be predicted by models that incorporate postzygotic survivorship alone.

The implication for the evolution of anisogamy is that conditions of sperm limitation resulted in selection for numerous smaller sperm that had an increased probability of finding an egg in a diffuse medium. Sperm limitation, along with factors associated with postzygotic survivorship, influenced selection on females for enlarged eggs. This modification of Parker *et al.*'s 1972 model assumes that sperm limitation, rather than sperm competition, along with selection for postzygotic success, was the mechanism driving the evolution of anisogamy (Levitan 1996a).

Taken at face value, and on the assumption that isogamy and external fertilization are the ancestral state, this result implies a modification in the sequence of events resulting in the evolution of sexual dimorphism and sexual selection. Protists that became colonial and larger began to experience increased selective pressure for increased zygote size to reduce the risks associated with growth to maturity (Parker *et al.* 1972; Knowlton 1974). Once the advantages of large zygote size became established, fertilization limitation would select for increased sperm numbers (at a cost of reduced size) and egg size (to a size larger than predicted solely by postfertilization survivorship), enhancing collision frequency and resulting in anisogamy. In addition, fertilization limitation would also select for behavioural modifications for increased aggregation and synchrony and for the maintenance of reduced sexual dimorphism with equal investment in gonads. Close-pair spawning, pseudocopulation, and copu-

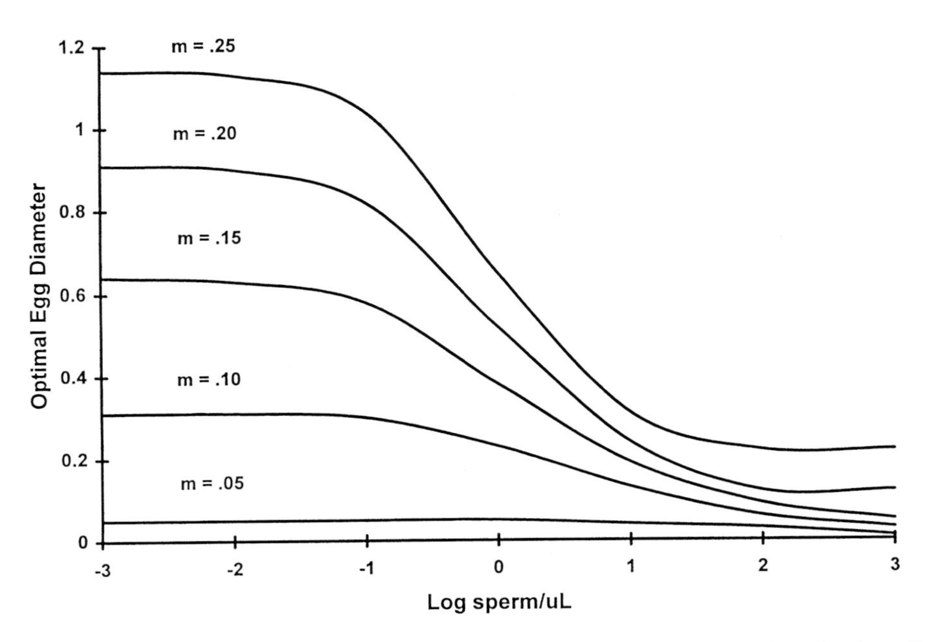

Fig. 6.6. *The optimal egg size that maximizes maternal fitness is predicted to be a function of sperm concentration and the daily larval mortality rate (m) in free-spawning echinoids. See text for model. The total amount of egg material is 1 ml, egg concentration is 0.01* μl^{-1}*, sperm–egg interaction time is 600 s, sperm velocity is 0.130 mm* s^{-1} *(Levitan 1993),* b_0 *is 0.0 000 952* $mm^3\ s^{-1}$ *(Levitan 1993). The lines represent egg sizes that maximize the number of settling offspring for a given sperm mortality rate and sperm concentration. Figure from Levitan (1996a).*

lation evolved as escapes, at least at first, from sperm limitation (rather than from sperm competition; Parker 1984). With the evolution of internal fertilization and brooding of postzygotic stages by females, two changes in the selective environment occurred. First, variation in female fertilization success drastically decreased, whereas male fertilization success remained highly variable (or increased in variability because single males could more easily monopolize females), setting the stage for sexual asymmetries in the intensity of sexual selection and male–male competition (including its special case, sperm competition). Second, females began holding and controlling the fate of the offspring, setting the stage for female choice. Sexual dimorphism then becomes a prominent feature of internally fertilizing taxa.

In this scenario, internal fertilization is selectively advantageous to both males and females. Males benefit because the average success of males increases, and males best at copulation gain greatly. Females also benefit greatly because fertilization ceases to be a limiting factor, and choice becomes an option. Because significant anatomical change in females is associated with internal fertilization, it seems reasonable to assume that, for internal fertilization to evolve, it must benefit females.

Alternately, if sperm competition drove selection for internal fertilization, it is not clear how reducing sperm competition would benefit females (Knowlton and Greenwell 1984).

The sperm-limitation hypothesis implies not that sperm competition is absent among broadcast spawners but that it has reduced importance and may be balanced by egg competition. Choosing among this hypothesis, the more traditional view of the effect of sperm competition on sexual dimorphism, and some 'middle ground' will require much more new information on the distribution, abundance, and parentage of gametes in the sea (e.g. Figs 6.3 and 6.4), as well as accurate phylogenies to establish which traits are ancestral to others. Given such a wide array of taxa, environments, and strategies, studies of gamete interactions in external fertilizers have great potential. With the advent of molecular markers and DNA-amplification techniques, this potential can now be fulfilled.

ACKNOWLEDGEMENTS

I thank R. Babcock, T. Birkhead, G. Farley, M. Gage, N. Knowlton, T. McGovern, A. Møller, J. Pearse, J. Travis, and A. Thistle for making comments on the manuscript. This work is funded by the National Science Foundation and a visiting scientist award from the Bamfield Marine Station.

REFERENCES

Ahlfield TE (1977) A disparate seasonal study of reproduction of eight deep-sea macroinvertebrate species from the northwestern Atlantic Ocean. Ph.D. dissertation, Florida State University, Tallahassee, USA.

Amanze D & Iyengar A (1990) The micropyle: a sperm guidance system in the teleost fertilization. *Development* **109:** 495–500.

Arnold JM & Williams-Arnold LD (1977) Cephalopoda; Decapoda. In *Reproduction in Marine Invertebrates*, Vol. 4. AC Giese & JS Pearse (eds), pp. 243–289. Academic Press, New York.

Arnold SJ (1994a) Is there a unifying concept of sexual selection that applies to both plants and animals? *Am. Nat.* **144:** S1-S12.

Arnold SJ (1994b) Bateman's principles and the measurement of sexual selection in plants and animals. *Am. Nat.* **144:** S126-S149.

Babcock RC & Mundy CN (1992) Reproductive biology, spawning and field fertilization rates of *Acanthaster planci*. *Aust. J. Mar. Freshw. Res.* **43:** 525–534.

Babcock RC, Bull GD, Harrison PL, Heyward AJ. Oliver JK, Wallace CC & Willis

BL (1986) Synchronous spawnings of 105 scleractinian coral species on the Great Barrier Reef. *Mar. Biol.* **90:** 379–394.

Babcock R, Mundy C, Keesing J & Oliver J (1992) Predictable and unpredictable spawning events: *in situ* behavioural data from free-spawning coral reef invertebrates. *Invert. Reprod. Develop.* **22:** 213–228.

Babcock RC, Mundy CN & Whitehead D (1994) Sperm diffusion models and *in situ* confirmation of long distance fertilization in the free-spawning asteroid *Acanthaster planci*. *Biol. Bull.* **186:** 17–28.

Bateman AJ (1948) Intra-sexual selection in *Drosophila*. *Heredity* **2:** 349–368.

Battle H (1932) Rhythmic sexual maturity and spawning of certain bivalve molluscs. *Contrib. Can. Biol. Fish. NS* **7:** 255–276.

Beeman RD (1977) Gastropoda: Opisthobranchia. In *Reproduction in Marine Invertebrates*, Vol. 4. AC Giese & JS Pearse (eds), pp. 115–179. Academic Press, New York.

Bennett J & Giese AC (1955) The annual reproductive and nutritional cycles in two western sea urchins. *Biol. Bull.* **109:** 226–237.

Benzie JAH & Dixon P (1994) The effects of sperm concentration, sperm : egg ratio, and gamete age on fertilization success in the crown-of-thorns starfish (*Acanthaster planci*) in the laboratory. *Biol. Bull.* **186:** 139–152.

Benzie JAH, Black KP, Moran PJ & Dixon P (1994) Small scale dispersion of eggs and sperm of the crown-of-thorns starfish (*Acanthaster planci*) in a shallow reef habitat. *Biol. Bull.* **186:** 153–167.

Berrill NJ (1975) Chordata: Tunicata. In *Reproduction in Marine Invertebrates*, Vol. 2. AC Giese & JS Pearse (eds), pp. 241–282. Academic Press, New York.

Berry AJ (1977) Gastropoda: Pulmonata. In *Reproduction in Marine Invertebrates*, Vol. 4. AC Giese & JS Pearse (eds), pp. 181–225. Academic Press, New York.

Bolton TF & Havenhand JN (1996) Chemical mediation of sperm activity and longevity in the solitary ascidians *Ciona intestinalis* and *Ascidiella aspersa*. *Biol. Bull.* **190:** 329–335.

Branham JM (1972) Comparative fertility of gametes from six species of sea urchins. *Biol. Bull.* **142:** 385–396.

Brazeau DA & Lasker HR (1992) Reproductive success in the Caribbean octocoral *Briareum asbestinum*. *Mar. Biol.* **114:** 157–163.

Brockmann HJ, Colsen T & Potts W (1994) Sperm competition in horseshoe crabs (*Limulus polyphemus*). *Behav. Ecol. Sociobiol.* **35:** 153–160.

Byrne M (1990) Annual reproductive cycles of the commercial sea urchin *Paracentrotus lividus* from an exposed intertidal and sheltered subtidal habitat on the west coast of Ireland. *Mar. Biol.* **104:** 275–289.

Byrne M (1992) Reproduction of sympatric populations of *Patiriella gunnii*, *P. calcar* and *P. exigua* in New South Wales, asterinid seastars with direct development. *Mar. Biol.* **114:** 297–316.

Cambell RD (1974) Cnidaria. In *Reproduction in Marine Invertebrates*, Vol. 1. AC Giese, JS Pearse & VB Pearse (eds), pp. 133–199. Academic Press, New York.

Caspers H (1984) Spawning periodicity and habitat of the palolo worm *Eunice viridis* (Polychaeta: Eunicidae) in the Samoan Islands. *Mar. Biol.* **79:** 229–236.

Charniaux-Cotton H, Payen GG & Ginsburger-Vogel T (1992) Arthropoda-Crustacea: Sexual differentiation. In *Reproductive Biology of Invertebrates*, Vol. V. KG Adiyodi & RG Adiyodi (eds), pp. 281–324. Wiley, New York.

Charnov EL (1982) *The Theory of Sex Allocation*. Princeton University Press, Princeton.

Chen B-Y & Chen C-P (1992) Reproductive cycle, larval development, juvenile growth and population dynamics of *Patiriella pseudoexigua* (Echinodermata: Asteroidea) in Taiwan. *Mar. Biol.* **113:** 271–280.

Chesher RH (1969) Contributions to the biology of *Meoma ventricosa* (Echinoidea: Spatangoida) *Bull. Mar. Sci.* **19:** 72–110.

Chia F-S & Bickell LR (1983) Echinodermata. In *Reproductive Biology of Invertebrates*, Vol. II. KG Adiyodi & RG Adiyodi (eds), pp. 545–620. Wiley, New York.

Chia F-S & Walker CW (1991) Asteroidea. In *Reproduction in Marine Invertebrates*, Vol. 6. AC Giese, JS Pearse & VB Pearse (eds), pp. 301–353. Boxwood Press, Pacific Grove.

Choe S (1962) *Biology of the Japanese Common Sea Cucumber* Stichopus japonicus *Selenka*. Pussan Fisheries College, Pussan National University.

Choi KS, Lewis DH, Powell EN & Ray SM (1993) Quantitative measurement of reproductive output in the American oyster, *Crassostrea virginica* (Gmelin), using an enzyme-linked immunosorbent assay (ELISA). *Aquacult. Fish. Manag.* **24:** 299–322.

Chuang H (1959) The breeding season of the brachiopod *Lingula unguis* (L.). *Biol. Bull.* **117:** 202–207.

Cocanour BA & Allen K (1967) The breeding cycles of a sand dollar and a sea urchin. *Comp. Biochem. Physiol.* **20:** 327–331.

Coe WR (1947) Nutrition, growth and sexuality in the pismo clam (*Tivela stultorum*). *J. Exp. Zool.* **4:** 1–24.

Conand C (1981) Sexual cycle of three commercially important holothurian species (Echinodermata) from the lagoon of New Caledonia. *Bull. Mar. Sci.* **31:** 532–543.

Conand C (1982) Reproductive cycle and biometric relations in a population of *Actinopyga echinites* (Echinodermata: Holothuroidea) from the lagoon of New Caledonia, western tropical Pacific. In *Echinoderms, Proc. Int. Conf., Tampa Bay*. JM Lawrence (ed.), pp. 437–442. Balkema, Rotterdam.

Conand C (1985) Distribution, reproductive cycle and morphometric relationships of *Acanthaster planci* (Echinodermata: Asteroidea) in New Caledonia, western tropical Pacific. In *Echinodermata*. BF Keegan & BDS O'Conner (eds), pp. 499–506. Balkema, Rotterdam.

Costelloe J (1985) The annual reproductive cycle of the holothurian *Aslia lefevrei* (Dendrochirota: Echinodermata). *Mar. Biol.* **88:** 155–165.

Denny MW & Shibata MF (1989) Consequences of surf-zone turbulence for settlement and external fertilization. *Am. Nat.* **134:** 859–889.

Denny MW, Dairiki J & Distefano S (1992) Biological consequences of topography on wave-swept rocky shores: I. Enhancement of external fertilization. *Biol. Bull.* **183:** 220–232.

Dix TG (1970) Biology of *Evechinus chloroticus* (Echinoidea: Echinometridea) from different localities. 3. Reproduction. *NZ J. Mar. Freshw. Res.* **4:** 385–405.

Eckelbarger KJ, Young CM & Cameron JL (1989) Modified sperm in echinoderms from the bathyal and abyssal zones of the deep sea. In *Reproduction, Genetics and Distribution of Marine Organisms*. JS Ryland & PA Tyler (eds), pp. 67–74. Olsen and Olsen, Fredensborg.

Emlet RB, McEdward LR & Strathmann RR (1987) Echinoderm larval ecology viewed from the egg. In *Echinoderm Studies*, Vol. 2. M Jangoux & JM Lawrence (eds), pp. 55–136. Balkema, Rotterdam.

Engstrom NA (1974) Population dynamics and prey-predation relations of a dendrochirote holothurian, *Cucumaria lubrica*, and sea stars of the genus *Solaster*. Ph.D. dissertation, University of Washington, Seattle, USA.

Engstrom NA (1980) Reproductive cycles of *Holothuria (Halodeima) floridana H. (H.) mexicana* and their hybrids (Echinodermata: Holothuroidea) in southern Florida, USA. *Int. J. Invert. Reprod.* **2:** 237–244.

Epel D (1978) Mechanisms of activation of sperm and egg during fertilization of sea urchin gametes. *Curr. Top. Develop. Biol.* **12:** 185–246.

Epel D (1991) How successful is the fertilization process of the sea urchin egg? In *Proc. 7th Int. Echinoderm Conf., Atami, 1990.* T Yanagisawa, I Yasumasu, C Oguro, N Suzuki & T Matukawa (eds), pp. 51–54. Balkema, Rotterdam.

Farmanfarmaian A, Giese AC, Boolootian RA & Bennett J (1958) Annual reproductive cycles in four species of west coast starfishes. *J. Exp. Zool.* **138:** 355–367.

Fell PE (1974) Porifera. In *Reproduction in Marine Invertebrates*, Vol. 1. AC Giese, JS Pearse & VB Pearse (eds), pp. 51–132. Academic Press, New York.

Ferguson JC (1974) Growth and reproduction of *Echinaster echinophorus. Florida Sci.* **37:** 57–60.

Ferguson JC (1975) The role of free amino acids in nitrogen storage during the annual cycle in a starfish. *Comp. Biochem. Physiol.* **52A:** 341–350.

Franzén A (1987) Spermatogenesis. In *Reproduction in Marine Invertebrates*, Vol. 9. AC Giese, JS Pearse & VB Pearse (eds), pp. 1–47. Blackwell Scientific/Boxwood Press, Palo Alto/Pacific Grove.

Fuji A (1960a) Studies on the biology of the sea urchin. I. Superficial and histological changes in gametogenic process of two sea urchins, *Strongylocentrotus nudus* and *S. intermedius. Bull Fac. Fish. Hokk. Univ.* **11:** 1–14.

Fuji A (1960b) Studies on the biology of the sea urchin. III. Reproductive cycle of two sea urchins, *Strongylocentrotus nudus* and *S. intermedius*, in southern Hokkaido. *Bull. Fac. Fish. Hokk. Univ.* **11:** 49–57.

Giese AC & Kanatani H (1987) Maturation and spawning. In *Reproduction of Marine Invertebrates*, Vol. 9. AC Giese, JS Pearse & VB Pearse (eds), pp. 251–329. Blackwell Scientific/Boxwood Press, Palo Alto/Pacific Grove.

Giese AC, Greenfield L, Huang H, Farmanfarmaian A, Boolootian R & Lasker R (1958) Organic productivity in the reproductive cycle of the purple sea urchin. *Biol. Bull.* **116:** 49–58.

Gillot C, Mathad SB & Nair VSK (1992) Arthropoda-Insecta. In *Reproductive Biology of Invertebrates*, Vol. V. KG Adiyodi & RG Adiyodi (eds), pp. 345–400. Wiley, New York.

Gonor JJ (1979) Monoplacophora. In *Reproduction in Marine Invertebrates*, Vol. 5. AC Giese, JS Pearse & VB Pearse (eds), pp. 87–93. Academic Press, New York.

Gould-Somero M (1975) Echiura. In *Reproduction in Marine Invertebrates*, Vol. 3. AC Giese & JS Pearse (eds), p. 277. Academic Press, New York.

Gray J (1955) The movement of sea-urchin spermatozoa. *J. Exp. Biol.* **32:** 775–801.

Grosberg RK (1991) Sperm-mediated gene flow and the genetic structure of a population of the colonial ascidian *Botryllus schlosseri. Evolution* **45:** 130–142.

Gutsell JS (1930) Natural history of the bay scallop (*Pectin irradians*). *Bull. US Bur. Fish.* **46:** 569–632.

Hadfield MG (1975) Hemichordata. In *Reproduction in Marine Invertebrates*, Vol. 2. AC Giese & JS Pearse (eds), pp. 185–240. Academic Press, New York.

Levitan DR & Grosberg RK (1993) The analysis of paternity and maternity in the marine hydrozoan *Hydractinia symbiolongicarpus* using randomly amplified polymorphic DNA (RAPD) markers. *Molec. Ecol.* **2:** 315–326.

Levitan DR & Petersen C (1995) Sperm limitation in the sea. *Trends Ecol. Evol.* **10:** 228–231.

Levitan DR & Young CM (1995) Reproductive success in large populations: empirical measures and theoretical predictions of fertilization in the sea biscuit *Clypeaster rosaceus*. *J. Exp. Mar. Biol. Ecol.* **190:** 221–241.

Levitan DR, Sewell MA & Chia FS (1991) Kinetics of fertilization in the sea urchin *Strongylocentrotus franciscanus*: interaction of gamete dilution age and contact time. *Biol. Bull.* **181:** 371–378.

Levitan DR, Sewell MA & Chia FS (1992) How distribution and abundance influences fertilization success in the sea urchin *Strongylocentrotus franciscanus*. *Ecology* **73:** 248–254.

Long JA & Strickler SA (1991) Brachiopoda. In *Reproduction in Marine Invertebrates*, Vol. 6. AC Giese, JS Pearse & VB Pearse (eds), pp. 47–84. Boxwood Press, Pacific Grove.

Lowe EF (1978) Relationships between biochemical and caloric composition and reproduction cycle in *Asterias vulgaris* (Echinodermata: Asteroidea) from the Gulf of Maine. Ph.D. dissertation University of Maine, Orono, USA.

Mariscal RN (1975) Entoprocta. In *Reproduction in Marine Invertebrates*, Vol. 2. AC Giese & JS Pearse (eds), pp. 1–41. Academic Press, New York.

McClintock JB, Watts SA, Marion KR & Hopkins TS (1995) Gonadal cycle, gametogenesis and energy allocation in two sympatric mid shelf sea stars with contrasting modes of reproduction. *Bull. Mar. Sci.* **57:** 442–452.

McEuen FS (1988) Spawning behaviors of northeast Pacific sea cucumbers (Holothuroidea: Echinodermata). *Mar. Biol.* **98:** 565–585.

McFadien-Carter M (1979) Scaphopoda. In *Reproduction in Marine Invertebrates*, Vol. 5. AC Giese, JS Pearse & VB Pearse (eds), pp. 95–111. Academic Press, New York.

McPherson BF (1968a) Contributions to the biology of the sea urchin *Eucidaris tribuloides* (Lamarck). *Bull. Mar. Sci.* **18:** 400–443.

McPherson BF (1968b) The ecology of the tropical sea urchin *Eucidaris tribuloides*. Ph.D. dissertation, University of Miami, Coral Gables, Florida, USA.

McPherson BF (1969) Studies on the biology of the tropical sea urchins *Echinometra lucunter* and *Echinometra viridis*. *Bull. Mar. Sci.* **19:** 194–213.

Mead KS & Denny MW (1995) The effects of hydrodynamic shear stress on fertilization and early development of the purple sea urchin *Strongylocentrotus purpuratus*. *Biol. Bull.* **188:** 46–56.

Menge BA (1975) Brood or broadcast? The adaptive significance of different reproductive strategies in the two intertidal sea stars *Leptasterias hexactis* and *Pisaster ochraceus*. *Mar. Biol.* **131:** 87–100.

Miller RL (1985) Demonstration of sperm chemotaxis in Echinodermata: Asteroidea, Holothuroidea, Ophiuroidea. *J. Exp. Zool.* **234:** 383–414.

Minchin D (1987) Sea-water temperature and spawning behaviour in the seastar *Marthasterias glacialis*. *Mar. Biol.* **95:** 139–143.

Minchin D (1992) Multiple species, mass spawning events in an Irish sea lough: the effect of temperatures on spawning and recruitment of invertebrates. *Invert. Reprod. Develop.* **22:** 229–238.

Møller AP (1994) *Sexual Selection and the Barn Swallow*. Oxford University Press, Oxford, UK.

Moore HB (1934) A comparison of the biology of *Echinus esculentus* in different habitats. Part I. *J. Mar. Biol. Assoc. UK* **19:** 869–881.

Moore HB (1936) The biology of *Echinocardium cordatum*. *J. Mar. Biol. Assoc. UK* **20:** 655–672.

Moore HB & Lopez NN (1966) The ecology and productivity of *Moira atropos* (Lamark). *Bull. Mar. Sci.* **16:** 648–667.

Moore HB & McPherson BF (1965) A contribution to the productivity of the urchins *Tripneustes esculentus* and *Lytechinus variegatus*. *Bull. Mar. Sci.* **15:** 855–871.

Moore HB, Jutare T, Bauer JC & Jones JA (1963) The biology of *Lytechinus variegatus*. *Bull. Mar. Sci.* **13:** 23–53.

Newell RIE & Bayne BL (1980) Seasonal changes in the physiology, reproductive condition and carbohydrate content of the cockle *Cardium* (= *Cerastoderma*) *edule* (Bivalvia: Cardiidae). *Mar. Biol.* **56:** 11–19.

Newell RIE, Hilbish TJ, Koehn RK & Newell CJ (1982) Temporal variation in the reproductive cycle of *Mytilus edulis* (Bivalvia, Mytilidae) from localities on the east coast of the United States. *Biol. Bull.* **162:** 299–310.

Nojima S (1979) Ecological studies of the sea star, *Astropecten latespinosus* Meissner. I. Survivorship curve and life history. *Publ. Amakusa Mar. Biol. Lab. Kyushu Univ.* **5:** 45–65.

O'Conner C, Riley G & Bloom D (1976) Reproductive periodicities of the echinoids of the Solitary Islands in the light of some ecological variables. II. Superficial and histological changes in the gonads of *Centrostephanus rodgersii* (Clark), *Phyllacanthus parvispinus* (Tenison-Woods), *Heliocidaris tuberculata* (Clark), and *Tripneustes gratilla* (Linnaeus), and their relevance to aquaculture. *Thalas. Yugoslav.* **12:** 245–267.

Oliver J & Babcock R (1992) Aspects of the fertilization ecology of broadcast spawning corals: sperm dilution effects and *in situ* measurements of fertilization. *Biol. Bull.* **183:** 409–418.

Oliver J & Willis B (1987) Coral-spawn slicks in the Great Barrier Reef: preliminary observations. *Mar. Biol.* **94:** 521–529.

Parker GA (1984) Sperm competition and the evolution of animal mating strategies. In *Sperm Competition and the Evolution of Animal Mating Systems*. RL Smith (ed.), pp. 1–60. Academic Press, Orlando, FL.

Parker GA, Baker RR & Smith VGF (1972) The origin and evolution of gamete dimorphism and the male–female phenomenon. *J. Theor. Biol.* **36:** 529–553.

Pearse JS (1979) Polyplacophora. In *Reproduction in Marine Invertebrates*, Vol. 5. AC Giese, JS Pearse & VB Pearse (eds), pp. 27–85. Academic Press, New York.

Pearse JS & Cameron RA (1991) Echinodermata: Echinoidea. In *Reproduction of Marine Invertebrates*, Vol. 6, *Echinoderms* and *Lophophorates*. AC Giese, JS Pearse & VB Pearse (eds), pp. 513–662. Boxwood Press, Pacific Grove.

Pearse JS, McClary DJ, Sewell MA, Austin WC, Perez-Ruzafa A & Byrne M (1988) Simultaneous spawning of six species of echinoderms in Barkley Sound, British Columbia. *Invert. Reprod. Develop.* **14:** 279–288.

Pennington JT (1985) The ecology of fertilization of echinoid eggs: the consequence of sperm dilution, adult aggregation, and synchronous spawning. *Biol. Bull.* **169:** 417–430.

Petersen CW (1991) Sex allocation in hermaphroditic sea basses. *Am. Nat.* **138:** 650–667.

Petersen CW, Warner RR, Cohen S, Hess HC & Sewell AT (1992) Variable pelagic fertilization success: implications for mate choice and spatial patterns of mating. *Ecology* **73:** 391–401.

Pianka HD (1974) Ctenophora. In *Reproduction in Marine Invertebrates*, Vol. 1. AC Giese, JS Pearse & VB Pearse (eds), pp. 201–265. Academic Press, New York.

Pitnick S, Markow TA & Spicer GS (1995) Delayed male maturity is a cost of producing large sperm in *Drosophila. Proc. Natl Acad. Sci. USA* **92:** 10614–10618.

Podolsky RD (1995) Effects of the echinoid egg jelly coat on fertilization through changes in effective egg size. *Am. Zool.* **35:** 54A.

Podolsky RD & Strathmann RR (1996) Evolution of egg size in free-spawners: consequences of the fertilization-fecundity tradeoff. *Am. Nat.* **148:** 160–173.

Pollock LW (1975) Tardigrada. In *Reproduction in Marine Invertebrates*, Vol. 2. AC Giese & JS Pearse (eds), pp. 43–54. Academic Press, New York.

Randall JE, Schroeder RE & Stark WA, II (1964) Notes on the biology of the echinoid *Diadema antillarum. Carib. J. Sci.* **4:** 421–433.

Rao GC (1965) Reproductive cycle of *Oreaster* (*Pentaceros*) *hedemanni* in relation to chemical composition of gonads. *Curr. Sci.* **5:** 87–88.

Rappaport R & Rappaport BN (1994) Cleavage in conical sand dollar eggs. *Develop. Biol.* **164:** 258–266.

Reed CG (1991) Bryozoa. In *Reproduction in Marine Invertebrates*, Vol. 6. AC Giese, JS Pearse & VB Pearse (eds), pp. 86–245. Boxwood Press, Pacific Grove.

Reeve MR & Cosper JC (1975) Chaetognatha. In *Reproduction in Marine Invertebrates*, Vol. 2. AC Giese & JS Pearse (eds), pp. 157–184. Academic Press, New York.

Reiswig HM (1970) Porifera: sudden sperm release by tropical demospongiae. *Science* **170:** 538–539.

Rice ME (1975) Sipunculida. In *Reproduction in Marine Invertebrates*, Vol. 2. AC Giese & JS Pearse (eds), pp. 67–127. Academic Press, New York.

Riser NW (1974) Nemertina. In *Reproduction in Marine Invertebrates*, Vol. 1. AC Giese, JS Pearse & VB Pearse (eds), pp. 359–389. Academic Press, New York.

Robertson DR & Hoffman SG (1977) The roles of female mate choice and predation in the mating systems of some tropical labroid fishes. *Z. Tierpsychol.* **45:** 298–320.

Rothschild L & Swann MM (1951) The fertilization reaction in the sea urchin. The probability of a successful sperm-egg collision. *J. Exp. Biol.* **28:** 403–416.

Rouse G & Fitzhugh K (1994) Broadcasting fables: is external fertilization really primitive? Sex, size, and larvae in sabellid polychaetes. *Zool. Scripta* **23:** 271–312.

Rowe FWE, Anderson DT & Healy JM (1991) Concentricycloidea. In *Reproduction in Marine Invertebrates*, Vol. 6. AC Giese, JS Pearse & VB Pearse (eds), pp. 751–760. Boxwood Press, Pacific Grove.

Rumrill SS (1990) Natural mortality of marine invertebrate larvae. *Ophelia* **32:** 163–198.

Run J-Q, Chen C-P, Chang K-H & Chia F-S (1988) Mating behaviour and reproductive cycle of *Archaster typicus* (Echinodermata: Asteroidea). *Mar. Biol.* **99:** 247–253.

Sastry AN (1979) Pelecypoda (excluding Ostridea). In *Reproduction in Marine Invertebrates*, Vol. 5. AC Giese, JS Pearse & VB Pearse (eds), pp. 113–292. Academic Press, New York.

Scheibling RE (1979) The ecology of *Oreaster reticulatus* (L) (Echinodermata:

Asteroidea) in the Caribbean. Ph.D. dissertation, McGill University, Montreal, Canada.

Scheibling RE & Lawrence JM (1982) Differences in reproductive strategies of morphs of the genus *Echinaster* (Echinodermata: Asteroidea) from the eastern Gulf of Mexico. *Mar. Biol.* **70:** 51–62.

Schroeder PC & Hermans CO (1975) Annelida. In *Reproduction in Marine Invertebrates*, Vol. 3. AC Giese & JS Pearse (eds), pp. 1–213. Academic Press, New York.

Schuster P & Sigmund K (1982) A note on the evolution of sexual dimorphism. *J. Theor. Biol.* **94:** 107–110.

Scudo FM (1967) The adaptive value of sexual dimorphism. I. Anisogamy. *Evolution* **21:** 285–291.

Sewell MA (1994) Small size, brooding and protandry in the apodid sea cucumber *Leptosynapta clarki*. *Biol. Bull.* **187:** 112–123.

Sewell MA & Levitan DR (1992) Fertilization success in a natural spawning of the dendrochirote sea cucumber *Cucumaria miniata*. *Bull. Mar. Sci.* **51:** 161–166.

Shlesinger Y & Loya Y (1985) Coral community reproductive patterns: Red Sea versus the Great Barrier Reef. *Science* **228:** 1333–1335.

Sinervo B & McEdward LR (1988) Developmental consequences of an evolutionary change in egg size: an experimental test. *Evolution* **42:** 885–899.

Smiley S, McEuen FS, Chaffee C & Krishnan S (1991) Holothuroidea. In *Reproduction in Marine Invertebrates*, Vol. 6. AC Giese, JS Pearse & VB Pearse (eds), pp. 663–750. Boxwood Press, Pacific Grove.

Smith CC & Fretwell SD (1974) The optimal balance between size and number offspring. *Am. Nat.* **108:** 499–506.

Smith RH (1971) Reproductive biology of the brooding sea-star, *Leptasterias pusilla* (Fisher (eds)), in the Monterey Bay region. Ph.D. dissertation, Stanford University, Stanford, USA.

Smith RI (1989) Observations on spawning behavior of *Eupolymnia nebulosa*, and comparisons with *Lanice conchilega* (Annelida, Polychaeta, Terebellidae). *Bull. Mar. Sci.* **45:** 406–414.

Southward EC (1975) Pogonophora. In *Reproduction in Marine Invertebrates*, Vol. 2. AC Giese & JS Pearse (eds), pp. 129–156. Academic Press, New York.

Stekoll MS & Shirley TC (1993) *In situ* spawning behavior of an Alaskan population of pinto abalone, *Haliotis kamtschatkana* Jonas, 1845. *Veliger* **36:** 95–97.

Sterrer W (1974) Gnathostomulida. In *Reproduction in Marine Invertebrates*, Vol. 1. AC Giese, JS Pearse & VB Pearse (eds), pp. 345–357. Academic Press, New York.

Strathmann RR (1990) Why life histories evolve differently in the sea. *Am. Zool.* **30:** 197–207.

Thane A (1974) Rotifera. In *Reproduction in Marine Invertebrates*, Vol. 1. AC Giese, JS Pearse & VB Pearse (eds), pp. 471–484. Academic Press, New York.

Thomas FIM (1994a) Transport and mixing of gametes in three free-spawning polychaete annelids, *Phragmatopoma californica* (Fewkes (eds), *Sabellaria cementarium* (Moore (eds), and *Schizobranchia insignis* (Bush). *J. Exp. Mar. Biol. Ecol.* **197:** 11–27.

Thomas FIM (1994b) Physical properties of gametes in three sea urchin species. *J. Exp. Mar. Biol. Ecol.* **194:** 263–284.

Thorson G (1950) Reproductive and larval ecology of marine bottom invertebrates. *Biol. Rev.* **25:** 1–45.

Thresher RE (1984) *Reproduction in Reef Fishes*. TFH Publications, Neptune City.

Tominga H, Komatsu M & Oguro C (1994) Aggregation for spawning in the breeding season of the sea-star, *Asterina minor* Hayashi. In *Echinoderms Through Time*. B David, J-P Féral & M Roux (eds), pp. 369–373. Balkema, Rotterdam.

Town JC (1980) Movement, morphology, reproductive periodicity, and some factors affecting gonad production in the seastar *Astrotole scabra* (Hutton). *J. Exp. Mar. Biol. Ecol.* **44:** 111–132.

Tyler PA, Young CM, Billet DSM & Giles LA (1992) Pairing behaviour, reproduction and diet in the deep-sea holothurian genus *Paroriza* (Holothuroidea: Synallactidae). *J. Mar. Biol. Assoc. UK* **72:** 447–462.

Uchida T & Yamada M (1968) Cnidaria. In *Invertebrate Embryology*. M Kume & K Dan (eds), pp. 86–116 (Published for the US National Library of Medicine, Washington, DC). Nolit, Belgrade.

Unger B & Lott C (1994) *In-situ* studies on the aggregation behaviour of the sea urchin *Sphaerechinus granularis* Lam (Echinodermata: Echinoidea). In *Echinoderms Through Time*. B David, J-P Féral & M Roux (eds), pp. 913–919. Balkema, Rotterdam.

Vance RR (1973) On reproductive strategies in marine bottom invertebrates. *Am. Nat.* **107:** 339–352.

van der Land J (1975) Priapulida. In *Reproduction in Marine Invertebrates*, Vol. 2. AC Giese & JS Pearse (eds), pp. 55–65. Academic Press, New York.

Vogel H, Czihak G, Chang P & Wolf W (1982) Fertilization kinetics of sea urchin eggs. *Math. Biosci.* **58:** 189–216.

von Bismark O (1959) Versuch einer Analyse der die Stoffwechsetintenstat ('Ruheumsatz') von *Asterias rubens* L. beeinflussesden Faktoren. *Kiel. Meeresforsch.* **15:** 164–186.

Warner RR, Robertson DR & Leigh EG (1975) Sex change and sexual selection. *Science* **190:** 633–638.

Webber HH (1977) Gastropoda: Prosobranchia. In *Reproduction in Marine Invertebrates*, Vol. 4. AC Giese & JS Pearse (eds), pp. 1–93. Academic Press, New York.

Wells MJ & Wells J (1977) Cephalopoda: Octopoda. In *Reproduction in Marine Invertebrates*, Vol. 4. AC Giese & JS Pearse (eds), pp. 291–335. Academic Press, New York.

Wickstead JH (1975) Chordata: Acrania. In *Reproduction in Marine Invertebrates*, Vol. 2. AC Giese & JS Pearse (eds), pp. 283–319. Academic Press, New York.

Wilborg K (1946) Undersokelser over oskjellet. *Rep. Norweg. Fish. Mar. Invest.* **8:** 1–85.

Willis BL, Babcock RC, Harrison PL & Wallace CC (in press). Hybridization and breeding incompatibilities within the mating systems of mass spawning reef corals. *8th Int. Coral Reef Symp., Panama.*

Wilson BC (1900) The habits and early development of *Cerebratulus lacteus*. *Quart. J. Microsc. Sci. Ser. 2* **43:** 97–198.

Wray GA (1995) Evolution of larvae and developmental modes. In *Ecology of Marine Invertebrate Larvae*. L McEdward (ed.), pp. 412–448. CRC Press, Boca Raton.

Young CM, Tyler PA, Cameron JL & Rumrill SG (1992) Seasonal breeding aggregations in low-density populations of a bathyal echinoid, *Styocidaris lineata*. *Mar. Biol.* **113:** 603–612.

Yund PO (1990) An *in situ* measurement of sperm dispersal in a colonial marine hydroid. *J. Exp. Zool.* **253:** 102–106.
Yund PO (1995) Gene flow via the dispersal of fertilizing sperm in a colonial ascidian (*Botryllus schlosseri*): the effect of male density. *Mar. Biol.* **122:** 649–654.
Yund PO & McCartney MA (1994) Male reproductive success in sessile invertebrates: competition for fertilizations. *Ecology* **75:** 2151–2167.
Zimmer RL (1991) Phoronida. In *Reproduction in Marine Invertebrates*, Vol. 6. AC Giese, JS Pearse & VB Pearse (eds), pp. 2–45. Boxwood Press, Pacific Grove.

7 Mating Conflicts and Sperm Competition in Simultaneous Hermaphrodites

Nico K. Michiels

Max-Planck-Institut für Verhaltensphysiologie, PO Box 1564, D-82305 Starnberg, Germany

'In the lowest classes the two sexes are not rarely united in the same individual, and therefore secondary sexual characters cannot be developed. . . . Moreover, it is almost certain that these animals have too imperfect senses and much too low mental powers to feel mutual rivalry, or to appreciate each other's beauty or other attractions.' (Charles Darwin 1871, Part II, p. 321).

I. INTRODUCTION

Although it is now generally accepted that sexual selection applies to all types of gender expression (Arnold 1994a,b; Morgan 1994), hermaphroditic animals are still largely absent from the sexual selection literature (see Andersson 1994). This is not due to a lack of conceptual basis – many theoretical contributions have been made (Ghiselin 1969;

Sperm Competition and Sexual Selection
ISBN 0-12-100543-7

Williams 1975; Heath 1977; Charnov *et al.* 1976; Charnov 1979, 1982; Bell 1982). In this chapter I speculate on the consequences of sexual selection and sperm competition in an hermaphroditic mating system. First, I show that hermaphrodites have the unique property that they can optimize their allocation to the male and female function. Second, I explain how conflicts between mating partners arise in hermaphrodites, and why they are particularly strong during copulation because of different mating interests. This is expected to result in mechanisms such as elaborate *conditional reciprocity* or aggressive *hypodermic impregnation*, both of which will be subject to ongoing adaptation and counter-adaptation. I shall show that sperm competition is expected to be common and that the adaptations resulting from it are sometimes drastic and unique to hermaphrodites.

Unfortunately, quantitative data that relate to the theme of this chapter, such as mating frequency, sperm transfer, sperm usage and fertilization success, are very rare for hermaphrodites. As a result, many of the views presented here are speculative and I have no doubt that some might eventually be proved wrong once appropriate data are available. Yet, I want to be thought-provoking and make clear that the animal taxa that Darwin deliberately ignored deserve a more central place in the evolutionary biological literature.

II. WHAT ARE HERMAPHRODITES?

Hermaphrodites are individuals that possess a functional male and female reproductive system during at least part of their lives. This type of gender expression is ubiquitous among plants (see Chapter 5) and widespread in the animal kingdom (Ghiselin 1969). Table 7.1 shows that 20 out of the 28 phyla listed have at least some hermaphroditic representatives and seven are exclusively hermaphroditic. These are the sponges, the entoprocts, the bryozoans, the free-living and parasitic flatworms, the arrowworms, the gastrotrichs, and the comb jellies. Three further phyla contain major classes or orders that are also almost entirely hermaphroditic: (1) anemones and corals, (2) sea slugs and pulmonate snails (see Chapter 8) and (3) gnathostomulids, leeches and earthworms. Although hermaphroditism is less common in other groups, some of these exceptions are well-studied: the nematode *Caenorhabditis elegans*, for example, is without doubt the best-studied hermaphrodite (Wood 1988). Hermaphroditism is also known as an aberrant condition in mammals (Bunch *et al.* 1991; De Guise *et al.* 1994), including humans (Akin *et al.* 1993; Krob *et al.* 1994; Spurdle *et al.* 1995). It is obvious that not only is this mode of gender expression widespread, but that it must have evolved repeatedly.

A. Sequential and simultaneous hermaphrodites

Two major types of hermaphroditism will be distinguished in this chapter. Sex-changing or *sequential hermaphrodites* (e.g. corals, certain fish and polychaetes) are well studied. These start out as one sex and change into the other later in life. In some polychaetes and fish, repeated, alternating changes are possible (Teuchert 1968; Kuwamura *et al.* 1994; Nakashima *et al.* 1995). Inspired by the size-advantage hypothesis proposed by Ghiselin (1969), many studies of sequential hermaphrodites have focused on the question of when to change sex (Warner 1975, 1988; Warner *et al.* 1975; Leigh *et al.* 1976; Policansky 1982; Berglund 1986, 1991; Charnov 1982).

The second type, and focus of this chapter, are *simultaneous hermaphrodites*. These have functional male and female genitalia simultaneously present for most of their lives and reproductive acts usually involve both the male and female function in each individual. Throughout this chapter, I use the term hermaphrodite to refer to simultaneous hermaphrodites. The term gonochorist refers to species where individuals are either male or female. In order to avoid confusion about the source of sperm, autosperm will be used for the self sperm an individual donates to its partner, whereas allosperm is used for the sperm an individual receives from its partner. With the exception of a few important case studies, I will not elaborate on hermaphrodite systems in which gametes are spawned into the (aquatic) environment (see Chapter 6), but concentrate on hermaphroditic groups that copulate or at least transfer a spermatophore. Gastropods are the best studied examples in this respect and are reviewed in a separate chapter (see Chapter 8).

III. WHY ARE HERMAPHRODITES HERMAPHRODITIC?

A. The resource allocation model

Hermaphroditism is favoured whenever the overall reproductive success achieved by a hermaphrodite is greater than that of a pure male or a pure female (Charnov *et al.* 1976; Charnov 1982). Although this may appear trivial at first, this leads to the more interesting question of under what sorts of conditions this may be expected. The resource allocation model (Charnov *et al.* 1976) predicts that, whenever offspring produced by one sex function are increasingly expensive as their number increases (diminishing returns), it is more efficient to limit investment in this function and to become hermaphroditic; the remaining resources can then be relocated to the other sex function, within the same individual. A limitation is that the additional costs arising from having two sexual functions

Table 7.1. A limited overview of the presence of hermaphroditism in the animal kingdom, as well as a rough indication of the presence of gamete exchange via spawning or copulation and external versus internal fertilization. Taxa where hermaphroditism is the dominant mode of gender expression are in **bold**. (Int.: internal fertilization; Ext.: external fertilization.)

Phylum	Lower taxon	Popular name	Hermaphroditism	Type of sperm transfer	Fertilization	N species
Porifera		**Sponges**	**Ubiquitous**	Spawning	Int.	5000
Cnidaria	Hydrozoa	Hydras and hydroids	Rare	Spawning	Int.	2700
Cnidaria	Scyphozoa and Cubozoa	Jellyfish	Rare	Spawning	Int.	215
Cnidaria	**Anthozoa**	**Anemones and corals**	**Ubiquitous**	Spawning	Ext. and Int.	6000
Sipunculida		Peanut worms	Rare	Spawning	Ext.	320
Annelida*	Polychaeta		Rare	Spawning	Ext.	8000
Annelida*	**Gnathostomulida**		**Ubiquitous**	Copulation	Int.	80
Annelida*	Pogonophora		Rare	Spawning	Int.	80
Annelida*	Echiura	Spoonworms	Absent	Spawning	Ext.	140
Annelida*	**Oligochaeta**	**Earthworms, freshwater oligochaetes**	**Ubiquitous**	Copulation	Int.	3100
Annelida*	**Hirudinea**	**Leeches**	**Ubiquitous**	Copulation or spermatophore	Int.	500
Onychophora		Velvet worms	Absent	Spermatophore	Int.	70
Arthropoda	Chelicerata	Horseshoe crabs, scorpions, spiders, mites, harvestmen, sea spiders	Absent	Copulation or spermatophore	Int.	72 000
Arthropoda	Crustacea	Crabs, shrimps, crayfish, woodlice, copepods, barnacles	Rare	Copulation	Int.	38 000
Arthropoda	Myriopoda	Centipedes, millipedes	Absent	Spermatophore	Int.	10 500
Arthropoda	Insecta	Insects	Rare	Copulation	Int.	750 000
Tardigrada		Water bears	Rare	Copulation	Int.	600
Entoprocta			**Ubiquitous**	Spawning	Int.	150
Ectoprocta		**Bryozoans**	**Ubiquitous**	Spawning	Int.	5000

Platyhelminthes	**Cestoidea**	**Tapeworms**	**Ubiquitous**	Copulation	Int.	3400
Platyhelminthes	**Trematoda**	**Flukes**	**Ubiquitous**	Copulation	Int.	11 000
Platyhelminthes	**Monogenea**	**Monogenean ectoparasites**	**Ubiquitous**	Copulation	Int.	1100
Platyhelminthes	**Turbellaria***	**Free-living flatworms**	**Ubiquitous**	Copulation	Int.	3000
Nemertini		Ribbon worms	Rare	Spawning	Ext.	900
Rotifera			Absent	Copulation	Int.	1500
Acanthocephala			Absent	Copulation	Int.	1150
Chaetognatha		**Arrowworms**	**Ubiquitous**	Copulation	Ext. and Int.	70
Gastrotricha			**Ubiquitous**	Copulation	Int.	430
Nematoda		Roundworms	Present	Copulation	Int.	12 000
Nematomorpha			Absent	Copulation or spermatophore	Int.	320
Priapulida			Absent	Spawning	Ext. and Int.	16
Kinorhyncha			Absent	Spermatophore	Int.	150
Ctenophora		**Sea walnuts, comb jellies**	**Ubiquitous**	Spawning	Ext.	50
Phoronidae			Present	Spermatophore	Int.	14
Brachiopoda		Lamp shells	Rare	Spawning	Ext.	325
Pterobranchia			Male and Female zoids	Spawning	Int.	21
Echinodermata		Starfish, brittle stars, sea urchins, sea cucumbers, sea lilies	Rare	Spawning	Ext. and Int.	6000
Enteropneusta		Acorn worms	Absent	Spawning	Ext.	70
Urochrodata		**Sea squirts, salps, ...**	**Ubiquitous**	Spawning	Ext.	1250
Cephalochordata		Lancelets	Absent	Spawning	Ext.	25
Vertebrata	Pisces	Fish	Present	Spawning (copulation)	Ext. and Int.	20 500
Vertebrata	Tetrapoda	Amphibians, reptiles, birds, mammals	Absent	Copulation (spawning)	Ext. and Int.	21 500
						1 032 995

* Polyphyletic groups for which the systematics have not been resolved satisfactorily. Molluscs are reviewed separately by Baur (Chapter 8). The taxonomic subdivision was taken from Nielsen (1995), and the other data from Barnes and Ruppert (1994).

Table 7.2. A summary of possible differences between gonochorists (species with separate sexes) and hermaphrodites.

Typical gonochorists (males and females)	Simultaneous hermaphrodites
No self-fertilization	Self-fertilization possible
Traits of only one sex expressed per individual	Traits of both genders always expressed
Sexual specialization less constrained	Sexual specialization constrained
Coevolution of males and females	Coevolution of male and female structures within one individual
Cost of paternal offspring lower than that of maternal offspring	Cost of paternal and maternal offspring balanced
Allocation to males and females 1 : 1: low flexibility to adjust sex allocation to conditions	Allocation to sperm and eggs: opportunistic resource utilization possible
Sexual conflict *before* copulation	Mating conflict *during* copulation
Nuptial gifts possible	Nuptial gifts *not* expected
Reluctance to donate sperm rare in males	Reluctance to donate sperm not uncommon?
Mating not necessarily assortative	When mate choice, then assortative
Females may 'allow' access to sperm stores	Access to sperm stores *not* expected
Manipulation of partner's sex allocation not possible	Manipulation of partner's sex allocation expected?

should not exceed the benefits that can be achieved (Heath 1977; Charnov 1982). Differences between hermaphrodites and gonochorists are summarized in Table 7.2.

B. Diminishing returns for the male and/or female sex

Diminishing returns for the male function may arise under conditions where few mating opportunities exist, so that inseminating a few partners may be easy, whereas inseminating many is very costly or simply impossible owing to low mobility or density (Charnov 1979). That optimal male allocation is reduced when the mating group is small, as predicted by the Local Mate Competition model adjusted for hermaphrodites (Charnov 1980, 1982), has been shown for serranid fish (Fischer 1984b; Petersen 1990, 1991), polychaetes (Sella 1990) and barnacles (Raimondi and Martin 1991). Strong mate choice could result in diminishing returns in dense populations.

The female function could show a saturating gain curve when offspring dispersal is limited and results in increased local sib competition, or when the number of young that can be produced per unit time is limited. Brooding, which is common in many hermaphroditic taxa, may exemplify such limitation (Ghiselin 1969; Charnov 1982).

C. Selfing

Hermaphrodites can, in principle, self-fertilize their eggs. Selfing is common among plants (Jarne and Charlesworth 1993) and well known from parasitic flatworms (Joyeux and Baer 1961), oligochaetes (Needham 1990), arrowworms (Jägertsen 1940; Reeve and Walter 1972; Alvariño 1990) and pulmonate snails (Jarne *et al.* 1993). Initially, selfing may appear to serve as an 'emergency exit' for reproduction when a partner cannot be found (Ghiselin 1969). However, if inbreeding is not too costly, a mixture of selfing and outbreeding may allow combination of the advantages of increased genetic propagation with those of outcrossing (Jarne and Charlesworth 1993). Extending the resource allocation model to selfers, it can be shown that they should reduce male allocation and produce more ova instead (Charlesworth and Charlesworth 1981; Charnov 1982). An extreme example is the nematode *C. elegans*, where selfing hermaphrodites have minimized autosperm production to such an extent that autosperm depletion is the rule later in life (Barker 1992).

D. Does hermaphroditism equal opportunism?

When a population is outcrossing and panmictic and neither function is limited in its access to common resources, hermaphrodites should spend, on average, equal amounts on the male and female function in accordance with the basic 1:1 sex ratio rule (Fisher 1958; Williams 1975). Under the resource allocation model, however, most situations that result in stable hermaphroditism are characterized by sex allocation that diverges from this ratio (Charnov 1982). Such conditions may arise when one sex function is limited by resources that are unimportant for the other. At first, this suggests that there is strong selection for one particular, optimal sex allocation.

What is somewhat misleading in these models is that, in both cases, the optimal sex allocation is the predicted average for the population. For the individual hermaphrodite, a major advantage is that it can diverge from this ratio. Whenever the two sex functions require different resources (which is probably the rule rather than the exception), and the presence of these resources changes unexpectedly or rapidly in time or space, a hermaphrodite may exploit such fluctuations opportunistically. For example, it can produce more sperm when mating opportunities are plentiful, or produce more eggs when certain food types are abundant. This view implies that they are (to some extent) able to shift resources from one function to the other, depending on which one produces offspring more cheaply at the time; note that this automatically results in the expected equal overall expenditure on paternal and maternal offspring.

It may be expected that opportunistic allocation is particularly important when competition for food is severe, as occurs in populations of planarian flatworms (Reynoldson and Young 1965; Reynoldson and Bellamy 1973) and when the local habitat is characterized by fluctuations in opportunities or limitations with regard to food and mate availability. Owing to low mobility in many hermaphroditic taxa (e.g. flatworms, snails, earthworms, leeches) such fluctuations cannot be averaged out by mobility alone, leaving sex allocation as an intra-individual solution to local perturbations.

In gonochorists, instant sex ratio adjustments are not possible: a pure female has only very limited possibilities to adjust the number of sons or daughters she produces. An important exception, however, are haplodiploid insects in which females can optimize the sex ratio of their clutch according to the mating arena in which their offspring find themselves when hatching (Werren 1980).

In hermaphrodites, the relative cost of paternally and maternally produced offspring must be equal (Williams 1975). When maternal offspring become cheaper to produce than paternal offspring, animals with higher maternal allocation would be favoured. This increase in egg production would, in turn, make paternal offspring cheaper, resulting in a new equilibrium at which paternal and maternal offspring cost the same again. If this prediction holds, classic sexual conflict does not exist in hermaphrodites: a cheap (and therefore) preferred sexual role cannot persist. However, offspring produced by one sex function can remain cheaper than those produced by the other when reproduction by the first is strictly constrained by external factors (e.g. brief mating period per day or brooding space).

Note that, in gonochorists, typical males have the potential to produce offspring at a lower cost than females, resulting in sexual conflict over matings. One could consider males as individuals with 'spare' resources relative to females, which they may invest in courtship or ornaments rather than directly into individual offspring.

IV. SEXUAL SELECTION IN HERMAPHRODITES

A. Evolutionary constraints

Before speculating on what traits may be favoured by sexual selection, it is important to stress four constraints resulting from hermaphroditism (see also Table 7.2). First, in hermaphrodites, selection on male traits cannot be independent from selection on female traits of the same individual (Morgan 1994). This means that in a hermaphroditic population, it is impossible to find the optimal male and female strategy combined in one individual, but one may find the optimal compromise between both. A second constraint exists on the evolution of genitalia: internally fertilizing hermaphrodites with reciprocal penis insertion must evolve compatible, yet identical genitalia. Simultaneous penis insertion may require special postures or even asymmetrical genital structures with the same skew towards left- or right-handedness, allowing a better 'handshake' between genitalia (e.g. *Polycelis tenuis*, Ball and Reynoldson 1981; gastropods, see Chapter 8). A third limitation involves mate choice. Whenever one trait is favoured by most individuals in the population (i.e. large body size), hermaphrodites are almost automatically expected to mate assortatively. Large individuals will prefer to mate with a large partner, leaving only small partners for small individuals (see also Ridley 1983). This reduces the impact of mate choice since skewed mating rates (known from species in which one successful male can mate with many females of all ranks) are prevented. In fact, assortative mating may actually stabilize selection on the trait involved, since average individuals may find a partner more easily. A fourth limitation follows from the fact that, in hermaphrodites, sexual preferences and traits are expressed simultaneously in each individual. No traits remain 'hidden', as female traits are in pure males and male traits are in pure females. The fact that in gonochorists, sex-related traits can jump one or more generations, allows them to recombine several times without being exposed to selection. Such a source of new variation is absent from hermaphrodites. However, despite these limitations, I shall show why I believe that sexual selection is not only strong in hermaphrodites, but may also favour rather peculiar traits.

B. Sex roles in hermaphrodites?

Assuming that the female function of a hermaphrodite rarely runs short of sperm (but see Chapter 6), and that the benefits of receiving multiple ejaculates are limited, one could expect hermaphrodites to mate mainly in order to inseminate their partner rather than to receive allosperm (Charnov 1979). This view stems from gonochorists, where females usually limit male mating opportunities and, as a result, males will rarely refuse to donate sperm when given an opportunity by a female partner (Bateman 1948). In hermaphrodites, however, mating opportunities are not limited in this way. Here, all conspecifics may want to donate autosperm and, as a result, copulate readily and accept allosperm in order to have an opportunity to inseminate a partner. Hence, a donor may not be limited by the number of mating partners, but by its own capacity to inseminate them. Consequently, it may be the donor who is 'choosy' about who to donate sperm to. This is particularly likely when individuals vary in quality (e.g. fecundity), when copulations are costly and when choice is possible owing to high density (see also Ridley 1983). For the reasons listed above, such sex-role reversal may be more common among hermaphrodites than among gonochorists.

Evidence suggestive of a 'choosy' donor function was found in the planarian flatworm *Dugesia gonocephala*. In this species, size is positively related to fecundity (Vreys and Michiels 1995). Under such conditions, size-assortative mating is expected (see earlier). Partners of this species engage in frequent precopulatory contacts, during which both lie on top of each other and flatten out completely. Genital contact is not possible in this position (Vreys *et al.* 1997a; Fig. 7.1). Experimental investigation indicated that partners use this peculiar behaviour to size each other up and mate only when both are of a similar size, resulting in very pronounced size-assortative mating in the field (Vreys and Michiels 1997).

If sperm donors can be choosy, this may eventually lead to a situation where the receiving function must advertise its receptivity to seduce reluctant sperm donors. Although this may be a logical expectation, it appears very difficult to distinguish whether partners stimulate each other to donate or to receive. The fact that up to one-third of all matings in the planarian *Dugesia polychroa* end without sperm transfer in either direction strongly suggests that matings are not always driven by an

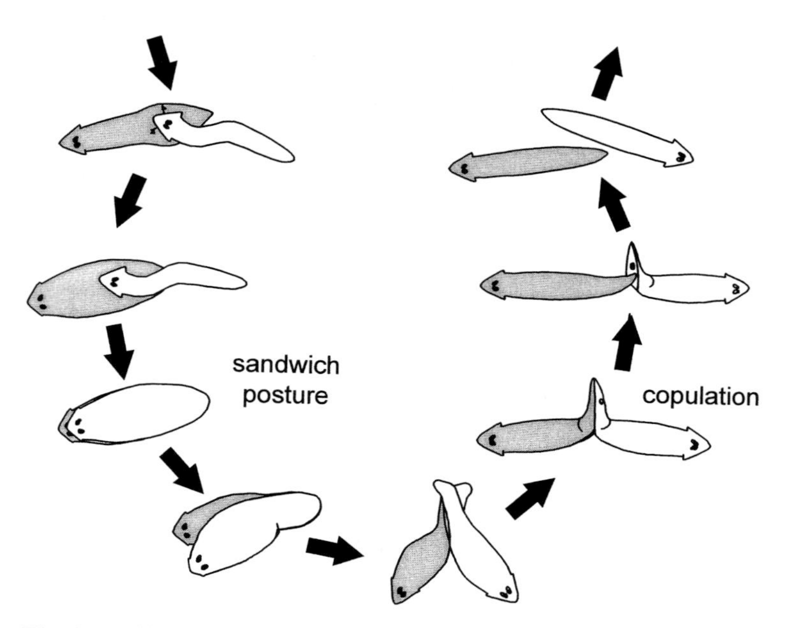

Fig. 7.1. *Mating sequence in the planarian flatworm* Dugesia gonocephala. *In this species, a typical 'sandwich' posture always precedes copulation. During this phase, individuals are assumed to measure each other's size (modified after Vreys* et al. *1997a). This behaviour is not known from any other planarian.*

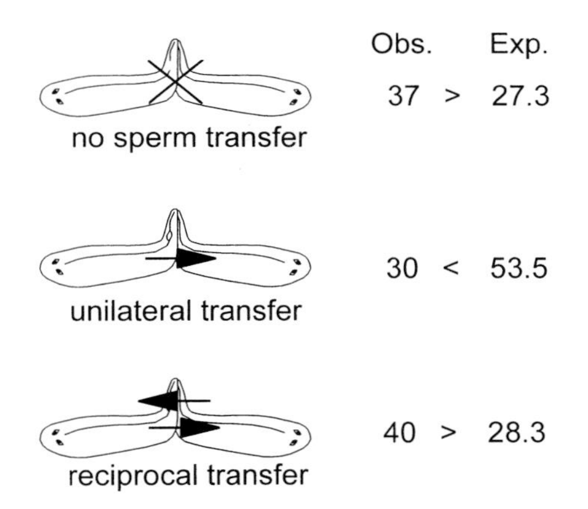

Fig. 7.2. *Summary of sperm transfer in 107 copulating pairs of the planarian* Dugesia polychroa. *One-third of all copulations ended without sperm transfer in either direction. A comparison with a binomial distribution showed that 'no exchange' or 'bilateral exchange' were more common than expected whereas 'unilateral sperm transfer' was less common than expected. This result is indicative of sperm trading (see text) (N. K. Michiels and B. Bakovsky, unpublished data). Obs., observed; Exp., expected.*

intention to inseminate (Fig. 7.2). Particularly interesting are observations of unilateral insemination when reciprocity is the rule. Reise (1995) observed that, after occasional unilateral sperm transfer in the slug *Deroceras rodnae*, sperm donors became aggressive when their partner (who had not donated sperm) tried to escape. Ghirardelli (1968) observed that in the arrowworm *Spadella cephaloptera* unilateral matings are prolonged until reciprocity occurs.

V. MATING CONFLICT

Confusion about the role that each partner plays during a copulation does not exist in gonochorists: when a male and a female agree to mate, their sexual role is implied. When two hermaphroditic partners intend to mate, however, they may not know what the interests of the partner are. Because both individuals can have three different interests (donate autosperm, receive allosperm, or both), their combined interests will vary from totally compatible to totally incompatible (Fig. 7.3). Mating interests will be incompatible when both partners want to donate but not receive sperm or vice versa (Fig. 7.3, areas 1 and 4). Only when both want to give and receive are their interests compatible (Fig. 7.3, centre). Compatibility also prevails when one partner wants to receive and the other to donate sperm (Fig. 7.3, areas 2 and 3). However, since it can be

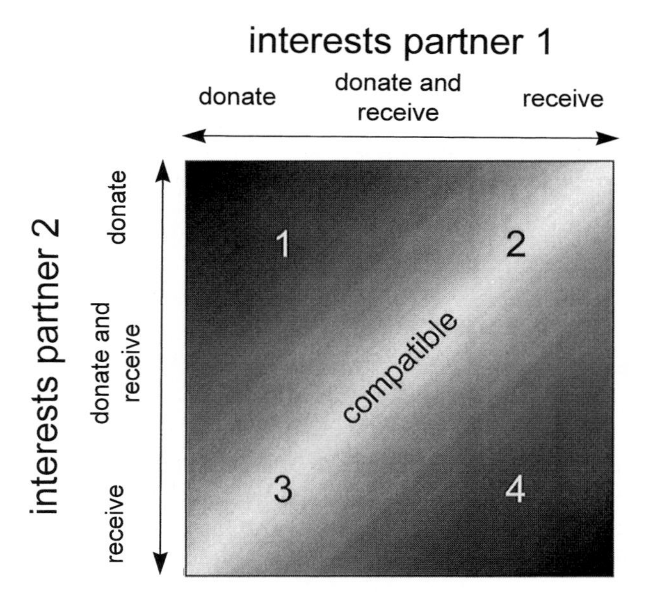

Fig. 7.3. *Interest matrix of all situations (likely or not) that can arise when two hermaphroditic partners with internal fertilization meet with the intention to mate. Each partner can have different interests, indicated as a gradient on the horizontal and vertical axis, ranging from 'donate only' to 'donate and receive' to 'receive only'. Black areas are where interests are incompatible because they are unilateral and identical. White indicates the zone of complete compatibility, either because of a difference in interest (away from the middle) or because both animals want to donate and receive sperm (in the centre). Numbers indicate four possible situations explained in the text: 1, both partners want to donate, but not receive – they trade sperm; 2 and 3, both partners have different, but compatible interests – mating can proceed; 4, both partners want to receive, and trade sperm; 5 (the centre), partners will insist on donating and receiving and trade this if necessary. Note that, for species with separate sexes (gonochorists), the situation is very simple and limited to the upper right or lower left corner of the matrix.*

expected that mating interests will be identical more often than they are different, it is also likely that hermaphrodites have a high likelihood of encountering mating conflicts. One way for hermaphrodites to resolve these is conditional, reciprocal exchange of gametes.

A. Reciprocity of gamete exchange

In the black hamlet *Hypoplectrus nigricans*, a hermaphroditic, coral reef fish, Fischer (1980) observed that two partners alternate the release of a small part of their clutch for fertilization by their partner. He coined the term 'egg trading' for this (Fischer 1984a, 1987) and interpreted the system as one in which individuals prefer to donate sperm in order to fertilize the partner's eggs (Fig. 7.3, area 1), but are allowed to do so only when they release some eggs themselves. This trade goes on for as long as both partners have some eggs left to give. This behaviour has now also been described in another serranid (Petersen 1995). Egg trading has also been observed in the hermaphroditic polychaete *Ophryotrocha diadema* and *Ophryotrocha gracilis*, in which pairs live together for several reproductive cycles (Sella 1985, 1988; Sella *et al.* 1997). Mating events consist of one partner spawning eggs and the other fertilizing them. Roles are alternated several times in a regular pattern, usually with the same partner. Both partners care for all the eggs. The advantage for traders is clear: by being 'monogamous' and sharing brood care, they can minimize sperm production (Sella 1990) and thus maximize the overall number of offspring (Premoli and Sella 1995).

The only known example of alternating sperm exchange in hermaphrodites with internal fertilization is in the opisthobranch sea slug *Navanax inermis*. In this species, penis insertion is unilateral but alternated repeatedly in the course of a copulatory bout, which Leonard and Lukowiak (1984, 1985, 1991) interpreted as sperm trading. They argued that in internal fertilizers the female role is preferred (Fig. 7.3, area 4) because it controls fertilization, in contrast to the male role, which has no guarantee that it will gain fertilizations in its partner. As a result, they expect mating partners to parcel their autosperm and alternate the exchange of small quantities in order to reduce the risk of donating a whole ejaculate without receiving allosperm or fertilizations in the partner. It implies that partners are reluctant to donate sperm and receivers risk running short of allosperm. Although I agree that a reluctance to donate sperm may evolve readily in hermaphrodites (see above), the 'risk' argument used by Leonard and Lukowiak (1984, 1985, 1991) is not a good explanation because the risk that autosperm are discarded by the partner simply adds to the total paternal cost, which ought to be compared with the total maternal costs before deciding which role may be cheaper. Since the number of paternally and maternally produced offspring is always equal in a sexual population, this 'risk' may actually average out as a low, rather than a high cost.

Whatever the primary mating interest of an hermaphroditic individual, it remains a fact that internally fertilizing hermaphrodites almost always show reciprocal insemination. This is true for free-living flatworms (Costello and Costello 1938; Hyman 1951; Apelt 1969; Peters *et al.* 1996; Vreys *et al.* 1997a,b; Figs 7.1 and 7.2), parasitic flatworms (Joyeux and Baer 1961; Williams and McVicar 1968; Kearn 1992; Kearn and Whittington 1992), oligochaetes (Grove 1925; Grove and Cowley 1926; Bahl 1928; Needham 1990), leeches (Brumpt 1900; Hoffman 1956; Wilkialis 1970; Wilkialis and Davies 1980; Kutschera 1989; Fig. 7.4), gastrotrichs (Ruppert 1978), arrowworms (Ghirardelli 1968; Reeve and Walter 1972; Alvariño 1990; Fig. 7.5), snails (Chapter 8). Therefore, I

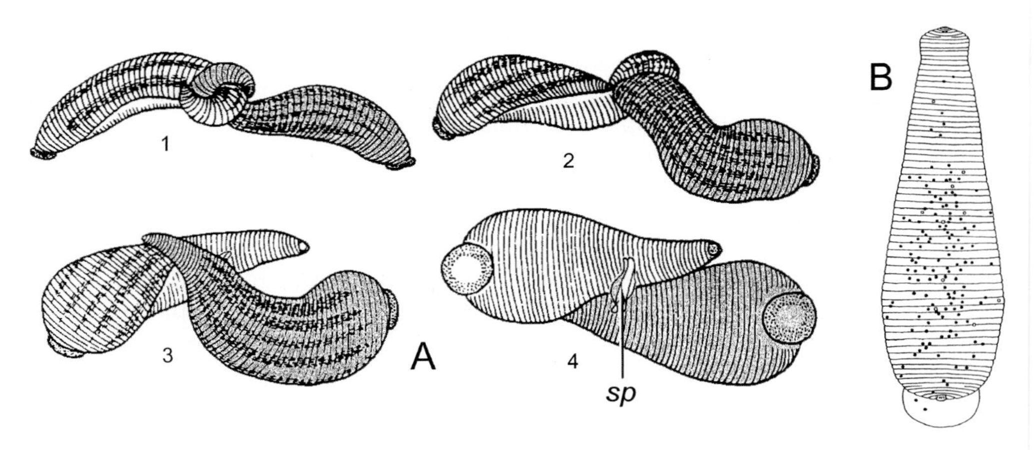

Fig. 7.4. **A.** *Reciprocal spermatophore (*sp.*) exchange in the leech* Glossiphonia lata *and deposition in the genital region.* **B.** *Summary chart of the distribution of spermatophore implantations observed in* Placobdella parasitica, *a species with random, unilateral spermatophore implantation. Filled dots are dorsal, open circles ventral (modified after Myers 1935).*

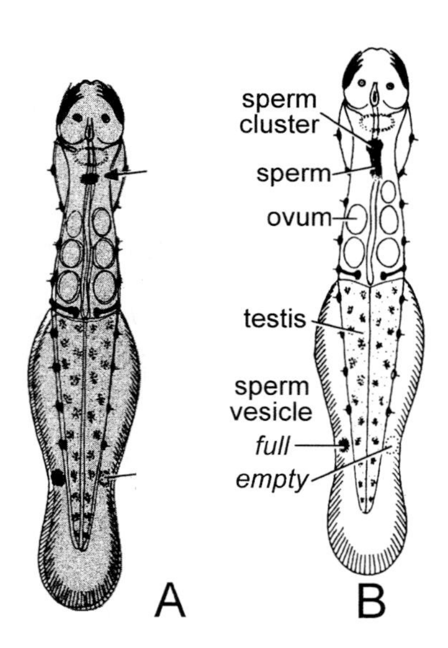

Fig. 7.5. *Sperm exchange in the arrowworm* Spadella cephaloptera. **A.** *Partners align head-to-tail and reciprocally transfer a cluster of sperm from their sperm vesicles to a specific spot under the head of the partner.* **B.** *After copulation, sperm (*sp.*) stream towards the vaginal opening (modified after Ghirardelli 1968, with permission).*

explore the ultimate causes of reciprocity and look more specifically for conditions under which reciprocity might be conditional, i.e. whether sperm donation depends on sperm receipt. If it does, it may solve conflicts arising from incompatible interests.

I limit the usage of the term 'trading' to systems where partners alternate roles such as egg trading in serranid fish and polychaetes, and sperm trading in *Navanax*. The more general term 'conditional reciprocity' is better suited to include all other copulatory mechanisms where sperm donation depends on sperm receipt, including those with synchronous sperm exchange. Although internal fertilizers with simultaneous, mutual penis insertion or spermatophore exchange are a large group among hermaphrodites, there are only very few studies that address the question of conditional reciprocity in this group (Michiels and Streng 1998; Vreys and Michiels 1998; Chapter 8).

B. *Unconditional reciprocity*

If sperm donation is the main reason why hermaphrodites mate (Charnov 1979) and partners do not refuse sperm donations, reciprocal insemination could simply be a coincidental side-effect of the mutual willingness to donate. This may be a trivial explanation for why reciprocity is so common and may make it difficult to recognize truly conditional reciprocity.

C. *When is conditional reciprocity expected?*

For conditional reciprocity to evolve, one important initial condition must be met: it must be possible to give up a partner at low cost when there are signs that it will not reciprocate. This implies that a new partner can be found easily, which may be true in high-density populations. Once this assumption is fulfilled, three situations should be distinguished.

First, partners may insist on receiving as well as donating (centre of Fig. 7.3). It is clear that donating frequently may be beneficial at high density. But why insist on sperm receipt when matings are common? If ejaculates are not infinitely small, nutrients derived from the ejaculate may be a reason to insist on reciprocity. As I shall show later, sperm digestion is widespread among hermaphrodites. A sperm donor may have no choice other than to accept that the majority of its sperm will be digested: it is part of the paternal costs that the donor may attempt, but fail, to minimize and it should therefore not be seen as a nuptial gift.

Second, both partners may wish to receive, but not donate sperm (Fig. 7.3, area 4). Such hermaphrodites may stimulate each other to donate but may not themselves donate before reciprocity has been assured. This may escalate in elaborate stimulation and assessment bouts before sperm transfer takes place. The donor function can be expected to have a simple 'come-and-get-it' sperm donation mechanism that may not even insert the sperm into the receiving function of the partner, as occurs in some flatworms (Ullyot and Beauchamp 1931) and slugs (see Chapter 8). Many planarian flatworms possess penis-like musculo-glandular organs associated with the genitalia that have been interpreted as sexual stimulators (Hyman 1951; see Fig. 7.9), and that may induce the partner to donate. It is difficult to force a partner to donate sperm. Therefore, escalation to physically damaging strategies appear unlikely.

Finally, both partners may want to donate, but not to receive (Fig. 7.3, area 1) because receiving is costly (e.g. manipulation by the donor or parasite transmission). Such situation may favour the evolution of hit-and-run mechanisms similar to hypodermic injection (see below). In response to these, the receiver may evolve successful avoidance tactics

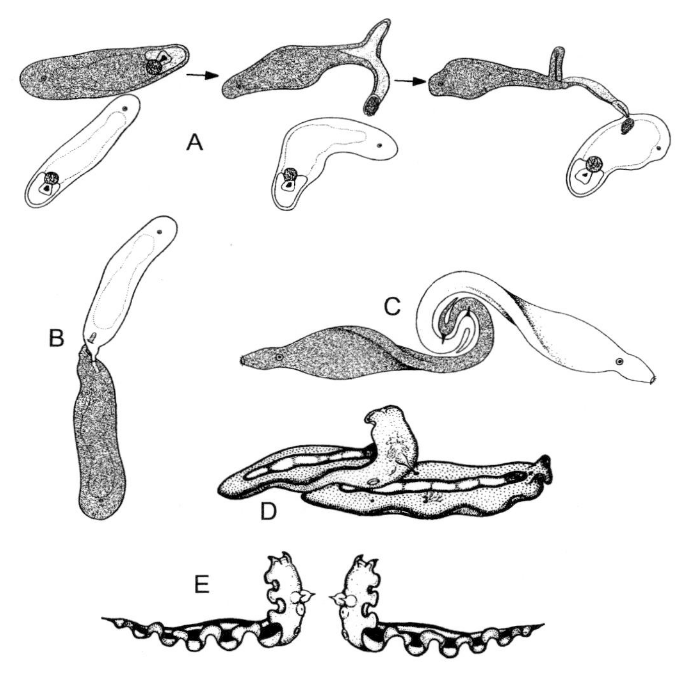

Fig. 7.6. *Several types of hypodermic impregnation in free-living flatworms.* **A.** *'Hit-and-run' injection of sperm in* Pseudophanostoma psammophilum *(modified after Apelt 1969, with permission).* **B.** *Reciprocal injection in* Archaphanostoma agile *(modified after Apelt 1969, with permission).* **C.** *Mutual tail holding and injection in* Monocelis fusca *(modified after Giesa 1966, with permission).* **D, E.** *Unilateral and bilateral stabbing in the polyclad flatworms* Pseudoceros *and* Pseudobiceros, *respectively. Note the double, everted penises in the latter (modified after Newman and Cannon 1994, with permission from the Queensland Museum).*

resulting in matings that can be best described as ritualized, reciprocal stabbing, suggestive of conditional reciprocity. This type of mating is common among leeches (Brumpt 1900; Hoffmann 1956; Nagao 1958; Kutschera and Wirtz 1986; Kutschera 1989; Fig. 7.4) and free-living and parasitic flatworms (Hyman 1951; Apelt 1969; Kearn 1992; Kearn and Whittington 1992; Newman and Canon 1994; Michiels and Newmann 1998; Fig. 7.6).

D. Evidence for conditional reciprocity

Except for the black hamlet fish, the polychaete *Ophryotrocha* and the sea slug *Navanax*, published data that suggest conditional reciprocity are lacking. Possible candidates appear to be oligochaetes. In the species *Pheretima communissima*, for example, the receiver function has three pairs of spermathecae which are reciprocally filled during one copulation using only one pair of penises. In order to achieve this, spermathecae are filled pair by pair, which takes 1.5 h per set, for 4–5 h in total (Avel 1959). It is clear that, since sperm exchange is simultaneous, interruption results in identical amounts of sperm exchanged. It is systems like these that may offer the best opportunities to collect quantitative data on conditional reciprocity in internal fertilizers with mutual penis insertion. In order to demonstrate conditionality in systems where ejaculates are exchanged in one single action rather than as a series of parcels, one would have to demonstrate that copulations with a symmetrical outcome (both or neither donate sperm) occur more often than expected by chance relative to unilateral inseminations. Indications that this may be true were found in the planarian *D. polychroa* (Michiels and Streng 1998) (Fig. 7.2). Alternatively, partners can trade by volume and donate as much sperm as they receive, as shown recently for *D. gonocephalia* (Vreys and Michiels 1998).

Conditional reciprocity is likely to be associated with pre-copula or in copula stimulation or assessment mechanisms whereby interruptions before or during early copulation should be relatively frequent. Of particular interest are cases where animals copulate in such a way that one partner keeps the other physically under control, thus preventing it from leaving before reciprocity is completed. The physical contact between earthworms during copulation is, for example, extreme and probably allows accurate control over what the partner is doing (Grove 1925; Grove and Cowley 1926). Many leeches intertwine during copulation as if to hold each other and it has been observed that this lasts longer when one partner does not reciprocate (Hoffmann 1956). In gastrotrichs, both partners intertwine in a tight 'knot' and first transfer sperm externally to a special sperm transfer organ, then to the partner (Ruppert 1978). The first step of this two-step process may reliably signal to the partner that sperm transfer will take place. In the flatworm *Amphiscolops langerhansi*

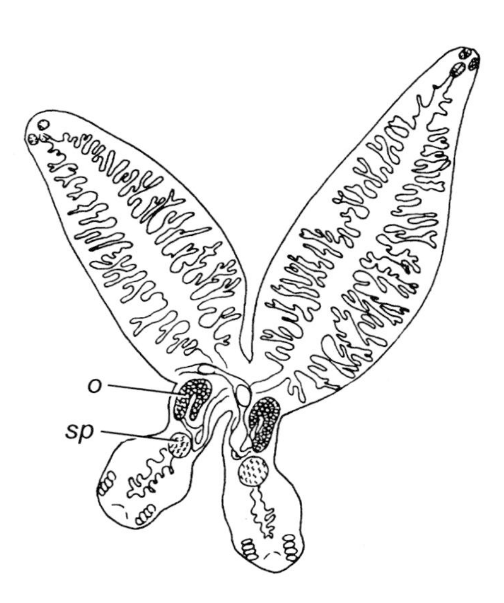

Fig. 7.7. *Extreme monogamy in the parasitic flatworm* Diplozoon paradoxum, *where partners fuse and stay together for life;* o, *ova;* sp, *sperm. Yolk glands are not indicated but if their size is similar to that seen in other monogeneans, sperm investment can be considered low (modified after Baer and Euzet 1961, with permission).*

partners use their penis only to hold on to that of their mate in a 'handshake' posture (Hyman 1951). Sperm exchange takes place on the outside of the interlocked penises and copulation takes a total of 40–60 min, which is unusually long. In another flatworm species *Monocelis fusca*, partners lock their tail ends into each other (Giesa 1966; Fig. 7.6C). The most extreme case is that of the fish parasite *Diplozoon paradoxum* (Fig. 7.7). In this species, larvae that have found a host do not mature until they pair up with another larva (Baer and Euzet 1961). Both individuals connect and fuse in the middle. Only then do they become sexually mature and stay together for life (which may be years).

Further detailed studies of copulatory mechanisms and behaviour should reveal how common conditional reciprocity really is. Unfortunately, descriptions of genitalia are rarely made *in copula* and data on reproductive behaviour are usually anecdotal and too vague to be useful in this context. If proven, conditional reciprocity will have important consequences for the evolution of assessment and signalling mechanisms in hermaphrodites. It may actually stabilize hermaphroditism because it sets additional limits to reproduction by one or both sexual functions. Conditional reciprocity may also result in the rejection of gonochoristic partners. In populations of the polychaete *Ophryotrocha diadema*, young individuals are male (protandry) and are rejected as partners by (older) hermaphrodites in favour of other hermaphrodites (Sella 1988).

VI. SPERM COMPETITION IN HERMAPHRODITES

A. A Contradiction?

As explained above, low population density is traditionally seen as one important explanation for the origin of hermaphroditism, suggesting a low importance of sperm competition. However, although many taxa indeed occur at low densities, population densities are so high in many others, that multiple matings are common (Brumpt 1900; Pearse and Wharton 1938; Nagao 1958; Ghirardelli 1968; Apelt 1969; Kutschera 1984; Nagasawa and Marumo 1984; Goto and Yoshida 1985; Kutschera and Wirtz 1986; Kearn 1992; Peters and Michiels 1996a,b; Peters *et al.* 1996; Vreys *et al.* 1997a). In leeches (Wilkialis 1970; Wilkialis and Davies 1980) and snails (see Chapter 8) multiple mating sometimes occurs in aggregations of several individuals during which partners are exchanged. Mating events whereby autosperm are donated to one partner while receiving allosperm from a third individual are known from sea slugs (see Chapter 8) and a monogenean flatworm (MacDonald and Caley 1975). In free-living flatworms one mating partner can also be actively displaced by a third individual (Darlington 1959; own unpublished observation). All this indicates that sperm competition may be as important in hermaphrodites as in any other animal species.

Even if low density were the general rule, it may actually have resulted in four preadaptations to sperm competition: (1) low density promotes promiscuous mating, since partners may be too rare to be selective; (2) long-term sperm storage is needed to assure continued fertility and is known from all hermaphrodites with internal fertilization (Adiyodi and Adiyodi 1983, 1988, 1990); (3) when new mating opportunities do occur, accurate control of the fertilization chances of old and new ejaculates may be favoured; (4) low density may also favour selfing. In order to control the optimal degree of inbreeding and outbreeding, mechanisms similar to selective storage will evolve in order to control the overlap between autosperm and allosperm, which then may also be used to distinguish between allosperm from different partners. An alternative is to have less fit autosperm. This may be disadvantageous in an outcrossing hermaphrodite, but is known from *C. elegans*, in which hermaphrodites use autosperm only to self-fertilize and where (rare) males inseminate hermaphrodites and subsequently outcompete their sperm (Dix *et al.* 1994; LaMunyon and Ward 1994, 1995). I shall now survey some of the possible responses to sperm competition, and consider the consequences for the sperm donating as well as the sperm receiving function (Table 7.3).

Table 7.3 Overview of possible adaptative responses to sperm competition, subdivided in adaptations related to the role of the receiver and that of the donor. Note that both roles are always combined in one hermaphroditic individual.

Donor		Receiver
Advertise own quality		*Selectively use allosperm*
• Genetical or phenotypical	↔	• Store and use allosperm selectively
• Sperm surface signals	↔	• Compatibility mechanisms
Improve chances of autosperm		*Counter-measures*
• Increase number of autosperm per ejaculate	↔	• Digest excess allosperm
• Optimize sperm packaging	↔	• Control allosperm movement
Reduce competition with allosperm stored by partner		*Counter-measures*
• Spermicidal ejaculatory fluid	↔	• Neutralize spermicide
• Evolve access to sperm stores	↔	• Prevent access to sperm stores
Reduce likelihood of future matings of partner		*Counter-measures*
• Induce refractory period	↔	• Recover mechanisms
• Mate guarding	↔	• Ignore
Increase immediate fecundity of partner		*Counter-measures*
• Suppress partner's male function	↔	• Stimulate own male function
• Stimulate partner's female function	↔	• Suppress own female function
Circumvent partner's counter-measures		*Counter-measures*
• Hypodermic impregnation	↔	• Behavioural avoidance
Reluctant to donate sperm (when sex-role reversed)		*Stimulate partner to donate*

(↔ indicates that the inscriptions in both columns are responses and counter responses to each other.)

B. *Postcopulatory sperm selection*

When multiple ejaculates overlap, the receiver function may develop post-copulatory mechanisms for sperm selection. This is particularly important in hermaphrodites since precopulatory mate choice may reflect the preference of the individual as a donor, rather than its preference as a receiver. It may therefore be essential to possess mechanisms that discriminate less favoured from preferred allosperm after a copulation. However, data suggestive of sperm selection in hermaphrodites are very rare. A preliminary survey of female reproductive organs across taxa suggests that a wide variety of mechanisms exist that have the potential of controlling which sperm are used for fertilization. Sperm resorption in the female genital tract is known from free-living flatworms (Cernosvitov 1931, 1932; Fischlschweiger and Clausnitzer 1984; Sluys 1989; Fischlschweiger 1991, 1994), oligochaetes (Grove 1925; Lasserre 1975; Adiyodi 1988) and snails (Chapter 8). In leeches, the behaviour of a spermatophore receiver indicates that it tries to rub it off immediately after receipt (Myers

1935), and particularly when in a poor condition, receivers may actually consume it (Brumpt 1900). Eating allosperm before they had a chance to enter the sperm receptacle has also been observed in arrowworms (John 1933). In leeches, hypodermically injected sperm are intensively phagocytosed in the coelomic sinuses, particularly in the vicinity of the ovaries (Brumpt 1900). A duct through which excess allosperm are transported from the sperm-receiving bursa copulatrix into the gut is characteristic of many free-living flatworms (Bock 1927; Hyman 1951; Ball and Reynoldson 1981). Others have 'nozzles' or 'mouthpieces' in the wall of the bursa that may function as sperm filters. Allosperm apparently have to move through their extremely narrow lumen to leave the bursa and reach the eggs (Costello and Costello 1938; Henley 1974). Selective usage may also be acquired through separate storage, as suggested by the presence of multiple bursae in some flatworms (up to 40 in *Oligochoerus limnophilus*; Henley 1974) and the large number of spermathecae seen in some oligochaetes (Adiyodi 1988; Fig. 7.8). In planarian flatworms, sperm are received in the caudally situated bursa (Fig. 7.9), and subsequently have to migrate from there, via the oviducts, to the sperm receptacles in the head region (Ball and Reynoldson 1981). The sheer length of this trajectory, plus the fact that sperm can be resorbed everywhere along it (Sluys 1989) suggests that the female system may function as a 'race track' for

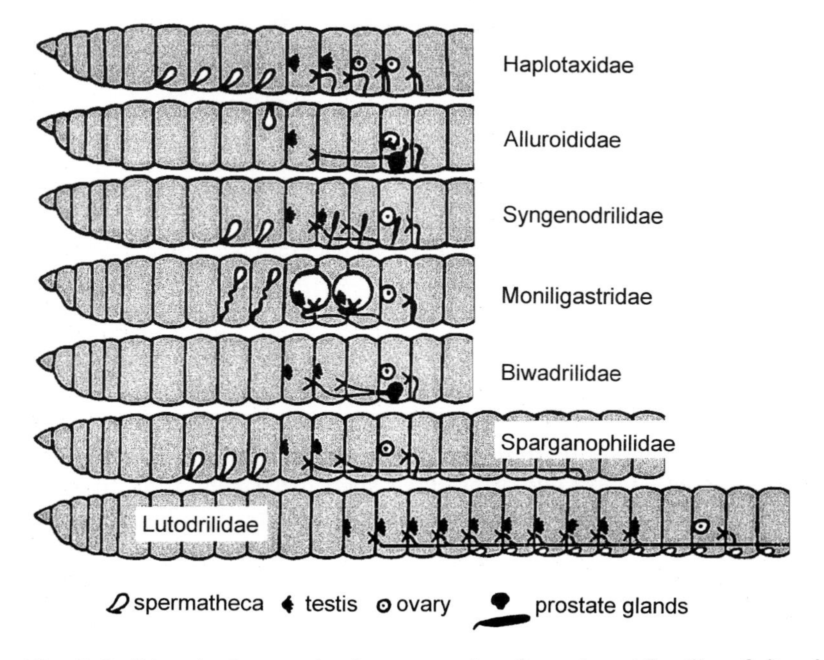

Fig. 7.8. *Diversity in reproductive systems in a few selected families of oligochaetes. Note the occasionally high number of testes and spermathecae (modified after Adiyodi 1988).*

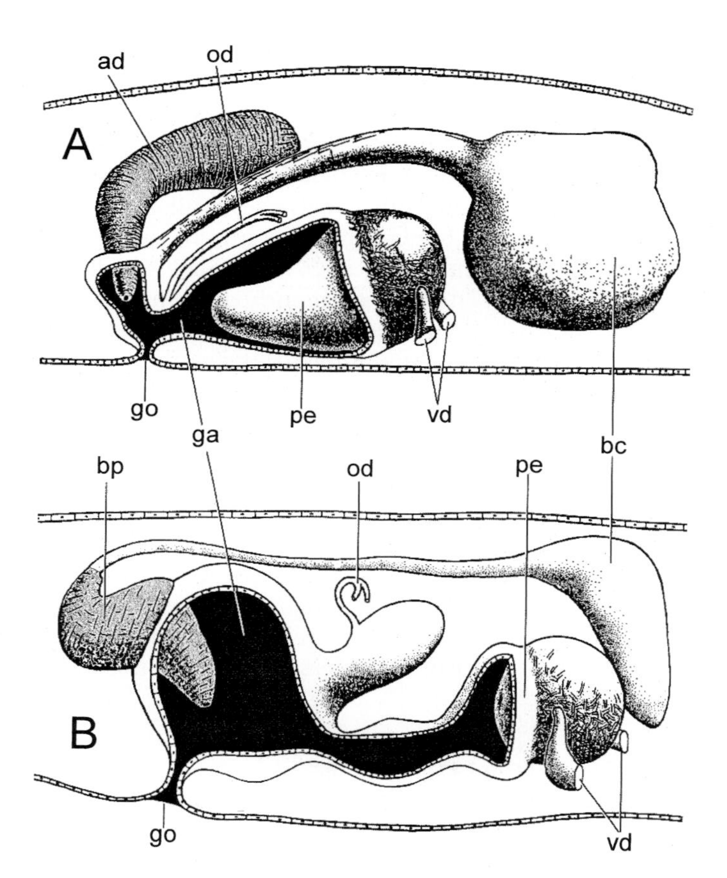

Fig. 7.9. *Two examples of peculiar genital morphologies in planarian flatworms.* **A.** Planaria torva *possesses a musculo-glandular organ or adenodactyl (*ad*) that may function as a sexual stimulator during copulation (Hyman 1951).* **B.** Bdellocephala punctata *(lower) has a penis (*pe*) that is not evertable. Instead, sperm are collected in the partner using the papilla-like extension (*bp*) of the bursal canal, a 'female penis'. Other abbreviations:* bc, *bursa copulatrix;* ga, *common atrium;* go, *gonopore;* od, *oviduct;* vd, *vasa deferentia (modified after Ball and Reynoldson 1981).*

sperm, similar to that in mammals (see Chapter 16) or in flowers (see Chapter 5). Arrowworms have specialized cells that connect the sperm receptacle directly to the cytoplasm of the ripening eggs (Kapp 1991), thus allowing individual sperm to migrate straight into eggs (Reeve and Lester 1974). This morphology suggests a sperm sieving function. The only known example of sperm incompatibility in hermaphrolitic animals comes from ascidians. In *Diplosoma listerianum*, autosperm and allosperm of the same type are blocked at the entrance of the female reproductive system (Bishop 1996; Bishop *et al.* 1996). This system bears a clear resemblance to the pollen–pistil interactions described by Delph (Chapter 5). In the ascidian *Ciona intestinalis*, eggs are enveloped by follicle cells which can selectively block autosperm (De Santis and Pinto 1991).

Sperm selection by the receiver poses a problem to the donor, who now has to convince its partner to use its sperm to fertilize eggs. Costly nuptial gifts are an unlikely solution since both functions of the hermaphroditic partner will benefit from them. A paternal offspring produced via such gift is therefore likely to be more expensive than using the gift oneself to produce additional maternal offspring. Hence, sperm selection in hermaphrodites must be based on less costly indicators of a donor's quality. Less expensive ornaments that enhance the attractiveness of an individual as a receiver and as a donor may be a cheaper strategy. The prediction of Eberhard (1986, 1993) and Eberhard and Cordero (1995) that genitalia and prostate products may be used in mate choice may very well apply to hermaphrodites. In free-living flatworms genitalia, as well as prostate glands, are complex (Hyman 1951). In the acoel flatworm *Convoluta convoluta*, for example, the penis is everted like a glove with finger-like glandular extensions that anchor the penis within the vagina (Apelt 1969). Because most hermaphroditic animals have poor eyesight, it is very likely that they require tactile or chemical signals that can only be detected over short distances, or while already in contact, or even in copula.

C. *Maximize success of autosperm in receiver*

A classic response to sperm competition is to increase the number of autosperm transferred per mating (see Chapter 1). Using principles from sperm displacement mechanisms in insects, Charnov (1996) investigated how sex allocation should change in response to sperm competition, and developed an ESS model that predicts that allocation to the male function is directly related to the number of sperm the donor can produce maximally, divided by the number of sperm that can be stored by the receiver. Although the simplicity of this result is appealing, data to test both the model and its assumptions may prove hard to obtain.

A rough comparison of anatomical drawings of the relative size of testes plus prostate glands vs. ovaries plus yolk glands (Grassé 1959, 1960, 1961) suggests that male allocation is high in most hermaphrodites (Figs 7.4, 7.8 and 7.10). Such measures may, however, represent an unreliable indicator of sperm competition. It may be necessary to limit comparisons to cases where the degree of sperm competition is the most important difference between a few related species (Petersen 1990). In addition, differential costs to maintain prostate glands, male genitalia and mating behaviour may make differences between species misleading. Finally, many hermaphrodites show a slight temporal separation in sex development. If, for example, young hermaphrodites start out as a male

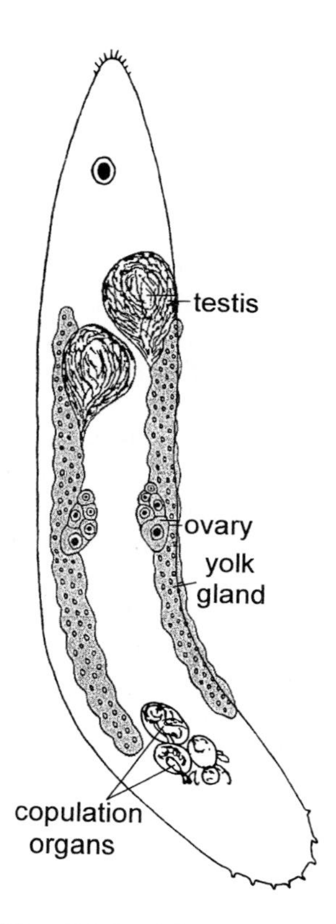

Fig. 7.10. *A free-living flatworm,* Monotoplana diorchis *with multiple copulatory complexes. It is assumed that copulatory complexes are lost during copulation and replaced (modified after Ax 1958, with permission).*

(protandry), this may reduce the apparent male allocation in 'adult' hermaphrodites.

When the receiver can obtain amounts of sperm that considerably exceed the amounts needed for fertilization, it may further extend the previously mentioned ability to digest sperm. Allosperm digestion by the receiver may, in return, result in the evolution of compensating traits in the donor such as chemicals that neutralize digestive enzymes or protective sperm packaging methods, such as a spermatophore. In the planarian *D. polychroa*, the transfer of a sperm clump is preceded and followed by the injection of a heterogeneous, sperm-free seminal fluid which might function as an inhibitor for digestion by the bursa copulatrix (A. Streng pers. comm.). In the related species *D. gonocephala*, however, sperm are transferred in a spermatophore and most of these are able to leave the bursa before the spermatophore is digested (Vreys *et al.* 1997b). Interestingly, the latter species also produces fewer sperm than the former (personal observation). But even when a spermatophore is used, the receiver function may still control the movement and survival of sperm further down the genital tract by employing many of the mechanisms mentioned above.

Prolonged copulation may also increase the fertilization chances of autosperm. This is particularly so when sperm need time to migrate away from a dangerous area. Arrowworms, for example, deposit a sperm cluster on each other's skin, and sperm subsequently move actively to the gonopores (Ghirardelli 1968). John (1933) observed that arrowworm partners held each other for up to 2 h (although sperm donation takes only seconds) and that the receiver ate the allosperm at the outside of its vagina after separation, suggesting that partners may hold each other to minimize the amount eaten. In the leech *Piscicola geometra* copulations last 5–6 h, even though spermatophores are exchanged at the beginning (Brumpt 1900). In this case, copulation may prevent the receiver from removing the spermatophore before it has discharged its content.

D. Reduce competition with allosperm

The ability to physically remove sperm directly from the sperm storage organs, as occurs in some insects (see Chapter 10), appears to be absent from hermaphrodites. Although hooked or spined penises are common among parasitic and free-living flatworms (Hyman 1951; Joyeux and Baer 1961; Williams and McVicar 1968), this armament appears to function as an anchor rather than a sperm displacement device. The only data suggestive of a sperm removal mechanism in hermaphrodites is from a study of the oligochaete *Euthyphoeus waltoni*, in which the paired penises carry a set of setae (Bahl 1928). Two of them are used during a single copulation and subsequently replaced. They are 5.5 mm long and can extend at least 1 mm beyond the tip of the penis. As the spermathecal duct is only 0.09 mm long it seems likely that the setae are driven deep into the spermathecae. Whether they are able to displace sperm is, however, not known. The rare occurrence of active sperm displacement among hermaphrodites may be due to the fact that they may not benefit from providing access to their sperm stores. In gonochorists, females can offer males increased paternity in this way in return for benefits such as guarding against other males and access to resources (see Chapter 10). In hermaphrodites, however, there appears to be nothing in terms of access or protection that a partner can offer that the receiver does not have already. Hermaphroditic partners could, in principle, 'trade' certainty of paternity with each other, but in internal fertilizers this possibility seems wide open for cheating.

E. Induction of a refractory period in the sperm receiver?

A donor would benefit if it could prolong the remating interval of the partner. Prostate fluids or other secretions may have an, as yet unknown, function here. Alternatively, deliberate injury of the partner could switch on an 'emergency physiology' and result in lower mobility and relocation of resources from storage to reproduction because of reduced survival perspectives. Hypodermic impregnation, although described in a different context below, may have such a function. Even if the receiver pays high long-term costs for such matings, the donor will be selected to inflict a wound on its partner if it can increase its short-term fertilization success in this way. This could explain the often aggressive way in which many hermaphrodites copulate. Some oligochaetes, for example, possess specialized, grooved chaetae that carry the secretion of specialized glands deep into the partner's body and might even transverse it completely (Needham 1990). It also offers a speculative explanation for dart-shooting and aggressive biting in mating snails (see Chapter 8).

A more direct reduction of the partner's maleness can be achieved by mutilating its male function so that it can no longer donate sperm, thus reducing its mating rate. In the slug *Ariolimax columbianus* (Mead 1943; see Chapter 8), individuals frequently lose their penis during copulation. It becomes stuck in the female genital tract and is bitten off either by the receiver or by its possessor (Leonard 1991). The presence of multiple copulation organs in some free-living flatworms (e.g. Hyman 1951; Fig. 7.10) has been interpreted as a way to replace copulatory complexes lost during copulation (Ax 1958).

Guarding against rival conspecifics is another possible way of forcing a refractory period onto the partner, but seems to be rare or even absent in hermaphrodites. One explanation may be that when a third individual interferes, the original partners may start to compete for the new, possibly better, individual rather than to protect each other. Another explanation may be that inducing a refractory period in the partner is likely also to breach the own optimal mating interval. However, an indication that mutual, 'traded' guarding might take place, comes from the flatworm *D. gonocephala*. When copulating, sperm transfer is completed after precisely 4.5 h, but partners remain in copula for up to 10 h (average 5 h) (Vreys *et al.* 1997b). Because in this species copulations take place mainly at night, a prolongation may reduce the chances that a partner mates again that same night, leaving the day for sperm migration within the female tract.

F. Manipulation of the receiver's sex allocation

Hermaphrodites might not only attempt to prolong the refractory period of their partner, but also unbalance its sex allocation to make it more female and less male. This not only increases the number of eggs it will produce, but may also reduce its mating rate. Although such a manipulation has not yet been documented, selection on the donor to do so must be high. This may be yet another explanation for the importance of prostate products. Hermaphrodites could defend themselves against feminizing products from the partner by a priori biasing allocation to the male function so that, when being feminized by the partner, the resultant sex allocation is still close to optimal for the manipulated individual. This could explain the widespread occurrence of protandry among hermaphrodites. Eventually, feminizing agents may result in hermaphrodites that are masculine as long as they are virgin and must receive an ejaculate in order to feminize. Arrowworms, for example, have to mate in order to produce eggs for the first time (Alvariño 1990). In order to keep feminizing factors separate from their own reproductive system, donors could exploit ejaculate resorption by the receiver (see earlier) and transfer products that become biochemically active when metabolized. Data for snails suggest that young individuals indeed need to metabolize an ejaculate from a partner in order to induce egg development (Kruglov 1980). A major difference with gonochorists is that, when males induce females to produce more eggs, they may only be able to do so by offering her additional resources, which is not in conflict with her interests. In a hermaphrodite, however, a donor is not interested in topping up the resources of its partner (see above), but may attempt to free resources by reallocation within its receiver instead, resulting in a conflict that is unique to hermaphrodites.

G. *Circumvention of the partner's defences: hypodermic impregnation*

Copulatory mechanisms may eventually become so complicated and costly for the donor function, particularly if the receiving function has managed to keep pace and controls fertilization completely, that it may pay the donor to circumvent completely the receiver's control. Injecting sperm hypodermically rather than into a female sperm-receiving organ is common among hermaphrodites but not in gonochorists. It is ubiquitous in large groups of free-living flatworms (Hyman 1951; Michiels and Newman 1998; Fig. 7.6), and is also known from parasitic flatworms (Joyeux and Baer 1961; Baer and Euzet 1961; Ubelaker 1983; Kearn 1992), polychaetes (Westheide 1979; Olive 1983; Adiyodi 1988) and sea slugs (see Chapter 8). Such species usually have a stylet-shaped penis. Leeches, however, glue a highly specialized spermatophore onto the skin of the partner (Myers 1935; Lasserre 1975; Kutschera 1989; Sawyer 1986; Fig. 7.5). Enzymes first dissolve the skin, after which sperm are injected into the body cavity by a spontaneous contraction of the spermatophore, which continues to inject sperm until long after the donor has gone.

In addition to circumventing the receiver's control, hypodermic impregnation saves time by being one order of magnitude shorter. Unilateral hit-and-run usually takes only seconds (Myers 1935; Apelt 1969; Borkott 1970; Kutschera 1984) whereas reciprocal hypodermic impregnation ranges from a few seconds up to 60 min (Brumpt 1900; Hyman 1951; Harrison 1953; Hoffman 1956; Nagao 1958; Apelt 1969; Kutschera and Wirtz 1986; Kutschera 1989; Kearn 1992; Kearn and Whittington 1992; Michiels and Newman 1998). Although normal copulation can be short in some species (Costello and Costello 1938; Hyman 1951; Darlington 1959; Apelt 1969) it often takes hours (Kato 1940; Yanagita 1964; Wilkialis and Davies 1980; Peters *et al.* 1996; Vreys *et al.* 1997a) or even days (Brumpt 1900). An unusual, unilateral mechanism of hypodermic impregnation has been described for the parasitic flatworm *Diclidophora merlangi* in which the penis applies suction to the epidermis of the partner and dissolves the skin, after which spines hold the wound open while sperm are injected (MacDonald and Caley 1975). This process takes 1 h.

One explanation for the relatively low occurrence of hypodermic impregnation in gonochorists may be that unreceptive females can simply hide or avoid males. Only when they are receptive will they seek a partner. As a result, males will not need forced sperm donation to inseminate them or could even lose attractiveness. In hermaphrodites, however, animals that do not want to receive still have to copulate in order to donate. Under such conditions, it may be advantageous to develop forced sperm transfer.

H. Sperm competition and sex-role reversal

Thus far, I have assumed that hermaphrodites mate because they want to donate sperm and that sperm receipt still leaves the option to accept or reject it. However, as explained earlier, sex-role reversal might be common in hermaphrodites. Sperm competition will be less significant as a result, and the receiving function may even become sperm limited and start to develop advertisement tactics to get its eggs fertilized. This may explain why, for example, some planarian flatworms have no evertible penis, but instead have a female intromittant organ that is used to suck up the donor's sperm (e.g. *Bdellocephala*; Ullyott and Beauchamp 1931; Fig. 7.9). An even more extreme case is offered by the suggestion that certain turbellarians obtain allosperm by cannibalizing their partner (Sterrer and Rieger 1974).

VII. CONCLUDING REMARKS

Although constrained in their evolutionary response to sexual selection, it is clear that there is scope for evolutionary arms races in hermaphrodites that may result in adaptations that are at least as diverse as those seen among gonochorists. Comparisons between gonochorists and hermaphrodites also show that the dominance of the former in sexual selection theory might actually hamper our understanding of the latter. Certain aspects of hermaphrodite biology, such as the extreme diversity in sperm morphology, the importance of cannibalism or the many specialized glands and receptors have been ignored, but would further enrich this image. Although I have (artificially) separated sex allocation, reciprocity of sperm transfer and sperm competition, it is obvious that the selection pressures that result from each of these are not independent, but will often result in similar adaptations. This may make it difficult to find out whether, for example, assessment occurs to measure the partners quality as a donor, as a receiver, or as a reciprocating partner. As a result, many of the examples that I have mentioned in one section, may equally serve as 'suggestive evidence' in another. In order to disentangle these phenomena, it is very important to have accurate data on sperm production, transfer, depletion, storage and selection, as well as life history data for the species. Events that diverge from the typical pattern for a given species may be of particular importance (e.g. unilateral transfer in an otherwise reciprocally inseminating species).

It is unfortunate that I have to confirm what Eric Charnov wrote 15 years ago: although the theoretical basis for the understanding of hermaphroditic mating systems is available, far too few examples have been worked out sufficiently to put the theory to the test (Charnov 1982). I hope that my (speculative) attempts to further elaborate existing ideas will prompt new students in behavioural ecology to consider the unexplored aspects of hermaphroditic systems and share my excitement. In the end, I hope it may also lead to enhanced communication between botanists (the professionals in this context) and zoologists.

ACKNOWLEDGEMENTS

This manuscript has benefited from frequent discussions on hermaphrodite mating behaviour with Rolf Weinzierl. I gratefully acknowledge helpful comments from Leo Beukeboom, Angél Martín Algánza, Bruno Baur, Tim Birkhead, Philippe Jarne, Anders Møller, Anne Peters, Tim Sharbel, Andrea Streng, Claus Wedekind, Jack Werren and Julie Zeitlinger. Many of the ideas presented have been positively influenced by the presentations and discussions at a workshop of the European Science

Foundation on the evolution of hermaphrodite mating systems (Jarne and Charlesworth 1996).

REFERENCES

Adiyodi KG (1988) Annelida. In *Reproductive Biology of Invertebrates. III. Accessory Sex Glands.* KG Adiyodi & RG Adiyodi (eds), pp. 189–250. Wiley, Chichester.

Adiyodi KG & Adiyodi RG (1983) *Reproductive Biology of Invertebrates. Vol. I. Oogenesis, Oviposition and Oosorption.* Wiley, Chichester.

Adiyodi KG & Adiyodi RG (1988) *Reproductive Biology of Invertebrates. Vol. III. Accessory Sex Glands.* Wiley, Chichester.

Adiyodi KG & Adiyodi RG (1990) *Reproductive Biology of Invertebrates. Vol. IV, Parts A and B. Fertilization, Development, and Parental Care.* Wiley, Chichester.

Akin JW, Tho SPT & Mcdonough PG (1993) Reconsidering the difference between mixed gonadal-dysgenesis and true hermaphroditism. *Adolesc. Pediat. Gynecol.* **6:** 102–104.

Alvariño A (1990) Chaetognatha. In *Reproductive Biology of Invertebrates. IV (Part B) Fertilization, Development, and Parental Care.* KG Adiyodi & RG Adiyodi (eds), pp. 255–282. Wiley, Chichester.

Andersson M (1994) *Sexual Selection.* Princeton University Press, Princeton, NJ.

Apelt G (1969) Fortpflanzungsbiologie, Entwicklungszyklen und vergleichende Frühentwicklung acoeler Turbellarien. *Mar. Biol.* **4:** 267–325.

Arnold SJ (1994a) Is there a unifying concept of sexual selection that applies to both plants and animals? *Am. Nat.* **144:** 1–12.

Arnold SJ (1994b) Bateman principles and the measurement of sexual selection in plants and animals. *Am. Nat.* **144:** 126–149.

Avel M (1959) Classe des Annélides Oligochètes (Oligochaeta Huxley, 1875) In *Traité de Zoologie. Anatomie, Systématique, Biologie. IV, Fasc. I. Annélides, Myzostomides, Sipunculiens, Echiuriens, Priapuliens, Endoproctes, Phoronidiens.* P-P Grassé (ed.), pp. 224–470. Masson et Cie, Paris.

Ax P (1958) Vervielfachung des männlichen Kopulationsapparates bei Turbellarien. *Verhandl. Deutsch. Zoologisch. Gesellsch. Graz* 227–249.

Baer J-G & Euzet L (1961) Classe de Monogènes. In *Traité de Zoologie. Anatomie, Systématique et Biologie. Tome IV. Plathelminthes, Mézozoaires, Acanthocéphales, Némertiens.* P-P Grassé (ed.), pp. 243–325. Masson et Cie, Paris.

Bahl KN (1928) On the reproductive processes of earthworms, Part 1: the process of copulation and exchange of sperms in *Eutyphoeus waltoni* Mich. *Quart. J. Microsc. Sci.* **71:** 479–502.

Ball IR & Reynoldson TB (1981) *British Planarians.* Cambridge University Press, Cambridge.

Barker DM (1992) Evolution of sperm shortage in a selfing hermaphrodite. *Evolution* **46:** 1951–1955.

Barnes RD & Ruppert EE (1994) *Invertebrate Zoology,* 6th edn. Saunders, Philadelphia.

Bateman AJ (1948) Intra-sexual selection in *Drosophila. Heredity* **2:** 349–368.

Bell G (1982) *The Masterpiece of Nature. The Evolution and Genetics of Sexuality.* Croom Helm, London.

Berglund A (1986) Sex change by a polychaete: effects of social and reproductive costs. *Ecology* **67:** 837–845.

Berglund A (1991) To change or not to change sex: a comparison between two *Ophryotrocha* species (Polychaeta). *Evol. Ecol.* **5:** 128–135.

Bishop JDD (1996) Female control of paternity in the internally fertilizing compound ascidian *Diplosoma listerianum.* I. Autoradiographic investigation of sperm movements in the female reproductive tract. *Proc. Roy. Soc. Lond. Biol. Ser.* **263:** 369–376.

Bishop JDD, Jones CS & Noble LR (1996) Female control of paternity in the internally fertilizing compound ascidian *Diplosoma listerianum.* II. Investigation of male mating success using RAPD markers. *Proc. Roy. Soc. Lond. Biol. Ser.* **263:** 401–407.

Bock S (1927) Ductus genito-intestinalis in the polyclads. *Arkiv Zoologi* **19:** 1–15.

Borkott H (1970) Geschlechtliche Organisation, Fortpflanzungsverhalten und Ursachen der sexuellen Vermehrung von *Stenostomum sthenum* nov. spec (Turbellaria Catenulida). *Zeitschr. Morphol. Tiere* **67:** 183–262.

Brumpt E (1900) Réproduction des Hirudinées. *Mém. Soc. Zool. Fr.* **13:** 286–430.

Bunch TD, Callan RJ, Maciulis A, Dalton JC, Figueroa MR, Kunzler R & Olson RE (1991) True hermaphroditism in a wild sheep: a clinical report. *Theriogenology* **36**(2): 185–190.

Cernosvitov L (1931) Studien über die Spermaresorption – III – Die Samenresorption bei den Tricladen. *Zoologisch. Jahrb. Abt. System.* **54:** 296–332.

Cernosvitov L (1932) Studien über die Spermaresorption – IV – Verbreitung der Samenresorption bei den Turbellarien. *Zoologisch. Jahrb. Abt. System.* **55:** 137–172.

Charlesworth D & Charlesworth B (1981) Allocation of resources to male and female functions in hermaphrodites. *Biol. J. Linn. Soc.* **15:** 57–74.

Charnov EL (1979) Simultaneous hermaphroditism and sexual selection. *Proc. Natl Acad. Sci. USA* **76:** 2480–2484.

Charnov EL (1980) Sex allocation and local mate competition in barnacles. *Mar. Biol. Lett.* **1:** 269–272.

Charnov EL (1982) *The Theory of Sex Allocation.* Princeton University Press, Princeton, NJ.

Charnov EL (1996) Sperm competition and sex allocation in simultaneous hermaphrodites. *Evol. Ecol.* **10:** 457–462.

Charnov EL, Maynard Smith J & Bull JJ (1976) Why be an hermaphrodite? *Nature* **263:** 125–126.

Costello HM & Costello DP (1938) Copulation in the acoelus turbellarian *Polychoerus carmelensis. Biol. Bull.* **75:** 85–98.

Darlington JD (1959) The Turbellaria of two granite outcrops in Georgia. *Am. Midl. Nat.* **61:** 257–294.

Darwin C (1871) *The Descent of Man and Selection in Relation to Sex.* J. Murray, London.

De Guise S, Lagace A & Beland P (1994) True hermaphroditism in a St. Lawrence beluga whale (*Delphinapterus leucas*). *J. Wildl. Dis.* **30**(2): 287–290.

De Santis R & Pinto MR (1991) Gamete self-discrimination in ascidians: a role for the follicle cells. *Molec. Reprod. Devel.* **29**(1): 47–50.

Dix I, Koltai H, Glazer I & Burnell AM (1994) Sperm competition in mated 1st generation hermaphrodite females of the hp-88 strain of heterorhabditis (nematoda, heterorhabditidae) and progeny sex-ratios in mated and unmated females. *Fund. Appl. Nematol.* **17:** 17–27.

lated sperm by artificial insemination of *Caenorhabditis elegans*. *Genetics* **138:** 689–692.

LaMunyon CW & Ward S (1995) Sperm precedence in a hermaphroditic nematode (*Caenorhabditis elegans*) is due to competitive superiority of male sperm. *Experientia* **51**(8)**:** 17–23.

Lasserre P (1975) Clitellata. In *Reproduction of Marine Invertebrates*, Vol. III. AS Giese & JS Pierse (eds), pp. 215–275. Academic Press, New York.

Leigh EG, Charnov EL & Warner RR (1976) Sex ratio, sex change and natural selection. *Proc. Natl Acad. Sci. USA* **73:** 3655–3660.

Leonard JL (1991) Sexual conflict and the mating systems of simultaneously hermaphroditic gastropods. *Am. Malac. Bull.* **9:** 45–58.

Leonard JL & Lukowiak K (1984) Male-female conflict in a simultaneous hermaphrodite resolved by sperm trading. *Am. Nat.* **124:** 282–286.

Leonard JL & Lukowiak K (1985) Courtship, copulation, and sperm trading in the sea slug, *Navanax inermis* (Opisthobranchia: Cephalaspidea). *Can. J. Zool.* **63:** 2719–2729.

Leonard JL & Lukowiak K (1991) Sex and the simultaneous hermaphrodite: testing models of male–female conflict in a sea slug, *Navanax inermis* (Opisthobranchia). *Anim. Behav.* **41:** 255–266.

MacDonald S & Caley J (1975) Sexual reproduction in the monogenean *Diclidophora merlangi*: tissue penetration by sperms. *Zeitschr. Parasiten.* **45:** 323–334.

Mead AR (1943) Revision of the giant West Coast land slugs of the genus *Ariolimax* Moerch (Pulmonata: Arionidae). *Am. Midl. Nat.* **30:** 675–717.

Michiels NK & Newman LJ (1998) Peru's fencing in flatworms. *Nature* (in press).

Michiels NK & Streng A (1998) Sperm exchange in a simultaneous hermaphrodite. *Behav. Ecol. Sociobiol.* (in press).

Morgan MT (1994) Models of sexual selection in hermaphrodites, especially plants. *Am. Nat.* **144:** 100–125.

Myers RJ (1935) Behaviour and morphological changes in the leech *Placobdella parasitica* during hypodermic insemination. *J. Morphol.* **57:** 617–653.

Nagao Z (1958) Some observations on the breeding habits of the freshwater leech *Glossiphonia lata* Oka. *Jpn J. Zool.* **12:** 219–228.

Nagasawa S & Marumo R (1984) Feeding habits and copulation of the chaetognath *Sagitta crassa*. *Mer (Tokyo)* **22**(1)**:** 8–14.

Nakashima Y, Kuwamura T & Yogo Y (1995) Why be a both-ways sex changer. *Ethology* **101:** 301–307.

Needham AE (1990) Annelida – Clitellata. In *Reproductive Biology of Invertebrates. IV (Part B) Fertilization, Development and Parental Care.* KG Adiyodi & RG Adiyodi (eds), pp. 1–36. Wiley, Chichester.

Newman LJ & Cannon LRG (1994) *Pseudoceros* and *Pseudobiceros* (Platyhelminthes, Polycladida, Pseudocerotidae) from eastern Australia and Papua New Guinea. *Mem. Queensl. Mus.* **37**(1)**:** 205–266.

Nielsen C (1995) *Animal Evolution: Interrelationships of the Living Phyla.* Oxford University Press, Oxford.

Olive PJW (1983) Annelida – Polychaeta. In *Reproductive Biology of Invertebrates. II. Spermatogenesis and Sperm Function.* KG Adiyodi & RG Adiyodi (eds), pp. 321–342. Wiley, Chichester.

Pearse AS & Wharton GW (1938) The oyster 'leech', *Stylochus inimicus* Palombi associated with oysters on the coasts of Florida. *Ecol. Monogr.* **8:** 605–655.

Peters A & Michiels NK (1996a) Evidence for a lack of inbreeding avoidance by former (personal observation). But even when a spermatophore is used, the receiver function may still control the movement and survival of sperm further down the genital tract by employing many of the mechanisms mentioned above.

Prolonged copulation may also increase the fertilization chances of autosperm. This is particularly so when sperm need time to migrate away from a dangerous area. Arrowworms, for example, deposit a sperm cluster on each other's skin, and sperm subsequently move actively to the gonopores (Ghirardelli 1968). John (1933) observed that arrowworm partners held each other for up to 2 h (although sperm donation takes only seconds) and that the receiver ate the allosperm at the outside of its vagina after separation, suggesting that partners may hold each other to minimize the amount eaten. In the leech *Piscicola geometra* copulations last 5–6 h, even though spermatophores are exchanged at the beginning (Brumpt 1900). In this case, copulation may prevent the receiver from removing the spermatophore before it has discharged its content.

D. Reduce competition with allosperm

The ability to physically remove sperm directly from the sperm storage organs, as occurs in some insects (see Chapter 10), appears to be absent from hermaphrodites. Although hooked or spined penises are common among parasitic and free-living flatworms (Hyman 1951; Joyeux and Baer 1961; Williams and McVicar 1968), this armament appears to function as an anchor rather than a sperm displacement device. The only data suggestive of a sperm removal mechanism in hermaphrodites is from a study of the oligochaete *Euthyphoeus waltoni*, in which the paired penises carry a set of setae (Bahl 1928). Two of them are used during a single copulation and subsequently replaced. They are 5.5 mm long and can extend at least 1 mm beyond the tip of the penis. As the spermathecal duct is only 0.09 mm long it seems likely that the setae are driven deep into the spermathecae. Whether they are able to displace sperm is, however, not known. The rare occurrence of active sperm displacement among hermaphrodites may be due to the fact that they may not benefit from providing access to their sperm stores. In gonochorists, females can offer males increased paternity in this way in return for benefits such as guarding against other males and access to resources (see Chapter 10). In hermaphrodites, however, there appears to be nothing in terms of access or protection that a partner can offer that the receiver does not have already. Hermaphroditic partners could, in principle, 'trade' certainty of paternity with each other, but in internal fertilizers this possibility seems wide open for cheating.

E. Induction of a refractory period in the sperm receiver?

A donor would benefit if it could prolong the remating interval of the partner. Prostate fluids or other secretions may have an, as yet unknown, function here. Alternatively, deliberate injury of the partner could switch on an 'emergency physiology' and result in lower mobility and relocation of resources from storage to reproduction because of reduced survival perspectives. Hypodermic impregnation, although described in a different context below, may have such a function. Even if the receiver pays high long-term costs for such matings, the donor will be selected to inflict a wound on its partner if it can increase its short-term fertilization success in this way. This could explain the often aggressive way in which many hermaphrodites copulate. Some oligochaetes, for example, possess specialized, grooved chaetae that carry the secretion of specialized glands deep into the partner's body and might even transverse it completely (Needham 1990). It also offers a speculative explanation for dart-shooting and aggressive biting in mating snails (see Chapter 8).

A more direct reduction of the partner's maleness can be achieved by mutilating its male function so that it can no longer donate sperm, thus reducing its mating rate. In the slug *Ariolimax columbianus* (Mead 1943; see Chapter 8), individuals frequently lose their penis during copulation. It becomes stuck in the female genital tract and is bitten off either by the receiver or by its possessor (Leonard 1991). The presence of multiple copulation organs in some free-living flatworms (e.g. Hyman 1951; Fig. 7.10) has been interpreted as a way to replace copulatory complexes lost during copulation (Ax 1958).

Guarding against rival conspecifics is another possible way of forcing a refractory period onto the partner, but seems to be rare or even absent in hermaphrodites. One explanation may be that when a third individual interferes, the original partners may start to compete for the new, possibly better, individual rather than to protect each other. Another explanation may be that inducing a refractory period in the partner is likely also to breach the own optimal mating interval. However, an indication that mutual, 'traded' guarding might take place, comes from the flatworm *D. gonocephala*. When copulating, sperm transfer is completed after precisely 4.5 h, but partners remain in copula for up to 10 h (average 5 h) (Vreys *et al.* 1997b). Because in this species copulations take place mainly at night, a prolongation may reduce the chances that a partner mates again that same night, leaving the day for sperm migration within the female tract.

F. Manipulation of the receiver's sex allocation

Hermaphrodites might not only attempt to prolong the refractory period of their partner, but also unbalance its sex allocation to make it more selective mating in a simultaneous hermaphrodite. *Invert. Biol.* **115**(2): 99–103.

Peters A & Michiels NK (1996b) Do simultaneous hermaphrodites choose their mates? Effects of body size in a planarian flatworm. *Freshw. Biol.* **36:** 623–630.

Peters A, Streng A & Michiels NK (1996) Mating behaviour in a hermaphroditic flatworm with reciprocal insemination: do they assess their mates during copulation? *Ethology* **102:** 236–251.

Petersen CW (1990) Sex allocation in simultaneous hermaphrodites: testing local mate competition theory. *Lect. Mathemat. Sci. Am. Mathemat. Soc.* **22**: 183–204.

Petersen CW (1991) Sex allocation in hermaphroditic sea basses. *Am. Nat.* **138**: 650–667.

Petersen CW (1995) Reproductive behavior, egg trading, and correlates of male mating success in the simultaneous hermaphrodite, *Serranus tabacarius*. *Environ. Biol. Fish.* **43:** 351–361.

Policansky D (1982) Sex change in plants and animals. *Annu. Rev. Ecol. System.* **13:** 471–495.

Premoli MC & Sella G (1995) Sex economy in benthic polychaetes. *Ethol. Ecol. Evol.* **7:** 27–48.

Raimondi PT & Martin JE (1991) Evidence that mating group size affects allocation of reproductive resources in a simultaneous hermaphrodite. *Am. Nat.* **138:** 1206–1217.

Reeve MR & Lester B (1974) The process of egg-laying in the chaetognath *Sagitta hispida*. *Biol. Bull.* **147:** 247–256.

Reeve MR & Walter MA (1972) Observations and experiments on methods of fertilization in the chaetognath *Sagitta hispida*. *Biol. Bull.* **143:** 207–214.

Reise H (1995) Mating-behavior of *Deroceras rodnae* Grossu and Lupu, 1965 and *Deroceras praecox* Wiktor, 1966 (Pulmonata, Agriolimacidae). *J. Mollusc. Stud.* **61:** 325–330.

Reynoldson TB & Bellamy LS (1973) Interspecific competition in lake-dwelling triclads – a laboratory study. *Oikos* **24:** 301–313.

Reynoldson TB & Young JO (1965) Food supply as a factor regulating population size in freshwater triclads. *Mitt. Int. Verein. Limnol.* **13:** 3–20.

Ridley M (1983) *The Explanation of Organic Diversity. The Comparative Method and Adaptions for Mating*. Clarendon Press, Oxford.

Ruppert EE (1978) The reproductive system of gastrotrichs. II. Insemination in *Macrodasys*: a unique mode of sperm transfer in Metazoa. *Zoomorphologie* **89:** 207–228.

Sawyer RT (1986) *Leech Biology and Behaviour*, Vols I–III. Clarendon Press, Oxford.

Sella G (1985) Reciprocal egg trading and brood care in a hermaphroditic polychaete worm. *Anim. Behav.* **33:** 938–944.

Sella G (1988) Reciprocation, reproductive success, and safeguards against cheating in a hermaphroditic polychaete worm, *Ophyotrocha diadema* Akesson. *Biol. Bull.* **175:** 212–217.

Sella G (1990) Sex allocation in the simultaneously hermaphroditic polychaete worm *Ophryotrocha diadema*. *Ecology* **71:** 27–32.

Sella G, Premoli MC & Turri F (1997) Egg trawling in the simultaneously hermaphroditic polychaete worm *Orphryotrocha gracilis*. *Behav. Ecol.* **8:** 83–86.

Sluys R (1989) Sperm resorption in triclads (Platyhelminthes, Tricladida). *Invert. Reprod. Devel.* **15**(2): 89–95.

Spurdle AB, Shankman S & Ramsay M (1995) XX true hermaphroditism in southern African blacks: exclusion of SRY sequences and uniparental disomy of the X chromosome. *Am. J. Med. Genet.* **55:** 53–56.

Sterrer W & Rieger R (1974) Retronectidae – a new cosmopolitan marine family of Catenulida (Turbellaria). In *Biology of the Turbellaria.* NW Riser & MP Morse (eds), pp. 63–92. McGraw-Hill, New York.

Teuchert G (1968) Zur Fortpflanzung und Entwicklung der Macrodasyoidea (Gastrotricha). *Zeitschr. Morphol. Tiere* **63:** 343–418.

Ubelaker JE (1983) The morphology, development and evolution of tapeworm larvae. In *Biology of the Eucestoda,* Vol. I. C Arme & PW Pappas (eds), pp. 235–296. Academic Press, London.

Ullyott P & Beauchamp RSA (1931) Mechanisms for the prevention of self-fertilization in some species of fresh-water triclads. *Quart. J. Microsc. Sci.* **74:** 477–490.

Vreys C & Michiels NK (1995) The influence of body size on immediate reproductive success in *Dugesia gonocephala* (Tricladida, Paludicola). *Hydrobiologia* **305**(1–3): 113–117.

Vreys C & Michiels NK (1997) Flatworms flatten to size up each other. *Proc. Roy. Soc. Lond. B* **264:** 1559–1564.

Vreys C & Michiels NK (1998) Sperm trading by volume in the hermaphroditic flatworm *Dugesia gonocephala. Anim. Behav.* (in press).

Vreys C, Schockaert ER & Michiels NK (1997a) Unusual pre-copulatory behaviour in the planarian flatworm *Dugesia gonocephala* (Tricladida: Paludicola). *Ethology* **103**(3): 208–221.

Vreys C, Schockaert ER & Michiels NK (1997b) Formation, transfer and assimilation of the spermatophore of the hermaphroditic flatworm *Dugesia gonocephala* (Tricladida, Paludicola). *Can. J. Zool.* **75:** 1479–1486.

Warner RR (1975) The adaptive significance of sequential hermaphroditism in animals. *Am. Nat.* **109**: 61–82.

Warner RR (1988) Sex change and the size-advantage model. *Trends Ecol. Evol.* **3**: 133–136.

Warner RR, Robertson DR & Leigh, EG Jr (1975) Sex change and sexual selection. *Science* **190:** 6333–6338.

Werren JH (1980) Sex ratio adaptations to local mate competition in a parasitic wasp. *Science* **208:** 1157–1158.

Westheide W (1979) Ultrastruktur der Genitalorgane interstitieller polychaeten – II – Männliche Kopulationsorgane mit intrazellulären Stilettstäben in einer *Microphthalmus*-Art. *Zoologica Scripta* **8:** 111–118.

Wilkialis J (1970) Investigations on the biology of leeches of the Glossiphoniidae family. *Zool. Polon.* **20**(1): 29–54.

Wilkialis J & Davies RW (1980) The reproductive biology of *Theromyzon tessulatum* (Glossiphoniidae: Hirudinoidea), with comments on *Theromyzon rude. J. Zool.* **192:** 421–429.

Williams GC (1975) *Sex and Evolution.* Princeton University Press, Princeton, NJ.

Williams HH & McVicar A (1968) Sperm transfer in Tetraphyllidea (Platyhelminths: Cestoda). *Nytt. Mag. Zool.* **16:** 60–71.

Wood WB (1988) *The Nematode* Caenorhabditis elegans. Cold Spring Harbor Laboratory Press, Cold Spring Harbor, NY.

Yanagita Y (1964) Observations on the copulation of a freshwater planarian, *Polycelis sapporo. J. Fac. Sci. Hokk. Univ. Ser. VI Zool.* **15:** 449–457.

8 Sperm Competition in Molluscs

Bruno Baur

Department of Integrative Biology, Section of Conservation Biology (NLU), University of Basel, St Johanns-Vorstadt 10, CH–4056 Basel, Switzerland

I. INTRODUCTION

Molluscs are numerically the second largest phylum in the animal kingdom with more than 120 000 living species, which are divided into eight classes (see Table 8.1). The visual differences between a snail, a clam, and a squid, each an example of a major class of the molluscs, belie the closeness of their relationship. Sexual behaviour is also extremely variable in this phylum. Many species are promiscuous and there are different forms of sperm storage, providing the potential for sperm competition. However, with a few exceptions, evolutionary and behavioural aspects of sperm competition have not been examined in this phylum. Most of the available evidence for sperm competition in molluscs is published in studies with other aims and therefore is indirect.

Sperm Competition and Sexual Selection
ISBN 0-12-100543-7

This chapter presents different lines of evidence which suggest that sperm competition might be important in the evolution of the reproductive behaviour in some groups of molluscs. It also shows that there are huge gaps in our understanding of the reproductive behaviour of molluscs. Theoretical aspects of sperm competition are dealt with in Chapter 1, and more specifically, with respect to hermaphrodites in Chapter 7 and by Charnov (1996) and are therefore not repeated here. General principles of mollusc reproduction have been reviewed by Duncan (1975), Thompson (1976), Beeman (1977), Giese and Pearse (1977, 1979), Tompa *et al.* (1984), Boyle (1983, 1987), Runham (1992) and Harrison and Kohn (1994). The following provides a taxonomic account of reproductive morphology, physiology and behaviour that potentially have implications for sperm competition.

II. CAUDOFOVEATA

Caudofoveata are shell-less molluscs 2–140 mm long (Salvini-Plawen 1985). Their worm-like body is totally covered by a chitinous cuticle with embedded scales (Scheltema *et al.* 1994). They are inhabitants of marine sediments and can be found at depths of up to 3540 m (Salvini-Plawen 1985). In this group of molluscs, sexes are separate, but external morphological distinctions are lacking (Hadfield 1979). Indirect evidence suggests that they are external fertilizers. No information on sperm competition is available in this class.

III. SOLENOGASTERS

Solenogasters are shell-less molluscs 0.3–300 mm long (Salvini-Plawen 1985). The laterally narrowed body is covered by a chitinous cuticle with embedded spicules, except on the narrowed foot or pedal groove. They live exclusively in the benthos at depths of 10–4000 m (Hyman 1967). Solenogasters are hermaphrodites, but little is known about their mating system (Hadfield 1979). Heath (1918) observed that two individuals of *Anamenia* (*Strophomenia*) *agassizi* copulated by coiling around each other with cloacal openings in contact. Similar copulation behaviour has been recorded in other species (Hyman 1967). No data are available on sperm competition in this group.

IV. POLYPLACOPHORA (CHITONS)

Chitons are untorted, bilaterally symmetrical molluscs with a distinct head and a shell composed of a longitudinal series of eight shingle-like, overlapping plates. The species range from 3 to 430 mm in body length (Salvini-Plawen 1985). The majority of chiton species inhabit marine rocky shores, where they graze on encrusting algae and sessile animals. Most chitons are gonochorists, with the exceptions of the two hermaphroditic species *Lepidochitona fernaldi* and *Lepidochitona caverna* (Eernisse 1988). Sexual dimorphism in the colour of the plates has been described in a variety of species. In many species the sex ratio is male-biased, and in some species it varies with respect to the size (age) of the individuals of a population (Pearse 1979). It has been suggested that the male-biased sex-ratio results from a higher mortality rate in females.

Chitons have external fertilization. Free spawning may last for more than 1 h. Males of *Mopalia lignosa* have been observed to release sperm in short periods lasting 3–5 min at 5–15-min intervals, while females release a steady stream of eggs (Watanabe and Cox 1975). A male *Cryptochiton stelleri* lost 16% of his wet weight over a 1-week spawning period, whereas a female that spawned for more than 1 week lost only 5% of her wet weight (Tucker and Giese 1962). The high reproductive investment of males suggests either sperm limitation in free spawners (cf. Levitan and Petersen 1995; see Chapter 6) or a high level of sperm competition.

In a variety of species males spawn some hours before females (Pearse 1979). It is assumed that males release substances in their seminal fluid that stimulate females nearby to spawn. In about 30 species (including the two hermaphroditic ones) females brood their eggs by retaining them in the pallial groove until the larvae or juveniles are released (Buckland-Nicks and Eernisse 1993). This may increase the potential for female control of paternity. Aspects of sperm competition have not been examined in chiton species.

V. MONOPLACOPHORA (SEGMENTED LIMPETS)

The Monoplacophora are uncoiled, univalve, bilaterally symmetrical molluscs that live in the abyssal zone (Hyman 1967). As Palaeozoic relics they are important because their bauplan and reproductive biology may show the original molluscan mode and serve as a standard for judging the degree of specialization found in other molluscs (Gonor 1979). The majority of the monoplacophorans are gonochoric with external fertilization (Healy *et al.* 1995). An exception is the phylogenetically more advanced species *Micropilina arntzi*, which is hermaphroditic, has internal fertilization and broods its eggs and young in the mantle cavity (Warén and Hain 1992; Haszprunar 1992). No information on sperm competition is available in this class.

VI. GASTROPODA (SNAILS AND SLUGS)

Gastropods constitute by far the largest and most diverse class of Mollusca, with an estimated 105 000 species (Götting 1974; Salvini-Plawen 1985). They are characterized by a distinct head with eyes and tentacles, and a (usually) coiled shell. Slugs are snails in which the shell is reduced to a remnant or has been lost. Gastropods occupy a wide variety of habitats, including intertidal rocks, the deep sea floor, mountain streams, interstitial ground water, grassland, tree stems, leaf litter and stone deserts (for reviews on gastropod ecology see Fretter and Graham (1962), Hyman (1967), Solem (1974), Cain (1983), Calow (1983), and South (1992)). Gastropods are divided into three subclasses: the prosobranchs, opisthobranchs and pulmonates (Table 8.1).

A. Mating systems

1. Type of sexuality and mode of fertilization

Most prosobranchs are gonochoric (97% of the 2100 genera; Heller 1993). However, only few data are available on the sex ratio in natural populations. A male-biased sex ratio (62% and 66% males in two successive years) was recorded in a *Turritella communis* population on the west coast of Ireland (Kennedy 1995).

In diotocardian prosobranchs (the most primitive level), gametes are broadcast into the surrounding water, and fertilization is external (Graham 1985). In monotocardian prosobranchs, gametes are released through a glandular genital duct and fertilization is internal (Fretter 1984). Parthenogenesis occurs in several prosobranch species. In *Potamorpyrgus antipodarum* (*P. jenkinsi*) and *Melanoides tuberculata*, males occur in some populations, but usually in low numbers (Heller and Farstey 1990; Wallace 1992). The possibility of self-fertilization in hermaphroditic prosobranchs has been examined in two species, but neither of them reproduced when kept isolated throughout their life (Fretter 1984).

Sex change has been found in a variety of prosobranch species (reviews in Fretter 1984; Wright 1988). In most species individuals are first male and then female, with a brief intervening hermaphrodite stage.

Table 8.1. Mode of sexuality, type of gamete release and fertilization, occurrence of courtship behaviour and sperm competition in different classes and subclasses of molluscs.[a]

Class/subclass	Common name	Number of species[b]	Habitat[c]	Mode of sexuality	Type of fertilization	Type of gamete release	Courtship behaviour	Occurrence of sperm competition
Caudofoveata	—	70	m	Gonochoric	External	?	?	?
Solenogasters	—	185	m	Hermaphroditic	Internal	Pairwise copulation	?	?
Polyplacophora	Chitons	830	m	Gonochoric (two species hermaphroditic)	External (internal)	Spawning	—	?
Monoplacophora	—	28	m	Gonochoric (one species hermaphroditic)	External (internal)	Spawning	?	?
Gastropoda	Snails, slugs	105 000						
Prosobranchia			m/f/t	Mainly gonochoric, some hermaphroditic	External or internal	Spawning, pairwise and chain copulation	+	+
Opisthobranchia			m/f	Mainly hermaphroditic, a few gonochoric	Internal	Unilateral, pairwise and chain copulation	+	+
Pulmonata			m/f/t	Hermaphroditic	Internal	Unilateral and pairwise copulation	+	+
Scaphopoda	Tooth shells	350	m	Mainly gonochoric, some hermaphroditic	External	Spawning	—	?
Bivalvia	Clams	20 000	m/f	Mainly gonochoric, some hermaphroditic	Mainly external	Spawning	—	(+)
Cephalopoda	Cephalopods, octopuses, squids	730	m	Gonochoric	Internal	Pairwise copulation	+	+

[a] For a systematic overview see Haszprunar (1988).
[b] Salvini-Plawen (1985), Eernisse and Reynolds (1994) and Haszprunar and Schaefer (1996).
[c] Habitat: m=marine, f=freshwater, t=terrestrial.

The timing of sex reversal in some genera is primarily genetic and unaffected by external stimuli; it typically takes place at the end of the first breeding season (e.g. *Trichotropis*; Yonge 1962). In other genera, the timing of sex change depends on external stimuli. Species of *Crepidula* exhibit labile sex determination and have socially influenced sex ratios (Hoagland 1978). Their gregarious behaviour and female-induced delay of sex change appear to be mediated by pheromones. In *Crepidula fornicata*, which lives in chains of up to 12 individuals, the lowest members (the largest) are females, the apical are males, and those between are changing sex. However, size at sex change is highly variable, depending on population structure and proximate factors (Collin 1995).

The vast majority of opisthobranchs are functional simultaneous hermaphrodites for most of their reproductive life (Hadfield and Switzer-Dunlap 1984). In some orders, however, there is a tendency towards protandrous hermaphroditism (e.g. in the Thecosomata; Lalli and Wells 1978). Fertilization is internal, and cross-fertilization is the rule in most species. Self-fertilization occurs in only a few opisthobranch species (e.g. the nudibranch *Trinchesia granosa*; Schönenberger 1969). Hypodermic insemination occurs in a variety of opisthobranch slugs (see below).

Pulmonates are simultaneous hermaphrodites (Heller 1993). However, there is a tendency towards protandry among the more primitive forms (Duncan 1975). For example, young individuals of *Bulinus globosus* are capable of acting as males before they are mature as females (Rudolph 1983). Pulmonates exhibit a variety of mating systems. Some species reproduce predominantly by cross-fertilization, others predominantly by self-fertilization and others have mixed-mating systems (for reviews see Selander and Ochman 1983; Jarne *et al.* 1993; Jarne and Städler 1995).

2. Mating patterns and multiple paternity

Evidence for promiscuity and multiple paternity in broods is available for several gastropod species. Multiple mating with different partners has been recorded in some prosobranch species, although most observations are restricted to animals kept in the laboratory (e.g. a female of *Kelletia kelletii* copulated six times with five different males within 30 days; Rosenthal 1970). Coe (1936) and Hoagland (1978) observed copulation with multiple individuals in *Crepidula fornicata*, which resulted in multiple-sired broods (Gaffney and McGee 1992). Individuals of *C. fornicata* are capable of storing sperm for more than 1 year (Hoagland 1978).

Most opisthobranchs copulate with many different partners (Hadfield and Switzer-Dunlap 1984). Sea hares are hermaphrodites which exhibit one sex role when they copulate in pairs (see below). In a natural *Aplysia kurodai* population, individuals mated on average 1.2 times per day as each sex (Yusa 1996a). Within 3 days, individual sea hares copulated with, on average, 2.6 different partners as each sex, and the

variance was slightly larger in the male role than in the female role. The presence of numerous scars in the skin of *Ercolania felina* (a species with hypodermic insemination) indicates multiple mating (Trowbridge 1995). In contrast, pair bonding for the entire lifespan has been observed in the sea slugs *Phestilla lugubris* (*sibogae*) and *Phestilla minor* (Rudman 1981).

Multiple mating with different partners has frequently been observed in freshwater pulmonates. Using enzyme electrophoresis, multiple paternity has been demonstrated in *Biomphalaria obstructa* (Mulvey and Vrijenhoek 1981), *Bulinus africanus* (Rudolph and Bailey 1985), *Bulinus cernicus* (Rollinson *et al.* 1989), and *Physa heterostropha pomilia* (Wethington and Dillon 1991). The freshwater pulmonate *Ancylus fluviatilis* reproduces predominantly by self-fertilization, although a few broods with multiple paternity have been observed (Städler *et al.* 1995).

Individuals of the pulmonate land snails *Helix pomatia*, *Cepaea nemoralis*, and *Arianta arbustorum* have been observed to mate repeatedly with different partners in the course of a reproductive season, resulting in multiple-sired broods (Wolda 1963; Murray 1964; Baur 1988; Lind 1988). *Helix pomatia* copulated two to six times per year in a Danish population (Lind 1988), two to four times in a German population (Tischler 1973), and *Helix aspersa* on average three times (maximum seven times) in a British population (Fearnley 1993, 1996). Paternity analysis in egg batches of *A. arbustorum* indicated that at least 63% of the snails used sperm from two or more mates for the fertilization of their eggs (Baur 1994a).

The few data available on mating frequency in gastropods suggest that marine and freshwater gastropods copulate more frequently than terrestrial gastropods. In intertidal and terrestrial gastropods the reproductive activity is limited by favourable environmental conditions (the high risk of desiccation may incur a significant cost of mating; see below).

3. Random or assortative mating?

Size-assortative mating has been reported in the intertidal, gonochoric prosobranchs *Olivella biplicata* (Edwards 1968) and *Littorina littorea* (in one of three populations; Erlandsson and Johannesson 1994), in the freshwater prosobranch *Viviparus ater* (Staub and Ribi 1995), and in the simultaneously hermaphroditic nudibranch *Chromodoris zebra* (Crozier 1918). In *O. biplicata*, size-assortative mating can be explained by a zonation of individuals of different size: larger animals live further up the shoreline and smaller ones lower (Edwards 1969). In the slug *C. zebra*, size-assortative mating can be explained by a physical constraint. Sexually mature slugs range in body length from 4 to 18 cm and two individuals that differ greatly in size are unable to bring their reproductive organs together. *Chromodoris zebra* can copulate at all hours of the day throughout the year; individuals have plenty of time to match the size of a potential mate and if they are unequal in size they move apart to search for another mate (Crozier 1918).

The terrestrial pulmonate *Achatina fulica* is a protandrous hermaphrodite. 'Young adults', which produce only sperm, continue to grow for a further 3–6 months to became true hermaphrodites ('old adults'), which produce both sperm and eggs (Tomiyama 1993). In a natural population in Japan, 72% protandric and 28% hermaphroditic snails were recorded; copulations between hermaphroditic individuals occurred more frequently than would be expected under random mating (Tomiyama 1996).

Random mating with respect to size has been observed in the intertidal prosobranchs *Littorina rudis* and *Littorina nigrolineata* (Raffaelli 1977), and in a natural population of the hermaphroditic sea hare *Aplysia californica* (Pennings 1991). In terrestrial pulmonates, mating has been reported to be random with respect to shell size (*Cepaea nemoralis*; Wolda 1963; *Helix pomatia* and *Arianta arbustorum*; Baur 1992a), shell colour and banding pattern (*C. nemoralis*; Schilder 1950; Schnetter 1950; Lamotte 1951; Wolda 1963), and degree of relatedness (*A. arbustorum*; Baur and Baur 1997). In contrast to most benthic marine and freshwater gastropods, courtship and copulation in intertidal and terrestrial gastropods is restricted to periods of favourable environmental conditions. It has been suggested that because of the time-constrained activity and high costs for locomotion, the best strategy for a snail is to mate with the first mating partner available to minimize the risk of either ending up without any mating at all or drying up during mating (Baur 1992a). The resulting random mating pattern does not imply random fertilization of eggs, because multiple mating and sperm storage offer opportunities for sperm competition (see below). Furthermore, the structure and morphology of the sperm storage site (spermatheca), fertilization chamber and sperm-digesting organ offer opportunities for sperm selection by the female function of the hermaphrodite (cryptic female choice; Eberhard 1996).

B. Sexual dimorphism, anatomy and reproductive physiology

I. Sexual dimorphism

Sexual dimorphism in shell size, radula structure and soft tissue has been reported in a variety of gonochoric prosobranch species: see reviews by Fretter and Graham (1962) and Fretter (1984). In most of the species, females are larger than males (Webber 1977). An exception is *Olivella biplicata*, in which males are larger than females (Edwards 1968). Size dimorphism is marked in parasitic prosobranchs (e.g. *Asterophila japonica*; Grusov 1965); the dwarf males are associated with the pseudopallial cavity of the female and consist almost entirely of gonadial tissue (Fretter

1984). Colour dimorphism of mantle, cephalic tentacles and sides of the foot of *Cypraea gracilis* are sex linked – red animals are females and brown are males (Griffiths 1961). It is, however, not known whether the sex-specific coloration has any significance in the reproductive behaviour of the animals.

2. Phally polymorphism

Phally polymorphism is defined as the co-occurrence of regular, hermaphrodite individuals (euphallics) and individuals lacking the distal part of the male reproductive tract (aphallics) in natural populations (de Larambergue 1939; Jarne and Städler 1995). Euphallics can reproduce by outcrossing as male or female, as well as by selfing. Aphallics, however, cannot transfer sperm to mating partners; they can only reproduce by outcrossing as female or by selfing. Aphally has been reported in numerous species of freshwater and terrestrial pulmonates (Watson 1923; Pokryszko 1987; Jarne *et al.* 1992; Schrag and Read 1996). The proportion of aphallic individuals varies widely among populations. In the freshwater snail *Bulinus truncatus*, pure aphallic, pure euphallic, and mixed populations have been found (Jarne and Städler 1995). In the rock-dwelling land snail *Chondrina clienta*, the frequency of aphally varied from 52 to 99% in 23 natural populations, and from 1 to 89% in 21 populations of *Chondrina avenacea* (Baur *et al.* 1993; Baur and Chen 1993).

The determination of aphally vs. euphally is still unclear. Genetic components are indicated by the results of some studies (de Larambergue 1939; Schrag *et al.* 1992; Schrag and Rollinson 1994; Doums *et al.* 1996). Conversely, the proportion of aphallic individuals can vary with temperature in populations of *B. truncatus*, suggesting that the expression of phally is at least partly mediated by environmental conditions (Schrag *et al.* 1994a; Jarne and Städler 1995).

An increase in the selfing rate has been assumed in populations with large proportions of aphallic individuals (Schrag and Read 1992; Schrag *et al.* 1994b). Furthermore, with increasing number of aphallics in a population the extent of sperm competition may decrease.

3. Functional anatomy of the female reproductive tract

The terminology of the morphology of the gastropod reproductive tract is often confusing, partly as a result of the use of the same term for different structures. For simplicity, I use descriptive terms throughout this chapter.

There is considerable variability in the reproductive anatomy among and within families in each of the three subclasses. In prosobranchs with external fertilization there is little or no elaboration of the genital duct or modification for copulation. In prosobranchs with internal fertilization, gametes are released through a glandular genital duct (Fretter 1984). In these species the reproductive system becomes complex with copulatory organs, sperm pouches (for storage and digestion), and glands secreting albumen and egg coverings (Graham 1985).

In simultaneously hermaphroditic opisthobranchs and pulmonates, the ovotestis produces both spermatozoa and ova, sometimes, but not always, simultaneously (Duncan 1975). When released, male and female gametes pass along the hermaphrodite duct (Beeman 1970). Thereafter, they follow separate paths. In *Helix pomatia*, autosperm are stored in the seminal vesicle of the hermaphrodite duct throughout the year (Lind 1973). Phagocytosis of autosperm by the hermaphrodite duct epithelium has been reported in *H. pomatia* and *Oxychilus cellarius* (Rigby 1963). Sperm can be expelled from the hermaphrodite duct at times other than copulation, eventually to be digested (as are foreign sperm) in the bursa copulatrix.

The role of the female duct is to receive sperm from a copulating partner, to store the sperm and provide a site for fertilization, to form the egg capsule and digest sperm and remnants of the spermatophore and to absorb nutritional fluids received with the ejaculate. In the context of sperm competition, the sites of sperm storage (spermatheca), fertilization and sperm digestion (bursa copulatrix) are of major interest.

In terrestrial pulmonates, the enormous variation in structure and morphology of the spermatheca, fertilization chamber, and sperm-digesting organ could have evolved in response to different levels of sperm competition or cryptic female choice. For example, the fertilization pouch is not divided into a separate spermatheca and fertilization chamber in *Trigonephrus gypsinus* (Brinders and Sirgel 1992). *Oxychilus draparnaudi* has a single spermathecal tubule beside the fertilization chamber (Flasar 1967). In *Succinea putris* two spermathecal tubules occur (Rigby 1965), and 34 tubules have been recorded in the spermatheca of *Drymaeus papyraceus* (van Mol 1971). There is also a considerable within-species variation in spermathecal tubules (three to five in *Helix pomatia*; Lind 1973; two to seven and three to eight in two populations of *Arianta arbustorum*; Haase and Baur 1995; Fig. 8.1). As a consequence of the large intraspecific variation in the number of spermathecal tubules, different individuals might have different possibilities to store allosperm from more than one mating partner (Fig. 8.1B). Mixing of sperm from different mates would be more likely in less structured spermatheca, whereas a large number of tubules would allow better separation of spermatozoa from different mates. Almost nothing is known about patterns of sperm storage in the spermatheca, except for *A. arbustorum*, in which individuals showed a nonrandom uptake of sperm in single tubules of the spermatheca (Haase and Baur 1995; see below).

In terrestrial pulmonates, the stalk of the bursa copulatrix can exhibit strong peristaltic waves (Lind 1973). The bursa generally contains dissolved or digested spermatozoa, which are probably both allosperm and

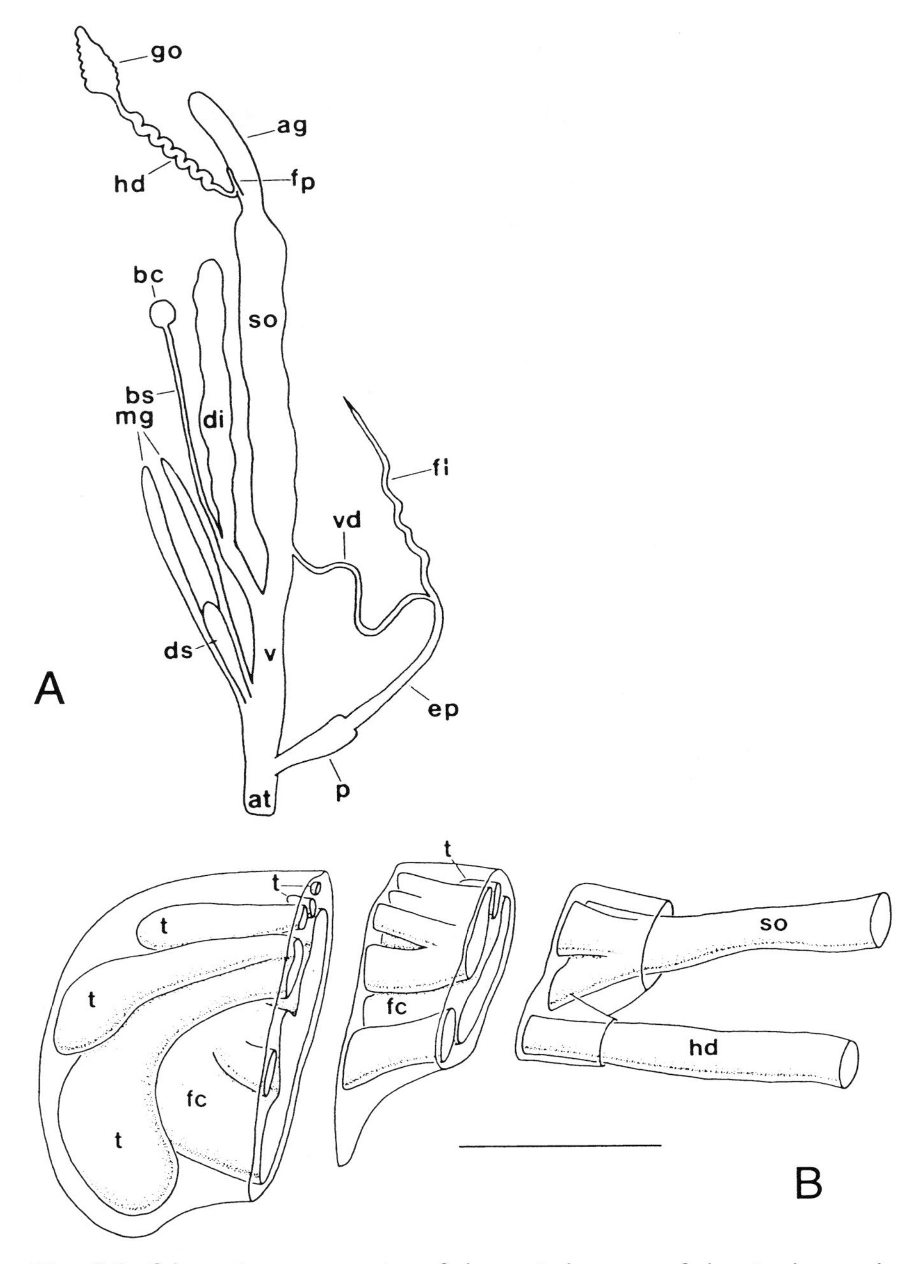

Fig. 8.1. *Schematic representation of the genital system of the simultaneously hermaphroditic land snail* Arianta arbustorum *(A), and reconstruction of the fertilization pouch (B; scale bar = 500 μm). ag, albumen gland; at, genital atrium; bc, bursa copulatrix; bs, bursa stalk; di, diverticulum; ds, dart sac; ep, epiphallus; fc, fertilization chamber; fl, flagellum; fp, fertilization pouch; go, gonad; hd, hermaphroditic duct; mg, mucous glands; p, penis; so, spermoviduct; t, spermathecal tubules; v, vagina; vd, vas deferens (from Haase and Baur 1995, with permission).*

autosperm. The bursa also digests remnants of the spermatophore and excess female secretions (Els 1978). In cases in which the sperm-digesting organ is connected to the female part of the reproductive tract there is – at least theoretically – an opportunity for sperm selection by the female function of the hermaphrodite (cryptic female choice; Thornhill 1983; Eberhard 1991, 1996). However, whether or not and to what extent the female function of hermaphroditic gastropods can control the fertilization of its eggs remains to be investigated.

4. Penis morphology

The sperm-transferring structure is not homologous in gastropods. Several families of gonochoric prosobranchs have a cephalic penis (Fretter 1984). For example, in neritids it lies on the head near the base of the right tentacle. In other prosobranchs the penis is a modified cephalic tentacle; the enlarged left in *Neomphalus*, the right in *Cocculinella*, and in viviparids a finger-like lobe that is normally folded back into a pouch at the tip of the right tentacle (Fretter 1984).

In a variety of prosobranchs and in hermaphroditic opisthobranchs, the penis may be simple, glandular, or armed with hooks and spines (Ghiselin 1965; Schmekel 1970; Rankin 1979; Gosliner 1994; Jensen 1996). In species of the order Sacoglossa, the type of penis appears to be related to the mode of insemination (Gascoigne 1974). The penes are either unarmed or equipped with spines, rodlets, or a cuticular style at the tip of the penis (Jensen 1992, 1993; Gosliner 1994; see Fig. 8.2). Gascoigne (1974) classified styles as either penetrant or coupling. Hypodermic insemination (see below) occurs mainly in species with penetrant styles, whereas species with coupling styles copulate reciprocally. However, hypodermic insemination can also occur in species with unarmed penis (e.g. *Elysia maoria*; Reid 1964). The chitinous spine on the penis of the pelagic sea slug *Glaucus atlanticus* may help to prolong contact with the mate in the face of wave action, and thus might be a morphological adaptation to long copulation duration (up to 1 h; Ross and Quetin 1990).

In terrestrial pulmonates, the penis is a muscular organ that is everted at copulation and is typically inserted into the genital atrium and vagina of the mate. It is assumed that a variety of structures at the penis tip, such as plates (*Ventridens suppressus*; Webb 1948), pilasters and hooks (*Striatura* sp.), suckers (*Parmacella* sp.), pads and calcareous spurs (*Phrixolestes* sp.), and spines all over the outer surface of the penis (*Ariophanta ligulata*; Dasen 1933) may have holdfast function (Tompa 1984). Some of these structures resemble those observed on the penes of insects (e.g. dragonflies; Eberhard 1985). In gastropods, the significance of these structures has not been examined with respect to sperm competition. Individuals of *Ellobium* have a long penis, which is evertible during copulation (Berry 1977). In *Ellobium aurisjudae* some 40 mm of the penis enters the coiled female reproductive tract of the partner. However, it is

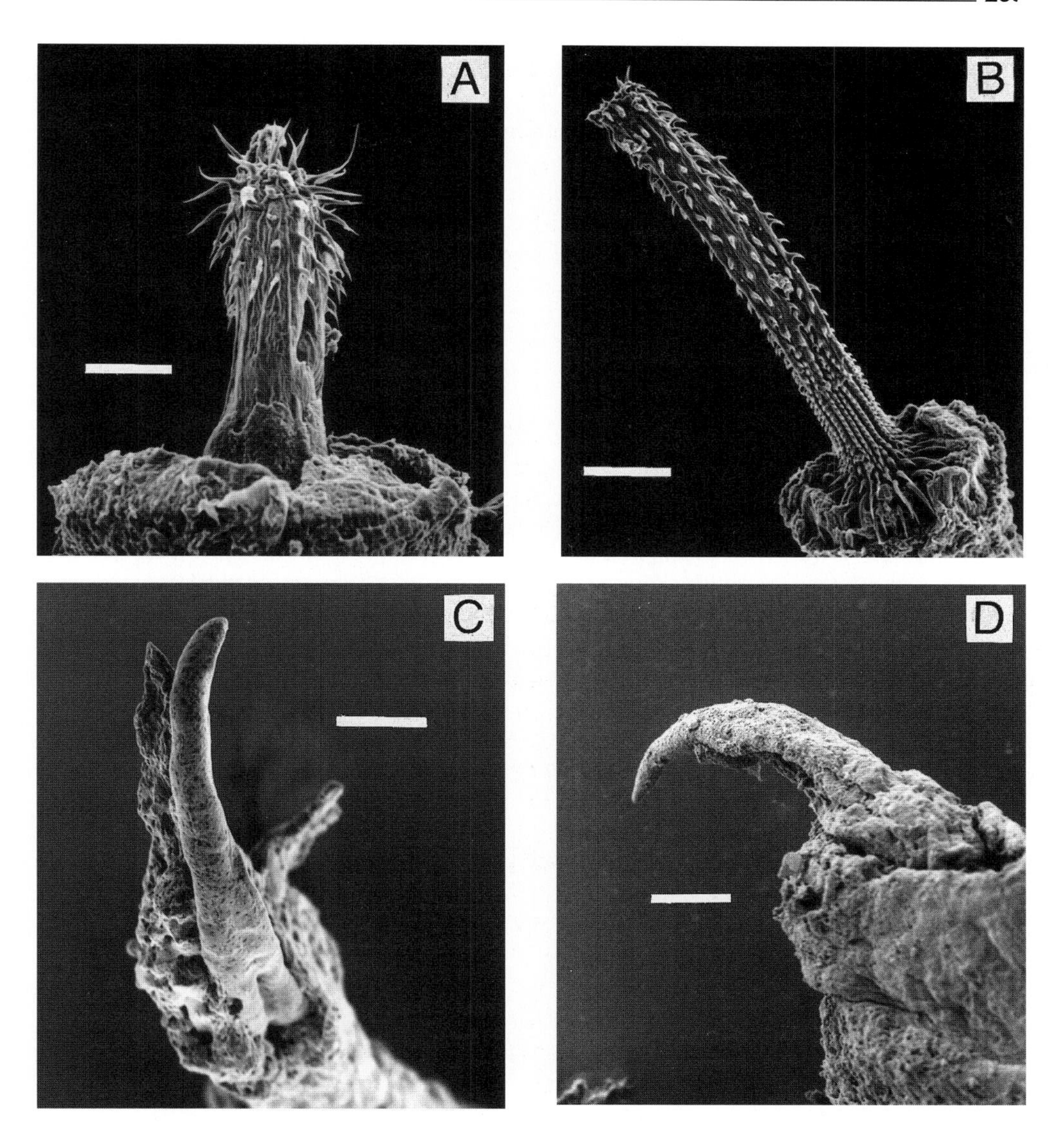

Fig. 8.2. *Penile anatomy of hermaphroditic opisthobranchs gastropods. (A, B) Everted penes of two undescribed species of* Nembrotha, *showing penile spines; (C) penile stylets of* Ascobulla fischeri *and (D)* Diaphana minuta. *Scale bars: A = 75 μm; B = 100 μm; C, D = 10 μm. (A and B, photographs courtesy of T.M. Gosliner; C and D, photographs from Jensen 1996, with permission).*

not known whether the enlarged penis of these snails is able to manipulate previously deposited sperm.

In those pulmonate species that have a penis, the animals insert it simultaneously or sequentially into the partner's genital pore during mating. However, in some slug species (e.g. *Limax maximus*; Gerhardt 1933; Falkner 1992), the penes entwine and exchange sperm at their

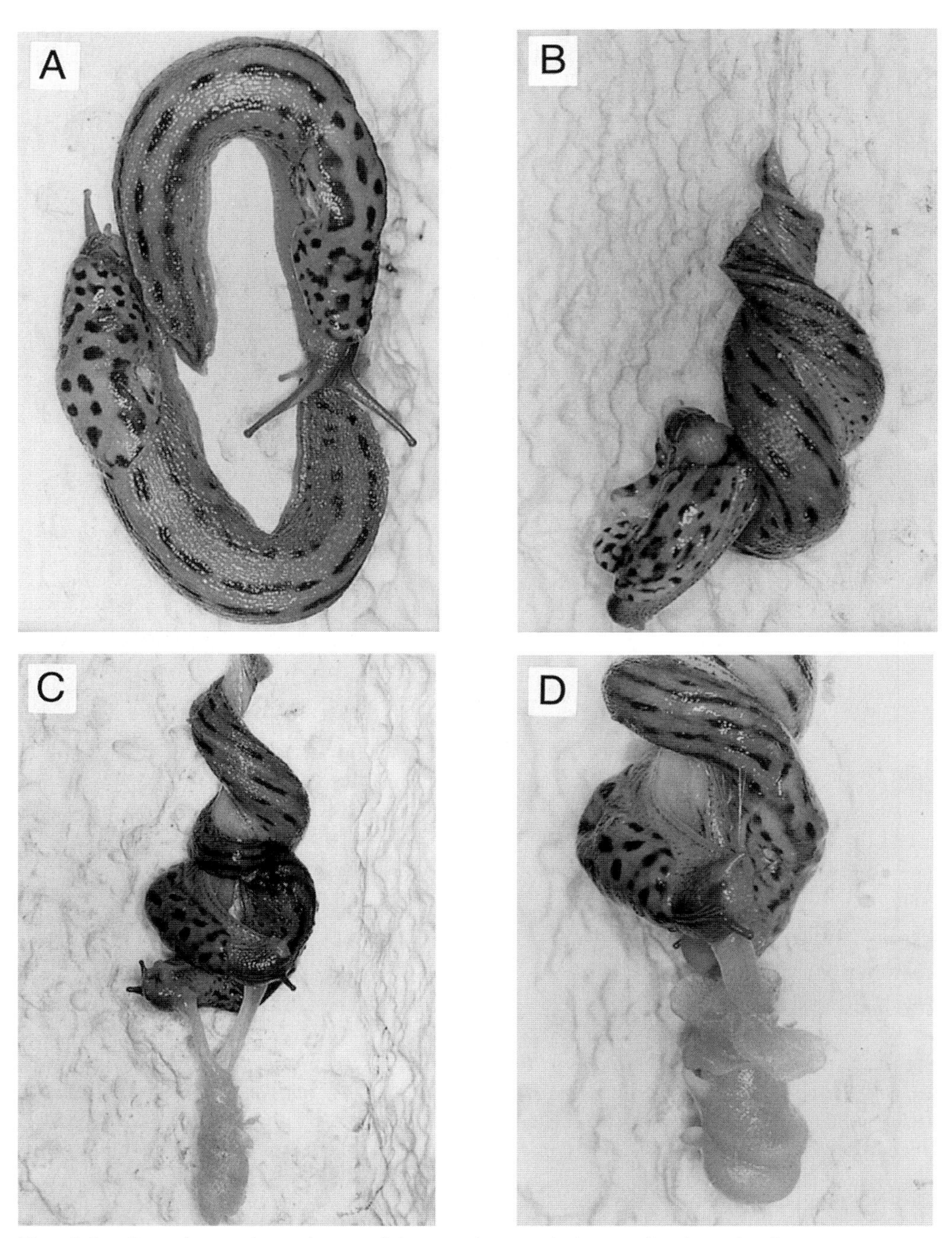

Fig. 8.3. *Courtship and copulation of the simultaneously hermaphroditic slug* Limax maximus *on a vertical wall. (A) courtship behaviour with clockwise-turning slugs; (B) the slugs mate with their bodies entwined hanging down from a mucous thread; (C) their penes are elongated and hang below the entwined pair; (D) exchange of sperm masses on the tip of the intertwined penes. (Photographs A. Limbrunner; courtesy G. Falkner).*

tips without intromission, completely outside of the body (Fig. 8.3). In these species the penis is often remarkably long in relation to body length. For example, in the 12–15 cm long slug *Limax corsicus*, which exhibits aerial mating (see below), the uncoiled penis reaches a length of 60 cm and in the 13–15 cm long *Limax redii* 85 cm (Falkner 1990; see Fig. 8.4).

Many species of terrestrial gastropods, especially slugs, normally lack penes; mating in these species may be accomplished by pressing the genital pores together directly, or by using nonhomologous penis-like structures (derived from other parts of the terminal genitalia) for sperm transfer (Tompa 1984).

C. Sperm polymorphism

The structure of the molluscan spermatozoa has been reviewed by Thompson (1973), Anderson and Personne (1976), Healy (1988, 1996) and Giusti *et al.* (1992). Prosobranchs have two basic types of sperm: typical or eupyrene sperm that fertilize the eggs and nonfertile, atypical sperm (Fretter and Graham 1964; Giusti and Selmi 1982). Atypical sperm include oligopyrene (chromatin deficient), apyrene (chromatin absent) and other abnormal (lack of acrosomes, modified centrioles, and multiple flagella) sperm (Fretter 1984). Atypical sperm have been described in more than 20 families of prosobranch snails (Nishikawi 1964; Giusti and Selmi 1982). These sperm may be vermiform and motile (e.g. Viviparacea), spindle-shaped and virtually immobile (e.g. Buccinacea), or consist of a large, fibrous, undulating plate and long tail to which numerous eupyrene sperm attach by their acrosome (Fretter 1984). Males of the freshwater prosobranch *Viviparus* produce eupyrene and oligopyrene sperm. Oligopyrene sperm are infertile but rich in polysaccharides (Hanson *et al.* 1952). In *Viviparus contectus*, oligopyrene sperm are ingested by epithelial cells of the female reproductive tract and probably serve as a source of nutrition (Dembski 1968).

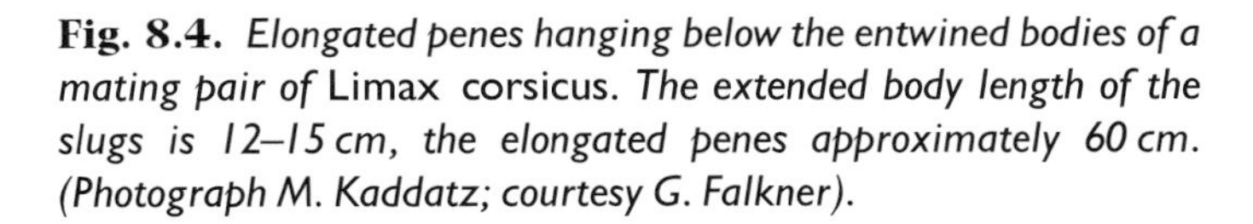

Fig. 8.4. *Elongated penes hanging below the entwined bodies of a mating pair of* Limax corsicus. *The extended body length of the slugs is 12–15 cm, the elongated penes approximately 60 cm. (Photograph M. Kaddatz; courtesy G. Falkner).*

Size dimorphism in eupyrene sperm has been found in several species (e.g. *Epitonium eusculptum* and *Epitonium eximium*; Nishikawi and Tochimoto 1969). It is assumed that both sizes of sperm are functional. In the prosobranch snail *Modulus modulus*, eupyrene spermatozoa are about 36 μm long (Houbrick 1980). Apyrene spermatozoa are larger, about 48 μm long, and bear six flagella. In males, the vas deferens is packed with eupyrene and apyrene sperm, in about equal numbers, tangled together by the numerous flagella into dense masses. Apyrene sperm move in a slow sinuous manner, while eupyrene sperm are fast moving (Houbrick 1980). Size dimorphism in apyrene sperm has been found in the freshwater prosobranch *Theodoxus fluviatilis*; the longer spermatozoa measure 120–140 μm and the shorter ones 60–80 μm (Selmi and Giusti 1983).

Eupyrene and atypical sperm have been reported in the opisthobranch *Haminoea navicula* (Dupouy 1964). It is not clear whether pulmonates have eupyrene and apyrene sperm. However, abnormal sperm (lack of acrosomes, modified centrioles, multiple flagella) occur in large numbers in several terrestrial slug species (e.g. *Milax gagates*, *Agriolimax agrestis* and *Arion ater*; South 1992). Despite much research, the function of abnormal sperm remains unclear (see also paternity guards).

Information on the size of molluscan spermatozoa is summarized by Thompson (1973). Free-spawning prosobranchs have simple and relatively short sperm. In opisthobranchs sperm length varies from 106 to 115 μm in *Bulla ampulla* to 440 μm in *Berthella plumula*. Spermatozoa of freshwater and terrestrial pulmonates are among the longest of the molluscs, e.g. 365 μm in *Physa acuta* (Brackenbury and Appleton 1991), 850 μm in *Helix pomatia* and 1140–1400 μm (of which the head accounts for only 10 μm) in *Hedleyella falconeri* (Thompson 1973).

In general, gastropod species with internal fertilization have larger and more complex sperm than species with external fertilization.

D. Modes of sperm transfer and sperm number

Hypodermic insemination occurs in a variety of opisthobranch slugs (e.g. Haase and Wawra 1996). In some species, the penis pierces the body wall of the mating partner above a closed vaginal bursa and injects sperm directly into the bursa (Gascoigne 1956, 1975). In other species hypodermic injection of sperm occurs randomly over the surface of the body (Hadfield and Switzer-Dunlap 1984). However, it is not known how injected sperm reach sperm-storage organs or unfertilized eggs. In *Ercolania felina*, hypodermic insemination can be unilateral (nonreciprocal) or reciprocal (Trowbridge 1995). Epidermal scars indicate that an individual has received sperm. The costs and benefits of hypodermic insemination are unknown. Hypodermic insemination may reduce or even eliminate courtship; this may be relevant to species inhabiting unpredictable or transient environments. Moreover, it may allow individuals to mate with many conspecifics within a short period. However, there is little information available about sperm competition in species with hypodermic insemination.

In molluscs with internal fertilization, sperm are transferred to the partner in the form of free sperm (i.e. as sperm suspension in seminal plasma) or the sperm are either aggregated into loosely assembled naked conglomerates (spermatozeugmata) or encapsulated into spermatophores (Mann 1984). In *Aplysia parvula*, the number of sperm transferred is positively correlated with duration of copulation. When mating duration increased from 2 to 47 min., the number of sperm transferred increased from 4×10^6 to 6×10^6 (Yusa 1994). In *Aplysia kurodai* and *Aplysia juliana*, the number of fertilized eggs laid by an individual, which was allowed to mate only once, was positively correlated with duration of copulation (Yusa 1996b). The ratio of transferred sperm to fertilized eggs is approximately 30:1.

Most freshwater pulmonates transfer a seminal fluid in which sperm is embedded (Geraerts and Joosse 1984). During one copulation the freshwater pulmonate *Bulinus globosus* ejaculates at least 350 000 sperm (Rudolph 1983). *Bulinus globosus* is able to copulate as male once per day for up to eight consecutive days. Following a single copulation after 1 week of isolation, the hermaphroditic duct of male-acting individuals contained an average of $87\,000 \pm 42\,000$ (mean $\pm$ SD) sperm. In the 10 days following the initial copulation, the snails produced approximately 50 000 sperm per day. The production of sperm will vary depending on the environment, the age, size and nutritional state of the snail and, probably, on the level of sperm competition.

There are some prosobranchs which produce spermatozeugmata. The very large 'giant spermatozeugmata', which can be occasionally as much as 1 mm diameter, consist of normal and abnormal (oligopyrene and apyrene) sperm. While some authors are of the opinion that the prime function of abnormal sperm within a spermatozeugma is to help with the transport of the eupyrene sperm to the storage site, others favour the view that their main role is to provide nutrients for the eupyrene spermatozoa (reviewed by Mann 1984).

Spermatophores occur in more than 40 families distributed over all three subclasses (Mann 1984; Tompa 1984; Robertson 1989). Sperm of sessile vermetids (gonochoric prosobranchs) are encapsulated in complex, inflated spermatophores that are released into the sea, where they are subsequently trapped in the mucous feeding-nets of neighbouring females (Scheuwimmer 1979; Hadfield and Hopper 1980; Fig. 8.5A). Spermatophores of the marine snail *Neritina reclivata* are 2.4–3 cm long and 1 mm wide – the length of the female's body is only 2 cm (Andrews 1936). The spermatophore looks like a 'white turgid cylinder bent in a loop and with both attenuated ends stuck together'. Spicules (chitinoid secretions) at the wider of the two end portions are supposed to provide an anchoring device that keeps the spermatophore in position while the

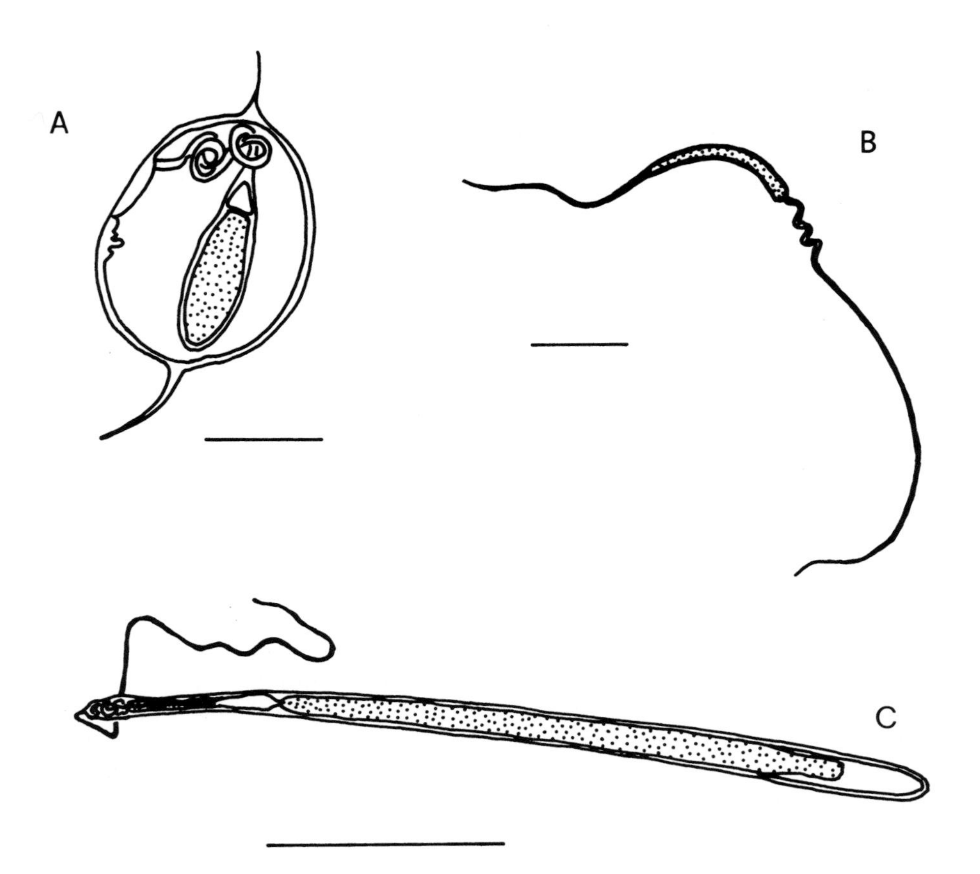

Fig. 8.5. *Diagrams of molluscan spermatophores. (A) The vermitid prosobranch* Petaloconchus montereyensis*: males release spermatophores into the sea water; some become entangled in the mucous feeding nets of females (redrawn after Hadfield and Hopper 1980). (B) The simultaneously hermaphroditic land snail* Arianta arbustorum*: copulating individuals exchange one spermatophore each (redrawn after Hofmann 1923). (C) The cephalopod* Loligo pealli*: males transfer spermatophores with their hectocotylus to the females (redrawn after Austin* et al. *1964). Scale bars: A = 0.5 mm; B, C = 3 mm.*

spermatozoa are being discharged and moved to the storage site of the female. Two to four newly received spermatophores and remnants of another 2–20 spermatophores were found in single females by Andrews (1936).

Some opisthobranch species produce spermatophores, which are morphologically simple thin-walled sacs (Hadfield and Switzer-Dunlap 1984; Kress 1985). Spermatophores of *Limacina inflata* (and most probably many other species) also contain varying amounts of prostatic secretion which may be used as nutrients by the recipient (Lalli and Wells 1978). In three species of hermaphroditic *Boonea*, spermatophores are placed into the mantle cavity of the recipient during mating, and in three species of *Fargoa* onto the partner's shell (Robertson 1978). In a sample of 10 *Boonea bisuturalis* collected in the wild, three snails lacked spermatophores in their mantle cavity, two had one spermatophore each, three had two spermatophores, one had four spermatophores and one had five spermatophores. In some animals the spermatophores received were of different ages and therefore may have been received from different partners. In a few interstitial opisthobranchs one or more spermatophores are attached to the body wall of the recipient. The spermatozoa penetrate the skin to enter the haemocoel directly, with no copulation (Swedmark 1968). The spindle-shaped spermatophores of *Microhedyle cryptophthalma* (body length of adults 1–1.6 mm) are 80–187 μm long and contain 200–300 spermatozoa (Westheide and Wawra 1974). In *Haminoea arachis* and *Metaruncina setoensis*, spermatophores are deposited into the reproductive openings of the mating partner (Hadfield and Switzer-Dunlap 1984).

Pulmonate spermatophores are not homologous with those of prosobranchs and opisthobranchs because they are formed by the epiphallus and flagellum, organs that are not present in the other two subclasses (Robertson 1989). In terrestrial pulmonates, the spermatophore has a species-specific morphology, which is of taxonomic significance. Spermatophores may be smooth, elaborated spined, calcified or uncalcified (Tompa 1984). In general, snails form a single spermatophore at each mating and exchange it reciprocal. In *Helix pomatia*, the spermatophore is 6–8 cm long and consists of a distinctive tip, a body (sperm container), and a long tail (Lind 1973). No estimates of the number of sperm in a spermatophore are available, except in the terrestrial pulmonate *Arianta arbustorum* (see below).

The adaptive significance of the spermatophore in terrestrial pulmonates with well-developed copulation organs and internal fertilization is unclear. In *H. pomatia*, sperm leave the spermatophore body through the spermatophore tail in the stalk of the bursa copulatrix and migrate into the spermatheca (Lind 1973). The spermatophore and any remaining sperm are digested later in the bursa copulatrix. Lind (1973) suggested that the function of the spermatophore is to ensure that a number of sperm can migrate into the oviduct and reach the spermatheca without coming into contact with the digesting bursa copulatrix. Thus, the significance of the peculiar method of transferring sperm in terrestrial pulmonates may be to allow only the most active sperm to pass to the spermatheca and thus can be considered as a means to mitigate sperm selection of the recipient (which still might occur in the spermatheca).

I. Viability of sperm stored

In most gastropod species long-term storage of sperm occurs in the spermatheca. In the gonochoric prosobranch *Lacuna* (*Epheria*) *variegata* sperm are stored in the ovary (Buckland-Nicks and Darling 1993).

Females of the freshwater prosobranch *Viviparus ater* reproduced successfully 2 years after the last copulation (Trüb 1990). In freshwater pulmonates maximum longevity of stored allosperm was 150 days in *Helisoma duryi* (Madsen *et al.* 1983), 111 days (mean 48 days) in *Biomphalaria glabrata* (Vianey-Liaud 1992), 116 days in *Lymnaea stagnalis* (Cain 1956) and 70 days in *Bulinus cornicus* (Rollinson and Wright 1984). In terrestrial pulmonates, viable allosperm have been found 108 days after the last copulation in the tropical snails *Limicolaria flammea* (Egonmwan 1990), 520 days in *Limicolaria martensiana* (Owiny 1974), 341 days in *Achatina fulica* and 476 days in *Macrochlamys indica* (Raut and Ghose 1979) and 4 years in *Cepaea nemoralis* (Duncan 1975).

Sperm viability in terrestrial gastropods may not be a simple function of time. In *A. fulica* and *M. indica*, the viability of sperm stored is influenced by the length of the aestivation period. Sperm viability decreased to 105 days in aestivating *A. fulica* and to 150 days in *M. indica* (Raut and Ghose 1982).

E. Reproductive behaviour

I. Mating aggregations

Mating aggregations occur in numerous prosobranch species. Males and females aggregate and breed either on their feeding grounds or on sites suitable for egg deposition. In *Olivella biplicata* and *Mitra idae*, males are attracted by pheromones in mucus trails of females (Cate 1968; Edwards 1968). Mating aggregations allow individuals to find mating partners, but at the same time they may increase the level of sperm competition. In broadcast spawners different male behaviours have been reported that may increase the fertilization chances of the sperm. In *Patella lusitanica* spawning pairs remain in close proximity, and in *Nacella concinna* the spawning male mounts the shell of the female (von Medem 1945; Picken 1980).

Breeding aggregations have also been observed in numerous opisthobranch species (e.g. *Aplysia californica*; Audesirk 1979; *Onchidoris bilamellata*; Todd 1979). In *A. californica*, large mating aggregations could persist for a full week, but turnover of individuals within aggregations was very high (there was a 33% chance that an individual stayed in the same aggregation for 24 h; Pennings 1991).

In terrestrial pulmonates, pheromones are used to attract and stimulate potential mating partners. In several species of the families Bradybaenidae, Hygromiidae and Helicidae, a pheromone is released from the 'head-wart', which protrudes between the two optical tentacles (Takeda and Tsuruoka 1979; Falkner 1993). In *Arianta arbustorum*, a protrusion of the 'head-wart' could only be observed in a small proportion of courting individuals (G. Baumgartner, unpublished data). In this species, aggregation behaviour of adults is most pronounced during the main mating period (Andreassen 1981; Baur 1986). The significance of this 'head-wart' with respect to sexual selection and sperm competition is unclear.

Mating aggregations can also be triggered by environmental cues. *Melampus bidentatus* aggregates during the day before full or new moon and copulates during the following day (Russell-Hunter *et al.* 1972; Price 1979).

2. Courtship and copulation

Gastropods show diverse mating behaviour and sperm transfer methods (Duncan 1975; Tompa 1984). Sexual behaviour can be relatively simple, as in the forms with external fertilization, or complex, as in forms with internal fertilization where courtship and mating may last many hours.

The marine slug *Navanax inermis* is a simultaneous hermaphrodite that normally mates in pairs (Leonard and Lukowiak 1984a,b, 1985). Copulations usually occur in bouts, with active alternation of sexual roles and functions. A mating bout is initiated when one individual assumes the male role. This snail then follows a conspecific mucous trail and courts and copulates as a male, after which the roles are changed several times. Leonard and Lukowiak (1984b, 1985) called this mating system 'sperm-trading' (for a discussion of different mating systems in hermaphrodites see Chapter 7).

In simultaneously hermaphroditic opisthobranchs of the genus *Aplysia* (sea hares), copulation is usually nonreciprocal between two individuals, with one individual (acting as a male) mounting the other (acting as a female) (Carefoot 1987). Animals acting as males (sperm donors) were found to be larger than animals acting as females (sperm recipients) in mating pairs of *Aplysia punctata* in the wild (Otsuka *et al.* 1980). A large proportion of both *Aplysia dactyomela* and *Aplysia juliana* was found to specialize in one sexual role in the laboratory, acting predominantly either as sperm donors or receivers (Lederhendler and Tobach 1977; Switzer-Dunlap *et al.* 1984). In contrast, individuals of *Aplysia californica* did not specialize in either sexual role (Pennings 1991). Furthermore, the sexual role chosen by an individual of *A. californica* was neither a function of its mass nor of its relative size compared with the partner. When more than two individuals participate in mating, the gastropods often form a copulatory chain (e.g. three to six individuals of *Aplysia kurodai*; Yusa 1993); the bottom-most individual acts only as a female, the top-most as a male, and those between as males and females. Copulation chains can last several days, although a large turnover of individuals may occur. These patterns have not been investigated in the light of sperm competition. In three species of sea hares, *A. kurodai*, *A. juliana* and *Aplysia parvula*, one successful copulation is sufficient to produce at least one entire batch of viable eggs (Yusa 1996b). However, there is a probability of 36% that a copulation as female in *A. kurodai* will not lead to fertilization of eggs. Corresponding figures for *A. juliana* and *A. parvula* were 32% and 10% (Yusa 1996b).

In all three subclasses, courtship and copulation duration is extremely variable, even within genera. For example, the time spent in copulation varies from 5 to 10 min in the prosobranch *Littorina pintado* to 1.5–15 h in *Littorina obtusata* (Struhsaker 1966), and from 1 to 3 h in the prosobranch apple snail *Pomacea haustrum* to 10–18 h in *Pomacea canaliculata* (Albrecht *et al.* 1996). In the carnivorous opisthobranch *Hermissenda crassicornis* reciprocal copulation is very rapid, lasting only about 6 s Rutowski 1983). Individuals of *H. crassicornis* are cannibalistic. The brief duration of sexual contact might be an adaptation to minimize the time spent with a mate that also is a potential predator (Longley and Longley 1982; Baur 1992b). In other opisthobranchs the duration of copulation varies from 10 to 30 s (e.g. *Acteonia cocksi*) to more than 24 h (e.g. *Hexabranchus sanguineus* and *Adalaria proxima*; Todd 1979).

Mating is rarely simultaneously reciprocal in freshwater pulmonates (van Duivenboden and ter Maat 1988; Wethington and Dillon 1996). Courtship begins when the 'male' (active snail) mounts the shell of the 'female'. It then moves around until it adopts the copulatory position and inserts the everted penis into the female aperture. Copulation may last from a few minutes to 12 h. Reversal of roles may frequently occur, but unilateral (e.g. *Siphonaria obliquata* and *Melampus bidentatus*; Berry 1977; Price 1979) and chain mating (e.g. *Ancylus fluviatilis*; Städler *et al.* 1995) have been observed in several families of freshwater pulmonates.

Mating behaviour in terrestrial gastropods is strongly influenced by environmental conditions. For example, a rise in temperature stimulates mating behaviour in *Helix pomatia* (Meisenheimer 1907; Lind 1988). Terrestrial gastropods locate potential mates by chemical cues (Croll 1983; Chase 1986). Courting *Arianta arbustorum* attract other conspecifics, which then may interfere with the courting pair (Baur and Baur 1992a). Courting groups involving 3–4 *A. arbustorum* have been observed in the field. In a natural population of *H. pomatia*, Lind (1988) recorded that more than two individuals were involved in courtship behaviour in 86 of 981 cases (9%), the maximum being five. In several cases two of the three courting *H. pomatia* copulated reciprocally, and a few hours later one of them copulated with the third. A laboratory experiment showed that *A. arbustorum* with larger shells were not capable of displacing individuals with smaller shells during courtship, indicating that courting tenacity rather than shell size may be important for mating success (Baur 1992a).

Courtship behaviour of terrestrial pulmonates is extremely variable. In most species, the two partners court reciprocally (e.g. turning dance in slugs; see Fig. 8.3). In some families there is a reversal of roles during mating (e.g. in *Partula*; Lipton and Murray 1979). In other species an asymmetry of courting behaviour can be observed. For example, in *Achatina fulica* one snail initiates courtship by approaching a potential partner from the back of the body and mounting on its shell (Tomiyama 1994). If the second snail accepts the courtship, it bends its head backwards and waves the head and shell. The upper snail bites into the soft body of the partner and rubs its everted penis against the partner's penis. Finally, both snails insert their penis into the vagina of each other. The copulating pair falls on the side and remains in this posture until copulation is finished. Rejection of courtship may frequently occur. In a natural population of *A. fulica*, only 11% of 223 courtship attempts resulted in a successful copulation; in all other cases one of the snails rejected its partner.

Courtship and mating duration ranges from a few hours to more than 36 h in terrestrial gastropods and thus often exceeds the period favourable for locomotory activity (conditions of high air humidity; Meisenheimer 1907; Lind 1973, 1976; Jeppesen 1976; Giusti and Lepri 1980; Chung 1987; Giusti and Andreini 1988). During courtship and copulation terrestrial gastropods are exposed to severe water loss and are more susceptible to predation than single adults (Pollard 1975). In most species intromission and sperm transfer is rather short compared with the extended courtship (Tompa 1984). Spermatophore formation is very rapid, taking only a few minutes during copulation. The adaptive significance of the elaborate courtship behaviour in simultaneously hermaphroditic gastropods is unclear; courtship behaviour probably has some function for the assessment of potential mates.

External sperm exchange in simultaneous hermaphrodites, by which sperm is deposited on the mate's everted penis without intromission, is unique to pulmonate gastropods (Emberton 1994). External sperm exchange is best known in the common garden slug *Limax maximus*, which exhibits a spectacular aerial mating (Gerhardt 1933; Chace 1952; Falkner 1992). Copulating pairs hang on thick mucus ropes suspended from trees or vertical walls with everted penes longer than their bodies (Fig. 8.3). Reciprocal sperm exchange occurs on the tip of the intertwined penes. The animals climb up the mucus thread or fall to the ground after copulation is finished.

Following sperm transfer the partners may quickly separate (e.g. in *Deroceras reticulatum*). In a variety of species, however, there is a period of immobility (e.g. lasting 0.5–9 h in *Helix pomatia*; Lind 1976), during which the spermatophore is transported in the reproductive tract of the recipient towards the bursa copulatrix, where it is eventually digested. During this period sperm leave the spermatophore. Depending on the location of the spermatophore in the female reproductive tract, sperm may reach the spermatheca (sperm storage site) or they may be transported into the bursa copulatrix where they are eventually digested.

A bizarre behaviour is apophallation, i.e. the gnawing off of the entire penis at the end of copulation, which occurs frequently in the simultaneously hermaphroditic slug *Ariolimax columbianus* and other congeners (Mead 1943). The penis in these species has a very attenuated tip which may become knotted on copulation. Separation is then impossible and, after several attempts, the owner cuts through the base of the penis using the radula. Individuals that have lost their penis cannot regenerate another, but they can continue to function as females.

Sperm exchange is usually reciprocal in terrestrial pulmonates. However, unidirectional sperm transfer may occur, although little information is available about its frequency. Reise (1995) observed that during three out of 15 apparently normal copulations of *Deroceras rodnae* only one of the partners transferred a sperm mass (slugs of this species exchange sperm on the tips of the everted penes). Unilateral sperm transfer in mating systems with normally simultaneously reciprocal copulation has been interpreted as 'cheating' in gamete trading models (Leonard 1990, 1991; see Chapter 7).

3. Dart shooting

In marine opisthobranchs and at least 10 families of terrestrial pulmonates, the animals form a sharp, hard, calcified or chitinous structure (the so-called love dart) in the female part of the reproductive organ (Pruvot-Fol 1960; Tompa 1980; Fig. 8.6). The dart is used to pierce the body of the mating partner during courtship. This peculiar behaviour has been hotly discussed for more than 250 years, but its adaptive significance is still unknown (Kothbauer 1988; Leonard 1992). Even though darts may wound or even kill a partner, the elaborate structure of the dart apparatus suggests that it serves some adaptive function.

Dart shooting is best studied in *Helix pomatia* and *Helix aspersa* (Jeppesen 1976; Lind 1976; Chung 1987; Adamo and Chase 1988). In both species, dart shooting is a facultative element of courtship behaviour; it

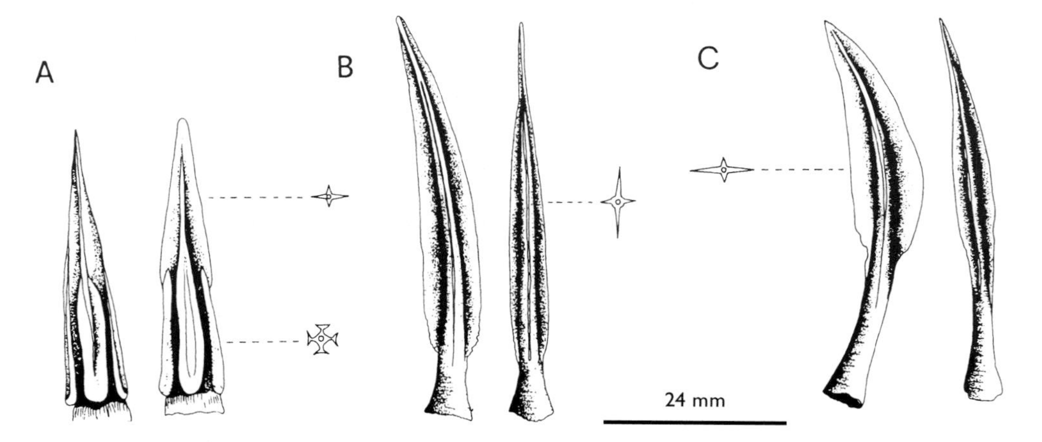

Fig. 8.6. *Love darts (with sections) of three simultaneously hermaphroditic land snails. (A)* Eobania vermiculata *(adult shell breadth 2.6 cm); (B)* Helix aspersa *(3.5 cm); (C)* Chilostoma tacheoides *(2.5 cm). Scale bar: 24 mm (after Giusti and Lepri 1980, with permission).*

occurs when a snail quickly everts the basal tubercle of the dart sac out of its everted genitals (Chung 1987). The dart is never propelled through the air, because it is firmly attached by its base to the tubercle until it is lodged in the partner's tissue. Occasionally, the dart does not hit the partner. A new dart is produced within 5–6 days after dart shooting (Dillaman 1981; Tompa 1982). During the expulsion of the dart, a globule of whitish mucus from the digitiform glands adheres to the dart. It has been shown that the mucus transferred with the dart decreases the duration of the courtship (Chung 1986; Adamo and Chase 1990; see also below). It is assumed that a pheromone in the mucus may cause a behavioural synchrony between the mating partners to assure a faster courtship and subsequent copulation, which might decrease both the chances of failure to complete mating and of predation (Adamo and Chase 1990). However, the fact that dart shooting results in 15% of the recipients breaking off courtship contradicts the stimulatory function of this behaviour (Lind 1976).

Virgin individuals of *H. aspersa* possess no dart (Chung 1987). Thus, mating sequences without dart shooting occur naturally in *H. aspersa*. Furthermore, in 10 of 23 observed copulations in *Helix aperta* the dart fell on the ground (Giusti and Lepri 1980).

Other hypotheses have been proposed to explain the adaptive significance of dart shooting (Leonard 1992; Adamo and Chase 1996). The hypothesis that the dart is a gift of calcium to the partner for the production of eggs (calcium is a limiting factor in the reproduction of most land snails) has recently been rejected (J. M. Koene, personal communication). Simultaneous hermaphrodites that mate in pairs are thought to experience a conflict over which individual will assume which sexual role (Fischer 1980, 1984; Leonard and Lukowiak 1984b, 1985). This conflict can be resolved by some sort of parcelling behaviour, where individuals repeatedly alternate sexual roles (Leonard 1991). According to Leonard (1992) the dart may serve to induce the partner to act as a male, by acting as an honest signal of an individual's willingness to reciprocate in a sperm-trading mating system (for a discussion of gamete-trading models see Chapter 7). An alternative hypothesis suggests that dart-shooting behaviour might have evolved in order to manipulate sperm utilization and/or oviposition in the mating partners (Charnov 1979; Tompa 1980). In other animals from different phyla, males influence female reproduction indirectly by stimulating the female's sensory systems (Adamo and Chase 1996). In a variety of terrestrial gastropods, copulation *per se* and/or the transfer of the spermatophore stimulates egg production via hormones in both partners (Griffond and Gomot 1989; Saleuddin *et al.* 1991; Baur and Baur 1992b). Thus, a copulation might be beneficial even if the sperm delivered are not used by the partner for the fertilization of eggs. Adamo and Chase (1996) suggest that dart shooting is an additional ability to influence fertilization directly by manipulating the endocrine system of the partner. The dart serves to increase the likelihood that the recipient will use the shooter's sperm to fertilize its eggs. Adamo and

Chase (1996) reviewed features of snails reproduction which are consistent with this hypothesis. However, detailed experimental studies that consider anatomical, physiological and behavioural specializations are needed to understand the significance of dart shooting.

F. Sperm competition

Direct evidence for sperm competition in gastropods is scarce, except for the two simultaneously hermaphroditic pulmonates *Biomphalaria glabrata* and *Arianta arbustorum*. Double-mated individuals of the freshwater snail *B. glabrata* were found preferentially to use sperm from a pigmented partner and not those from an albino partner (Vianey-Liaud 1995). However, it is possible that pigmented and albino snails differ in the competitive ability of their sperm, since pigmented and albino strains of *B. glabrata* differ in shell size and fecundity (Vianey-Liaud 1989).

More detailed information on sperm competition is available for the land snail *A. arbustorum*. The snail is common in moist habitats of northwestern and central Europe and has determinate growth. Individuals become sexually mature at an age of 2–4 years, and adults live for a further 3–4 years (maximum 14 years; Baur and Raboud 1988). In the field, snails deposit one to three clutches, each containing 20–50 eggs, per reproductive season. Female fecundity (i.e. the number of clutches produced per season, clutch size and egg size) is positively correlated with adult shell size (Baur 1994b). Breeding experiments showed that 12 out of 44 virgin individuals (27%) prevented from mating produced a few hatchlings by self-fertilization in the second and third reproductive year (Chen 1993). However, the reproductive success of selfing individuals was less than 2% of that of mated snails, suggesting high costs of selfing (Chen 1994).

Mating of *A. arbustorum* includes elaborate courtship behaviour with dart shooting and lasts 2–18 h (Hofmann 1923). Mating was found to be random with respect to shell size in natural populations (Baur 1992a): mate-choice tests revealed that individuals of *A. arbustorum* discriminate neither between potential mates of different size nor between snails of different degrees of relatedness (Baur 1992a; Baur and Baur 1997).

Copulation is reciprocal in *A. arbustorum*; spermatophores are transferred after simultaneous intromission (Fig. 8.5B). The snails mate repeatedly in the course of a reproductive season and viable sperm can be stored for more than 1 year (Baur 1988). The sperm storage organ (spermatheca) of *A. arbustorum* consists of 2–8 tubules with a common entrance into the duct of the fertilization pouch (Fig. 8.1; Haase and Baur 1995). Paternity analysis in broods of wild-caught *A. arbustorum* showed a high frequency of multiple insemination (Baur 1994a). A controlled laboratory experiment showed that one successful copulation per reproductive season is sufficient to fertilize all the eggs produced by an individual (Chen and Baur 1993). However, there is a probability of 5–8% that a copulation will not lead to fertilization of eggs (either no sperm are transferred or nonfertile sperm are transferred).

Sperm precedence (P_2) was estimated in double-mated *A. arbustorum* using shell colour as a genetic marker (Baur 1994a). P_2 was influenced by the time between the two matings when the mating delay exceeded 70 days (one reproductive season). P_2 averaged 0.34 in the first brood of snails that mated twice within 70 days, indicating first mate sperm precedence. In contrast, P_2 averaged 0.76 in broods of snails that remated in the following season, indicating either a partial loss of sperm or a loss of sperm viability from the first mate. The size of sperm-donating individuals had no effect on the fertilization success of their sperm in the first brood produced after the second copulation.

Analysis of long-term sperm utilization in 23 double-mated snails that laid three to nine batches over 2 years revealed striking differences among individuals (Baur 1994a). Five snails (22%) exhibited first mate sperm precedence throughout, eight snails (35%) showed second mate sperm precedence throughout, and 10 snails (43%) exhibited sperm mixing in successive batches. The individual variation in sperm precedence in *A. arbustorum* could have arisen for several reasons, including selective sperm storage by the female function, differences in the amount of sperm transferred, and differential fertilizing capacity of sperm from different 'males'.

In *A. arbustorum*, the spermatophore is formed and filled with sperm during copulation (Hofmann 1923). The spermatophore has a distinctive form consisting of head, body (sperm container) and a 2–3 cm long tail (Fig. 8.5B). The number of sperm in the spermatophores of *A. arbustorum* vary considerably, from 802 600 to 3 968 800 (median: 2 048 650; n = 73), in snails living in a subalpine population (R. Locher and B. Baur, unpublished data).

The sites of sperm storage were examined in *A. arbustorum* that remated successfully. In some snails a part of the spermathecal tubules was filled with spermatozoa, while in other animals no sperm were found in the spermatheca (Haase and Baur 1995). In these snails sperm were found exclusively in the sperm-digesting bursa copulatrix. This suggests that the female reproductive system of *A. arbustorum* may be able to control fertilization by a selective storage of sperm from different mating partners (cf. Eberhard 1991, 1996).

G. Paternity guards

Different behavioural patterns and mechanisms have evolved to increase the certainty of paternity of a particular male. Mate guarding, i.e. the close following of females by their mates during the female's period of fertility, is an efficient way for males to increase their certainty of paternity. Precopulatory mate guarding has been reported in the marine prosobranch *Littorina planaxis* (Gibson 1964). A male exploring the shell of a female becomes aggressive if another male tries to mount the shell of that female. In combats lasting 30–180 s the potential mates push each other with the head ends of their shells until one is dislodged. Post-copulatory mate guarding has been observed in natural populations of the marine prosobranch *Strombus pugilis* (Bradshaw-Hawkins and Sander 1981). After a successful copulation, the male guarded the female while she was laying eggs. Males that actively attempted to copulate with a guarded female were displaced. The observed associations between an ovipositing female and a male guard were maintained either until the female completed egg-laying or until the male was displaced by another male.

The production of copulatory plugs is another paternity guard. A copulatory plug, composed of secretions of the male accessory glands, is formed in the female duct of freshwater pulmonates (*Stagnicola elodes*; Rudolph 1979a; *Lymnaea stagnalis*; Horstmann 1955; *Bulinus contortus*; de Larambergue 1939; *Bulinus globosus*; Rudolph 1979b). The plug could prevent sperm leakage, inhibit successive copulation by another snail, and/or prevent sperm from the second partner entering the spermatheca. In general, plugs are not effective in preventing further matings (Rudolph 1979b; van Duivenboden and ter Maat 1988). However, it is not known whether the plugs block the passage of sperm from the second mate.

It has been suggested that some sperm (the kamikaze sperm) may be specifically designed not to fertilize eggs but to destroy or block the passage of sperm from other males (Silberglied *et al.* 1984; Baker and Bellis 1988); this kamikaze sperm hypothesis is still controversial. In prosobranch gastropods apyrene and oligopyrene sperm could potentially reduce the passage of sperm received from a second partner in the female reproductive tract. However, this hypothesis needs to be tested.

H. Costs of copulation

In gastropods the costs of copulation may include an increased risk of predation (e.g. in *Helix pomatia*; Pollard 1975), physical damage by courtship behaviour (e.g. dart shooting) and/or the act of copulation itself, parasite infection, and interference of the long-lasting courtship and copulation with other activities such as feeding. After copulation, the protozoon *Cryptobia helicis* was found in the sperm-digesting organ of *H. pomatia* (Lind 1973) and *Cryptobia* spp. was found in *Triodopsis multilineata* (Current 1980). Transmission of these protozoa is venereal, but any costs to the host have not been examined. The mite *Riccardoella limacis* is a common parasite of several pulmonate land snails. It feeds on the blood of the host while in the lung, resulting in pathological changes in the lung epithelium. Heavy mite infections result in a 40% reduction in growth rate of *Helix aspersa* and retardation of reproductive development (Graham *et al.* 1995). Transmission of mites occurs during courtship and copulation as well as among individuals living under crowded conditions (e.g. *Arianta arbustorum*; B. Baur, unpublished data).

Sexual transmission of nematodes has been documented in several terrestrial pulmonates (*Agfa flexilis* in the slug *Limax cinereoniger*; Morand and Hommay 1990; *Nemhelix lamottei* and *Nemhelix ludesensis* in *Cepaea nemoralis* and *Cepaea hortensis*; Morand 1988a, 1989). In *H. aspersa*, new hosts are infested during mating by transmitting the nematode *Nemhelix bakeri* along with the spermatophore (Morand 1988b). Nematodes do not affect the survivorship of the snails, but significantly reduce egg production (Morand 1989).

VII. BIVALVIA

Bivalves, or pelecypods, are untorted, bilaterally symmetrical molluscs with two valves continuously joined by a mid-dorsal mantle isthmus that secretes a largely uncalcified ligament (Allen 1985). About two-thirds of the 10 000 species are marine, the others are freshwater species. Bivalves vary in adult shell length from 0.05 cm to 140 cm (e.g. the giant clam *Tridacna gigas*; Solem 1974).

A. Mating systems

The majority of bivalves are gonochoric, with broadcast spawning and external fertilization. Of the approximately 10 000 recent species, about 400 are hermaphroditic (Moore 1969). However, the type of sexuality may vary among species belonging to the same genus and also within populations of the same species (Heller 1993). Occasionally, hermaphroditic individuals have been found in species that are considered strictly gonochoric (e.g. *Mya arenaria*; Coe and Turner 1938). For example, hermaphroditic individuals in otherwise predominantly gonochoristic species have been encountered in 30 out of 101 species of the family Unionidae examined (Kat 1983). This suggests that the intensity of sperm competition may vary between populations of the same species.

In most gonochoric bivalves, the sex ratio is close to 1:1 (Mackie 1984). In a few species it is slightly female biased (reviewed by Mackie 1984). In some species with parasitic dwarf males, the sex ratio is male biased (see below).

The majority of hermaphroditic bivalves are functional hermaphrodites, producing simultaneously ova and spermatozoa. In these species, self-fertilization may occur under appropriate circumstances (e.g. *Lasaea subviridis*; Ó Foighil 1987). Some species are protandric hermaphrodites (e.g. *Teredo diegensis*), other species, although less common, are protogynous hermaphrodites (e.g. *Kellia suborbicularis* and *Montacuta ferruginosa*; Oldfield 1961). In oysters of the genus *Ostrea*, individuals usually function either as males or females for one reproductive season (Mackie 1984). The change of sex is repeated, either annually or at closer intervals, leading to a rhythmical alteration between sexes. Alternative sexuality occurs in species with separate sexes, but the gender may or may not be reversed in the next reproductive season (e.g. *Crassostrea* sp.; Andrews 1979). In some protandrous hermaphrodites, a small number of young males never change sex. Factors that influence sex change in bivalves have been reviewed by Fretter and Graham (1964), Sastry (1979) and Mackie (1984). Patterns of sex change might influence the intensity of sexual selection and sperm competition.

B. Sexual dimorphism, anatomy and reproductive physiology

Sexual dimorphism is rare in bivalves (Mackie 1984). Females are larger than males in some species of *Anodonta* (Heard 1975). In the commensal bivalve *Montacuta percompressa*, females are larger than the parasitic males (Chanley and Chanley 1970). Females of *Ephippodonta oedipus* incubate two dwarf males in a pair of pallial pouches (Morton 1976). These dwarf males fertilize the eggs of their host female. There may be competition among sperm from the two males, but differences in fertilization success have not been examined.

Coe and Turner (1938) found that the number of eggs in females of *Macoma balthica* is positively correlated with animal size. Heard (1975) recorded a predominance of either testicular or ovarian tissue in some simultaneously hermaphroditic species of *Anodonta*. This pattern of asymmetrical sex allocation in functional hermaphrodites could be a response to a variable intensity of sperm competition. Except in a few species, there are no copulatory organs or accessory sex glands (Sastry 1979; Mackie 1984).

As in other external fertilizers, the spermatozoa of bivalves are very simple in structure, and compared with other molluscs, relatively short (39 μm in *Unio pictorum*; Franzén 1955; Thompson 1973). Dimorphic spermatozoa occur in bivalve species (e.g. *Kellia symmetros*, *Mysella bidentata* and *Montacuta tenella*; Ockelmann 1965). The significance of atypical spermatozoa is not clear, but it has been suggested that they have a secretory function concerned either with the formation of typical spermatozoa or with the spawning process (Sastry 1979).

In bivalves spermatophore production occurs in a small number of species, whose individuals are normally aggregated on or around a host animal (Ó Foighil 1985). Individuals of the hermaphroditic *Mysella tumida* release several transparent spermatophores, up to 0.6 mm long, into the water. Floating spermatophores that contact the extended foot of these highly mobile bivalves adhere to the shell surface and sperm are released into the suprabranchial chamber. Sperm storage is achieved by the mass attachment of sperm to the surface of gill filaments (Ó Foighil 1985). Other forms of sperm storage occur in a number of brooding bivalves (Sastry 1979). Sperm morulae are multinucleated structures of atypical spermatogenesis that occur in members of the superfamily Unioacea and some marine species (Mackie 1984). Their function is not clearly understood. In *Pseudopythina subsinuata* sperm can be stored as sperm morulae in the suprabranchial chamber for short periods of time (Morton 1972). Sperm competition might occur in bivalve species with sperm storage.

A variety of protozoan and trematode parasites are reported in bivalves, some of which are found in gonadal tissue, but their effects on gametogenesis in the host is unclear (Cheng 1967; Sastry 1979). Infection by trematode parasites has been reported to cause parasitic castration in some bivalves (e.g. the trematode *Bucephalus haimeanus* in tissues of *Cerastoderma* (*Cardium*) *tuberculum* and *Syndosmya alba*; Cheng, 1967).

C. Reproductive behaviour and sperm competition

Copulatory behaviour is absent in bivalves because they lack copulatory organs (Mackie 1984). However, individuals of numerous clams and oyster species actively aggregate (e.g. *Modiolus demissus*; Lent 1969). Several authors argue that the aggregation behaviour might be important for the synchronization of spawning and fertilization of the gametes released (Sastry 1979). Spawning in aggregated individuals will also promote sperm competition and result in multiple paternity in broods from single individuals.

In most bivalves the timing and duration of reproductive activity may be determined through an interaction between endogenous and exogenous factors. Spawning can be induced by a variety of abiotic factors, including changes in temperature, salinity and light, lunar periodicity, depth, food abundance, chemicals, and mechanical factors (Sastry 1979; Mackie 1984). In some species, gametes of the opposite sex also stimulate spawning. Young (1946) showed that testicular tissue of *Mytilus californianus* contains a substance which causes spawning in females. In *Mytilus edulis* and *Mytilus galloprovincialis*, males can be stimulated to release gametes by eggs or egg extracts, but fresh sperm or sperm extracts are not effective in stimulating the release of eggs by the female (Lubet 1951, 1955). Bivalves that broadcast their eggs produce numerous, small eggs (approximately 50 μm diameter) that develop into planktotrophic larvae (e.g. *Crassostrea virginica* 100×10^6, and *Crassostrea gigas*, 55 × 10^6; Galtsoff 1964). Species that lay their eggs in mucous strings or cases produce only a few eggs (e.g. *Nucula delphinodonta*, 2–70; Fretter and Graham 1964).

I am not aware of any study that provides direct evidence for sperm competition in bivalves.

VIII. SCAPHOPODA (TOOTH SHELLS)

Scaphopods are molluscs with a curved, tooth-like shell open at both ends. Scaphopods occur in the benthos of both shallow and deep waters of all major seas (McFadien-Carter 1979). The narrow end of the shell often protrudes above the mud or sand in which the animal lives. Scaphopods are gonochoric, but hermaphroditic individuals of *Dentalium* sp. are occasionally found (McFadien-Carter 1979). Fertilization is external; gametes are released either through the posterior or anterior aperture (Steiner 1993). No information on sperm competition is available in this class.

IX. CEPHALOPODA

A. Mating systems

All cephalopods are gonochoric (Heller 1993), and cross-fertilization is the rule. In general, a sex ratio of 1:1 is assumed in cephalopods, as has been observed in *Loligo pealei* (Summers 1971). However, deviations from an even sex ratio have been recorded in several species (review by Mangold-Wirz 1963). These deviations may reflect sex-specific biases in the sampling methods (Wells and Wells 1977).

Both sexes of all cephalopods appear to be promiscuous. Females can copulate several months before they lay eggs. Viable sperm can be stored for up to 10 months (Wells 1978). In most cephalopods, mating behaviour and egg laying is followed by death, an exception being *Nautilus* (Arnold 1984).

B. Sexual dimorphism, anatomy and reproductive physiology

Sexual dimorphism is evident in a majority of cephalopods (Arnold 1984). Characters of external sexual dimorphism include size, body patterns, enlarged suckers, gonad shape or colour and the presence of a specialized arm in males (the hectocotylus, which transfers spermatophores to the females) (Mangold 1987). Body proportions may be slightly different between sexes and the size of the adult animals may also differ. Males are larger than females in loligonid squids and *Nautilus*, but in other pelagic octopods (*Argonauta*, *Ocythoe*, and *Tremoctopus*), the males are very small and have a relatively large hectocotylus, which detaches and remains for some time in the mantle of its mate (Arnold 1984). During mating, the sexes can be easily distinguished because the males are aggressive; both sexes show behavioural differences.

In *Octopus*, fertilization occurs in the oviduct as the eggs are shed (Wells 1978). Female fecundity is size-related in many cephalopod species (e.g. females of *Loligo pealei* lay between 3500 and 6000 eggs; Summers 1971).

Spermatozoa can be stored 'externally', in the buccal pouch (as in many teuthoids, sepiids), the oral pits (*Vampyroteuthis*) or 'internally', with a series of possible sites ranging from the inside of the mantle cavity (*Illex*) to the spermatheca (*Sepiola*, *Sepietta*), the oviducts and oviductal glands (most *Octopus* species) to the ovary (*Eledone*) (Mangold 1987).

I. Sperm size, sperm number and spermatophores

There is a great variation in the size and structure of cephalopod spermatozoa, which is of taxonomic significance (Arnold 1984). Sperm length varies from 50 μm in *Loligo pealei* (Austin *et al.* 1964) to 500 μm in *Octopus dofleini* (Mann *et al.* 1970). Similarly, there is an enormous variation in size and structure of cephalopod spermatophores (Mann 1984). The 1-cm long spermatophore of *L. pealei* consists of a central coiled mass of sperm ($7.2–9.6 \times 10^6$ spermatozoa), an associated cement gland, and the membranes and fluid-filled spaces which encase the whole structure (Austin *et al.* 1964; Fig. 8.6C). In *Nautilus pompilius*, the spermatophores average 14 cm long and vary in diameter from 2 to 5 mm, those of *Nautilus belauensis* measure 35 cm long and about 5 mm diameter (Arnold 1984). *Octopus dofleini* is a large octopus (20 kg or more) which produces 1-m long spermatophores (Mann *et al.* 1970). In *Loligo opalescens*, the length of the spermatophore is proportional to the size of the male producing it (Fields 1965). Up to several dozens of spermatophores may be formed each day and a few hundred may be stored in Needham's sac. The number of spermatophores varies from species to species (about six in *O. dofleini* and several hundreds in *Octopus cyanea*; Wells and Wells 1977). When the male transfers the spermatophores to the female, his hectocotylus pulls the cap thread at the tip of one end. The spermatophore suddenly bursts, the sperm mass is pushed down through the ejaculatory apparatus, and the cement gland breaks and fixes the sperm reservoir to the female's buccal seminal receptacle, oviduct, or body wall, depending on the position of copulation (Arnold 1984). In *L. pealei*, this process takes a few seconds, but in the 1-m long spermatophores of *O. dofleini*, the reaction takes 1–2 h (Mann *et al.* 1966, 1970).

C. Reproductive behaviour

There are distinct differences in reproductive behaviour, mating and site of sperm storage among genera or groups of cephalopods. Boyle (1983, 1987) and Hanlon and Messenger (1996) reviewed courtship and mating behaviour in a variety of cephalopods.

Shoaling squids (decapods) have ample opportunity for social interactions with conspecifics and some species have developed elaborate agonistic and courtship behaviour. Octopuses are mostly solitary, exhibit little courtship behaviour and male agonistic behaviour, but mating is prolonged compared with squids.

Courtship behaviour is relatively well studied in *Sepia* (Tinbergen 1939; Corner and Moore 1980), and *Loligo pealei* (Drew 1911; Arnold 1984). In *L. pealei*, courtship behaviour can be initiated by the visual stimulus of an egg mass or occurs spontaneously. Copulation occurs in two positions, either head-to-head or side-by-side. Female choice may be involved in the decision of which mating position is used. However, it is not known how the mating position is determined, or by which sex. In *Loligo*, spermatophores transferred in the head-to-head position appear to be passed only to the seminal receptacle below the mouth for storage by the female, whereas in the side-by-side position the spermatophores are placed inside the mantle cavity near the opening of the oviduct. Arnold and Williams-Arnold (1977) assume that the head-to-head position is used offshore, possibly before the animals have reached sexual maturity, to ensure fertility of eggs, and the side-to-side position is used when fertilization is imminent. Different positions of the spermatophores could result in differential fertilization success of the sperm.

Loligo pealei may copulate repeatedly. All females appear to arrive on spawning grounds with sperm stored in the receptacle below the mouth (Drew 1911). Theoretically, these females need not mate again. However, large males compete vigorously in agonistic contests, primarily of visual displays, that can escalate to mild fin beating (Hanlon 1996). Large males mate the females in the side-by-side position (Arnold 1962; Hanlon and Messenger 1996), and guard them as they approach the egg mass and lay each egg capsule (Hanlon 1996). 'Sneaker' males, which are smaller than the typical large males, use an alternative tactic. They do not engage large males in agonistic contests; instead they jet forward suddenly onto the arms of a female as the large male and female pair approaches the egg mass (Hanlon 1996). They swiftly deposit spermatophores directly on the egg capsule, which is presumably amid the female's arms. Thus, there are several potential sources of sperm for each egg capsule: stored sperm, the large male and the sneaker male. However, methods of paternity assessment must be used to determine the winner (or winners) of these sperm competition games. In *L. pealei*, this behavioural pattern results in the formation of a rather permanent social hierarchy in which the males establish and maintain pairs on the basis of their size, aggressiveness, and persistence (Arnold and Williams-Arnold 1977). In *L. opalescens*, however, males frequently switch females after a period of copulation (Fields 1965).

Female octopuses brood the eggs by continually flushing them with water from the funnel and running the arm tips over them (Lane 1960; Wells and Wells 1977; Wells 1978).

D. Sperm competition

There is so far little direct evidence for sperm competition in cephalopods (except in pygmy octopuses; see below). However, several features suggest that sperm competition may occur: the sperm are packed in spermatophores, sperm are stored by females, the morphology of the oviduct and spermatheca are appropriate for sperm competition, mating systems are polygamous, there are different types of mating, and there are delays between mating and egg laying.

Male pygmy octopuses are able to assess the mating history of females, as indicated by the increase in duration when mating with recently mated females (Cigliano 1995). Male octopuses possibly use the presence/absence of sperm to determine whether females have recently mated. Males copulating with recently mated females did not transfer more spermatophores, but increased the duration of the first phase of mating most probably to remove or displace previously deposited sperm. This can be done by placing the 'spoon'-shaped lingula in the distal portion of the oviduct and scooping out previously deposited sperm. Cigliano (1995) suggested that the lingula's function may be to scoop out sperm instead of placing the spermatophores into the oviduct.

Mate guarding has been observed in several cephalopod species. In *L. opalescens*, the male may retain his grasp on the female after he has withdrawn his hectocotylus and stay with her as she deposits the first egg strings (Fields 1965). In *Sepia officinalis*, the male tends to stay with the female, copulating repeatedly with her, and defends his mate from other approaching males with intense colour displays, extended arms, and by attacking and biting (Tinbergen 1939).

X. CONCLUSIONS AND FUTURE RESEARCH DIRECTIONS

Sperm competition may occur more frequently among molluscs than commonly assumed. There exists a huge literature on the anatomy and morphology of reproductive systems in molluscs, although not all groups have received the same attention. However, the significance of these structures with respect to sexual selection and sperm competition is, in most cases, unclear. Several mollusc species may be well-suited for studies on sperm competition, but their potential as experimental organisms has not yet begun to be exploited by behavioural and evolutionary biologists.

Studies on sperm competition in gastropods and cephalopods should prove to be particularly rewarding. Both gastropods and cephalopods are unique because of several important features. Their elaborate mating behaviour may rival the complexity of those of various vertebrates. Do individuals assess the quality of potential mates during courtship? Is there a displacement of previous sperm in gastropods and cephalopods that show protracted copulation? Sperm selection (either in the form of a selective storage or digestion) might occur in both groups. This could provide a unique possibility of female choice in sperm competition, which deserves attention. Careful studies of the morphology of the female reproductive tract with respect to sperm storage and mating experiments using molecular techniques for paternity analyses could begin to answer some of these questions. Furthermore, the adaptive significance of sperm polymorphism is still not known. There is much to be learned in these most interesting groups of animals.

ACKNOWLEDGEMENTS

I thank Anette Baur, Tim R. Birkhead, Martin Haase, Hans Kothbauer, Nico K. Michiels and Anders P. Møller for helpful discussions and comments. I am especially indebted to Gerhard Falkner, Terrence M. Gosliner and Kathe R. Jensen for providing negatives and photos. Financial support for my own research was from the Swiss National Science Foundation.

REFERENCES

Adamo SA & Chase R (1988) Courtship and copulation in the terrestrial snail, *Helix aspersa*. *Can. J. Zool.* **66:** 1446–1453.

Adamo SA & Chase R (1990) The 'love dart' of the snail *Helix aspersa* injects a pheromone that decreases courtship duration. *J. Exp. Zool.* **255:** 80–87.

Adamo SA & Chase R (1996) Dart shooting in helicid snails: an 'honest' signal or an instrument of manipulation? *J. Theor. Biol.* **180:** 77–80.

Albrecht EA, Carreno NB & Castro-Vazquez A (1996) A quantitative study of copulation and spawning in the South American apple-snail, *Pomacea canaliculata* (Prosobranchia: Ampullariidae). *Veliger* **39:** 142–147.

Allen JA (1985) The recent Bivalvia: their form and evolution. In *The Mollusca, Vol. 10, Evolution.* ER Trueman & MR Clarke (eds), pp. 337–403. Academic Press, London.

Anderson AW & Personne P (1976) The molluscan spermatozoon: dynamic aspects of its structure and function. *Am. Zool.* **16:** 293–313.

Andreassen EM (1981) Population dynamics of *Arianta arbustorum* and *Cepaea hortensis* in western Norway. *Fauna Norv. Ser. A* **2:** 1–13.

Andrews EA (1936) Spermatophores of the snail *Neritina reclivata. J. Morphol.* **60:** 191–209.

Andrews EB (1964) The functional anatomy and histology of the reproductive system of some pilid gastropod molluscs. *Proc. Malacol. Soc. Lond.* **36:** 121–140.

Andrews JD (1979) Pelecypoda: Ostreidae. In *Reproduction of Marine Invertebrates, Vol. 5, Molluscs: Pelecypods and Lesser Classes.* AC Giese & JS Pearse (eds), pp. 293–341. Academic Press, London.

Arnold JM (1962) Mating behavior and the social structure in *Loligo pealii. Biol. Bull.* **123:** 53–57.

Arnold JM (1984) Cephalopods. In *The Mollusca, Vol. 7, Reproduction.* AS Tompa, NH Verdonk & JAM van den Biggelaar (eds), pp. 419–454. Academic Press, London.

Arnold JM & Williams-Arnold LD (1977) Cephalopoda: Decapoda. In *Reproduction of Marine Invertebrates, Vol. 4, Molluscs: Gastropods and Cephalopods.* AC Giese & JS Pearse (eds), pp. 243–290. Academic Press, London.

Audesirk TE (1979) Chemoreception in *Aplysia californica* III. Evidence for pheromones influencing reproductive behavior. *Behav. Biol.* **20:** 235–243.

Austin CR, Lutwak-Mann C & Mann T (1964) Spermatophores and spermatozoa of the squid *Loligo pealii. Proc. Roy. Soc. Lond. Ser. B* **161:** 143–152.

Baker RR & Bellis MA (1988) 'Kamikaze' sperm in mammals? *Anim. Behav.* **37:** 867–869.

Baur B (1986) Patterns of dispersion, density and dispersal in alpine populations of the land snail *Arianta arbustorum. Holarctic Ecol.* **9:** 117–125.

Baur B (1988) Repeated mating and female fecundity in the simultaneously hermaphroditic land snail *Arianta arbustorum. Invert. Reprod. Devel.* **14:** 197–204.

Baur B (1992a) Random mating by size in the simultaneously hermaphroditic land snail *Arianta arbustorum*: experiments and an explanation. *Anim. Behav.* **43:** 511–518.

Baur B (1992b) Cannibalism in gastropods. In *Cannibalism: Ecology and Evolution Among Diverse Taxa.* MA Elgar & BJ Crespi (eds), pp. 102–127. Oxford University Press, Oxford.

Baur B (1994a) Multiple paternity and individual variation in sperm precedence in the simultaneously hermaphroditic land snail *Arianta arbustorum. Behav. Ecol. Sociobiol.* **35:** 413–421.

Baur B (1994b) Parental care in terrestrial gastropods. *Experientia* **50:** 5–14.

Baur B & Baur A (1992a) Reduced reproductive compatibility in the land snail *Arianta arbustorum* from distant populations. *Heredity* **69:** 65–72.

Baur B & Baur A (1992b) Effect of courtship and repeated copulation on egg production in the simultaneously hermaphroditic land snail *Arianta arbustorum*. *Invert. Reprod. Devel.* **21:** 201–206.

Baur B & Baur A (1997) Random mating with respect to relatedness in the simultaneously hermaphroditic land snail *Arianta arbustorum*. *Invert. Biol.* **116:** 294–298.

Baur B & Chen X (1993) Genital dimorphism in the land snail *Chondrina avenacea*: frequency of aphally in natural populations and morph-specific allocation to reproductive organs. *Veliger* **36:** 252–258.

Baur B & Raboud C (1988) Life history of the land snail *Arianta arbustorum* along an altitudinal gradient. *J. Anim. Ecol.* **57:** 71–87.

Baur B, Chen X & Baur A (1993) Genital dimorphism in natural populations of the land snail *Chondrina clienta* and the influence of the environment on its expression. *J. Zool.* **231:** 275–284.

Beeman RD (1970) An autoradiographic study of sperm exchange and storage in a sea hare, *Phyllaplysia taylori*, a hermaphroditic gastropod (Opisthobranchia: Anaspidea). *J. Exp. Zool.* **175:** 125–132.

Beeman RD (1977) Gastropoda: Opisthobranchia. In *Reproduction of Marine Invertebrates, Vol. 4, Molluscs: Gastropods and Cephalopods.* AC Giese & JP Pearse (eds), pp. 115–179. Academic Press, London.

Berry AJ (1977) Gastropoda: Pulmonata. In *Reproduction of Marine Invertebrates, Vol. 4, Molluscs: Gastropods and Cephalopods.* AC Giese & JP Pearse (eds), pp. 181–226. Academic Press, London.

Boyle PR (ed.) (1983) *Cephalopod Life Cycles, Vol. I. Species Accounts.* Academic Press, London.

Boyle PR (ed.) (1987) *Cephalopod Life Cycles, Vol. II. Comparative Reviews.* Academic Press, London.

Brackenbury TD & Appleton CC (1991) Morphology of the mature spermatozoon of *Physa acuta* (Draparnaud, 1801) (Gastropoda: Physidae). *J. Mollusc. Stud.* **57:** 211–218.

Bradshaw-Hawkins VI & Sander F (1981) Notes on the reproductive biology and behavior of the West Indian fighting conch, *Strombus pugilis* Linnaeus in Barbados, with evidence of mate guarding. *Veliger* **24:** 159–164.

Brinders EM & Sirgel WF (1992) The morphology and histology of the genital system of *Trigonephrus gypsinus* and *Trigonephrus latezonatus* (Gastropoda: Pulmonata). *Ann. Univ. Stellenbosch* **3:** 1–27.

Buckland-Nicks J & Darling P (1993) Sperm are stored in the ovary of *Lacuna (Epheria) variegata* (Carpenter, 1864) (Gastropoda: Littorinidae). *J. Exp. Zool.* **267:** 624–627.

Buckland-Nicks JA & Eernisse DJ (1993) Ultrastructure of mature sperm and eggs of the brooding hermaphroditic chiton, *Lepidochitona fernaldi* Eernisse 1986, with special reference to the mechanism of fertilization. *J. Exp. Zool.* **265:** 567–574.

Cain AJ (1983) Ecology and ecogenetics of terrestrial molluscan populations. In *The Mollusca, Vol. 6, Ecology*. WD Russell-Hunter (ed.), pp. 597–647. Academic Press, London.

Cain GL (1956) Studies on cross-fertilization and self-fertilization in *Lymnaea stagnalis appressa* Say. *Biol. Bull.* **111:** 45–52.

Calow P (1983) Life-cycle patterns and evolution. In *The Mollusca, Vol. 6, Ecology*. WD Russell-Hunter (ed.), pp. 649–678. Academic Press, London.

Carefoot TH (1987) *Aplysia*: its biology and ecology. *Oceanogr. Mar. Biol. Annu. Rev.* **25:** 167–284.

Cate JM (1968) Mating behavior in *Mitra idae* Melvill, 1893. *Veliger* **10:** 247–252.

Chace LM (1952) The aerial mating of the great slug. *Discovery* **13:** 356–359.

Chanley PE & Chanley MH (1970) Larval development of the commensal clam, *Montacuta percompressa* Dall. *Proc. Malacol. Soc. Lond.* **39:** 59–67.

Charnov EL (1979) Simultaneous hermaphroditism and sexual selection. *Proc. Natl Acad. Sci. USA* **76:** 2480–2484.

Charnov EL (1996) Sperm competition and sex allocation in simultaneous hermaphrodites. *Evol. Ecol.* **10:** 457–462.

Chase R (1986) Lessons from snail tentacles. *Chem. Senses* **11:** 411–426.

Chen X (1993) Comparison of inbreeding and outbreeding in hermaphroditic *Arianta arbustorum* (land snail) (L.). *Heredity* **71:** 456–461.

Chen X (1994) Self-fertilization and cross-fertilization in the land snail *Arianta arbustorum* (Mollusca, Pulmonata: Helicidae). *J. Zool.* **232:** 465–471.

Chen X & Baur B (1993) The effect of multiple mating on female reproductive success in the simultaneously hermaphroditic land snail *Arianta arbustorum. Can. J. Zool.* **71:** 2431–2436.

Cheng TC (1967) Marine molluscs as hosts for symbioses with a review of known parasites of commercially important species. *Adv. Mar. Biol.* **5:** 1–424.

Chung DJD (1986) Stimulation of genital eversion in the land snail *Helix aspersa* by extracts of the glands of the dart apparatus. *J. Exp. Zool.* **238:** 129–139.

Chung DJD (1987) Courtship and dart shooting behavior of the land snail *Helix aspersa. Veliger* **30:** 24–39.

Cigliano JA (1995) Assessment of the mating history of female pygmy octopuses and a possible sperm competition mechanism. *Anim. Behav.* **49:** 849–851.

Coe WR (1936) Sexual phases in *Crepidula. J. Exp. Zool.* **77:** 401–424.

Coe WR & Turner HJ (1938) Development of the gonads and gametes in the soft-shell clam (*Mya arenaria*). *J. Morphol.* **62:** 91–111.

Collin R (1995) Sex, size, and position: a test of models predicting size at sex change in the protandrous gastropod *Crepidula fornicata. Am. Nat.* **146:** 815–831.

Corner BD & Moore HT (1980) Field observations on reproductive behavior of *Sepia latimanus. Micronesica* **16:** 235–260.

Croll RP (1983) Gastropod chemoreception. *Biol. Rev.* **58:** 293–319.

Crozier WJ (1918) Assortative mating in a nudibranch, *Chromodoris zebra* Heilprin. *J. Exp. Zool.* **27:** 247–292.

Current WL (1980) *Cryptobia* sp. in the snail *Triodopsis multilineata* (Say): fine stucture of attached flagellates and their mode of attachment to the spermatheca. *J. Protozool.* **27:** 278–287.

Dasen DD (1933) Structure and function of the reproductive system in *Ariophanta ligulata. Proc. Zool. Soc. Lond.* 1933, 97–118.

Dembski WJ (1968) Histochemische Untersuchungen über Funktion und Verbleib eu- und oligopyrener Spermien von *Viviparus contectus* (Millet 1813) (Gastropoda, Prosobranchia). *Z. Zellforsch.* **89:** 151–179.

Dillaman RM (1981) Dart formation in *Helix aspersa* (Mollusca, Gastropoda). *Zoomorphology* **97:** 247–261.

Doums C, Bremond P, Delay B & Jarne P (1996) The genetical and environmental determination of phally polymorphism in the freshwater snail *Bulinus truncatus*. *Genetics* **142:** 217–225.

Drew GA (1911) Sexual activities in the squid *Loligo pealei* (Les.) I. Copulation, egg-laying and fertilization. *J. Morphol.* **22:** 327–360.

Duncan CJ (1975) Reproduction. In *Pulmonates, Vol. 1.* V Fretter & J Peake (eds), pp. 309–365. Academic Press, London.

Dupouy J (1964) La tératogénèse germinale male des Gastéropodes et ses rapports avec l'oogénèse atypique et la formation des oeufs nourriciers. *Arch. Zool. Exp. Génér.* **103:** 217–368.

Eberhard WG (1985) *Sexual Selection and Animal Genitalia.* Harvard University Press, Cambridge, MA.

Eberhard WG (1991) Copulatory courtship and cryptic female choice in insects. *Biol. Rev.* **66:** 1–31.

Eberhard WG (1996) *Female Control: Sexual Selection by Cryptic Female Choice.* Princeton University Press, Princeton.

Edwards DC (1968) Reproduction in *Olivella biplicata. Veliger* **10:** 297–304.

Edwards DC (1969) Zonation by size as an adaptation for intertidal life in *Olivella biplicata. Am. Zool.* **9:** 399–417.

Eernisse DJ (1988) Reproductive patterns in six species of *Lepidochitona* (Mollusca: Polyplacophora) from the Pacific Coast of North America. *Biol. Bull.* **174:** 287–302.

Eernisse DJ & Reynolds PD (1994) Polyplacophora. In *Microscopic Anatomy of Invertebrates, Vol. 5, Mollusca I.* FW Harrison & AJ Kohn (eds), pp. 55–110. Wiley-Liss, New York.

Egonmwan RI (1990) Viability of allosperm in the garden snail *Limicolaria flammea*, Muller (Gastropoda: Pulmonata). *Biosci. Res. Commun.* **2:** 87–92.

Els WJ (1978) Histochemical studies on the maturation of the genital system of the slug *Deroceras laeve* (Pulmonata, Limacidae), with special reference to the identification of mucosubstances secreted by the genital tract. *Ann. Univ. Stellenbosch (A2)* **1:** 1–116.

Emberton KC (1994) Polygyrid land snail phylogeny: external sperm exchange, early North American biogeography, iterative shell evolution. *Biol. J. Linn. Soc.* **52:** 241–271.

Erlandsson J & Johannesson K (1994) Sexual selection on female size in a marine snail, *Littorina littorea* (L.). *J. Exp. Mar. Biol. Ecol.* **181:** 145–157.

Falkner G (1990) Binnenmollusken. In *Weichtiere. Europäische Meeres- und Binnenmollusken.* R Fechter & G Falkner (eds), pp. 112–278. Mosaik Verlag, Munich.

Falkner G (1992) Grandioser Seilakt zu nächtlicher Stunde: Paarung des Tigerschnegels. In *Die grosse Bertelsmann Lexikothek, Naturenzyklopädie Europas, Vol. 6.* JH Reichholf & G Steinbach (eds), pp. 282–283. Mosaik Verlag, Munich.

Falkner G (1993) Lockspiel und Lockstoffdrüsen bei Hygromiiden und Heliciden (Gastropoda: Stylommatophora). *Heldia* **2:** 15–20.

Fearnley RH (1993) Sexual selection, dispersal and reproductive behaviour in hermaphrodite land snails, with particular reference to *Helix aspersa* Müller (Pulmonata, Gastropoda). Ph.D. Thesis, University of Manchester, UK.

Fearnley RH (1996) Heterogenic copulatory behaviour produces non-random mating in laboratory trials in the land snail *Helix aspersa* Müller. *J. Mollusc. Stud.* **62:** 159–164.

Fields WG (1965) The structure, development, food relations, reproduction and life history of the squid *Loligo opalescens* Berry. *Fish. Bull. Calif. Fish Game* **131:** 1–108.

Fischer EA (1980) The relationship between mating system and simultaneous hermaphroditism in the coral reef fish *Hypoplectus nigricans. Anim. Behav.* **28:** 620–633.

Fischer EA (1984) Egg trading in the chalk bass *Serranus tortugarum*, a simultaneous hermaphrodite. *Z. Tierpsychol.* **66:** 143–151.

Flasar I (1967) Der innere Bau der Befruchtungstasche bei *Oxychilus draparnaudi* (Beck) und die Geschichte ihrer Entdeckung und Erforschung bei anderen Pulmonaten. *Acta Soc. Zool. Bohem.* **31:** 150–158.

Franzén A (1955) Comparative morphological investigations into the spermiogenesis among Mollusca. *Zool. bidr. Uppsala* **30:** 399–456.

Fretter V (1984) Prosobranchs. In *The Mollusca, Vol. 7, Reproduction.* AS Tompa, NH Verdonk & JAM van den Biggelaar (eds), pp. 1–45. Academic Press, London.

Fretter V & Graham A (1962) *British Prosobranch Molluscs. Their Functional Anatomy and Ecology.* Ray Society, London.

Fretter V & Graham A (1964) Reproduction. In *Physiology of Mollusca, Vol. 1.* KM Wilbur & CM Yonge (eds), pp. 127–164. Academic Press, London.

Gaffney PM & McGee B (1992) Multiple paternity in *Crepidula fornicata* (Linnaeus). *Veliger* **35:** 12–15.

Galtsoff PS (1964) The American oyster *Crassostrea virginica* Gmelin. *Fish. Bull.* **64:** 1–480.

Gascoigne T (1956) Feeding and reproduction in the Limapontiidae. *Trans. Roy. Soc. Edin.* **63:** 129–151.

Gascoigne T (1974) A note on some sacoglossan penial styles (Gastropoda: Opisthobranchia). *Zool. J. Linn. Soc.* **55:** 53–59.

Gascoigne T (1975) The radula and reproductive system of *Olea hansineensis* Agersborg, 1923 (Gastropoda: Opisthobranchia: Sacoglossa). *Veliger* **17:** 313–317.

Geraerts WPM & Joosse J (1984) Freshwater snails (Basommatophora). In *The Mollusca, Vol. 7, Reproduction.* AS Tompa, NH Verdonk & JAM van den Biggelaar (eds), pp. 141–207. Academic Press, London.

Gerhardt U (1933) Zur Kopulation der Limaciden. I. Mitteilung. *Z. Morphol. Ökol. Tiere* **27:** 401–450.

Ghiselin MT (1965) Reproductive function and the phylogeny of opisthobranch gastropods. *Malacologia* **3:** 327–378.

Gibson DG (1964) Mating behavior in *Littorina planaxis* Philippi (Gastropoda: Prosobranchiata). *Veliger* **7:** 134–139.

Giese AC & Pearse JS (eds) (1977) *Reproduction of Marine Invertebrates, Vol. 4, Molluscs: Gastropods and Cephalopods.* Academic Press, New York.

Giese AC & Pearse JS (eds) (1979) *Reproduction of Marine Invertebrates, Vol. 5, Molluscs: Pelecypods and Lesser Classes.* Academic Press, New York.

Giusti F & Andreini S (1988) Morphological and ethological aspects of mating in two species of the family Helicidae (Gastropoda: Pulmonata): *Theba pisana* (Müller) and *Helix aperta* Born. *Monit. Zool. Ital.* **22:** 331–363.

Giusti F & Lepri A (1980) Aspetti morfologici ed etologici dell'accoppiamento in alcune specie della famiglia Helicidae (Gastropoda: Pulmonata). *Accad. Sci. Siena Fisiocr.* **1980**, 11–71.

Giusti F & Selmi MG (1982) The atypical sperm in the prosobranch molluscs. *Malacologia* **22:** 171–181.

Giusti F, Manganelli G & Selmi MG (1992) Spermatozoon fine structure in the phylogenetic study of the Helicoidea (Gastropoda, Pulmonata). Proc. 10th Int. Malacol. Congr., Tübingen 1989, pp. 611–616. Unitas, Tübingen.

Gonor JJ (1979) Monoplacophora. In *Reproduction of Marine Invertebrates, Vol. 5, Molluscs: Pelecypods and Lesser Classes.* AC Giese & JS Pearse (eds), pp. 87–93. Academic Press, London.

Gosliner TC (1994) Gastropoda: Opisthobranchia. In *Microscopic Anatomy of Invertebrates, Vol. 5, Mollusca I.* FW Harrison & AJ Kohn (eds), pp. 253–355. Wiley-Liss, New York.

Götting K-J (1974) *Malakozoologie.* Gustav Fischer Verlag, Stuttgart.

Graham A (1985) Evolution within the Gastropoda: Prosobranchia. In *The Mollusca, Vol. 10: Evolution.* ER Trueman & MR Clarke (eds), pp. 151–186. Academic Press, London.

Graham F, Runham NW & Ford JB (1995) *Riccardoella limacis* a parasite of *Helix aspersa.* In *Abstracts of the Twelfth International Malacological Congress, Vigo, Spain.* A Guerra, E Rolán & F Rocha (eds), pp. 189. Feito, Vigo.

Griffiths RJ (1961) Sexual dimorphism in Cypraeidae. *Proc. Malacol. Soc. Lond.* **34:** 203–206.

Griffond B & Gomot L (1989) Endocrinology of reproduction in stylommatophoran pulmonate gastropods with special reference to *Helix. Comp. Endocrinol.* **8:** 23–32.

Grusov EN (1965) The endoparasite mollusk *Asterophila japonica* Randall & Heath (Prosobranchia: Melanellidae) and its relation to the parasitic gastropods. *Malacologia* **3:** 111–181.

Haase M & Baur B (1995) Variation in spermathecal morphology and storage of spermatozoa in the simultaneously hermaphroditic land snail *Arianta arbustorum* (Gastropoda: Pulmonata: Stylommatophora). *Invert. Reprod. Devel.* **28:** 33–41.

Haase M & Wawra E (1996) The genital system of *Acochlidium fijiense* (Opisthobranchia: Acochlidioidea) and its inferred function. *Malacologia* **38:** 143–151.

Hadfield MG (1979) Aplacophora. In *Reproduction of Marine Invertebrates, Vol. 5, Molluscs: Pelecypods and Lesser Classes.* AC Giese & JS Pearse (eds), pp. 1–25. Academic Press, London.

Hadfield MG & Hopper CN (1980) Ecological and evolutionary significance of pelagic spermatophores of vermetid gastropods. *Mar. Biol.* **57:** 315–325.

Hadfield MG & Switzer-Dunlap M (1984) Opisthobranchs. In *The Mollusca, Vol. 7, Reproduction.* AS Tompa, NH Verdonk & JAM van den Biggelaar (eds), pp. 209–350. Academic Press, London.

Hanlon RT (1996) Evolutionary games that squids play: fighting, courting, sneaking, and mating behaviors used for sexual selection in *Loligo pealei. Biol. Bull.* **191:** 309–310.

Hanlon RT & Messenger JB (1996) *Cephalopod Behaviour.* Cambridge University Press, Cambridge.

Hanson J, Randall JT & Bayley ST (1952) The microstructure of the spermatozoa of the snail *Viviparus. Exp. Cell Res.* **3:** 65–78.

Harrison FW & Kohn AJ (eds) (1994) *Microscopic Anatomy of Invertebrates, Vol. 5, Mollusca I.* Wiley-Liss, New York.

Haszprunar G (1988) On the origin and evolution of major gastropod groups, with special reference to the Streptoneura. *J. Mollusc. Stud.* **54:** 367–441.

Haszprunar G (1992) Preliminary anatomical data on a new neopilinid (Monoplacophora) from the Antartic waters. In *Abstracts of the 11th International*

Malacological Congress in Siena. F Giusti & G Manganelli (eds), pp. 307–308. University of Siena, Siena.

Haszprunar G & Schaefer K (1996) Monoplacophora. In *Microscopic Anatomy of Invertebrates, Vol. 6, Monoplacophora, Bivalvia, Scaphopoda and Cephalopoda*. FW Harrison & AJ Kohn (eds), pp. 387–426. Wiley-Liss, New York.

Healy JM (1988) Sperm morphology and its systematic importance in the Gastropoda. *Malacol. Rev. Suppl.* **4:** 251–266.

Healy JM (1996) Molluscan sperm ultrastructure: correlation with taxonomic units within the Gastropoda, Cephalopoda and Bivalvia. In *Origin and Evolutionary Radiation of the Mollusca*. JD Taylor (ed.), pp. 99–113. Oxford University Press, Oxford.

Healy JM, Schaefer K & Haszprunar G (1995) Spermatozoa and spermatogenesis in a monoplacophoran mollusc, *Laevipilina antarctica*: ultrastructure and comparison with other Mollusca. *Mar. Biol.* **122:** 53–65.

Heard WH (1975) Sexuality and other aspects of reproduction in *Anodonta* (Pelecypoda: Unionidae). *Malacologia* **15:** 81–103.

Heath H (1918) Solenogastres from the eastern coast of North America. *Mem. Mus. Comp. Zool. Harv.* **45:** 187–260.

Heller J (1993) Hermaphroditism in molluscs. *Biol. J. Linn. Soc.* **48:** 19–42.

Heller J & Farstey V (1990) Sexual and parthenogenetic populations of the freshwater snail *Melanoides tuberculata* in Israel. *Israel J. Zool.* **37:** 75–87.

Hoagland KE (1978) Protandry and the evolution of environmentally mediated sex change: a study of the Mollusca. *Malacologia* **17:** 365–391.

Hofmann E (1923) Ueber den Begattungsvorgang von *Arianta arbustorum* (L.). *Jen. Z. Naturwiss.* **59:** 363–400.

Horstmann H-J (1955) Untersuchungen zur Physiologie der Begattung und Befruchtung der Schlammschnecke *Lymnaea stagnalis* L. *Z. Morphol. Ökol. Tiere* **44:** 222–268.

Houbrick RS (1980) Observations on the anatomy and life history of *Modulus modulus* (Prosobranchia: Modulidae). *Malacologia* **20:** 117–142.

Hyman LH (1967) *The Invertebrates, Vol. 6, Mollusca 1*. McGraw-Hill Book Company, New York.

Jarne P & Städler T (1995) Population genetic structure and mating system evolution in freshwater pulmonates. *Experientia* **51:** 482–497.

Jarne P, Finot L, Bellec C & Delay B (1992) Aphally vs. euphally in self-fertile hermaphrodite snails from the species *Bulinus truncatus* (Pulmonata: Planorbidae). *Am. Nat.* **139:** 424–432.

Jarne P, Vianey-Liaud M & Delay B (1993) Selfing and outcrossing in hermaphrodite freshwater gastropods (Basommatophora): where, when and why? *Biol. J. Linn. Soc.* **49:** 99–125.

Jensen KR (1992) Anatomy of some Indo-Pacific Elysiidae (Opisthobranchia: Sacoglossa (= Ascoglossa)), with a discussion of the generic division and phylogeny. *J. Mollusc. Stud.* **58:** 257–296.

Jensen KR (1993) Sacoglossa (Mollusca, Opisthobranchia) from Rottnest Island and central Western Australia. In *Proc. Fifth International Marine Biological Workshop: The Marine Flora and Fauna of Rottnest Island, Western Australia*. FE Wells, DI Walker, H Kirkman & R. Lethbridge (eds), pp. 207–253. Western Australia Museum, Perth.

Jensen KR (1996) The Diaphanidae as a possible sister group of the Sacoglossa (Gastropoda, Opisthobranchia). In *Origin and Evolutionary Radiation of the Mollusca*. JD Taylor (ed.), pp. 231–247. Oxford University Press, Oxford.

Jeppesen LL (1976) The control of mating behaviour in *Helix pomatia* L (Gastropoda: Pulmonata). *Anim. Behav.* **24:** 275–290.

Kat PW (1983) Sexual selection and simultaneous hermaphroditism among the Unionidae (Bivalvia: Mollusca). *J. Zool.* **201:** 395–416.

Kennedy JJ (1995) The courtship, pseudocopulation behaviour and spermatophore of *Turritella communis* Risso 1826 (Prosobranchia: Turritellidae). *J. Mollusc. Stud.* **61:** 421–434.

Kothbauer H (1988) Ueber Liebespfeile, Schnecken und Weltbilder. *Ann. Naturhist. Mus. Wien* **90B:** 163–169.

Kress A (1985) A structural analysis of the spermatophore of *Runcina ferruginea* Kress (Opisthobranchia: Cephalaspidea). *J. Mar. Biol. Assoc. UK* **65:** 337–342.

Lalli CM (1978) Reproduction in the genus *Limacina* (Opisthobranchia: Thecosomata). *J. Zool.* **186:** 95–108.

Lalli, CM & Wells FE (1978) Reproduction in the genus *Limacina* (Opisthobranchia: Thecosomata). *J. Zool.* **186:** 95–108.

Lamotte M (1951) Recherches sur la structure génétique des populations naturelles de *Cepaea nemoralis*. *Bull. Biol. Fr. Belg. Suppl.* **35:** 1–239.

Lane FW (1960) *Kingdom of the Octopus: The Life History of the Cephalopoda*. Sheridan House, New York.

de Larambergue, M. (1939) Étude de l'autofécondation chez les gastéropodes pulmoné. Recherche sur l'aphallie et la fécondation *Bulinus (Isidora) contortus* Michaud. *Bull. Biol. Fr. Belg.* **73:** 19–231.

Lederhendler II & Tobach E (1977) Reproductive roles in the simultaneous hermaphrodite *Aplysia dactyomela*. *Nature* **270:** 238–239.

Lent CM (1969) Adaptations of the ribbed mussel, *Modiolus demissus* (Dillwyn), to the intertidal habitat. *Am. Zool.* **9:** 283–292.

Leonard JL (1990) The hermaphrodite's dilemma. *J. Theor. Biol.* **147:** 361–372.

Leonard JL (1991) Sexual conflict and the mating systems of simultaneously hermaphroditic gastropods. *Am. Malacol. Bull.* **9:** 45–58.

Leonard JL (1992) The 'love-dart' in helicid snails: a gift of calcium or a firm commitment? *J. Theor. Biol.* **159:** 513–521.

Leonard JL & Lukowiak K (1984a) An ethogram of the sea slug, *Navanax inermis* (Gastropoda, Opisthobranchia). *Z. Tierpsychol.* **65:** 327–345.

Leonard JL & Lukowiak K (1984b) Male–female conflict in a simultaneous hermaphrodite resolved by sperm trading. *Am. Nat.* **124:** 282–286.

Leonard JL & Lukowiak K (1985) Courtship, copulation, and sperm trading in the sea slug, *Navanax inermis* (Opisthobranchia: Cephalaspidea). *Can. J. Zool.* **63:** 2719–2729.

Levitan DR & Petersen C (1995) Sperm limitation in the sea. *Trends Ecol. Evol.* **10:** 228–231.

Lind H (1973) The functional significance of the spermatophore and the fate of spermatozoa in the genital tract of *Helix pomatia* (Gastropoda: Stylommatophora). *J. Zool.* **169:** 39–64.

Lind H (1976) Causal and functional organization of the mating behaviour sequence in *Helix pomatia* L (Pulmonata, Gastropoda). *Behaviour* **59:** 162–202.

Lind H (1988) The behaviour of *Helix pomatia* L (Gastropoda, Pulmonata) in a natural habitat. *Videnskab. Meddel. Dansk Naturhist. For. Køben.* **147:** 67–92.

Lipton CS & Murray J (1979) Courtship of land snails of the genus *Partula*. *Malacologia* **19:** 129–146.

Longley RD & Longley AJ (1982) Hermissenda: agonistic behavior or mating behavior? *Veliger* **24:** 230–231.

Lubet P (1951) Sur l'émission des gamètes chez *Chlamys varia* L (Mollusques, Lamellibranches). *CR Acad. Sci.* **235:** 1680–1681.

Lubet P (1955) Cycle neurosécrétoire chez *Chlamys varia* et *Mytilus edulis* L (Mollusques, Lamellibranches). *CR Acad. Sci.* **241:** 119–121.

Mackie GL (1984) Bivalves. In *The Mollusca, Vol. 7: Reproduction.* AS Tompa, NH Verdonk & JAM van den Biggelaar (eds), pp. 351–418. Academic Press, London.

Madsen H, Thiongo FW & Ouma JH (1983) Egg laying and growth in *Helisoma duryi* (Wetherby) (Pulmonata: Planorbidae): effect of population density and mode of fertilization. *Hydrobiologia* **106:** 185–191.

Mangold K (1987) Reproduction. In *Cephalopod Life Cycles, Vol. 2, Comparative Reviews.* PR Boyle (ed.), pp. 157–200. Academic Press, London.

Mangold-Wirz K (1963) Biologie des cephalopodes benthiques et nectonique del la Mer Catalane. *Vie Milieu Suppl.* **13:** 1–285.

Mann T (1984) *Spermatophores. Development, Structure, Biochemical Attributes and Role in the Transfer of Spermatozoa.* Springer-Verlag, Berlin.

Mann T, Martin AW & Thiersch JB (1966) Spermatophores and spermatophoric reaction in the giant octopus of the North Pacific, *Octopus dofleini martini. Nature* **211:** 1279–1282.

Mann T, Martin AW & Thiersch JB (1970) Male reproductive tract, spermatophores and spermatophoric reaction in the giant octopus of the North Pacific, *Octopus dofleini martini. Proc. Roy. Soc. Lond. Ser. B* **175:** 31–61.

McFadien-Carter M (1979) Scaphopoda. In *Reproduction of Marine Invertebrates, Vol. 5, Molluscs: Pelecypods and Lesser Classes.* AC Giese & JS Pearse (eds), pp. 95–111. Academic Press, London.

Mead AR (1943) Revision of the giant west coast land slugs of the genus *Ariolimax* Moerch (Pulmonata: Arionidae). *Am. Midl. Nat.* **30:** 675–717.

Meisenheimer J (1907) Biologie, Morphologie und Physiologie des Begattungvorganges und der Eiablage von *Helix pomatia. Zool. Jahrb. Abt. Syst. Oekol.* **25:** 461–502.

Moore RC (1969) *Treatise on Invertebrate Paleontology, Vol. 6.* Geological Society of America, Boulder.

Morand S (1988a) Cycle évolutif de *Nemhelix bakeri* Morand et Petter (Nematoda, Cosmocercidae) parasite de l'appareil génital de *Helix aspersa* (Gastropoda, Helicidae). *Can. J. Zool.* **66:** 1796–1802.

Morand S (1988b) Eléments d'épidémiologie de *Nemhelix bakeri* (Nematoda, Cosmocercidae), parasite de l'appareil génital de *Helix aspersa* (Gastropoda, Helicidae). *Haliotis* **18:** 297–304.

Morand S (1989) Deux nouveaux nématodes Cosmocercidae parasites des escargots terrestres *Cepaea nemoralis* L. et *Cepaea hortensis* Müller. *Bull. Mus. Natl Histoire Nat.* **4:** 563–570.

Morand S & Hommay G (1990) Redescription de *Agfa flexilis* Dujardin, 1845 (Nematoda, Agfidae) parasite de l'appareil génital de *Limax cinereoniger* Wolf (Gastropoda, Limacidae). *System. Parasitol.* **15:** 127–132.

Morton B (1972) Some aspects of the functional morphology and biology of *Pseudopythina subsinuata* (Bivalvia: Leptonacea) commensal on stomatopod crustaceans. *J. Zool.* **166:** 79–96.

Morton BS (1976) Secondary brooding of temporary dwarf males in *Ephippodonta* (*Ephippodontina*) *oedipus* sp. nov. (Bivalvia: Leptonacea). *J. Conchol.* **29:** 31–39.

Mulvey M & Vrijenhoek RC (1981) Multiple paternity in the hermaphroditic snail, *Biomphalaria obstructa. J. Hered.* **72:** 308–312.

Murray J (1964) Multiple mating and effective population size in *Cepaea nemoralis. Evolution* **18:** 283–291.

Nishikawi S (1964) Phylogenetic study on the type of the dimorphic spermatozoa in Prosobranchia. *Sci. Rep. Tokyo Kyoiku Daigaku Sec. B* **11:** 237–275.

Nishikawi S & Tochimoto T (1969) Dimorphism in typical and atypical spermatozoa forming two types of spermatozeugmata in two epitoniid prosobranchs. *Venus* **28:** 37–46.

Ó Foighil D (1985) Sperm transfer and storage in the brooding bivalve *Mysella tumida. Biol. Bull.* **169:** 602–614.

Ó Foighil D (1987) Cytological evidence for self-fertilization in *Lasea subviridis* (Galeommatacea: Bivalvia). *Invert. Reprod. Devel.* **12:** 83–90.

Ockelmann KW (1965) Redescription, distribution, biology, and dimorphous sperm of *Montacuta tenella* Lovén (Mollusca, Leptonacea). *Ophelia* **2:** 211–221.

Oldfield E (1961) The functional morphology of *Kellia suborbicularis* (Montagu), *Montacuta ferruginosa* (Montagu) and *M. substriata* (Montagu) (Mollusca, Lamellibranchiata). *Proc. Malacol. Soc. Lond.* **34:** 255–295.

Otsuka C, Rouger Y & Tobach E (1980) A possible relationship between size and reproductive behavior in a population of *Aplysia punctata* Cuvier, 1803. *Veliger* **23:** 159–162.

Owiny AM (1974) Some aspects of the breeding biology of the equatorial land snail *Limicolaria martensiana* (Achatinidae: Pulmonata). *J. Zool.* **172:** 191–206.

Pearse JS (1979) Polyplacophora. In *Reproduction of Marine Invertebrates, Vol. 5, Molluscs: Pelecypods and Lesser Classes.* AC Giese & JS Pearse (eds), pp. 27–85. Academic Press, London.

Pennings SC (1991) Reproductive behavior of *Aplysia californica* Cooper: diel patterns, sexual roles and mating aggregations. *J. Exp. Mar. Biol. Ecol.* **149:** 249–266.

Picken GB (1980) The distribution, growth, and reproduction of the Antarctic limpet *Nacella (Patinigera) concinna* (Strebel 1908). *J. Exp. Mar. Biol. Ecol.* **42:** 71–85.

Pokryszko BM (1987) On the aphally in the Vertiginidae (Gastropoda: Pulmonata: Orthurethra). *J. Conchol.* **32:** 365–375.

Pollard E (1975) Aspects of the ecology of *Helix pomatia* L. *J. Anim. Ecol.* **44:** 305–329.

Price CH (1979) Physical factors and neurosecretion in the control of reproduction in *Melampus* (Mollusca: Pulmonata). *J. Exp. Zool.* **207:** 269–282.

Pruvot-Fol A (1960) Les organes géniteaux des opisthobranches. *Arch. Zool. Exp. Gén.* **99:** 135–224.

Raffaelli DG (1977) Observations on the copulatory behavior of *Littorina rudis* (Maton) and *Littorina nigrolineata* Gray (Gastropoda: Prosobranchia). *Veliger* **20:** 75–77.

Rankin JJ (1979) A freshwater shell-less mollusc from the Caribbean: structure, biotics, and contribution to a new understanding of the Acochlidioidea. *Life Sci. Contrib. Roy. Ont. Mus.* **116:** 1–123.

Raut SK & Ghose KC (1979) Viability of sperm in two land snails, *Achatina fulica* Bodwich and *Macrochlamys indica* Godwin-Austen. *Veliger* **21:** 486–487.

Raut SK & Ghose KC (1982) Viability of sperms in aestivating *Achatina fulica* Bodwich and *Macrochlamys indica* Godwin-Austen. *J. Mollusc. Stud.* **48:** 84–86.

Reid JD (1964) The reproduction of the sacoglossan opisthobranch *Elysia maoria. Proc. Zool. Soc. Lond.* **143:** 365–393.

Reise H (1995) Mating behaviour of *Deroceras rodnae* Grossu & Lupu, 1965 and

D. praecox Wiktor, 1966 (Pulmonata: Agriolimacidae). *J. Mollusc. Stud.* **61:** 325–330.

Rigby JE (1963) Alimentary and reproductive systems of *Oxychilus cellarius*. *Proc. Zool. Soc. Lond.* **141:** 311–359.

Rigby JE (1965) *Succinea putris*, a terrestrial opisthobranch mollusc. *Proc. Zool. Soc. Lond.* **144:** 445–487.

Robertson R (1978) Spermatophores of six Eastern North American pyramidellid gastropods and their systematic significance (with the new genus *Boonea*). *Biol. Bull.* **155:** 360–382.

Robertson R (1989) Spermatophores of aquatic non-stylommatophoran gastropods: a review with new data on *Heliacus* (Architectonicidae). *Malacologia* **30:** 341–364.

Rollinson D & Wright CA (1984) Population studies on *Bulinus cernicus* from Mauritius. *Malacologia* **25:** 447–464.

Rollinson D, Kane RA & Lines JRL (1989) An analysis of fertilization in *Bulinus cernicus* (Gastropoda: Planorbidae). *J. Zool.* **217:** 295–310.

Rosenthal RJ (1970) Observations on the reproductive biology of the Kellet's whelk, *Kelletia kelletii*. *Veliger* **12:** 319–324.

Ross RM & Quetin LB (1990) Mating behavior and spawning in two neustonic nudibranchs in the family Glaucidae. *Am. Malacol. Bull.* **8:** 61–66.

Rudman WB (1981) Further studies of the anatomy and ecology of opisthobranch molluscs feeding on the scleractinian coral *Porites*. *Biol. J. Linn. Soc.* **71:** 373–412.

Rudolph PH (1979a) The strategy of copulation in *Stagnicola elodes* (Say) (Basommatophora: Lymnaeidae). *Malacologia* **18:** 381–389.

Rudolph PH (1979b) An analysis of copulation in *Bulinus (Physopsis) globosus* (Gastropoda: Planorbidae). *Malacologia* **19:** 147–155.

Rudolph PH (1983) Copulatory activity and sperm production in *Bulinus* (*Physopsis*) *globosus* (Gastropoda: Planobidae). *J. Mollusc. Stud.* **49:** 125–132.

Rudolph PH & Bailey JB (1985) Copulation as females and use of allosperm in the freshwater snail genus *Bulinus* (Gastropoda: Planorbiae). *J. Mollusc. Stud.* **51:** 267–275.

Runham NW (1992) Mollusca. In *Reproductive Biology of Invertebrates, Vol. 5, Sexual Differentiation and Behaviour.* KG Adiyodi & RG Adiyodi (eds), pp. 193–229. John Wiley & Sons, New York.

Russell-Hunter WD, Apley ML & Hunter RD (1972) Early life-history of *Melampus* and the significance of semilunar synchrony. *Biol. Bull.* **143:** 623–656.

Rutowski RL (1983) Mating and egg mass production in the aeolid nudibranch *Hermissenda crassicornis* (Opisthobranchia). *Veliger* **24:** 227–229.

Saleuddin ASM, Griffond B & Ashton ML (1991) An ultrastructural study of the activation of the endocrine dorsal bodies in the snail *Helix aspersa* by mating. *Can. J. Zool.* **69:** 1203–1215.

Salvini-Plawen Lv (1985) Early evolution and the primitive groups. In *The Mollusca, Vol. 10, Evolution.* ER Trueman & MR Clarke (eds), pp. 59–150. Academic Press, London.

Sastry AN (1979) Pelecypoda (excluding Ostreidae). In *Reproduction of Marine Invertebrates, Vol. 5, Molluscs: Pelecypods and Lesser Classes.* AC Giese & JS Pearse (eds), pp. 113–292. Academic Press, London.

Scheltema AH, Tscherkassky M & Kuzirian AM (1994) Aplacophora. In *Microscopic Anatomy of Invertebrates, Vol. 5, Mollusca I.* FW Harrison & AJ Kohn (eds), pp. 13–54. Wiley-Liss, New York.

Scheuwimmer A (1979) Sperm transfer in the sessile gastropod *Serpulorbis* (Prosobranchia: Vermetidae). *Mar. Ecol. Progr. Ser.* **1:** 65–70.

Schilder A (1950) Die Ursachen der Variabilität bei *Cepaea. Biol. Zentralbl.* **69:** 79–103.

Schmekel L (1970) Anatomie der Genitalorgane von Nudibranchiern (Gastropoda, Euthyneura). *Pubbl. Staz. Zool. Napoli* **38:** 120–217.

Schnetter M (1950) Veränderungen der genetischen Konstitution in natürlichen Populationen der polymorphen Bänderschnecken. *Verhand. Deutsch. Zool. Gesellsch.* **13:** 192–206.

Schönenberger N (1969) Beiträge zur Entwicklung und Morphologie von *Trinchesia granosa* Schmekel (Gastropoda, Opisthobranchia). *Pubbl. Staz. Zool. Napoli* **37:** 236–292.

Schrag SJ & Read AF (1992) Temperature determination of male outcrossing ability in a simultaneous hermaphrodite. *Evolution* **46:** 1698–1707.

Schrag SJ & Read AF (1996) Loss of male outcrossing ability in simultaneous hermaphrodites: phylogenetic analyses of pulmonate snails. *J. Zool.* **238:** 287–299.

Schrag SJ & Rollinson D (1994) Effects of *Schistosoma haematobium* infection on reproductive success and male out-crossing ability in the simultaneous hermaphrodite, *Bulinus truncatus* (Gastropoda: Planorbidae). *Parasitology* **108:** 27–34.

Schrag SJ, Rollinson D, Keymer AE & Read AF (1992) Heritability of male outcrossing ability in the simultaneous hermaphrodite, *Bulinus truncatus* (Gastropoda, Planorbidae). *J. Zool.* **226:** 311–319.

Schrag SJ, Mooers AØ, Ndifon GT & Read AF (1994a) Ecological correlates of male outcrossing ability in a simultaneous hermaphrodite snail. *Am. Nat.* **143:** 636–655.

Schrag SJ, Ndifon GT & Read AF (1994b) Temperature-determined outcrossing ability in wild populations of a simultaneous hermaphrodite snail. *Ecology* **75:** 2066–2077.

Selander RK & Ochman H (1983) The genetic structure of populations as illustrated by molluscs. *Isozymes* **10:** 93–123.

Selmi MG & Giusti F (1983) The atypical spermatozoon of *Theodoxus fluviatilis* (L.) (Gastropoda, Prosobranchia). *J. Ultrastruct. Res.* **84:** 173–181.

Silberglied RE, Shepherd JG & Dickinson JL (1984) Eunuchs: the role of apyrene sperm in Lepidoptera? *Am. Nat.* **123:** 255–265.

Solem GA (1974) *The Shell Makers. Introducing Molluscs.* John Wiley and Sons, New York.

South A (1992) *Terrestrial Slugs: Biology, Ecology and Control.* Chapman and Hall, London.

Städler T, Weisner S & Streit B (1995) Outcrossing rates and correlated matings in a predominantly selfing freshwater snail. *Proc. Roy. Soc. Lond. Ser. B* **262:** 119–125.

Staub R & Ribi G (1995) Size-assortative mating in a natural population of *Viviparus ater* (Gastropoda: Prosobranchia) in Lake Zürich, Switzerland. *J. Mollusc. Stud.* **61:** 237–247.

Steiner G (1993) Spawning behaviour of *Pulsellum lofotensis* (M. Sars) and *Cadulus subfusiformis* (M. Sars) (Scaphopoda, Mollusca). *Sarsia* **78:** 31–33.

Struhsaker JW (1966) Breeding, spawning periodicity and early development in the Hawaiian *Littorina: L. pintado* (Wood), *L. picta* (Philippi) and *L. scabra* (Linné). *Proc. Malacol. Soc. Lond.* **37:** 137–166.

Summers WC (1971) Age and growth of *Loligo pealei*, a population study of the common Atlantic coast squid. *Biol. Bull.* **141:** 189–201.

Swedmark B (1968) The biology of interstitial Mollusca. In *Studies in the Structure, Physiology and Ecology of Molluscs. Symposia of the Zoological Society of London 22.* V Fretter (ed.), pp. 135–149. Zoological Society of London, London.

Switzer-Dunlap M, Meyer-Schulte K & Gardner EA (1984) The effect of size, age, and recent egg-laying on copulatory choice of the hermaphroditic mollusc *Aplysia juliana. Invert. Reprod. Devel.* **7:** 217–225.

Takeda N & Tsuruoka (1979) A sex pheromone secreting gland in the terrestrial snail, *Euhadra peliomphala. J. Exp. Zool.* **207:** 17–26.

Thompson TE (1973) Euthyneuran and other molluscan spermatozoa. *Malacologia* **14:** 167–206.

Thompson TE (1976) *Biology of Opisthobranch Molluscs, Vol. 1.* Ray Society, London.

Thornhill R (1983) Cryptic female choice and its implications in the scorpionfly *Harpobittacus nigriceps. Am. Nat.* **122:** 765–788.

Tinbergen L (1939) Zur Fortpflanzungsethologie von *Sepia officinalis* L. *Arch. Néer. Zool.* **7:** 213–286.

Tischler W (1973) Zur Biologie und Ökologie der Weinbergschnecke (*Helix pomatia*). *Faun.-ökol. Mitteil.* **4:** 283–298.

Todd CD (1979) Reproductive energetics of two species of dorid nudibranchs with planktotrophic and lecithotrophic larval strategies. *Mar. Biol.* **53:** 57–68.

Tomiyama K (1993) Growth and maturation pattern in the giant African snail, *Achatina fulica* (Férussac) (Stylommatophora: Achatinidae) in the field. *Venus* **52:** 87–100.

Tomiyama K (1994) Courtship behaviour of the giant African snail, *Achatina fulica* (Férussac) (Stylommatophora: Achatinidae) in the field. *J. Mollusc. Stud.* **60:** 47–54.

Tomiyama K (1996) Mate-choice criteria in a protandrous simultaneously hermaphroditic land snail *Achatina fulica* (Férussac) (Stylommatophora: Achatinidae). *J. Mollusc. Stud.* **62:** 101–111.

Tompa AS (1980) The ultrastructure and mineralogy of the dart from *Philomycus carolinianus* (Pulmonata: Gastropoda) with a brief survey of the occurrence of darts in land snails. *Veliger* **23:** 35–42.

Tompa AS (1982) X-ray radiographic examination of dart formation in *Helix aspersa. Neth. J. Zool.* **32:** 63–71.

Tompa AS (1984) Land Snails (Stylommatophora). In *The Mollusca, Vol. 7, Reproduction.* AS Tompa, NH Verdonk & JAM van den Biggelaar (eds), pp. 47–140. Academic Press, London.

Tompa AS, Verdonk NH & van den Biggelaar JAM (eds) (1984) *The Mollusca, Vol. 7, Reproduction.* Academic Press, London.

Trowbridge CD (1995) Hypodermic insemination, oviposition, and embryonic development of a pool-dwelling ascoglossan (= sacoglossan) opisthobranch: *Ercolania felina* (Hutton, 1882) on New Zealand shores. *Veliger* **38:** 203–211.

Trüb H (1990) Züchtung von Hybriden zwischen *Viviparus ater* und *V. contectus* (Mollusca, Prosobranchia) im Zürichsee und ökologische Untersuchungen in einer gemischten Population im Gardasee. Ph.D. Thesis, University of Zurich, Switzerland.

Tucker JS & Giese AC (1962) Reproductive cycle of *Cryptochiton stelleri* (Middendorff). *J. Exp. Zool.* **150:** 33–43.

van Duivenboden YA & ter Maat A (1988) Mating behaviour of *Lymnaea stagnalis*. *Malacologia* **28:** 53–64.

van Mol JJ (1971) Notes anatomiques sur les Bulimulidae (Mollusques Gastéropodes Pulmonés). *Ann. Soc. Roy. Zool. Belg.* **101:** 183–225.

Vianey-Liaud M (1989) Growth and fecundity in a black-pigmented and an albino strain of *Biomphalaria glabrata* (Gastropoda: Pulmonata). *Malacol. Rev.* **22:** 25–32.

Vianey-Liaud M (1992) Sperm storage time in *Biomphalaria glabrata* (Gastropoda: Planorbidae): fertilization of albino snails by surgically castrated pigmented snails. *J. Med. Appl. Malacol.* **4:** 99–101.

Vianey-Liaud M (1995) Bias in the productions of heterozygous pigmented embryos from successively mated *Biomphalaria glabrata* (Gastropoda: Planorbidae) albino snails. *Malacol. Rev.* **28:** 97–106.

von Medem F (1945) Untersuchungen über die Ei- und Spermawirkstoffe bei marinen Mollusken. *Zool Jahrb. Abt. Allg. Zool. Physiol.* **61:** 1–44.

Wallace C (1992) Parthenogenesis, sex and chromosomes in *Potamopyrgus*. *J. Mollusc. Stud.* **58:** 93–107.

Warén A & Hain S (1992) *Laevipilina antarctica* and *Micropilina arntzi*, two new monoplacophorans from the Antarctic. *Veliger* **35:** 165–176.

Watanabe JM & Cox LR (1975) Spawning behavior and larval development in *Mopalia lignosa* and *Mopalia muscosa* (Mollusca: Polyplacophora) in Central California. *Veliger* **18**(Suppl.): 18–27.

Watson H (1923) Masculine deficiencies in the British Vertigininae. *Proc. Malacol. Soc. Lond.* **15:** 270–280.

Webb GR (1948) Notes on the mating of some Zonitoides (Ventridens) species of land snails. *Am. Midl. Nat.* **40:** 453–461.

Webber HH (1977) Gastropoda: Prosobranchia. In *Reproduction of Marine Invertebrates, Vol. 4, Molluscs: Gastropods and Cephalopods*. AC Giese & JP Pearse (eds), pp. 1–97. Academic Press, London.

Wells MJ (1978) *Octopus. Physiology and Behaviour of an Advanced Invertebrate*. Chapman and Hall, London.

Wells MJ & Wells J (1977) Cephalopoda: Octapoda. In *Reproduction of Marine Invertebrates, Vol. 4*. AC Giese & JS Pearse (eds), pp. 291–336. Academic Press, London.

Westheide W & Wawra E (1974) Organisation, Systematik und Biologie von *Microhedyle cryptophthalma* nov. spec. (Gastropoda, Opisthobranchia) aus dem Brandungsstrand des Mittelmeeres. *Helgo. Wiss. Meeresuntersuch.* **26:** 27–41.

Wethington AR & Dillon RT (1991) Sperm storage and evidence for multiple insemination in a natural population of the freshwater snail, *Physa*. *Am. Malacol. Bull.* **9:** 99–102.

Wethington AR & Dillon RT (1996) Gender choice and gender conflict in a non-reciprocally mating simultaneous hermaphrodite, the freshwater snail, *Physa*. *Anim. Behav.* **51:** 1107–1118.

Wolda H (1963) Natural populations of the polymorphic landsnail *Cepaea nemoralis* (L.). *Arch. Néerl. Zool.* **15:** 381–471.

Wright WG (1988) Sex change in the Mollusca. *Trends Ecol. Evol.* **3:** 137–140.

Yonge CM (1962) On the biology of the mesogastropod *Trichotropis cancellata* Hinds, a benthic indicator species. *Biol. Bull.* **122:** 160–181.

Young RT (1946) Stimulation of spawning in the mussel, *Mytilus californianus*. *Ecology* **26:** 58–69.

Yusa Y (1993) Copulatory load in a simultaneous hermaphrodite *Aplysia kurodai* Baba, 1937 (Mollusca: Opisthobranchia). *Publ. Seto Mar. Biol. Lab.* **36:** 79–84.

Yusa Y (1994) Factors regulating sperm transfer in an hermaphroditic sea hare, *Aplysia parvula* Mörch, 1863 (Gastropoda: Opisthobranchia). *J. Exp. Mar. Biol. Ecol.* **181:** 213–221.

Yusa Y (1996a) The effects of body size on mating features in a field population of the hermaphroditic sea hare *Aplysia kurodai* Baba, 1937 (Gastropoda: Opisthobranchia). *J. Mollusc. Stud.* **62:** 381–386.

Yusa Y (1996b) Utilization and degree of depletion of exogenous sperm in three hermaphroditic sea hares of the genus *Aplysia* (Gastropoda: Opisthobranchia). *J. Mollusc. Stud.* **62:** 113–120.

drumming rates by males suggests that this behaviour has been selected by female choice (Parri *et al.* 1997). Further experiments by Mappes *et al.* (1996) reveal that this behaviour provides a reliable indication of male viability, and represents a condition-dependent sexual display (Andersson 1994). Males that experienced higher levels of food intake maintained drumming rates at a higher level than males provided with less food. Furthermore, drumming behaviour is costly; males induced to sustain longer periods of drumming activity, by being artificially exposed to females, suffered higher mortality and lost more weight than other males. These experiments suggest that drumming in *H. rubrofasciata* is an honest signal of male quality, and they supply among the most comprehensive demonstration of sexual selection of a male trait by female choice in a spider (but see also Watson 1990, 1994).

B. Copulation and cryptic female choice

Females may continue to exercise some choice in the sire of their offspring even after copulation has commenced (Thornhill 1983; Birkhead and Møller 1987; Eberhard 1991, 1996). Sperm are rarely deposited by the male at the site of fertilization, and must therefore be transported, stored and nourished – processes that may be under female control. Thus, a female may be able to exercise 'cryptic' choice (Thornhill 1983) by influencing whether the sperm of a male that has achieved intromission actually fertilizes her eggs. Eberhard (1985, 1996) suggests that cryptic female choice is widespread and may have an important influence on the many aspects of sexual reproduction. Presumably, females could exercise cryptic choice only if they could also discriminate between males during copulation and manipulate the sperm between intromission and fertilization accordingly.

The transfer of sperm from male to female spiders involves two stages. First, the male transfers sperm from his gonopore to each of his modified palpi by a process called sperm induction (Montgomery 1903). The male presses his abdomen against a specially constructed silk sperm web and releases onto it a drop of sperm from his genital pore. He then dips each of his palpi into the drop of sperm, which are taken up into the palpae and temporarily stored. Following the successful courtship of a sexually receptive female, the male copulates with the female by inserting the tips of the palpi, the emboli, into the female genital opening and ejaculating the sperm. Sperm induction typically occurs shortly after the male has attained sexual maturity and subsequently after each mating, depending upon the quantity of sperm in his palpae. In some families, including theridiids and linyphiids, sperm induction takes place after the male has located the female and during his subsequent courtship (Austad 1984).

What evidence is there for copulatory courtship and sperm manipulation in spiders? Eberhard (1994, 1996) provides numerous examples of activities during copulation that may represent courtship behaviour. However, the extent to which females may manipulate sperm is less clear. Watson (1991b) argues that the 'sharp elbow within a minute fleshy portion of the pre-spermathecal sperm duct' may provide female filmy dome spiders *Linyphia marginata* (Linyphiidae) with a mechanism for controlling sperm utilization. More compelling, but still indirect evidence of sperm manipulation, is provided by Gunnarsson and Andersson (1996), who demonstrated that the position of female sheet web spiders *Pityohyphantes phrygianus* (Linyphiidae) during the 24 h period following copulation significantly influenced the sex ratio of their offspring. In particular, the primary sex ratio (proportion of males) of clutches from females that remained in the typical, ventral side up position was 0.28, compared with 0.41 for females that remained dorsal side up. Gunnarsson and Andersson (1996) argue that females may manipulate the sex ratio of their clutches by changing position immediately after copulation, and thereby control where the sperm are stored.

The duration of copulation in spiders is frequently longer than might be expected for only intromission to take place (Elgar 1995), and this may provide the female with the opportunity to assess the suitability of her mate according to the structure of his genitalia (Eberhard 1985). For example, the retrolateral tibial apophysis (rta) is found on the palpae of adult male spiders of some 38 families of spiders. Histological sections of several species reveal that the rta is used to fix the male palpi to the female epigyne, thereby facilitating intromission of the embolus (Huber 1995a). The structure of the rta is highly species-specific and, like other morphological structures of the male genitalia, may have evolved through cryptic female choice (see Eberhard 1985).

While cryptic female choice of variants in male genitalia remains a possibility, two lines of evidence are still required. First, the area of physical contact between the female and male genitalia of several spiders appears to be free of mechanoreceptors, despite their abundance elsewhere on the body surface of the female (Huber 1993, 1995b; Uhl *et al.* 1995). Thus, we lack evidence of a mechanism by which females can discriminate between males according to different sensory stimulation (Huber 1995b). Second, cryptic female choice in this particular context implies a correlation between a male's genital morphology and his fertilization success, which may be a function of the number of sperm transferred during intromission, and/or the number retained by the female until fertilization of her eggs. No such data are currently available.

Interestingly, the males of some spiders copulate with the female before sperm induction has taken place, and hence these males are, at first, unable to transfer sperm (van Helsdingen 1965). The function of this 'pseudocopulation' is not fully understood, but appears to provide a mechanism for sexual selection by female choice (Watson 1991b, 1994; Willey Robertson and Adler 1994).

C. Sperm transfer and fertilization

Spiders can be classified into one of two groups, according to the morphology of the reproductive tract of the female (Fig. 9.1). Generally, the reproductive tract of female entelegyne spiders is bilateral, with two genital pores covered by a sclerotized plate called the epigynum, two

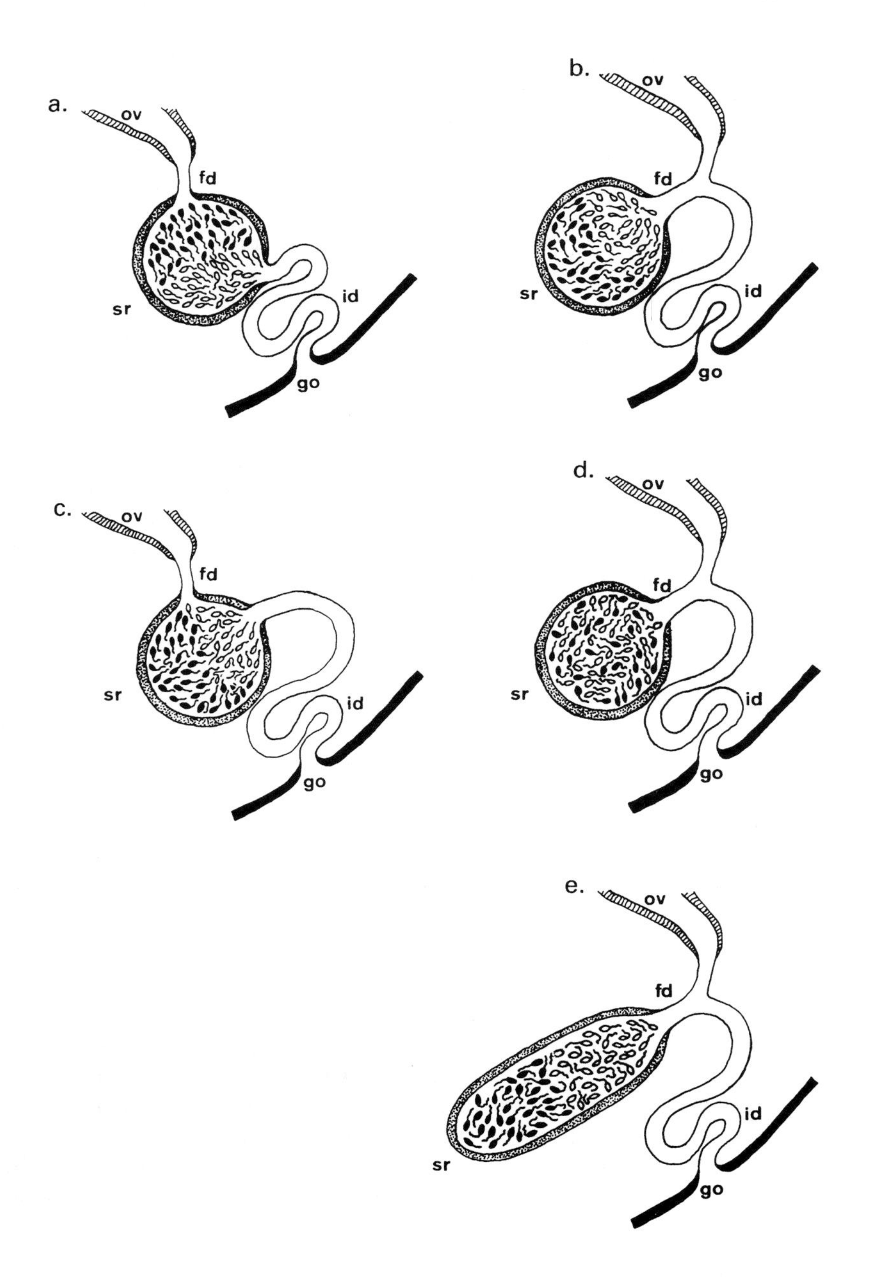

sperm ducts and two receptacula seminis where the seminal fluid is stored. Sperm are transferred into both sides, apparently from the corresponding palpi (but see Yoward and Oxford 1996). In some species, the copulating male alternates between left and right palpal insertion without releasing the female, while in others the male may insert one palpi then release the female, and court her again before inserting his other palpi. In contrast, the reproductive tract of female haplogyne spiders is typically rather less complicated, comprising a single genital opening that leads into the uterus externus, and males use both palpae to transfer sperm. Until recently, it was not clear how sperm were stored in haplogyne spiders, but Uhl (1994a,b) has demonstrated that females of the haplogyne daddy long-legs *Pholcus phalangioides* (Pholcidae) release glandular secretions into the uterus externus that bind the sperm mass into a specific position.

Despite the different reproductive morphologies of entelegyne and haplogyne spiders, females of both types are capable of storing sperm until oviposition takes place. The process of sperm capacitation is not well understood in spiders, but sperm change from an encysted to a flagellate state prior to oviposition in the orb-weaving spider *Nephila clavipes* (Brown 1985). The time between insemination and subsequent fertilization and oviposition of the first clutch of eggs varies both within and between species, but some spiders are able to store sperm for many months before fertilizing successfully several clutches of eggs (e.g. Miyashita 1987; Uhl 1993a). Long-lived species, on the other hand, may need to mate every year (e.g. Miyashita 1992), particularly those that moult annually, thereby removing the lining of the spermathecae and any stored sperm. Females of some species also discard at least some

Fig. 9.1. *Schematic, hypothetical representation of the reproductive anatomy of entelegyne and haplogyne spiders. ov, oviposition duct; fd, fertilization duct; sr, seminal receptical; id, insemination duct; go, genital opening. Black sperms are those from the first male to mate, while white sperms are those from the second male. According to Austad (1984), in (a) the sperm from the first male to mate with an entelegyne, 'conduit-design' female will be closer to the point of fertilization and thus have sperm priority, while in (b) the sperm from the second male to mate with the haplogyne 'cul-de-sac design' female will be closer and hence have priority. However, the relative position of the oviposition duct and insemination duct may influence patterns of sperm precedence. In (c) the ducts are close together and the reproductive tract may be more 'cul-de-sac design' than 'conduit design' (Eberhard, personal communication). The degree of sperm stratification and the shape of the seminal receptical may also affect patterns of sperm precedence. In (d) the seminal receptical is spherical and there is no stratification, leading to sperm mixing, while in (e) the seminal receptical is ovoid and the sperm are stratified, leading to stronger last-male sperm priority (Yoward 1996). Finally, the pattern of sperm capacitation may also affect precedence patterns; although the sperms in the figure are depicted in an essentially flagellate state, spiders always transmit encapsulated sperm, and all transmitted sperm may not be decapsulated at any given moment.*

sperm following mating. Discarding sperm, or 'sperm dumping' may represent a mechanism of cryptic female choice (Eberhard 1996).

III. MULTIPLE MATING BY FEMALES

Although a substantial body of evidence indicates that multiple mating by females is widespread in spiders (see reviews by Austad 1984; Yoward 1996), it is not true of all species. Females of some species apparently rarely mate more than once (e.g. *H. rubrofasciata*; R. Alatalo, personal communication) while in others their receptivity is very considerably reduced after mating (see below).

Females may face a variety of potential costs of courting and copulation, including increased risk of predation, transmission of pathogens and lost foraging opportunities (Lewis 1987). Thus, there must be benefits to mating with more than one mate if the behaviour is to be maintained by selection. Polyandry may provide both genetic and material benefits, although the former are likely to have greater taxonomic generality. Males may provide material benefits in the form of nuptial gifts or other care of the young.

There are no records of paternal care in spiders; indeed, the tropical harvestman *Zygopachylus albomarginis* is the only arachnid in which male care of offspring occurs (Mora 1990). Provisioning the female with prey items or other male-synthesized material, while widespread among insects (see Chapter 10) is unusual in spiders. Female *Pisaura mirabilis* (Pisauridae) will not mate with a male unless he bears a nuptial gift in the form of a prey item (Lang 1996), although providing a gift does not necessarily guarantee that the female will mate with the male (Austad and Thornhill 1986). Female fecundity is a function of feeding rate in this species, and thus multiple mating may allow females to produce more eggs.

Substances other than sperm may also be transferred during courtship and copulation. In some species, females ingest through their mouthparts secretions that are released from the male during courtship and/or copulation (Blest and Taylor 1977; Schaible *et al.* 1986; Blest 1987; Lopez 1987; Huber 1997). Although the secretion may be haemolymph in the linyphiid *Baryphyma pratense* (Blest 1987), the amount seems unlikely to provide a measurable increase in fecundity for polyandrous females. In contrast, females of some sexually cannibalistic species (Elgar 1992) may benefit from multiple mating if the consumption of males increases their fecundity. There is little evidence of a fecundity advantage in consuming a single male (Elgar and Nash 1988; Andrade 1996; Arnqvist and Henriksson 1997; Fahey and Elgar 1997), although it may be more apparent if the female consumes several males.

Multiple mating in the linyphiid *L. litigiosa* provides the female with a mechanism of reducing the cost of sharing her web with male suitors, rather than providing a direct benefit (Watson 1993). Courting males remain on the web of females and are able to steal substantial amounts of prey caught on the web because, in these circumstances, they are the dominant sex. As a consequence, females obtain less prey when sharing their webs with males than when alone. However, their foraging success is improved if they copulate with the male, because he then leaves shortly afterwards (Watson 1993). Females of other web-building spiders that cohabit with dominant males may benefit similarly, although this may not always be true (see Eberhard and Briceño 1983; Blanchong *et al.* 1995).

The genetic benefits of polyandry are essentially of two kinds (see Chapter 2). For species in which females are typically faced with a sequential choice of partners, polyandry may allow females to mate with a male of higher quality than her previous choice, either in terms of offspring viability or arbitrary attractiveness of sons. However, this benefit depends critically on the order of sperm precedence, and is unlikely to apply to those species with absolute first-male sperm priority (see below). Females may also increase the probability of producing some offspring with rare genotypes, which may ensure that some progeny survive against rapidly changing environmental challenges, such as pathogens or parasites (Austad 1984; Hamilton *et al.* 1990). Polyandry may also provide an advantage to females because it results in progeny that are genetically more variable. However, this advantage lies in reducing the variance in reproductive success, rather than increasing the mean.

Watson (1991b, 1997) provides evidence that polyandry in *L. litigiosa* represents bet-hedging against a poor initial choice of partner. Females invariably mate with the first male they encounter after sexual maturation; these males are likely to be of superior fighting ability because competition between males for mating opportunities is intense (Watson 1986, 1990). While some females reject opportunities to remate, most mate more than once and the estimated paternity of these secondary mating males is negatively correlated with the duration of courtship and positively correlated with factors (such as body size) that are associated with superior fighting ability (Watson 1991b). Thus, mated females are more willing to mate again with relatively superior males, and appear to adjust the fertilization success of these secondary males accordingly (Watson 1991b). Significantly, the growth rate of the offspring of polyandrous females was greater than that of singly mated females (Watson 1997). Such an argument assumes that females exert control over patterns of sperm precedence (see Watson 1991a), and that these mechanisms of control have evolved in concert with male mating strategies.

Whatever the advantages of polyandry, multiple mating by females establishes the possibility that male reproductive success may be compromised as a consequence of sperm competition from other males (Parker 1970, 1984, 1990; see Chapter 1). Male spiders have a variety of physiological and behavioural characteristics that appear to have evolved as a mechanism for winning this competition. However, the effectiveness of these mechanisms in spiders critically depends on the patterns of sperm precedence.

IV. SPERM COMPETITION AND THE REPRODUCTIVE ANATOMY OF SPIDERS

In his influential paper, Austad (1984) argued that the differences in the reproductive tract between entelegyne and haplogyne spiders may have important implications for patterns of sperm precedence and, consequently, male and female reproductive strategies. The reproductive tract of entelegyne spiders consists of an insemination duct that opens near the vaginal opening and into which the male intromittent organ dispenses seminal fluid, and the fertilization duct from which sperm issue when the eggs are fertilized. The reproductive tract of haplogyne spiders consists of a single duct into which sperm enter following intromission and from which they must return when the eggs are fertilized. Austad (1984) suggested that these reproductive anatomies predispose the spiders to different patterns of sperm precedence. The 'conduit' reproductive tract of entelegyne spiders might favour first-male sperm priority because the last sperm to enter the female is likely to be furthest from the fertilization duct (Fig. 9.1a). In contrast, the 'cul-de-sac' reproductive tract of haplogyne spiders should favour last-male sperm priority because the last sperm to enter the female will be closest to the eggs as they are released for fertilization (Fig. 9.1b).

It has become clear that while Austad's (1984) original suggestion may prove correct in principle, it suffers in detail. In principle, the pattern of sperm priority is predicted to be a function of the relative position within the spermathecae of the sperm of each male with respect to the point of fertilization. However, the prediction is not strongly supported by the available comparative data of the patterns of sperm precedence of both entelegyne and haplogyne spiders (Table 9.1). Although this survey comprises rather few species from even less families, there is no obvious difference in the patterns of P_2 values (the proportion of eggs fertilized by the second male) between haplogyne and entelegyne spiders. Indeed, the most striking observation is the extensive range of P_2 values between and within species, suggesting that complete sperm priority is not typical for spiders.

The deviations from Austad's (1984) original predictions may arise through differences in spermathecal morphology (Yoward 1996; W. Eberhard, personal communication). For example, the entelegyne design of some species could be not so much a conduit as a cul-de-sac, depending upon the position of the fertilization and insemination ducts (Fig. 9.1c). The P_2 values of species with cul-de-sac designs may depend upon

Table 9.1. Sperm priority patterns, P_2 (proportion of eggs fertilized by the second male) and timing of mate guarding in spiders.

Taxa	P_2 Mean	P_2 Variation[a]	Mate Guarding[b]	Source
Haplogynes				
Pholcidae				
Holocnemus pluchei	0.74	0.17–1.00	None	Kaster and Jakob (1997)
Pholcus phalangoides	0.63	0.00–1.00	None	Yoward (1996)
Physocyclus globosus	0.55	±0.38	None	Eberhard *et al.* (1993)
Entelegyne				
Agelenidae				
Agelena limbata[a]	0.63	±0.11	None	Masumoto (1993)
Tegenaria saeva[c]	0.50	—	—	Oxford (1993)
Araneoidea				
Nephila clavipes	0.18	0.15–0.21	Before	Brown (1985)
Eresidae				
Stegodyphus lineatus	0.49	—	None	Schneider and Lubin (1996, 1997)
Linyphiidae				
Frontinella pyramitela	0.05	—	Before	Austad (1982)
Linyphia hortensis	0.21	±0.23	Before	Stumpf (1990)
Neriene litigiosa[e]	0.35	0.30–0.40	Before	Watson (1991a)
Salticidae				
Phidippus johnsoni	0.50	—	Before	Jackson (1980), Austad (1984)
Theridiidae				
Latrodectus hasselti	0.56	0.00–1.00	None	Andrade (1996)

[a] Variation expressed as either absolute range or ±SE.
[b] Before: guarding before copulation takes place; none: no guarding.
[c] Inferred from enzyme electrophoretic analysis of a single female.
[d] When the copulatory plug of the first male is incomplete, otherwise it is always 0.00.
[e] Inferred from enzyme electrophoretic analyses.

the shape of the spermathecae and the degree of stratification, or mixing, of sperm (Fig. 9.1d,e). Intraspecific variation in P_2 values may be explained by the duration of copulation (Andrade 1996) or the time between copulations (Christenson and Cohn 1988). Clearly, more comparative data are required before we can examine the causes of interspecific variation in P_2 values and, in particular, determine whether the two different reproductive tracts predispose the species to higher or lower P_2 values. Nevertheless, two comparative studies of patterns of cohabitation (Jackson 1986; Eberhard *et al.* 1993) and the duration of copulation (Elgar 1995) reveal interspecific patterns that are consistent with Austad's (1984) suggestion (see below).

Studies of sperm competition in spiders have examined the paternity of secondary males using sperm sterilization techniques involving two males (but see Watson 1991a). Recently, Zeh and Zeh (1994) have shown that in the harlequin beetle-riding pseudoscorpion *Cordylochernes scorpioides*, the pattern of last-male sperm precedence in matings with two males is less pronounced when the interval between matings increases. More significantly, this pattern of sperm precedence disappears when females mate with more than two males. However, Watson (1991a) found that under natural conditions, the pattern of first-male sperm precedence in *L. litigiosa* was not influenced by the number of subsequent mating partners. Nevertheless, the experimental data provided by Zeh and Zeh (1994) are important because they suggest that the opportunity for postcopulation sexual selection through sperm competition may be more widespread than is suggested by the results of two-male experiments that yield low, or negligible, P_2 values.

V. MALE ADAPTATIONS FOR FERTILIZATION SUCCESS

Like other organisms, spiders exhibit an extraordinary diversity of mechanisms that have apparently evolved in response to polyandry, including mate guarding by physical combat, reducing female sexual attractiveness or receptivity, adjusting the duration of copulation, and infanticide. The kinds of mechanisms used by males to secure fertilization success in competition with other males will depend, at least in part, on the pattern of sperm priority. The following section describes these mechanisms and, where possible, examines their prevalence in conduit-design compared with cul-de-sac-design spiders.

A. Contest competition and mate-guarding

Contests between males over mating opportunities have been documented for a large number of spiders (e.g. Bristowe 1958; Robinson and Robinson 1980; Table 9.2). This behaviour has been the subject of numerous studies that attempt to explain the variation in the outcome and degree of escalation in these contests (Table 9.2). In many instances, but particularly in wandering spiders such as salticids and lycosids, this behaviour may reflect a general antipathy between males that occurs even in the absence of females (Wells 1988; Faber and Baylis 1993). However, this behaviour among web- or nest-building spiders (Jackson 1986) appears to represent more conventional mate guarding (*sensu* Parker 1974; Ridley 1983). Presumably, the females of these species are

Table 9.2. Female sexual status and male attributes as determinants of escalation and outcome of male–male contests over females in spiders.

	Factors determining		
Taxa	**Outcome**	**Escalation**	**Source**
Agelenidae			
Agelenopsis aperta	Body size	—	Singer and Riechert (1995)
Araneoidea			
Metellina segmentata	Body size	Female weight	Rubenstein (1987), Prenter *et al.* (1994), Hack *et al.* (1997)
Nephila clavata	Body size	—	Miyashita (1993)
Nephila clavipes	Body size	Not virginity	Christenson and Goist (1979), Christenson *et al.* (1985), Vollrath (1980), Cohn *et al.* (1988)
Nephila plumipes	Body size	—	Elgar and Fahey (1996)
Phonognatha graeffei	Not body size	Immaturity	Fahey and Elgar (1997)
Gasteracantha minax	Not body size	Virginity	Elgar and Bathgate (1996)
Ctenidae			
Cupiennius getazi	Body size	—	Schmitt *et al.* (1992)
Linyphiidae			
Linyphia triangularis	Body size	—	Rovner (1968), Neilsen and Toft (1990)
Neriene litigiosa	Body size	Virginity	Watson (1990, 1991b)
Frontinella pyramitela	Body size	Virginity	Austad (1983)
Lycosidae			
Lycosa tarentula	Not body size	—	Fernandez-Montraveta and Ortega (1993)
Salticidae			
Zygoballus rufipes	Body size	Size difference	Faber and Baylis (1993)
Euophrys parvula	Body size	—	Wells (1988)
Theridiidae			
Argyrodes antipodiana	Body size	—	Whitehouse (1991)
Thomisidae			
Misumenoides formosipes	Body size	—	Dodson and Beck (1993)

more sedentary and hence more easily defended than their cursorial counterparts.

In theory, precopulation mate guarding should be more prevalent among entelegyne than haplogyne spiders, assuming first-male sperm priority in the former and last-male sperm priority in the latter. Although the available experimental evidence suggests that this may not be the prevailing pattern (Table 9.1), interspecific comparative studies lend some support to this idea: in a broad survey of largely anecdotal data, Jackson (1986) documented 156 entelegyne and five haplogyne species that, at least sometimes, cohabit with immature females; Eberhard *et al.* (1993) found that males of two out of three entelegyne species prefer immature to mature females, while males of three haplogyne species were generally indifferent. Several other studies of single species have demonstrated that males of entelegyne spiders cohabit more readily with immature than mature females (Miller and Miller 1986; Toft 1989; Watson 1990; Fahey and Elgar 1997). While postcopulation mate guarding might also be expected among haplogynes, it has not been widely reported (Yoward 1996). One reason may be that effective postcopulation mate-guarding may be impossible because the length of time between mating and oviposition can be considerable, and may even exceed the life expectancy of the male (Austad 1984). Males may also find it difficult to remain with a travelling female.

Several studies show that male scorpions (Benton 1992), mites (Enders 1993) and spiders (Table 9.2) alter their fighting behaviour according to the mating status of the female. The level to which contests may escalate should reflect both the similarity of contestants and the relative value of the resource (Enquist and Leimar 1990; Leimar *et al.* 1991). Thus, males in contests over females should risk higher levels of escalation if they expect to fertilize more eggs (Austad 1983). Male *Gasteracantha minax* (Araneidae) defend females from rival males shortly after mating has taken place and while mated females are still receptive, but cease to do so the next day when the female will not remate (Elgar and Bathgate 1996). Males of the entelgyne linyphiid *Frontinella pyramitela* compete over mating opportunities for females, and these potentially fatal contests are more likely to escalate if the female is immature than if she has mated (Austad 1983). There is strong first-male sperm priority in this species (Austad 1982) and thus the reproductive benefit of defending an immature, virgin female are much greater than that of defending a mature female. A similar pattern occurs in the leaf-curling, orb-web spider *Phonognatha graeffei*: males defend immature females more vigorously than mature females (Fahey and Elgar 1997). *Phonognatha graeffei* is an entelegyne spider and presumably this pattern of male fighting behaviour similarly reflects the patterns of sperm precedence.

B. Reducing female sexual attractiveness and receptivity

An alternative to guarding a female after mating has taken place is to reduce the probability that she is inseminated by rival males. This can be achieved in several ways: by reducing her attractiveness; her receptivity; or even the chance that sperm is transferred during copulation. It could be argued that these mechanisms are more likely found in spiders with last-male than first-male sperm priority, since rival males of the latter species may be unlikely to fertilize many eggs. However, there is considerable intraspecific variation in sperm precedence patterns, and absolute first-male sperm precedence is rare (Table 9.1). Consequently, selection will usually favour males that reduce the likelihood that their mate will copulate again.

The release of sex-attracting pheromones is a widespread mechanism that enables females to attract males (Mayer and McLaughlin 1991) that has been documented in many spiders (Schulz and Toft 1993). Clearly, a male that interferes with the broadcast of these pheromones will reduce the chance that rival males will locate his mate. Males of several linyphiid spiders destroy the web of the female after locating her and before courting her, (Watson 1986; Willey Robertson and Adler 1994), thereby removing the source of the sex-attracting pheromone. Elegant experiments by Watson (1986) showed that web reduction dramatically reduces the probability the female is located by rival males (Fig. 9.2). Interestingly, competition between males is sufficiently intense that the male destroys the web as soon as he arrives, since it is possible a rival male would otherwise interfere with his courtship and mating. It is

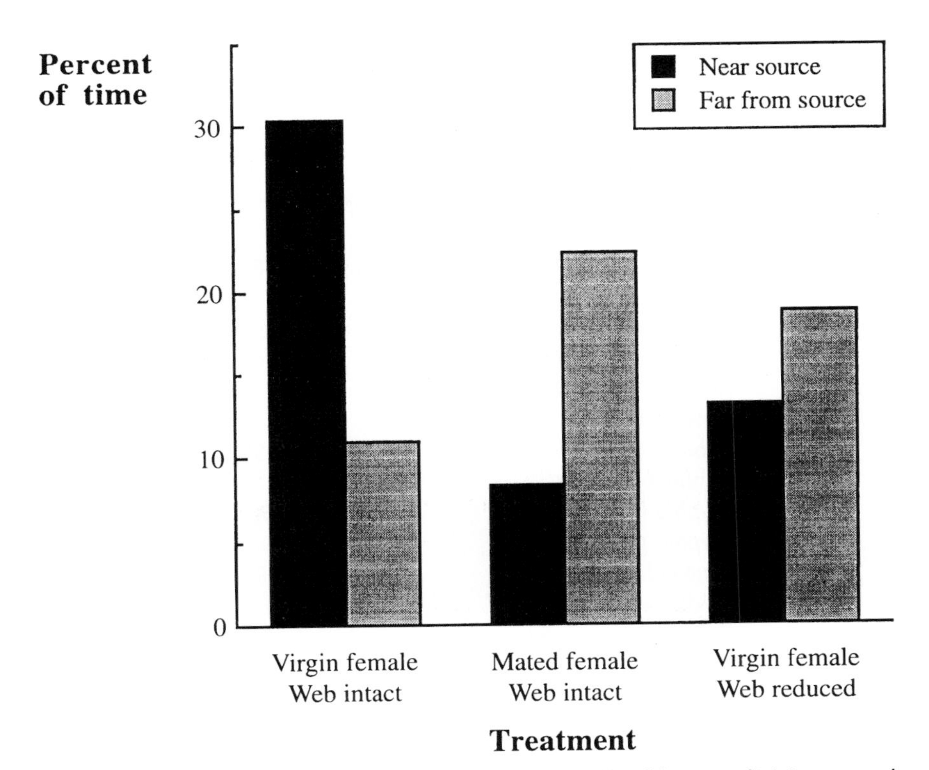

Fig. 9.2. *The proportion of time male sierra dome spiders* Neriene litigiosa *spend either close to or distant from virgin or mated females with or without intact webs (redrawn from Watson 1986).*

curious that web-reduction behaviour has not been reported in other web-building spiders, but perhaps the pheromones in these species are not found on the silk, or an undamaged web is required for mating.

Males may reduce the possibility of sperm competition by manipulating female receptivity after mating has taken place, and thus the possibility that she will mate with another male. A marked decrease or even absence of sexual receptivity of mated females, common in insects (Ringo 1996), has also been recorded in spiders from several families (Table 9.3). The reasons for this reduction in sexual receptivity following mating are not clear. It is tempting to argue that it is induced by the male and represents an outcome of the conflict between the sexes that has favoured males. For example, male spiders may, like other invertebrates (Thornhill and Alcock 1983), transfer substances in the seminal fluid that interfere with female receptivity. The reduction in

Table 9.3. Reduction in female receptivity following mating in spiders.

Taxa	Source
Haplogyne	
Pholcidae	
Pholcus phalangioides	Uhl *et al.* (1995)
Entelegynes	
Anyphaenidae	
Anyphaena accentuata	Huber (1995b)
Agelenidae	
Agelenopsis aperta	Singer and Riechert (1995)
Araneidae	
Metallina segmentata	Prenter *et al.* (1994)
Gasteracantha minax	Elgar and Bathgate (1996)
'Araneus' bradleyi	Elgar and Clode (unpublished data)
Clubionidae	
Clubiona pallidula	Huber (1995b)
Linyphiidae	
Linyphia hortensis	Stumpf (1990)
Linyphia triangularis	Stumpf (1990)
Neriene litigiosa	Watson (1986)
Lycosidae	
Hygrolycosa rubrofasciata	R. Alatalo (personal communication)
Nesticidae	
Nesticus cellulanus	Huber (1993)
Salticidae	
Phidippus johnsoni	Jackson (1980)
Thomisidae	
Misumenoides formosipes	Dodson and Beck (1993)

female receptivity may not take place immediately, and thus the male guards the female for a while after mating. This explanation is consistent with the behaviour of male and female *G. minax* (Elgar and Bathgate 1996).

However, it is possible that the costs of polyandry to females may be greater than the benefits, and thus selection will not favour female multiple mating, irrespective of the behaviour of males. The behaviour of females of species in which the female attempts to cannibalize the male before mating has taken place may also change according to her mating status; mated females may be more aggressive to subsequent males because the costs of sexual cannibalism (in terms of remaining unmated) now no longer exceed the nutritional benefits of eating a male suitor (Elgar 1992; M. A. Elgar and D. Clode, unpublished data).

If absentee males are unable to eliminate completely female attractiveness or receptivity, they may also reduce the potential for sperm competition by preventing the sperm of rival males being transferred into the female reproductive tract. This can be achieved by placing a physical barrier, such as a sperm plug, over the genital opening of the female. Such sperm plugs occur in a wide range of taxa, including spiders (Yoward 1996). Amorphous secretion-like substances blocking the epigynum of just-mated females is quite common and the embolus tip or its covering cap, which often breaks off inside the female during copulation, may also prevent fertilization (Jackson 1980; Lopez 1987; Matsumoto 1993). However, the effectiveness of these plugs at preventing sperm transmission by rival males has only been evaluated in one species, *Agelena limbata* (Agelenidae) (Matsumoto 1993), which has a conduit-design spermatheca. Male *A. limbata* cannot always place a complete plug over the female, and thus Matsumoto (1993) was able to compare the P_2 values of males that mated with females with either complete or incomplete plugs. The data showed that a complete sperm plug prevented the second male from achieving any fertilizations, but an incomplete plug was ineffective. Following the reasoning of Austad (1984), sperm plugs should be more common in spiders with cul-de-sac spermathecae because they are predicted to have last-male sperm precedence, and a sperm plug can effectively reverse this pattern. However, they may also evolve in entelegyne spiders whose spermathecae may be less conduit and more cul-de-sac in design (see Fig. 9.1c).

C. Copulation and sperm competition

The duration of copulation in spiders varies considerably, both within and between species (Elgar 1995; Table 9.4), and may reflect the outcome of conflicts of interest between the sexes over the choice and number of mating partners. Several studies have demonstrated that the duration of copulation in spiders is longer than is necessary to transfer sufficient sperm to fertilize the eggs of the female, which suggests that copulation may have additional functions (see Elgar 1995). These might include cryptic female choice (e.g. Eberhard 1985) or manipulating the sperm already transferred by previous males (see Chapter 10). Physical manipulation of the sperm of rival males has not been documented for arachnids, although the high level of last-male sperm priority in deer ticks *Ixodes dammini* (Acari; Ixodidae) may be due to males removing or manipulating the sperm of rival males (Yuval and Spielman 1990). The mouthparts of copulating male *I. dammini* may be inserted in the genital aperture of the female while she is obtaining her blood meal.

It is tempting to speculate that male spiders might remove sperm of rival males, perhaps by attaching them to the embolus before releasing their own sperm. Such activities might explain why males of many species copulate for longer with mated than virgin females (Table 9.4).

Table 9.4. Duration of copulation (in minutes ±SE) and female mating status in spiders.

	Female status		
Taxa	**Virgin**	**Mated**	**Source**
Haplogyne			
Pholcidae			
Pholcus phalangioides	74.2 (19.2–129)	25.3 (1.5–105.8)	Yoward (1996)
Physocyclus globosus	36.8 ± 5.6	21.4 ± 11.5	Huber and Eberhard (1997)
Holocnemus pluchei	31.7 ± 2.1	40.9 ± 2.1	Kaster and Jacob (1997)
Entelegyne			
Araneoidea			
Gasteracantha minax[a]	95.8 ± 6.8	100.2 ± 7.1	Elgar and Bathgate (1996)
Leucauge mariana	18.5 ± 1.6	9.9 ± 3.7	Eberhard and Huber (unpublished data)
Nephila clavipes[b]	4.8 ± 0.6	10.2 ± 0.3	Christenson and Cohn (1988)
Phonognatha graeffei	100.8 ± 8.4	136.6 ± 5.1	Fahey and Elgar (1997)
Tetragnatha montana	3.5–37.4	5.3–40.5	Yoward (1996)
Zygiella x-notata	14.0–73.0	9.0–70.0	Yoward (1996)
Linyphiidae			
Frontinella pyramitela[c]	36.6	75.0	Suter (1990)
Salticidae			
Phidippus johnsoni	51.4 (2–1200)	261.5 (0.3–1920)	Jackson (1980)
Theridiidae			
Achaearanea wau[c]	1.0 (0.2–9.0)	0.3 (0.1–2.1)	Lubin (1986)
Thomisidae			
Misumenoides formosipes	4.4 ± 0.8	2.8 ± 0.7	Dodson and Beck (1993)

[a] Duration of copulation with one palp.
[b] Frequency of copulations observed during point sampling.
[c] Values are medians with range in parentheses.

However, the male genitalia of many entelegyne spiders do not reach the spermathecae, and in those species that do, the thin thread-like embolus must pass through a narrow insemination duct, making it mechanically unrealistic that sperm could be removed by moving the male genitalia (Huber and Eberhard 1997).

In general, haplogyne spiders copulate for a shorter time than entelegyne spiders, which may reflect differences in male mating strategies (Fahey and Elgar 1996). Male entelegyne spiders should prefer to mate with virgin females because they may then fertilize most of the clutch. These males may further maximize their reproductive success by ejaculating larger quantities of sperm. In contrast, male haplogyne spiders may be able to obtain high levels of paternity only by preventing rival males from subsequently mating with the female. In the absence of such mechanisms (such as sperm plugs), males of these spiders may maximize their reproductive success by ejaculating relatively smaller quantities of sperm, but mating with many females. Elgar (1995) speculates that if the number of sperm transferred is positively correlated with the duration of copulation (Austad 1982; Cohn 1990; Andrade 1996; Fahey and Elgar 1996), then selection may favour longer copulations in entelegyne spiders than in haplogyne spiders (Elgar 1995). An important underlying assumption of this argument is that there is a finite supply of sperm, otherwise an increase in sperm volume (and hence longer copulations) is predicted under conditions of greater risk of sperm competition (Parker 1990). Resolution of these ideas requires more data on the costs and benefits to males of adjusting the quantity of sperm transferred during copulation.

The duration of copulation may reflect a conflict between the sexes: females may prefer shorter copulations if that facilitates equitable sperm mixing, while males may prefer longer copulations if that increases the number of sperm that are transferred and hence their fertilization success. This conflict is manifested in an extraordinary way by the Australian redback spider *Latrodectus hasselti*. Male *L. hasselti* always perform a somersault while copulating, which results in the abdomen of the male resting directly on the mouthparts of the female, where it remains throughout copulation (Forster 1992). Depending upon her hunger level, the female may then attempt to consume the male (Andrade 1996). Sexual cannibalism in this species therefore appears to involve male complicity, and may represent a form of paternal investment (Buskirk *et al.* 1984; Elgar 1992). However, this paternal investment function of sexual cannibalism is unlikely for *L. hasselti* because the consumption of a single male does not significantly increase female fecundity (Andrade 1996).

Andrade (1996) demonstrated that the behaviour provides the male with two ways of increasing his fertilization success with polyandrous females (Fig. 9.3). First, cannibalized males copulate longer than males that are not cannibalized, and longer copulations increase the proportion of eggs the male fertilizes when in competition with another male.

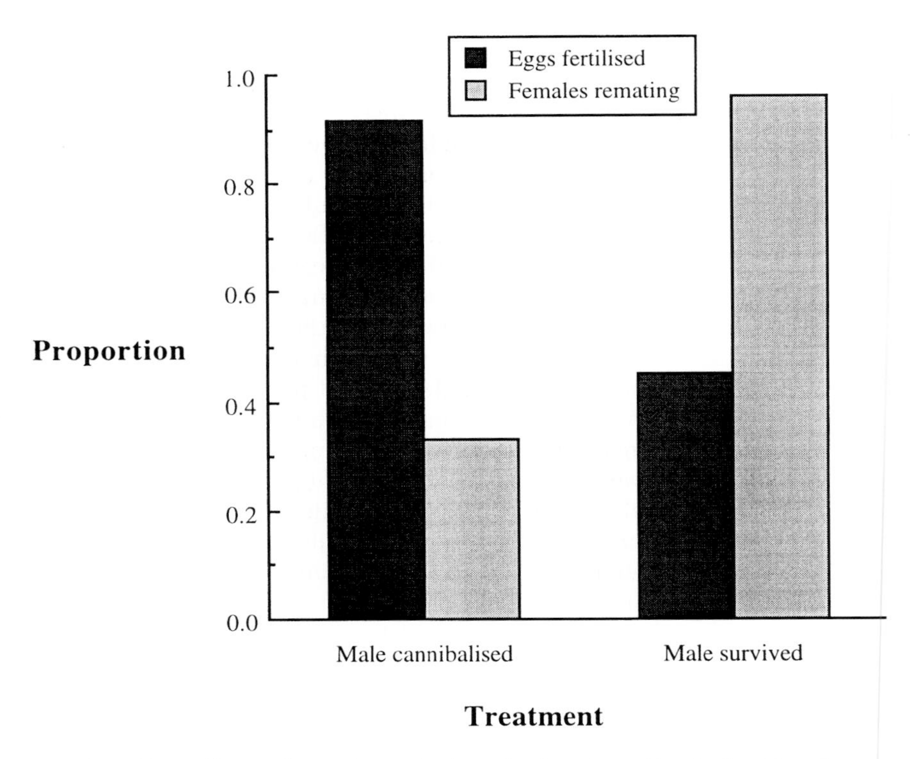

Fig. 9.3. *The effects of sexual cannibalism by female red-backed spiders* Latrodectus hasselti *on the proportion of eggs a male may fertilize and the probability that the female will subsequently mate with another male (redrawn from Andrade 1996).*

Second, males that are sexually cannibalized by virgin females increase their paternity because cannibalism decreases the probability (by over 60%) that the female mates with another male. This paternity security function of sexual cannibalism may also explain similar occurrences of sexual cannibalism during copulation in other species, such as *Argiope* and *Nephila* (Sasaki and Iwahashi 1995; Elgar and Fahey 1996; but see also Christenson and Cohn 1988; Elgar 1992).

Explanations of post-insemination sexual cannibalism in spiders show a remarkable parallel with those of the function of nuptial gifts by male bushcrickets. Originally, nuptial gifts were attributed a paternal investment function, but subsequent studies indicate that they more likely evolved as a paternity protection function (see Wedell 1991, 1993). Andrade's (1996) study may have implications for the provision of nuptial gifts in *P. mirabilis* (Austad and Thornhill 1986). Males that provide females with larger prey items copulate for longer (Austad and Thornhill 1986), perhaps thereby decreasing the fertilization success of rival males. A similar pattern may occur in those spiders in which the male secretes substances that the female imbibes during copulation (Blest 1987; Huber 1997); paternity protection may also explain why male spiders sometimes restrain their potential female mates with silk (Bruce and Carico 1988) if that increases the duration of copulation.

Finally, males may increase their success in sperm competition by producing larger sperm. Both male and female bulb mites *Rhizoglyphus robini* (Acari: Astigmata) mate promiscuously (Radwan 1996), and the proportion of eggs fertilized by the second male is on average 54%, but can range from 0 to 93% (Radwan and Siva-Jothy 1996). Radwan (1996) found that sperm size was significantly positively correlated with fertilization success, while male body size, duration of copulation, and sperm number had no effect. Further, there was no correlation between sperm size and sperm number. It is not clear why larger sperm are better at fertilizing eggs than smaller sperm, although Radwan (1996) suggests that they may be able to survive longer in the seminal receptacle. It would be interesting to examine similar patterns in spiders; perhaps large sperm are particularly favourable in species in which oviposition takes place a long time after fertilization.

D. Infanticide

The conflict between the sexes over the frequency of mating is perhaps most evident in the infanticidal behaviour of male *Stegodyphus lineatus* spiders (Eresidae; Schneider and Lubin 1996, 1997). This behaviour also illustrates how competition between males may persist even after the female has oviposited. Female *S. lineatus* provide suicidal care to matriphagous hatchlings of her single clutch (see also Evans *et al.* 1995). Typically, females produce a single clutch only, although they can lay replacement clutches if the original eggsac is removed (Schneider and Lubin 1997). Therefore, a female protecting her clutch does not benefit by remating because she will not oviposit. Unlike many other annual, semiparous spiders, late-maturing male *S. lineatus* may encounter females that have already oviposited and are guarding their eggsacs. Late-maturing males that encounter these guarding females attempt to overpower the female and remove her eggs. If the male is successful, he will mate with the female, who will then produce a second clutch.

The benefits to the infanticidal male are straightforward, since he cannot sire offspring with a female that is guarding her eggsac. However, the female sustains considerable costs from infanticidal males: the clutch size of the second brood is significantly smaller, the additional time required to form a new clutch substantially decreases her expected survivorship (Schneider and Lubin 1997), and she may not achieve the same body mass before her matriphagous offspring consume her, which may decrease their chances of survival (Evans *et al.* 1995). It is therefore surprising that the first male to mate with the female does not stay to guard both the female and their offspring. Sterile male mating experiments revealed that there is complete sperm mixing (Schneider and Lubin 1996) and thus the reproductive success of a male whose mate is subsequently overpowered by a rival, infanticidal male would be more than halved. Perhaps it pays the male to leave the female, since she would expect to encounter and mate with less than two males in her lifetime (Schneider and Lubin 1996), and the male may find a second female.

VI. SEXUAL SELECTION AND SIZE DIMORPHISM

Both overt competition between males for mating opportunities and covert competition for fertilization success have interesting implications for sexual size dimorphism in spiders. Typically, larger males are favoured by sexual selection through competition between males for mating opportunities or mate guarding (Table 9.2). However, the reproductive success of male spiders may also be a function of the number of receptive females they encounter. Like other invertebrates, male spiders may encounter more females by reaching sexual maturity earlier (protandry), perhaps by having faster developmental rates or smaller body sizes (Nylin and Wedell 1994). The strength of selection for protandry may be increased in spiders with first-male sperm precedence. In contrast, fecundity selection may favour larger female size at sexual maturity or further growth as an adult, since both will allow females to produce more, or larger eggs (Head 1995; but see Prenter *et al.* 1995).

Fecundity selection may explain why female spiders are typically larger than males (Elgar 1992), but the magnitude of this dimorphism may be explained by the conflicting selection pressures on male size (but see Head 1995; Hormiga *et al.* 1995; Prenter *et al.* 1997), which may be modified by patterns of sperm precedence. For example, spiders belonging to the genus *Argyrodes* are kleptoparasitic, feeding on prey caught in the webs of their host. In some species, a large number of individuals of both sexes may be found on the web of a host. There is often intense competition between males for access to receptive females, and large males have an advantage (Whitehouse 1991). The intensity of male–male competition may explain why the typical pattern of sexual size dimorphism in spiders is reversed among these kleptoparasitic species of *Argyrodes*, in which males are larger than females (Elgar 1993).

Vollrath and Parker (1992) suggested that the variation in sexual size dimorphism may be due to differences in the sex ratio bias that arises from differences in male and female lifestyles (see also Piel 1996). A female bias may occur in species with relatively sedentary adult females and more mobile adult males that suffer higher mortality as a result of their searching behaviour. According to Vollrath and Parker (1992), the female biased sex ratio relaxes selection by male–male competition for larger male size and thus selection for protandry will favour diminutive males. Although Vollrath and Parker (1992) claim support for their idea from an analysis of comparative data, subsequent comparative studies indicate that it may apply only to species with long breeding seasons (Prenter *et al.* 1997; see also Elgar 1991; Head 1995; Hormiga *et al.* 1995). A more thorough phylogenetic analysis suggests that selection favours larger female size, rather than diminutive males (Coddington *et al.* 1997; see also Elgar 1991).

VII. CONCLUSIONS AND PROSPECTS

Polyandry, which sets the stage for the potential for sperm competition, may provide female spiders with both material and genetic benefits. The material benefits include a reduction in the costs of male harassment and, perhaps, an increase in fecundity through the provision of nuptial gifts, including the body of the male. The genetic benefits may be modified by the patterns of sperm precedence, and may be less apparent for females of species with a predominantly first-male sperm priority. For male spiders, most of the opportunities for securing paternity appear to be related to the timing of mating, physical guarding females from rival males, blocking the genital region of the female with a plug, or increasing the duration of copulation. The latter may reduce female receptivity and/or increase fertilization success in competition with other males, although the mechanisms remain unknown.

Although our understanding of the mating systems of spiders has advanced considerably over the last two decades, there are still several important gaps. We need more information on patterns of sperm priority because this has important implications for male and female mating strategies. Clearly, it is unwise to assume that there is a clear difference in the sperm priority patterns of entelgyne and haplogyne spiders, since exceptions have been found. In particular, it may be useful to examine the anatomy of the female reproductive tract in more detail, thereby establishing the degree to which it follows a cul-de-sac or conduit design, and whether this influences sperm precedence patterns. Studies of sperm competition may also benefit by examining precedence patterns involving more than two males. The clear sexual dimorphism and often elaborate courtship behaviour of many cursorial spiders may provide a rich seam of model systems with which to investigate the evolutionary significance of both overt and cryptic female choice.

ACKNOWLEDGEMENTS

I thank Rauno Alatalo, Tim Birkhead, Deborah Cole, Bill Eberhard, Bernhard Huber, Anders Pape Møller, John Prenter, Gabrielle Uhl, Jutta Schneider, Nina Wedell and Paul Yoward for their help, discussion or comments on this manuscript; Sharon Wragg for drawing Fig. 9.1; and the Australian Research Council for financial support.

REFERENCES

Alcock J (1994) Post-insemination associations between males and females in insects: the mate guarding hypothesis. *Annu. Rev. Entomol.* **39:** 1–21.

Andersson M (1994) *Sexual Selection.* Princeton University Press, Princeton.

Andrade MCB (1996) Sexual selection for male sacrifice in the Australian redback spider. *Science* **271:** 70–72.

Arnqvist G & Henriksson S (1997) Sexual cannibalism in the fishing spider and a model for the evolution of sexual cannibalism based on genetic constraints. *Evol. Ecol.* **11:** 255–273.

Austad SN (1982) First male sperm priority in the bowl and doily spider, *Frontinella pyramitela* (Walckenaer). *Evolution* **36:** 777–785.

Austad SN (1983) A game-theoretical interpretation of male combat in the bowl and doily spider (*Frontinella pyramitela*). *Anim. Behav.* **31:** 59–73.

Austad SN (1984) Evolution of sperm priority patterns in spiders. In *Sperm Competition* and *the Evolution of Animal Mating Systems.* RL Smith (ed.), pp. 233–249. Academic Press: London.

Austad SN & Thornhill R (1986) Female reproductive variation in a nuptial-feeding spider, *Pisaura mirabilis. Bull. Br. Arachnol. Soc.* **7:** 48–52.

Benton TG (1992) Determinants of male mating success in a scorpion. *Anim. Behav.* **43:** 125–135.

Birkhead TR & Møller AP (1987) *Sperm Competition in Birds: Evolutionary Causes and Consequences.* Academic Press, London.

Blanchong JA, Summerfield MS, Popson MA & Jakob EM (1995) Chivalry in pholcid spiders revisited. *J. Arachnol.* **23:** 165–170.

Blest AD (1987) The copulation of a linyphiid spider, *Baryphyma pratense*: does a female receive a blood-meal from her mate? *J. Zool.* **213:** 189–191.

Blest AD & Taylor HH (1977) The clypeal glands of *Mynoglenes* and of some other linyphiid spiders. *J. Zool.* **183:** 473–492.

Bristowe WS (1929) The mating habits of spiders, with special reference to the problems surrounding sex dimorphism. *Proc. Zool. Soc.* **100:** 309–358.

Bristowe WS (1958) *The World of Spiders.* London, Collins.

Brown SG (1985) Mating behavior of the golden orb-weaving spider, *Nephila clavipes*: II. Sperm capacitation, sperm competition, and fecundity. *J. Comp. Psychol.* **99:** 167–175.

Bruce JA & Carico JE (1988) Silk use during mating in *Pisaurina mira* (Walckenaer) (Araneae, Pisaurdiae). *J. Arachnol.* **16:** 1–4.

Buskirk RE, Frohlich C & Ross KG (1984) The natural selection of sexual cannibalism. *Am. Nat.* **123:** 612–625.

Christenson TE & Cohn J (1988) Male advantages for egg fertilisation in the golden orb-weaving spider (*Nephila clavipes*). *J. Comp. Psychol.* **102:** 312–318.

Christenson TE & Goist KC, Jr (1979) Costs and benefits of male–male competition in the orb-weaving spider *Nephila clavipes*. *Behav. Ecol. Sociobiol.* **5:** 87–92.

Christenson TE, Brown SG, Wenzl PA, Hill EM & Goist KC (1985) Mating behavior of the golden-orb-weaving spider, *Nephila clavipes*: I. Female receptivity and male courtship. *J. Comp. Psychol.* **99:** 160–166.

Clark DL & Uetz GW (1992) Morph-independent mate selection in a dimorphic jumping spider: demonstration of movement bias in female choice using video-controlled courtship behaviour. *Anim. Behav.* **43:** 247–254.

Coddington JA, Hormiga G & Scharff N (1997) Giant female or dwarf male spiders? *Nature* **385:** 687–688.

Cohn J (1990) Is it the size that counts? Palp morphology, sperm storage, and egg hatching frequency in *Nephila clavipes* (Araneae, Araneidae). *J. Arachnol.* **18:** 59–71.

Cohn J, Balding FV & Christenson TE (1988) In defense of *Nephila clavipes*: post-mate guarding by the male golden orb-weaving spider. *J. Comp. Psychol.* **102:** 319–325.

Crane J (1949) Comparative biology of salticid spiders at Rancho Grande, Venezuela, Part IV. An analysis of display. *Zoologica* **34:** 159–215.

Darwin C (1871) *The Descent of Man and Selection in Relation to Sex*. Murray, London.

Dodson GN & Beck MW (1993) Pre-copulatory guarding of penultimate females by male crab spiders *Misumenoides formosipes*. *Anim. Behav.* **46:** 951–959.

Eberhard WG (1985) *Sexual Selection and Animal Genitalia*. Harvard University Press, Cambridge, MA.

Eberhard WG (1991) Copulatory courtship and cryptic female choice in insects. *Biol. Rev.* **66:** 1–31.

Eberhard WG (1994) Evidence for widespread courtship during copulation in 131 species of insects and spiders, and implications for cryptic female choice. *Evolution* **48:** 711–733.

Eberhard WG (1996) *Female Control: Sexual Selection by Cryptic Female Choice*. Princeton University Press, Princeton.

Eberhard WG & Briceño RD (1983) Chivalry in pholcid spiders. *Behav. Ecol. Sociobiol.* **13:** 189–195.

Eberhard WG, Guzmàn-Gòmez S & Catley KM (1993) Correlation between spermathecal morphology and mating systems in spiders. *Biol. J. Linn. Soc.* **50:** 197–209.

Elgar MA (1991) Sexual cannibalism, size dimorphism and courtship behavior in orb-weaving spiders (Araneae). *Evolution* **45:** 444–448.

Elgar MA (1992) Sexual cannibalism in spiders and other invertebrates. In *Cannibalism: Ecology* and *Evolution Among Diverse Taxa*. MA Elgar & BJ Crespi (eds), pp. 128–155. Oxford University Press, Oxford.

Elgar MA (1993) Inter-specific associations involving spiders: kleptoparasitism, mimicry and mutualism. *Mem. Queensl. Mus.* **33:** 411–430.

Elgar MA (1995) The duration of copulation in spiders: comparative patterns. *Rec. West Aust. Mus. Suppl.* **52:** 1–11.

Elgar MA & Bathgate R (1996) Female receptivity and male mate-guarding in the jewel spider *Gasteracantha minax* Thorell (Araneidae). *J. Insect Behav.* **9:** 729–738.

Elgar MA & Fahey BF (1996) Sexual cannibalism, male-male competition and sexual size dimorphism in the orb-weaving spider *Nephila plumipes*. *Behav. Ecol.* **7:** 195–198.

Elgar MA & Nash DR (1988) Sexual cannibalism in the garden spider *Araneus diadematus*. *Anim. Behav.* **36:** 1511–1517.

Enders MM (1993) The effect of male size and operational sex ratio on male mating success in the common spider mite, *Tetranychus urticae* Koch (Acari: Tetranychidae). *Anim. Behav.* **46:** 835–846.

Enquist M & Leimar O (1990) The evolution of fatal fighting. *Anim. Behav.* **39:** 1–9.

Evans TA, Wallis EJ & Elgar MA (1995) Making a meal of mother. *Nature* **376:** 299.

Faber DB & Baylis JR (1993) Effects of body size on agonistic encounters between male jumping spiders (Araneae: Salticidae). *Anim. Behav.* **45:** 289–299.

Fahey, BF & Elgar MA (1997) Sexual cohabitation as mate-guarding in the leaf-curling spider *Phonognatha graeffei* Keyserling (Araneoidea, Araneae). *Behav. Ecol. Sociobiol.* **40:** 127–133.

Fernandez-Montraveta C & Ortega J (1993) Sex differences in the agonistic behaviour of a lycosid spider (Araneae Lycosidae). *Ethol. Ecol. Evol.* **5:** 293–301.

Forster LM (1992) The stereotyped behaviour of sexual cannibalism in *Latrodectus hasselti* Thorell (Araneae: Theridiidae): the Australian redback spider. *Aust. J. Zool.* **40:** 1–11.

Gaffin DD & Brownell PH (1992) Evidence of chemical signalling in the sand scorpion, *Paruroctonus mesaensis* (Scorpionida: Vaejovida). *Ethology* **91:** 59–69.

Gunnarsson B & Andersson A (1996) Sex ratio variation in sheet-web spiders: options for female control? *Proc. Roy. Soc. Lond. Ser. B* **263:** 1177–1182.

Hack MA, Thompson DJ & Fernandes DM (1997) Fighting in males of the autumn spider, *Metellina segmentata*: the effects of relative body size, prior residency and female value on contest outcome and duration. *Ethology* **103:** 488–498.

Hamilton WD, Axelrod R & Tanese R (1990) Sexual reproduction as an adaptation to resist parasites (a review). *Proc. Natl Acad. Sci. USA* **87:** 3566–3573.

Head G (1995) Selection on fecundity and variation in the degree of sexual size dimorphism among spider species (class Araneae). *Evolution* **49:** 776–781.

Hormiga G, Eberhard WG & Coddington JA (1995) Web-construction behaviour in Australian *Phonognatha* and the phylogeny of nephiline and tetragnathid spiders (Araneae: Tetragnathidae). *Aust. J. Zool.* **43:** 313–364.

Huber BA (1993) Genital mechanics and sexual selection in the spider *Nesticus cellulanus* (Araneae: Nesticidae). *Can. J. Zool.* **71:** 2437–2447.

Huber BA (1995a) The retrolateral tibial apophysis in spiders – shaped by sexual selection? *Zool. J. Linn. Soc.* **113:** 151–163.

Huber BA (1995b) Genital morphology and copulatory mechanics in *Anyphaena accentuata* (Anyphaenidae) and *Clubiona pallidula* (Clubionidae: Araneae). *J. Zool.* **235:** 689–702.

Huber BA (1997) Evidence for gustatorial courtship in a haplogyne spider (*Hedypsilus culicinus:* Pholcidae: Araneae). *Netherlands J. Zool.* **47:** 95–98.

Huber BA & Eberhard WG (1997) Courtship, copulation and genital mechanics in *Physocyclus globosus* (Araneae, Pholcidae). *Can. J. Zool.* **74:** 905–918.

Jackson RR (1980) The mating strategy of *Phidippus johnsoni* (Araneae: Salticidae): II. Sperm competition and the function of copulation. *J. Arachnol.* **8:** 217–240.

Jackson RR (1982) The behavior of communicating in jumping spiders (Salticidae). In *Spider Communication: Mechanisms* and *Ecological Significance.* PN Witt & JS Rovner (eds), pp. 213–247. Stanford University Press, Stanford.

Jackson RR (1986) Cohabitation of males and juvenile females: a prevelant mating tactic of spiders. *J. Nat. Hist.* **20:** 1193–1210.

Jackson RR (1992) Eight-legged tricksters. *BioScience* **42:** 590–598.

Kaster JL & Jakob EM (1997) Last-male sperm priority in a haplogyne spider (Araneae, Pholcidae): correlations between female morphology and patterns of sperm usage. *Ann. Entomol. Soc. Am.* **90:** 254–259.

Kotiaho J, Alatalo RV, Mappes J & Parri S (1997) Sexual selection in a wolf spider: male drumming activity, body size and viability. *Evolution* **50:** 1977–1981.

Lang A (1996) Silk investment in gifts by males of the nuptial feeding spider *Pisaura mirabilis* (Araneae: Pisauridae). *Behaviour* **133:** 697–716.

Leimar O, Austad SN & Enquist M (1991) A test of the sequential assessment game: fighting in the bowl and doily spider *Frontinella pyramitela. Evolution* **45:** 862–874.

Lewis WM, Jr (1987) The costs of sex. In *The Evolution of Sex and its Consequences.* SC Stearns (ed.), pp. 33–57. Birkhuser Verlag, Basel.

Lopez A (1987) Glandular aspects of sexual biology. In *Ecophysiology of Spiders.* W Nentwig (ed.), pp. 121–132. Springer-Verlag, Heidelberg.

Lubin YD (1986) Courtship and alternative mating tactics in a social spider. *J. Arachnol.* **14:** 239–257.

Mappes J, Alatalo RV, Kotiaho J & Parri S (1996) Viability costs of condition-dependent sexual male display in a drumming wolf spider. *Proc. Roy. Soc. Lond. Ser. B* **263:** 785–789.

Masumoto T (1993) The effect of the copulatory plug in the funnel-web spider *Agelena limbata* (Araneae: Agelenidae). *J. Arachnol.* **21:** 55–59.

Mayer MS & McLaughlin JR (1991) *Handbook of Insect Pheromones and Sex Attractants.* CRC, Boca Raton.

Miller GJ & Ramey Miller P (1986) Pre-courtship cohabitation of mature male and penultimate female *Geolycosa turricola. J. Arachnol.* **14:** 133–134.

Miyashita K (1987) Development and egg sac production of *Achaearanea tepidariorum* (C. L. Koch) (Araneae, Theridiidae) under long and short photoperiods. *J. Arachnol.* **15:** 51–58.

Miyashita K (1992) Postembryonic development and life cycle of *Atypus karschi* Dönitz (Araneae: Atypidae). *Acta Arachnol.* **41:** 177–186.

Miyashita T (1993) Male–male competition and mating success in the orb-web spider *Nephila clavata*, with reference to temporal factors. *Ecol. Res.* **8:** 93–102.

Miyashita T & Hayashi H (1996) Volatile chemical cue elicits mating behavior of cohabiting males of *Nephila clavata* (Araneae, Tetragnathidae). *J. Arachnol.* **24:** 9–15.

Montgomery TH, Jr (1903) Studies on the habits of spiders, particularly those of the mating period. *Proc. Acad. Nat. Sci. Philadel.* **55:** 59–149.

Montgomery TH, Jr (1910) The significance of the courtship and secondary sexual characters of araneads. *Am. Nat.* **44:** 151–177.

Mora G (1990) Paternal care in a neotropical harvestman, *Zygopachylus albomarginis* (Arachnida, Opiliones: Gonyleptidae). *Anim. Behav.* **39:** 582–593.

Nielsen N & Toft S (1990) Alternative male mating strategies in *Lyniphia triangularis* (Araneae, Lyniphiidae). *Acta Zool. Fenn.* **190:** 293–297.

Nylin S & Wedell N (1994) Sexual size dimorphism and comparative methods. In *Phylogenetics and Ecology*. P Eggleton & R Vane-Wright (eds), pp. 253–280. Academic Press, London.

Oxford GS (1993) Patterns of sperm usage in large house spiders (*Tegenaria* spp): genetics of esterase markers. *Heredity* **70:** 413–419.

Parker GA (1970) Sperm competition and its evolutionary consequences in insects. *Biol. Rev.* **45:** 525–567.

Parker GA (1974) Courtship persistence and female-guarding as male time investment strategies. *Behaviour* **48:** 157–184.

Parker GA (1984) Sperm competition and the evolution of animal mating strategies. In *Sperm Competition* and *the Evolution of Animal Mating Systems*. RL Smith (ed.), pp. 1–60. Academic Press, London.

Parker GA (1990) Sperm competition: raffles and roles. *Proc. Roy. Soc. Lond. Ser. B* **242:** 120–126.

Parri S, Alatalo RV, Kotiaho J & Mappes J (1997) Female preference for male drumming in the wolf spider *Hygrolycosa rubrofasciata*. *Anim. Behav.* **53:** 305–312.

Peckham GW & Peckham EG (1889) Observations on sexual selection in spiders of the family Attidae. *Occas. Pap. Nat. Hist. Soc. Wiscon.* **1:** 1–60.

Piel WH (1996) Ecology of sexual dimorphism in spiders of the genus *Metepeira* (Araneae: Araneidae). *Rev. Suisse Zool.* Vol. hors série: 523–529.

Platnick N (1971) The evolution of courtship behaviour in spiders. *Bull. Br. Arachnol. Soc.* **2:** 40–47.

Polis G & Sisson WD (1990) Life history. In *The Biology of Scorpions*. GA Polis (ed.), pp. 161–223. Stanford University Press, Stanford.

Pollard SD, Macnab AM & Jackson JJ (1987) Communication with chemicals: pheromones and spiders. In *Ecophysiology of Spiders*. W Nentwig (ed.), pp. 133–141. Springer-Verlag, Heidelberg.

Prenter J, Elwood RW, Montgomery WI (1994) Assessments and decisions in *Metellina segmentata* (Araneae: Metidae): evidence of a pheromone involved in mate-guarding. *Behav. Ecol. Sociobiol.* **35:** 39–43.

Prenter J, Montgomery WI & Elwood RW (1995) Multivariate morphometrics and sexual dimorphism in the orb-web spider *Metellina segmentata* (Clerck, 1757) (Araneae, Metidae). *Biol. J. Linn. Soc.* **55:** 345–354.

Prenter J, Montgomery WI & Elwood RW (1997) Sexual dimorphism in northern temperate spiders: implications for the differential mortality model. *J. Zool.* **243:** 341–349.

Proctor HC (1992) Mating and spermatophore morphology of water mites (Acari: Parasitengona). *Zool. J. Linn. Soc.* **106:** 341–384.

Radwan J (1996) Intraspecific variation in sperm competition success in the bulb mite: a role for sperm size. *Proc. Roy. Soc. Lond. Ser. B* **263:** 855–859.

Radwan J & Siva-Jothy MT (1996) The function of the postinsemination mate association in the bulb mite *Rhizoglyphus robini*. *Anim. Behav.* **52:** 651–657.

Ridley M (1983) *The Explanation of Organic Diversity: the Comparative Method and Adaptations for Mating*. Clarendon Press, Oxford.

Ringo J (1996) Sexual receptivity in insects. *Annu. Rev. Entomol.* **41:** 473–494.

Robinson MH (1982) Courtship and mating behaviour in spiders. *Annu. Rev. Entomol.* **27:** 1–20.

Robinson MH & Robinson B (1980) Comparative studies on the courtship and mating behavior of tropical araneid spiders. *Pacif. Insects Monog.* **36:** 1–218.

Rovner JS (1968) Territoriality in the sheet-web spider *Linyphia triangularis* (Araneae, Linyphiidae). *Z. Tierpsychol.* **25:** 232–242.

Rubenstein DI (1987) Alternative reproductive tactics in the spider *Meta segmentata*. *Behav. Ecol. Sociobiol.* **20:** 229–237.

Sasaki T & Iwahashi O (1995) Sexual cannibalism in an orb-weaving spider *Argiope aemula*. *Anim. Behav.* **49:** 1119–1121.

Schaible U, Gack C & Paulus HF (1986) Zur Morphologie, Histologie und Biologischen Bedeutung der Kopfstruksturen männlicher Zwergspinnen (Linyphiidae: Erigoninae). *Zool. Jb (Syst)* **113:** 389–408.

Scheffer SJ, Uetz GW & Stratton GE (1996) Sexual selection, male morphology and the efficacy of courtship signalling in two wolf spiders (Araneae: Lycosidae). *Behav. Ecol. Sociobiol.* **38:** 17–23.

Schmitt A, Schuster M & Barth FG (1992) Male competition in a wandering spider (*Cupiennius getazi*, Ctendiae). *Ethology* **90:** 293–306.

Schneider JM & Lubin Y (1996) Infanticidal male eresid spiders. *Nature* **381:** 655–656.

Schneider JM & Lubin Y (1997) Infanticide by males in a spider with suicidal maternal care, *Stegodyphus lineatus* (Eresidae). *Anim. Behav.* **54:** 305–312.

Schulz S & Toft S (1993) Identification of a sex pheromone from a spider. *Science* **260:** 1635–1637.

Singer F & Riechert S (1995) Mating system and mating success of the desert spider *Agelenopsis aperta*. *Behav. Ecol. Sociobiol.* **36:** 313–322.

Smith RL (1984) *Sperm Competition and the Evolution of Animal Mating Systems.* Academic Press, New York.

Stratton GE & Uetz GW (1981) Acoustic communication and reproductive isolation in two species of wolf spiders. *Science* **214:** 575–577.

Stratton GE & Uetz GW (1986) The inheritance of courtship behaviour and its role as a reproductive isolating mechanism in two species of *Schizocosa* wolf spiders (Araneae: Lycosidae). *Evolution* **40:** 129–141.

Stumpf H (1990) Observations on the copulation behaviour of sheet-web spiders *Linyphia horlensis* Sundevall and *Linyphia triangularis* (Clerck)(Araneae: Linyphiidae). *Bull. Soc. Eur. Arachnol. HS* **1:** 340–345.

Suter RB (1990) Courtship and assessment of virginity by male bowl and doily spiders. *Anim. Behav.* **39:** 307–313.

Thomas RH & Zeh DW (1984) Sperm transfer and utilisation strategies in arachnids: ecological and morphological constraints. In *Sperm Competition* and *the Evolution of Animal Mating Systems.* RL Smith (ed.), pp. 180–222. Academic Press, London.

Thornhill R (1983) Cryptic female choice and its implications in the scorpionfly *Harpobittacus nigriceps*. *Am. Nat.* **122:** 763–788.

Thornhill R & Alcock J (1983) *The Evolution of Insect Mating Systems.* Harvard University Press, Cambridge, MA.

Tietjen WJ & Rovner JS (1982) Chemical communication in lycosids and other spiders. In *Communication in Spiders: Mechanisms and Ecological Significance.* PN Witt & JS Rovner (eds), pp. 249–279. Princeton University Press, Princeton.

Toft S (1989) Mate guarding in two *Linyphia* species (Araneae: Linyphiidae). *Bull. Br. Arachnol. Soc.* **8:** 33–37.

Trivers RL (1972) Parental investment and sexual selection. In *Sexual Selection and the Descent of Man.* B Campbell (ed.), pp. 136–179. Heinemann, London.

Uetz GW, McClintock WJ, Miller D, Smith EI & Cook KK (1996) Limb regeneration and subsequent asymmetry in a male secondary sexual character influences sexual selection in wolf spiders. *Behav. Ecol. Sociobiol.* **38:** 253–258.

Uhl G (1993a) Sperm storage and repeated egg production in female *Pholcus phalangioides* Fuesslin (Araneae). *Bull. Soc. Neuchâtel. Sci. Nat.* **116:** 245–252.

Uhl G (1993b) Mating behaviour and female sperm storage in *Pholcus phalangioides*. *Mem. Queensl. Mus.* **33:** 667–674.

Uhl G (1994a) Genital morphology and sperm storage in *Pholcus phalangioides* (Fuesslin, 1775) (Pholcidae; Araneae). *Acta Zoologica* **75:** 1–12.

Uhl G (1994b) Ultrastructure of the accessory glands in female genitalia of *Pholcus phalangioides* (Fuesslin, 1775) (Pholcidae; Araneae). *Acta Zoologica* **75:** 13–25.

Uhl G, Huber BA & Rose W (1995) Male pedipalp morphology and copulatory mechanism in *Pholcus phalangioides* (Fuesslin, 1775) (Araneae, Pholicidae). *Bull. Br. Arachnol. Soc.* **10:** 1–9.

van Helsdingen PJ (1965) Sexual behaviour of *Lepthyphantes leprosus* (Ohlert) (Araneida, Linyphiidae), with notes on the function of the genital organs. *Zool. Mededel.* **41:** 15–42.

Vollrath F (1980) Male body size and fitness in the web-building spider *Nephila clavipes*. *Z. Tierpsychol.* **53:** 61–78.

Vollrath F & Parker GA (1992) Sexual dimorphism and distorted sex ratios in spiders. *Nature* **360:** 156–159.

Watson PJ (1986) Transmission of a female sex pheromone thwarted by males in the spider *Linyphia litigiosa* Keyserling (Linyphiidae). *Science* **233:** 219–221.

Watson PJ (1990) Female-enhanced male competition determines the first mate and principle sire in the spider *Linyphia litigiosa* (Linyphiidae). *Behav. Ecol. Sociobiol.* **26:** 77–90.

Watson PJ (1991a) Multiple paternity and first mate sperm precedence in the sierra dome spiders, *Linyphia litigiosa* Kerserling (Linyphiidae). *Anim. Behav.* **41:** 135–148.

Watson PJ (1991b) Multiple paternity as genetic bet-hedging in female sierra dome spiders, *Linyphia litigiosa* (Linyphiidae). *Anim. Behav.* **41:** 343–360.

Watson PJ (1993) Foraging advantage of polyandry for female sierra dome spiders (*Linyphia litigiosa*: Linyphiidae) and assessment of alternative direct benefit hypotheses. *Am. Nat.* **141:** 440–465.

Watson PJ (1994) Sexual selection and the energetics of copulatory courtship in the sierra dome spider, *Linyphia litigiosa*. *Anim. Behav.* **48:** 615–626.

Watson PJ (1997) Preferential multi-mating by females increases offspring size and growth rates in the spider *Neriene litigiosa* (Linyphiidae). *Anim. Behav.*, in press.

Watson PJ & Thornhill R (1994) Fluctuating assymetries and sexual selection. *Trends Ecol. Evol.* **9:** 21–25.

Wedell N (1991) Sperm competition selects for nuptial feeding in a bushcricket. *Evolution* **45:** 1975–1978.

Wedell N (1993) Spermatophore size in bushcrickets: comparative evidence for nuptial gifts as a sperm protection device. *Evolution* **47:** 1203–1212.

Wells MS (1988) Effects of body size and resource value on fighting behaviour in a jumping spider. *Anim. Behav.* **36:** 321–326.

Whitehouse MEA (1991) To mate or fight? Male–male competition and alternative mating strategies in *Argyrodes antipodiana* (Theridiidae, Araneae). *Behav. Proces.* **23:** 163–172.

Willey Robertson M & Adler PH (1994) Mating behavior of *Florinda coccinea* (Hentz) (Araneae: Linyphiidae). *J. Insect Behav.* **7:** 313–326.

Yoward P (1996) Spider sperm competition: the conduit/cul-de-sac hypothesis – a route to understanding or a dead end? D.Phil. thesis, York University, UK.

Yoward P & Oxford G (1996) Single palp usage during copulation in spiders. *Newsl. Br. Arachnol. Soc.* **77:** 8–9.

Yuval B & Spielman A (1990) Sperm precedence in the deer tick *Ixodes dammini*. *Physiol. Entomol.* **15:** 123–128.

Zeh JA & Zeh DW (1994) Last-male sperm precedence breaks down when females mate with three males. *Proc. Roy. Soc. Lond. Ser. B* **257:** 287–292.

10 Sperm Competition in Insects: Mechanisms and the Potential for Selection

L. W. Simmons[1] *and M. T. Siva-Jothy*[2]

[1] Department of Zoology, The University of Western Australia, Nedlands, WA 6009, Australia; [2] Department of Animal and Plant Sciences, University of Sheffield, Sheffield S10 2UQ, UK

I. INTRODUCTION

Twenty-five years have passed since Parker (1970c) proposed that sexual selection would continue beyond the competitive struggle between males for access to females, because females often remate before sperm from their previous partners have been fully utilized. Males will thus compete for fertilizations after copulation. Parker (1970c) defined sperm competition as 'the competition within a single female between the sperm from two or more males for the fertilization of the ova' and recognized that sperm competition would generate opposing evolutionary pressures on males. On the one hand, selection should favour adaptations that enabled males to pre-empt the sperm of rivals stored inside the

Sperm Competition and Sexual Selection
ISBN 0-12-100543-7

female, while on the other, favour adaptations that enabled males to prevent remating by their mates and thereby resist sperm competition from future males.

Insects are predisposed to high levels of sperm competition because females show a propensity for multiple mating, and maintain sperm in specially adapted sperm-storage organs (usually termed spermathecae) (Parker 1970c; Ridley 1988). Parker (1970c) thus used the insects as a model to demonstrate both the occurrence of sperm competition, and its evolutionary consequences for reproductive morphology, physiology and behaviour. Subsequently, the evolutionary significance of sperm competition has been recognized in most sexually reproducing organisms, including plants (Smith 1984; Birkhead and Møller 1992; see Chapter 5). Despite the ubiquity of sperm competition and the wealth of information on its fitness consequences, the proximate mechanisms underlying sperm competition in insects, and most other taxa, are poorly understood. In this review we provide a conceptual base for approaching studies of sperm competition mechanisms in insects and illustrate why an understanding of the mechanisms involved is the key to revealing and understanding adaptations that arise under selection from sperm competition.

II. MEASUREMENT AND INTERPRETATION OF SPERM UTILIZATION PATTERNS

Sperm utilization by females after insemination by more than one male is expressed empirically as the proportion of offspring sired by the last male to mate. Because of the experimental protocol that has gained currency in this branch of science (Boorman and Parker 1976) this equates to the proportion of eggs fathered by the second male to mate in a controlled double-mating trial, hence P_2. P_2 values are often used to infer the mechanism underlying patterns of sperm utilization: intermediate values are taken as indicative of sperm mixing while high values of P_2 are taken as evidence for sperm precedence or sperm displacement. The terms sperm precedence and sperm displacement are often used synonymously, despite the fact that they imply very different mechanisms of sperm transfer, storage, and utilization. It is thus necessary to define terms explicitly.

Sperm precedence is simply the nonrandom utilization of sperm from one of several males to mate with a female. The last sperm to enter the female's sperm-storage organ(s) may be the first to leave, yielding a high value of P_2. However, sperm precedence can also occur when the sperm from the first male are the first to leave the spermatheca. In the latter case, sperm precedence would be characterized by a very low value of P_2. Precedence can result from a passive process, where sperm mixing is negligible, so that the sperm from different males become stratified within the spermatheca of the female. An extreme example of this situation may occur in insect species where the sperm from different males are delivered in 'spermatodoses' – packets of sperm bound within their own membranes – that overlay one another within the spermatheca (Boldyrev 1915). Alternatively, precedence may be achieved by an active process, as in some odonates, where the male packs sperm to the rear of the female's spermatheca before placing his own over the top (Waage 1984). Whatever the mechanism, females store all of the sperm received from different males. Consequently, sperm precedence can break down if mixing of the spermathecal contents occurs with time (Siva-Jothy and Tsubaki 1989).

In contrast, sperm displacement can involve the active removal of previously stored sperm by the copulating male, either by flushing the female's sperm-storage organ(s) with his own ejaculate (Ono *et al.* 1989), or by mechanically removing sperm prior to ejaculation (Waage 1979b). Gromko *et al.* (1984) recommended the use of the term sperm predominance because this did not infer any particular mechanism for nonrandom fertilization. However, it does assume that females retain the sperm of different males within their stores, which is not always the case. Apart from the obvious cases of sperm displacement, females may use up and/or lose sperm from their sperm-storage organ(s) prior to the second mating, resulting in a high value of P_2 after the second mating (Yamagishi *et al.* 1992). In short, inferring mechanisms of sperm competition from P_2 data can be misleading without a knowledge of the patterns of sperm transfer and storage. P_2 simply indicates the proportion of a female's eggs fertilized by each of two males after a double mating. It is not indicative of sperm displacement or any other mechanism of sperm competition. In general, a complex list of terms has built up in the sperm competition literature, terms that have rarely, if ever, been defined explicitly. Incorrect usage of these terms is rife and often leads to erroneous conclusions. We therefore offer a classification and clarification of terminology in Appendix A.

Sperm competition is a term that has, in our opinion, been frequently stretched beyond acceptable conceptual boundaries. Sperm competition is implicitly, and sometimes explicitly, linked to the value of P_2; when P_2 is high, sperm competition is said to be intense. However, P_2 will be high whenever sperm are lost or removed from the female's sperm-storage organ(s) prior to, or during the second mating. Clearly, in this situation the sperm of successive males are not present within the reproductive tract of the female and do not compete for fertilizations. Sperm competition will occur only when sperm from successive males mix within the female's sperm-storage organ(s), either immediately following mating, or after sperm precedence mechanisms have broken down. Thus, sperm competition is likely to be at its most intense when intermediate values of P_2 are observed. This is an important point because adaptations, such as increased testis size and/or ejaculate size, have been predicted to arise when sperm are in direct competition (Parker 1990a,b, 1993; Parker and Begon 1993). However, selection on increased testis size may occur for other reasons, for example if the ejaculate is used directly to displace previously stored sperm and thus avoid sperm competition. Large testes could thus be characteristic of species with intermediate values of P_2 (where sperm competition occurs) and very high values of P_2 (where there may be little or no sperm competition). Large testis size could even be characteristic of very low values of P_2, if ejaculate size functioned to prevent insemination by future males. Thus, for comparative analyses it is important to distinguish species on the basis of adaptations that have arisen to avoid or reduce sperm competition, such as mechanisms of sperm displacement and precedence, and those that have arisen to increase success in sperm competition when it occurs, such as sperm numbers, survival or motility. Values of P_2 alone provide little information on the intensity of sperm competition acting within a species because P_2 *per se* is not always indicative of the coexistence of sperm from two males within the female's sperm stores. This problem is perhaps most pertinent in the insects, where a diversity of mechanisms for the avoidance of sperm competition may confound general predictions concerning adaptive responses to sperm competition across species. Any measure of sperm competition intensity would have to take into account the degree of multiple mating by females coupled with the extent to which multiple ejaculates coexist within the females' sperm stores.

Since Parker (1970c), there have been four reviews of insect sperm competition, each attempting to draw associations between morphological or behavioural adaptations thought to be subject to selection via sperm competition, and the patterns of sperm utilization (expressed as P_2). Boorman and Parker (1976) argued that mating plugs are an adaptation for the avoidance of sperm displacement. Thus, they reviewed the literature, predicting that P_2 should be low in species with effective mating plugs. Walker (1980) attempted to establish a mechanism for sperm precedence based on the idea that the degree of sperm mixing was dependent on spermathecal morphology; species with tubular spermathecae were predicted to have high values of P_2 because of low mixing potential, while those with spherical spermathecae were predicted to have intermediate values. Gwynne (1984) examined the patterns of sperm utilization in relation to male parental investment. Because parental investment is predicted to arise only in species where confidence of paternity is high, Gwynne (1984) predicted that species with male parental investment should be characterized by high values of P_2. Each of these reviews claimed to provide evidence in favour of their relevant hypotheses. However, Ridley (1989) pointed out that none had performed appropriate comparative analyses, and tested each hypothesis using outgroup comparisons. P_2 values were higher in species without mating plugs, with tubular spermathecae, and with male investment, although none of the comparisons were significant. Ridley (1989) also tested the hypothesis that selection for sperm displacement mechanisms is predicted in polyandrous species and/or that polyandry is more likely to evolve in species with sperm displacement. In support of his hypothesis

Ridley (1989) found a significantly higher P_2 value in polyandrous species. Each of these comparative studies suffer from, among other things, a lack of understanding of the mechanisms of sperm competition and the interpretation of P_2 values. Where mating plugs function to prevent sperm pre-emption by future males, the observed values for P_2 are unlikely to be low. Rather, within any given species observed P_2 values may be either very high, when the plug's protection is breached, or very low, in cases where the plug remains intact. Thus, the average P_2 value for species with mating plugs is likely to be intermediate. Although adaptations for sperm precedence and/or displacement may generate opposing selection for effective mating plugs, this is unlikely to be reflected by the mean P_2 values across species. Walker's (1980) hypothesis of sperm precedence assumes that females store all of the sperm from successive copulations and ignores the fact that some species have evolved mechanisms of sperm displacement, irrespective of the shape of the female's spermathecae. For example, the spermathecae of some odonates are spherical yet sperm displacement by males results in high levels of P_2 (Waage 1984). Gwynne's (1984) hypothesis was too simplistic. Although we might expect parental investment to arise only in species with confidence of paternity, this may be associated either with very high values of P_2 as suggested, or with very low values of P_2. That is, males that invest parentally may do so because there is little risk that females will utilize sperm from future males. The expected pattern of sperm utilization in parentally investing species will depend on additional reproductive variables, such as the nature of male investment, the interval between matings, or the time taken for females to utilize male contributions. Further, Ridley's (1989) comparative analysis included male investments such as postcopulatory guarding that are clearly not parental investment. Like mating plugs, mate-guarding strategies are counter-adaptations to sperm pre-emption by future males. High values of P_2 might be predicted for species with mate guarding (but see below), although they can occur for a variety of reasons which have already been discussed. Combining predictions based on adaptations for the avoidance of sperm pre-emption with predictions based on parental investment theory must confound Ridley's (1989) analysis. Finally, although Ridley (1989) found high P_2 values in polyandrous species, this may have little to do with sperm displacement. High P_2 values are expected in species where sperm are lost and/or used up from the female's sperm-storage organ(s) prior to the second copulation. Polyandry is also predicted in these species in order for females to maintain their fertility. Thus, Ridley's (1989) result can be explained without the involvement of sperm competition. Indeed, Ridley (1988) had previously shown that mating frequency in insects was associated with the maintenance of fertility, evidence that is counter to his sperm displacement hypothesis. Clearly, while P_2 values may be useful indicators of the patterns of sperm utilization their use for predicting selection arising from sperm competition should be made with extreme caution.

III. VARIATION IN SPERM UTILIZATION PATTERNS

An extraordinary number of insect studies have examined the patterns of sperm utilization following double matings, reporting the species-specific mean P_2 value. However, Lewis and Austad (1990) drew attention to the fact that almost all studies have revealed considerable variation about the mean value of P_2. Despite its ubiquity, intraspecific variation in P_2 has been largely ignored. Yet variation in P_2 can provide greater insight into the mechanisms involved in sperm utilization than the mean value itself.

Table 10.1 contains the data on the mean and variance in P_2 values for all available studies of sperm utilization in insects. Many studies provide some estimate of the variance about the mean P_2, either in the form of a standard deviation or standard error, and/or simply as maximum and minimum values. Others provide the raw data. We have given the full range of P_2 values for each species (*sensu* Lewis and Austad 1990) and, where sufficient information was available, we have given the standard deviation in P_2 as a standard measure of variance. The standard deviation is a more realistic measure of variability than the range because single values of 0 or 1 may often represent outliers in an otherwise invariant distribution. For example, although P_2 has a full range of 1.00 in the dragon fly *Sympetrum danae* the variance is only 0.05 compared with the fruitfly *Drosophila hydei*, where the variance is an order of magnitude higher even though the range in P_2 is only 0.75 (Table 10.1).

Two main techniques have been used to measure sperm utilization in insects. The irradiated male technique relies on the induction of chromosomal abnormalities in sperm by exposing males to sublethal doses of γ- or X-irradiation. Eggs fertilized by irradiated sperm suffer early embryonic mortality, making the proportion of eggs fertilized by an irradiated male competing with a normal male easy to determine from the proportion of eggs that hatch. Genetic markers, such as colour variants or enzyme polymorphisms with known mechanisms of inheritance, have also been used in sperm competition studies. For both techniques, reciprocal crosses are often performed to control for any differences in the competitive ability of sperm from the males in each treatment. Where two estimates for the variance in P_2 were available from reciprocal crosses, we calculated the mean variance from the two values. It should be noted that different techniques will have different sampling errors and some of the variation in P_2 may come from the technique adopted. Many studies also provide a range of mean values for P_2 and a corresponding range of variances, determined under different experimental regimes. These are provided, together with information on some of the sources of variation in P_2 where they have been identified.

The interspecific distribution of mean P_2 values in Fig. 10.1a shows, as is commonly claimed, that in general the last male to mate with a female tends to gain higher paternity than the first; the distribution is skewed to

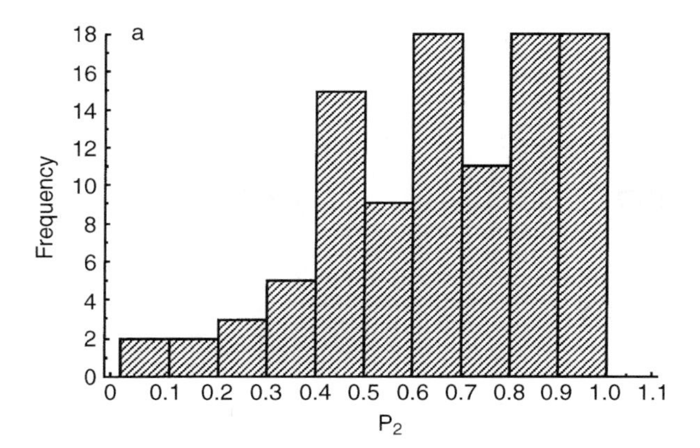

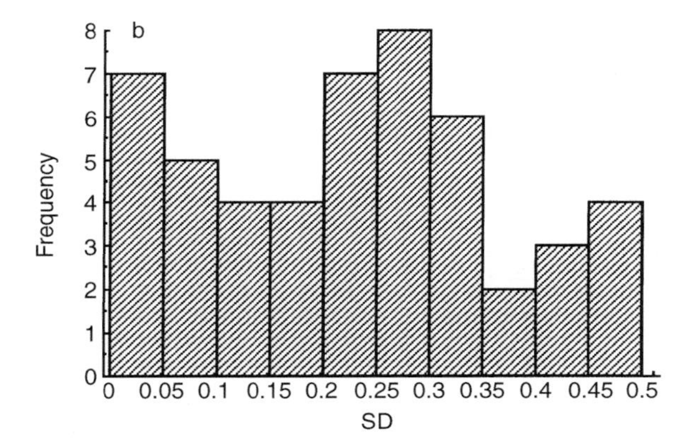

Fig. 10.1. *Interspecific (a) and intraspecific (b) variation in the proportion of offspring sired by the second male to mate, P_2. In general, mixed sperm utilization and/or varying levels of second male priority are the most common patterns observed, while almost all species show moderate to high variability in P_2 (data from Table 10.1).*

the right. Nevertheless, a considerable proportion of species show mixed paternity to varying degrees. In reality, less than half (45%) of the species studied show appreciable second male advantages at fertilization ($P_2 > 0.75$) while only 17% show near complete second-male paternity ($P_2 > 0.90$). Cases of first male advantage are comparatively rare. The data in Fig. 10.1b shows that high variance in P_2 is the norm rather than the exception. It is clear from Table 10.1 that the intraspecific variance in P_2 is as large as the interspecific variance; across 109 species from 10 orders of insect, the standard deviation in P_2 is 0.23, while the mean intraspecific standard deviation is 0.23 ± 0.14 ($n = 52$ species). One intuitive pattern to emerge from the data is that, in general, variation in P_2 is negatively associated with the species-specific mean; species with strong second male paternity have a low variance in paternity

Table 10.1. Patterns of sperm utilization seen in insects, expressed as the proportion of offspring fathered by the second male to mate, P_2, together with estimates of variation in sperm utilization in the form of the range and/or standard deviation (SD) in P_2.

Species	Mean P_2	Range	SD	Source of variance
Odonata				
Calopteryx maculata[1]	0.90^s			
Calopteryx dimidiata[2]	0.98^s			
Calopteryx spendens[3]	0.98			
Ischnura ramburi[4]	0.82^s			
Ischnura graellsi[5]	0.99	0.92–1.00	0.03	Time since final copulation (−)
Ischnura elegans[6]	0.82	0.44–1.00	0.19	
Ischnura senegalensis[7]	1.00^s			
Enellagma hageni[8]	0.78	0.44–0.95	0.21	
Erythemis simplicollis[9]	0.99–0.65		0.00–0.13	
Lestes vigilax[10]	0.75^s			
Argia moesta[4]	0.93^s			
Argia sedula[4]	0.71^s			
Leucorrhina intacta[11]	0.90	0.00–1.00	0.21	Copula duration of second male (+asymptotic)
Nanophya pygmaea[12]	0.98–0.72		0.02–0.10	Time since final copulation (−)
Sympetrum danae[13]	0.96	0.00–1.00	0.05	Copulation duration of second male (+asymptotic)
Anax parthenope[14]	1.00		0.00	
Orthetrum coerulescens[14]	1.00–0.77	0.19–1.00	0.00–0.34	Copulation duration of second male (+); time since final copulation (−)
Mnais pruinosa pruinosa[15]	1.00–0.50		0.00–0.40	Time since final copulation (−); copula duration of second male (+)
Orthoptera				
Gryllus bimaculatus[16]	0.33–0.45–0.68	0.00–0.80	0.06–0.03–0.06	Relative spermatophore attachment times; relative number of copulations
Gryllus integer[17]	0.72	0.08–1.00		
Grylloides sigillatus[18]	0.42	0.04–0.88	0.20	Relative spermatophore attachment times

Allenomobius fasciatus[19]	0.62			
Allenomobius socius[19]	0.43			
Truljalia hibinonis[20]	0.88[s]		0.12	
Decticus verrucivorous[21]	0.50	0.03–1.00	0.31	Relative spermatophore size
Requena verticalis[22]	0.00–0.19	0.00–0.94	0.17	Spermatophore size of first male (−); remating interval (+)
Kawanaphila nartee[23]	0.69	0.13–1.00	0.33	
Poecilimon veluchianus[24]	0.90	0.87–0.93	0.03	
Metaplastes ornatus[25]	0.85[s]			
Schistocerca gregaria[26]	1.00			
Locusta migratoria[27]	0.38–0.86	0.00–1.00		Remating interval (+)
Paratettix texanus[28]	0.58	0.00–0.91	0.37	
Podisma pedestris[29]	0.39	0.00–1.00		Homogamy
Eyprepocnemis plorans[30]	0.92	0.40–1.00		Relative number of copulations
Chorthippus parallelus[31]	0.60	0.35–0.84		Homogamy
Blattodea				
Blatella germanica[32]	<0.43[a]	0.00–1.00		
Diploptera punctata[33]	0.68	0.00–1.00	0.28	
Phasmatoidea				
Extatosoma tiaratum[34]	0.98			
Baculum sp. 1[35]	0.99	0.80–1.00	0.04	
Hemiptera				
Abedus herberti[36]	0.99	0.97–1.00	0.01	
Oncopeltus fasciatus[37]	0.50	0.03–0.63	0.23	
Dysdercus koenigii[38]	0.66			
Nezara viridula[39]	0.51	0.00–1.00	0.31	Size of first male (−); remating interval (+)
Gerris lateralis[40]	0.81			
Gerris remigis[41]	0.65	0.28–1.00		Relative copula duration
Neacoryphus bicrucis[42]	0.78		0.01	Copula duration of second male (+asymptotic)

Continued

Table 10.1. Continued

Species	Mean P_2	Range	SD	Source of variance
Jadera haematoloma[43]	0.62	0.05–0.95	0.30	
Lygaeus equestris[44]	0.92			
Coleoptera				
Popillia japonica[45]	0.85			
Epilachna varivestis[46]	0.70			
Harmonia axyridis[47]	0.55	0.16–1.00	0.53	Copula duration of second male (+)
Henosepilachna pustulosa[48]	0.60	0.00–0.97	0.31	
Tribolium confusum[49]	0.82			
Tribolium castaneum[50]	0.62	0.40–0.86	0.27	Relative male size; females; time since final copulation[51] (−); homogamy[52]; parasites[53]
Trogoderma inclusum[54]	0.52			
Lasioderma serricorne[55]	0.84			
Tetraopes tetraophthalmus[56]	0.72	0.33–1.00	0.22	
Labidomera clivicollis[57]	0.65–0.78–0.87	0.29–1.00	0.27–0.32–0.26	Copula duration of second male (+asymptotic)
Anthonomus grandis[58]	0.52–0.90	0.10–0.90		Remating interval (+)
Conotrachelus nenuphar[59]	0.50	0.02–0.90	0.28	
Leptinotarsa decemlineata[60]	0.32–0.53			Relative copulation frequency
Necrophorus vespilloides[61]	0.11–0.92	0.00–1.00		Number of copulations by second male (+)
Necrophorus orbicollis[62]	0.94			
Callosobruchus maculatus[63]	0.83	0.55–1.00	0.34	Number of sperm transferred by second male (+)
Adelia bipunctata[64]	0.60	0.00–1.00		
Onymacris unguicularis[65]	0.82			
Tenebrio molitor[66]	0.91			Time since final copulation (−)
Chelymorpha alternans[67]	0.50			Copula duration (+); genital flagellum length (+)
Aleochara curtula[68]	c. 1.0[s]			

Mecoptera				
Panorpa vulgaris[69]	0.46	0.02–0.95	0.27	Relative copula duration
Diptera				
Aedes aegypti[70]	0.15	0.08–0.83		
Anopheles gambiae[71]	0.02			
Culex pipiens[72]	0.11–1.00			Remating interval (+)
Culicoides mellitus[73]	0.29			
Scatophaga stercoraria[74]	0.88	0.02–1.00	0.26	Copula duration of second male (+asymptotic), size of second male (+), male nutrient reserves
Drosophila melanogaster[75]	0.93	0.31–1.00	0.13	Number of sperm transferred by second male (+)[76], resistance by first male (−), genetic[77], remating interval (+)
Drosophila pseudoobscura[78]	0.82	0.00–1.00	0.07	Resistance by first male (−)
Drosophila mojavensis[79]	0.66	0.52–0.81	0.09	Remating interval (+); time since final copulation (−)
Drosophila teissieri[80]	0.77	0.30–1.00	0.18	Homogamy; sperm polymorphism
Drosophila hydei[81]	0.48	0.25–1.00	0.28	Remating interval (+)
Glossina austeni[82]	0.29			
Glossina morsitans[83]	0.45–0.11			Remating interval (−)
Ceratitis capitata[84]	0.68	0.16–0.99	0.27	
Rhagoletis pomenella[85]	0.83	0.31–1.00	0.20	
Dacus oleae[86]	0.46–0.52			Age at irradiation
Dacus cucurbitae[87]	0.42–0.74	0.35–1.00	0.19	Relative copula duration, remating interval (+)
Dryomyza anilis[88]	0.33 0.18–0.75 0.45–0.85	0.05–0.73	0.17	Size of second male (+); eggs laid (−); mating situation; number of tapping sequences (asymptotic +); time since final copulation (−); number of copulations by second male (+)

Continued

Table 10.1. Continued

Species	Mean P_2	Range	SD	Source of variance
Hymenoptera				
Dahlbominus fuscipennis[89]	0.32	0.20–0.76	0.11	Time since final copulation (+)
Nasonia vitripennis[90]	0.67	0.09–0.99	0.35	
Apis mellifera[91]	0.50			
Diachasmimorpha longicaudata[92]	0.49			Remating interval (+)
Aphytis melinus[93]	0.14	0.00–0.50		Remating interval (−)
Lepidoptera				
Choristoneura fumiferana[94]	0.46	0.00–1.00[b]	0.48	Remating interval (−)
Carpocapsa pomonella[95]	0.56			
Plodia interpunctella[96]	0.57–0.82	0.00–1.00[b]	0.55–0.29	Relative spermatophore size
Papilio dardanus[97]	0.86			
Colias erytheme[98]	1.00		0.00	
Danaus plexippus[99]	0.67	0.00–1.00	0.30	Male size (+); male mating history (+); time since final copulation (+/−)
Euphydryas editha[100]	0.72	0.00–1.00[b]	0.44	
Bombyx mori[101]	0.95–0.06	0.00–1.00[b]	0.05–0.05	Remating interval (−)
Heliothis virescens[102]	0.47	0.00–1.00[b]		
Trichoplusia ni[103]	0.92			
Spodoptera frugiperda[104]	0.54	0.00–1.00[b]	0.44	
Spodoptera litura[105]	1.00[s]			
Pseudoplusia includens[106]	0.27	0.00–1.00[b]	0.43	
Helicoverpa zea[107]	0.71	0.00–1.00		
Utetheisa ornatrix[108]	0.52	0.00–1.00[b]	0.45	Relative spermatophore size
Pseudolutia unipuncta[109]	0.47	0.00–1.00[b]		
Pseudaletia separata[110]	0.83			

[a] Value calculated from mixed broods only; 22% of females had mixed broods, 78% of females had $P_2 = 0$.
[s] Values based on direct observation of sperm removal and the assumption of random mixing of remaining sperm.
[b] Bimodally distributed with modes at zero and one.
+/− Indicate direction of change in P_2 value with described variable.
Note: although many authors choose to exclude values of $P_2 = 0$ in their analyses, they are included in the calculation of means and variances here since they are genuine observations of P_2 after double matings.

Notes to Table 10.1
[1] Waage 1979b; [2] Waage 1984; [3] Hooper and Siva-Jothy 1996; [4] Waage 1986; [5] Cordero and Miller 1992; [6] Cooper 1995; [7] Sawada 1995; [8] Fincke 1984; [9] McVey and Smittle 1984; [10] Waage 1982; [11] Wolf *et al.* 1989; [12] Siva-Jothy and Tsubaki 1994; [13] Michiels 1992; [14] Hadrys *et al.* 1993; [15] Siva-Jothy and Tsubaki 1989; [16] Simmons 1987b; [17] Backus and Cade 1986; [18] Sakaluk and Eggert 1996; [19] Gregory and Howard 1994; [20] Ono *et al.* 1989; [21] Wedell 1991; [22] Gwynne and Snedden 1995; [23] Simmons 1995b; [24] Achmann *et al.* 1992; [25] Helversen and Helversen 1991; [26] Hunter-Jones 1960; [27] Parker and Smith 1975; [28] Nabours 1927; [29] Hewitt *et al.* 1989; [30] Lopez-Leon *et al.* 1993; [31] Bella *et al.* 1992; [32] Cochran 1979; [33] Woodhead 1985; [34] Carlberg 1987b; [35] Carlberg 1987a; [36] Smith 1979; [37] Economorpoulos and Gordon 1972; [38] Harwalker and Rahalkar 1973; [39] McLain 1985; [40] Arnqvist 1988; [41] Rubenstein 1989; [42] McLain 1989; [43] Carroll 1991; [44] Sillen-Tullberg 1981; [45] Ladd 1966; [46] Webb and Smith 1968; [47] Ueno 1994; [48] Nakano 1985; [49] Vardell and Brower 1978; [50] Lewis and Austad 1990; [51] Schlager 1960; [52] Robinson *et al.* 1994; [53] Yan and Stevens 1995; [54] Vick *et al.* 1972; [55] Coffelt 1975; [56] McCauley and Reilly 1984; [57] Dickinson 1986, 1988; [58] Bartlett *et al.* 1968; [59] Huettel *et al.* 1972; [60] Boiteau 1988; [61] Muller and Eggert 1989; [62] Trumbo and Fiore 1991; [63] Eady 1994a; [64] Jong *et al.* 1993; [65] De Villiers and Hanrahan 1991; [66] Siva-Jothy *et al.* 1996; [67] V. Rodriguez, W. G. Eberhard and D. Windsor unpublished data; [68] Gack and Peschke 1994; [69] Thornhill and Sauer 1991; [70] George 1967; [71] Bryan 1968; [72] Bullini *et al.* 1976; [73] Linley 1975; [74] Parker 1970d, Simmons and Parker 1992; [75] Gromko *et al.* 1984; [76] Letsinger and Gromko 1985; [77] Clark *et al.* 1995; [78] Turner and Anderson 1984; [79] Markow 1988; [80] Joly *et al.* 1991; [81] Markow 1985; [82] Curtis 1968; [83] Dame and Ford 1968; [84] Saul *et al.* 1988, Saul and McCombs 1993; [85] Opp *et al.* 1990; [86] Cavalloro and Delrio 1974; [87] Tsubaki and Sokei 1988, Yamagishi *et al.* 1992; [88] Otronen 1990, 1994a,b; [89] Wilkes 1966; [90] Beukeboom 1994; [91] Page and Metcalf 1982; [92] Martinez *et al.* 1993; [93] Allen *et al.* 1994; [94] Retnakaran 1974; [95] Proverbs and Newton 1962; [96] Cook 1996; [97] Clarke and Shepard 1962; [98] Boggs and Watt 1981; [99] K. Oberhauser, R. Hampton, B. Jenson and S. Weisberg unpublished data; [100] Labine 1966; [101] Omura 1939, Suzuki *et al.* 1996; [102] Flint and Kressin 1968; [103] North and Holt 1968; [104] Snow *et al.* 1970; [105] Etman and Hooper 1979; [106] Mason and Pashley 1991; [107] Carpenter 1992; [108] LaMunyon and Eisner 1993, 1994; [109] Svard and McNeil 1994; [110] He *et al.* 1995.

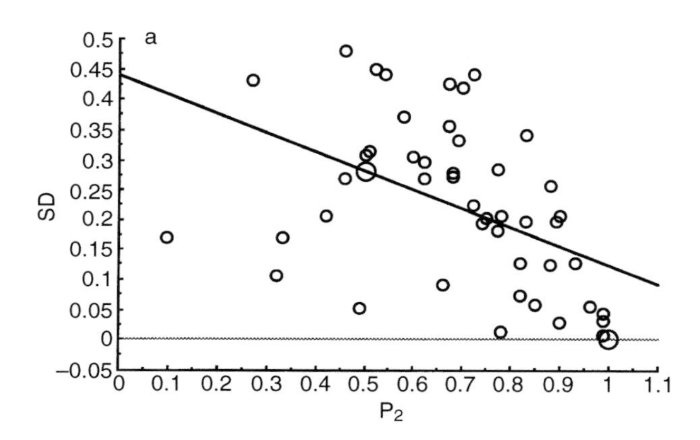

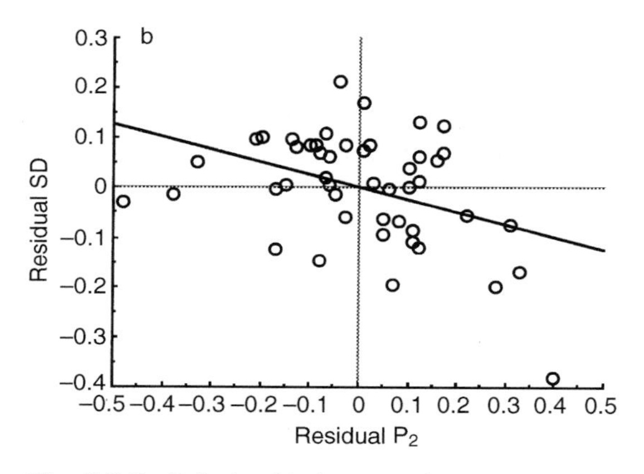

Fig. 10.2. *Relationship between the mean species value of P_2 and the intraspecific variability expressed as the standard deviation in P_2 (a) across species, $r^2 = 0.237$, $F_{(1,48)} = 14.92$, $P < 0.001$, and (b) after phylogenetic subtraction using order means, $r^2 = 0.159$, $F_{(1,45)} = 8.51$, $P = 0.006$.*

outcome (Fig. 10.2a). This association may be confounded by common ancestry within the insects because it is apparent when analysed using the mean P_2 values and standard deviations in P_2 at the Order level ($r^2 = 0.522$, $F_{(1,9)} = 8.73$, $P = 0.018$). We therefore performed a comparative analysis using Stearns (1983) method of phylogenetic subtraction. We calculated the mean P_2 and standard deviation in P_2 for each Order and subtracted these values from each species value within each Order. The relationship between P_2 and the variance in P_2 remained (Fig. 10.2b). Furthermore, even within species, when P_2 is low, variance is high but when P_2 is high, variance is low (Table 10.1). Because of the mathematical properties of P_2 data, values are restricted between 0 and 1, extreme values of P_2 might be expected to have restricted variances so that the relationship observed may be a mathematical, rather than a biological one. However, identical results are obtained using the range in observed P_2 values, an estimate of the variance in P_2 that is not restricted by the magnitude of the mean.

A. Variance and mechanisms

A low variance in P_2 might be expected whenever the second male to mate pre-empts the sperm from previous males through sperm displacement or precedence, or when the female's sperm-storage organ(s) are depleted prior to the second mating. Thus, in general, the sperm displacement mechanisms of odonates result in both a high value of P_2 and a low variance. A moderate variance in P_2 might be expected whenever sperm from both males are present within the female's spermatheca. However, the magnitude of variation will depend on the degree of sperm mixing; as mixing becomes more homogeneous so the variation in P_2 should increase. In dragonflies where the male packs sperm to the rear of the female's spermathecae an initially high P_2 with low variance is achieved. However, as sperm mixing proceeds, P_2 declines and the variation in P_2 increases (McVey and Smittle 1984; Siva-Jothy and Tsubaki 1989, 1994; Hadrys *et al.* 1993). The occurrence of homogeneous mixing of ejaculates has been established in gryllids, *Gryllus bimaculatus* and *Grylloides supplicans,* which also have moderate variation in P_2 (Table 10.1). Finally, high levels of variation in P_2 might be expected where first males install a mating plug to guard against sperm preemption by future males. Some second males will succeed in circumventing the plugs of first males while others will not, leading to high variation in P_2. In Parker and Smith's (1975) study of *Locusta migratorioides,* the variation in P_2 was greatly reduced when oviposition occurred between rematings. Prior to oviposition the first male's spermatophore remains in the genital tract and appears to act as a mating plug; double matings prior to oviposition were characterized by P_2 values that were either very high or very low (Parker and Smith 1975). In general, high variance in P_2 appears to be a characteristic of species that show, on average, very low levels of second-male sperm utilization; high variance may result from the breakdown of mechanisms that prevent sperm from the second male entering the sperm-storage organ(s) (see below).

The distributions of P_2 may similarly provide insight into potential mechanisms of sperm utilization. One striking consistency seen in the Lepidoptera are distributions that are bimodal, the two modes occurring at 0 and 1 (Table 10.1). Lepidopteran spermatophores are transferred to the bursa copulatrix where the spermatophore tube must be aligned with the ductus seminalis to enable sperm to migrate to the spermatheca (Drummond 1984). Spermatophores remain in the bursa copulatrix of the female and, as in *L. migratorioides,* have the potential to act as mating plugs. Second males must dislodge the previous spermatophore and align their own for successful insemination. Spermatophore alignment is obviously a difficult process – failed insemination occurs in 5–30% of matings with virgin females (Drummond 1984). Thus, alignment in the presence of another spermatophore is likely to be even more difficult. Indeed, in her study of *Euphydryas editha*, Labine (1966) noted that in the case where P_2 was zero, the first spermatophore had prevented proper orientation of the second so that second-male sperm remained in the bursa and never reached the spermatheca. Brower (1975) made similar observations with *Plodia interpunctella*. In LaMunyon and Eisner's (1993) study of *Utetheisa ornatrix*, second males that were very much larger than the first fell into the mode where P_2 approached 1, while second males that were smaller than the first fell into the mode where P_2 approached 0. Since larger males produce larger spermatophores (LaMunyon and Eisner 1994), this pattern could reflect the relative efficiencies of large and small spermatophores as mating plugs. Manipulation of spermatophore size by varying male mating history has since shown that males transferring large spermatophores relative to the first are more likely to achieve high paternity than those transferring relatively small spermatophores (LaMunyon and Eisner 1994). Retnakaran (1974) offered an alternative hypothesis to explain the apparent 'all-or-nothing' sperm utilization patterns seen in Lepidoptera. He suggested that when the female's sperm-storage organ(s) were filled by the first ejaculate, there would be no room for sperm from the second spermatophore: P_2 would consequently be zero. Because sperm migration from the spermatophore to the spermatheca occurs over an extended period of time, a short remating interval could result in the first spermatophore being displaced before sperm have had the opportunity to migrate to the sperm-storage organ(s), resulting in almost complete paternity for the second male. Intermediate values of P_2 would result when the sperm-storage organs were only partially filled by the first spermatophore, leaving room for sperm from the second male. The magnitude of P_2 would depend on the time the first male's spermatophore remained in place prior to the second copulation. The proposed mechanism relies on the sperm-storage organ(s) having a fixed storage capacity. In support of his hypothesis, Retnakaran (1974) found that increasing the interval between copulations resulted in complete first male paternity, suggesting (1) that the second male's sperm were unable to enter the spermatheca and (2) that spermathecal filling may be a mechanism by which males avoid sperm competition from future males. Similarly, in their study of *Bombyx mori* Suzuki *et al.* (1996) found complete second-male paternity when second copulations followed immediately after first copulations, and complete first-male paternity when copulations occurred 2 h later. Moreover, they examined sperm movement and found that the spermatheca did have a finite sperm-carrying capacity that was reached after a single copulation, and that the immediate placement of a second spermatophore in the bursa copulatrix hindered sperm migration from the first spermatophore, supporting Retnakaran's (1974) hypothesis. Nevertheless, variation in the time available for sperm movement from the spermatheca to the fertilization canal may also be important (Suzuki *et al.* 1996). The effect of spermatophore size noted by LaMunyon and Eisner (1994) could also result from a fixed sperm-storage capacity. Partial filling of the sperm-storage organs by a small first spermatophore could allow the second male's sperm to enter the spermatheca and gain precedence over the first male's sperm. In contrast, a large first spermatophore would fill the spermatheca and thereby not allow sperm from a second spermatophore to enter the storage organ. Using the moth *Plodia interpunctella*, Cook (1996) has recently shown that when two small ejaculates (spermatophores) are received by the female, P_2 approximates to 0.5. However, when a large second ejaculate competes with a small first ejaculate, P_2 is increased to 0.83, supporting the notion that relative sperm numbers may be an important component of sperm utilization patterns in Lepidoptera. Because larger spermatophores contain more sperm, LaMunyon and Eisner's (1994) result could equally be due to the numerical superiority of sperm from males producing larger spermatophores, even in the absence of a limit to the female's sperm-storage capacity. Finally, a bimodal distribution of P_2 values might be expected because of the high insemination failure rates seen in Lepidoptera (Drummond 1984), irrespective of the mechanisms of sperm transfer and utilization.

Spermathecal filling as a mechanism for paternity assurance is suggested by the work of Gwynne and Snedden (1995). Male *Requena verticalis* donate nutrient gifts to females at mating that serve as male parental investment (Gwynne 1988a). Males have a high confidence of paternity since P_2 approaches zero (Gwynne 1988b). However, Gwynne and Snedden (1995) found that by increasing the interval between matings, P_2 was increased to 20%. Further, P_2 was negatively related to the size of the first male's spermatophore. These data are consistent with the spermathecal filling hypothesis for the avoidance of sperm competition. First, large spermatophores contain more sperm and should therefore represent a more effective block to future sperm, and second, as females utilize first-male sperm for fertilization, room becomes available for the second male's sperm resulting in a deterioration of the first male's paternity assurance mechanism and an increased variance in P_2. By dissecting female *Dahlbominus fuscipennis* after copulation, Wilkes (1966) demonstrated that female sperm-storage organs reached 70% of their storage capacity during the first mating and become completely filled during the second mating. The fact that P_2 in this species is 0.30 provides strong support for the spermathecal filling hypothesis for paternity assurance.

B. Determinants of variation

It is clear from Table 10.1 that an important parameter generating variation in sperm utilization patterns is the relative number of sperm stored from different males at the time of fertilization. Information on variation in P_2 comes from 41 studies of insects from eight orders. Of these, 15 species show an effect of remating interval on P_2. With just four exceptions, the longer the interval between successive matings the greater the P_2. This is consistent with the notion that sperm from the first mating are used up by the female or lost from her sperm-storage organ(s) such that after the second mating, a decreasing proportion of sperm in the female's sperm-storage organ(s) will have been derived from the first male. Several studies provide detailed evidence for the effect of sperm longevity on P_2. When female melon flies (*Dacus cucurbitae*) mate with a second male within 24 h, P_2 does not differ significantly from 0.5, indicating that sperm mix randomly in storage (Yamagishi *et al.* 1992). However, there is an exponential loss of sperm from the spermathecae with time (Tsubaki and Yamagishi 1991), so that P_2 increases to 1.0 as the interval between matings is increased (Yamagishi *et al.* 1992). The observed variation in P_2 fits a model in which the mechanism of sperm utilization involves random mixing of ejaculates, with the size of the first male's ejaculate decreasing with time in storage. Thus, high values of P_2 in this species do not arise from sperm precedence or displacement. Rather, they reflect an absence of sperm from the first male. The depletion of stored sperm has also been implicated as an important determinant of sperm utilization patterns in *Drosophila melanogaster* (Newport and Gromko 1984; Letsinger and Gromko 1985). The four exceptions to the general pattern are informative because the opposite, negative relationship between remating interval and P_2, are associated with strong first-male advantages at fertilization. Allen *et al.*'s (1994) study of the parasitoid wasp *Aphytis melinus* showed that the duration of postcopulatory mate guarding by first males increased their expectation of paternity, or decreased the expectation of second males. Allen *et al.*'s (1994) interpretation was similar to that offered by Retnakaran (1974). If females have a fixed sperm-storage capacity, then males able to delay remating until their ejaculate has been fully stored will avoid sperm competition from future males. Conversely, males that can dislodge first males before sperm storage is complete will achieve some fertilization success. Spermathecal filling during the first copulation seems a reasonable explanation for the negative effect of remating interval on P_2 in the short term. However, increases in remating intervals over a greater time scale may show the reverse pattern because the sperm from a previously filled spermatheca will be depleted, making room for additional sperm from second males (see above).

Another temporal factor that generates variance in P_2 is the period over which sperm are stored and utilized. In general, mechanisms of sperm precedence result in initially high values of P_2 that decline as sperm mix while in the female's sperm-storage organs (Schlager 1960; Siva-Jothy and Tsubaki 1989, 1994) or initially low values of P_2 that increase as sperm mix (Wilkes 1966; Ueno and Ito 1992).

A second factor generating variation in P_2 that can be seen from the data in Table 10.1 is the number of sperm transferred. There are 23 cases in Table 10.1 where either the number of sperm transferred, or a variable associated with sperm transfer (such as copulation duration, copulation frequency, spermatophore size and attachment time) affects P_2. There are two ways in which the amount of sperm transferred can influence P_2. If sperm mix randomly in storage and there is no sperm displacement, there will be an advantage to the male that has the greatest proportional representation of sperm in the female's sperm-storage organ(s) at the time of fertilization. Thus, there are eight cases where the numbers of sperm transferred, duration of copula, spermatophore size or attachment durations of second males relative to first males have a positive linear effect on P_2. The mean value of P_2 across these eight species is 0.57 ± 0.07, suggesting that a mechanism of random sperm mixing generates a selective advantage on males that can transfer greater quantities of sperm.

The second way in which the amount of sperm transferred can influence P_2 is where a mechanism of sperm displacement involves a fixed sperm-storage capacity in females and the use of the second male's own ejaculate to flush out sperm stored from the first mating. Assuming that the female's sperm-storage organ(s) are filled by the first male, P_2 should be dependent only on the amount of ejaculate transferred by the second male and further, should increase asymptotically with the amount of sperm transferred because an increasing proportion of the second male's own ejaculate will be displaced as insemination proceeds (see below). There are five cases in which the copula duration of the second male has a positive and asymptotic effect on P_2, indicative of sperm displacement by flushing. The mean P_2 across these species is correspondingly high (0.86 ± 0.04).

Recent work with *D. melanogaster* has revealed considerable genetic variation in the ability of males both to achieve high values of P_2, termed sperm offence, and to withstand the sperm displacement attempts of further males, termed sperm defence (Clark *et al.* 1995). Interestingly, the two processes are not correlated, suggesting that they are achieved by different mechanisms. Clark *et al.* (1995) were able to identify four candidate genes associated with accessory gland protein loci that encode proteins transmitted to the female during insemination. These genes were significantly associated with a male's ability to resist sperm displacement. However, a total of seven accessory gland protein genes were implicated in the sperm competition mechanism of *Drosophila* and it remains to be established which, if any, are responsible for sperm offence. That these genes respond to selection is clear from the work of Service and Fales (1993) who found that selection for delayed senescence showed a correlated response in the ability of males to withstand sperm displacement; males from selected lines were superior to males from control, rapid senescence, lines. There were no correlated responses in sperm offence, which supports the finding of Clark *et al.* (1995) that defensive and offensive mechanisms of sperm competition differ. The ability to resist sperm displacement also accounts for variation in P_2 in *Drosophila pseudoobscura* (Turner and Anderson 1984).

Finally, five studies have shown an effect of male body size on P_2. In the case of *Tribolium castaneum*, the size of the second male relative to the first has a positive effect on P_2, which could be explained by, among other things, a relationship between body size and the relative numbers of sperm transferred (Table 10.1). The mean and variance in P_2 suggest that sperm mix randomly in storage so that larger males might be expected to gain more fertilizations if they transferred more sperm. The rate of sperm transfer is related to male size in *Scatophaga stercoraria* such that large second males have a higher rate of fertilization gain, P_2, during copulation (Simmons and Parker 1992; Parker and Simmons 1994). Otronen (1994a) and K. Oberhauser, R. Hampton, B. Jenson and S. Weisberg (unpublished data) similarly found that P_2 increased with the size of the second male. McLain (1985) found that it was the size of the first male that was related to P_2, the larger the first male the lower the value of P_2. This pattern is reminiscent of that seen in *Requena verticalis* where the size of the first spermatophore adversely affects P_2 (Gwynne and Snedden 1995). McLain (1985) also found that P_2 increased with the remating interval. Together, these patterns again suggest a mechanism of spermathecal filling, with large first males being able to fill the female's spermatheca to a greater extent than small males, leaving less room for second male sperm immediately after copulation. Utilization of sperm between copulations increases the storage space available for second males. Consistent with spermathecal filling by the first male was the result of Itou (1992) who obtained a mean value of P_2 for *Nezara viridula* of just 0.15. However, Itou (1992) found that it was the size of the second male, rather than the first, that explained a significant proportion of the variance in P_2.

IV. MECHANISMS OF SPERM COMPETITION

As with most studies of sperm competition, the preceding sections rely heavily on verbal arguments for mechanisms that may generate the observed sperm utilization patterns. In general, few studies have examined in detail the mechanisms of sperm competition. The exceptions are those species which practice sperm displacement. Waage's (1979b) early work with odonates provided the first documentation of sperm displacement. The penis of the male damselfly is covered with proximally oriented spines that entrap stored sperm and remove them from the female's sperm-storage organ(s) (Fig. 10.3a). Further, the shape of the penis appears precisely matched to the structure of the female's bursa copulatrix and, at least in some species, the spermatheca, so that the male gains almost complete access to the female's sperm-storage organ(s). Estimates of sperm volume at various stages during copulation have shown that up

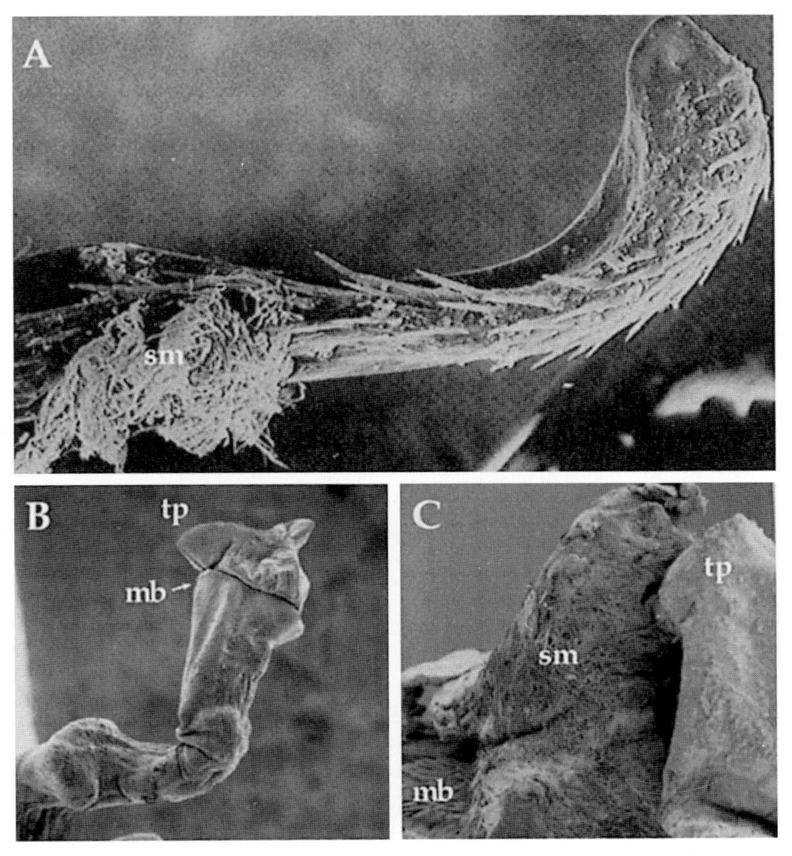

Fig. 10.3. *Adaptation in male genitalia for removing sperm from the female's reproductive tract. (a) Scanning electron micrograph of the penis of* Calopteryx maculata, *showing one of the lateral horns which enter the spermathecae and remove sperm (sm), seen trapped under the spines on the horn (from Waage 1979b, with permisssion. Copyright 1979 by the American Association for the Advancement of Science). (b) Scanning electron micrograph of the penis of* Psacothea hilaris *showing the triangular projection (tp) and microbristles (mb) used to remove sperm and (c) the removed sperm mass (sm) attached to the penis (micrographs supplied by Naoto Yokoi).*

to 99% of the female's sperm stores are removed before the male delivers his own ejaculate. Studies of sperm utilization show high values of P_2 because first male sperm are unavailable for competition (Table 10.1).

Recently, two species of Orthoptera and two species of Coleoptera have also been shown to exercise novel mechanisms of sperm displacement. In the bushcricket *Metaplastes ornatus* the male everts the females reproductive tract using a modified subgenital plate and the female consumes any stored sperm. In this way 85% of stored sperm are removed before the male transfers his spermatophore (Helversen and Helversen 1991). In tree cricket *Truljalia hibinonis* the male's penis enters the spermatheca and sperm are ejaculated into its anterior region, forcing previously stored sperm posteriorly onto the shaft of the penis from which they are consumed after copulation. Using males with dyed semen Ono *et al.* (1989) determined that 88% of the previous male's ejaculate was displaced from the female's spermatheca during copulation. Like odonates, the penis of the longicorn beetle *Psacothea hilaris* has adaptations for sperm removal (Yokoi 1990) (Fig. 10.3b,c). Male rove beetles *Aleochara curtula* use the spermatophore, rather than the penis to displace stored sperm (Gack and Peschke 1994). The spermatophore is positioned at the anterior end of the spermatheca and expands, forcing previously stored sperm from the sperm-storage organ(s) before delivering its own contents.

What the above studies have in common are detailed quantification of the processes of sperm transfer and storage. This information is essential for an understanding of the mechanisms of sperm competition. Estimates of P_2 can only support proposed mechanisms of sperm competition that are based on observations of sperm transfer and storage. They cannot be used to predict mechanisms of sperm competition without such information (see above). For example, in his study of the wasp *Dahlbominus fuscipennis*, Wilkes (1966) used quantification of sperm transfer and storage, coupled with measures of P_2, to show that spermathecal filling was the mechanism behind the first-male advantage at fertilization.

Parker *et al.* (1990) advocated a theoretical approach to quantifying the mechanisms behind sperm utilization patterns, based on a knowledge of sperm transfer and storage, and on observed values of P_2. They developed prospective models based on two principle sperm competition mechanisms: the 'raffle' (random sperm mixing) whereby sperm are equivalent to tickets in a lottery, and sperm displacement where the male flushes sperm from the sperm-storage organ(s) using his own ejaculate. In the fair raffle, each sperm from each male has an equal chance of entering the fertilization set. Females store all of the sperm transferred by both males. Considering just two males, the total number of sperm transferred by male 1 is S_1 and the number transferred by male 2 is S_2. The probability of paternity for the second male is simply:

$$P_2 = S_2/(S_1 + S_2) \qquad (1a)$$

Thus, given information on the relative numbers of sperm transferred by males 1 and 2, it is easy to determine whether the observed P_2 fits that predicted from the sperm mixing model. Parker *et al.* (1990) tested the data of Simmons (1987b), using a linearized version of equation 1:

$$1/P_2 = (S_1/S_2) + 1 \qquad (1b)$$

plotting the observed $1/P_2$ against S_1/S_2 yielded an intercept of +1.0 and a slope of +1.0 supporting a mechanism of random sperm mixing in the cricket, *Gryllus bimaculatus* (Fig. 10.4a). Parker *et al.* (1990) used a linearized version of the model for simplicity. However, a statistically

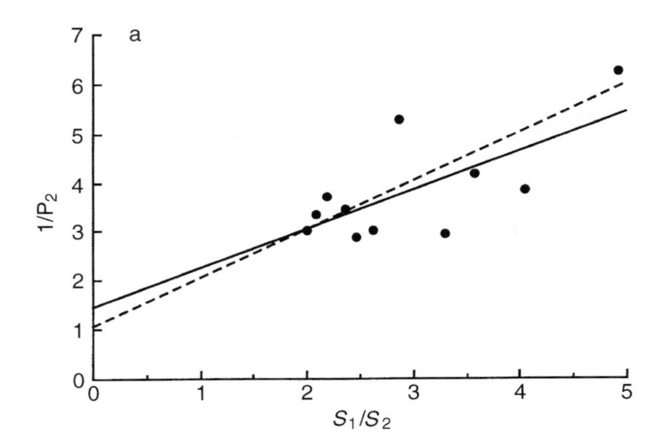

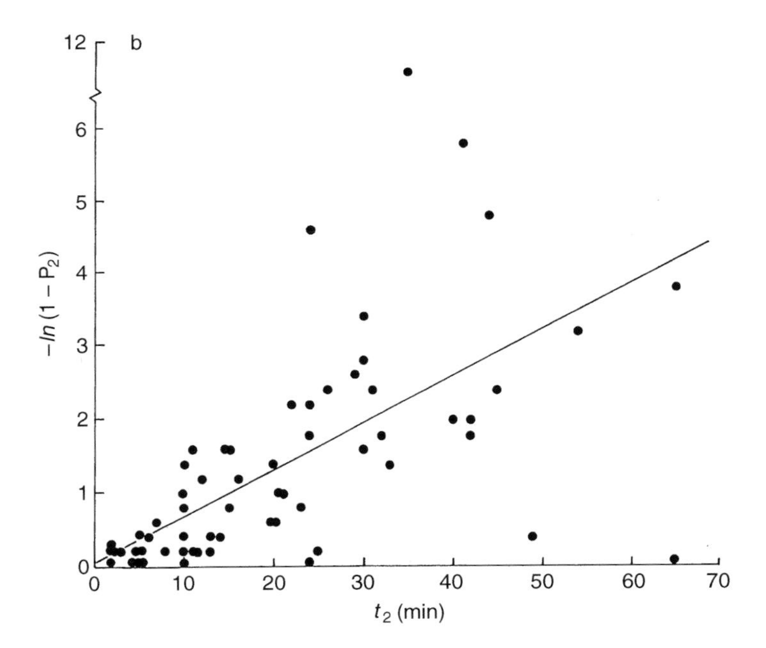

Fig. 10.4. *P_2 values obtained from sperm competition experiments with (a) field crickets fitted to a predictive model of sperm competition based on the assumptions of a fair raffle (equation 1b, dashed line), and (b) yellow dung flies, fitted to a model of volumetric sperm displacement (equation 3b). Both data sets provide significant fits to their respective models of sperm competition mechanisms (from Parker* et al. *1990; Parker and Simmons 1991, with permission. Copyright (a) Springer-Verlag 1990, (b) The Royal Society 1991.)*

better approach may be to fit the data using a nonlinear regression model.

In the sperm displacement model, Parker *et al.* (1990) assumed instantaneous random mixing of sperm during displacement such that:

$$P_2 = 1 - \exp(-pS_2/S) \qquad (2a)$$

where pS_2 is the proportion of sperm transferred by the second male that have entered the fertilization set, and S is the total carrying capacity of the female's sperm-storage organ(s) which, at the beginning of the second copulation are all derived from male 1. Thus, the model assumes a fixed sperm-storage capacity. As copulation proceeds, each sperm that enters the fertilization set displaces one sperm already stored so that as pS_2 increases, there is an increasing probability that male 2 will displace his own sperm such that pS_2/S increases with diminishing gradient. In its linear form:

$$-\ln(P_1) = pS_2/S \qquad (2b)$$

Parker *et al.* (1990) used data from *Scatophaga stercoraria* and obtained a significant fit, but one that was qualitatively poor. Parker and Simmons (1991) later refined the sperm displacement model, assuming that sperm were not displaced on a one-for-one basis but rather on a volumetric basis such that insemination involved a change in the density of sperm in the sperm-storage organ(s). Here:

$$P_2 = [1 - \exp(c_2t_2)]/[1 - \exp(-c_1t_1 - c_2t_2)] \qquad (3a)$$

where c is equal to the constant rate of sperm transfer and displacement and t is the time spent copulating such that a total input volume of $(c_1t_1 + c_2t_2)$ has been ejaculated by the two males, causing the sperm density to rise to $1 - \exp(-c_1t_1 - c_2t_2)$ of its maximum value. Of this, a density equal to $1 - \exp(-c_2t_2)$ will be attributable to male 2. When the total input volume approaches the carrying capacity of the female's sperm-storage organ(s), the denominator of equation 3a approaches unity, so that

$$P_2 \approx 1 - \exp(-c_2t_2) \qquad (3b)$$

P_2 is independent of the copula duration of male 1. A plot of $-\ln(1 - P_2)$ against t_2 should yield an intercept of +1.0 and a positive slope, equal to the constant rate of sperm displacement, c. The data for *S. stercoraria* gave a much improved fit to the model of volumetric sperm displacement (Fig. 10.4b) compared with the original displacement model (equation 2a; see Parker *et al.* 1990).

Parker *et al.* (1990) showed how both sperm mixing and sperm displacement models could be modified to account for different patterns of sperm mixing or for raffles in which the sperm from different males were not of equal value. These examples show how observed values of P_2 can be used to examine the underlying mechanisms of sperm competition. Economorpoulos and Gordon (1972) also developed a model that could explain the observed patterns of P_2 in *Oncopeltus fasciatus* but were unable to test their model because they had no knowledge of the patterns of sperm transfer and storage. Eady (1994b) adopted the approach of Parker *et al.* (1990) in his study of *Callosobruchus maculatus.* He fitted his data on sperm transfer, storage and utilization to three alternative models developed by Parker *et al.* (1990) and modified them to incorporate the observation that sperm were lost from the bursa copulatrix at a constant, exponential rate after insemination. However, none of the models could provide an adequate fit to the data. The models assume that all of the sperm ejaculated by the male enter the female's sperm-storage organ(s). However, Eady (1994b) showed empirically that only 14% of the sperm ejaculated into the bursa copulatrix enter the spermatheca; the remaining sperm are digested and lost. Including a parameter *p*, the proportion of sperm stored, into the models and solving for *p*, using empirically determined values for sperm utilization and sperm numbers ejaculated, Eady (1994b) found that, as in *S. stercoraria,* the model of volumetric sperm displacement provided the best fit, requiring *p* to be exactly 14%.

The sperm mixing and sperm displacement models outlined above represent two ends of a spectrum, the former assuming that the sperm-storage organ(s) are perfectly elastic so that the number of sperm stored after *i* matings, $S = \Sigma S_i$. The sperm displacement models assume that the sperm-storage organ(s) have a fixed capacity, *S*, so that when the organ(s) are full, further insemination causes sperm to be displaced. In some species the sperm-storage organ(s) may be filled after a single copulation (Parker *et al.* 1990), while in others multiple copulations may be necessary to fill the stores so that the patterns of sperm utilization could change with copulation frequency. In many species, the sperm-storage organ(s) may be less than perfectly elastic or have a more flexible storage capacity so that the patterns of sperm storage and displacement reflect variation in the degree of sperm displacement, which is itself determined by the morphology of the female's sperm-storage organ(s). Sakaluk and Eggert (1996) extended Parker *et al.*'s (1990) models to include a parameter *b*, the displacement efficacy. They used data from *Grylloides sigillatus* and found that a model of sperm displacement with *b* set to 0.5 provided a significant fit to the observed P_2 data. However, *b* was assessed by linear measurements of spermathecae from singly and doubly mated females (doubly mated females had spermathecae 1.49 times the volume of singly mated females) and Sakaluk and Eggert (1996) thus assumed an isometric relationship between spermathecal volume and number of sperm stored. Further, the number of sperm transferred by male *G. sigillatus* is now known to be highly variable and dependent on the perceived risks of sperm competition (Gage and Barnard 1996), yet Sakaluk and Eggert (1996) assumed that spermatophore attachment time was an accurate predictor of sperm numbers. Deviations in P_2 values from those predicted from a model of random mixing may have resulted from higher sperm numbers within the spermatophores of second males. Without knowledge of the numbers of sperm stored from rival males, results based on P_2 data alone must remain equivocal.

V. AVOIDANCE OF SPERM COMPETITION

Sexual selection will favour any adaptation in males that will avoid competition between their own sperm and those of other males (Parker 1970c). By manipulating the stored sperm of rivals within the female's sperm-storage organ(s) and consequently placing those sperm at a competitive disadvantage, a copulating male can reduce, or even avoid, sperm competition. The time scale over which sperm competition avoidance occurs will vary from species to species, and will depend on the cost:benefit ratio of implementing the avoidance mechanism. Most sperm competition avoidance mechanisms that operate during copulation and insemination function to provide the sperm of the copulating male with a competitive advantage over the sperm of rivals already stored within the female. In contrast, those sperm competition avoidance mechanisms that operate after copulation tend to function to avoid sperm competition with future males. Both types of adaptations can arise within a single species owing to the opposing selection pressures acting upon males (Parker 1970c). Alternatively, adaptations for the avoidance of sperm competition may favour one or other of the above mechanisms, depending on the relative intensity of sexual selection generated by past and future rivals. Here, we examine adaptations for the avoidance of sperm competition and their general relation to sperm utilization patterns.

A. Avoiding competition with stored rival sperm

A variety of male traits operate during, and sometimes prior to, mating that reduce the paternity of males that have already mated with the female. All involve the copulating male manipulating the stored sperm of rivals within the female's sperm-storage organ(s) with the result that the manipulator's sperm are used preferentially in the subsequent oviposition bout(s). Males can avoid competition with the stored sperm of rivals by any of the following mechanisms.

1. Anatomical displacement of rival sperm

The Odonata provide the best examples of anatomical traits that operate during copulation to avoid sperm competition. Despite major structural differences in the male genitalia of damselflies and true dragonflies (Pfau 1971) the distal part of the intromittent organ of both groups enters the female's sperm-storage organ(s) during copulation. Males utilize this ability to displace directly the stored sperm of their rivals before they inseminate their mate. By displacing sperm from the exit of the female's sperm-storage organ (i.e. the site from where sperm are used during fertilization) and placing their own sperm into that competitively favourable site (Miller 1984; Siva-Jothy and Tsubaki 1989) males ensure they achieve high, invariant fertilization success in the eggs the female lays immediately after copulation (see Table 10.1). The paradigm for this kind of displacement mechanism occurs in calopterygid damselflies where the copulating male removes all stored rival sperm from the female before insemination (Waage 1979b) thereby completely avoiding sperm competition (see Fig. 10.3a). However, other odonate species achieve equally high levels of sperm precedence without removing all rival sperm from the female (McVey and Smittle 1984). Males of these species only remove sperm from the exit to the bursa copulatrix and inseminate into the same area to ensure their sperm are used for fertilization. Because rival sperm are still present in the female's sperm-storage organs this high initial P_2 declines with time after copula, until the female remates (Miller 1984; Siva-Jothy and Tsubaki 1989). Whether or not sperm removal is complete, males avoid sperm competition for fertilization of the eggs the female lays immediately after copula. Because female odonates usually mate only when they have mature eggs, and because they take several days to mature a new clutch of eggs, copulating males need only avoid sperm competition in the oviposition bout following copulation. Whether a species opts for complete or partial sperm removal will depend on ecological variables, the female's sperm-storage organ anatomy (and therefore the ease with which males can remove stored sperm) and the time before female remating and periodicity of oviposition. For example, if males can defend all available oviposition resources, the frequency of remating will be high. If males cannot defend all resources then females can potentially avoid males, giving previous ejaculates an opportunity to fertilize eggs. Under the latter condition, males may be selected to remove all rival sperm and thereby avoid sperm competition until the female remates. Some odonates do not remove sperm from the female's sperm-storage organs but reposition it away from the exit, towards its distal regions (Waage 1984; Siva-Jothy 1988). This is achieved by lobes on the male's intromittent organ which undergo a large volume increase when they are in the female's sperm-storage organ(s) (Fig. 10.5). Because the male moves sperm away from the fertilization bottle-neck at the exit to the bursa, and inseminates into that position, he avoids sperm competition during the subsequent oviposition bout, and thereby achieves high immediate fertilization success (Siva-Jothy and Tsubaki 1994). However, sperm repositioning only avoids short-term spatial overlap between ejaculates. Last-male sperm precedence declines with time since last copula, and sperm competition is the consequence.

Sperm removal and repositioning occur in species with essentially

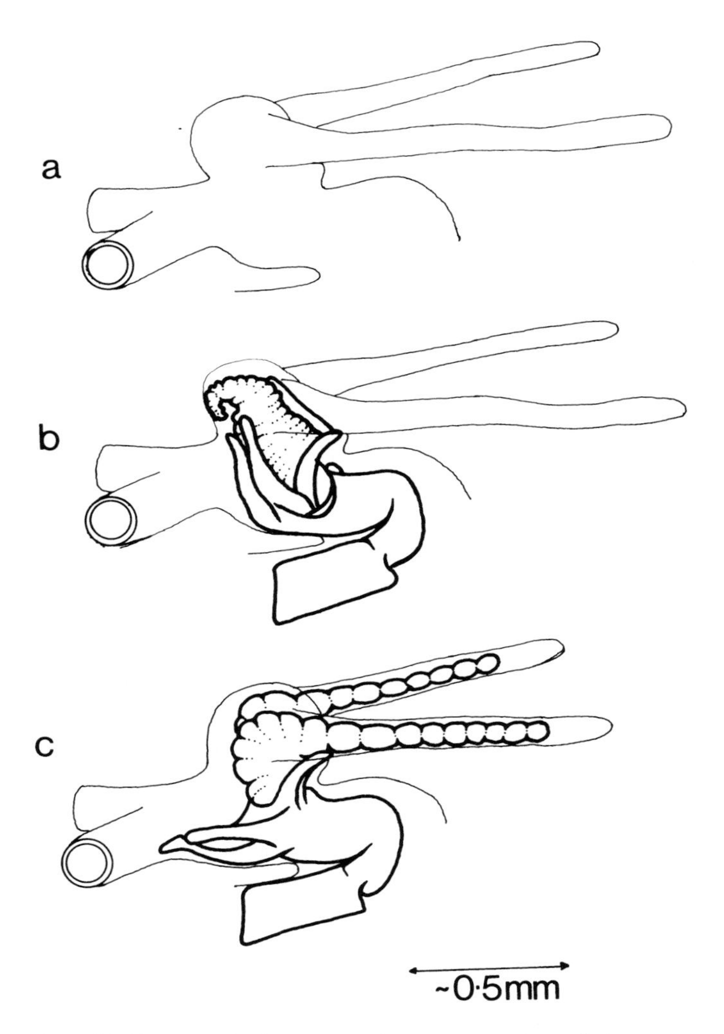

Fig. 10.5. *a–c. A diagrammatic representation of the sequence of events that occur during the copulation of* Hadrothemis defecta pseudodefecta. *(a) The female sperm-storage organs in three-quarter latero-dorsal view; (b) the fourth and third segment of the male's intromittent organ just after introduction, but before inflation; (c) the fourth and third segment of the male's intromittent organ after inflation, showing the extension of the dorsal lobes into the spermathecae. This last action is believed to reposition rival sperm to the ends of the spermathecae, away from the exit to the sperm-storage organs. At the end of copulation the male ejaculates into the region of the female's sperm-storage organ from where sperm are used to fertilize eggs. Modified from Miller (1991).*

similar mating systems: females of both type are highly likely to remate when returning to oviposit the next clutch of eggs, and their new mates prevent the female from remating during oviposition. Why such different mechanisms have evolved is not clear. An obvious economic consequence is the duration of copulation. Sperm removers tend to have long copulations (1–20 min) while sperm repositioners tend to have short copula durations (7–20 s) (Waage 1984). Reduced time costs may be important in species where the operational sex ratio (fertilizable females to sexually active males, Emlen and Oring 1977) is less biased towards males because males will have both a higher potential mating rate and a reduced risk of competition. It is also important to bear in mind that female sperm-storage organ morphology will play an important part in determining the type of sperm competition avoidance mechanisms that evolves. The females of odonate species, where the male repositions sperm, have large sac-like sperm-storage organs (Siva-Jothy 1984; Miller 1991). Complete sperm removal from such a structure would be impractical compared with the rapid repositioning solution employed. Likewise, female sperm-storage organs in sperm-removing species tend to store a relatively small volume of sperm in narrow, tubular sperm-storage organs (Miller 1984, 1990; Siva-Jothy 1984). Repositioning sperm in these structures may be just as time-consuming as removing, and does not have the long-term advantages. Ensuring that rival sperm cannot fertilize eggs by physically removing them from the competitive arena is an elegant solution to avoiding sperm competition. It is likely to be a mechanism that is identified frequently, not necessarily because it is ubiquitous but because it is easy to establish empirically. In all species that use this mechanism of sperm competition avoidance, ejaculation occurs after rival sperm have been removed thereby avoiding the removal of 'self' sperm (Pfau 1971; Miller 1991). The anatomical features associated with direct sperm manipulation in odonates are well known. Spines on the region of the intromittent organ that enters the female's sperm-storage organ(s) are responsible for trapping and removing rival sperm (Fig. 10.3a). However, similar spines are often associated with structures that do not enter the female's sperm-storage organ: the presence of spines on a male's intromittent organ should not be taken as an indication of sperm removal ability *per se*: spines may, for example, function to maintain genital contact.

Finally, sperm can be removed from the competitive arena within the female by means other than displacement with the male's intromittent organ. The spermatophore of some insect species functions as a surrogate intromittent organ and can be used to displace sperm (Gack and Peschke 1994). In other species, males use their own ejaculate to flush rival sperm from storage (Ono *et al.* 1989; Parker and Simmons 1991). Chemicals in the seminal fluids of male *D. melanogaster* have been implicated in sperm displacement. Males incorporate into their ejaculate chemicals that facilitate the mobilization of sperm from the female's sperm-storage organ(s) (Gilbert *et al.* 1981). When mated fertile females are remated using *tudor* males (a genetic strain in which the male produces the normal accessory fluids but fail to produce sperm) the number of progeny produced is greatly reduced, implying that the seminal fluid of *tudor* males may be responsible for loss of sperm from the female's sperm stores (Scott and Richmond 1990; Scott and Williams 1993; Harshman and Prout 1994; Gilchrist and Partridge 1995). Moreover, using a genetic strain that lacks main cell products in the seminal fluids, Harshman and Prout (1994) showed that the reduction in progeny production was only 67% of that seen with *tudor* males, suggesting that sperm displacement may be mediated by chemical products derived from the main cells. Unfortunately, sperm were never observed directly in any of these experiments so it is not yet clear whether sperm displacement actually occurs. Chapman *et al.* (1995) have since found that main cell products that facilitate sperm displacement by males, impose a cost of reproduction on females in terms of reduced lifespan.

2. Repeated copulations

Frequent copulation is a tactic commonly adopted by the males of socially monogamous birds to assure paternity (Birkhead and Møller 1992). Because pairs remain in close proximity throughout the female's fertile period, potential rivals may copulate only once, or at best infrequently. However, it still pays pair-males to ensure their sperm are present in the female's tract in larger numbers because of the inferred dynamics of sperm storage, release and utilization (see Chapter 14). This type of sperm competition avoidance mechanism also appears to occur in insects. Male and female *Nicrophorus* burying beetles form pair associations over an extended period during the construction of a nest. During this period the pair are usually isolated from rivals, but the female is rarely a virgin. Her current mate achieves high levels of paternity (i.e. minimizes sperm competition) only by repeatedly mating with the female (Muller and Eggert 1989). It may be that repeated copulation increases the male's numerical representation in the spermatheca, thereby improving his success in sperm competition, as is the case for field crickets (Simmons 1987b). Alternatively it is possible that, like carrion flies, repeated copulations are required to displace rival sperm and avoid sperm competition (Otronen 1994b). Paternally caring water bugs (*Abedus herberti*) also copulate repeatedly during the period that the female oviposits on the male's back (Smith 1979). Here, increased sperm numbers do not appear to be important since a single copulation results in a high and invariant P_2 (Table 10.1). It is likely that this behaviour evolved because of the relatively high cost to males of paternal care: by interrupting egg laying with bouts of copulation males ensure that their effort is directed only towards their offspring. Presumably, if males were prevented from repeated copulation the egg-laying female would fertilize eggs with sperm received from previous matings.

3. Physiological incapacitation/killing of rival sperm

Rival sperm could be 'removed' from the competitive arena if they were physiologically incapacitated. In other words, males could avoid sperm competition if they killed or disabled the sperm of rivals in the female's sperm-storage organ (see Silberglied *et al.* 1984). This sperm competition avoidance mechanism could probably only be achieved in a few restricted ways in insects. If the sperm of rivals were killed by nonspecific agents then we would expect to see the initial transfer of sperm-free ejaculate containing the agent, followed by a period of inactivity (while the agent took effect and then became neutralized) only then should self-sperm arrive at the site of storage (a process that mirrors the events that occur in sperm removal in odonates). It is unlikely that sperm killing in insects could be affected by specific 'immunological' agents since insects have relatively poor xeno- and allo- graft recognition abilities (Lackie 1986). Specific agents are more likely to act via molecular markers on sperm associated with being in storage, such as age-related acquisition of surface molecules or the acquisition, or modification by, female-secreted molecules. An incapacitating agent which acted on sperm bearing such markers would selectively neutralize stored (and therefore rival) sperm, without affecting the mating male's ejaculate (which would not bear these markers). Accessory gland products in the seminal fluids have been implicated as agents for the incapacitation of sperm stored in the female's sperm-storage organs in *D. melanogaster* (Harshman and Prout 1994; Clark *et al.* 1995). Alternatively, these compounds may facilitate displacement because of their influence on the release of sperm from the sperm-storage organs (Gilbert 1981). It is known that there is a delay of several hours between ejaculation (when the agent is introduced into the female) and the migration of sperm into the spermathecae (Gilbert 1981), which is compatible with the idea that incapacitation of rival sperm should precede storage of self sperm. However, Fowler (1973) showed that a male's sperm is present in the uterus of the female together with his own accessory gland fluids for all but the first few minutes of copulation. Moreover, sperm stored in the seminal receptacle appear to be those used by the female to fertilize eggs during the first few days following mating (Gromko *et al.* 1984) and are themselves the target of displacement (Fowler 1973; Gilchrist and Partridge 1995). Accessory gland products are unlikely to function as incapacitating agents in this context because a male's accessory gland fluid is unable to distinguish between self and rival sperm (Gilchrist and Partridge 1995).

B. *Avoiding sperm competition with future ejaculates*

When sexual selection results in the evolution of traits that ensure last-male sperm precedence immediately after copulation, it will also favour counter-adaptations that prevent rivals from mating with the female (Parker 1970c, 1984). Male insects prevent competition with future ejaculates by either staying with their mate after copulation to reduce the chance of take-overs by other males, or they change the female's ability to remate when she leaves the site of copulation. We term the first type of sperm competition avoidance behaviour 'proximate' mate guarding because the male stays in physical and/or sensory contact with his mate while attempting to prevent her from remating. We term the second form of sperm competition avoidance behaviour 'remote' mate guarding because males rely on a transferred structure, and/or chemical, that reduces or prevents their mate's ability to remate.

1. Proximate mate guarding

Proximate mate-guarding can occur before, during and/or after copulation. Perhaps the most important male precopulatory trait that avoids sperm competition is mate concealment, especially when local competition for mates is high. Male flies in the empid genus *Ramphomyia* leave the swarm as soon as they find a mate (Downes 1970) and so reduce the chances that rivals will come into contact with their mate. Similar behaviour is shown by some species of dragonflies and dungflies: males usually encounter females at oviposition sites, and once a male has grasped a female and is assured of copulating, the pair flies into nearby vegetation where they are less likely to be discovered and disturbed by rival males (Parker 1970b; Rowe 1987). Males of other species conceal their mates by more devious means. In species where the female releases a mate-attracting pheromone, males quickly release a pheromone of their own when they find a receptive mate. This male pheromone appears to mask, or diminish, the effectiveness of the female's attractant (Happ 1969; Happ and Wheeler 1969; Happ *et al.* 1970) making the receptive female less conspicuous to rivals. Many male insects avoid attracting interest from potential rivals by using short-range courtship signals. Some crickets switch from using long, loud chirps (to attract mates) to shorter duration, low-volume courtship songs when a female appears (Boake 1983). Precopulatory guarding (i.e. a passive phase where the male remains with, or near the female prior to copulation) usually functions to assure a male of an opportunity to copulate when female receptivity is difficult for males to predict, and females are receptive only for a short time. For example, the spermatophore of male locusts *Locusta migratorioides* plugs the female's genital tract from the time of copulation until oviposition, when it is ejected. The last male to mate achieves a significantly greater paternity if he copulates after the previous male's spermatophore has been ejected: consequently, males show precopulatory mate-guarding of females that are about to oviposit (Parker and Smith 1975). Precopulatory mate guarding is also selected when the male's sperm competition avoidance mechanism and the female's reproductive anatomy/physiology combine to produce first-male sperm precedence and/or when virgin females are defensible until receptive. The males of many species of parasitoid wasp defend parasitized hosts because virgin females will emerge from them (Thornhill and Alcock 1983); first-male sperm precedence is often high in these species (Table 10.1). A similar phenomenon occurs in heliconid butterflies, where males guard, and fight for possession of pupae containing females (Gilbert 1976; Ehrlich and Ehrlich 1978; Deinert *et al.* 1994). Female *Heliconius hewitsoni* mate only once, thereby selecting for males that mate as soon as the female becomes adult (Deinert *et al.* 1994). Defending sites where virgin females will emerge is clearly an adaptation to monopolize females of high reproductive value in this species.

Proximate mate guarding can also occur during copula, when it manifests as prolonged mating. Such copulations are characterized by periods of physiological inactivity during full genital contact. The most dramatic example of this kind of adaptation comes from phasmids, where genital contact can last 79 days (Gangrade 1963). Prolonged copulation generally functions to reduce or avoid sperm competition primarily by excluding potential rivals until the female is unreceptive and/or she is ready to oviposit. Many insect species that show prolonged copulation, rather than postcopulatory guarding, have specially adapted male external genitalia that can maintain a strong grasp on the female and/or block her genital entrance. Such traits enable males to prevent rivals from gaining access to the female's genitalia for as long as he is locked on. The consequences of not maintaining genital contact while guarding are well illustrated in the weevil *Brenthus anchorago*. Small males often manage to copulate with the mate of a large male, who is 'guarding' (Johnson 1982). Maintaining genital contact pre-empts this possibility. The damselfly *Ischnura elegans* extends copulation to 5 h by pausing, but maintaining genital contact, during the standard coenagrionid copulatory repertoire (Miller 1987). The last male to mate gains the majority of fertilizations in this species (Cooper 1995; Cooper *et al.* 1996) which explains why males guard their ovipositing mates. Female *I. elegans* do not oviposit until late in the afternoon and males pair up with females early in the day. Only a fraction of female *I. elegans* are gravid and consequently receptive to copulation on any one day. The only way males may have of gauging a female's reproductive potential is to attempt copulation. Because females in this, and other, zygopteran species can resist copulation, males presumably use receptivity as an indication of the likelihood that the female will oviposit on that day. Once engaged with a female a male will achieve fertilization success provided that he completes copulation successfully and prevents the female from remating until, and while, she oviposits. Because the oviposition period is restricted to later in the day the male must guard the receptive female he has just entered copula with. Entering copulation early in the day therefore enables males to find receptive females; prolonging copulation enables the successful male to

Fig. 10.6. *(a) Contact guarding in* Enallagma cyathigerum*: the male remains attached to his recent mate's prothorax by his anal appendages throughout the female's oviposition period. (b) Noncontact guarding as typically displayed in calopterygid damselflies. A male* Calopteryx splendens xanthostoma *defends his ovipositing female and his territory from a perch on the oviposition substrate.*

monopolize the receptive female until she is ready to oviposit. A similar situation occurs in monarch butterflies *Danaus plexippus* where the male remains in copula until nightfall, when he transfers sperm; females oviposit the following morning (Svärd and Wiklund 1988).

It is generally assumed that in species where males obtain the majority of fertilizations after copulation, selection should favour strong postcopulatory guarding of the female until she has finished ovipositing. By excluding potential mates the male avoids sperm competition with future ejaculates. During proximate postcopulatory guarding males either remain in direct physical contact with the female or they release the female after copulation but remain in her vicinity. This difference in behaviour is likely to be the result of the balance between potential reproductive pay-offs (in terms of the likelihood of encountering and mating with other females) and the likelihood of the current mate being captured and copulated by a rival. The outcome is well demonstrated by considering odonates, which show both contact and non-contact postcopulatory guarding (Fig. 10.6). The difference in the form of postcopulatory guarding in this order does not appear to be related to P_2 (which is high in both cases). Species that show noncontact guarding tend to be territorial. By releasing their mate after copulation males can protect their paternity and their territory as well as mate with other females arriving at their territory. In contrast, species that show contact guarding tend to occur at high density and tend not to be territorial. Operational sex ratios tend to be highly male biased in these species. Since paired males do not need to defend a territory and gravid females are relatively scarce, it pays a male to remain attached to his mate and completely prevent take-overs until she has laid the eggs he will father. The importance of ecology, rather than sperm utilization patterns, in determining which type of mate guarding prevails is well illustrated by the libellulid *Sympetrum parvulum*. Territorial males only show noncontact guarding. In contrast nonterritorial males show contact guarding at high population densities (when there are more rivals) and noncontact guarding at low population densities (Ueda 1979). In general, males will avoid sperm competition by contact guarding if (1) they do not have to defend a resource, (2) they can repossess a vacated territory easily and/or (3) they encounter receptive females at a low rate. In contrast paternity assurance by noncontact guarding may be favoured when males (1) defend a valuable resource that is therefore likely to be costly to acquire and repossess, and (2) are likely to encounter receptive females at a relatively high rate.

Postcopulatory guarding can function to provide benefits in addition to paternity assurance for males (Alcock 1994). The females of some odonate species oviposit underwater (e.g. *Enallagma* spp., Bick and Bick 1963; *Hetaerina vulnerata*, Alcock 1982). Because males never submerge to pursue females, an ovipositing female is effectively inaccessible to rivals. We might expect males to abandon their mates once they have submerged, but they do not. Some even remain attached to, and submerge with, their mate (Robert 1958). P_2 is probably high in all these species, but guarding a mate that is effectively inaccessible appears to be taking sperm competition avoidance to the extreme. The reason for this fidelity is that females often reject their current oviposition site, which they can only assess once they are underwater. It therefore pays males to be on hand in case his mate resurfaces. In addition, this behaviour enables males to rescue females who became trapped at the surface of the water when returning from submerged ovipositing (Fincke 1986). Recent work by Tsubaki *et al.* (1994) shows that postcopulatory mate guarding is associated with subtle effects on female reproductive output. The number of eggs laid by guarded females is higher than by solitary females, independent of the effects of reduced harassment. The reasons why female reproductive output increases in the presence of a guardian are unclear, however, in *D. melanogaster* an accessory gland peptide produced by males induces females to increase oviposition (Chen *et al.* 1988). Alternative explanations for sperm competition may often account for adaptations in male reproductive behaviour (Thornhill 1984).

In a recent study of the milkweed beetle *Chrysochus cobaltinus*, Dickinson (1995) examined the costs of postcopulatory mate guarding, in terms of lost mating opportunities, and its fitness benefits in terms of offspring production. Dickinson (1995) modelled the fitness consequences of guarding under varying levels of last-male sperm utilization. She found that guarding would confer fitness advantages on males even when P_2 was as low as 0.4. This study shows that we should not expect to find postcopulatory mate guarding restricted to species that have a high P_2, and illustrates the necessity for studies of mate guarding that take into account the relative costs and benefits of the behaviour. Indeed, postcopulatory mate guarding occurs in many species of water striders (Rowe *et al.* 1994) where estimates of P_2 show that the last male can fertilize as few as 65% offspring (Rubenstein 1989) and where mate guarding is known to have the nonsexual function of increasing female foraging efficiency (Rubenstein 1989; Wilcox 1984). A case in point is Allen *et al.*'s (1994) study of postcopulatory mate guarding in *Aphytis melinus*, which shows that guarding functions to ensure paternity of the first male (average $P_2 = 0.14$). Thus, the mean species value of P_2 may provide little predictive power for patterns of mate guarding, and vice versa.

2. Remote postcopulatory mate guarding

Males do not have to remain with their current mate in order to prevent them from remating. Proximate mate guarding may be costly in time and/or energy which may be better spent searching for other mates. If a male transfers a chemical, or a physical barrier that reduces the female's propensity to remate he can also reduce competition from future ejaculates. Male house flies (*Musca domestica*, Reimann *et al.* 1967) and hanging flies (*Bittacus apicalis*, Thornhill 1976) transfer a compound(s) that reduces the receptivity of the female; prolonged copulation in both of these species is a form of mate guarding during the period over which the chemical becomes effective. These compounds are probably in the seminal fluid, and so are likely to be manufactured by the accessory glands. In *D. melanogaster*, the seminal fluid also induces nonreceptivity in the female (Kalb *et al.* 1993) and a 36 amino acid peptide is known to be partly responsible for this effect, as well as increasing egg-laying rates after mating (Chen *et al.* 1988). This type of remote mate guarding is not confined to the Diptera. Prostaglandins in the male's seminal fluid induce nonreceptivity in some female Orthoptera (Stanley-Samuelson and Loher 1986), and materials passed in the spermatophore of moths in the genus *Cecropia* act to trigger the female's corpora cardiacum to release another hormone which inhibits the release of sex pheromones (Riddiford and Ashenhurst 1973), thereby reducing the likelihood that the female will remate. In addition to transferring compounds that act on the female's central nervous system to change behaviour, males can exploit the sensory systems that females use to monitor how much sperm they have in their sperm-storage organs. Some female insects remate when they become sperm limited (Fukui and Gromko 1989). By completely filling the female's sperm-storage organ a male may induce nonreceptivity (Nakagawa *et al.* 1971; Thibout 1975). The selection pressures that result in this kind of sperm competition avoidance mechanism may also have resulted in morphological adaptation in sperm. The anucleate sperm of Lepidoptera may act as 'cheap fillers' that enable the male to fill the female's spermatheca and thereby delay female remating (Silberglied *et al.* 1984; Cook and Gage 1995). In taxa where the female's sperm-storage organ is of a fixed capacity and is inaccessible to the male's intromittent organ during copula, selection may favour males that produce ejaculates large enough to prevent the sperm from subsequent matings gaining access to the sperm-storage organ(s) (Retnakaran 1974). Sperm gigantism may have evolved for this reason. Dybas and Dybas (1981) studied featherwing beetles from the genus *Bambara* and showed that the size of the sperm enables the first male's sperm to completely fill the female's spermatheca and preclude further sperm from gaining access. Large spermatophores can also influence female receptivity but it is important to show experimentally that the change in female behaviour is due to the spermatophore and not a compound transferred with it. In the cockroach *Nauphoeta cinerea* the male passes a spermatophore into the female's bursa copulatrix. Once in place, the female will refuse to mate, but if the spermatophore is removed experimentally, the female will readily remate (Roth 1962). By injecting silicone oil into the bursa of *Pereis rapae*, Sugawara (1979) showed that bursal stretching, rather than male-transferred material, was responsible for inducing female nonreceptivity.

A male can also prevent a female from remating by blocking her genital tract. Because the entrance to the female's genital tract is usually also the exit, such blocking devices are only effective until the female oviposits (Parker and Smith 1975). However, in the ditrysian Lepidoptera (where the female has a copulatory tract and an oviposition tract) males may be able effectively to prevent the female from remating over several oviposition bouts. In many ditrysian Lepidoptera the spermatophore functions to block future access to the female's sperm-storage organs by potential rivals (Labine 1966). Because spermatophores often occupy positions that block the entrance to the duct which enables sperm to get to the spermatheca, they may hinder the successful evacuation of subsequent spermatophores (Labine 1966; Brower 1975). Purpose-built mating plugs are often formed from reactive components in the seminal fluid which harden upon contact with air. In some Lepidoptera this produces a highly visible, and hard, sphragis which blocks the female's genital orifice and the parts of her external genitalia (Drummond 1984; Pierre 1985). Experiments by Dickinson and Rutowski (1989) have shown that the sphragis functions to prevent intromission in *Euphydryas chaledon*, while Orr and Rutowski (1991) have shown it is a highly effective visual signal (male *Cressida cressida* make no attempt to mate with females that have a visible sphragis, presumably because its presence signals female nonreceptivity, and/or the presence of an effective mating plug). Some Diptera also use mating plugs derived from seminal fluids, for example ceratopogonids (Linley and Adams 1972), mosquitoes (Lum 1961), and *Drosophila* (Markow and Ankney 1988; Alonso-Pimentel *et al.* 1994). Although often considered to be a mating plug, the detached genitalia, or mating sign, of the male honey bee seems particularly ineffective since genetic analysis of offspring indicate that queens mate successfully with several males on their nuptial flight (Peer 1956; Moritz *et al.* 1991), although it is possible that levels of polyandry might be higher if the mating sign had not evolved. Thornhill and Alcock (1983) have suggested that the female can decide when to remove the mating sign, thereby enabling her to exercise some degree of choice over her partners. However, as Koeniger (1986) points out, queens returning to their natal nest after their nuptial flight are particularly inept at removing the last male's mating sign, which has to be pulled out by the workers. Moreover, Koeniger's (1986) detailed study of the male's endophallus shows that it can easily remove the 'mating plug'. The actual function of the detached genitalia of drone bees is far from clear.

Finally, by physically damaging the female's genital tract males may prevent their mate from successfully achieving later copulations. The problem with this strategy is that males may also be reducing the chances that their mate is able to lay eggs, since the eggs must pass down the damaged tract. Damage accrued to the female's genital tract as a consequence of mating has been documented in the bushcricket *Metaplastes ornatus* (Helverson and Helverson 1991) and the bruchid beetle *Callosobruchus maculata*, where it may reduce the probability of female remating (Tufton 1993).

VI. SPERM IN COMPETITION: A MALE PERSPECTIVE

A. Ejaculate size

When the sperm of two or more males are present within the reproductive tract of the same female, sexual selection will favour any adaptation in males, or their sperm, that increases the probability that their sperm are used for fertilization (Parker 1970c). Thus, in species without sperm displacement or precedence we might expect to find adaptations that enhance a male's success in sperm competition. The most obvious adaptation to sperm competition is selection on males for increased sperm numbers (Parker 1982). When sperm mix in storage there will be a selective advantage to the male that ejaculates more sperm than his competitor(s) since he will gain proportionally more fertilizations. In its extreme, increasing sperm numbers could also be viewed as an adaptation for the avoidance of sperm competition, for example if a male could ejaculate enough sperm to completely overwhelm previously stored sperm there will be no competition for available ova. Thus, the functional significance of increased sperm numbers might best be viewed as a continuum between the avoidance of, and engagement in, sperm competition. One consequence of selection for increased sperm production will be an increase in spermatogenic tissue, and comparative studies across a variety of vertebrate taxa have shown that increases in the degree of multiple mating by females are associated with increased testis size in males (Harcourt *et al.* 1981; Ginsberg and Rubenstein 1990; Møller 1991; Jennions and Passmore 1993). In general, sperm mixing appears to be the rule in vertebrate systems (Dewsbury 1984; Birkhead *et al.* 1995; Colegrave *et al.* 1995). However, in insects, a simple association between female mating frequency and testis size might not be expected because of the diversity of sperm competition mechanisms. Where sperm displacement is achieved by mechanical means, sperm numbers will be unimportant in determining the fertilization success of second males so that selection via female mating frequency is unlikely to favour increased testis size. On the other hand, where males use their own ejaculate to flush rival sperm from the sperm-storage organs of females, sperm numbers will be unimportant at the time of fertilization, yet selection could favour increased testis size because of the necessity for males to use their ejaculates for reasons beyond fertilization *per se*. Of course, selection may act on accessory glands rather than testes if it is seminal secretions rather than sperm that are involved in sperm displacement. These types of confounding factors need to be considered in any comparative analysis across insects, and when drawing conclusions from such analyses.

The best evidence for an effect of sperm competition on ejaculate size comes from the Lepidoptera. Svärd and Wiklund (1989) examined the influence of polyandry on the ejaculation features of two families of butterflies, the pierids and satyrids. Because empty spermatophores remain in the female's bursa copulatrix, it is possible to assess the degree of polyandry, and thus potential for sperm competition, by performing spermatophore counts on wild caught females. Svärd and Wiklund (1989) found that the pierids were highly polyandrous while the satyrids were relatively monandrous. Accordingly, they found that the pierids had significantly greater ejaculate weights and, further, within the peirids there was a significant positive association between the degree of polyandry and both ejaculate weight, and the rate at which males could produce sperm and accessory secretions. Svärd and Wiklund's (1989) study did not control for the effects of common ancestry on the variables considered. In a more detailed study involving 74 species of butterflies from five families, Gage (1994) found that after controlling for body size and phylogeny, there was a positive association between testis size and degree of polyandry, again estimated from spermatophore counts of wild-caught females. Superficially, these data could be interpreted as supporting the hypothesis that increased potential for sperm competition is associated with increased male investment in sperm production. However, the data for Lepidoptera show that sperm utilization favours either the first or the second male (Table 10.1); within species, cases of mixed sperm utilization are rare. What then could be the selective advantage to males of increased sperm production if sperm do not compete numerically? The answer may lie in the mechanism that generates the bimodal distribution of sperm utilization. If Retnakaran's (1974) hypothesis of spermathecal filling is correct, selection for increased ejaculate size may arise because of the advantage it bestows on males as a guard against sperm competition from future males, rather than as a response that enhances success during sperm competition. Males that are able to completely fill the female's sperm-storage organ(s) will prevent future males from gaining access to the sperm stores and competing for fertilizations, thus favouring the evolution of large ejaculates. One readily testable prediction of the spermathecal filling hypothesis is that, after controlling for body size and phylogeny, across species there should be a positive association between the volume of female sperm-storage organ(s) and ejaculate (testis) size. Further, within species, where sperm-store volume is correlated with body size, males should transfer larger ejaculates to larger females.

Pitnick and Markow (1994a) found a positive association between testis size and female remating frequency in the *nannoptera* species group of the Drosophilidae. Again, this relationship could be taken as evidence for selection via sperm competition. The Drosophilidae are an interesting group in that sperm gigantism is common. Across species, there is a positive association between sperm length and testis size (Pitnick 1996). Increased testis size appears to be associated with the production of giant sperm but not with increases in the number of sperm produced (Pitnick 1996). Further, there is a trade-off between sperm size and number such that sperm gigantism results in sperm limitation; males partition their limited supply of sperm among multiple females so that after a single mating, females are submaximally inseminated (Pitnick 1993, 1996; Pitnick and Markow 1994a). What role sperm competition plays in the evolution of sperm gigantism is unknown (but see below). However, since females are submaximally inseminated, the tendency to remate is increased and, hence, so is the risk of sperm competition. In *D. melanogaster* and *Drosophila pseudoobscura*, sperm are relatively small and males maximally inseminate females (Gilbert 1981; Snook *et al.* 1994). Because of sperm displacement P_2 values are high in these species so that direct competition between the sperm from two males is low (Table 10.1). In contrast, like *Drosophila pachea* (Pitnick 1993), males of the giant sperm-producing *Drosophila hydei* may submaximally inseminate, so that sperm from successive males are stored by the female (Markow 1985). Sperm mixing means that selection via sperm competition is greater for *D. hydei* than for *D. melanogaster* or *D. pseudoobscura* (Table 10.1). Thus, while sperm competition is associated with increased testis size across drosophilids, the relationship is not causal. Rather, increased sperm competition may be the inevitable consequence of increased sperm length. This example clearly demonstrates how the often complex mechanisms of sperm transfer, storage and utilization found in insects can confound predictions based on models that assume a simple mechanism of sperm competition by numerical superiority.

B. Sperm morphology

Parker (1970c) noted that there is every reason to expect selection through sperm competition to act on individual sperm. Whenever sperm are in competition, any trait of an individual sperm that enhances its success in fertilization over its competitors will be favoured in the male producing it. Insect sperm are characterized by a remarkable array of morphological features that may arise via selection through sperm competition (Sivinski 1980, 1984). Sperm morphologies range from the simple disc like structures with no means of locomotion, seen in the proturan *Eosentonon transitorium*, to the multiflagellate spermatozoa of the termite *Mastotermes darwiniensis*, and the aggregations of spermatozoa attached to a central spermatostyle seen in gyrinid beetles (Fig. 10.7). The rapid and divergent nature of spermatozoon evolution can, in some cases, make sperm structure a useful means for constructing phylogenies (Jamieson 1987) and suggests that they are under considerable selection pressure. Sperm gigantism is common in *Drosophila* spp. with sperm lengths ranging from 0.32 mm in *Drosophila persimilis* to 58.29 mm in *Drosophila bifurca*, some 20 times the length of the male producing them (Pitnick *et al.* 1995a,b). In some species, sperm also exhibit polymorphisms in size. Thirteen species in the *obscura* group have both long and short spermatozoa whose lengths differ from 180% in *Drosophila obscura*

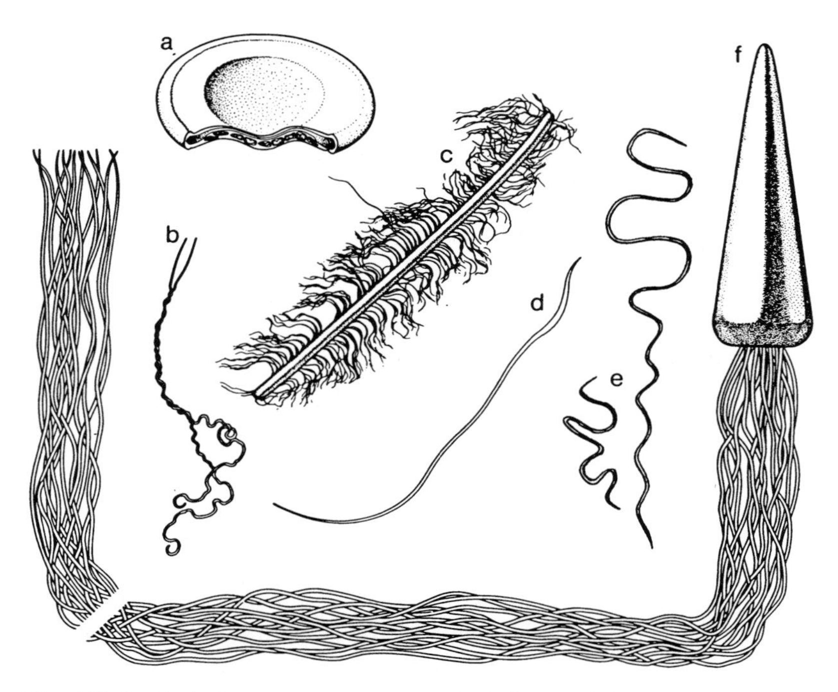

Fig. 10.7. *Variation in insect sperm morphology. (a) The proturan* Eosentonon transitorium; *(b) paired sperm of the firebrat* Thermobia domestica; *(c) spermatostyle and associated spermatozoa of a gyrinid beetle,* Dineutus *sp.; (d) typical sperm from the firefly* Pyractomena barberi; *(e) apyrene (small) and eupyrene (large) sperm of the moth* Plodia interpunctella; *and (f) giant, multiflagellate sperm of the Australian termite* Mastotermes darwiniensis *(modified from Sivinski 1980).*

to as much as 900% in the case of *Drosophila azteca* (Joly and Lachaise 1994). The pentatomid bug *Arvelius albopunctatus* has three sperm lengths (Schrader and Leuchtenberger 1950), while the hymenopteran wasp *Dahlbominus fuscipennis* has as many as five different sperm morphs that vary in size and shape (Lee and Wilkes 1965). Sperm dimorphism is an almost universal phenomenon in the Lepidoptera. One sperm morph, the eupyrene sperm, is the usual type of nucleated sperm that fertilize eggs. The other, the apyrene morph is smaller and lacks any genetic material. Apyrene sperm are produced in large numbers, often comprising in excess of 90% of the total sperm numbers ejaculated (Gage and Cook 1994; Silberglied *et al.* 1984).

1. Inter-ejaculate sperm competition

Relatively few data are available on the functional significance of sperm polymorphisms or their role in sperm competition. Silberglied *et al.*

(1984) reviewed and dismissed three proposed functions for the apyrene sperm of Lepidoptera that were based on the ideas that apyrene sperm aid the movement of eupyrene sperm from the testis and through the female's reproductive tract, or that they provided nourishment to the eupyrene sperm, the female, or the zygote. They presented two alternative hypotheses based on sperm competition theory. They argued that apyrene sperm may represent a type of worker morph that is dedicated to an offensive or defensive role in sperm competition, allowing their eupyrene counterparts to achieve fertilization. They envisage two ways in which this could be achieved: by displacing sperm stored by the female from previous matings and/or by preventing further matings.

Silberglied *et al.* (1984) argued that in Lepidoptera the last male to mate with a female fathers the majority of offspring and suggested that apyrene sperm may be responsible for the removal, inactivation, or destruction of sperm from previous males, generating a last-male advantage. However, it is clear from the studies in Table 10.1 that last-male sperm priority is not the norm in Lepidoptera; as frequently it is the first male that fathers the majority offspring. Across Lepidoptera, the mean value of P_2 is only 0.65 ± 0.05 and within species a bimodal distribution with P_2 values with peaks at the extremes of zero and one is common. These data do not provide general support for a role for apyrene sperm in displacement. Nevertheless, sperm displacement may occur in some species.

Etman and Hooper (1979) dissected females at various intervals after copulation and examined their spermathecae for the presence of sperm. After initial matings, sperm first appeared in the spermatheca 45 min after spermatophore transfer. Following the second mating, however, sperm were lost from the spermatheca during the first 45 min after spermatophore transfer and then reappeared 60 min later. These data support the notion of sperm displacement and led Etman and Hooper (1979) to the conclusion that last male sperm precedence occurred (Table 10.1). However, that the spermatheca is devoid of sperm for over 30 min shows that the sperm of the second male are not directly responsible for this displacement. Etman and Hooper (1979) concluded that some unknown physiological mechanism of sperm expulsion was responsible. Pair *et al.*'s (1977) work with *Heliothis* spp. provides similar evidence for sperm displacement. Backcrosses between hybrids of *Heliothis subflexa* and *Heliothis viriscens* produce sterile males owing to morphogenic defects in eupyrene sperm: eupyrene sperm from hybrid males fail to reach the spermatheca, hence male sterility (Proshold and LaChance 1974; Proshold *et al.* 1975). Apyrene sperm, however, appear normal (Richard *et al.* 1975). Pair *et al.* (1977) used these males in competitive situations with normal males to assess sperm utilization patterns. Normal males had an estimated P_2 of 0.76, while there was a reduction in fertility of females remated to backcross hybrids that predicted a P_2 of 0.83, had these males also transferred normal eupyrene sperm. Sperm counts from spermathecae of doubly mated females showed that after backcross with hybrid second males, 88% of the eupyrene sperm previously stored was lost from the female's spermatheca after the second mating. These data might suggest a role of apyrene sperm in sperm displacement. However, given the results of Etman and Hooper (1979), those obtained by Pair *et al.* (1977) could be explained by some physiological process in the female, triggered by the transfer of a complete spermatophore, regardless of the type of sperm contained within it, or even receipt of seminal fluids that are similarly unaffected by hybrid crossing. Seminal fluids are thought to play a major role in sperm displacement in *D. melanogaster* (Harshman and Prout 1994; Chapman *et al.* 1995).

Silberglied *et al.*'s (1984) second hypothesis for apyrene sperm function proposed that apyrene sperm served the role of avoiding sperm competition by preventing the female from remating. Female Lepidoptera become unreceptive to further males after mating and the duration of the refractory period has been linked to the presence of a spermatophore in the bursa copulatrix (Sugawara 1979) and the presence of motile sperm in the bursa and/or spermatheca (Thibout 1975). Thus, Silberglied *et al.* (1984) suggested that apyrene sperm may represent a 'cheap filler' that delays remating by the female and thereby reduces the risk of future sperm competition. Apyrene sperm are transferred in large numbers and are highly mobile in the reproductive tract, so that females may be fooled into believing that they have received a larger ejaculate than the male has actually provided. Males might benefit by producing apyrene rather than eupyrene sperm because the costs of cell growth and synapsis necessary for the production of effective nucleated cells are bypassed. The idea of 'cheap filler' is also consistent with the hypothesis that spermathecal filling by the first male serves to reduce the risks of sperm competition from future males (see above). By topping up the female's reproductive tract with relatively inexpensive apyrene sperm, males could prevent the sperm of future males gaining access to the female's sperm-storage organ(s).

Variation in ejaculate features within species may provide insight into the functional significance of apyrene sperm. In Cook and Gage's (1995) study of *Plodia interpunctella*, the numbers of sperm in storage had no influence on the number of apyrene sperm ejaculated. If apyrene sperm were involved in sperm displacement, the number of apyrene sperm ejaculated might be predicted to increase with the number of previously stored sperm that must be displaced. Males are capable of adjusting ejaculate features since the number of eupyrene sperm ejaculated does increase with the number of sperm in the female's sperm-storage organ(s). Further, males do transfer larger quantities of apyrene sperm to young virgin females compared with old mated females. Cook and Gage (1995) interpret this result as evidence for a role of apyrene sperm in delaying the onset of sexual receptivity – the 'cheap filler' hypothesis of Silberglied *et al.* (1984). Old females are unlikely to remate so that the transfer of apyrene sperm to old females would be unnecessary. Similarly, males would not need to fill the sperm-storage organ(s) of old females if the likelihood of them remating was low.

Studies of sperm form and function have been made in the sperm heteromorphic drosophilids (Joly *et al.* 1991). Dimorphism in sperm length occurs in *Drosophila teissieri*. Unlike the discrete sperm morphs seen in members of the subgroup *obscura*, dimorphism in *D. teissieri* is characterized by a continuous distribution of sperm lengths with two major peaks. Moreover, there is remarkable geographic variation, with some populations showing monomorphic sperm and others, dimorphic sperm. Joly *et al.* (1991) used this variation in experiments to investigate the influence of sperm dimorphism on patterns of sperm utilization. They crossed females from sperm monomorphic populations with both a sperm monomorphic male from their own population and a sperm dimorphic male from a different population. The experiment was repeated using females from dimorphic populations. They found no significant female effect on P_2 but a significant affect due to males and a significant male $\times$ female interaction (Fig. 10.8). When females from sperm dimorphic populations were tested there was no difference in the proportion offspring sired by the second male when he was either sperm monomorphic or sperm dimorphic. However, when females from sperm monomorphic populations were used, sperm dimorphic males had a significant disadvantage. These data thus raise the question of how sperm dimorphism can be maintained in a population when a mutant individual with monomorphic sperm would be at a selective advantage? Joly *et al.* (1991) followed the arguments of Sivinski (1980, 1984) that long and short sperm morphs may be favoured at different times; abundant short sperm may be favoured early after ejaculation while long sperm may be favoured later if they are better able to resist sperm displacement by future males. This argument relies on the assumption that both sperm morphs are involved in fertilization. However, Snook *et al.* (1994) showed that long and short sperm were functionally nonequivalent in *D. pseudoobscura*. Only long sperm persist in significant numbers in the female's sperm-storage organs and further, only long sperm participate in fertilization. The same is true for *Drosophila subobscura* (Bressac and Hauschteck-Jungen 1996). Thus, the reduced paternity of sperm dimorphic males in competition with monomorphic males may arise because sperm dimorphic males have relatively fewer fertilizing sperm. Bressac *et al.* (1991) found differences in the behaviour of long and short sperm morphs of the sperm dimorphic *obscura* group; within species, long sperm have higher beat frequencies and beat frequency is positively correlated with wave propagation velocity. Further, the kinetics of long sperm change after transfer to the female, with a burst of activity towards high wave propagation velocities for long sperm. These data suggest that the short sperm morphs of *Drosophila* spp. may function in a manner similar to the apyrene sperm morphs of Lepidoptera. One possibility is that sperm movement within the bursa copulatrix works in conjunction with seminal fluid molecules to mobilize previously stored sperm, thereby facilitating displacement. Males may produce cheap small sperm to aid in sperm displacement and long sperm destined for storage in the female's

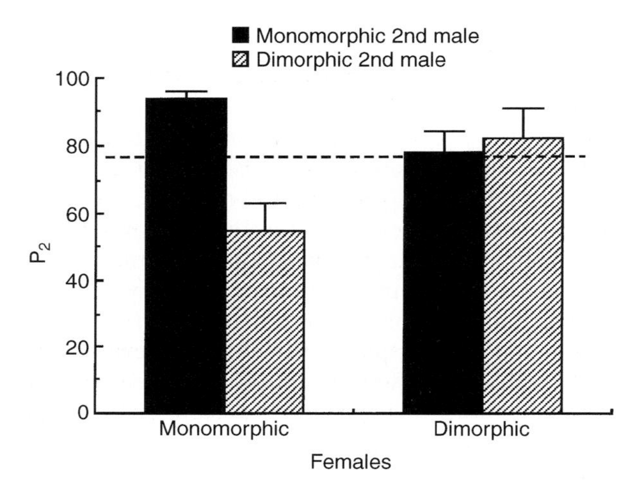

Fig. 10.8. *Outcome of sperm competition experiments in which both a sperm monomorphic and sperm dimorphic male* Drosophila teissieri *were allowed to copulate with a single female. The experiment was repeated using females from each of the sperm monomorphic and dimorphic populations. Sperm dimorphism in this species is characterized by a continuous distribution of sperm lengths but with major modes characterizing short and long sperm types. The dotted line indicates the mean value of P_2 across all treatments, error bars represent 1 SE (data from Joly* et al. *1991).*

spermatheca. Unfortunately, the sperm utilization patterns seen in *D. teissieri* could equally be due to homogamy, that is the tendency for sperm from the female's own population to be more successful in fertilization. Thus, across all possible combinations, there is a tendency for the second male to father the majority of offspring (Fig. 10.8). However, homogamy results in second males from the female's own population gaining more fertilizations than would be expected from sperm competition, while second males from different populations obtain fewer fertilizations than expected. It is not possible to disentangle these effects from the experiments performed by Joly *et al.* (1991).

Parker's (1982) theoretical analysis suggested that sperm competition should maintain small and numerous sperm. Assuming a trade-off between sperm size and number, and that sperm competition conforms to the simplest mechanism of numerical superiority, any increase in sperm size would have an immediate selective disadvantage. Parker's analysis was aimed at understanding why males should not increase their investment in sperm so as to contribute parentally to the zygote. Parker (1993) later examined possible selection via sperm competition on sperm size. He concluded that sperm size should be optimized in relation to competitive weight (a measure of its success in a fertilization raffle) and survivorship, but that its size should be independent of the risk of sperm competition.

However, Pitnick and Markow (1994a) and Pitnick (1996) provide clear evidence that drosophilid sperm can evolve to considerable sizes, even at the expense of sperm numbers.

Parker's (1982) analysis concluded that in the absence of sperm competition an increase in sperm size at the expense of sperm number could be favoured. Competition between the sperm of different males is reduced in *D. pseudoobscura* and *D. melanogaster* due to displacement of previously stored sperm (Table 10.1). The phylogeny of drosophilids places *D. pseudoobscura* and *D. melanogaster* ancestral to the giant sperm-producing *nannoptera* and *hydei* species groups (Karr and Pitnick 1996) so that reduced sperm competition could perhaps favour increased sperm provisioning. Bressac *et al.* (1995) favour this scenario, based on the observation that the tails of giant sperm enter the egg, and that for individual sperm, paternity assurance increases with sperm length. However, Pitnick *et al.* (1995b) show that in many species only a fraction of the sperm tail enters the egg, for *D. bifurca* less than 5%. In a phylogenetic analysis of sperm–egg interactions, Karr and Pitnick (1996) showed that sperm gigantism has evolved independently of sperm tail entry into the egg. The costs of giant sperm production appear considerable. Across species, increasing sperm length results in an increase in the energetic investment in spermatogenic tissue (Pitnick 1996) and delayed male maturity (Pitnick *et al.* 1995a). Further, increased sperm length is associated with a reduction in male fitness because the number of progeny produced per copulation is negatively associated with sperm length in singly-mated females (Pitnick 1993; Pitnick and Markow 1994). Increased sperm length may also be associated with increased sperm competition in multiple mated females: P_2 is 0.93 in *D. melanogaster* (sperm length 1.8 mm) and only 0.50 in *D. hydei* (sperm length 23.3 mm). Across 11 species of *Drosophila* Pitnick (1996) notes a general trade-off between sperm length and sperm number that is predicted to generate an increase in sperm competition because of submaximal insemination (Markow 1985; Pitnick 1993). Parker's analysis shows that even an extremely low level of sperm competition, such as that seen in *D. melanogaster* and *D. pseudoobscura*, should counter any increase in sperm size.

Parker's (1993) analysis of the affects of sperm competition on sperm size noted four special circumstances under which sperm competition could favour an increase in sperm size: (1) ejaculate mass can only increase by increasing sperm size; (2) the competitive benefits of sperm size increase with increasing numbers of sperm in competition; (3) size affects sperm survival and sperm competition risk increases with female remating interval; and (4) size increases competitive ability at the expense of survivorship, but sperm competition risk decreases with female remating interval. We know little of how sperm size affects their competitive ability or survivorship, but it is generally assumed that larger sperm should have greater swimming speeds and be more successful in fertilization. In sperm dimorphic *Drosophila*, the long sperm morphs do have a higher wave propagation once transferred to females (Bressac *et al.* 1991) and only long sperm participate in fertilization (Snook *et al.* 1994; Bressac and Hauschteck-Jungen 1996). If it is the case that the competitive benefits of sperm size increase with increasing numbers of sperm in competition, we should only expect to see sperm gigantism in species with high female mating frequency and, most importantly, where females store sperm from all of their mating partners and sperm mix while in storage. That is, increased sperm size would not be predicted for species with mechanisms of sperm displacement. These criteria appear to be met in the giant sperm-producing drosophilids. However, sperm lengths appear to be much longer than the distances they need to travel within the female so that increased swimming speed of longer sperm is unlikely to represent a significant advantage (Pitnick and Markow 1994). Species with sperm precedence could be selected to have increased sperm size under Parker's (1993) third exception. Because sperm competition increases as sperm precedence mechanisms break down, increased sperm survival, and thus sperm size, could be favoured. A number of studies show that actual competition between sperm becomes increasingly important (P_2 approaches 0.5) with time since the final copulation (Table 10.1). Nevertheless, until we know more about the effects of sperm size on survival and competitive ability, we can only speculate on its evolutionary significance.

Sivinski (1980, 1984) suggested that sperm size may play a role in the avoidance of sperm competition, giant sperm effectively filling the female's reproductive tract so that future males are unable to transfer their sperm to the female's sperm-storage organ(s). Filling of the female's spermatheca may be an important mechanism for avoiding sperm competition in Lepidoptera (Retnakaran 1974) and Gage's (1994) comparative study showed that increased risk of sperm competition was associated with an increase in the length of eupyrene sperm but not apyrene sperm (Fig. 10.9). That only eupyrene sperm persist in the spermatheca (Silberglied *et al.* 1984) provides compelling support for the idea that increased eupyrene sperm size has been selected as a mechanism to resist sperm competition from future males. Further, when subject to nutritional stress, male *P. interpunctella* maintain the size of their sperm while sperm numbers are significantly reduced (Gage and Cook 1994), suggesting that sperm size may have a more significant role to play in male reproductive success than sperm numbers. Similarly, nutritional stress imposed on developing male *D. hydei* was found to influence testis size and the number of sperm produced, but to have no effect on the length of their giant sperm (Pitnick and Markow 1994b).

The production of large sperm is associated with increased testis size in the genus *Drosophila* (Pitnick 1996). Gage (1994) also found that increased sperm competition was associated with increased testis size across Lepidoptera and this relationship may be due to selection for increased sperm size rather than increased sperm numbers. Selection via spermathecal filling would predict that eupyrene sperm lengths, although not necessarily apyrene sperm lengths, should be positively associated

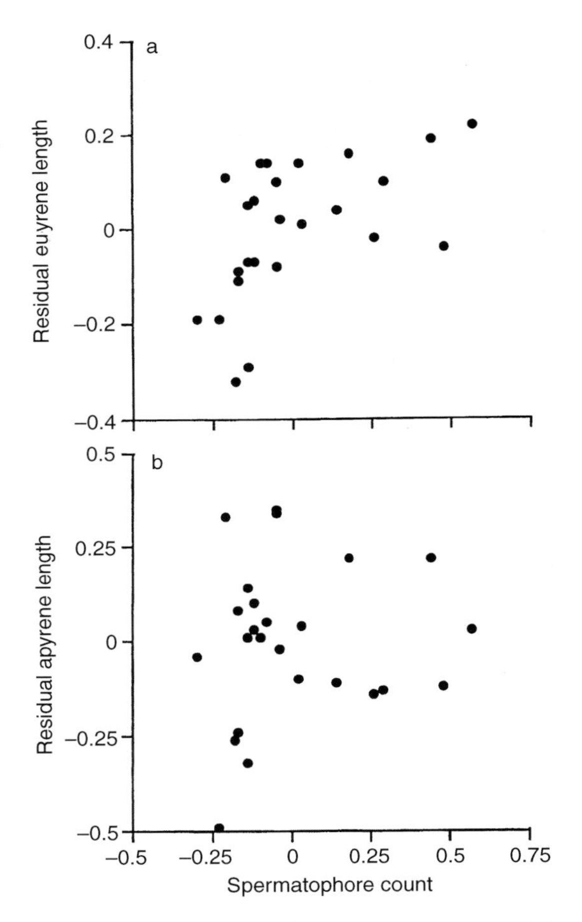

Fig. 10.9. *Associations between sperm competition risk (spermatophore count in wild-caught females) and the lengths of (a) eupyrene and (b) apyrene sperm across 25 butterfly species after phylogenetic subtraction using family means (from Gage 1994, with permission. Copyright 1994 by The Royal Society.)*

with the dimensions of the female's spermatheca. Dybas and Dybas (1981) found such a relationship in featherwing beetles (*Bambara* spp.) where the first male apparently fills the female's spermatheca to capacity, precluding access by future males. However, Pitnick and Markow's (1994) work with the giant sperm-producing drosophilids does not support the spermathecal filling hypothesis.

2. Intra-ejaculate sperm competition

Sivinski (1980, 1984) also noted that intra-ejaculate sperm competition could impose selection on sperm morphology. However, sperm in competition with their siblings could result in a reduction in the effectiveness of the ejaculate as a whole, and Sivinski (1984) suggested that the absence of phenotypic expression (haploid control) may be an adaptation to suppress the potential deleterious effects of intra-ejaculate competition, a suggestion supported by Haig and Bergstrom's (1995) theoretical analysis. Nevertheless, studies of vertebrates show that haploid expression can be an important component of spermatogenesis and sperm function (Erickson 1990), so that it should not be dismissed entirely. Parker and Begon's (1993) models predict a conflict between parental and gametic interests over sperm size and number that is dependent on the degree of inter-ejaculate sperm competition. They found that sperm size could increase at the expense of sperm numbers under intra-ejaculate sperm competition but that this effect decreases with increasing inter-ejaculate sperm competition. Mutations affecting sperm number result in a decrease in sperm size and this effect is increased by inter-ejaculate sperm competition. Parker and Begon (1993) concluded that sperm morphology is likely to be a reflection of the resolution of conflict between haploid and diploid interests. Because the haploid genomes of sperm can vary considerably in their genetic relatedness, Haig and Bergstrom (1995) suggest that intra-ejaculate sperm competition may be a significant obstacle to the evolution of sperm polymorphisms where different morphs have different functions. They suggest that because the sperm produced by haplodiploid species are genetically identical, haplodiploid species might be more likely to exhibit sperm polymorphisms because of the greatly reduced intra-ejaculate sperm competition. Interestingly, the haplodiploid hymenopteran, *Dahlbominus fuscipennis*, has a total of five sperm morphs (Lee and Wilkes 1965). A detailed comparative study focusing on the degree of sperm polymorphism in diploid vs. haplodiploid species is required to test Haig and Bergstrom's (1995) hypothesis.

C. Strategic ejaculation

Parker (1974) suggested that when the risk of remating by females is high, males should invest more heavily in mate guarding. Male insects have been shown to exhibit plasticity in their mate-guarding strategies, increasing their investment when the risks of local mate competition are increased (McLain 1980; Sillén-Tullberg 1981; Sivinski 1983). Parker (1990a,b) also produced a series of sperm competition game models in which he argued that within species, males should vary their ejaculate expenditure, depending on the risks of sperm competition. Because ejaculates can be costly to produce (Dewsbury 1982) males should transfer the minimum necessary to ensure fertilization when there is no sperm competition. But it will obviously pay males to increase sperm numbers to outcompete rivals. Thus, when sperm compete on a numerical basis, males should ejaculate strategically, dependent on the risks of sperm competition. Parker (1990a) showed that the evolutionarily stable ejaculation strategy depended on the information available to males. When males 'know' that they are the first or second male to mate but these roles are assigned randomly, the two males are predicted to ejaculate the same numbers of sperm into the female. However, when roles are not random, so that one male is consistently the first or second male to mate, then the male whose sperm are less likely to achieve fertilization is predicted to ejaculate more sperm than the male in the favoured role. If there is no bias in paternity toward either male then both males are again predicted to ejaculate the same number of sperm. Parker's (1990a) models are based on a mechanism of sperm competition with numerical superiority, hence the advantage of ejaculating more sperm. The best place to look for evidence for strategic ejaculation then is species without sperm displacement. When males mechanically remove previous sperm prior to ejaculation they will effectively avoid sperm competition and ejaculate size should be the minimum required to ensure fertilization. When males use their ejaculates to displace previously stored sperm, ejaculate size is always expected to be high although the number of sperm contained in the ejaculate will depend on whether sperm *per se*, or the accessory secretions are involved in displacement.

A number of studies support the idea of strategic ejaculation by male insects, although few have examined directly the predictions of Parker's (1990a) models. Gage (1991) and Gage and Baker's (1991) work with *Ceratitis capitata* and *Tenebrio molitor* show that when males copulate within the presence of rivals they ejaculate two to three times the number of sperm than when copulating alone. The mechanism of sperm competition is unknown in *T. molitor*, although Gage (1992) later suggested that males may be able to remove sperm from the female using spines on the shaft of the penis when they copulate soon after the first mating. This would appear to negate the necessity, or adaptive advantage, of ejaculating more sperm in the presence of rivals unless the total number of sperm removed remained constant. Siva-Jothy *et al.* (1996) have since shown that sperm removed by the penile spines of *T. molitor* probably come from the male's own ejaculate that leaks into the posterior region of the female's reproductive tract during copulation. The last male to mate achieves the majority of fertilizations immediately after copulation but P_2 declines with time (Siva-Jothy *et al.* 1996) so that numerical sperm competition becomes increasingly important. Increasing ejaculate size when the risks of sperm competition are high may thus give advantage to the copulating male in one of two ways: (1) it may extend the period over which sperm precedence is achieved and/or (2) once precedence has broken down it may increase the male's chances of success in numerical competition. *Ceratitis capitata* shows sperm mixing immediately after copulation (Table 10.1) so that males could similarly increase their fertilization success by ejaculating more sperm. The same appears true for the crickets *Acheta domesticus* and *Grylloides sigillatus* where, within the limits of the experiment, there was a linear increase in the number of sperm transferred by males under increasing male density (Gage and Barnard 1996). Sperm mixing appears to be a common feature of all gryllids studied to date (Table 10.1).

Parker's (1990a) models predict that only the male in the disfavoured role should increase ejaculate expenditure. The above studies do not examine the relative number of sperm ejaculated by first or second males. For *T. molitor* and *C. capitata* P_2 is significantly greater than 0.5 while for *G. sigillatus*, P_2 is significantly lower than 0.5. Thus, for *T. molitor* and *C. capitata*, Parker's (1990a) models predict that it should be only the first male that increases its ejaculate expenditure, while for *G. sigillatus* it should be only the second male that increases ejaculate expenditure. Further, strategic ejaculation is expected only when roles are nonrandom (Parker 1990a). Thus, it is difficult to evaluate Parker's (1990a) models from these studies. It is possible that all males in the population have the same ejaculate expenditure, which is dependent on the current risk of sperm competition for the population. That is, both first and second males have the same ejaculate expenditures which are both elevated with the perceived risk of sperm competition. Such a scenario would fit Parker's (1990a) model where roles are assigned randomly. It is difficult to envisage a general situation where roles would not be random, except perhaps in the case where some males adopt an alternative sneak strategy (see Parker 1990b) or where some ecological factor determines a male's role (see below). Evidence that all males in the population respond equally to sperm competition risk comes from Gage's (1995) study of *P. interpunctella*. Increases in the population density of developing larvae result in increased testis size, and the number of apyrene and eupyrene sperm ejaculated by males when adult (see also He and Tsubaki 1992). Female mating frequency is a positive function of larval population density so that the risk of sperm competition for males increases. Nevertheless, within-population variation is also apparent. Cook and Gage (1995) showed that the number of eupyrene but not apyrene sperm ejaculated by second males increases with the size of the ejaculate transferred by the first male. The adaptive significance of within-population variation in eupyrene numbers will depend on the mechanism of sperm competition. In common with most Lepidoptera, P_2 values for *P. interpunctella* are bimodally distributed about 0 and 1. Between-population variance in ejaculate expenditure (Gage 1995) seems logical if spermathecal filling functions to resist future sperm competition (Retnakaran 1974); it would pay males to transfer larger ejaculates to females in order to maximally fill the sperm-storage organ(s) when the risk that females will remate is high and/or where increased sperm numbers result in increased delays in female remating (Thibout 1975). However, it is difficult to see how such a mechanism could favour increased ejaculate expenditure by second males within populations because sperm competition would not be numerical where second male sperm are prevented from entering the spermatheca by first male sperm. Conversely, if the bimodal distributions of P_2 values seen in

Lepidoptera arise because of the plugging effect of previous spermatophores, second males may be selected to increase ejaculate expenditure if larger spermatophores are more successful in breaching the barrier of first spermatophores and, once breached, success in sperm competition is dependent on the number of sperm transferred from the spermatophore. Again, we need to understand the mechanisms of sperm competition before the adaptive significance of variation in ejaculate expenditure can be assessed.

Evidence for an effect of male roles on ejaculate expenditure comes from work on bushcricket spermatophores. For *R. verticalis* the first male to mate with a female is most often in the favoured role since P_2 is close to 0 (Table 10.1). Thus, males prefer virgin females as mates and selection has favoured extreme protandry (Simmons *et al.* 1994). However, as the mating season progresses, the probability of encountering a virgin female declines so that males are increasingly likely to be in the disfavoured role of second mate. Although males appear unable to assess virginity directly, they transfer around twice as many sperm to 15-day-old females as they do to 9-day-old females (Simmons *et al.* 1993). Field observations show that sexually active 15-day-old females will have mated once and recovered from their first refractory period while 9-day-old females are unlikely to have mated previously (Simmons *et al.* 1994). Wedell (1992) found that male *Decticus verrucivorous* respond to female virginity directly, transferring larger spermatophores when in the role of first mate. However, the mean P_2 for *D. verrucivorous* does not differ significantly from 0.5 so that there is no disfavoured role. In this situation, males are predicted to transfer equal numbers of sperm to females (Parker 1990a).

Roles may be nonrandom when males adopt different mate-finding strategies. For a number of insects males can obtain females either by actively searching or by take-overs from other males. Take-overs occur in two species of Diptera, *Scatophaga stercoraria* (Parker 1970d) and *Dryomyza anilis* (Otronen 1989). In both species take-overs are size dependent in that larger males are more successful in take-over attempts and more likely to engage in them, while small males are both less successful in resisting take-overs and subject to a greater number of attempts (Sigurjónsdóttir and Parker 1981; Otronen 1993, 1995). Thus, small males should be consistently in the role of first male because they are subject to a higher probability of take-over. For *S. stercoraria* neither copulation duration (Parker 1970a) nor resultant P_2 (Simmons *et al.* 1996) are influenced by whether the mating results from searching for a newly arrived female or from a take-over from a previous male. However, for *D. anilis*, P_2 is up to 10% lower for matings following take-overs (Otronen 1994a). These differences may illustrate the importance of different sperm competition mechanisms. Male *S. stercoraria* displace 80% of the females stored sperm with their own ejaculate so that the number of sperm transferred by the first male has little or no affect on sperm utilization (Parker and Simmons 1991; see above). The ejaculation strategy for *S. stercoraria* is thus subject to a simple optimization with respect to the costs of mate searching (Parker and Stuart 1976; Parker and Simmons 1994). In contrast, second males of *D. anilis* only gain an advantage at fertilization through repeated copulations (Otronen 1994b) and tapping (Otronen 1990) during which first-male sperm are mobilized from the spermathecae to the bursa were they mix with the second male's sperm (Otronen and Siva-Jothy 1991). Numerical sperm competition, required by Parker's (1990a) evolutionarily stable strategy models is thus a prominent feature of sperm utilization in *D. anilis* so that variation in male ejaculation strategies may well evolve. Unfortunately, Otronen (1994a) did not count the numbers of sperm transferred by males in different roles. Nevertheless, Parker's (1990a) models predict that the male in the disfavoured role should increase ejaculate investment. For *D. anilis*, the second male is disfavoured since P_2 is significantly lower than 0.5 (Otronen 1994a). Males attempting take-overs are thus predicted to increase the number of tapping sequences and/or copulation bouts. Although the sample sizes in Otronen's (1994a) study are small, second males during take-overs do appear to have higher numbers of tapping sequences compared with ordinary matings, but the number of copulation bouts are similar.

These studies lend support to the notion that within species, males should invest in sperm in relation to the perceived risks of sperm competition. However, future studies need to consider the mechanisms of sperm competition and should also test the assumptions and predictions of Parker's (1990a) models more explicitly. In particular, the prediction that roles are an important determinant of selection requires study because in many of the examples cited above, males seem to respond to sperm competition risk, regardless of whether roles are random or nonrandom. The basis for the effect of roles is the trade-off between ejaculate expenditure and future mating opportunities. When roles are random, a male that trades future opportunities for increased current investment should not have a higher average fitness than a male who maintains constant investment. It would be useful to examine directly this assumed trade-off.

D. Seminal fluids

Sperm are not the only constituent of ejaculates. The males of many species incorporate nutrients and chemical substances within the ejaculate that may be subjected to selection through sperm competition (Leopold 1976; Chen 1984). Displacement of previously stored sperm in *D. melanogaster* may involve chemical substances in the seminal fluids produced by the main cells (Harshman and Prout 1994). Other substances in the *Drosophila* ejaculate are involved in suppressing female receptivity to remating and increasing oviposition (Baumann 1974). Prostaglandins appear to be an active constituent of the ejaculate of orthopterans that increases oviposition and suppresses remating (Stanley-Samuelson and Loher 1986). Sperm competition is likely to favour the evolution of substances that suppress female remating because these will reduce the risks of sperm competition or displacement from future males. Increased oviposition during the refractory period will ensure paternity of a greater proportion of the female's lifetime egg production. The induction of a sexual refractory period is common in insects. In some species the chemically induced refractory period lasts for the female's entire lifespan, thereby removing all future risk of sperm competition (Reimann *et al.* 1967). In others, the length of the refractory period is more variable (Smith *et al.* 1990). In bushcrickets, the length of the refractory period is positively associated with the amount of ejaculate transferred, both within (Gwynne 1986; Simmons and Gwynne 1991) and between species (Wedell 1993). Ejaculate volume in Lepidoptera appears to be a major determinant of the female's remating interval (Labine 1964; Sugawara 1979; Wiklund and Kaitala 1995). Thus, selection for increased ejaculate size, independent of sperm number, may arise via the effect ejaculate size has on reducing the risks of future sperm competition. Karlsson (1995) has shown that across 21 species of butterfly, ejaculate weight increases with the degree of polyandry and Bissoondath and Wiklund (1995) show that males of polyandrous species incorporate greater quantities of protein into the ejaculate than those of monandrous species. Within species, the degree of polyandry is negatively associated with ejaculate mass (Kaitala and Wiklund 1994). These data are consistent with the hypothesis that substances contained within the seminal fluids have been selected in the context of sperm competition. This effect may arise independent of increases in sperm numbers, although no studies have yet examined how sperm numbers vary in relation to ejaculate volume. These data are also consistent with the idea that, in general, Lepidoptera have been subject to selection to resist sperm competition, rather than to increase success in competition when it occurs (see above). Selection acting on ejaculate substances is predicted to generate a positive association across species between the size of male accessory glands and sperm competition risk.

VII. SPERM MANIPULATION: A FEMALE PERSPECTIVE

Sperm competition was viewed by Parker (1970c) as an extension of male–male competition, favouring adaptation in males for achieving greater numbers of fertilizations in the competitive arena of the female's reproductive tract. However, females are unlikely to be passive arenas for male combat. Sperm storage and utilization is ultimately under the control of the female and, as Lloyd (1979) pointed out, female choice might also be expected to continue during and after copulation, especially where males have evolved mechanisms that bypass female attempts to choose among potential mates prior to copulation. Thus, Lloyd (1979) suggested that females may manipulate ejaculates, selectively storing, using, or digesting them, dependent on the characteristics of the copulating male. Indeed, multiple mating, coupled with an ability of females to store and maintain sperm in often complex sperm-storage organs, has been interpreted as an adaptation in females for mate choice (Lloyd 1979; Walker 1980; Sivinski 1984; Simmons 1986).

Knowlton and Greenwell (1984) provided a theoretical analysis of the influence of female interests on the evolution of sperm competition avoidance mechanisms in males. They assumed that mechanisms of sperm competition avoidance were costly to females, an assumption that now has empirical support (Chapman *et al.* 1995). Knowlton and Greenwell's (1984) models recognized that selection on females could prevent the evolution of sperm competition avoidance mechanisms in males. Nevertheless, selection on males for the avoidance of sperm competition is likely to be intense and the evolutionary outcome of sexual conflict over sperm competition avoidance is likely to reflect the relative costs for females and the benefits for males. In many cases the costs for females will be typically less than the benefits for males (Knowlton and Greenwell 1984). Further, the costs of sperm competition avoidance for females may even be outweighed by female benefits. Such benefits could include the potential for females to determine paternity of their offspring. Females could store sperm from the first male they encounter to ensure fertility, and then allow sperm displacement only from males that are of higher quality than their first mate (Thornhill and Alcock 1983). Alternatively, if sperm ageing resulted in a depression in fertility and/or zygote fitness, sperm displacement would constitute a significant direct advantage for females so that their interests need not conflict with those of males. Proximate mechanisms for the avoidance of sperm competition may also provide significant benefits to females (Waage 1984; Wilcox 1984; Fincke 1986; Tsubaki *et al.* 1994). Thus Knowlton and Greenwell (1984) concluded that the resolution of sexual conflict would probably favour the evolution of sperm competition avoidance mechanisms in males. Nevertheless, recent work strongly supports the notion that an arms race equal in intensity to any host–parasite tussle exists between the sexes in determining the nature of traits selected through sperm competition (Rice 1996). The main cell products involved in sperm displacement in *D. melanogaster* decrease female survival in a dose-dependent manner (Chapman *et al.* 1995). In an elegant experiment, Rice (1996) showed that when female *D. melanogaster* were prevented from coevolving with males, the male sperm displacement mechanism rapidly evolved to a state where it was highly detrimental to the static female line. These experiments show that the evolutionary interests of females can be in conflict with those of males, and play an important part in moulding the traits we see.

The sexual conflict over insemination envisaged by Lloyd (1979) led Eberhard (1985) to propose that, in general, the rapid and divergent evolution of animal genitalia, may be due to runaway female choice. The female reproductive tracts consist of complex networks of pipes and spermathecal sacs, while the morphology of male genitalia often reflects the internal structure of the female. Thus, Eberhard (1985) envisaged female tracts evolving in such a way as to make it increasingly difficult for males to achieve insemination while male genitalia evolve to match any shift in females. Eberhard (1985, 1991, 1994) also suggested that females may assess males on the basis of their ability to stimulate them during copulation, and recognized a widespread occurrence of copulatory courtship in insects; 81% of 131 species exhibited behaviour patterns during copulation that he interpreted as copulatory courtship. He argued that females exercised cryptic female choice and that copulatory courtship by males functions to persuade females to transport and retain ejaculates within their sperm-storage organs, thus imposing sexual selection on males.

Eberhard (1993) considered and rejected the notion that selection on male genitalia could operate via a 'good genes' model of sexual selection. However, we see no reason why good genes models should not predict the same elaboration in genitalia as an arbitrary trait, or Fisherian model of sexual selection. If there was covariation between genital elaboration and male quality, by using sperm from males with elaborate genitalia females would bias fertilization toward males of greater quality and gain fitness benefits for their offspring. Eberhard (1993) claimed that within species variation in genital complexity was too small to allow selection via good gene mechanisms and genitalia are unlikely to be costly to produce. However, runaway selection also relies on variation in genital elaboration. Variation in male genitalia has not been studied widely, although Cordero and Miller (1992) noted a relationship between male size and genital horn length in the damselfly *Ischnura graellsii* that may be responsible for variation in the ability of males to remove sperm. The costs of producing complex genital structures have not been examined but we see no reason why they should not be similar to those associated with the production of other sexual traits (Arnqvist 1994).

The genital environment of the female could also play a role in shaping sperm morphology. The female insect genital tract is lined with cuticle and is effectively an externalized structure (albeit involuted). It has several potentially conflicting physiological functions: (1) it provides a physical conduit during copulation, and must therefore be resilient enough to withstand the stresses and strains of copula; (2) part of it is developed into structures that store and maintain sperm and seminal fluid in a viable condition; (3) it must exclude pathogens from the stored sperm and the oviducts (by physiological and/or anatomical mechanisms); and (4) it must coordinate the transport, fertilization and coating (i.e. application of accessory gland secretions) of eggs, from the oviducts, past the sperm-storage organs to the exterior. Because of the different selection pressures acting on each of these functions in different species, it is likely that even the females of closely related species will have very different physiological (and possibly even anatomical) conditions within their genital tracts. Evidence that such differences might occur and, more importantly, influence the migration, storage and utilization of sperm within the female's genital tract comes from studies of heterospecific matings (Katakura 1986; Howard and Gregory 1993). Katakura (1986) showed that although heterospecific matings appeared to proceed normally, and that sperm were transferred, these sperm failed to migrate to the female's sperm-storage organs. It is therefore likely that conditions in the female's genital tract will select for efficient sperm function in the context of that environment. If we accept that sperm function partly manifests as form, then sperm morphology will, to some extent, be selected by these variables.

Eberhard (1996) speculated on 20 different means by which female behaviour, morphology and/or physiology could directly or indirectly contribute to variation in nonrandom paternity. Of Eberhard's (1996) suggested mechanisms, those relevant to our discussion can be broadly categorized as occurring at one of two stages in the reproductive process. Females could directly influence insemination so that only sperm from desirable males gained access to their sperm-storage organs, following which sperm are utilized at random. Alternatively, females could exercise sperm selection after insemination so that female choice of sires occurred via the nonrandom utilization of sperm from a pool that contained the sperm of all potential sires. The ability of females to exercise control over paternity, and the mechanisms by which this can be achieved are likely to be strongly dependent on the mechanisms of sperm competition. Control over insemination could be an effective means of female choice in species with a mechanism of numerical sperm competition, because females could determine the relative numbers of sperm transferred by each of their mating partners. Thus, species that might yield evidence for female influence over paternity would be those with intermediate values of P_2. The high variance in P_2 that is associated with such cases (Table 10.1) may reflect the influence of females. For a mechanism of sperm selection to be an effective means of female choice requires the storage of sperm from a diversity of males so that females have at least the potential to choose among sperm genotypes. Sperm selection would be unlikely where sperm displacement mechanisms have evolved in males because almost all of the sperm available to females at oviposition would come from the last male to copulate. Sperm selection by females is thus less likely to be found in species with high values of P_2 and low variance. Females of such species may be constrained to exercise choice either prior to copulation or, if they can control insemination, by influencing the degree of sperm displacement.

While the arguments for a role of female influence over sperm utilization are compelling, identifying variation in P_2 that results from female choice is difficult. Female choice as a mechanism of nonrandom mating was an issue of great controversy (Partridge and Halliday 1984). The combination of intense male–male competition and often subtle or cryptic mechanisms of female choice make it difficult to disentangle the causal mechanisms underlying observed nonrandom mating, particularly since the traits that enhance male success in competition are often also those subject to female choice. The problems of identifying female influences over nonrandom paternity are even more daunting because the processes involved in sperm transfer, storage and utilization within the female's reproductive tract are often impossible to observe. Further, sperm competition, the equivalent of male–male competition at the gametic level, is very intense and may easily obscure female effects, or even pre-empt female control where sexual conflict over the evolution of sperm displacement mechanisms has been resolved in favour of males (Knowlton and Greenwell 1984). An understanding of the mechanisms of sperm utilization are thus essential for progress to be made in this area.

Eberhard (1996) looks at current literature from the female's perspective to illustrate potential mechanisms by which females could influence paternity. Many arguments put forward by Eberhard for female control can be countered by arguments for male control; the differing views are rather like the faces of Dawkins' (1982) Necker Cubes. In reality, the observed outcome of fertilization is likely to represent a resolution of conflict between males and females, a conflict that will be resolved in favour of the sex whose costs of random paternity (or benefits of nonrandom paternity) are greatest. The work with *Drosophila* provides perhaps the clearest illustration of sexual conflict; last males obtain some 90% of fertilizations even though the cost to females associated with sperm displacement favours adaptations in females that counter further evolution of male displacement ability (Rice 1996). Here, we concentrate our attention on the few examples in which both male and female influences on sperm transfer, storage or utilization have been empirically examined.

A. Control over insemination

The most obvious way in which females can influence the paternity of their offspring is to be selective about which males they copulate with. In many odonates females have two distinct oviposition strategies, they either copulate with the male resident on the territory before oviposition or they oviposit without copulation (Waage 1979a; Koenig 1991). In *Calopteryx splendens xanthostoma* females that oviposit without copulation do so by actively rejecting the copulation attempts of the territory owner (Siva-Jothy and Hooper 1995). DNA analysis of the sperm within the sperm-storage organs of the female reproductive tract revealed that there was a greater genetic diversity of sperm in both the spermathecae and bursa of females that oviposit without copulating (Siva-Jothy and Hooper 1995) so that they are likely to produce offspring sired by a greater number of males than copulating females. The lower genetic diversity of sperm in copulating females arises because territory owners displace sperm from the bursa copulatrix prior to delivering their own ejaculate (Siva-Jothy and Hooper 1995) thereby gaining 98% of fertilizations (Hooper and Siva-Jothy 1996). Thus, females may indeed exploit the mechanism of sperm displacement to bestow paternity on males of higher quality than their previous mates, simply by allowing them to displace sperm (Thornhill and Alcock 1983). When males are less attractive they can deny them paternity and utilize sperm stored by males that were presumably deemed attractive on previous visits to the oviposition site.

Thornhill's (1976, 1983) work with hanging flies *Hylobittacus apicalis* and *Harpobittacus apicalis*, was the first to demonstrate the influence females can have over insemination. Males offer their mates a prey item on which to feed during copulation. Females prefer males with larger prey items, rejecting males offering small prey items prior to copulation. However, females also determine the duration of copulation, terminating copulations with males offering small prey items before insemination is completed (Fig. 10.10a). Female control over copulation duration appears to be a general phenomenon in insects where males provide prey items at copulation (see also Thornhill 1979; Svensson *et al.* 1990). In at least one of these species, *Panorpa vulgaris*, the outcome of sperm competition conforms to a mechanism of random sperm mixing so that copulation duration, and thus the number of sperm transferred, is directly related to a male's paternity expectation (Thornhill and Sauer 1991). Thus, females in gift-giving species may trade paternity for immediate nutritional benefits offered by males.

Female control over insemination is also a common feature of orthopteran species where the male's ejaculate is transferred from an externally attached spermatophore. Transfer of sperm from the spermatophore takes approximately 60–70 min in the field cricket *Gryllus bimaculatus* (Simmons 1986). After copulation, females remove and consume the externally attached spermatophore, often before insemination is complete. Females appear to adopt spermatophore removal as a mechanism of mate choice, removing the spermatophores of some males before insemination, yet leaving those of others attached often longer than necessary for insemination (Simmons 1986, 1991a). Moreover, females remate with preferred males, accepting multiple ejaculates. Females store all of the sperm received from multiple matings and the mechanism of sperm competition is one of numerical superiority (Simmons 1987b; Parker *et al.* 1990). Thus, by allowing spermatophores of preferred males to remain attached for full insemination, and by copulating repeatedly with the same male, females bias the paternity of their offspring in favour of preferred mates. Female choice in this species may contribute to the competitive fitness of the female's offspring (Simmons 1987a).

Sexual conflict over insemination is evident in most field crickets, where males enter a period of postcopulatory guarding during which they try to subvert female attempts to remove the spermatophore or move away in search of other males (Fig. 10.11) (Simmons 1991b; Zuk

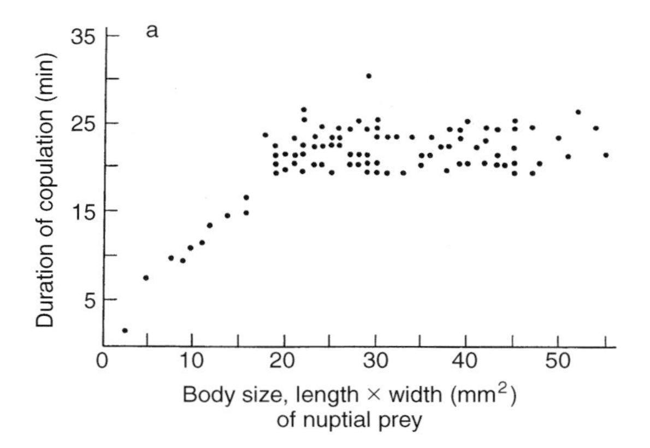

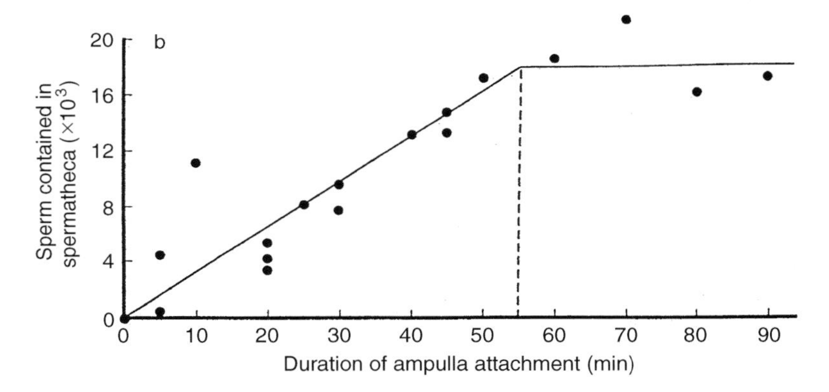

Fig. 10.10. *Female control over the number of sperm inseminated. (a) male hanging flies* Hylobittacus apicalis *offering larger nuptial prey items are allowed to copulate for longer and thereby inseminate more sperm (from Thornhill 1976); (b) male decorated crickets* Grylloides sigillatus *attach a gelatinous spermatophylax to the ampulla of the spermatophore. Females remove the ampulla, and thereby terminate insemination after completing the spermatophylax meal. Males feed females to ensure complete insemination (from Sakaluk 1984, with permission. Copyright by the University of Chicago Press.)*

and Simmons 1997). Males' interests will always be served by successful insemination. However, females' interests may best be served by insemination only from males of high quality. Post-copulatory mate guarding by male *G. bimaculatus* does not appear sufficient to overcome female interests and a male's success in postcopulatory guarding has been suggested as a means by which females assess male quality (Thornhill and Alcock 1983). Males of the decorated cricket *Grylloides sigillatus* have

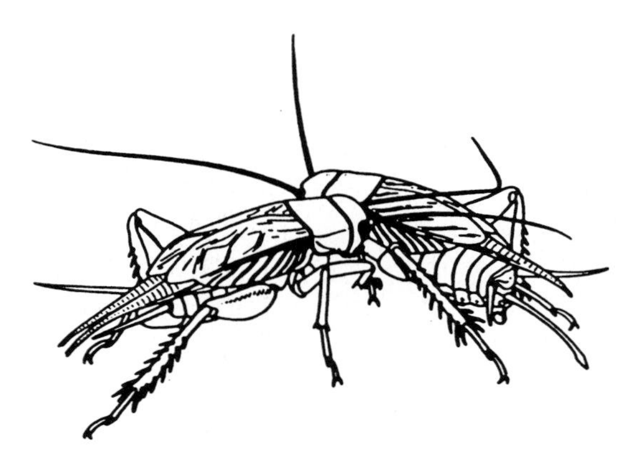

Fig. 10.11. *Male field crickets guard females after spermatophore transfer by maintaining constant antenna contact and responding aggressively to restless females. Females often manage to evade male guarding attempts and remove spermatophores, thereby avoiding insemination.*

adopted an additional means by which to gain control over insemination. Attached to the spermatophore is a gelatinous mass, known as the spermatophylax, which is grasped by the female when she attempts to remove the spermatophore after copulation. Consumption of the spermatophylax delays spermatophore removal until the ejaculate has been transferred (Sakaluk 1984), suggesting that in *G. sigillatus*, males have regained control over insemination (Fig. 10.10b). Nevertheless, Sakaluk and Eggert (1996) argue that the spermatophylax may represent an honest signal of male quality that facilitates female choice. Sakaluk (1985) noted that there was considerable variation in the mass of spermatophylax material transferred by males and concluded that because small spermatophylaxes were consumed before the mean time at which complete sperm transfer occurred (roughly 55 min) that females would, by removing the sperm containing ampulla of the spermatophore early, reduce a male's paternity. Sakaluk and Eggert (1996) used experimental manipulations of ampulla attachment duration to show that, as in *G. bimaculatus*, the number of sperm transferred by the male is an important determinant of paternity. Like *G. bimaculatus*, the mechanism of sperm competition in *G. sigillatus* is predominantly one of random sperm mixing; sperm displacement is weak. Sakaluk and Eggert (1996) argue that females exercise choice by controlling sperm transfer and paternity, favouring males of high quality, as indicated by the size of their nuptial gift. However, the basic assumption of their argument is that variance in spermatophylax size is not associated with variance in sperm numbers contained in the ampulla. Indeed, they assume that sperm numbers do not vary. If spermatophylax size and sperm content of the ampulla covary, females may not be actively influencing insemination at all. Gage and Barnard (1996) have found that spermatophylax size and sperm

Fig. 10.12. *Nuptial feeding is widespread in the tettigoniids. Male* Requena verticalis *invests just enough material to ensure complete sperm transfer when the potential mating rate is high but invests extra material, as paternal investment, when potential mating rate is low (Simmons 1995a; photograph by L. W. Simmons).*

number do covary in *G. sigillatus*, suggesting that males may be in control of the numbers of sperm transferred to the female, varying the size of the spermatophylax with regard to the number of sperm they choose to ejaculate. Ejaculate size and spermatophylax size appears to be adjusted in relation to the male's current risk of sperm competition (Gage and Barnard 1996). The transfer of a protective spermatophylax is widespread in the tettigoniids (Fig. 10.12) and Wedell (1993) suggests that because spermatophylax size and ejaculate size covary across species, sperm competition may have been responsible for its evolutionary origin (see also Vahed and Gilbert 1996). Because of sperm mixing, relative spermatophore size, and thus sperm transferred, is the principal determinant of paternity in *D. verrucivorous* (Wedell 1991). Female choice, implemented through interference with insemination, may be generally responsible for the evolution of counter adaptations in males, such as nuptial feeding and coercive mating, that subvert female interests (Thornhill 1986; Thornhill and Sauer 1991).

B. Control over sperm storage

In some species, males ejaculate directly into the female's sperm-storage organs (Ono *et al.* 1989; Gack and Peschke 1994) while in others sperm transportation to the storage organs may be directly influenced by muscular contractions of the female's reproductive tract (Drummond 1984; Heming-Van Battum and Heming 1986; LaMunyon and Eisner 1993). Female control over sperm transport and storage in such cases may be an avenue by which they could choose among males (Birkhead and Møller 1993; Birkhead *et al.* 1993). Evidence that females can control sperm usage comes from cases of haplodiploid hymenoptera, where females facultatively choose whether to fertilize their eggs and produce males or females (King 1962; Werren 1980).

Female control over sperm entry and exit from the spermatheca has been examined in two species of beetle in which the spermatheca has associated musculature, implying female control over sperm storage and release (Villavaso 1975; Rodriguez 1994). By cutting the spermathecal muscle of female boll weevils *Anthonomus grandis*, Villavaso (1975) showed that a functional spermathecal muscle was not necessary for sperm storage; sperm either enter the spermatheca owing to the fluid pressures involved in ejaculation and/or they swim there under their own control. The spermathecal muscle in Chrysomelid beetles *Chelymorpha alternans* does influence the uptake of sperm (Rodriguez 1994) and in both species, it controls the exit of sperm from the spermatheca. Females with cut spermathecal muscles were unable to utilize sperm stored in their spermathecae and thus, laid infertile eggs. Moreover, in *A. grandis* sperm displacement by second males was reduced from 66% to 22% by cutting the spermathecal muscle. These data thus show that females assist in sperm displacement by second males. However, whether females show selective cooperation with males, dependent on mate phenotype has not been established. It is possible that males exploit a sensory system in females that triggers sperm release. In most insects the female reproductive tract acts as an egg-processing conduit in which eggs must be shunted from the oviducts to the site of fertilization, positioned correctly for fertilization and then shunted out of the female (often while having accessory gland secretions applied to them). Given the logistics of the second role it is not surprising that the females of some insect species have a sophisticated proprioreception system associated with their genital tract (Miller 1987; Siva-Jothy 1987). Copulating males may exploit this sensory system to elicit sperm ejection by the female. Miller (1987, 1990, 1991) suggested that copulating male damselflies and libellulid dragonflies might stimulate the campaniform sensillae that line the female's genital tract, and which control the fertilization reflex, in order to effect the ejection of stored sperm. He showed that artificial stimulation of these sensillae resulted in the reflex contraction of the spermathecal muscle (Miller 1990). Helversen and Helversen (1991) argued that male *Metaplastes ornatus* similarly mimic egg passage through the females reproductive tract to affect sperm removal. Alternatively, males may chemically stimulate the reproductive tract of the female to release sperm, which is implicit in the work on *D. melanogaster* (Harshman and Prout 1994). A lack of spermathecal muscle function in Villavaso's (1975) study would have prevented the female from responding to male signals. The evolution of signals in males for sperm release may thus arise because of pre-existing sensory biases in females (Ryan *et al.* 1990). In general, the evolution of seminal secretions that contain chemicals which influence female sexual receptivity, rates of oviposition, or sperm movement, may have evolved under selection imposed via female physiology, in that males better equipped to elicit responses in females will be at a selective advantage (Cordero 1995; Eberhard and Cordero 1995; Eberhard 1996).

The influence of copulatory courtship (*sensu* Eberhard 1994) on ejaculate distribution and sperm storage is perhaps best illustrated by the work of Otronen and Siva-Jothy (1991). Between multiple copulations with the same female, male *Dryomyza anilis* tap the female's external genitalia with their genital claspers and females emit a droplet of sperm prior to remating. Tapping results in the mobilization of sperm stored in the spermatheca from previous matings, into the bursa copulatrix where they mix with the second male's ejaculate. Sperm then re-enter the spermatheca and about 50% of the sperm remaining in the bursa are emitted in the droplet. The more tapping sequences performed by the male the lower the representation of his sperm in the emitted droplet (Otronen and Siva-Jothy 1991). Thus, a male's success in fertilization increases with the number of tapping sequences performed (Otronen 1990). If females were able to control the number of tapping sequences they would be able to determine the success of their mating partners in gaining fertilization. Females do actively resist males during tapping sequences and the strength of female resistance increases with successive copulation bouts (Otronen 1989). Paternity is not related to female resistance *per se* (Otronen 1990), but large males are able to perform a greater number of tapping sequences (Otronen 1990). Thus, the inevitable consequence of sexual conflict over the number of tapping sequences and resultant distribution of sperm within the female's tract, is that males most successful in exerting their control over the female are likely to achieve higher paternity. By resisting all males, females may effectively exercise a form of passive female choice for those males vigorous enough to subdue them.

Finally, a recent study by V. Rodriguez, W. G. Eberhard and D. Windsor (unpublished data) supports the notion that variation in male genitalia can affect sperm storage. Male *C. alternans* with longer genital flagella father more offspring when females mate with either two or three different males. Males that copulate for longer also father more offspring. Rodriguez and colleagues also show that females are more likely to emit sperm, and therefore less likely to store sperm, when mating with males that have short flagella. A mechanism of sperm mixing appears to facilitate an advantage to the male with the greater number of sperm in the females sperm-storage organ(s) (Table 10.1). The flagella is inserted to the full length of the spermathecal duct and often into the spermatheca itself (Eberhard 1996). The sperm migrate from a spermatophore placed in the female's bursa along the length of the spermathecal duct containing the flagella, and into the spermatheca. Experimentally cutting the flagellum increased the likelihood of sperm emission, demonstrating that an intact flagellum is essential for successful transfer of sperm to the sperm-storage organ(s). It is possible that females respond in a selective manner to males that can stimulate deep inside their reproductive tract (Eberhard 1985). Alternatively, males with longer flagella may simply be more efficient at delivering their ejaculate to the site of storage, irrespective of female behaviour. Distinguishing between male and female influences on paternity are likely to prove difficult (Simmons *et al.* 1996).

C. Sperm selection

The final avenue open to females for mate choice occurs at fertilization; females have the potential to exercise choice by using only the sperm from particular males to fertilize their eggs (Sivinski 1984). Sperm selection may be possible since, in a number of heterogametic organisms, sex ratios can be varied facultatively, suggesting that females at least have the ability to distinguish between male-producing and female-producing sperm (Werren and Charnov 1978). Nevertheless, distinguishing between competitive superiority of sperm and female sperm selection in generating nonrandom paternity is likely to prove difficult (Simmons *et al.* 1996).

There are a number of sperm competition studies of insects where variation in P_2 is indicative of homogamy (see Table 10.1; Hewitt *et al.* 1989; Bella *et al.* 1992; Gregory and Howard 1994; Robinson *et al.* 1994). When females are mated to a conspecific and heterospecific male, a significant tendency for eggs to be fertilized by the conspecific male overlies the basic pattern of sperm utilization. In all of these studies P_2 is intermediate, suggesting extensive sperm mixing. The bias towards conspecific sperm is akin to the loaded raffle mechanism of sperm competition proposed by Parker *et al.* (1990) in that the conspecific male's sperm have an advantage over and above their numerical representation. There are two mechanisms by which loading could occur: females may actively select conspecific sperm to fertilize their eggs or conspecific sperm may be competitively superior to heterospecific sperm. An interaction between these mechanisms is also possible in that a competitive disadvantage for heterospecific sperm may arise because of a lack of adaptation to the environment in the female's reproductive tract. In evolutionary terms, females can ensure fertilization by genetically compatible mates by keeping the environment in the reproductive tract within narrow physiological limits. In their hybridization studies of *Allonemobius fasciatus* and *Allonemobius socius*, Gregory and Howard (1994) found that there was almost total superiority of conspecific sperm when in competition with heterospecific sperm, irrespective of mating order. Heterospecific sperm are capable of fertilizing eggs – females do produce hybrid offspring when multiply mated to heterospecific males. However, offspring production by female *A. fasciatus* is greatly reduced when females mate only twice with heterospecific males (Howard and Gregory 1993). Thus, postinsemination barriers to fertilization exist in these species but are dependent on the numbers of sperm present in the reproductive tract. The sperm number dependency might imply that the barrier is determined by the competitive abilities of sperm since increasing sperm numbers of heterospecific males can overcome the homospecific advantage. The proximate mechanisms responsible for sperm selection may include species-specific molecules associated with gamete recognition during fertilization interactions (Howard and Gregory 1993). Alternatively, the physiological environment of the female's reproductive tract may selectively incapacitate sperm from incompatible males, a phenomenon recently demonstrated in the compound ascidian *Diplosom listerianum* (Bishop 1996; Bishop *et al.* 1996).

The above examples use heterospecific matings in which there exist major genetic differences between the sperm transferred by males. Sperm selection, if it occurs, acts as a species-isolating mechanism so the examples do not illustrate sexual selection. For females to choose between males of their own species would require much finer levels of discrimination. Childress and Hartl (1972) described a phenomenon in *D. melanogaster* that provides some evidence for an ability for females to exercise sperm selection, even within a single ejaculate. Using a strain of *D. melanogaster* carrying a chromosomal translocation, T(1;4)B^s, they found that females produced a decreasing proportion of offspring that were the products of B^s + 4-bearing sperm with time since mating. Females were inseminated by only a single male in these experiments. Childress and Hartl (1972) interpreted this result as selective use of sperm that is dependent on exposure of the reproductive tract to B^s + 4-bearing sperm – females essentially learn to discriminate against B^s + 4 sperm. They consider various alternative arguments for the observed 'brood' effects, including sperm competition. Different sperm genotypes may also have different viabilities and/or motilities (Gromko *et al.* 1984). Childress and Hartl (1972) argued that if B^s + 4 sperm were less competitive than other sperm types, their representation in offspring should increase with time as the more competitive genotypes are used up. However, the reverse result was obtained. The result could also be explained if there were nonrandom loss of sperm from the female's reproductive tract, for example if B^s + 4 sperm had a shorter life-span than other sperm genotypes, or their competitive ability declined with age. If this were the case, identical patterns should be obtained when females are remated after exhausting their current sperm supply. However, when Childress and Hartl (1972) remated females the discrimination against $B^s + 4$ sperm was even greater than that seen in females mated once, suggesting that females 'remembered' the disfavoured genotype and supporting their conclusion that females were capable of sperm selection.

Evidence that females may influence nonrandom paternity after multiple matings comes from the carefully designed experiments of Lewis and Austad (1990). They assessed the levels of P_2 attained for 11 pairs of male *Tribolium castaneum* that were allowed to copulate within eight replicate females. Thus, they were able to partition variance in P_2 that was due to differences between males and that due to differences between females. They found significant male and female effects; 17.8% of the variance in P_2 was due to differences among male pairs while 58% of the variance was due to differences among females. This constitutes strong evidence for a female influence over paternity. Because *T. castaneum* appears to exhibit numerical sperm competition (Table 10.1), much of the variance among males may have been due to differences in ejaculate size. Large males attained a higher P_2 and may have transferred greater quantities of sperm, either through longer or repeated copulations. Second males were found to exhibit greater mounting frequencies and longer mount durations, suggesting that males are able to recognize and respond to the risks of sperm competition by increasing ejaculate expenditure (Parker 1990a). This could account for the fact that second males tend, on average, to attain more than 50% of fertilizations. Lewis and Austad (1994) have since found an association between male olfactory attractiveness in two-choice trials and P_2, and Yan and Stevens (1995) found that parasitic infection reduces P_2. These data could be interpreted as evidence for female choice based on sperm selection, in that males of superior quality appear to be favoured at fertilization. However, it is important to interpret them within the context of the mechanisms of sperm competition. With sperm mixing, the results are equally compatible with an interpretation of sperm competition; where males of superior quality are able to produce and transfer greater quantities of sperm they will be expected to attain higher paternity. Until sperm numbers in the ejaculates of males of different qualities are counted and controlled it will be impossible to assign nonrandom paternity to female choice via active sperm selection or intermale competition via sperm competition.

A further problem arises due to differing competitive abilities of sperm. If male quality and sperm quality are positively correlated then nonrandom fertilization is expected as a product of intermale competition via sperm competition, without intervention of the female. Variation in sperm competitiveness could arise because of differences in motility and/or longevity or because of meiotic drive, a phenomena not uncommon in *Drosophila* (Zimmering *et al.* 1970). If sperm quality and male quality were positively correlated, the females of species in which sperm mixing was extensive could ensure the production of offspring of high quality by multiple mating, allowing sperm competition to filter out inferior mates

(Sivinski 1984; Curtsinger 1991; Keller and Reeve 1995). However, the most competitive sperm may not always come from males of superior quality. In *Tribolium confusum*, males infected with the intracellular parasite *Wolbachia pipientis* have a competitive advantage over uninfected males during sperm competition (Wade and Chang 1995). Increased fertility of infected males enhances the parasite's own fitness at the expense of female *T. confusum*; females produce few sterile offspring. Despite the deleterious effects on uninfected females, they do not select against sperm from infected males. The data suggest some form of meiotic drive generated by *W. pipientis* present in the sperm of infected males.

LaMunyon and Eisner (1993) found that large males of the moth *Utetheisa ornatrix* are more likely to be principal sires than small males and concluded that sperm selection by females was responsible for nonrandom paternity. However, interpreted within a framework of sperm competition mechanisms these data provide little support for the notion of sperm selection. Sperm utilization patterns of *U. ornatrix* show the typical lepidopteran bimodal distribution of P_2 values (Table 10.1). Thus, P_2 values about the mode of 0 indicate the first male as principal sire, while those about the mode of 1.0 indicate the second male as principal sire. These patterns might be expected from the high rate of failed matings seen in the Lepidoptera (Drummond 1984), if large males were better able to position the spermatophore within the reproductive tract. Male size influences spermatophore attachment ability in crickets (Simmons 1988). Of course, that female reproductive tracts have evolved in such a way as to make it difficult for males to position spermatophores for insemination could be interpreted as a consequence of female choice (*sensu* Eberhard 1990). However, the mechanism of choice in this instance would be pre-insemination, not sperm selection. The patterns observed by LaMunyon and Eisner (1993) are also consistent with a mechanism of sperm transfer and utilization in which the first spermatophore acts as a mating plug (see above); when the second male is able to breach the plug he becomes principal sire. Large males transfer larger spermatophores (LaMunyon and Eisner 1994) and the observation that principal sires are larger than nonprincipal sires is consistent with a hypothesis in which large spermatophores are both more effective plugs and more effective breachers of existing plugs. Indeed, LaMunyon and Eisner (1994) later found that spermatophore size was the important variable. By competing nonvirgin large males against virgin small males they were able to reverse the size-dependent patterns of principal sire; recently mated males produce spermatophores *c.* 40% smaller than virgin males. Thus, males with small spermatophores may have been nonprincipal sires because they failed to inseminate females, rather than because of sperm selection by females. Equally, if spermatophores do not function as mating plugs, smaller spermatophores contain fewer sperm (Cook and Gage 1995) so that paternity disadvantages for males with small spermatophores would be expected without sperm selection by females.

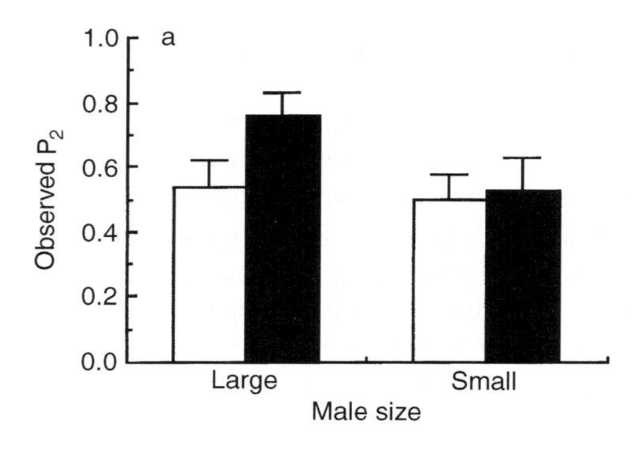

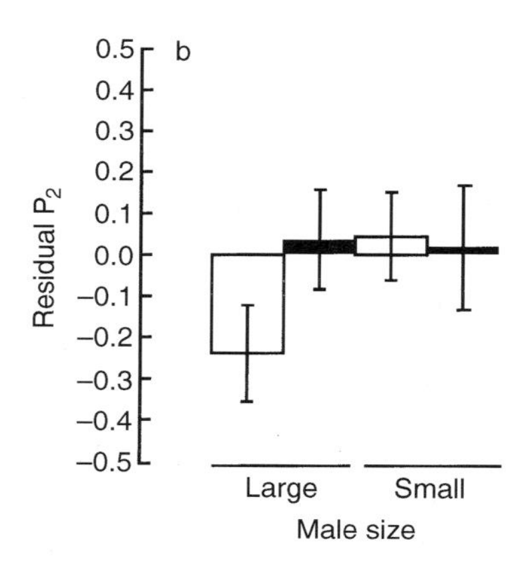

Fig. 10.13. *The outcome of double matings in* Scatophaga stercoraria *for large and small males assessed with replicate females that differed in their previous mating experience: females mated first with a large male (open bars) or first with a small male (closed bars). (a) Large males achieve a higher* P_2 *than small males. However, after controlling for differences in the rates of sperm transfer for large and small males (b) there was no significant influence of male size on* P_2. *There was no significant variance due to females either before or after controlling for differences in sperm transfer rates of males (Simmons* et al. *1996, with permission. Copyright 1996 by Springer-Verlag.)*

A recent attempt to disentangle the effects of sperm competition mechanisms and sperm selection in determining nonrandom paternity has been made using dung flies, *S. stercoraria*, as an experimental model (Simmons *et al.* 1996). Ward (1993) performed an experiment in which he allowed two males to copulate with a single female for a fixed time and assessed their paternity using enzyme polymorphic loci. He found that large males attained a higher P_2 than did small males and concluded that females were exercising sperm selection, favouring large males as mates. However, Simmons and Parker (1992), Ward (1993), and Parker and Simmons (1994) have shown that large males have a higher constant rate of sperm transfer and displacement than small males. Parker and Simmons (1991) have shown that the mechanism of sperm transfer is one in which there is volumetric displacement of previously stored sperm and instantaneous random sperm mixing so that as copulation proceeds a male increasingly displaces his own sperm (see equation 3a above). Following sperm displacement, sperm are assumed to be used at random from the female's sperm-storage organ(s) so that there is no sperm selection. Because large males have a higher rate of sperm displacement they are expected to attain a higher P_2, even without sperm selection on the basis of male size. Simmons *et al.* (1996) adopted the design of Lewis and Austad's (1990) experiments in that they assessed a given male's success in fertilization as a second male with replicate females that had experienced first males of different sizes. They found no significant variance in P_2 among females but significant variance among males that was related to male size; as in Ward's (1993) study, large males gained a higher P_2 than small males (Fig. 10.13a). After calculating the P_2 expected on the basis of size-dependent rates of sperm displacement (using equation 3a) and subtracting this from the observed value of P_2, variance in residual P_2 was unrelated to male size and showed no significant variation among males or females (Fig. 10.13b). Thus, sperm utilization from the female's sperm-storage organ(s) was random with respect to male size, once the numbers of sperm transferred was controlled. In general these types of experimental approaches, coupled with an understanding of the knowledge of sperm competition mechanisms are essential for interpreting nonrandom patterns of sperm utilization. It should be noted that Ward (1993) did find brood effects in females mated with small second males; P_2 increased across three successive clutches when females were not allowed to remate between clutches. Such brood effects are at least consistent with the notion that increasing numbers of sperm from small males were utilized, perhaps as sperm from large first males become exhausted. Implicit in this argument is the notion of nonrandom utilization of large male sperm. Sperm from large and small males do not differ in size (Ward and Hauschteck-Jungen 1993) ruling out size-related advantage in sperm competitive ability. The biological significance of this result is unclear however, since, in natural populations, females are invariably captured and mated on arrival at the oviposition site (Parker 1970a) and 30% of females copulate more than once owing to take-overs (Parker 1970d). Further, the sperm-storage organs of females in Ward's (1993) study were not filled during the first copulation because of experimental interruptions of copulation, a situation that will affect sperm displacement by the second male (see equations 3a and 3b; Simmons *et al.* 1996) and one which would not arise in nature.

VIII. SPERM COMPETITION AND SEXUAL SELECTION

The aim of this volume is to provide a comprehensive overview of how sperm competition gives rise to sexual selection. In his general overview, Møller (Chapter 2) provides evidence from the study of birds, that extra-pair paternity can favour the evolution of elaborate secondary sexual ornaments. This is because female birds often engage in extra-pair matings with males having more exaggerated ornaments than their current nest mates. Although not due to sperm competition *per se* (there is no evidence to suggest that males with more exaggerated ornaments have competitively superior sperm) the timing and/or frequency of extra pair matings performed by females may promote nonrandom fertilization by extra pair males. Behavioural mechanisms that favour nonrandom paternity thus drive sexual selection.

With few exceptions insect taxa rarely exhibit ornamental secondary sexual traits that function as visual signals. Rather, insect signalling is often in the olfactory and/or auditory channels. Thus, sexual selection has favoured the evolution of elaborate pheromonal cocktails and acoustic displays (Eisner and Meinwald 1995; Bailey 1991; Thornhill and Alcock 1983; Andersson 1994). The products of sexual selection via sperm competition are clearly apparent in insects. We have shown in this chapter how sexual selection has favoured the evolution of behavioural, physiological and morphological traits in males, or their sperm, that increase reproductive success in the face of competition from other males and female choice. For example, the spermatophylax of many bushcrickets, crickets and dobsonflies is one obvious adaptation (Sakaluk 1984; Hayashi 1992; Wedell 1993; Vahed and Gilbert 1996). Variation in the number of sperm inseminated occurs because females remove spermatophores before sperm are transferred. The spermatophylax delays spermatophore removal and so prolongs sperm transfer, thereby generating variation in male reproductive success. Males with relatively larger spermatophylaxes transfer more sperm and are thus favoured at fertilization, thereby generating directional sexual selection for further elaboration of the spermatophylax. Other examples of sexual traits that are subject to sexual selection arising from variation in male success at insemination, and thus fertilization, include the nuptial feeding behaviour and salivary mass production of scorpion flies (Thornhill 1979), and the clamping devices of scorpion flies, water striders and sagebrush crickets (Thornhill and Sauer 1991; Arnqvist 1989; Sakaluk *et al.* 1995). Although intromittent organs are classified as primary sexual traits, their obvious elaboration in many insect taxa is clearly the product of sexual selection via sperm competition. The horns and spines on the intromittent organs of odonates, for example, have been shown to be adaptations for the removal of stored sperm and avoidance of sperm competition (Waage 1984); males with relatively more elaborately armed intromittent organs would presumably remove more sperm, gain more fertilizations, and thus be favoured by sexual selection. Furthermore, it has been suggested that males better equipped to provide appropriate stimuli to females are likely to be favoured by sexual selection via cryptic female choice (Eberhard 1996). Mate-guarding behaviours are also subject to directional sexual selection; males better able to guard their mates will avoid sperm displacement by future males (Alcock 1994) and/or be more likely to achieve complete sperm transfer and therefore paternity (Zuk and Simmons 1997).

Male body size is a trait often subject to sexual selection in insects (Thornhill and Alcock 1983; Choe and Crespi 1997). The recent move to examine causes of intraspecific variation in the outcome of sperm competition has revealed that large male size is similarly favoured via sperm competition (Table 10.1) so that the net action of sexual selection on male size may comprise components of both pre- and post-mating selection. McLain (1991) partitioned size-dependent variance in male reproductive success in stinkbugs *Nezara viridula* between different components of fitness (Table 10.2). Standardized selection differentials suggest that the intensity of directional sexual selection on male size derives more from the effects of male size on the outcome of sperm competition than from success in acquiring mates, although both components impose directional selection for increasing male size. In contrast, the selective advantage of male size during sperm displacement in *S. stercoraria* appears to be confined to the precopulatory phase of reproduction. Although small males have both a lower mating success and rate of sperm displacement, selection appears to have favoured a size-dependent duration of copula so that small males copulate for longer to achieve the same fertilization gain per female as do large males (Simmons and Parker 1992; Parker and Simmons 1994). Lewis and Austad's (1994) work with *T. castaneum* provides further evidence for the combined action of pre- and post-copulatory sexual selection. Males produce a pheromone that attracts sexually receptive females. Lewis and Austad (1994) found a positive association between male olfactory attractiveness and P_2. It is known that P_2 is positively associated with male size in this species (Lewis and Austad 1990) and there was a tendency for heavier males to attract more females, so that the association may derive from the joint

Table 10.2. The intensity of selection (standardized selection differentials) on body size partitioned between components of fitness in the southern green stink bug, *N. viridula* (from McLain 1991).

Fitness parameter	Male	Female
Mating success	0.011	—
Sperm precedence	0.125	—
Development rate	0.096	0.096
Fecundity	—	0.120

action of male size on pheromone and sperm production. Sexual selection via female choice of males with stronger pheromone titres is similarly coincident with selection via sperm competition in the moth *U. ornatrix*. Females prefer males offering larger spermatophores, as shown from their pheromone titre (Dussourd *et al.* 1991) and males with larger spermatophores are favoured in sperm competition (LaMunyon and Eisner 1994).

Thus, sexual selection via sperm competition is intense in the insects although the products of selection may differ from those seen in other taxa such as birds. Many of the traits discussed above could arguably be viewed as primary sexual traits. Nevertheless, some insects are characterized by secondary sexual ornaments that are used in visual displays. In the stork-eyed fly *Cyrtodiopsis whitei* male eye-span width is a sexually selected ornament currently subject to selection via female choice (Wilkinson and Reillo 1994). Burkhardt *et al.* (1994) showed that large males with wider eye-spans fathered more offspring than small males with narrow eye-spans. This result could be explained by a greater mating frequency for wide-eyed males given that females prefer such males, although Burkhardt *et al.* (1994) dismiss this possibility; preliminary studies suggested that large and small males have the same daily mating frequency when in competition for females. The alternative explanation is that large males are more successful in sperm competition, transferring a relatively larger number of sperm and/or being more successful in avoiding sperm competition. Lorch *et al.* (1993) find that patterns of P_2 are bimodally distributed with peaks at 0 and 1. Insemination is via an internally positioned spermatophore that may block the female's reproductive tract to future males. Thus, if larger males are more successful in avoiding sperm displacement from future males because of their larger spermatophore plugs, sperm competition will favour males with larger eye stalks, compounding premating sexual selection in this species.

IX. CONCLUSION

Despite 25 years of research into insect sperm competition we are still largely ignorant of the mechanisms by which nonrandom paternity is generated. Throughout this chapter we have argued that an understanding of the mechanisms of sperm competition are essential for interpreting nonrandom patterns of paternity and for predicting the types of adaptations that sperm competition can generate. Researchers in sperm competition have only recently begun to look at within-species variation in sperm utilization patterns which, in many instances, can provide more information than the species mean value of P_2. The bimodal distribution of P_2 values seen in Lepidoptera is an excellent example of how little the mean value of P_2 can tell us about the patterns of sperm competition. Sperm competition studies almost always use just two males mated to a single female, a situation that is highly unlikely to reflect what happens in natural populations of promiscuous species. Sperm utilization patterns can be very different when more than two males copulate. In some cases the patterns remain the same regardless of the numbers of males in competition (Markow 1985; Moritz 1986) while in others, variation in the numbers of sperm inseminated into the female prior to the final mating can have marked effects on the patterns of sperm utilization (Dickinson 1988; Zeh and Zeh 1994). Without knowledge of the natural mating frequency of insects it is difficult to determine how observed patterns of nonrandom paternity relate to selection in the wild. Patterns of sperm competition can change markedly with changes in the intervals between copulations (Table 10.1) so that many laboratory studies of P_2 may be poor estimates of the actual patterns of sperm utilization in nature. With few exceptions (Cobbs 1977; Allen *et al.* 1994; LaMunyon 1994) sperm utilization patterns in nature have not been examined. It is now becoming increasingly clear that males can adjust the numbers of sperm ejaculated into females, depending on female quality and/or the risks of sperm competition. Thus, patterns of sperm utilization may depend on the behaviour of males. Alternatively, they may depend on the responses of females to males that vary in their quality as mates. We have come a long way since Parker (1970c) conceptualized the phenomenon of sperm competition, but we still have a long way to go before we can appreciate all the evolutionary implications. In this field, perhaps more than any other, a bridge between evolution, behaviour and physiology needs to be built.

ACKNOWLEDGEMENTS

We thank Yoshitaka Tsubaki for kindly sending reprints of papers not available to us and Janis Dickinson, Scott Pitnick, Matt Gage, Tim Birkhead, and Anders Møller for insightful comments on the manuscript. William Eberhard and Karen Oberhauser allowed us to use their unpublished data.

REFERENCES

Achmann R, Heller K-G & Epplen JT (1992) Last-male sperm precedence in the bushcricket *Poecilimon veluchianus* (Orthoptera, Tettigoniidae) demonstrated by DNA fingerprinting. *Molec. Biol.* **1:** 47–54.

Alcock J (1982) Post-copulatory mate guarding by males of the damselfly *Hetaerina vulnerata* Selys (Odonata: Calopterygidae). *Anim. Behav.* **30:** 99–107.

Alcock J (1994) Postinsemination associations between males and females in insects: the mate-guarding hypothesis. *Annu. Rev. Entomol.* **39:** 1–21.

Allen GR, Kazmer DJ & Luck RF (1994) Post-copulatory male behaviour, sperm precedence and multiple mating in a solitary parasitoid wasp. *Anim. Behav.* **48:** 635–644.

Alonso-Pimentel H, Tolbert LP & Heed WB (1994) Ultrastructural examination of the insemination reaction in *Drosophila. Cell Tiss. Res.* **275:** 467–479.

Andersson M (1994) *Sexual Selection.* Princeton University Press, Princeton.

Arnqvist G (1988) Mate guarding and sperm displacement in the water strider *Gerris lateralis* Schumm (Heteroptera: Gerridae). *Freshw. Biol.* **19:** 269–274.

Arnqvist G (1989) Sexual selection in a water strider: the function, mechanism of selection and heritability of a male grasping apparatus. *Oikos* **56:** 344–350.

Arnqvist G (1994) The cost of male secondary sexual traits: developmental constraints during ontogeny in a sexually dimorphic water strider. *Am. Nat.* **144:** 119–132.

Backus VL & Cade WH (1986) Sperm competition in the field cricket *Gryllus integer* (Orthoptera: Gryllidae). *Florida Entomol.* **69:** 722–728.

Bailey WJ (1991) *Acoustic Behaviour of Insects: An Evolutionary Perspective.* Chapman and Hall, London.

Bartlett AC, Mattix EB & Wilson NM (1968) Multiple matings and use of sperm in the boll weevil, *Anthonomus grandis. Ann. Entomol. Soc. Am.* **61:** 1148–1155.

Baumann H (1974) Biological effects of paragonial substances PS1 and PS2 in female *Drosophila funebris. J. Insect Physiol.* **20:** 2347–2363.

Bella JL, Butlin RK, Ferris C & Hewitt GM (1992) Asymmetrical homogamy and unequal sex ratio from reciprocal mating-order crosses between *Chorthippus parallelus* subspecies. *Heredity* **68:** 345–352.

Beukeboom LW (1994) Phenotypic fitness effects of the selfish B chromosome, paternal sex ratio (PSR) in the parasitic wasp *Nasonia vitripennis. Evol. Ecol.* **8:** 1–24.

Bick GH & Bick JC (1963) Behavior and population structure of the damselfly *Enallagma civile* (Hagen) (Odonata: Coenagrionidae). *SW Nat.* **8:** 57–84.

Birkhead TR & Møller AP (1992) *Sperm Competition in Birds: Evolutionary Causes and Consequences.* Academic Press, London.

Birkhead TR & Møller AP (1993) Female control of paternity. *Trends Ecol. Evol.* **8:** 100–104.

Birkhead TR, Møller AP & Sutherland W (1993) Why do females make it so difficult for males to fertilise their eggs? *J. Theor. Biol.* **161:** 51–60.

Birkhead TR, Wishart GJ & Biggins JD (1995) Sperm precedence in the domestic fowl. *Proc. Roy. Soc. Lond. Ser. B* **261:** 285–292.

Bishop JDD (1996) Female control of paternity in the internally fertilizing compound ascidian *Diplosoma listerianum.* I. Autoradiographic investigation of sperm movements in the female reproductive tract. *Proc. Roy. Soc. Lond. Ser. B* **263:** 369–376.

Bishop JDD, Jones CS & Noble LR (1996) Female control of paternity in the internally fertilizing compound ascidian *Diplosoma listerianum.* II. Investigation of male mating success using RAPD markers. *Proc. Roy. Soc. Lond. Ser. B* **263:** 401–407.

Bissoondath CJ & Wiklund C (1995) Protein content of spermatophores in relation to monandry/polyandry in butterflies. *Behav. Ecol. Sociobiol.* **37:** 365–371.

Boake CRB (1983) Mating systems and signals in crickets. In *Orthopteran Mating Systems Sexual Competition in a Diverse Group of Insects.* DT Gwynne & GK Morris (eds), pp. 28–44. Westview Press, Boulder.

Boggs CL & Watt WB (1981) Population structure of pierid butterflies IV. Genetic and physiological investment in offspring by male *Colias*. *Oecologia* **50:** 320–324.

Boiteau G (1988) Sperm utilization and post-copulatory female-guarding in the Colorado potato beetle, *Leptinotarsa decemlineata*. *Entomol. Exp. Appl.* **47:** 183–188.

Boldyrev BT (1915) Contributions a l'étude de la structure des spermatophores et des particularites de la copulation chez Locustodea et Gryllidea. *Horae. Soc. Ent. Ross.* **41:** 1–245.

Boorman E & Parker GA (1976) Sperm (ejaculate) competition in *Drosophila melanogaster*, and the reproductive value of females to males in relation to female age and mating status. *Ecol. Entomol.* **1:** 145–155.

Bressac C, Fleury A & Lachaise D (1995) Another way of being anisogamous in *Drosophila* subgenus species: giant sperm, one-to-one gamete ratio, and high zygote provisioning. *Proc. Natl Acad. Sci. USA* **91:** 10399–10402.

Bressac C & Hauschteck-Jungen E (1996) *Drosophila subobscura* females preferentially select long sperm for storage and use. *J. Insect Physiol.* **42:** 323–328.

Bressac C, Joly D, Devaux J, Serres C, Feneux D & Lachaise D (1991) Comparative kinetics of short and long sperm in sperm dimorphic *Drosophila* spp. *Cell Motil. Cytoskel.* **19:** 269–274.

Brower JH (1975) Sperm precedence in the indian meal moth, *Plodia interpunctella*. *Ann. Entomol. Soc. Am.* **68:** 78–80.

Bryan JH (1968) Results of consecutive matings of female *Anopheles gambiae* species *B* with fertile and sterile males. *Nature* **218:** 489.

Bullini L, Coluzzi M & Bianchi Bullini AP (1976) Biochemical variants in the study of multiple insemination in *Culex pipiens* L. (Diptera, Culicidae). *Bull. Ent. Res.* **65:** 683–685.

Burkhardt D, de la Motte I & Lunau K (1994) Signalling fitness: large males sire more offspring. Studies of the stalk-eyed fly *Cyrtodiopsis whitei* (Diopsidae, Diptera). *J. Comp. Physiol. A* **174:** 61–64.

Carlberg U (1987a) Mate choice, sperm competition and storage of sperm in *Baculum* sp. 1 (Insects: Phasmida). *Zool. Anz.* **219:** 182–196.

Carlberg U (1987b) Reproduction behavior of *Extatosoma tiaratum* (MacLeay) (Insecta: Phasmida). *Zool. Anz.* **219:** 331–336.

Carpenter JE (1992) Sperm precedence in *Helicoverpa zea* (Lepidoptera: Noctuidae): response to a substerilizing dose of radiation. *J. Econ. Entomol.* **85:** 779–782.

Carroll SP (1991) The adaptive significance of mate guarding in the soapberry bug, *Jadera haematoloma* (Hemiptera: Rhopalidae). *J. Insect Behav.* **4:** 509–530.

Cavalloro R & Delrio G (1974) Mating behavior and competitiveness of gamma-irradiated olive fruit flies. *J. Econ. Entomol.* **67:** 253–255.

Chapman T, Liddle LF, Kalb JM, Wolfner MF & Partridge L (1995) Cost of mating in *Drosophila melanogaster* females is mediated by male accessory gland products. *Nature* **373:** 241–244.

Chen PS (1984) The functional morphology and biochemistry of insect male accessory glands and their secretions. *Annu. Rev. Entomol.* **29:** 233–255.

Chen PS, Stumm-Zollinger E, Aigaki T, Balmer J, Bienz M & Bohlen P (1988) A male accessory gland peptide that regulates reproductive behaviour of female *D. melanogaster*. *Cell* **54:** 291–298.

Childress D & Hartl DL (1972) Sperm preference in *Drosophila melanogaster*. *Genetics* **71:** 417–427.

Choe JC & Crespi BJ (1997) *The Evolution of Mating Systems in Insects and Arachnids*. Cambridge University Press, Cambridge.

Clark AG, Aguade M, Prout T, Harshman LG & Langley CH (1995) Variation in sperm displacement and its association with accessory gland protein loci in *Drosophila melanogaster*. *Genetics* **139:** 189–201.

Clarke CA & Shepard PM (1962) Offspring from double matings in swallow butterflies. *Entomologist* **95:** 199–203.

Cobbs G (1977) Multiple insemination and male sexual selection in natural populations of *Drosophila pseudoobscura*. *Am. Nat.* **111:** 641–656.

Cochran DG (1979) A genetic determination of insemination frequency and sperm precedence in the german cockroach. *Entomol. Exp. Appl.* **26:** 259–266.

Coffelt JA (1975) Multiple mating by *Lasioderma serricorne* (F) – effects on fertility and fecundity. *Proceedings of the First International Working Conference on Stored Product Entomology*, pp. 549–553. Savannah, Georgia.

Colegrave N, Birkhead TR & Lessells CM (1995) Sperm precedence in zebra finches does not require special mechanisms of sperm competition. *Proc. Roy. Soc. Lond. Ser. B* **259:** 223–228.

Cook P (1996) Sperm competition in butterflies and moths (Lepidoptera). PhD thesis, University of Liverpool.

Cook PA & Gage MJG (1995) Effects of risks of sperm competition on the numbers of eupyrene and apyrene sperm ejaculated by the male moth *Plodia interpunctella* (Lepidoptera: Pyralidae). *Behav. Ecol. Sociobiol.* **36:** 261–268.

Cooper G (1995) Analysis of genetic variation and sperm competition in dragonflies. PhD thesis, University of Oxford.

Cooper G, Miller PL & Holland PWH (1996) Molecular genetic analysis of sperm competition in the damselfly *Ischnura elegans* (Vander Linden). *Proc. Roy. Soc. Lond. Ser. B* **263:** 1343–1349.

Cordero A & Miller PL (1992) Sperm transfer, displacement and precedence in *Ischnura graellsii* (Odonata: Coenagrionidae). *Behav. Ecol. Sociobiol.* **30:** 261–267.

Cordero C (1995) Ejaculate substances that affect female insect reproductive physiology and behavior – honest or arbitrary traits. *J. Theor. Biol.* **174:** 453–461.

Curtis CF (1968) Radiation sterilization and the effect of multiple mating of females in *Glossina austeni*. *J. Insect Physiol.* **14:** 1365–1380.

Curtsinger JW (1991) Sperm competition and the evolution of multiple mating. *Am. Nat.* **138:** 93–102.

Dame DA & Ford HR (1968) Multiple mating of *Glossina morsitans* Westw. and its potential effect on the sterile male technique. *Bull. Ent. Res.* **58:** 213–219.

Dawkins R (1982) *The Extended Phenotype*. Freeman, Oxford.

De Villiers PS & Hanrahan SA (1991) Sperm competition in the namib desert beetle, *Onymacris unguicularis*. *J. Insect Physiol.* **37:** 1–8.

Deinert EI, Longino JT & Gilbert LE (1994) Mate competition in butterflies. *Nature* **370:** 23–24.

Dewsbury DA (1982) Ejaculate cost and male choice. *Am. Nat.* **119:** 601–610.

Dewsbury DA (1984) Sperm competition in muroid rodents. In *Sperm Competition and the Evolution of Animal Mating Systems*. RL Smith (ed.), pp. 547–571. Academic Press, London.

Dickinson JL (1986) Prolonged mating in the milkweed leaf beetle *Labidomera clivicollis clivicollis* (Coleoptera: Chrysomelidae): a test of the 'sperm-loading' hypothesis. *Behav. Ecol. Sociobiol.* **18:** 331–338.

Dickinson JL (1988) Determinants of paternity in the milkweed leaf beetle. *Behav. Ecol. Sociobiol.* **23:** 9–19.

Dickinson JL (1995) Trade-offs between post-copulatory riding and mate location in the blue milkweed beetle. *Behav. Ecol.* **6:** 280–286.

Dickinson JL & Rutowski RL (1989) The function of the mating plug in the chalcedon checkerspot butterfly. *Anim. Behav.* **38:** 154–162.

Downes JA (1970) The feeding and mating behaviour of the specialised Empidinae (Diptera); observations on four species of *Rhamphomyia* in the high arctic and general discussion. *Can. Entomol.* **102:** 769–791.

Drummond BA (1984) Multiple mating and sperm competition in the lepidoptera. In *Sperm Competition and the Evolution of Animal Mating Systems.* RL Smith (ed.), pp. 547–572. Academic Press, London.

Dussourd DE, Harvis CA, Meinwald J & Eisner T (1991) Pheromonal advertisement of a nuptial gift by a male moth (*Utetheisa ornatrix*). *Proc. Natl Acad. Sci. USA* **88:** 9224–9227.

Dybas LK & Dybas HS (1981) Coadaptation and taxanomic differentiation of sperm and spermathecae in featherwing beetles. *Evolution* **35:** 168–174.

Eady P (1994a) Intraspecific variation in sperm precedence in the bruchid beetle *Callosobruchus maculatus. Ecol. Entomol.* **19:** 11–16.

Eady P (1994b) Sperm transfer and storage in relation to sperm competition in *Callosobruchus maculatus. Behav. Ecol. Sociobiol.* **35:** 123–129.

Eberhard WG (1985) *Sexual Selection and Animal Genitalia.* Harvard University Press, Cambridge, MA.

Eberhard WG (1990) Animal genitalia and female choice. *Am. Sci.* **78:** 134–141.

Eberhard WG (1991) Copulatory courtship and cryptic female choice in insects. *Biol. Rev.* **66:** 1–31.

Eberhard WG (1993) Evaluating models of sexual selection: genitalia as a test case. *Am. Nat.* **142:** 564–571.

Eberhard WG (1994) Evidence for widespread courtship during copulation in 131 species of insects and spiders, and implications for cryptic female choice. *Evolution* **48:** 711–733.

Eberhard WG (1996) *Female Control: Sexual Selection by Cryptic Female Choice.* Princeton University Press, Princeton.

Eberhard WG & Cordero C (1995) Sexual selection by cryptic female choice on male seminal products – a bridge between sexual selection and reproductive physiology. *Trends Ecol. Evol.* **10:** 493–496.

Economorpoulos AP & Gordon HT (1972) Sperm replacement and depletion in the spermatheca of the s and cs strains of *Oncopeltus fasciatus. Entomol. Exp. Appl.* **15:** 1–12.

Ehrlich AH & Ehrlich PR (1978) Reproductive strategies in butterflies. I. Mating frequency, plugging and egg number. *J. Kan. Entomol. Soc.* **51:** 666–697.

Eisner T & Meinwald J (1995) The chemistry of sexual selection. *Proc. Natl Acad. Sci. USA* **92:** 50–55.

Emlen ST & Oring LW (1977) Ecology, sexual selection, and the evolution of mating systems. *Science* **197:** 215–223.

Erickson RP (1990) Post-meiotic gene expression. *Trends Genet.* **6:** 264–269.

Etman AAM & Hooper GHS (1979) Sperm precedence of the last mating in *Spodoptera litura. Ann. Entomol. Soc. Am.* **72:** 119–120.

Fincke OM (1984) Sperm competition in the damselfly *Enallagma hageni* Walsh (Odonata: Coenagrionidae): benefits of multiple mating to males and females. *Behav. Ecol. Sociobiol.* **14:** 235–240.

Fincke OM (1986) Underwater oviposition in a damselfly (Odonata: Coenagrionidae) favors male vigilance, and multiple mating by females. *Behav. Ecol. Sociobiol.* **18:** 405–412.

Flint HM & Kressin EL (1968) Gamma irradiation of the tobacco budworm: sterilization, competitiveness, and observations on reproductive biology. *J. Econ. Entomol.* **61:** 477–483.

Fowler GL (1973) Some aspects of the reproductive biology of *Drosophila melanogaster*: sperm transfer, sperm storage, and sperm utilisation. *Adv. Genet.* **17:** 293–360.

Fukui HH & Gromko MH (1989) Female receptivity to remating and early fecundity in *Drosophila melanogaster*. *Evolution* **43:** 1311–1315.

Gack C & Peschke K (1994) Spermathecal morphology, sperm transfer and a novel mechanism of sperm displacement in the rove beetle, *Aleochara curtula* (Coleoptera, Staphylinidae). *Zoomorphology* **114:** 227–237.

Gage AR & Barnard CJ (1996) Male crickets increase sperm number in relation to competition and female size. *Behav. Ecol. Sociobiol.* **38:** 349–353.

Gage MJG (1991) Risk of sperm competition directly affects ejaculate size in the Mediterranean fruit fly. *Anim. Behav.* **42:** 1036–1037.

Gage MJG (1992) Removal of rival sperm during copulation in a beetle, *Tenebrio molitor*. *Anim. Behav.* **44:** 587–589.

Gage MJG (1994) Associations between body size, mating pattern, testis size and sperm lengths across butterflies. *Proc. Roy. Soc. Lond. Ser. B* **258:** 247–254.

Gage MJG (1995) Continuous variation in reproductive strategy as an adaptive response to population density in the moth *Plodia interpunctella*. *Proc. Roy. Soc. Lond. Ser. B* **261:** 25–30.

Gage MJG & Baker RR (1991) Ejaculate size varies with socio-sexual situation in an insect. *Ecol. Entomol.* **16:** 331–337.

Gage MJG & Cook PA (1994) Sperm size or numbers? Effects of nutritional stress upon eupyrene and apyrene sperm production strategies in the moth *Plodia interpunctella* (Lepidoptera: Pyralidae). *Funct. Ecol.* **8:** 594–599.

Gangrade GA (1963) A contribution to the biology of *Necroscia sparaxes* Westwood (Phasmidae: Phasmida). *Entomologist* **96:** 83–93.

George JA (1967) Effect of mating sequence on egg-hatch from female *Aedes aegypti* (L) mated with irradiated and normal males. *Mosquito News* **27:** 82–86.

Gilbert DG (1981) Ejaculate esterase 6 and initial sperm use by female *Drosophila melanogaster*. *J. Insect Physiol.* **27:** 641–650.

Gilbert DG, Richmond RC & Sheenhan KB (1981) Studies of esterase 6 in *Drosophila melanogaster* V. Progeny production and sperm use in females inseminated by males having active or null alleles. *Evolution* **35:** 21–37.

Gilbert LE (1976) Postmating female odour in *Heliconius* butterflies: a male contributed antiaphrodisiac? *Science* **193:** 419–420.

Gilchrist AS & Partridge L (1995) Male identity and sperm displacement in *Drosophila melanogaster*. *J. Insect Physiol.* **41:** 1087–1092.

Ginsberg JR & Rubenstein DJ (1990) Sperm competition and variation in zebra mating behavior. *Behav. Ecol. Sociobiol.* **26:** 427–434.

Gregory PG & Howard DJ (1994) A postinsemination barrier to fertilization isolates two closely related ground crickets. *Evolution* **48:** 705–710.

Gromko MH, Gilbert DG & Richmond RC (1984) Sperm transfer and use in the multiple mating system of *Drosophila*. In *Sperm Competition and the Evolution of Animal Mating Systems*. RL Smith (ed.), pp. 371–426. Academic Press, London.

Gwynne DT (1984) Male mating effort, confidence of paternity, and insect sperm competition. In *Sperm Competition and the Evolution of Animal Mating Systems*. RL Smith (ed.), pp. 117–149. Academic Press, London.

Gwynne DT (1986) Courtship feeding in katydids (Orthoptera: Tettigoniidae): investment in offspring or in obtaining fertilizations? *Am. Nat.* **128:** 342–352.

Gwynne DT (1988a) Courtship feeding and the fitness of female katydids (Orthoptera: Tettigoniidae). *Evolution* **42:** 545–555.

Gwynne DT (1988b) Courtship feeding in katydids benefits the mating male's offspring. *Behav. Ecol. Sociobiol.* **23:** 373–377.

Gwynne DT & Snedden AW (1995) Paternity and female remating in *Requena verticalis* (Orthoptera: Tettigoniidae). *Ecol. Entomol.* **20:** 191–194.

Hadrys H, Schierwater B, Dellaporta SL, Desalle R & Buss LW (1993) Determination of paternity in dragonflies by random amplified polymorphic DNA fingerprints. *Molec. Ecol.* **2:** 79–87.

Haig D & Bergstrom CT (1995) Multiple mating, sperm competition and meiotic drive. *J. Evol. Biol.* **8:** 265–282.

Happ GH, Schroeder ME & Wang JCH (1970) Effects of male and female scent on reproductive maturation in young female *Tenebrio molitor*. *J. Insect Physiol.* **16:** 1543–1548.

Happ GM (1969) Multiple sex pheromones of the mealworm beetle *Tenebrio molitor*. *Nature* **222:** 180–181.

Happ GM & Wheeler J (1969) Bioassay, preliminary purification and effect of age, crowding and mating on the release of sex pheromone by female *Telebrio molitor*. *Ann. Ent. Soc. Am.* **62:** 846–851.

Harcourt AH, Harvey PH, Larson SG & Short RV (1981) Testis weight, body weight and breeding system in primates. *Nature* **293:** 55–57.

Harshman L & Prout T (1994) Sperm displacement without sperm transfer in *Drosophila* melanogaster. *Evolution* **48:** 758–766.

Harwalker MR & Rahalkar GW (1973) Sperm utilization in the female red cotton bug. *J. Econ. Entomol.* **66:** 805–806.

Hayashi F (1992) Large spermatophore production and consumption in dobsonflies *Protohermes* (Megaloptera, Corydalidae). *Jpn. J. Entomol.* **60:** 59–66.

He Y & Tsubaki Y (1992) Variation in spermatophore size in the armyworm, *Pseudaletia separata* (Lepidoptera: Noctuidae) in relation to rearing density. *Appl. Entomol. Zool.* **27:** 39–45.

He YB, Tsubaki Y, Itou K & Miyata T (1995) Gamma radiation effects on reproductive potential and sperm use patterns in *Pseudoletia seporata* (Lepidoptera: Noctuidae). *J. Econ. Entomol.* **88:** 1626–1630.

Helversen DV & Helversen OV (1991) Pre-mating sperm removal in the bushcricket *Metaplastes ornatus* Ramme 1931 (Orthoptera, Tettigoniidae, Phaneropteridae). *Behav. Ecol. Sociobiol.* **28:** 391–396.

Heming-van Battum KE & Heming BS (1986) Structure, function and evolution of the reproductive system in females of *Hebrus pusillus* and *H. ruficeps* (Hemiptera, Gerromorpha, Hebridae). *J. Morphol.* **190:** 121–167.

Hewitt GM, Mason P & Nichols RA (1989) Sperm precedence and homogamy across a hybrid zone in the alpine grasshopper *Podisma pedestris*. *Heredity* **62:** 343–354.

Hooper RE & Siva-Jothy MT (1996) Last male sperm precedence in a damselfly demonstrated by RAPD profiling. *Molec. Ecol.* **5:** 449–453.

Howard DJ & Gregory PG (1993) Post-insemination signalling systems and reinforcement. *Philosophical Transactions of the Royal Society of London B* **340:** 231–236.

Huettel MD, Calkins CO & Hill AJ (1972) Allozyme markers in the study of sperm precedence in the plum curculio, *Conotrachelus nenuphar. Ann. Entomol. Soc. Am.* **69:** 465–468.

Hunter-Jones P (1960) Fertilization of eggs of the desert locust by spermatozoa from successive copulations. *Nature* **185:** 336.

Itou K (1992) Studies on the prolonged copulation of the southern stink bug, *Nezara viridula* L. (Heteroptera: Pentatomidae). MSc thesis, Nagoya University, Japan.

Jamieson BGM (1987) *The Ultrastructure and Phylogeny of Insect Spermatozoa.* Cambridge University Press, Cambridge.

Jennions MD & Passmore NI (1993) Sperm competition in frogs: testis size and a 'sterile male' experiment on *Chiromantis xerampelina* (Rhacophoridae). *Biol. J. Linn. Soc.* **50:** 211–220.

Johnson LK (1982) Sexual selection in a tropical brentid weevil. *Evolution* **36:** 251–262.

Joly D, Cariou ML & Lachaise D (1991) Can sperm competition explain sperm polymorphism in *Drosophila teissieri? Evol. Biol.* **5:** 25–44.

Jong PW de, Verhoog MD & Brakefield PM (1993) Sperm competition and melanic polymorphism in the 2-spot ladybird, *Adalia bipunctata* (Coleoptera, Coccinellidae). *Heredity* **70:** 172–178.

Kaitala A & Wiklund C (1994) Polyandrous female butterflies forage for matings. *Behav. Ecol. Sociobiol.* **35:** 385–388.

Kalb JM, DiBenedetto AJD & Wolfner MF (1993) Probing the function of *Drosophila* accessory glands by directed cell ablation. *Proc. Natl Acad. Sci. USA* **90:** 8093–8097.

Karlsson B (1995) Resource allocation and mating systems in butterflies. *Evolution* **49:** 955–961.

Karr TL & Pitnick S (1996) The ins and outs of fertilization. *Nature* **379:** 405–406.

Katakura H (1986) Evidence for the incapacitation of heterospecific sperm in the female genital tract in a pair of closely related ladybirds (Insecta, Coleoptera, Coccinellidae). *Zool. Sci.* **3:** 115–121.

Keller L & Reeve HK (1995) Why do females mate with multiple males? The sexually selected sperm hypothesis. *Adv. Stud. Behav*, **24:** 291–315.

King PE (1962) The structure and action of the spermatheca in *Nasonia vitripennis* (Walker) (Hymenotera: Pteromalidae). *Proc. Roy. Soc. Lond.* **37:** 73–75.

Knowlton N & Greenwell SR (1984) Male sperm competition avoidance mechanisms: the influence of female interests. In *Sperm Competition and the Evolution of Animal Mating Systems.* RL Smith (ed.), pp. 62–85. Academic Press, London.

Koenig WD (1991) Levels of female choice in the white-tailed skimmer *Plathemis lydia* (Odonata: Libellulidae). *Behaviour* **119:** 193–224.

Koeniger G (1986) Mating sign and multiple mating in the honeybee. *Bee World* **67:** 141–150.

Labine PA (1964) Population biology of the butterfly *Euphydryas editha.* I. Barriers to multiple inseminations. *Evolution* **18:** 335–336.

Labine PA (1966) The population biology of the butterfly, *Euphydryas editha*. IV. Sperm precedence – a preliminary report. *Evolution* **20:** 580–586.

Lackie AM (1986) Transplantation: the limits of recognition. In *Hemocytic and Humoral Immunity in Arthropods*. AP Gupta (ed.), pp. 191–223. CRC Press, Boca Raton.

Ladd TL (1966) Egg viability and longevity of Japanese beetles treated with tepa, apholate, and metepa. *J. Econ. Entomol*, **59:** 422–425.

LaMunyon CW (1994) Paternity in naturally occurring *Utetheisa ornatrix* (Lepidoptera: Arctiidae) as estimated using enzyme polymorphism. *Behav. Ecol. Sociobiol.* **34:** 403–408.

LaMunyon CW & Eisner T (1993) Postcopulatory sexual selection in an arctiid moth (*Utetheisa ornatrix*). *Proc. Natl Acad. Sci. USA* **90:** 4689–4692.

LaMunyon CW & Eisner T (1994) Spermatophore mass as determinant of paternity in an arctiid moth (*Utetheisa ornatrix*). *Proc. Natl Acad. Sci. USA* **91:** 7081–7084.

Lee PE & Wilkes A (1965) Polymorphic spermatozoa in the hymenopterous wasp *Dahlbominous*. *Science* **147:** 1445–1446.

Leopold RA (1976) The role of male accessory glands in insect reproduction. *Annu. Rev. Entomol.* **21:** 199–221.

Letsinger JT & Gromko MH (1985) The role of sperm numbers in sperm competition and female remating in *Drosophila melanogaster*. *Genetica* **66:** 195–202.

Lewis SM & Austad SN (1990) Sources of intraspecific variation in sperm precedence in red flour beetles. *Am. Nat.* **135:** 351–359.

Lewis SM & Austad SN (1994) Sexual selection in flour beetles: the relationship between sperm precedence and male olfactory attractiveness. *Behav. Ecol.* **5:** 219–224.

Linley JR (1975) Sperm supply and its utilization in doubly inseminated flies, *Culicoides melleus*. *J. Insect Physiol.* **21:** 1785–1788.

Linley JR & Adams GM (1972) A study of the mating behaviour of *Culicoides mellus* (Coquillet) (Diptera: Ceratopogonidae). *Trans. Roy. Entomol. Soc. Lond.* **126:** 279–303.

Lloyd JE (1979) Mating behavior and natural selection. *Florida Entomol.* **62:** 17–34.

Lopez-Leon MD, Cabrero J, Pardo MC, Viseras E & Camacho JPM (1993) Paternity displacement in the grasshopper *Eyprepocnemis plorans*. *Heredity* **71:** 539–545.

Lorch PD, Wilkinson GS & Reillo PR (1993) Copulation duration and sperm precedence in the stalk-eyed fly *Cyrtodiopsis whitei* (Diptera: Diopsidae). *Behav. Ecol. Sociobiol.* **32:** 303–311.

Lum PT (1961) The reproductive system of some Florida mosquitoes II. The male accessory glands and their roles. *Ann. Entomol. Soc. Am.* **54:** 430–433.

Markow TA (1985) A comparative investigation of the mating system of *Drosophila hydei*. *Anim. Behav.* **33:** 775–781.

Markow TA (1988) *Drosophila* males provide a material contribution to offspring sired by other males. *Funct. Ecol.* **2:** 77–79.

Markow TA & Ankney PF (1988) Insemination reaction in *Drosophila*: found in species whose males contribute material to oocytes before fertilization. *Evolution* **42:** 1097–1101.

Martinez L, Leyvavazquez JL & Mojica HB (1993) Sperm competition in the female *Diachasmimorpha longicaudata* (Hymenotera, Braconidae). *SW Entomol.* **18:** 293–299.

Mason LJ & Pashley DP (1991) Sperm competition in the soybean looper (Lepidoptera: Noctuidae). *Ann. Entomol. Soc. Am.* **84:** 268–271.

McCauley DE & Reilly LM (1984) Sperm storage and sperm precedence in the milkweed beetle *Tetraopes tetraophthalmus* (Forster) (Coleoptera: Cerambycidae). *Ann. Entomol. Soc. Am.* **77:** 526–530.

McLain DK (1980) Female choice and the adaptive significance of prolonged copulation in *Nezara viridula* (Hemiptera: Pentatomaidae). *Psyche* **87:** 325–336.

McLain DK (1985) Male size, sperm competition, and the intensity of sexual selection in the southern green stink bug, *Nezara viridula* (Hemiptera: Pentatomidae). *Ann. Entomol. Soc. Am.* **78:** 86–89.

McLain DK (1989) Prolonged copulation as a post-insemination guarding tactic in a natural population of the ragwort seed bug. *Anim. Behav.* **38:** 659–664.

McLain DK (1991) Heritability of size: a positive correlate of multiple fitness components in the southern green stink bug (Hemiptera: Pentatomidae). *Ann. Entomol. Soc. Am.* **84:** 174–178.

McVey M & Smittle BJ (1984) Sperm precedence in the dragonfly *Erythemis simplicollis. J. Insect Physiol.* **30:** 619–628.

Michiels NK (1992) Consequences and adaptive significance of variation in copulation duration in the dragonfly *sympetrum danae. Behav. Ecol. Sociobiol.* **29:** 429–435.

Miller PL (1984) The structure of the genitalia and the volumes of sperm stored in male and female *Nesciothemis farinosa* (Foerster) and *Orthetrum chrysostigma* (Burmeister) (Anisoptera: Libellulidae). *Odonatologica* **13:** 415–428.

Miller PL (1987) Sperm competition in *Ischnura elegans* (Vander Linden) (Zygoptera: Coenagrionidae). *Odonatologica* **16:** 201–208.

Miller PL (1990) Mechanisms of sperm removal and sperm transfer in *Orthetrum coerulescens* (Fabricus) (Odonata: Libellulidae). *Physiol. Entomol.* **15:** 199–209.

Miller PL (1991) The structure and function of the genitalia in the Libellulidae (Odonata). *Zool. J. Linn. Soc.* **102:** 43–74.

Møller AP (1991) Sperm competition, sperm depletion, paternal care, and relative testis size in birds. *Am. Nat.* **137:** 882–906.

Moritz RFA (1986) Intracolonial worker relationship and sperm competition in the honeybee (*Apis mellifera* L.). *Experientia* **42:** 445–448.

Moritz RFA, Meusel MS & Haberl M (1991) Oligonucleotide DNA fingerprinting discriminates super- and half-sisters in honeybee colonies (*Apis melifera* L). *Naturwissenschaften* **78:** 422–424.

Muller JK & Eggert A-K (1989) Paternity assurance by 'helpful' males: adaptations to sperm competition in burying beetles. *Behav. Ecol. Sociobiol.* **24:** 245–249.

Nabours RK (1927) Polyandry in the grouse locust, *Paratettix texanus* Hancock, with notes on inheritance of acquired characters and telegony. *Am. Nat.* **61:** 531–538.

Nakagawa S, Farias GJ, Suda D, Cunningham RT & Chambers DL (1971) Reproduction of the mediterranean fruit fly: frequency of mating in the laboratory. *Ann. Entomol. Soc. Am.* **69:** 949–950.

Nakano S (1985) Sperm displacement in *Henosepilachna pustulosa* (Coleoptera, Coccinellidae). *Kotyu* **53:** 516–519.

Newport MEA & Gromko MH (1984) The effect of experimental design on female receptivity to remating and its impact on reproductive success in *Drosophila melanogaster. Evolution* **38:** 1261–1272.

North DT & Holt GG (1968) Genetic and cytogenetic basis of radiation-induced sterility in the adult male cabbage looper *Trichoplusia ni*. In *IAEA/FAO Symposium, Isotopes and Radiation in Entomology (1967)*, pp. 391–403. IAEA, Vienna.

Omura S (1939) Selective fertilization in *Bombyx mori*. *Jpn. J. Genet.* **15:** 29–35.

Ono T, Siva-Jothy MT & Kato A (1989) Removal and subsequent ingestion of rivals' semen during copulation in a tree cricket. *Physiol. Entomol.* **14:** 195-202.

Opp SB, Ziegner J, Bui N & Prokopy RJ (1990) Factors influencing estimates of sperm competition in *Rhagoletis pomonella* (Walsh) (Diptera: Tephritidae). *Ann. Entomol. Soc. Am.* **83:** 521–526.

Orr AG & Rutowski RL (1991) The function of the shpragis in *Cressida cressida* (Fab) (Lepidoptera, Papilionidea): a visual deterent to copulation attempts. *J. Nat. Hist.* **25:** 703–710.

Otronen M (1989) Female mating behaviour and multiple matings in the fly, *Dryomyza anilis*. *Behaviour* **111:** 77–97.

Otronen M (1990) Mating behavior and sperm competition in the fly, *Dryomyza anilis*. *Behav. Ecol. Sociobiol.* **26:** 349–356.

Otronen M (1993) Male distribution and interactions at female oviposition sites as factors affecting mating success in the fly *Dryomyza anilis* (Dryomyzidae). *Evol. Ecol.* **7:** 127–141.

Otronen M (1994a) Fertilisation success in the fly *Dryomyza anilis* (Dryomyzidae): effects of male size and the mating situation. *Behav. Ecol. Sociobiol.* **35:** 33–38.

Otronen M (1994b) Repeated copulations as a strategy to maximize fertilization in the fly, *Dryomyza anilis* (Dryomyzidae). *Behav. Ecol.* **5:** 51–56.

Otronen M (1995) Male distribution and mate searching in the yellow dung fly *Scathophaga stercoraria*: comparison between paired and unpaired males. *Ethology* **100:** 265–276.

Otronen M & Siva-Jothy MT (1991) The effect of postcopulatory male behaviour on ejaculate distribution within the female sperm storage organs of the fly, *Dryomyza anilis* (Diptera: Dryomyzidae). *Behav. Ecol. Sociobiol.* **29:** 33–37.

Page RE & Metcalf RA (1982) Multiple mating, sperm utilization, and social evolution. *Am. Nat.* **119:** 263–281.

Pair SD, Laster ML & Martin DF (1977) Hybrid sterility of the tobacco budworm: effects of alternate sterile and normal matings on fecundity and fertility. *Ann. Entomol. Soc. Am.* **70:** 952–954.

Parker GA (1970a) The reproductive behaviour and the nature of sexual selection in *Scatophaga stercoraria* L (Diptera: Scatophagidae) V. The females's behaviour at the oviposition site. *Behaviour* **37:** 140–168.

Parker GA (1970b) The reproductive behaviour and the nature of sexual selection in *Scatophaga stercoraria* L (Diptera: Scatophagidae) VI. The adaptive significance of emigration from the oviposition site during the phase of genital contact. *J. Anim. Ecol.* **40:** 215–233.

Parker GA (1970c) Sperm competition and its evolutionary consequences in the insects. *Biol. Rev.* **45:** 525–567.

Parker GA (1970d) Sperm competition and its evolutionary effect on copula duration in the fly *Scatophaga stercoraria*. *J. Insect Physiol.* **16:** 1301–1328.

Parker GA (1974) Courtship persistence and female-guarding as male time investment strategies. *Behaviour* **48:** 157–184.

Parker GA (1982) Why are there so many tiny sperm? Sperm competition and the maintenance of two sexes. *J. Theor. Biol.* **96:** 281–294.

Parker GA (1984) Sperm competition and the evolution of animal mating strategies. In *Sperm Competition and the Evolution of Animal Mating Systems.* RL Smith (ed.), pp. 2–60. Academic Press, London.

Parker GA (1990a) Sperm competition games: raffles and roles. *Proc. Roy. Soc. Lond. Ser. B* **242:** 120–126.

Parker GA (1990b) Sperm competition games: sneaks and extra-pair copulations. *Proc. Roy. Soc. Lond. Ser. B* **242:** 127–133.

Parker GA (1993) Sperm competition games: sperm size and sperm number under adult control. *Proc. Roy. Soc. Lond. Ser. B* **253:** 245–254.

Parker GA & Begon M (1993) Sperm competition games: sperm size and number under gametic control. *Proc. Roy. Soc. Lond. Ser. B* **253:** 255–262.

Parker GA & Simmons LW (1991) A model of constant random sperm displacement during mating: evidence from *Scatophaga. Proc. Roy. Soc. Lond. Ser. B* **246:** 107–115.

Parker GA & Simmons LW (1994) Evolution of phenotypic optima and copula duration in dungflies. *Nature* **370:** 53–56.

Parker GA & Smith JL (1975) Sperm competition and the evolution of the precopulatory passive phase behaviour in *Locusta migratoria migratorioides. J. Entomol. (A)* **49:** 155–171.

Parker GA & Stuart RA (1976) Animal behavior as a strategy optimizer: evolution of resource assessment strategies and optimal emigration thresholds. *Am. Nat.* **110:** 1055–1076.

Parker GA, Simmons LW & Kirk H (1990) Analysing sperm competition data: simple models for predicting mechanisms. *Behav. Ecol. Sociobiol.* **27:** 55–65.

Parker GA, Ball MA, Stockley P & Gage MJG (1996) Sperm competition games: individual assessment of sperm competition intensity by group spawners. *Proc. Roy. Soc. Lond. Ser. B* **263:** 1291–1297.

Partridge L & Halliday T (1984) Mating patterns and mate choice. In *Behavioural Ecology An Evolutionary Approach.* JR Krebs & NB Davies (eds), pp. 222–250. Blackwell Scientific Publications, Oxford.

Peer DF (1956) Multiple mating of queen honey bees. *J. Econ. Entomol.* **49:** 741–743.

Pfau HK (1971) Struktur und funktion des sekundaren kopulationsapparates der Odonaten (Insecta, Palaeoptera), ihre wandlung in der stammesgeschichte und Bedeutung fur die adaptive entfaltung der ordnung. *Zeit. Morphol. Tierre* **70:** 281–371.

Pierre J (1985) Le Sphragis chez les Acraeinae (Lepidoptera Nymphalidae). *Ann. Soc. Ent. Fr. (NS)* **21:** 393–398.

Pitnick S (1993) Operational sex ratios and sperm limitation in populations of *Drosophila pachea. Behav. Ecol. Sociobiol.* **33:** 383–391.

Pitnick S (1996) Investment in testes and the cost of making long sperm in *Drosophila. Am. Nat.* **148:** 57–80.

Pitnick S & Markow TA (1994a) Male gametic strategies: sperm size, testes size, and the allocation of ejaculate among successive mates by the sperm-limited fly *Drosophila pachea* and its relatives. *Am. Nat.* **143:** 785–819.

Pitnick S & Markow TA (1994b) Large-male advantages associated with costs of sperm production in *Drosophila hydei*, a species with giant sperm. *Proc. Natl Acad. Sci. USA* **91:** 9277–9281.

Pitnick S, Markow TA & Spicer GS (1995a) Delayed male maturity is a cost of producing large sperm in *Drosophila. Proc. Natl Acad. Sci. USA* **92:** 10614–10618.

Pitnick S, Spicer GS & Markow TA (1995b) How long is a giant sperm? *Nature* **375:** 109.

Proshold FI & LaChance LE (1974) Analysis of sterility in hybrids from interspecific crosses between *Heliothis virescens* and *H. subflexa*. *Ann. Entomol. Soc. Am.* **67:** 445–449.

Proshold FI, LaChance LE & Richard RD (1975) Sperm production and transfer by *Heliothis virescens, H. subflexa* and the sterile hybrid males. *Ann. Entomol. Soc. Am.* **68:** 31–34.

Proverbs MD & Newton JR (1962) Some effects of gamma radiation on the potential of the codling moth, *Carpocapsa pomonella* (L.) (Lepidoptera: Olethreutidae). *Can. J. Zool.* **94:** 1162–1170.

Reimann JG, Moen DO & Thorson BJ (1967) Female monogamy and its control in the housefly, *Musca domestica* L. *J. Insect Physiol.* **13:** 407–418.

Retnakaran A (1974) The mechanism of sperm precedence in the spruce budworm, *Choristoneura fumiferana* (Lepidoptera: Tortricidae). *Can. Entomol.* **106:** 1189–1194.

Rice WR (1996) Sexually antagonistic male adaptation triggered by experimental arrest of female evolution. *Nature* **381:** 232–234.

Richard RD, LaChance LE & Proshold FI (1975) An ultrastructural study of sperm in sterile-hybrids from crosses of *Heleothis virescens* and *Heleothis subflexa*. *Ann. Entomol. Soc. Am.* **68:** 35–39.

Riddiford LM & Ashenhurst J (1973) The switchover from virgin to mated behaviour in female *Cercropia* moths: the role of the bursa copulatrix. *Biol. Bull.* **144:** 162–171.

Ridley M (1988) Mating frequency and fecundity in insects. *Biol. Rev.* **63:** 509–549.

Ridley M (1989) The incidence of sperm displacement in insects: four conjectures, one corroboration. *Biol. J. Linn. Soc.* **38:** 349–367.

Riemann JG, Moen DO & Thorson BJ (1967) Female monogamy and its control in the housefly, *Musca domestica* L. *J. Insect Physiol.* **13:** 407–418.

Robert P-A (1958) *Les Libellules (Odonates)*. Delachaux and Niestlé, Paris.

Robinson T, Johnson NA & Wade MJ (1994) Postcopulatory, prezygotic isolation: intraspecific and interspecific sperm precedence in *Tribolium* spp., flour beetles. *Heredity* **73:** 155–159.

Rodriguez V (1994) Function of the spermathecal muscle in *Chelymorpha alternana* Boheman (Coleoptera: Chrysomelidae: Cassidinae). *Physiol. Entomol.* **19:** 198–202.

Roth LM (1962) Hypersexual activity induced in females of the cockroach *Nauphoeta cinerea*. *Science* **138:** 1267–1269.

Rowe L, Arnqvist G, Sih A & Krupa JJ (1994) Sexual conflict and the evolutionary ecology of mating patterns: water striders as a model system. *Trends Ecol. Evol.* **9:** 289–293.

Rowe R (1987) *The Dragonflies of New Zealand*. Auckland University Press, Auckland.

Rubenstein DI (1984) Resource acquisition and alternative mating strategies in water striders. *Am. Zool.* **24:** 345–353.

Rubenstein DI (1989) Sperm competition in the water strider, *Gerris remigis*. *Anim. Behav.* **38:** 631–636.

Ryan MJ, Fox JH, Wilczynski W & Rand AS (1990) Sexual selection for sensory exploitation in the frog *Physalaemus pustulosus*. *Nature* **343:** 66–67.

Sakaluk SK (1984) Male crickets feed females to ensure complete sperm transfer. *Science* **223:** 609–610.

Sakaluk SK (1985) Spermatophore size and its role in the reproductive behaviour of the cricket, *Gryllodes supplicans* (Orthoptera: Gryllidae). *Can. J. Zool.* **63:** 1652–1656.

Sakaluk SK & Eggert A-K (1996) Female control of sperm transfer and intraspecific variation in sperm precedence: antecedents to the evolution of a courtship food gift. *Evolution* **50:** 694–703.

Sakaluk SK, Bangert PJ, Eggert A-K, Gack C & Swanson LV (1995) The gin trap as a device facilitating coercive mating in sagebrush crickets. *Proc. Roy. Soc. Lond. Ser. B* **261:** 65–71.

Saul SH & McCombs SD (1993) Dynamics of sperm use in the Mediterranean fruit fly (Diptera: Tephritidae): reproductive fitness of multiple-mated females and sequentially mated males. *Ann. Entomol. Soc. Am.* **86:** 198–202.

Saul SH, Tam SYT & McInnis DO (1988) Relationship between sperm competition and copulation duration in the Mediterranean fruit fly (Diptera: Tephritidae). *Ann. Entomol. Soc. Am.* **81:** 498–502.

Sawada K (1995) Male's ability of sperm displacement during prolonged copulations in *Ischnura senegalensis* (Rambur) (Zygoptera: Coenagrionidae). *Odonatologica* **24:** 237–244.

Schlager G (1960) Sperm precedence in the fertilization of eggs in *Tribolium castaneum*. *Ann. Entomol. Soc. Am.* **53:** 557–560.

Schrader F & Leuchtenberger C (1950) A cytochemical analysis of the functional interrelationships of various cell structures in *Arvelius albopunctatus* (De Geer). *Exp. Cell Res.* **1:** 421–452.

Scott D & Richmond RC (1990) Sperm loss by remating *Drosophila melanogaster* females. *J. Insect Physiol.* **36:** 451–456.

Scott D & Williams E (1993) Sperm displacement after remating in *Drosophila melanogaster*. *J. Insect Physiol.* **39:** 201–206.

Service PM & Fales AJ (1993) Evolution of delayed reproductive senescence in male fruit flies: sperm competition. *Genetica* **91:** 111–125.

Sigurjónsdóttir H & Parker GA (1981) Dung fly struggles: evidence for assessment strategy. *Behav. Ecol. Sociobiol.* **8:** 219–230.

Silberglied RE, Sheperd JG & Dickinson JL (1984) Eunuchs: the role of apyrene sperm in lepidoptera? *Am. Nat.* **123:** 255–265.

Sillén-Tullberg B (1981) Prolonged copulation: a male 'postcopulatory' strategy in a promiscuous species, *Lygaeus equestris* (Heteroptera: Lygaenidae). *Behav. Ecol. Sociobiol.* **9:** 283–289.

Simmons LW (1986) Female choice in the field cricket, *Gryllus bimaculatus* (De Geer). *Anim. Behav.* **34:** 1463–1470.

Simmons LW (1987a) Female choice contributes to offspring fitness in the field cricket, *Gryllus bimaculatus* (De Geer). *Behav. Ecol. Sociobiol.* **21:** 313–321.

Simmons LW (1987b) Sperm competition as a mechanism of female choice in the field cricket, *Gryllus bimaculatus*. *Behav. Ecol. Sociobiol.* **21:** 197–202.

Simmons LW (1988) Male size, mating potential and lifetime reproductive success in the field cricket, *Gryllus bimaculatus* (De Geer). *Anim. Behav.* **36:** 372–379.

Simmons LW (1991a) Female choice and the relatedness of mates in the field cricket, *Gryllus bimaculatus*. *Anim. Behav.* **41:** 493–501.

Simmons LW (1991b) On the post-copulatory guarding behaviour of male field crickets. *Anim. Behav.* **42:** 504–505.

Simmons LW (1995a) Courtship feeding in katydids (Orthoptera: Tettigoniidae):

investment in offspring and in obtaining fertilizations. *Am. Nat.* **146:** 307–315.

Simmons LW (1995b) Relative parental investment, potential reproductive rates, and the control of sexual selection in katydids. *Am. Nat.* **145:** 797–808.

Simmons LW & Gwynne DT (1991) The refractory period of female katydids (Orthoptera: Tettigoniidae): sexual conflict over the remating interval? *Behav. Ecol.* **2:** 276–282.

Simmons LW & Parker GA (1992) Individual variation in sperm competition success of yellow dung flies, *Scatophaga stercoraria. Evolution* **46:** 366–375.

Simmons LW, Craig M, Llorens T, Schinzig M & Hosken D (1993) Bushcricket spermatophores vary in accord with sperm competition and parental investment theory. *Proc. Roy. Soc. Lond. Ser. B* **251:** 183–186.

Simmons LW, Llorens T, Schinzig M, Hosken D & Craig M (1994) Sperm competition selects for male mate choice and protandry in the bushcricket, *Requena verticalis* (Orthoptera: Tettigoniidae). *Anim. Behav.* **47:** 117–122.

Simmons LW, Stockley P, Jackson RL & Parker GA (1996) Sperm competition or sperm selection: no evidence for female influence over paternity in yellow dung flies *Scatophaga stercoraria. Behav. Ecol. Sociobiol.* **38:** 199–206.

Siva-Jothy MT (1984) Sperm competition in the family Libellulidae (Anisoptera) with special reference to *Crocothemis erythraea* (Brulle) and *Orthetrum cancellatum* (L.). *Adv. Odonatol.* **2:** 195–207.

Siva-Jothy MT (1987) The structure and function of the female sperm-storage organs in libellulid dragonflies. *J. Insect Physiol.* **33:** 559–568.

Siva-Jothy MT (1988) Sperm 'repositioning' in *Crocothemis erythraea*, a libellulid dragonfly with a brief copulation. *J. Insect Behav.* **1:** 235–245.

Siva-Jothy MT & Hooper RE (1995) The disposition and genetic diversity of stored sperm in females of the damselfly *Calopteryx splendens xanthostoma* (Charpentier). *Proc. Roy. Soc. Lond. Ser. B* **259:** 313–318.

Siva-Jothy MT & Tsubaki Y (1989) Variation in copula duration in *Mnais pruinosa pruinosa Selys* (Odonata: Calopterygidae). *Behav. Ecol. Sociobiol.* **24:** 39–45.

Siva-Jothy MT & Tsubaki Y (1994) Sperm competition and sperm precedence in the dragonfly *Nanophya pygmaea. Physiol. Entomol.* **19:** 363–366.

Siva-Jothy MT, Earle Blake D, Thompson J & Ryder JJ (1996) Short- and long-term sperm precedence in the beetle *Tenebrio molitor:* a test of the 'adaptive sperm removal' hypothesis. *Physiol. Entomol.* **21:** 313–316.

Sivinski J (1980) Sexual selection and insect sperm. *Florida Entomol.* **63:** 99–111.

Sivinski J (1983) Predation and sperm competition in the evolution of coupling durations, particularly in the stick insect *Diapheromera veliei.* In *Orthopteran Mating Systems: Sexual Competition in a Diverse Group of Insects.* DT Gwynne & MGK (eds), pp. 147–162. Westview Press, Boulder, Colorado.

Sivinski J (1984) Sperm in competition. In *Sperm Competition and the Evolution of Animal Mating Systems.* RL Smith (ed.), pp. 86–115. Academic Press, London.

Smith PH, Gillott C, Barton Browne L & Gerwen ACMV (1990) The mating-induced refractoriness of *Lucilia cuprina* females: manipulating the male contribution. *Physiol. Entomol.* **15:** 469–481.

Smith RL (1979) Repeated copulation and sperm precedence: paternity assurance for a male brooding water bug. *Science* **205:** 1029–1031.

Smith RL (1984) *Sperm Competition and the Evolution of Animal Mating Systems.* Academic Press, London.

Snook RR, Markow TA & Karr TL (1994) Functional nonequivalence of sperm in *Drosophila pseudoobscura*. *Proc. Natl Acad. Sci. USA* **91:** 11222–11226.

Snow JW, Young JR & Jones RL (1970) Competitiveness of sperm in female fall armyworms mating with normal and chemosterilized males. *J. Econ. Entomol.* **63:** 1799–1802.

Stanley-Samuelson DW & Loher W (1986) Prostaglandins in insect reproduction. *Ann. Entomol. Soc. Am.* **79:** 841–853.

Stearns SC (1983) The influence of size and phylogeny on patterns of covariation among life-history traits in mammals. *Oikos* **41:** 173–187.

Sugawara P (1979) Stretch reception in the bursa copulatrix of the butterfly, *Pieris rapae crucivora*, and its role in behaviour. *J. Comp Physiol.* **130:** 191–199.

Suzuki N, Okuda T & Shinbo H (1996) Sperm precedence and sperm movement under different copulation intervals in the silkworm, *Bombyx mori*. *J. Insect Physiol.* **42:** 199–204.

Svärd L & McNeil JN (1994) Female benefit, male risk: polyandry in the true armyworm *Pseudaletia unipuncta*. *Behav. Ecol. Sociobiol.* **35:** 319–326.

Svärd L & Wiklund C (1988) Prolonged mating in the monarch butterfly *Danaus plexippus* and nightfall as a cue for sperm transfer. *Oikos* **52:** 351–354.

Svärd L & Wiklund C (1989) Mass and production rate of ejaculates in relation to monandry/polyandry in butterflies. *Behav. Ecol. Sociobiol.* **24:** 395–402.

Svensson BG, Petersson E & Frisk M (1990) Nuptial gift size prolongs copulation duration in the dance fly *Empis borealis*. *Ecol. Entomol.* **15:** 225–229.

Thibout E (1975) Analyse des causes de l'inhibition de la receptivite sexualle et de l'influence d'une eventualle seconde copulation sur la reproduction chez la Teigne du poireau, *Acrolepia assectella* (Lepidoptera: Plutellidae). *Entomol. Exp. Appl.* **18:** 105–116.

Thornhill R (1976) Sexual selection and nuptial feeding behavior in *Bittacus apicalis* (Insecta: Mecoptera). *Am. Nat.* **110:** 529–548.

Thornhill R (1979) Male and female sexual selection and the evolution of mating systems in insects. In *Sexual Selection and Reproductive Competition in Insects*. MS Blum & NA Blum (eds), pp. 81–122. Academic Press, New York.

Thornhill R (1983) Cryptic female choice and its implications in the scorpionfly *Harpobittacus nigriceps*. *Am. Nat.* **122:** 765–788.

Thornhill R (1984) Alternative hypotheses for traits believed to have evolved by sperm competition. In *Sperm Competition and the Evolution of Animal Mating Systems*. RL Smith (ed.), pp. 151–178. Academic Press, London.

Thornhill R (1986) Relative parental contribution of the sexes to their offspring and the operation of sexual selection. In *Evolution of Animal Behavior: Paleontological and Field Approaches*. MH Nitecki & JA Kitchell (eds), pp. 113–136. Oxford University Press, New York.

Thornhill R & Alcock J (1983) *The Evolution of Insect Mating Systems*. Harvard University Press, Cambridge, MA.

Thornhill R & Sauer KP (1991) The notal organ of the scorpionfly (*Panorpa vulgaris*): an adaptation to coerce mating duration. *Behav. Ecol.* **2:** 156–164.

Trumbo ST & Fiore AJ (1991) A genetic marker for investigating paternity and maternity in the burying beetle *Nicrophorus orbicollis* (Coleoptera: Silphidae). *J. NY Entomol. Soc.* **99:** 637–642.

Tsubaki Y & Sokei Y (1988) Prolonged mating in the melon fly, *Dacus cucurbitae* (Diptera: Tephritidae): competition for fertilization by sperm-loading. *Res. Popul. Biol.* **30:** 343–352.

Tsubaki Y & Yamagishi M (1991) 'Longevity' of sperm within the female of the melon fly, *Dacus cucurbitae* (Diptera: Tephritidae), and its relevence to sperm competition. *J. Insect Behav.* **4:** 243–250.

Tsubaki Y, Siva-Jothy MT & Ono T (1994) Re-copulation and post-copulatory mate guarding increase immediate female reproductive output in the dragonfly *Nannophya pygmaea* Rambur. *Behav. Ecol. Sociobiol.* **35:** 219–225.

Tufton TJ (1993) The cost of reproduction in *Callosobruchus maculatus*. University of Sheffield.

Turner ME & Anderson WW (1984) Sperm predominance among *Drosophila pseudoobscura* karyotypes. *Evolution* **38:** 983–995.

Ueda T (1979) Plasticity of the reproductive behaviour of dragonfly, *Sympetrum parvulum* Bartneff, with reference to the social relationships of males and the density of territories. *Res. Pop. Ecol.* **21:** 135–152.

Ueno H (1994) Intraspecific variation of P_2 value in a coccinellid beetle, *Harmonia axyridis*. *J. Ethol.* **12:** 169–174.

Vahed K & Gilbert FS (1996) Differences across taxa in nuptial gift size correlate with differences in sperm number and ejaculate volume in bushcrickets (Orthoptera: Tettigoniidae). *Proc. Roy. Soc. Lond. Ser. B* **263:** 1257–1265.

Vardell HH & Brower JH (1978) Sperm precedence in *Tribolium confusum* (Coleoptera: Tenebrionidae). *J. Kan. Entomol. Soc.* **51:** 187–190.

Vick KW, Burkholder WE & Smittle BJ (1972) Duration of mating refractory period and frequency of second matings in female *Trogoderma inclusum* (Coleoptera: Dermestidae). *Ann. Entomol. Soc. Am.* **65:** 790–793.

Villavaso EJ (1975) Functions of the spermathecal muscle of the boll weevil, *Anthonomis grandis*. *J. Insect Physiol.* **21:** 1275–1278.

Waage JK (1979a) Adaptive significance of postcopulatory guarding of mates and non-mates by male *Calopteryx maculata* (Odonata). *Behav. Ecol. Sociobiol.* **6:** 147–154.

Waage JK (1979b) Dual function of the damselfly penis: sperm removal and transfer. *Science* **203:** 916–918.

Waage JK (1982) Sperm displacement by male *Lestes vigilax* Hagen (Odonata: Zygoptera). *Odonatologica* **11:** 201–209.

Waage JK (1984) Sperm competition and the evolution of odonate mating systems. In *Sperm Competition and the Evolution of Animal Mating Systems*. RL Smith (ed.), pp. 251–290. Academic Press, London.

Waage JK (1986) Evidence for widespread sperm displacement ability among Zygoptera (Odonata) and the means for predicting its presence. *Biol. J. Linn. Soc.* **28:** 285–300.

Wade MJ & Chang NW (1995) Increased male fertility in *Tribolium confusum* beetles after infection with the intracellular parasite *Wolbachia*. *Nature* **373:** 72–74.

Walker WF (1980) Sperm utilization strategies in nonsocial insects. *Am. Nat.* **115:** 780–799.

Ward PI (1993) Females influence sperm storage and use in the yellow dung fly *Scathophaga stercoraria* (L.). *Behav. Ecol. Sociobiol.* **32:** 313–319.

Ward PI & Hauschteck-Jungen E (1993) Variation in sperm length in the yellow dung fly *Scathophaga stercoraria* (L.). *J. Insect Physiol.* **39:** 545–547.

Webb RE & Smith FF (1968) Fertility of eggs of mexican bean beetles from females mated alternately with normal and Apholate-treated males. *J. Econ. Entomol.* **61:** 521–523.

Wedell N (1991) Sperm competition selects for nuptial feeding in a bushcricket. *Evolution* **45:** 1975–1978.

Wedell N (1992) Protandry and mate assessment in the wartbiter *Decticus verrucivorus* (Orthoptera: Tettigoniidae). *Behav. Ecol. Sociobiol.* **31:** 301–308.
Wedell N (1993) Spermatophore size in bushcrickets: comparative evidence for nuptial gifts as a sperm protection device. *Evolution* **47:** 1203–1212.
Werren JH (1980) Sex ratio adaptations to local mate competition in a parasitic wasp. *Science* **208:** 1157–1159.
Werren JH & Charnov EL (1978) Facultative sex ratios and population dynamics. *Nature* **272:** 349–350.
Wiklund C & Kaitala A (1995) Sexual selection for large male size in a polyandrous butterfly: the effect of body size on male vs. female reproductive success in *Pieris napi. Behav. Ecol.* **6:** 6–13.
Wilcox RS (1984) Male copulatory guarding enhances female foraging in a water strider. *Behav. Ecol. Sociobiol.* **15:** 171–174.
Wilkes A (1966) Sperm utilization following multiple insemination in the wasp *Dahlbominus fuscipennis. Can. J. Genet. Cytol.* **8:** 451–461.
Wilkinson GS & Reillo PR (1994) Female choice response to artificial selection on an exaggerated male trait in a stalk-eyed fly. *Proc. Roy. Soc. Lond. B* **255:** 1–6.
Wolf LL, Waltz EC, Wakeley K and Klockowski D (1989) Copulation duration and sperm competition in white-faced dragonflies (*Leucorrhinia intacta*; Odonata: Libellulidae). *Behav. Ecol. Sociobiol.* **24:** 63–68.
Woodhead AP (1985) Sperm mixing in the cockroach *Diploptera punctata. Evolution* **39:** 159–164.
Yamagishi M, Ito Y & Tsubaki Y (1992) Sperm competition in the melon fly, *Bactrocera cucurbitae* (Diptera: Tephritidae): effects of sperm 'longevity' on sperm precedence. *J. Insect Behav.* **5:** 599–608.
Yan G & Stevens L (1995) Selection by parasites on components of fitness in *Tribolium* beetles: the effect of intraspecific competition. *Am. Nat.* **146:** 795–813.
Yokoi N (1990) The sperm removal behaviour of the yellow spotted longicorn beetle *Psacothea hilaris* (Coleoptera: Cerambycidae). *Appl. Entomol. Zool.* **25:** 383–388.
Zeh JA & Zeh DW (1994) Last-male sperm precedence breaks down when females mate with three males. *Proc. Roy. Soc. Lond. Ser. B* **257:** 287–292.
Zimmering S, Sandler L & Nicoletti B (1970) Mechanisms of meiotic drive. *Annu. Rev. Genet.* **4:** 409–436.
Zuk M & Simmons LW (1997) Reproductive strategies of the crickets. In *The Evolution of Mating Systems in Insects and Arachnids*. JC Choe & BJ Crespi (eds), pp. 89–109. Cambridge University Press, Cambridge.

APPENDIX A

Classification and definition of terms frequently encountered in sperm competition literature

Sperm competition

The competition within a single female between the sperm of two or more males for the fertilization of ova (Parker 1970c). For external fertilizers, competition would occur within the spawning media. The selection pressures that arise as a consequence of sperm competition should favour somatic adaptations that increase the likelihood of fertilization success for the copulating male.

Sperm mixing

The mixing of sperm from two or more males within a single female, usually within her sperm-storage organs, that results in the conditions for sperm competition.

Sperm pre-emption

The ability of males to gain fertilizations with previously mated females (Parker 1970c).

P_n

The proportion offspring sired by the *n*th male to mate when *n* males copulate with the same female (generally this is P_2 with two males; Boorman and Parker 1976).

Sperm displacement

The spatial displacement of sperm derived from a female's previous mate(s) by the copulating male with the consequence that self sperm is more likely to fertilize ova, while displaced sperm is less likely to do so. Mechanisms include the following:

(1) *Sperm removal* – the direct anatomical removal of previous sperm from the female's sperm-storage organ(s) prior to ejaculation (Waage 1979b).
(2) *Sperm flushing* – the indirect removal of previous sperm from the female's sperm-storage organ(s) by the incoming ejaculate of the copulating male.
(3) *Sperm repositioning* – the copulating male 'repositions' sperm derived from his mate's previous partner(s) to a position away from the site of fertilization (Waage 1984).

Sperm precedence

The nonrandom utilization of sperm from a particular male when the sperm of two or more males overlap within a single female. Also termed *Sperm predominance* (Gromko *et al.* 1984). It is important to note that nonrandom fertilization following multiple matings does not indicate the occurrence of sperm precedence. Sperm precedence can be a consequence of one of the following phenomena:

(1) *Sperm stratification* – the last sperm to enter the sperm-storage organ(s) have a positional advantage so that they are the first

out during fertilization. Sperm stratification may be a direct consequence of sperm repositioning or the passive consequence of mating order. It refers to the layered disposition of the sperm of several males within a female's sperm-storage organ(s).

(2) *Sperm loading* – increasing fertilization success by increasing the numerical representation of self sperm with respect to rival sperm already within the female's sperm-storage organ(s) (Dickinson 1986).

(3) *Sperm selection* – nonrandom usage of sperm derived from a particular male when the female has sperm stored from a variety of potential sires. It is thus the extension of female choice at the gametic level.

Sperm incapacitation

The killing and/or inhibition of function of sperm from a female's previous mate by physiological and/or anatomical adaptations in the copulating male.

Sperm competition avoidance

A host of adaptations in males that prevent the sperm from past and/or future males obtaining fertilizations. Also termed *paternity assurance* (Smith 1979). Mechanisms include male-controlled mechanisms of sperm displacement and precedence plus:

(1) *Proximate mate guarding* – males remain in genital contact or in close proximity to their mates to prevent female remating.

(2) *Remote mate guarding* – males install physical and/or chemical barriers to prevent female remating.

(3) *Repeated copulation* – males increase fertilization success by repeated copulations prior to and during oviposition.

Sperm competition risk

The probability (between zero and one) that females will engage in promiscuous mating activity that will result in the temporal and spatial overlap of the ejaculates from two or more males.

Sperm competition intensity

The extent of overlap between the ejaculates of different males once competition occurs. Intensity is determined by the relative numbers of sperm from different males and the absolute number of males engaged in competition for the ova of a single female (Parker *et al.* 1996).

11 Sperm Competition in Fishes

C. W. Petersen[1] *and R. R. Warner*[2]

[1] College of the Atlantic, 105 Eden Street, Bar Harbor, Maine 04609, USA and [2] Department of Ecology, Evolution and Marine Biology, University of California, Santa Barbara, California 93106, USA

I. INTRODUCTION

Sperm competition occurs when sperm from different males compete for fertilizations (Parker 1970). The degree of sperm competition ranges widely in fishes, both interspecifically and intraspecifically, from males that never experience sperm competition to males that are in frequent competition with several males. The relative ease of observing fishes in natural populations, combined with the high diversity of mating systems, make them excellent subjects for studying sperm competition and conducting intraspecific and interspecific tests of sperm competition theory. However, only recently have comparative studies on sperm competition been published (Stockley *et al.* 1996, 1997), and in Smith's (1984) volume the single chapter on fish focused on a single family, the livebearing Poeciliidae (Constantz 1984). In this chapter we review the theory and data on sperm competition in fishes, compare these data with predictions from sperm competition theory, and discuss the evolutionary consequences of sperm competition in this diverse group of vertebrates.

Sperm Competition and Sexual Selection
ISBN 0-12-100543-7

The diversity of mating systems and life histories in fishes is impressive. Fertilization can occur in the environment by either the release of gametes into the water column (pelagic spawning) or the deposition of eggs on the substrate (demersal spawning). Fertilization can also take some rather unexpected twists, as in the case of the catfish *Corydoras aeneus* (Callichthyidae), where females drink sperm by putting their mouth around the abdomen of the male. The sperm quickly pass through the digestive system, exit the anus, and externally fertilize the female's eggs in a protected space formed by the female's pelvic fins (Kohda *et al.* 1995). In internal fertilizers, sperm are introduced into the female's body with subsequent events ranging from internal development and viviparity to cases where fertilization and development are delayed until the eggs are released. Internal fertilization can also occur in males, as is the case of pipefishes and seahorses where the female transfers unfertilized eggs into the pouch of her mate. Where it occurs, parental care can consist of uniparental care of eggs, biparental care of eggs and mobile offspring, attachment of eggs to the body, mouth-brooding, and viviparity (Breder and Rosen 1966; Blumer 1979).

Are there special characteristics of fishes that set them apart from other animal groups when we consider sperm competition? Like all animals, sperm from fishes must access eggs through an aqueous medium, but because fishes themselves live in an aqueous environment, fertilization often occurs outside the female's body. Pelagic fertilization with highly mobile males and females is more characteristic of fishes than other animal groups and results in selective pressures on sperm and spawning activity unique to this group. The close alignment of the partners means that while eggs and sperm are shed into a three-dimensional medium and both are free to disperse, they are initially in very close proximity. This type of external fertilization offers a precise location and time where unfertilized eggs are ready for insemination, and the opportunity for further fertilization declines very rapidly. It is probably because of this limited window of opportunity that sperm in some fish species have extremely short lifespans (Petersen *et al.* 1992). As outlined below, rapid, repeated mating and the detailed information available on each successive partner appears to have led to the evolution of mechanisms allowing precise control of ejaculate amount in several fishes (e.g. Shapiro *et al.* 1994).

While external fertilization has been proposed as a fundamental factor affecting sperm competition (see Chapter 6), it is not clear just how the short-term availability of eggs characteristic of many fish matings should play out in terms of sperm competition. External fertilization provides an opportunity that can be exploited by several males simultaneously, resulting in high levels of sperm competition. Often, this takes the form of smaller peripheral males contributing their sperm to a mating between a larger male and a female (Taborsky 1994). Most of the documented sperm competition has indeed been noted in externally fertilizing fishes, and some males have gonads comprising more than 10% of their body weight. However, external fertilization may also make mate guarding easier, since it is not necessary to guard an inseminated female for prolonged periods until fertilization occurs. In fact, many fish species with external fertilization where males defend either mates or preferred spawning sites often do not experience sperm competition.

In addition to its effect on intrasexual competition, some forms of external fertilization may also limit the extent to which females can choose among potential mates. Group matings, in which many males simultaneously release sperm, effectively prevent choice of a particular male, although females may still control their matings according to some group composition criterion. In benthic or pelagic pair matings, even in the presence of peripheral male activity, a female may still exercise choice over the male who will fertilize at least the majority of her eggs. Because males are more often the sex providing parental care of benthic eggs in fishes, female allocation of paternity to a particular male may have profound implications for the subsequent care that the eggs receive (see Chapter 4). Thus, sperm competition (and the resultant dilution of paternity) can potentially reduce the direct benefits that a female might receive from a male of her choosing.

II. SPERM COMPETITION THEORY

The effect of sperm competition on gamete characteristics and relative gametic investment has been modelled, largely by Parker and his co-workers. All of these models use an Evolutionary Stable Strategy (ESS) approach, since the optimal strategy for a male depends on the strategy adopted by other individuals in the population. These models attempt to predict the effects of varying the level of sperm competition on the proportion of reproductive effort devoted to gamete production (Parker 1990a,b, 1993; Parker and Begon 1993; Parker *et al.* 1996; Ball and Parker 1996), the proportion of total gametic investment allocated to male gametes in simultaneous hermaphrodites (Lloyd 1984; Charnov 1980, 1982; Fischer 1984), or characteristics of individual sperm, such as size or longevity (Parker 1984, 1993; Parker and Begon 1993; Ball and Parker 1996). These models make predictions about both interspecific and intraspecific variation in male reproductive characteristics.

Recently Parker *et al.* (1996) made a useful distinction between the risk of, and the intensity of sperm competition, and their definitions will be used in this chapter. The risk of sperm competition refers to the probability that a male will spawn with at least one other male competing for a batch of eggs. The intensity of sperm competition refers to the number of males whose sperm are competing for a batch of eggs. These distinctions are important because they represent two ways of measuring the level of sperm competition, with risk being more useful when sperm competition levels are low, and intensity being more useful when sperm competition levels are high. They are also important because they often refer to levels of sperm competition that represent very different types of mating systems and may have differing predictions based on ESS models (Parker *et al.* 1996).

All of the models assume that males obtain more fertilizations by releasing more sperm (at least in the case of sperm competition), but that this is done at a cost of investment in other fitness-enhancing functions such as growth, finding additional mates, investment in female function (for simultaneous hermaphrodites), mate defence or territoriality, or releasing more sperm in other spawnings. As applied to sperm competition, ESS theory generates predictions in three areas: gonadal investment, sperm characteristics and male reproductive anatomy, and life history strategies. These areas are examined below.

III. MEASURING SPERM COMPETITION

In practice, the level of sperm competition in fishes has been estimated in two fundamental ways, either behaviourally or by the paternity of progeny. Behaviourally, the simplest way to estimate the level of sperm competition is to count the number of males that release sperm in a spawn or copulate with a female during a period when sperm would compete for the same batch of eggs. A batch of eggs here refers to a spawning event by an external fertilizer or a group of similarly receptive oocytes in an internal fertilizer.

Using behavioural criteria, the presence of sperm competition can be assessed much more easily in external vs. internal fertilizers. Several authors have used the number of males associated with a female during gamete release as an estimate of the level of sperm competition (Warner *et al.* 1975; Fischer 1984; Petersen 1991a). Sperm competition in internal fertilizers is more difficult to document because matings are temporally separated. In these species, assessing multiple paternity in a brood via either phenotypic (e.g. coloration, growth) or genotypic (e.g. allozyme, DNA) paternal markers is a typical method of assessing the presence of sperm competition. Genetic markers, and especially DNA markers that can be magnified using polymerase chain reaction (PCR), are currently under development in several laboratories to assess paternity and the degree of sperm competition in a variety of fish species.

A more accurate measure of the level of sperm competition in a spawn would be the proportion of sperm in a spawn that came from a particular male. Using this definition, different males can experience different levels of sperm competition in a spawn if the number of sperm they release differs. However, even knowing the relative gametic contribution of the males in a spawn may not give an accurate representation of the degree that sperm are competing with heterologous sperm for fertilizations. All of these estimates assume complete mixing and equal competitive ability of sperm from different males, but these assumptions may often be violated. The ramifications of violating these assumptions and their effect on our estimate on the level of sperm competition are discussed later in this chapter.

IV. GONADAL INVESTMENT

Patterns of sperm production and the effects of sperm competition on the evolution of number of sperm released and testes size have been the subject of more theoretical and empirical work than any other aspect of sperm competition in fishes. The theory and data address patterns of variability at several different levels, including comparisons among species, among populations within species, among individuals within species, and among ejaculates within an individual. In all but the last level, most studies relate the size of the testes to the degree of sperm competition.

The most common measurement of relative gametic investment in fishes is the weight of the gonad relative to the total body weight of the fish. This measurement, called the gonosomatic index or GSI (= 100 × gonad weight/somatic weight), expresses gonad weight as a percentage of body weight. It has been used extensively by fisheries biologists and ichthyologists to determine reproductive seasonality, especially for females. There can be problems comparing GSIs among species, since GSI may represent different levels of reproductive effort in different species depending on spawning frequency or adult size. These problems can be minimized by comparing species with similar reproductive biology. GSI appears to be the best proxy available for testing predictions of sperm competition theory that concern differences in reproductive effort devoted to gamete production in males, both because of its widespread use and its relationship with gametic investment. Using testes size as a measure of male gametic investment assumes that standing crop biomass devoted to testicular material is correlated with energy devoted to sperm production and to sperm production rates. This assumption, although rarely tested, seems reasonable. There is a strong correlation between the number of sperm stripped from males and the size of the testes in one species tested, the bucktooth parrotfish *Sparisoma radians* (Marconato and Shapiro 1996).

GSI may not provide a useful comparison of male allocation if similar GSIs represent different relative reproductive effort for individuals of different sizes. Stockley *et al.* (1997) examined patterns of sperm competition, testes size, and adult size, and found a slight interspecific tendency to decreasing GSI with adult size, and found a slight opposite tendency towards decreasing GSI with adult size. This led them to conclude that the testes size–body size relationship would tend to mask the effects of sperm competition on GSI, and that any effects of sperm competition on GSI were probably real and not an artefact of allometric patterns.

With the increase of field observations of fish behaviour in the 1970s, correlations between testicular investment and mating system were documented both interspecifically and intraspecifically (Robertson and Choat 1974; Choat and Robertson 1975; Robertson and Warner 1978; Warner and Robertson 1978). These studies were some of the first to note the correlation of increased testes size with increasing levels of sperm competition. The interpretation by those researchers of this pattern is consistent with our current understanding of sperm competition: males facing chronic sperm competition have proportionately larger testes and produce larger numbers of sperm in order to increase their share of available fertilizations. Predictions and data for sperm competition and gonadal investment are given for the different levels of comparison in the following sections.

A. Interspecific comparisons

In externally fertilizing species, when multiple males release sperm in spawns, producing additional sperm is expected to increase the number of eggs that a male fertilizes at the expense of the other males releasing sperm. In the absence of sperm competition, producing additional sperm will increase fertilization success only when unfertilized eggs remain, and the numbers of unfertilized eggs will show an exponential decline as sperm output increases (Denny and Shibata 1989; Petersen 1991a). The presence or absence of sperm competition results in different male-fitness gain for increased sperm release in a spawn and leads to a simple prediction: species with higher levels of sperm competition are expected to have greater investment in sperm production (Parker 1984, 1990a). This prediction has been elaborated for separate-sexed species in a series of papers by Parker and his coworkers (Parker 1984, 1990a,b, 1993; Parker and Begon 1993; Ball and Parker 1996; Parker *et al.* 1996) and for simultaneous hermaphrodites by several authors (Charnov 1980, 1982; Fischer 1981, 1984; Lloyd 1984; Petersen 1991a; Petersen and Fischer 1996).

Interspecific patterns of GSI and the degree of sperm competition have been recently reviewed by Stockley *et al.* (1997). Using techniques to control for phylogenetic effects (comparative analysis by independent contrasts), they found that species with more intense sperm competition tended to have higher GSIs. Their raw data from 24 species, plotted in Fig. 11.1, shows the same trend without the control for phylogeny: species with more intense sperm competition have larger relative testes mass.

An independent test of the pattern of interspecific variation in sperm competition and GSI can be shown by examining four families of coral-

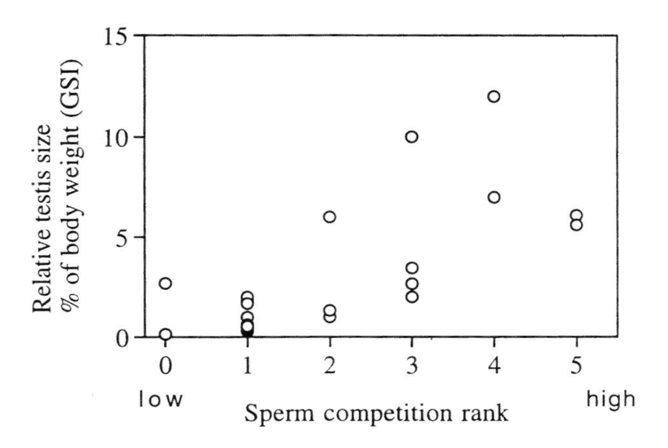

Fig. 11.1. *Relative testes size in relation to sperm competition intensity in 24 species of fishes from Stockley* et al. *(1997). Sperm competition rankings are also from Stockley* et al. *(1997), with increasing numbers representing higher levels of sperm competition. Sperm competition rank definitions: 0 = internal fertilization (including fertilization in the mouth), no evidence for communal spawning or polygamy; 1 = internal fertilization, low communal spawning or polygamy, or external fertilization, distinct pairing, no obvious communal spawning; 2 = internal fertilization, high communal spawning or polygamy, or external fertilization, distinct pairing, low communal spawning; 3 = external fertilization, distinct pairing, moderate communal spawning, or no pairing, low communal spawning; 4 = external fertilization, distinct pairing and high communal spawning, or no pairing, moderate communal spawning; 5 = no pairing, high communal spawning.*

reef fishes with external fertilization – a data set that was not examined by Stockley *et al.* (1997). This data set uses GSI values and sperm competition estimates from dominant males (or all males in species with only one male-mating strategy) in each species. All species in these families release pelagic eggs that are fertilized in a spawning rush, where a female and one or more males rise up the water column, releasing gametes at the apex of the rush. Dominant males are exposed to sperm competition when other males join pair spawns just as gametes are being released (streaking), or when they are one of a group of males with a female in a spawning rush (group spawning). Streaking rates range from zero in some haremic species to the majority of spawns in several high-density species. In many coral-reef fishes, multiple male-mating tactics exist within a population, and these tactics are associated with different levels of sperm competition. In all four families for which there are data, there is an increase in relative testes size with increasing levels of sperm competition (Fig. 11.2). The highest GSI values in Fig. 11.2 are recorded for species of surgeonfish where pair spawning does not occur and all males take part in group spawns in which several males simultaneously release sperm as the female releases eggs (Robertson 1983, 1985).

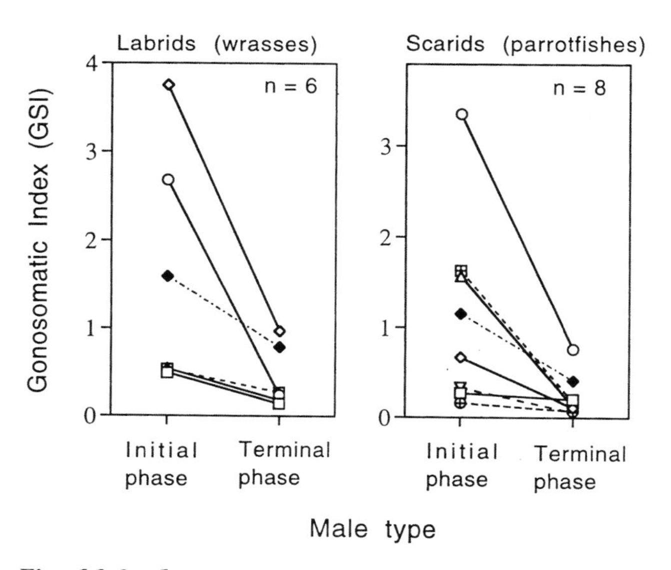

Fig. 11.3. *Comparison of relative testis size for initial phase (IP) and terminal phase (TP) males for scarids (parrotfishes) and labrids (wrasses). Each line connects data points from the same species. Data are from Robertson and Warner (1978), Warner and Robertson (1978), Warner (1982).*

proportion of matings with competing ejaculates, or having more males releasing sperm in an average spawn, these males may not be able to place their sperm as close to eggs as the dominant males. The sperm from one male may be at a competitive disadvantage compared with the sperm of another male, either because they are spatially more distant from the cloud of eggs or because the sperm are released at a time when they are less likely to reach unfertilized eggs. Parker calls this type of situation an unfair raffle (Parker 1990a), and it results in a male's sperm having a lower probability of fertilizing an egg. This is mathematically very similar to the male releasing his sperm into a much larger pool of sperm: it selects for males at a disadvantage to release more sperm, and males at an advantage to release less sperm, than would otherwise be predicted (Parker 1990a). It is difficult to find unambiguous support for this prediction because males that engage in alternative mating tactics are expected to be both in a poorer position to release sperm and be involved in more sperm competition. One possible way to test this prediction is to compare a mating tactic in which males are able to approach mating females closely for long periods of time to a tactic in which males dart into spawns and release sperm when spawning occurs (sneaking). In bluegill sunfish *Lepomis macrochirus*, sneakers differ from satellites in that satellites resemble females and may be able to approach females more closely by temporarily deceiving the dominant male (Gross 1982). Sneakers have higher GSIs than satellites, as would be predicted if satellite sperm is placed closer to females. However, these differences in GSI could also reflect simple differences in spawning rate, or even reflect selection for similar-sized gonads in the two types. In the latter case, the smaller sneaker-type males would automatically have a larger relative testis size.

In some species, local differences in sperm competition intensity may lead to differences in male allocation to testes mass. In the bucktooth parrotfish *S. radians* the GSI of individual territorial males was positively correlated with the proportion of their spawns in which additional males released sperm (Marconato and Shapiro 1996). However, in two other species, there was no evidence for differences in GSI levels among populations that exhibited different intensities of sperm competition. In the simultaneous hermaphrodite *Serranus psittacinus* (*Serranus fasciatus*), differences in testicular allocation between two populations were correlated with male mating success and not the estimated level of sperm competition among individuals or populations (Petersen 1990a). In the surgeonfish *Acanthurus nigrofuscus*, the mating system varied from one site with a predominantly haremic mating system and no sperm competition to a second site where group spawning and no mate defence was the rule. The GSI for males at the site with the higher intensity of sperm competition was not higher than the males from the population with low sperm competition (Robertson 1985).

C. Variation in sperm competition among spawns

In the previous two sections we examined mean differences in the level of sperm competition among species, populations within species, or males employing different reproductive tactics at either one point in their life or for their entire lives. In this section we examine cases where individual males encounter variable levels of sperm competition among a series of spawns. The predictions of how much sperm an individual should release in different spawns or with different females are not as straightforward as the comparisons among individuals or species. Parker *et al.* (1996) have recently modelled selection on an individual male that can alter its sperm release pattern with varying sperm competition levels and predicted different patterns for different absolute levels of sperm competition. These models were based on the biology of externally fertilizing fishes, where males can assess the intensity of sperm competition by the number of males present at a spawning event. At low levels of sperm competition, when males spawn either with no or sometimes one sperm competitor, the predictions match the previous sections, and males are predicted to release more sperm in cases of increased risk of sperm competition. However, in cases where individuals are exposed to chronic but variable sperm competition, the model predicted that males should release less sperm in cases of higher sperm competition. The reason, although not initially intuitive, is because a male will more substantially increase the number of eggs he fertilizes by increasing his sperm release in spawns with fewer competitors. In spawns with many competitors, the chances of his additional sperm encountering an unfertilized egg are too low to favour the production of additional sperm. When comparing two spawns with differing numbers of sperm competitors, the model predicts that under most circumstances he will obtain more fertilizations by releasing more sperm when he competes with fewer males. In other words, given a choice of where to increase sperm output, the same increase in output will profit a male more in situations of low competition.

There is no information on individual variation in male sperm release in response to sperm competition; it appears to be anatomically possible, since males in two species have been shown to alter their sperm release depending on the size or fecundity of a female in the absence of sperm competition (Shapiro *et al.* 1994; Marconato and Shapiro 1996). This prediction is difficult to test. Theoretically, it should be reflected in a negative relationship between estimated per capita sperm output and the estimated number of males in a spawning group. However, it is notoriously difficult to estimate the actual number of males releasing sperm in a spawn because the spawning rush happens so rapidly. Moreover, we currently have no means to estimate how many males are actually releasing sperm in group spawns. Error in this estimate will yield the predicted negative relationship (e.g. for a given total sperm amount, a high estimate of the number of participating males will simultaneously lower the per capita output estimate). More accurate tests would require the identification of sperm from a specific individual. One approach could involve a focal male spawning with gonadectomized males so that all the sperm in a spawn could be assigned to the focal male. Alternatively, genetic or other marking techniques will need to be developed in order to assign relative numbers of sperm released in a spawn to different males. Efforts are currently underway to use DNA microsatellite markers designed for bluehead wrasse *Thalassoma bifasciatum* and quantitative PCR to identify the relative contributions of males in matings with sperm competition (L. Wooninck, pers. comm.).

V. SPERM CHARACTERISTICS AND MALE REPRODUCTIVE ANATOMY

A. Sperm characteristics

In fishes, as in other organisms, morphological differences among sperm reflect differences in functional capabilities and phylogeny. In most bony fishes, sperm lack acrosomes, and enter the egg through a single micropyle located near the animal pole. In fishes there are differences in sperm morphology that are related to whether fertilization is internal or external (Jamieson 1991; Stockley *et al.* 1996). Sperm of internally fertilizing species, at least in the higher bony fishes, appear to be more morphologically complex, with longer heads, than externally fertilizing sperm in related species (Jamieson 1991).

During sperm competition sperm compete with unrelated sperm from other males for fertilizations, and there are more sperm competing for the fertilization of each egg. Selection will favour sperm that get to eggs faster because unfertilized eggs will be converted to fertilized eggs more quickly when more sperm are present. There will be differences in how selection acts on sperm number, morphology and behaviour, depending on whether sperm characteristics are controlled by the male or the sperm themselves (Parker and Begon 1993), but the general qualitative trends will be the same for both scenarios. Increased levels of sperm competition should increase selection on sperm number, and potentially sperm velocity, at the expense of sperm longevity (Parker 1993).

However, the predictions of sperm attributes from models of sperm competition are not well supported by the data (Stockley *et al.* 1997). Interspecifically, there is a negative correlation between the degree of sperm competition and sperm length (which may be positively correlated with sperm velocity), while two models of sperm attributes predict either no relationship or a positive correlation (Ball and Parker 1996; Stockley *et al.* 1997). Intraspecifically, in both sticklebacks (*Gasterosteus aculeatus*, de Fraipoint *et al.* 1993) and Atlantic salmon (*Salmo salar*, Kazakov 1981; Gage *et al.* 1995) smaller or younger males that are more likely to engage in sperm competition have a higher percentage of motile sperm and these sperm are motile for a longer period. This duration of motility is opposite to that predicted by sperm competition models.

The lack of support for the evolution of gamete characteristics based on sperm competition models could occur for several reasons. There is currently a poor understanding of trade-offs in sperm production and sperm life history. For example, with increasing sperm competition larger sperm may be selected for because of their increased tail length and velocity, or selected against because the greater investment per sperm reduces the total number of sperm that individuals can produce. The models of sperm investment and behaviour assume that increased investment per sperm is achieved at the cost of reduced numbers of sperm produced and, for a fixed energetic investment per sperm, increased activity is achieved at the cost of longevity. The main problems with moving from theory to data may be our lack of understanding of the functional and energetic significance of variation in sperm morphology.

In addition, there may be general selection for short lifespan in sperm and variation in this selection may outweigh selection from sperm competition. Sperm in externally fertilizing fishes is very short-lived, and often exhibits motility for less than a minute (Ginzburg 1968; Stockley *et al.* 1996). This lifespan is much shorter than that of other organisms in the same environment, particularly compared with marine invertebrates (Levitan 1995).

There are probably two reasons for the relatively short lifespan of fish sperm, and both are related to the close proximity of the sexes during gamete release compared with most other animals that release their gametes into the water. First, unfertilized eggs are at their greatest concentration in the first seconds after release and the density of unfertilized eggs in the water column decreases very quickly after spawning. This puts a premium on rapid access to eggs, even if this must be traded off against sperm longevity or characteristics that influence sperm longevity. Species that spawn in calmer water appear to have longer-lived sperm than those that spawn under more turbulent conditions (Ginzburg 1968; Billard 1987), suggesting that higher levels of water mixing might also disperse eggs more quickly and favour more active or numerous sperm over longer-lived sperm. Thus, water mixing and its effects on the dispersal of unfertilized eggs may help to mask any effect of sperm competition on sperm characteristics.

Second, unlike marine invertebrates, where fertilization rates may often be quite low (Levitan and Petersen 1995), in fishes there is evidence that fertilization success is typically over 90% (van den Berghe *et al.* 1989; Marconato *et al.* 1995; Warner *et al.* 1995; Marconato and Shapiro 1996). High fertilization rates imply more intense sperm competition in marine fishes than marine invertebrates, and selection for more or faster sperm that may have a cost in terms of reduced sperm energetic reserves and reduced sperm longevity. This second explanation is less clear, however, since many marine invertebrate species also spawn in synchrony (Levitan 1995), which should result in intense sperm competition if much mixing of sperm occurs.

B. Male reproductive anatomy

Although the testes of fishes have a simple anatomy, there are often associated accessory organs that appear to have multiple evolutionary origins, that have multiple names, and almost certainly do not perform the same function in all fishes (Fishelson 1991). The testicular gland can vary interspecifically from 15 to 250% of the testis weight (Lahnsteiner *et al.* 1990; Rasotto 1995); it is often unclear whether it has been included in measurements of GSI (Miller 1984). We include discussion of these accessory organs here because of some recent work suggesting that secretions from these glands may sometimes affect sperm longevity (see below).

The list of hypothesized functions of the accessory organs in males has included sperm storage or maturation (Blum 1972), production of mucins, which may be used in nest building, steroid synthesis (Seiwald and Patzner 1989), production of antibacterial or antifungal compounds, or the deposition of sperm in the nest (reviewed for gobies in Fishelson

1991; for blennies in Rasotto 1995). Some accessory organs have been shown to produce glycogen, which is believed to be important in germ cell nutrition, and nonsulphated mucins, which are part of the seminal fluid (Rasotto 1995). In other species these accessory organs contain abundant mature sperm and appear to function in sperm storage. In some gobies, males have two different accessory structures that contain visibly different but unknown contents (Miller 1984; Cole 1990). Given the variety of cell types and structure of the accessory glands it is likely that they are variable in function, both interspecifically and possibly intraspecifically.

Recently, another function has been hypothesized for male accessory organs: the production of sperm trails. These trails are a mixture of sperm and viscous material that allows for the slow release of active sperm (Marconato *et al.* 1996). Males may deposit these trails before eggs are deposited. Sperm trails may also prolong sperm life in the vicinity of eggs (Ruchon *et al.* 1995; Marconato *et al.* 1996), allowing nest-guarding males to continue to release sperm in a nest while defending the nest from possible sperm competitors, or allowing for 'sneaker' males to release sperm in a nest after departure. In some species with nest-guarding and sneaker males, the nest-guarding males appear to have much larger accessory organs (de Jonge *et al.* 1989; Ruchon *et al.* 1995). This supports the idea that sperm trails function as part of the parental male strategy (Marconato *et al.* 1996). Interspecific and intraspecific variation in the development of accessory organs and the production of sperm trails in species with distinctive oviposition patterns and male mating strategies is being examined.

VI. PATTERNS OF PATERNITY UNDER SPERM COMPETITION

There is much less information on paternity in fishes when compared with other groups such as birds (see Chapter 14) or mammals (see Chapter 16). Many fish species appear to have almost completely exclusive paternity, including all pipefishes and seahorses (Syngnathidae) where females transfer unfertilized eggs to individual males, and many examples of fishes with external fertilization and male parental care where additional males have never been observed in the proximity of spawning sites (e.g. most damselfishes, Pomacentridae).

A. Internal fertilizers

In species with internal fertilization, multiple paternity of broods has been shown to occur in livebearers (Poeciliidae; Winge 1937; Hildemann and Wagner 1954; Constantz 1984; Zimmerer and Kallman 1989), surfperches (Embiotocidae; Darling *et al.* 1980), and sculpins (Cottidae; Munehara *et al.* 1990). There are field observations of multiple matings in an embiotocid (Warner and Harlan 1982), and a large volume of data on multiple matings in poeciliids from aquaria studies (Constantz 1984).

In some fishes, although sperm enter ovaries, fertilization does not take place until eggs are released by the female (Hubbs 1966; Munehara *et al.* 1989, 1991). The delay in fertilization appears to be a result of a low concentration of calcium ions in the ovarian fluid (Munehara *et al.* 1994). In these species sperm associate with the micropyle but fertilization is delayed until eggs are released. In two species of cottid exhibiting this pattern of development, eggs are cared for by the male (Ragland and Fischer 1987; Munehara *et al.* 1990), and in one of these species paternity analysis revealed evidence of cuckoldry (Munehara *et al.* 1990). In the elkhorn sculpin *Alcichthys alcicornis*, females copulate with males after laying eggs in the male's site (Munehara 1988). This pattern of copulation means that the first clutch of a season will not be internally fertilized. Munehara (1988) believed that this clutch was externally fertilized by sperm emitted from the penis before and after copulation and by sperm leakage from the female's genital pore after copulation.

In guppies, there is evidence that the last male to spawn during the receptive period of the female fertilizes the most eggs (Hildemann and Wagner 1954). Sperm disappear slowly from the female's ovary after insemination, so that sperm that are deposited just before eggs become available for fertilization appear to have a higher probability of persisting in the ovary and fertilizing eggs.

Evidence of mate guarding in internal fertilizers is restricted to a brief mention of possible mate guarding in the viviparous shiner perch *Cymatogaster aggregata* (DeMartini 1988) and in the cardinalfish *Apogon imberbis*, where courtship lasts several days and males attempt to keep other males away from the female (Garnaud 1962, 1963). This cardinalfish is the only species of fish other than the sculpins mentioned earlier known to have both internal fertilization (or internal gamete association) and male parental care. In guppies, there is speculation that the gonopodium of the male may damage the female, cause bleeding and swelling around the genital pore, and may make subsequent inseminations more difficult (Hildemann and Wagner 1954). If true, male-induced damage has an effect similar to postcopulatory plugs in insects and could reduce the extent of sperm competition. However, there are no data to test this hypothesis.

B. External fertilizers

Paternity analysis has been conducted in several species of salmon. Typically, in salmonids one male is closest to the female while the redd, a depression dug in the gravel bottom, is excavated and eggs are released. The male closest to the female is larger than the other males that compete for fertilizations, but the difference in size between the dominant and other males varies from over a 100-fold difference in weight in the case of male parr in Atlantic salmon (*Salmo salar*; Gage *et al.* 1995) to near equality in size in chum salmon (*Oncorhynchus keta*; Schroder 1981). The relative success of these nondominant males appears to depend on their numbers and their size relative to the dominant male. In Atlantic salmon, females that spawned in artificial streams with varying number of parr had approximately 5–25% of their eggs fertilized by parr, as estimated from allozyme data from embryos and parents, with numbers increasing with higher numbers of parr (Hutchings and Myers 1988). However, the average parr fertilized fewer eggs as parr density increased, from 5% to approximately 1% at the highest density. This was substantially less than the fertilization achieved by a single satellite male chum salmon, which in one study averaged approximately 25% of the fertilizations (Schroder 1981). Hutchings and Myers (1988) examined interspecific salmonid data and noted that nondominant males tended to have higher paternity in those species where they were closer in size to the dominant male.

Additional data on paternity assessment using both allozyme (Phillip and Gross 1994) and DNA analysis (Rico *et al.* 1992; Colbourne *et al.* 1996) has revealed successful fertilization by males other than the nest-guarding parent in sticklebacks *Gasterosteus aculeatus* (Rico *et al.* 1992) and bluegill sunfish *Lepomis macrochirus* (Phillip and Gross 1994; Colbourne *et al.* 1996).

Without data on the relative amounts of sperm released or genetic analysis of offspring, many studies have estimated paternity either by dividing the potential fertilizations equally among all males releasing sperm in a spawn (pair-spawning equivalents of Warner *et al.* 1975) or have assigned all fertilizations to the 'parasitic' males, based on their larger testes (Gross and Charnov 1980; Gross 1982; Chan 1987 as cited in Taborsky). With the broader application of genetic paternity analysis, the practicality of these assumptions should be clarified in the near future.

VII. WHAT DETERMINES THE DEGREE OF SPERM COMPETITION IN A POPULATION?

Until this point, this chapter has dealt with how fishes respond to a relatively fixed level of sperm competition, and how they are expected to evolve changes in energetic allocation to sperm production and sperm characteristics in the face of that competition. In this section, we ask the question from a different perspective. Are there predictors of the degree of sperm competition in a population, and what types of life histories are expected to evolve under various scenarios of sperm competition?

Males often appear to fall within two categories of reproductive tactics in fishes, those that are involved in intense sperm competition and those that invest heavily in mate or spawning-site defence and are therefore exposed to low levels of sperm competition. If this dichotomy is accurate, it implies that there is selection for extremes in investment and specialization in either sperm competition or mate monopolization.

Several ecological and demographic features clearly influence the degree of sperm competition in fishes. The economic defendability of spawning sites, in both demersal and pelagic spawners, has been linked to several factors, including population density (Warner and Hoffman 1980a; Petersen 1990b), and the potential for cover for small males employing alternative mating tactics (Warner and Robertson 1978; Warner 1984a; Gross 1984).

The local population density and temporal and spatial predictability of spawning females appear to be important factors in the evolution of fish mating systems, including the success of males who engage in alternative mating tactics (Warner and Robertson 1978; Warner 1984a; Fischer and Petersen 1987; Hourigan 1989). As densities increase, dominant males may be less able to restrict the access of smaller males to females or mating sites. The importance of population density on the rate of sperm competition has been suggested from data on naturally occurring variation in mating systems at the interspecific level (Warner 1984a; Fischer and Petersen 1987; Hourigan 1989), the intraspecific level (Warner and Hoffman 1980b; Gross 1984; Phillip and Gross 1994; Petersen 1990b) and from the experimental manipulation of population density (Warner and Hoffman 1980a). In the bluehead wrasse, as population density increases, the mating success of dominant males at first increases as the local density of females increases but then decreases at higher densities as competition from other, smaller males increases (Warner and Hoffman 1980a). The exclusion of other males from preferred spawning sites appears costly and is probably the most important factor determining an upper limit for mating success in this species; territorial males given additional food were able to decrease the rate of intrusion by other males and increased their mating success (Warner *et al.* 1995). At high densities, bluehead wrasse become a predominantly group-spawning species, with most males experiencing sperm competition in all spawns (Hunt von Herbing and Hunte 1991; Warner 1995).

In species where females spawn and feed in the same areas, harems may evolve, where males can defend areas or the females within those areas. At low to moderate population densities, these males can effectively restrict other males from spawning with those females. In Fig. 11.2, many of the species with little or no sperm competition are haremic. In coral-reef fishes, harems are most likely to evolve in species where feeding territories are included within preferred spawning sites at the edges of reefs. Haremic species include low-density species that range widely, but over the same areas over time, and species with smaller home ranges whose habitat is at the edge of reefs (Warner 1984a; Baird 1988). In most small-bodied haremic species of coral reef fishes, females spawn within their territories, which they defend from other females (Petersen and Fischer 1986; Petersen 1987; Victor 1987; Baird 1988). In species that forage on the reef proper and migrate to the edge of the reef to spawn, resource defence polygyny of spawning sites by males is much more common. In the case of resource defence polygyny, small males exhibiting both streaking behaviour and group spawning are common and their abundance appears to depend on local population density (Warner and Robertson 1978; Warner 1984a).

VIII. LIFE HISTORY IMPLICATIONS

The success of alternative male mating tactics tends to reduce the variance in male mating success by lowering mate monopolization by large, dominant males. Specifically, such tactics create situations where smaller males can obtain some mating success by allocating energy to sperm production. Early maturation of individuals involved in sperm competition is common in fishes, with the best known freshwater examples being parr and jacks in salmon, sneakers and satellites in sunfish (reviewed by Gross 1984). Males that mature early grow more slowly and can have higher mortality rates than males that delay reproduction and grow to larger adult sizes before becoming reproductively active (Gross 1982). The marine equivalents are the small streakers and group spawners of many wrasses and parrotfishes, but the variation here appears to be ontogenetic (Warner 1984a). That is, individuals spend the first part of their lives as sperm-competitive spawners and the later part as mate-monopolizers. In the younger age classes, individuals that engage in daily reproduction have only about half the growth rate of those that do not (Warner 1984b).

IX. LOOKING AHEAD: FUTURE DIRECTIONS

A. The dynamics of fertilization in external fertilizers

Several models of fertilization in external fertilizers suggest that males will often release amounts of sperm that will not fertilize all the eggs in a spawn (Petersen 1991a; Ball and Parker 1996; Shapiro and Giraldeau 1996). Data from tropical reef fishes confirms that, although high, fertilization success is dependent on sperm released in a spawn and is less than 100% (Warner *et al.* 1995; Marconato *et al.* 1995; Marconato and

Shapiro 1996). Several factors in addition to the amount of sperm released in a spawn might also affect fertilization success, including water turbulence (ambient or created by the spawning fish themselves), the presence of heterospecific sperm, and the gametic compatibility of individuals releasing eggs and sperm. Water turbulence or water velocity may alter fertilization success by dispersing gametes more quickly and reduce fertilization success, or select for different amounts of sperm release, or different sperm characteristics (Billard 1987; Petersen *et al.* 1992). Many species use the same spawning sites, sometimes at the same times, and the presence of sperm from other species is a distinct possibility. This is very likely in the case when small males streak on spawns of other species that spawn at the same times and locations; this has been observed in the Caribbean (Petersen, pers. obs.). If these sperm interfere with fertilization or produce less fit offspring, this could constrain optimization of spawning times in reef fishes and lead to divergence in spawning times among species.

Several additional questions will need to be answered in the near future to advance our understanding of sperm competition in external fertilizers. In spawns involving multiple males, how well is paternity explained by the relative contribution of a male to the total sperm released in a spawn or, in the terminology of Parker (1990a), is it a fair lottery? Among the possible factors causing deviations from a fair lottery might be differential placement of sperm in a gamete cloud and sperm behaviour or longevity.

In species where there is a risk of sperm competition, dominant males probably release fewer sperm than sperm competitors exhibiting alternative reproductive tactics, but the dominant males appear to release their sperm closer to the female. For example, in one damselfish species, Gronell (1989) observed males outside the nest fanning with their pectoral fins, which may direct their sperm toward a nest where a territorial male was actively spawning with a female. Similar behaviour has been observed in other fish (Brantley and Bass 1994; R. Warner, pers. obs.). Although this is an extreme example of differential distances of males from females during spawning, a similar pattern, although to a lesser degree, probably results in most cases where males streak into nests and release sperm. One exception may be when males enter defended oviposition sites and appear to deceive the territorial male by looking and acting like spawning females. Thus, different alternative mating tactics are often lumped together under a general classification of sperm competition, but the success of sperm from males involved in different tactics may differ considerably.

One area where the dynamics of fertilization may affect male mate choice is in group-spawning tropical reef fishes. In many tropical reef fishes females mate at 'group spawning sites' where scores to hundreds of males congregate. However, most spawns contain only a few males that appear to release sperm, and spawning can be size assortative (van den Berghe and Warner 1989). The results of Parker's *et al.* (1996) model suggests that if the number of sperm competitors is high but variable, males should release fewer sperm when more competitors are present. At some point it may not be in a male's interest to join a large spawning group, especially if the male's sperm would be at a disadvantage to sperm from males closer to the female. If spawning has costs such as predation risk or a loss of time or energy to other activities (including other spawns), there should be a threshold group size or level of sperm competition above which males will choose not to join in spawning. Expanding the Parker *et al.* (1996) model to cases with unfair raffles and decisions about which spawns a male should join may help us to understand male mating decisions in this situation.

Even if sperm from multiple males have an equal chance of fertilizing eggs in a spawn, incomplete mixing of sperm will create biases in our current estimates of sperm competition intensity based on behavioural observations. As an illustration, imagine two males releasing sperm in a spawn, but only half of their sperm overlap spatially. Even if this half is completely mixed, and eggs are distributed equally throughout the total sperm cloud, these males would be under half the intensity of sperm competition that we would predict assuming complete mixing. The incomplete mixing of sperm among individuals in spawns with external fertilization seems highly likely, with the result that the level of sperm competition is typically overestimated when using current estimates based on complete mixing of gametes (cf. Fischer 1984; Petersen 1991a).

B. Sperm competition in internal fertilizers

In internal fertilizers, the ways that sperm compete, die, or are killed inside the reproductive system of the female are not well understood. Sperm competition appears to be common in internal fertilizers, based on the occurrence of mixed broods in groups such as poeciliids (Winge 1937; Hildemann and Wagner 1954; Constantz 1984). Understanding whether or not sperm precedence occurs in internal fertilizers, or if females are able to exercise choice by differentially destroying or limiting access of sperm to eggs may help us understand why behaviours such as postcopulatory mate guarding are seldom reported for fishes.

C. Sexual conflict and female choice

Sperm competition may change the benefits females receive during spawning, either by affecting fertilization rate or the paternity of offspring. In situations where sperm competition is nonexistent or minimal, males faced with many future matings may release amounts of sperm that do not maximize the fertilization rates of females (Warner *et al.*

1995). In cases where additional males join a spawning, any female choice of mate based on indirect benefits may be negated by sperm competition from nonselected males. Females may be able to choose between the quantity and quality of their offspring, but be unable to obtain both. Overall, females may find themselves in conflict with their mates in terms of reproductive choices and, depending on the circumstances, neither or only one of the individuals may 'win' the mating game.

Aggression towards potential sperm competitors is one way for dominant males to reduce the probability of sperm competition, while another is to attempt to prevent females from spawning during circumstances when sperm competition is likely. In the presence of potential streakers, dominant males may attempt to delay spawning of the female until these peripheral males can be displaced (van den Berghe *et al.* 1989).

In principle, females should be able to influence the degree of sperm competition that occurs in a population. Constantz (1984) hypothesized that multiple-sired guppy broods were an adaptation to produce genetically diverse young in a fluctuating environment. Although female guppies can allow mating to occur, it is not clear to what extent they are able to restrict unwanted matings by not cooperating with males. In externally fertilizing species, sperm competition may result in increased fertilization success for a spawn, so that situations that increase the fitness of females are precisely those that decrease fitness of individual males (Petersen 1991b; Marconato *et al.* in press). In some cases, females can choose between sites in terms of the degree of sperm competition their eggs might encounter at those sites, as in the bluehead wrasse where group and pair spawning occur on the same reef. In that species, it appears that larger females choose to spawn with single large males rather than in a group (Warner 1985), although group spawning sites have equal or higher fertilization rates (Marconato *et al.* in press). Where small males streak pair spawns, females may have less control over whether a male joins the pair spawn. Females have been observed to hesitate during spawning rushes, and this could be interpreted either as a mechanism to encourage additional males or, conversely, to alert the dominant male to the presence of sperm competitors (Petersen 1991b). Taborsky (1994) cites several studies where female behaviour was interpreted as a mating preference for dominant or nest-guarding males over satellite or sneaker males. Future studies directed at the consequences of sperm competition for female fitness would be valuable and could help to interpret the female mate-choice behaviours observed.

D. Functional morphology and life history of sperm

Sperm vary in morphology, behaviour (swimming speed and direction), and longevity in fishes, but the functional significance of this variation is not well understood (Stockley *et al.* 1997). There appear to be intraspecific differences that may correlate with male-mating type in some species (de Fraipoint *et al.* 1993; Gage *et al.* 1995). The degree of phenotypic plasticity in sperm behaviour is also not known. Under certain circumstances sperm appear capable of increasing swimming speed at a potential cost of longevity (Bolton and Havenhand 1996).

Until recently, very little effort was made to understand variation in sperm size, morphology, and behaviour among species. By treating sperm as individuals and applying current life-history models to sperm characteristics, we are in the position to create another data base for testing life history theory. The fitness effects of sperm size (head, tail, or entire body) and investment in sperm production (caloric investment or production rate) need to be better understood so that possible negative correlations among characteristics (implying trade-offs) are easier to detect.

Repeating a theme in many of the chapters of this book, one of the most important lessons that we have learned over the past decade is that sperm are not necessarily cheap to produce, and that sperm production must be included in fitness costs. Nakatsuru and Kramer (1982) noted that males in the lemon tetra *Hyphessobrycon pulchripinnis* (Characidae), which appear to produce only enough sperm to accommodate an expected number of matings, experience sperm depletion beyond this point, and (most importantly) females can detect males with insufficient sperm. This occurs in a species without apparent sperm competition, and it implies that even here sperm are not produced profligately. The careful allocation of available sperm supplies, or 'sperm economy', may be widespread among fishes (Shapiro *et al.* 1994; Warner *et al.* 1995). The costs of sperm production are magnified considerably among males in which sperm competition occurs because there is an even greater allocation of energy to gamete production. At present, we know only that large allocations to sperm production significantly restrict growth rate (Warner 1984b). There is a real need to explore the dynamics of allocation to sperm production and the subsequent allocation of sperm among spawns in males subject to sperm competition.

ACKNOWLEDGEMENTS

Paula Stockley let us read and liberally cite her work, in press, including the data in Fig. 11.1. Several other individuals helped us with the literature, including A. Kodric-Brown, R. Robertson, N. Stacey, A. Magurran, R. Liley, and J. Reynolds. We especially thank T. Birkhead, J. Godwin, H. Hess, A. Møller, M. Rasotto, E. Schultz, P. Stockley, and L. Wooninck for providing helpful comments on the manuscript. This work was supported by grants from NSF to both authors.

REFERENCES

Baird TA (1988) Female and male territoriality and mating system of the sand tilefish, *Malacanthus plumeri*. *Environ. Biol. Fish.* **22:** 101–116.

Ball MA & Parker GA (1996) Sperm competition games: external fertilization and 'adaptive' infertility. *J. Theor. Biol.* **180:** 141–150.

Bass AH (1996) Shaping brain sexuality. *Am. Sci.* **84:** 352–363.

Billard R (1987) Testis growth and spermatogenesis in teleost fish: the problem of the large interspecific variability in testis size. *Proc. 3rd Int. Symp. on Reproductive Physiology of Fish, St John's, Newfoundland*, pp. 183–186.

Blum V (1972) The influence of ovine follicle stimulating hormone (FSH) and luteinizing hormone (LH) on the male reproductive system and the skin of the Mediterranean blenniid fish *Blennius sphynx* (Valenciennes). *J. Exp. Zool.* **181:** 203–216.

Blumer LS (1979) Male parental care in the bony fishes. *Quart. Rev. Biol.* **54:** 149–161.

Bolton TF & Havenhand JN (1996) Chemical mediation of sperm activity and longevity in the solitary ascidians *Ciona intestinalis* and *Ascidiella aspersa*. *Biol. Bull.* **190:** 329–335.

Brantley RK & Bass AH (1994) Alternative male spawning tactics and acoustic signalling in the plainfin midshipman fish, *Porichthys notatus*. *Ethology* **96:** 213–232.

Breder CM & Rosen DE (1966) *Modes of Reproduction in Fishes*. Natural History Press, Garden City, NY.

Chan T-Y (1987) The role of male competition and female choice in the mating success of a lek-breeding southern African cichlid fish *Pseudocrenilabrus philander* (Pisces: Cichlidae). PhD thesis, Rhodes University, Grahamstown, South Africa.

Charnov EL (1980) Sex allocation and local mate competition in barnacles. *Mar. Biol. Lett.* **1:** 269–272.

Charnov EL (1982) *The Theory of Sex Allocation*. Princeton University Press, Princeton.

Choat JH & Robertson DR (1975) Protogynous hermaphroditism in fishes of the family Scaridae. In *Intersexuality in the Animal Kingdom*. R Reinboth (ed.), pp. 263–283. Springer-Verlag, Heidelberg.

Colbourne JK, Neff BD, Wright JM & Gross MR (1996) DNA fingerprinting of bluegill sunfish (*Lepomis macrochirus*) using $(GT)_n$ microsatellites and its potential for assessment of mating success. *Can. J. Fish. Aqua. Sci.* **53:** 342–349.

Cole K (1990) Patterns of gonad structure in hermaphroditic gobies (Teleostei: Gobiidae). *Environ. Biol. Fish.* **28:** 125–142.

Constantz GD (1984) Sperm competition in Poeciliid fishes. In *Sperm Competition and the Evolution of Animal Mating Systems*. RL Smith (ed.), pp. 465–485. Academic Press, Orlando.

Darling JDS, Noble ML & Shaw E (1980) Reproductive strategies in surfperches. I. Multiple insemination in natural populations of the shiner perch, *Cymatogaster aggregata*. *Evolution* **34:** 271–277.

de Fraipoint M, Fitzgerald GJ & Gurderley H (1993) Age related differences in reproductive tactics in the three-spined stickleback, *Gasterosteus aculeatus*. *Anim. Behav.* **46:** 961–968.

de Jonge J, de Ruiter AJH & van den Hurk R (1989) Testis-testicular gland complex of two *Tripterygion* species (Belnnioidei, Teleostei): differences between territorial and non-territorial males. *J. Fish Biol.* **35:** 497–508.

DeMartini EE (1988) Size-assortative courtship and competition in two embiotocid fishes. *Copeia* **1988:** 336–344.

Denny MW & Shibata MF (1989) Consequences of surf-zone turbulence for settlement and external fertilization. *Am. Nat.* **134:** 859–889.

Dominey WJ (1980) Female mimicry in male bluegill sunfish – a genetic polymorphism? *Nature* **284:** 546–548.

Fischer EA (1981) Sexual allocation in a simultaneously hermaphroditic reef fish. *Am. Nat.* **117:** 64–82.

Fischer EA (1984) Local mate competition and sex allocation in simultaneous hermaphrodites. *Am. Nat.* **124:** 590–596.

Fischer EA & Petersen CW (1987) The evolution of sexual patterns in the seabasses. *Bioscience* **37:** 482–489.

Fishelson L (1991) Comparative cytology and morphology of seminal vesicles in male gobiid fishes. *Jpn. J. Ichthyol.* **38:** 17–30.

Gage MJG, Stockley P & Parker GA (1995) Effects of alternative male mating strategies on characteristics of sperm production in the Atlantic salmon (*Salmo salar*): theoretical and empirical investigations. *Phil. Trans. Roy. Soc. Lond.* **350:** 391–399.

Garnaud J (1962) Monographie di l'*Apogon* mediterraneen, *Apogon imberbis* (Linne) 1758. *Bull. Inst. Oceanogr. (Monaco)* **1248:** 1–83.

Garnaud J (1963) Ethologie d'un poisson extraordinaire: *Apogon imberbis*. In *Congres International D'Aquariologie, Monaco (1960) Communications*, Vol. 1D, pp. 51–60. Musee Oceanographique, Monaco.

Ginzburg AS (1968) *Fertilization in Fishes and the Problem of Polyspermy*. Akademiya Nuak SSSR, Institut Biologii Razvitiya. Translated from Russian by Israel Program for Scientific Translations, Jerusalem 1972.

Grober MS, Fox S, Laughlin C & Bass AH (1994) GnRh cell size and number in a teleost fish with two male reproductive morphs: Sexual maturation, final sexual status and body size allometry. *Brain Behav. Evol.* **43:** 61–78.

Gronell AM (1989) Visiting behaviour by females of the sexually dichromatic damselfish, *Chrysiptera cyanea* (Teleostei: Pomacentridae): a probable method of assessing male quality. *Ethology* **81:** 89–122.

Gross MR (1979) Cuckholdry in sunfishes (*Lepomis*: Centrarchidae). *Can. J. Zool.* **57:** 1507–1509.

Gross MR (1982) Sneakers, satellites and parentals: polymorphic mating strategies in North American sunfishes. *Z. Tierpsychol.* **60:** 1–26.

Gross MR (1984) Sunfish, salmon, and the evolution of alternative reproductive strategies and tactics in fishes. In *Fish Reproduction: Strategies* and *Tactics*. R Wooton & G Potts (eds), pp. 55–75. Academic Press, London.

Gross MR & Charnov EL (1980) Alternative male life histories in bluegill sunfish. *Proc. Natl Acad. Sci. USA* **77:** 6937–6940.

Hildemann WH & Wagner ED (1954) Intraspecific sperm competition in *Lebistes reticulatus*. *Am. Nat.* **88:** 87–91.

Hourigan TF (1989) Environmental determinants of butterflyfish social systems. *Environ. Biol. Fish.* **25:** 61–78.

Hubbs C (1966) Fertilization, initiation of cleavage, and developmental temperature tolerance of the cottid fish, *Clinocottus analis*. *Copeia* **1966:** 29–42.

Hunt von Herbing I & Hunte W (1991) Spawning and recruitment of the bluehead wrasse *Thalassoma bifasciatum* in Barbados, WI. *Mar. Ecol. Progr. Ser.* **72:** 49–58.

Hutchings JA & Myers RA (1988) Mating success of alternative maturation phenotypes in male Atlantic salmon, *Salmo salar*. *Oecologia* **75:** 169–174.

Jamieson BGM (1991) *Fish Evolution and Systematics: Evidence from Spermatozoa.* Cambridge University Press, Cambridge.

Kazakov RV (1981) Peculiarities of sperm production by anadromous and parr Atlantic salmon (*Salmo salar* L.) and fish cultural characteristics of such sperm. *J. Fish Biol.* **18:** 1–8.

Kohda M, Tanimura M, Kikue-Nakamura M & Yamagishi S (1995) Sperm drinking by female catfishes: a novel mode of insemination. *Environ. Biol. Fish.* **42:** 1–6.

Lahnsteiner F, Richtarski U & Patzner RA (1990) Functions of the testicular gland in two blenniid fishes, *Salaria* (= *Blennius*) *pavo* and *Lipophrys* (= *Blennius*) *dalmatinus* (Blenniidae, Teleostei) as revealed by electron microscopy and enzyme histochemistry. *J. Fish Biol.* **37:** 85–97.

Levitan DR (1995) The ecology of fertilization in free-spawning invertebrates. In *Ecology of Marine Invertebrate Larvae.* L McEdward (ed.), pp. 123–156. CRC Press, Boca Raton.

Levitan DR & Petersen C (1995) Sperm limitation in the sea. *Trends Ecol. Evol.* **10:** 228–231.

Lloyd DG (1984) Gender allocations in outcrossing cosexual plants. In *Perspectives on Plant Population Ecology.* R Dirzo & J Sarukhan (eds), pp. 277–300. Sinauer, Sunderland, MA.

Marconato A & Shapiro DY (1996) Sperm allocation, sperm production and fertilization rates in the bucktooth parrotfish. *Anim. Behav.* **52:** 971–980.

Marconato A, Tessari V & Marin G (1995) The mating system of *Xyrichthys novacula*: sperm economy and fertilization success. *J. Fish Biol.* **47:** 292–301.

Marconato A, Rasotto MB & Mazzoldi C (1996) On the mechanism of sperm release in three gobiid fishes (Teleostei: Gobiidae). *Environ. Biol. Fish.* **46:** 321–327.

Marconato A, Shapiro DY, Petersen CW, Warner RR & Yoshikawa T (1998) Methodological analysis of fertilization rate in the bluehead wrasse, *Thalassoma bifasciatum*: pair vs. group spawns. *Mar. Ecol. Progr. Ser.*, in press.

Miller PJ (1984) The tokology of gobioid fishes. In *Fish Reproduction: Strategies and Tactics.* GW Potts & JR Wooton (eds), pp. 119–153. Academic Press, London.

Munehara H (1988) Spawning and subsequent copulating behavior of the elkhorn sculpin, *Alcichthys alcicornis* in an aquarium. *Jpn. J. Ichthyol.* **35:** 358–364.

Munehara H, Takano K & Koya Y (1989) Internal gametic association and external fertilization in the elkhorn sculpin, *Alcicththys alcicornis*. *Copeia* **1989:** 673–678.

Munehara H, Okamato H & Shimazaki K (1990) Paternity estimated by isozyme variation in the marine sculpin *Alcichthys alcicornis* (Pisces: Cottidae) exhibiting copulation and paternal care. *J. Ethol.* **8:** 21–24.

Munehara H, Takano K & Koya Y (1991) The little dragon sculpin *Blepsias cirrhosus*, another case of internal gametic association and external fertilization. *Jpn. J. Ichthyol.* **37:** 391–394.

Munehara H, Koya Y & Takano K (1994) Conditions for initiation of fertilization of eggs in the copulating elkhorn sculpin. *J. Fish Biol.* **45:** 1105–1111.

Nakatsuru K & Kramer DL (1982) Is sperm cheap? Limited male fertility and female choice in the lemon tetra (Pisces, Characidae). *Science* **216:** 753–755.

Parker GA (1970) Sperm competition and its evolutionary consequences in the insects. *Biol. Rev.* **45:** 425–467.

Parker GA (1984) Sperm competition and the evolution of animal mating strategies. In *Sperm Competition* and *the Evolution of Animal Mating Systems.* RL Smith (ed.), pp. 1–60. Academic Press, Orlando.

Parker GA (1990a) Sperm competition games: raffles and roles. *Proc. Roy. Soc. Lond. B* **242:** 120–126.

Parker GA (1990b) Sperm competition games: sneaks and extra-pair copulations. *Proc. Roy. Soc. Lond. B* **242:** 127–133.

Parker GA (1993) Sperm competition games: sperm size and sperm number under adult control. *Proc. Roy. Soc. Lond. B* **253:** 245–254.

Parker GA & Begon ME (1993) Sperm competition games: sperm size and number under gametic control. *Proc. Roy. Soc. Lond. B* **253:** 255–262.

Parker GA, Ball MA, Stockley P & Gage MJG (1996) Sperm competition games: individual assessment of sperm competition intensity by group spawners. *Proc. Roy. Soc. Lond. B* **263:** 1291–1297.

Petersen CW (1987) Reproductive behaviour and gender allocation in *Serranus fasciatus*, a hermaphroditic reef fish. *Anim. Behav.* **35:** 1601–1614.

Petersen CW (1990a) Variation in reproductive success and gonadal allocation in the simultaneous hermaphrodite, *Serranus fasciatus. Oecologia* **83:** 62–67.

Petersen CW (1990b) The relationships among population density, individual size, mating tactics, and reproductive success in a hermaphroditic fish, *Serranus fasciatus. Behaviour* **113:** 57–80.

Petersen CW (1991a) Sex allocation in hermaphroditic sea basses. *Am. Nat.* **138:** 650–667.

Petersen CW (1991b) Variation in fertilization rate in the tropical reef fish, *Halichoeres bivattatus*: correlates and implications. *Biol. Bull.* **181:** 232–237.

Petersen CW & Fischer EA (1986) Mating system of the hermaphroditic coral-reef fish, *Serranus baldwini. Behav. Ecol. Sociobiol.* **19:** 171–178.

Petersen CW & Fischer EA (1996) Intraspecific variation in sex allocation in a simultaneous hermaphrodite: the effect of individual size. *Evolution* **50:** 636–645.

Petersen CW, Warner RR, Cohen S, Hess HC & Sewell AT (1992) Variation in pelagic fertilization success: implications for production estimates, mate choice, and the spatial and temporal distribution of spawning. *Ecology* **73:** 391–401.

Philipp DP & Gross MR (1994) Genetic evidence for cuckoldry in bluegill *Lepomis macrochirus. Mol. Ecol.* **3:** 563–569.

Ragland HC & Fischer EA (1987) Internal fertilization and male parental care in the scalyhead sculpin, *Artedius harringtoni. Copeia* **1987:** 1059–1062.

Rasotto MB (1995) Male reproductive apparatus of some blennioidei (Pisces: Teleostei). *Copeia* **1995:** 907–914.

Rico C, Kuhnlein U & Fitzgerald GJ (1992) Male reproductive tactics in the three-spine stickleback: an evaluation by DNA fingerprinting. *Mol. Ecol.* **1:** 79–87.

Robertson DR (1983) On the spawning behavior and spawning cycles of eight surgeonfishes (Acanthuridae) from the Indo-Pacific. *Environ. Biol. Fish.* **9:** 193–223.

Robertson DR (1985) Sexual size dimorphism in surgeon fishes. *Proc. 5th Int. Coral Reef Congr.* **5:** 403–408.

Robertson DR & Choat JH (1974) Protogynous hermaphroditism and social systems in labrid fish. *Proc. 2nd Int. Coral Reef Symp.* **1:** 217–225.

Robertson DR & Warner RR (1978) Sexual patterns in the labroid fishes of the Western Caribbean. II. The parrotfishes (Scaridae). *Smiths. Contrib. Zool.* **255:** 1–26.

Ruchon F, Laugier T & Quignard JP (1995) Alternative male reproductive strategies in the peacock blenny. *J. Fish Biol.* **47:** 826–840.

Schroder SL (1981) The role of sexual selection in determining overall mating patterns and mate choice in chum salmon. PhD thesis, University of Washington, Seattle.

Seiwald M & Patzner RA (1989) Histological, fine-structural and histochemical differences in the testicular glands of gobiid and blenniid fishes. *J. Fish Biol.* **35:** 631–640.

Shapiro DY & Giraldeau A (1996) Mating tactics in external fertilizers when sperm is limited. *Behav. Ecol.* **7:** 19–23.

Shapiro DY, Marconato A & Yoshikawa T (1994) Sperm economy in a coral reef fish, *Thalassoma bifasciatum*. *Ecology* **75:** 1334–1344.

Smith RL (ed.) (1984) *Sperm Competition and the Evolution of Animal Mating Systems*. Academic Press, London.

Stockley P, Gage MJG, Parker GA & Møller AP (1996) Female reproductive biology and the coevolution of ejaculate characteristics in fish. *Proc. Roy. Soc. Lond. B* **263:** 451–458.

Stockley P, Gage MJG, Parker GA & Møller AP (1997) Sperm competition in fishes: the evolution of testis size and ejaculate characteristics. *Am. Nat.* **149:** 933–954.

Taborsky M (1994) Sneakers, satellites, and helpers: parasitic and cooperative behavior in fish reproduction. *Adv. Stud. Behav.* **23:** 1–100.

van den Berghe EP & Warner RR (1989) The effects of mating system on male mate choice in a coral reef fish. *Behav. Ecol. Sociobiol.* **24:** 409–415.

van den Berghe EP, Wernerus F & Warner RR (1989) Female choice and the mating cost of peripheral males. *Anim. Behav.* **38:** 875–884.

Victor BC (1987) The mating system of the Caribbean rosy razorfish, *Xyrichtys martinicensis*. *Bull. Mar. Sci.* **40:** 152–160.

Warner RR (1982) Mating systems, sex change and sexual demography in the rainbow wrasse, *Thalassoma lucasanum*. *Copeia* **1982:** 653–661.

Warner RR (1984a) Mating systems and hermaphroditism in coral reef fish. *Am. Sci.* **72:** 128–136.

Warner RR (1984b) Deferred reproduction as a response to sexual selection in a coral reef fish: a test of the life historical consequences. *Evolution* **38:** 148–162.

Warner RR (1985) Alternative mating behaviors in a coral reef fish: a life-history analysis. *Proc. 5th Int. Coral Reef Congr.* **4:** 145–150.

Warner RR (1995) Large mating aggregations and daily long-distance spawning migrations in the bluehead wrasse, *Thalassoma bifasciatum*. *Environ. Biol. Fish.* **44:** 337–345.

Warner RR & Harlan RK (1982) Sperm competition and sperm storage as determinants of sexual dimorphism in the dwarf surfperch, *Micrometrus minimus*. *Evolution* **36:** 44–55.

Warner RR & Hoffman SG (1980a) Population density and the economics of territorial defense in a coral reef fish. *Ecology* **61:** 772–780.

Warner RR & Hoffman SG (1980b) Local population size as a determinant of mating system and sexual composition in two tropical marine fishes (*Thalassoma* spp). *Evolution* **34:** 508–518.

Warner RR & Robertson DR (1978) Sexual patterns in the labroid fishes of the Western Caribbean. I. The wrasses (Labridae). *Smiths. Contrib. Zool.* **254:** 1–24.

Warner RR, Robertson DR & Leigh EG Jr (1975) Sex change and sexual selection. *Science* **190:** 633–638.

Warner RR, Shapiro DY, Marconato A & Petersen CW (1995) Sexual conflict: males with highest mating success convey the lowest fertilization benefits to females. *Proc. Roy. Soc. Lond. B* **262:** 135–139.

Winge O (1937) Succession of broods in *Lebistes*. *Nature* **140:** 467.

Zimmerer EJ & Kallman KD (1989) Genetic basis for alternative reproductive tactics in the pygmy swordtail, *Xiphophorus nigrensis*. *Evolution* **43:** 1298–1307.

12 Sperm Competition in Amphibians

T. Halliday

Department of Biology, The Open University, Milton Keynes, MK7 6AA, UK

I. INTRODUCTION

A previous review concluded that, while there had been no studies that explicitly addressed the role of sperm competition in amphibian mating patterns, amphibians offer rich opportunities for such studies (Halliday and Verrell 1984). Over the last 12 years, many new studies of the reproductive biology and sexual behaviour of amphibians have been published and it is now possible to provide a more comprehensive and less speculative review. It remains true, however, that our understanding of sperm competition in amphibians as a group is fragmentary and lags far behind that of some other taxa, notably insects and birds.

Among amphibians (Class Amphibia), there is a rich diversity of reproductive modes which does not correspond very closely to their taxonomy (Wake 1982, 1993; Jørgensen 1992). For example, internal fertilization has evolved independently in all three living amphibian orders, the anurans (Salientia; frogs and toads), the apodans (Gymnophonia; caecilians), and the urodeles (Caudata; salamanders and newts) (Wake 1993). While internal fertilization is universal in caecilians, it occurs in the majority of urodeles, but in only a very small minority of anurans. Associated with internal fertilization are various forms of viviparity and

Sperm Competition and Sexual Selection
ISBN 0-12-100543-7

ovoviviparity, spread across the three orders. True viviparity (involving oviductal gestation and maternal provision of embryos after the yolk has been exhausted) occurs in the majority of the caecilians, but in only a few species of anurans and urodeles, in which ovoviviparity is more common. Among amphibians with internal fertilization, sperm transfer mechanisms are similarly diverse. Male caecilians possess an intromittent organ; in anurans, sperm transfer is typically achieved by cloacal apposition; in urodeles, sperm transfer is indirect, by means of spermatophores, similar to those seen in many arthropods.

Sperm competition typically occurs within the reproductive tract of the female and is thus generally associated with the dual phenomena of internal fertilization and sperm storage by females, features that are found only in some amphibians. There is potential, however, for sperm to compete outside the female's body in species with external fertilization and some instances where this occurs in anurans are described below. The spermatophore system of sperm transfer that occurs in many urodeles is associated with a diverse array of male behaviour patterns, called sexual interference and sexual defence (Arnold 1976), in which males seek, respectively, to disrupt the transfer of their rivals' sperm, and to prevent them from doing so (Halliday 1990; Halliday and Tejedo 1995). Such behaviour, while not involving sperm competition as normally envisaged, can be regarded as a particular form of ejaculate competition and is examined in some detail.

II. ANURANS

In frogs and toads, the male typically grasps the female dorsally in a very tight embrace, called amplexus, which is maintained until she spawns, at which time the male sheds sperm onto the emerging eggs. In most species the operational sex ratio is skewed towards an excess of males (Wells 1977) and males in amplexus commonly have to guard females against the attempts of single males either to displace them or to gain a position alongside them on the female's back. Displacement during amplexus is quite common in the European toad *Bufo bufo*, with nearly 40% of mating males obtaining their partners by displacing rivals from the backs of females (Davies and Halliday 1979), but does not occur in all anuran species. It is very rare, for example, in *Bufo calamita* (Tejedo 1988) and *Bufo terrestris* (Lamb 1984). There is potential for sperm competition, and for more than one male to fertilize a female's eggs, when single males are very close to amplectant pairs at the moment of oviposition. Only one study, however, has demonstrated multiple paternity of egg masses in an anuran that breeds in open water, and that was conducted in the laboratory. Berger and Rybacki (1992, 1994) studied the European water frog species complex, in which *Rana esculenta* individuals (genome RL) are derived by hybridogenesis from matings between *Rana lessonae* (LL) and *Rana ridibunda* (RR). When R and L sperm are mixed in equal numbers in a suspension into which R and L eggs are introduced, 77–99% of eggs are fertilized by L sperm. To achieve equal numbers of eggs fertilized by the two kinds of sperm requires a suspension in which R sperm outnumber L sperm by 5:1. Berger and Rybacki (1994) suggest that this effect may be due to the larger genome of *R. ridibunda*, which may make R sperm heavier and slower than L sperm.

In several anuran species, single males may hold onto a female who is already in amplexus and, in that position, they may have an opportunity to shed sperm onto the emerging eggs at oviposition. Such behaviour would be analogous to the opportunistic 'sneaky' fertilizations achieved by males in several species of externally fertilizing fish, and has been described in the Mexican leaf frogs *Phyllomedusa* (*Agalychnis*) *callidryas* and *Pachymedusa dacnicolour* (Pyburn 1970). In these two species, eggs are laid on leaves overhanging water and are produced by the female in batches. Between each batch of eggs the female descends to the water to refill her bladder with water which she then releases onto the eggs. During these movements to and from the pond, an amplectant pair may encounter unpaired males who try to climb onto the female. If the amplectant male fails to resist these attempts, both males may shed sperm onto the eggs. In a related species, *Agalychnis saltator*, males parachute into dense mating aggregations that form on lianas, in which large communal egg masses are formed. Occasionally, females lay eggs with two males on their backs and it is likely that both fertilize some of the eggs (Roberts 1994).

The mating dynamics of *Agalychnis callidryas* have been described in detail by d'Orgeix (d'Orgeix and Turner 1995; d'Orgeix 1996). Among females already in amplexus with a primary male, 87% showed behaviour categorized as avoidance of would-be secondary males that attempted also to establish amplexus with them. When secondary males did succeed in establishing amplexus, females laid fewer eggs than when clasped only by a primary male and, among females that were clasped by three or more males, there was mortality of their eggs before they reached the oviposition site. These observations suggest that being in amplexus with more than one male is costly to females, and that they actively seek to avoid such a situation. d'Orgeix (1996) detected no difference in the mass or size of primary and secondary males.

The greatest potential for sperm competition arises in those species in which mating pairs spawn in a nest and are closely attended by satellite or peripheral males. Just such a situation has been studied in the African foam-nesting rhacophorid *Chiromantis xerampelina*. Foam nests are constructed over water by amplectant pairs, 90% of which are attended by one to seven peripheral males in addition to an amplectant male (Jennions *et al.* 1992). Those peripheral males that are closest to the pair compete to position their cloacae against the female's cloaca during oviposition; male participation in these groups, as an amplectant or a peripheral male, is not related to body size and individual males may participate in mating, on different occasions, in both amplectant and peripheral roles. To test the hypothesis that peripheral males obtain fertilizations, Jennions and Passmore (1993) performed a 'sterile male' experiment, in which the amplectant male was enclosed in a condom-like sheath, and found that the eggs were fertilized, demonstrating that peripheral males are fully capable of fertilizing eggs. Similar behaviour has been observed in the Japanese foam-nesting frog *Rhacophorus arboreus* (Maeda and Matsui 1990). In a related species, *Rhacophorus schlegelii*, a foam nest is made by a pair in a burrow in the soil and unpaired males sneak into the burrow after it has been constructed and before oviposition (Fukuyama 1991). Unpaired males were found in four out of nine nests in the field and in 10 out of 12 nests in the lab, suggesting that attempts by single males to intrude on matings are common in this species. All males can adopt the two alternative mating strategies observed in this species, calling and burrow-building and sneaking into another male's burrow, and there is no size difference between individuals adopting them (Fukuyama 1991).

Among males, sperm competition is predicted to select for greater sperm production, and thus large testes, a relationship that has been found in a number of taxa, such as birds (Birkhead and Møller 1992; Møller and Briskie 1995) and primates (Harcourt *et al.* 1981). Jennions and Passmore (1993) compared the testis size of three foam-nesting rhacophorids, *Chiromantis xerampelina*, *Rhacophorus arboreus* and *R. schlegelli*, and found that relative testis mass (correcting for body mass) is 3.8–14.6 times greater in these species than in 31 non-foam-nesting species. The relatively large testes of Japanese rhacophorid foam-nesters has also been noted by Kusano *et al.* (1991).

A consequence of the relatively small testes of frogs and toads that engage in scramble competition could be that their capacity for multiple ejaculations is limited. Evidence that this is the case has been obtained for *Rana sylvatica*, in which a male's sperm supply is quite quickly depleted by repeated ejaculations (Smith-Gill and Berven 1980). In the American toad *Bufo americanus*, however, the capacity of males to fertilize eggs showed no reduction over five matings carried out over five days, despite the fact that, in this species, males very rarely mate with more than one female in a season (Kruse and Mounce 1982).

Sperm competition is predicted to select for mate-guarding and sequestering of females by competing males (Parker 1974, 1984; Andersson 1994); amplexus can therefore be regarded as an adaptation that achieves these ends (Halliday and Tejedo 1995). It is important to note, however, that in his position on a female's back, an amplectant male may be able to prevent a rival male from clasping her but he cannot control her movements. The movements of an amplectant pair are entirely controlled by the female. It should also be noted that amplexus is a very effective and energy-saving method through which a male can maintain contact with a female between encountering her and when she lays her eggs, and so it should not be regarded solely as a form of mate-guarding. The effectiveness of amplexus is enhanced by a number of morphological adaptations in males, notably longer and more muscular forelimbs than are found in females, and nuptial pads that ensure a very tight grip (Halliday and Tejedo 1995). In some taxa, male nuptial pads also contain sexually dimorphic skin glands that may secrete adhesive secretions that enhance a male's ability to hold onto a female (Thomas *et al.* 1993). A bizarre use of adhesive secretions is shown by the African rain frogs (*Breviceps*), in which a rotund body shape and the small relative size of the male do not lend themselves to a secure amplexus; instead, the male becomes glued to the female's back (Passmore and Carruthers 1995). Successful defence of a female during amplexus is also related to male body size in *Bufo bufo*, with larger males being less likely to be displaced than smaller ones (Davies and Halliday 1979). Selection for larger body size clearly results from competition for mates among males in many anurans, but the fact that, in a majority of species, males are smaller than females, indicates that there are a number of other selection pressures that influence body size (for a full discussion, see Halliday and Tejedo 1995).

In some anuran species males prevent or reduce intrusion by rivals by establishing a territory around the spawning site. Anurans vary considerably in the extent to which they show territorial behaviour, but it generally occurs only in species that have prolonged breeding seasons (Wells 1977). In some species, males defend a territory from which they call to attract females and in which spawning occurs; examples include the North American bullfrog *Rana catesbeiana* (Howard 1978a,b), and the gladiator frog *Hyla rosenbergi* (Kluge 1981). In others, such as the natterjack toad *Bufo calamita* (Arak 1983) and the painted reed frog *Hyperolius marmoratus* (Dyson and Passmore 1992), males defend calling sites, but spawning occurs at an undefended site elsewhere in the pond.

In the majority of anurans males are smaller than, or similar in size to females, but in some species that maintain territories the males are larger than the females. For example, males of the African bullfrog *Pyxicephalus adspersus* are typically twice as large as females (Passmore and Carruthers 1995). (This species is unusual in that, although it has a short, explosive breeding season, males are territorial.) It is also in territorial species that specialized weapons, which are rare in anurans (Halliday and Tejedo 1995), are found. In a small, monophyletic clade of *Rana* species from Asia, including *R. blythi*, a suite of male characters, comprising fangs, hypertrophied jaw muscles, enlarged head, and body-size greater than the female's, has replaced the nuptial pads, advertisement call, vocal sacs and the smaller than female body size typical of the genus. This appears to be associated with a reproductive pattern that involves paternal care and male defence of a nest (Emerson and Inger 1992; Emerson and Voris 1992; Emerson *et al.* 1993; Emerson 1994).

Among male anurans, there is a rich diversity of morphological and behavioural adaptations that enhance an individual's ability to monopolize a female during mating. To what extent the function of such monopolization should be seen as preventing the female being totally lost to another male, as opposed to preventing sperm competition, is a question that has not been explored in any species.

In a few anurans fertilization is internal. In the tailed frog (*Ascaphus truei*) the male possesses a penis-like cloacal protuberance with which he inseminates the female (Slater 1931; Wernz 1969). This appears to be an adaptation to a fast-flowing mountain stream habitat, where external fertilization would be highly ineffective, although Jamieson *et al.* (1993) have suggested that internal fertilization is the primitive condition for all amphibians. In *Nectophrynoides occidentalis*, a viviparous, terrestrial species, male and female directly oppose their cloacae during mating (Noble 1931; Boisseau and Joly 1975). Internal fertilization has also been reported for *Nectophrynoides malcolmi* (Wake 1980), *Eleutherodactylus jasperi* (Wake 1978) and *Eleutherodactylus coqui* (Townsend *et al.* 1981). It is possible that sperm competition could occur within the female in such species but, at present, very little is known about their breeding biology.

III. APODANS

The reproductive biology of apodans has been described by Wake (1977, 1992), who has also made detailed morphological studies of their urogenital system (Wake 1968, 1970a,b, 1972). Fertilization is internal, sperm being transferred by means of an erectile intromittent organ, the phallodeum, consisting of the posterior part of the cloaca. Many caecilians give birth to live young. The Mullerian glands of the male produce a secretion that provides both a fluid medium and metabolites for the sperm (Wake 1981) but, in contrast to urodeles, the female possesses no sperm-storage organ (Wake 1972). Very little is known about their mating behaviour, in particular whether females are inseminated by more than one male, and so it is impossible to assess the potential for sperm competition in this group.

IV. URODELES

Fertilization is internal in 90% of urodele species; only the primitive cryptobranchids, sirenids and hynobiids have external fertilization. In those species with internal fertilization, sperm transfer is indirect, by means of a spermatophore deposited on the substrate (Salthe 1967; Arnold 1972,

1977; Halliday 1990). In some genera, such as *Ambystoma* (Arnold 1976), *Triturus* (Halliday 1974, 1977), *Notophthalmus* (Verrell 1982) and *Taricha* (Propper 1991), mating takes place in water; in others, such as *Plethodon* (Arnold 1976) and *Salamandra* (Joly 1966; Himstedt 1965), it occurs on land. There is considerable variation between genera, both in the manner in which males stimulate females, and in the extent to which males physically restrict female movements (Arnold 1972, 1977; Halliday 1977, 1990; Houck and Verrell 1993; Sullivan *et al.* 1995). In *Salamandra* (Joly 1966), *Chioglossa* (Arnold 1987) and *Euproctus* (Thiesmeier and Hornberg 1990; Brizzi *et al.* 1995a) the female is held in amplexus throughout the entire courtship and mating sequence; in *Taricha* (Davis and Twitty 1964; Propper 1991) and *Notophthalmus* (Arnold 1972; Verrell 1982) she is released just before spermatophore transfer, although in *Taricha* she is recaptured after spermatophore transfer (Propper 1991). In *Triturus* (Halliday 1974, 1977) and *Cynops* (Sparreboom 1994) she is not restrained at any stage during courtship. As discussed below, the degree to which males restrain females has an important impact on whether or not females mate with more than one male.

Parker (1970) listed four conditions which must exist if sperm competition is to occur. These are: (1) individual females are inseminated by more than one male; (2) females can store sperm, at least for the duration of the period over which matings with different males occur; (3) sperm remain viable during the storage period; (4) sperm are stored and used efficiently by females. Conditions 1–3 have all been studied in urodeles, but very little is known about condition 4.

A. Multiple mating

In many species of urodele, there are two ways in which an individual female can become inseminated by more than one male (Halliday and Verrell 1984; Verrell 1989). First, she may respond positively, on different occasions, to the courtship behaviour of different males. Second, she may become multiply inseminated during a single mating encounter, by both the male who initiates courtship (the courting male) and a male who sexually interferes in the encounter (the interfering male).

The extent to which female urodeles mate with more than one male depends on several factors including the duration of the breeding season, the frequency with which females encounter sexually active males, variation over time in female receptivity, variation over time in the ability of males to produce spermatophores, and the relative timing of mating and oviposition. Our knowledge of the dynamics of mating within urodele populations is very fragmentary and incomplete, primarily because the sexual behaviour of urodeles is very difficult to observe in natural situations. Much of what is known about mating dynamics is based on observations made in captivity, combined with inferences from limited field studies (Verrell 1989). Knowledge of urodele mating dynamics is particularly poor for those taxa, such as the plethodontids, which mate on land and which typically do not form mating assemblages. In contrast, aquatic breeding species generally migrate to ponds or streams, where they can form large and dense populations. A considerable amount of information can be gathered about the composition of such populations by intercepting animals at drift fences (Halliday 1996) on their way to their breeding sites (Verrell and Halliday 1985a,b). In aquatic mating assemblages, for example, drift fence studies reveal that the operational sex ratio is typically strongly male-biased, partly because males stay longer in the water than females (Halliday and Verrell 1984).

There is a great deal of variation in the frequency and duration of breeding activity among urodeles (Duellman and Trueb 1985; Jørgensen 1992). For example, whether individuals breed annually or biennially varies between species and between populations within a species. Much of this variation is related to latitude and altitude. At higher latitudes, breeding tends to be highly seasonal, while towards the equator it is more sporadic and prolonged. In a number of central American genera, such as *Bolitoglossa*, mating activity has been observed throughout the year (Houck 1977). In some species breeding in northern temperate habitats, such as some *Ambystoma* species (Garton 1972; Arnold 1972, 1976, 1977), breeding is explosive and is completed in a few days. In *Ambystoma macrodactylum*, breeding is completed in 3 weeks or less (Verrell and Pelton 1996). In the European newts, breeding in Britain is seasonal and prolonged. In both *Triturus cristatus* (Verrell and Halliday 1985a) and *Triturus vulgaris* (Verrell and Halliday 1985b; Verrell *et al.* 1986), there is great variation among individuals in terms of how long they spend in water, with some being present in a pond for as much as 9 months; mating activity is, however, confined to a 3-month period in the spring.

For the European smooth newt (*Triturus v. vulgaris*) breeding in Britain, data from a number of studies (Halliday 1974, 1976; Verrell and Halliday 1985b; Verrell *et al.* 1986; Hosie 1992; Waights 1996) enable us to build up a picture of how mating dynamics change over the course of the breeding season (Table 12.1). The primary determinant of the long duration of breeding activity in this species is the mode of egg-laying. Females lay their eggs individually, each carefully wrapped in a leaf; this is a time-consuming process and it takes a female up to 3 months to lay her entire clutch (Baker 1992). This prolonged oviposition period appears to have selected for an early initiation of breeding activity in female smooth newts who, unlike many amphibians, migrate to water at the same time as, and occasionally slightly before males (Verrell and Halliday 1985b). Females become highly receptive within a few days of their arrival and, in the laboratory, mate with several males over a few days (Hosie 1992). At this time, the secondary sexual characteristics of males,

ochrophaeus have more than one father. Labanick (1983), studying the same species, used variable phenotypic traits (red legs and cheeks) to establish parentage and estimated that 25% of females were multiply inseminated at one locality. In a laboratory study, Houck *et al.* (1985) mated females of *D. ochrophaeus* with a number of males obtained from different populations for which they had genetic markers. From data on the paternity of their progeny, there was no clear competitive advantage to either the first or the last male to have mated with a particular female, suggesting that sperm from different males become mixed. Rafinski (1981), using electrophoretic information, has reported high levels of multiple paternity in *Triturus alpestris*.

Sperm competition has been investigated extensively in *Triturus* by J. Rafinski and A. Pecio (pers. comm.). In *T. alpestris*, a paternity study, using males from different populations that can be identified by electrophoretic markers, revealed considerable variation, with the second male fathering 15–95% of a female's progeny but, overall, a second-male advantage in a majority of families (Rafinski, pers. comm.). Pecio (pers. comm.) has investigated paternity by mating females sequentially with males of different species, *T. vulgaris* and *T. montandoni*. There is a tendency for the second male to have a mating advantage, although this is tempered by a tendency for conspecific sperm to be more likely to fertilize a female's eggs. Overall, these data suggest that there is a last-male advantage in *Triturus*.

The prevailing view in the sexual selection literature is that sperm competition is a manifestation of male–male competition, and attention has focused on male adaptations related to it. However, given that sperm competition can occur only if females provide appropriate conditions, both by mating multiply and by storing sperm, more attention should be paid to possible adaptive consequences of sperm competition for females (Eberhart 1990). Halliday (1983) suggested that females might use multiple mating, combined with sperm competition in which there is a last-male advantage, to ensure that they mate with high-quality males. If a female mates with the first male she meets, she guarantees that her eggs will be fertilized. Thereafter, she can sample further males and, by mating only with those of higher quality than previous partners, she can maximize the quality of her progeny. This hypothesis has been tested by Gabor and Halliday (1997) in the smooth newt *T. v. vulgaris*; in this species females show a preference for picking up the spermatophores of males with larger crests (Green 1991; Hosie 1992). Females were presented with single males, varying in crest height, in two separate tests, separated by a period of 20 days, during which females could lay eggs. In the second test, a significant majority of females mated only if the male had a higher crest than the male with whom they mated in the first test. This result supports the hypothesis that females may combine multiple mating and sperm competition in an adaptive mating strategy, although multiple mating may serve other, additional functions (Halliday and Arnold 1987).

B. *Sperm storage*

The primitive condition among urodeles is for fertilization to be external and for females to lack sperm storage organs, as in cryptobranchids, hynobiids and sirenids (Sever 1991a; Sever *et al.* 1996a). Female storage organs in the other six families may be polyphyletic in origin (Sever and Kloepfer 1993; Sever 1994). The adaptive value of internal fertilization and sperm storage for female urodeles is that it facilitates the temporal and spatial separation of mating and egg-laying (Houck *et al.* 1985b; Brizzi *et al.* 1989; Houck and Verrell 1993; Trauth *et al.* 1994).

The spermatheca of urodeles opens into the roof of the cloaca. Reviews of spermathecal morphology by Kingsbury (1895), Noble (1931) and Boisseau and Joly (1975) distinguish two basic types of spermathecae. *Necturus*, a proteid (Sever 1992c), the amphiumids (Sever 1992a), and salamandrids, such as *Notophthalmus* and *Triturus*, have numerous short tubules opening separately into the roof of the female's cloaca. The spermatheca of *T. vulgaris* consists of 40–60 tubules (Verrell and Sever 1988). Most plethodontids, such as *Desmognathus*, *Plethodon* and *Hydromantes*, have a few short tubules, or diverticulae, opening into a common duct that leads to the cloaca (Sever 1992e, 1994). Noble (1931) suggested that *Ambystoma* has an intermediate structure, consisting of many tubules opening into a short duct, but extensive and detailed studies of several ambystomatids by Sever (1992d) revealed a multiple tubule pattern similar to that of salamandrids. The gross morphology and histology of the spermathecae of all the urodele families have been studied extensively by Sever (1991a,c, 1992a–e). The plethodontids are said to have a single 'complex' spermatheca consisting of compound tubuloalveolar glands with a common duct; in the 'simple' pattern found in other families, each tubule, opening independently into the cloaca, can be referred to as a spermatheca so that these groups have multiple spermathecae (Kingsbury 1895; Sever and Kloepfer 1993).

Some authors have suggested that stored sperm are nourished by spermathecal secretions (Dent 1970; Boisseau and Joly 1975) and Houck and Schwenk (1984) found that sperm were still apparently normal after oviposition in *Desmognathus ochrophaeus*. Sever and Kloepfer (1993), however, are sceptical about the nutritive function of spermathecal secretions and suggest that they serve simply to maintain stored sperm in a quiescent state.

There is considerable variation among urodeles in the interval between mating and oviposition, when fertilization typically occurs. In axolotls (*Ambystoma mexicanum*), sperm do not survive for more than 12 days (Humphrey 1977), but in *Ambystoma talpoideum* they are viable after being stored from November to February (Trauth *et al.* 1994). European newt (*Triturus*) females start to lay within 3–5 days of starting to mate (Hosie 1992), but in a related salamandrid, *Notophthalmus viridescens*, sperm are stored for up to 6 months (Sever *et al.* 1996b). In *Salamandra terdigitata*, sperm are stored from the autumn mating period until the next spring (Brizzi *et al.* 1989). In many plethodontids there is an interval of as much as 8 months between mating in spring or autumn and egg-laying during the summer or subsequent spring (Marynick 1971). Sperm remain viable for about 8 months in the plethodontid *Eurycea quadridigitatus* (Pool and Hoage 1973), and in the salamandrid *Salamandra salamandra*, it is reported that the interval between mating and fertilization may be as long as a year (Boisseau and Joly 1975). There are reports of very long periods of sperm storage in two salamandrids: 190 days in *Cynops pyrrhogaster* (Tsutsui 1931) and 2 years in *S. salamandra* (Boisseau and Joly 1975).

While some studies suggest that sperm can be stored over a long period in some urodeles, other data suggest that females of some species do not store sperm for very long. In the tiger salamander *Ambystoma tigrinum*, sperm are flushed out of the spermatheca within two days after mating (Sever 1995) and in *Ambystoma opacum* all sperm have been removed within 1 month post-mating (Sever *et al.* 1995). Sever (1995) suggests that removal of sperm soon after mating is the primitive condition in urodeles. In *Notophthalmus viridescens*, sperm are stored between December, when mating occurs, and May, when the female lays her eggs but, contrary to the suggestion of Massey (1990), no sperm are retained until the following breeding season (Sever *et al.* 1996b). In *Triturus vulgaris*, sperm do not appear to be stored from one season to the next (Verrell and Sever 1988). This is supported by Pecio's (1992) observation that the eggs laid by *T. vulgaris* females at the start of the breeding season, before they have encountered males, are not fertilized.

An intriguing aspect of recent studies of sperm storage in female urodeles is the discovery that sperm are actively destroyed in the spermatheca of some species. In the two-lined salamander *Eurycea cirrigera*, the spermatheca bathes stored sperm until oviposition is complete; thereafter, sperm become degraded and are phagocytosed (Sever 1991b). Spermiophagy occurs selectively in the distal bulbs of the storage tubules (Sever 1993; Sever and Brunette 1993). Some sperm are still present in the spermatheca at the beginning of the next breeding season, but do not appear to be viable (Sever 1992f). Evidence for the degradation and phagocytosis of sperm has also been obtained for *Ambystoma opacum* (Sever and Kloepfer 1993) and *Notophthalmus viridescens* (Dent 1970). In *Salamandrina terdigitata*, degradation and phagocytosis of sperm begins after oviposition and the spermatheca is clear of sperm by the end of the breeding season (Brizzi *et al.* 1995b). It is possible that the destruction of sperm is a selective process that eliminates only defective sperm, or it may be a more general process, the function of which is to ensure that females do not retain old sperm (Birkhead *et al.* 1993).

Among the conditions for sperm competition listed by Parker (1970) is the efficient use of stored sperm by females. The spermathecae of urodeles are rather small and it is possible that they are not large enough to store all the sperm that females receive from males. Rafinski (pers. comm.) believes that the spermatheca of *Triturus* spp. is large enough to store the contents of only one spermatophore cap so that when a female picks up more than one spermatophore in quick succession, either from the same male during a single mating encounter or from two males through sexual interference, much of the sperm transferred is lost.

An important difference between anurans and urodeles is that the anuran egg has a block to polyspermy whereas the urodele egg does not (Elinson 1986; Zug 1993). If, in urodeles, a large number of sperm enter each egg, it could have a profound effect on the quantity of sperm that females require to fertilize their eggs and therefore that males need to produce. For example, if an average of 10 sperm enter each egg, the sperm contained within a single spermatophore will have the potential to fertilize only one-tenth of the number of eggs than would be the case if urodeles were not polyspermic. Data for five species of internally fertilizing urodeles indicate that there is considerable variation, with between one and 30 sperm typically entering each egg (V. Waights pers. comm.). In *Triturus vulgaris*, a sample of 30 eggs studied by V. Waights (pers. comm.) revealed four that contained no sperm, one that contained 54 and another contained 100; in the remaining 24 eggs, there were one to 20 sperm (mean $\pm$ SD, 4.23 $\pm$ 2.37).

C. *Sexual interference and sexual defence*

Many urodeles gather to mate at confined breeding sites, such as ponds, and the resultant high population density provides frequent opportunities for mating competition to occur. In all but the most primitive urodele families, sperm transfer is achieved by means of spermatophores, a mechanism that is both intrinsically unreliable and open to a particular form of male–male competition called sexual interference (Arnold 1972, 1976, 1977; Halliday 1977, 1990; Halliday and Tejedo 1995). The most important opportunity for interference arises in the time interval between when the male deposits a spermatophore on the substrate and when the female picks it up a few seconds later. Successful transfer depends on the female behaving in a very precise and accurate way and she can be very easily disrupted by the intrusion of another male. Male urodeles interfere in the courtship of other males in a variety of ways, the simplest being attempts to display to a female that is already being courted. Other forms of sexual interference are summarized in Table 12.3. In descriptions of sexual interference, the male that is already engaged in sexual behaviour is called the courting male; the male that then intrudes is called the interfering male. The most sophisticated forms of interference are female mimicry and spermatophore covering. In female mimicry the interfering male causes the courting male to deposit a spermatophore by mimicking female behaviour, usually by a nudge with his snout against the courting male's tail. This spermatophore is not transferred to the female for one of three reasons, depending on the

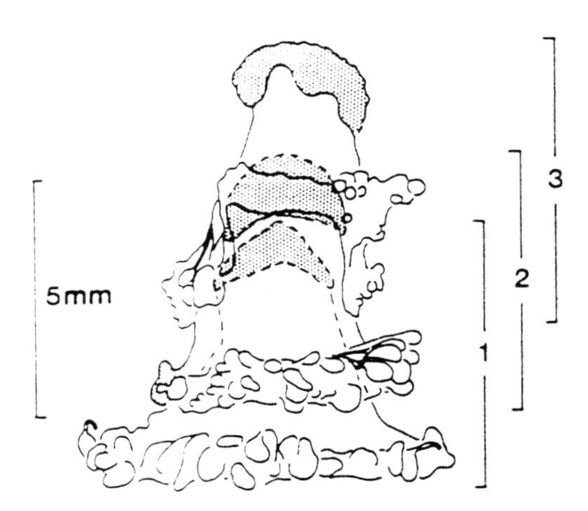

Fig. 12.2. *Stacked spermatophores of the spotted salamander* Ambystoma maculatum. *Three spermatophores have been deposited, one on top of another, so that the sperm masses (stippled) of the first two are covered and are thus inaccessible to a female. From Arnold (1976).*

species: the interfering male pushes the female out of the way, or he leads her away to initiate spermatophore transfer himself, or he deposits a spermatophore of his own on top of it. In spermatophore covering, only the sperm of the second male is accessible to the female's cloaca.

In many species in which interference occurs, male behaviour patterns can be identified that function either to counter the adverse effects of sexual interference, or to prevent it happening at all; these are called sexual defence (Arnold 1972, 1976). Like sexual interference, sexual defence occurs in a diversity of forms (Table 12.3). They include simply pushing the rival away from the female, increasing the rate of display directed at the female, and retaliatory spermatophore covering. The latter can lead to a situation in which several spermatophores, deposited alternately by two males, are stacked on top of one another (Fig. 12.2).

Because of the complexity of the behaviour involved, sexual interference and defence have mostly been described in the laboratory, making it difficult to assess the significance of such behaviour in the field. Such field data as are available suggest that, in nature, sexually active males often greatly outnumber receptive females and sexual interference is very common, leading to very low rates of successful spermatophore transfer. Such a situation has been reported in field studies of *Triturus vulgaris* (Verrell and McCabe 1988), *T. italicus* (Giacoma and Crusco 1987) and *Notophthalmus viridescens* (Massey 1988).

Sexual interference has so far been described in three urodele families, the plethodontids, the ambystomatids and the salamandrids; it is not known whether it occurs in other families. The taxonomic distribution of sexual interference suggests that it is an extremely ancient form of behaviour, and that it pre-dates the evolution of many of the diverse

Table 12.3. Sexual interference and sexual defence in salamanders and newts.

Species	Nature of interference shown by interfering male (IM) towards courting male (CM)	Nature of defence shown by courting male (CM)	References
Ambystoma maculatum	IM mimics female behaviour to elicit spermatophore deposition by CM, then deposits his spermatophore on top of that of CM	CM covers IM's spermatophore with one of his own	Arnold (1976)
Ambystoma texanum	IM deposits his spermatophore on top of that of CM	CM covers IM's spermatophore with one of his own. Increased spermatophore production	McWilliams (1992)
Ambystoma tigrinum	IM mimics female behaviour to elicit spermatophore deposition by CM, then deposits his spermatophore on top of that of CM IM pushes female away from rival male	CM carries female away from IM	Arnold (1976)
Cynops ensicauda	IM mimics female behaviour to elicit spermatophore deposition by CM		Sparreboom (1994)
Desmognathus ochrophaeus		CM forcibly removes female from the vicinity of rivals	Houck (1988)
Euproctus spp.	IM attacks CM and attempts to displace him from the female	CM captures female and holds her in amplexus with his tail, transferring his spermatophore directly into her cloaca	Thiesmeier and Hornberg (1990), Brizzi *et al.* (1995a)
Eurycea cirrigera	IM mimics female behaviour to elicit spermatophore deposition by CM and leads female away		Thomas (1989)

Notophthalmus viridescens	IM mimics female behaviour to elicit spermatophore deposition by CM and leads female away	CM prevents interference by clasping female in amplexus	Verrell (1982, 1983)
Plethodon jordani	IM mimics female behaviour to elicit spermatophore deposition by CM, then deposits his spermatophore on top of that of CM	CM chases potential IM away from female	Arnold (1976)
Taricha torosa	IM nudges female to one side just before spermatophore transfer	CM carries female away from potential rivals	Arnold (1977), T. R. Halliday and C. A. Hosie (unpublished data)
Triturus cristatus and *T. marmoratus*	IM mimics female behaviour to elicit spermatophore deposition by CM and leads female away	Males defend display sites, chase rivals away	Zuiderwijk and Sparreboom (1986)
Triturus italicus	IM pushes CM out of the way during display and displays to the female himself	CM chases and pushes IM away	Giacoma and Crusco (1987)
Triturus vulgaris	IM mimics female behaviour to elicit spermatophore deposition by CM and leads female away	CM increases his display rate. CM attempts to lead female away from IM by performing retreat display	Verrell (1984)

forms of courtship behaviour observed among living urodeles (Halliday 1990). Thus, many aspects of courtship behaviour, such as elaborate displays and leading the female prior to spermatophore deposition, may be forms of sexual defence, as well as serving other courtship functions, such as stimulation of the female and ensuring spermatophore transfer. For example, the 'retreat display' phase of courtship in *Triturus vulgaris* was interpreted by Halliday (1974) as a mechanism by which the male 'tests' the responsiveness of the female; later studies by Verrell (1984) suggest that it may also serve to lead the female away from a potential interfering male.

Sexual interference entails a number of costs for courting males. Interference generally increases the duration of courtship interactions and sexual defence involves an increase in energy expenditure. Such costs are probably trivial, but spermatophores that are not picked up by the female, or that are covered by a rival's spermatophores, represent a cost in terms of wasted reproductive effort. It is not known whether sperm *per se* are costly for males to produce, but, as discussed above, there is good evidence that costs involved in the production of spermatophores impose quite severe physiological constraints on males (Halliday 1987).

Another cost of sexual interference, in some species, arises from the observation that females become less sexually responsive when interference occurs (Verrell 1984; Sparreboom 1994). A consequence of this effect is that interference not only reduces the mating success of the courting male, but also yields a very low pay-off for an interfering male, because females rarely pick up their spermatophores (Verrell 1984).

Mechanisms of sexual defence can be divided into two categories, preemptive and retaliatory. Pre-emptive mechanisms include carrying, pushing or leading a female away from rivals prior to spermatophore deposition. In the great majority of urodeles, there is some form of amplexus prior to spermatophore transfer; amplexus in *Notophthalmus* and *Taricha* can be regarded as a form of mate guarding that reduces the risk of interference (Halliday and Tejedo 1995). Halliday (1990) has suggested that the enormous diversity of forms of amplexus observed among urodeles is the result of a number of independent evolutionary events in which taxa have evolved forms of female capture in response to a number of selection pressures, one of which is the risk of sexual interference. A number of urodeles are territorial, notably terrestrial-breeding plethodontids (Mathis *et al.* 1995); effective territoriality must reduce the possibility that intruders will interfere with a male's courtship attempts. It has been suggested that males of some of the aquatic European newts (*Triturus*) defend ephemeral territories during the breeding period and that these reduce the incidence of interference (Zuiderwijk and Sparreboom 1986; Raxworthy 1989; Hedlund 1990).

As discussed above, there is marked variation among urodeles in terms of the number of spermatophores that males produce during courtship encounters. Across three families, exemplified by the three genera *Ambystoma, Triturus* and *Plethodon*, there is a trend for the amount of male courtship activity invested in each spermatophore to increase, with a parallel increase in the rate at which spermatophores are successfully transferred to females (Arnold 1977; Halliday 1990). The longer courtship lasts, however, the more opportunities there are for rivals to interfere. The limited data available suggest that it is in species such as *Plethodon jordani* and *Desmognathus ochrophaeus* (Houck 1980), that have very lengthy courtship that sexual defence is pre-emptive, with males seeking to repel rivals before courtship begins. It is in the more explosive breeders, such as *Ambystoma maculatum*, that sexual defence generally takes the form of retaliation.

The most detailed and complete study of alternative mating strategies is that of *Notophthalmus viridescens* by Verrell (1982, 1983) (see also Halliday 1990). Males seek to obtain matings in one of four ways, depending both on the receptivity of the female and on whether rival males are present (Fig. 12.3):

1. 'Hula' display: a brief display preceding spermatophore transfer, performed only if the female is receptive and if no rival is present.
2. Amplexus: a lengthy activity, performed if the female is unreceptive and/or if a rival is present. Amplexus serves both to stimulate the female and to guard her against rivals.
3. Sexual interference: adopted if the female is already engaged in courtship with another male. There are two forms of interference:

   (a) If the pair are engaged in spermatophore transfer, the interfering male shows female mimicry and attempts to inseminate the female himself.
   (b) If the pair are in amplexus, the interfering male attempts to displace the courting male from the female's back, but is seldom successful.

These alternative strategies, and those shown by *Triturus* species, can be categorized as conditional strategies (Dunbar 1982); which strategy a male adopts on a particular occasion depends on the conditions prevailing at the time, not on any inherent property of the male, such as body size. This is well illustrated by *Ambystoma* species, in which two males take it in turns to be courter and interferer, covering each other's spermatophores. In the pre-emptive type of sexual defence shown by plethodontids, body size does appear to be an important factor. In the laboratory, Houck (1988) reported that larger males of *Desmognathus ochrophaeus* consistently drive smaller males away from females and, in the field, Mathis (1991) found that males of *Plethodon cinereus* that are close to females are significantly larger than those found alone. This effect may, however, reflect female choice, as Mathis has also found that, in the laboratory, females prefer to associate with larger males.

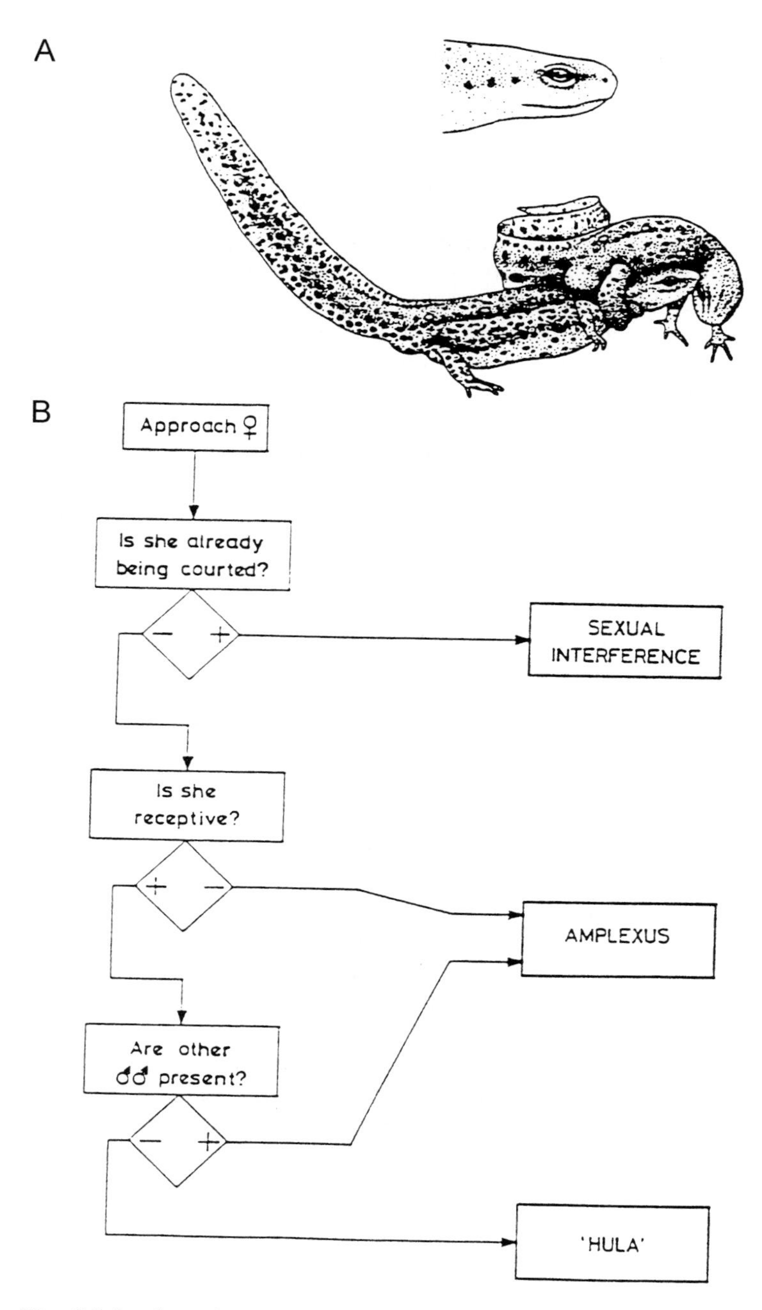

Fig. 12.3. *Courtship in the red-spotted newt* Notophthalmus viridescens. *(A) The male in amplexus with the female, rubbing his cheek glands (inset) against her nostrils. (B) Alternative male strategies in relation to the female's receptivity and the presence or absence of other males. From Halliday (1990).*

D. Competitive mating in urodeles with external fertilization

From descriptions of mating and mating competition in the Asian hynobiid salamanders, it is clear that there is the potential for sperm competition to occur in this primitive family. Among members of the genus, males of *Hynobius takedai* typically develop a larger tail and more massive head during the breeding season (Tanaka 1987). Males defend potential mating and oviposition sites on submerged twigs and males with larger heads enjoy a competitive advantage in defending such sites. Females approach males and the male assists in removing a large sac of eggs from the female's cloaca (*Hynobius retardatus*; Sato 1992), a behaviour pattern that Nussbaum (1985) describes as 'midwife' behaviour. Having pulled the egg sac out of the female, the male holds it and rubs it with his cloaca, apparently fertilizing the eggs (*Hynobius nigrescens*; Usuda 1993). Other males quickly start to scramble over the egg sac soon after its removal from the female; in *H. nigrescens* the mean number of males on each new egg sac is 5.6 (Usuda 1993). In *Hynobius nebulosus*, males are aggressive towards one another in defence of egg sacs, behaviour that appears to be adapted to ensure paternity (Kusano 1980). In the Korean salamander *Hynobius leechii*, the male wraps himself around the egg sac during and after ejaculation, a posture that appears to be adapted to protect his paternity of the eggs (Park *et al.* 1996). Hasumi (1994), describing the behaviour of *H. nigrescens*, differentiates between those males that first grasp the egg sac ('monopolist' males) and those that arrive later ('scrambler' males) and reports that scrambler males ejaculate sperm onto the eggs sacs. In marked contrast to the eggs of externally fertilizing anurans, whose eggs are typically fertilizable for only a few minutes, the eggs of this species remain fertilizable after 3 h in the water (Hasumi *et al.* 1993). This long interval of fertilizability is also much greater than that reported for urodeles with internal fertilization; in both *Notophthalmus viridescens* (McLaughlin and Humphries 1978) and *Cynops pyrrhogaster* (Matsuda and Onitake 1984), eggs are fertilizable for only 15 min once the ovum has passed the spermatheca. The eggs of hynobiids, unlike those of urodeles with internal fertilization, are not polyspermic (Iwao 1989).

The swollen heads of breeding males in *H. nigrescens* are attributable to the absorption of water (Hasumi and Iwasawa 1990; Hasumi 1994). Males with larger heads have an advantage in monopolizing egg sacs that is independent of their overall body size, as measured by snout-vent length. This genus thus appears to have evolved post-mating defence of eggs, together with an anatomical adaptation that enhances such defence, the function of which appears to be to reduce the impact of sperm competition. In the giant Japanese cryptobranchid *Andrias japonicus*, the male builds and defends a deep burrow with an entrance trench (Kuurabora *et al.* 1989). He continues to defend the nest until the eggs hatch and the larvae leave. Several females enter the nest and lay their eggs there; the nest is also entered by several males who ejaculate onto the eggs (Kuurabora *et al.* 1989).

E. *The sexual behaviour of* Triturus *and* Taricha *compared*

The sexual behaviour of urodeles is notable for its diversity and for the high degree of specialization shown by individual genera and species (Halliday 1990). As a consequence, there are few generalizations that can be made about their sexual behaviour. Comparative studies can, however, be very illuminating in revealing the selection pressures that have shaped the sexual behaviour of individual taxa. For example, recent studies of the western North American newts (*Taricha*) make a very interesting contrast with the European newts (*Triturus*); in particular, they reveal the potential significance of sperm competition as a factor shaping the evolution of newt reproductive biology.

Three species of *Taricha* are recognized, but recent genetic and biogeographic work on the genus suggests that further species should be recognized (Tan 1994; Tan and Wake 1995). The sexual behaviour of *Taricha granulosa* has been described in detail by Propper (1991) and some information is available also for *Taricha torosa* and *Taricha rivularis* (Davis and Twitty 1964; Twitty 1966; Brame 1968). In addition, the reproductive physiology of *T. granulosa* has been studied extensively (see Moore 1994 for an overview). Like *Triturus*, *Taricha* males have a dissociated reproductive cycle in which sperm are matured in late summer, just after the breeding season. The complex breeding dynamics of *Triturus vulgaris* (Table 12.1) show a number of contrasts with those of *Taricha torosa* (Table 12.4). While the breeding season of *Triturus vulgaris* lasts for several months in Britain, that of *Taricha torosa* in central California lasts for only a few weeks, depending on how long breeding ponds retain water after spring rains. *Taricha granulosa*, which generally inhabits less arid habitats, has a breeding season lasting several months (Moore 1994). In all *Triturus* species, females lay their eggs singly; the same is true of *Taricha granulosa*, but the eggs of *Taricha torosa* are laid in clusters (Twitty 1942; Brame 1968; Moore 1994). The most striking difference is that, whereas the female in *Triturus* is not restrained by the male at any point during a courtship and mating encounter lasting only a few minutes, male *Taricha granulosa* hold the female in amplexus throughout, an interaction lasting up to 5 days, releasing her only during the brief spermatophore transfer phase (Propper 1991; Moore 1994). In *Taricha torosa*, courtship and mating are less protracted, lasting only about 1 h (Davis and Twitty 1964). A further striking difference between these two genera is that, whereas in *Triturus* both sexes become fully physiologically adapted to life in water and remain aquatic for a comparable period (Halliday 1974; Verrell and Halliday 1985b), *Taricha* males undergo much more extensive physiological changes, notably in the texture of

Table 12.4. A comparison of the reproductive biology of the European smooth newt (*Triturus v. vulgaris*) in Britain with the California newt (*Taricha torosa*) in central California.

European smooth newt	California newt
Eggs laid singly	Eggs laid in clusters
Breeding season lasts several months	Breeding season typically lasts a few weeks, depending on permanence of breeding sites
Males and females arrive at breeding pond together (11)	Males arrive approximately 3 weeks before females (4, 9)
Females enter pond and stay for as long as males (11)	Females enter pond twice, staying briefly, first to mate, then to lay eggs (4, 9)
No amplexus or any form of male constraint on female's behaviour (3)	Male holds female in amplexus before and after insemination (1, 7)
Female choice favours males with larger crests (2, 6)	Little or no opportunity for female choice (7)
Males compete by sexual interference (10)	Males compete aggressively, seeking amplexus displacement, favouring large body bulk (1, 5, 7)
Male's tail same length as female's (4)	Male's tail much larger than female's – an adaptation for carrying the female away from rivals (4)
Low spermatophore transfer success (3)	High spermatophore transfer success (7)
Large testis (relative to body size) (12)	Small testis (relative to body size) (8)
Males allocate limited sperm supply to many sparsely-filled spermatophores	Males allocate limited sperm supply to few well-filled spermatophores
Sperm competition is a major feature of the mating system	Little or no opportunity for sperm competition

Sources: 1, Davis and Twitty (1964); 2, Green (1991); 3, Halliday (1974); 4, Halliday, personal observation; 5, T. R. Halliday and C. A. Hosie (unpublished data); 6, Hosie (1992); 7, Propper (1991); 8, Specker and Moore (1980); 9, Trenham (personal communication); 10, Verrell (1984); 11, Verrell and Halliday (1985b); 12, Verrell *et al.* (1986).

their skin, and remain aquatic for much longer than females (Deviche *et al.* 1990; Moore 1994).

Amplexus in *Taricha* is very similar to that of most frogs and toads: the male holds the female dorsally and defends her against attempts by rival males to displace him (Pimental 1960; Janzen and Brodie 1989; Propper 1991). Associated with the defence of females during amplexus, males in the breeding season develop three anatomical adaptations: a long, deep tail, keratinized nuptial pads and a marked increase in body bulk (Fig. 12.4) (Deviche *et al.* 1990). Janzen and Brodie (1989) showed that males with larger tails have a competitive advantage in competition for females and T. R. Halliday and C. A. Hosie (unpublished data) have shown that

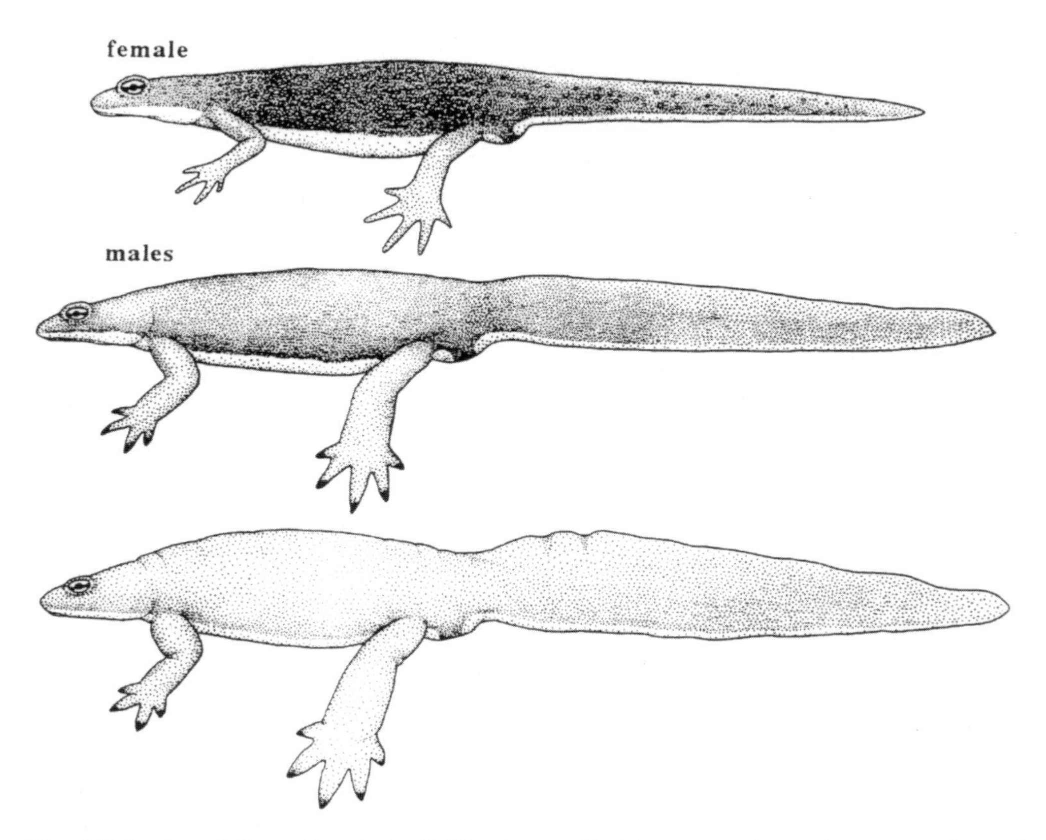

Fig. 12.4. *Sexual dimorphism in the California newt* Taricha torosa *during the aquatic phase of the breeding season. The female (top) retains the rough skin of the terrestrial phase and has a relatively short tail, lacking a fin. After arrival in the water, males (below) become increasingly smooth-skinned and puffy in appearance as they increase in bulk. Their relatively long tails also develop a deeper fin.*

bulkier males, when matched for snout-vent length, have a similar advantage. The large tail of the *Taricha* male also enables him, when approached by a rival male, to swim rapidly away, holding the female beneath him (personal observation). Because she is held in amplexus throughout courtship, the female *Taricha* has very limited opportunities to express mate choice, in contrast to the situation in *Triturus*, her only opportunity is when she is released during spermatophore transfer. Propper (1991) observed that 89.5% of first spermatophores are successfully transferred in *Taricha granulosa*, a much higher success rate than has been observed for any *Triturus* species (see Table 12.2). Observations by Propper (pers. comm.) suggest that females may be able to exercise some mate choice during initial amplexus. If a female remains in the unreceptive, head-down posture for more than 2 h, the male often releases her. Subsequently, such females may become receptive when clasped by a new male.

During amplexus in *Taricha*, the male stimulates the initially unresponsive female by rubbing her with glands on his head and by stroking her cloaca with his hindfeet (Propper 1991; Moore 1994). She is not released by the male until she signals her receptivity by lifting her head; this occurs at a median interval of 7 h, but may be as much as one or more days after capture in *T. granulosa* (Propper 1991). After spermatophore transfer, the male again clasps the female and resumes amplexus. If, in *T. granulosa*, spermatophore transfer is not successful, the female remains receptive and the pair attempt further spermatophore transfers until she is inseminated (Propper 1991; Moore 1994). When the female is inseminated the male maintains amplexus until she is once again unresponsive, a process that takes less than 24 h (Propper 1991). Post-insemination amplexus may, however, last for up to a few days if a pair have to make several attempts to achieve insemination (Propper 1991). This behaviour, which is interpreted as mate-guarding on the part of the male, severely limits the opportunity of the female to mate with other males. Such opportunities are further limited, at least in *T. torosa*, by the fact that females apparently leave breeding ponds after mating and do not return until they are ready to lay their eggs about 2 weeks later (P. Trenham, pers. comm.; own observations). At present, it is not clear whether all females leave the water between mating and egg-laying or whether it is a proportion of them that do so (P. Trenham, pers. comm.). This contrasts markedly with *Triturus*, in which females remain responsive and available to males over a period of several weeks. In *Triturus vulgaris*, females start to lay eggs within 5 days of being inseminated, but in female *Taricha granulosa* this interval is 2 weeks.

Data collected at two *Taricha torosa* breeding ponds in Carmel Valley, central California, USA, show that males arrive before females at the beginning of the breeding season (P. Trenham, pers. comm.; own observations). Males also arrive before females in *T. granulosa* (C. R. Propper, pers. comm.). This contrasts with *Triturus cristatus* and *Triturus vulgaris*, in which males and females arrive together (Verrell and Halliday 1985a,b). The adaptive significance of early arrival by males in *Taricha* appears to be that it enables males to absorb water and thus increase their body bulk, making themselves more competitive in contests for females, before females arrive. Also, because females mate only once, males that arrive late may miss the opportunity to mate. In *Triturus*, because females mate multiply through the season and because there is last-male advantage in sperm competition (see above), there is no apparent advantage to males from arriving at breeding sites before females.

Taken together, these various aspects of the reproductive behaviour of *Taricha* and *Triturus* suggest that, in *Taricha*, there is very limited opportunity for multiple mating by females and that sperm competition is therefore unlikely to be an important factor in its mating system. This is

Table 12.5. The testis weight, expressed as a proportion of body weight, in *Taricha* and *Triturus*.

Species	Minimum	Maximum	Reference
Taricha granulosa	0.3% (March)	2.6% (August)	Specker and Moore (1980)
Triturus v. vulgaris	1.2% (May)	6.0% (August)	Verrell *et al.* (1986)

reflected in the relative testis size of the two genera; correcting for the difference in body size between them, male *Taricha* have much smaller testes than male *Triturus* (Table 12.5). However, the spermatophores of *Taricha* contain many more sperm than those of *Triturus*. A sample of 300 *Triturus vulgaris* spermatophores analysed by V. Waights (pers. comm.) contained 30 000–120 000 sperm, whereas five *Taricha torosa* spermatophores contained 627 000–1 889 000 sperm. While sperm competition is precluded by the reproductive behaviour of males and females in *Taricha*, there can be little doubt that it has shaped the evolution of that behaviour in the past and that it is a factor in maintaining it in the present.

Despite the superficial similarity of their breeding seasons, *Triturus* and *Taricha* have sharply contrasting reproductive strategies. Both enter ponds in spring and have a long aquatic breeding season. However, the evolution of amplexus in *Taricha* (an event that appears to have occurred several times among urodeles; see Halliday 1990) has had far-reaching consequences for both males and females. The very effective mate-guarding exercised by males in *Taricha* seems to provide females little opportunity to choose their mates and even less opportunity to mate with more than one male (Propper 1991). In terms of time, male *Taricha* invest more in a single mating than any other urodele described, thus increasing their certainty of paternity but reducing their opportunity to mate with more than one female (Propper 1991). The large time investment shown by *Taricha* males also reflects the fact that, as a result of the dynamics of breeding populations, males always greatly outnumber receptive females; males have to make the most of limited opportunities to mate (C. R. Propper, pers. comm.). From the rather limited data on testis size and spermatophore contents for the two genera given above, it appears that *Taricha* males make relatively fewer sperm and allocate them, in larger 'ejaculates', to relatively fewer spermatophores than do *Triturus* males. These differences reflect the fact that sperm competition appears to be rare or nonexistent in *Taricha*, but is a major feature of the mating system in *Triturus*.

V. CONCLUSION

Comparison of the more recent studies covered by this review with those discussed over 12 years ago (Halliday and Verrell 1984) shows that sperm competition is now firmly on the agenda of researchers working on the reproductive biology of amphibians. Sperm competition is clearly a major factor in the sexual behaviour of many species and has led to the evolution in males of a variety of behaviour patterns, morphological features and physiological processes that serve to reduce its impact, such as mate-guarding and sexual dimorphism.

For most anurans, sperm competition is probably not a significant factor because external fertilization offers only limited and very brief opportunities for a female's eggs to be fertilized by more than one male. The very close proximity of males and females in amplexus also offers little space for 'sneaky' mating of the type seen among fish. However, the potential for sperm competition may well have shaped the evolution of anuran sexual behaviour. In particular, amplexus, by which males sequester and guard females, may have evolved, in part, as an adaptation that prevents sperm competition. There is good evidence, however, that sperm competition is important in those species that spawn in foam nests. There is also the potential for it to be important in species with internal fertilization and these require much more research.

Among urodeles, the necessary conditions for sperm competition, internal fertilization, multiple mating by females and sperm storage are widespread, and several recent studies provide evidence that it has been very important in the evolution of urodele sexual behaviour. An important and interesting aspect of the spermatophore mode of sperm transfer is that males generally cannot inseminate females forcibly. If females mate multiply, it is apparently because they are adapted to do so. This allows for a scenario in which, for example, females of the European newts may combine multiple mating and sperm competition adaptively to enhance the quality of the fathers of their progeny. Furthermore, the detailed work by Sever and others on the physiology of sperm storage suggests that females may have evolved sophisticated strategies for the use of stored sperm; females are clearly not passive recipients of sperm.

The existence of female strategies that exploit sperm competition set up a conflict of interest between the sexes and have fascinating implications for males. As suggested by the comparison between European and Californian newts, a female strategy that incorporates multiple mating and sperm competition must impose strong selection on how much males invest in sperm production and on the way that males allocate sperm to individual matings; it must also lead to a diverse array of behaviour patterns and other adaptations.

ACKNOWLEDGEMENTS

I thank Cathy Propper and David Sever for their detailed comments on parts of this chapter, and Tim Birkhead, Mandy Dyson and Anders Møller for reading and commenting on all of it. I also thank Juliet Kauffmann, Peter Trenham and Verina Waights for allowing me to quote their unpublished data.

REFERENCES

Andersson M (1994) *Sexual Selection.* Princeton University Press, Princeton.

Arak PA (1983) Sexual selection by male–male competition in natterjack toad choruses. *Nature* **306:** 261–262.

Arnold SJ (1972) The evolution of courtship behavior in salamanders. PhD dissertation, University of Michigan, Ann Arbor.

Arnold SJ (1976) Sexual behavior, sexual interference and sexual defense in the salamanders *Ambystoma maculatum, Ambystoma tigrinum* and *Plethodon jordani. Z. Tierpsychol.* **42:** 247–300.

Arnold SJ (1977) The evolution of courtship behavior in New World salamanders with some comments on Old World salamanders. In *The Reproductive Biology of Amphibians.* DH Taylor & SI Guttman (eds), pp. 141–183. Plenum Press, New York.

Arnold SJ (1987) The comparative ethology of courtship in salamandrid salamanders. 1. *Salamandra* and *Chioglossa. Ethology* **74:** 133–145.

Baker JMR (1992) Egg production in the smooth newt (*Triturus vulgaris*). *Herpetol. J.* **2:** 90–93.

Berger L & Rybacki M (1992) Sperm competition in European water frogs. *Alytes* **10:** 113–116.

Berger L & Rybacki M (1994) Sperm competition between two species of European water frogs (*Rana ridibunda* and *Rana lessonae*). *Zool. Polon.* **39:** 281–291.

Birkhead TR & Møller AP (1992) *Sperm Competition in Birds: Evolutionary Causes and Consequences.* Academic Press, London.

Birkhead TR, Møller AP & Sutherland WJ (1993) Why do females make it so difficult for males to fertilise their eggs? *J. Theor. Biol.* **161:** 51–60.

Boisseau C & Joly J (1975) Transport and survival of spermatozoa in female Amphibia. In *The Biology of Spermatozoa.* ESE Hafez & CG Thibault (eds), pp. 94–104. Karger, Basel.

Brame AH (1968) The number of egg masses and eggs laid by the California newt, *Taricha torosa. J. Herpetol.* **2:** 169–170.

Brizzi R, Delfino G & Calloni C (1989) Female cloacal anatomy in the spectacled salamander, *Salamandrina terdigitata* (Amphibia: Salamandridae). *Herpetologica* **45:** 310–322.

Brizzi R, Calloni C, Delfino G & Tanteri G (1995a) Notes on the male cloacal

anatomy and reproductive biology of *Euproctus montanus* (Amphibia: Salamandridae). *Herpetologica* **51:** 8–18.
Brizzi R, Delfino G, Selmi MG & Sever DM (1995b) Spermathecae of *Salamandrina terdigitata* (Amphibia: Salamandridae): patterns of sperm storage and degradation. *J. Morphol.* **223:** 21–33.
Crews D (1987) Diversity and evolution of behavioral controlling mechanisms. In *Psychobiology of Reproductive Behavior: an Evolutionary Perspective.* D Crews (ed.), pp. 88–119. Prentice Hall, Englewood Cliffs.
Davies NB & Halliday TR (1979) Competitive mate searching in male common toads, *Bufo bufo. Anim. Behav.* **27:** 1253–1267.
Davis WC & Twitty VC (1964) Courtship behavior and reproductive isolation in the species of *Taricha* (Amphibia, Caudata). *Copeia* **1964:** 601–609.
Dent JN (1970) The ultrastructure of the spermatheca in the red-spotted newt. *J. Morphol.* **132:** 397–424.
Deviche P, Propper CR & Moore FL (1990) Neuroendocrine, behavioral, and morphological changes associated with the termination of the reproductive period in a natural population of male rough-skinned newts (*Taricha granulosa*). *Horm. Behav.* **24:** 284–300.
Duellman WE & Trueb L (1985) *Biology of Amphibians.* McGraw-Hill, New York.
Dunbar RIM (1982) Intraspecific variations in mating strategy. In *Perspectives in Ethology*, Vol. 5. PPG Bateson & PH Klopfer (eds), pp. 385–431. Plenum Press, New York.
Dyson ML & Passmore NI (1992) Inter-male spacing and aggression in African painted reed frogs, *Hyperolius marmoratus. Ethology* **91:** 237–247.
Eberhard WG (1990) Inadvertent machismo? *Trends Ecol. Evol.* **5:** 263.
Elinson RP (1986) Fertilization in amphibians: the ancestry of the block to polyspermy. *Int. Rev. Cytol.* **111:** 59–100.
Emerson SB (1994) Testing pattern predictions of sexual selection: a frog example. *Am. Nat.* **143:** 848–869.
Emerson SB & Inger RF (1992) The comparative ethology of voiced and voiceless Bornean frogs. *J. Herpetol.* **26:** 482–490.
Emerson SB & Voris H (1992) Competing explanations for sexual dimorphism in a voiceless Bornean frog. *Funct. Ecol.* **6:** 654–660.
Emerson SB, Rowsemitt CN & Hess DL (1993) Androgen levels in a Bornean voiceless frog, *Rana blythi. Can. J. Zool.* **71:** 196–203.
Fukuyama K (1991) Spawning behaviour and male mating tactics of a foam-nesting treefrog, *Rhacophorus schlegelii. Anim. Behav.* **42:** 193–199.
Gabor CR & Halliday TR (1997) Sequential mate choice by smooth newts: females become more choosy. *Behav. Ecol.* **8:** 162–166.
Garton JS (1972) Courtship of the small-mouthed salamander, *Ambystoma texanum*, in southern Illinois. *Herpetologica* **28:** 41–45.
Giacoma C & Crusco N (1987) Courtship and male interference in the Italian newt: a field study. *Monit. Zool. Ital. (NS)* **21:** 190–191.
Green AJ (1991) Large male crests, an honest indicator of condition, are preferred by female smooth newts, *Triturus vulgaris* (Salamandridae) at the spermatophore transfer stage. *Anim. Behav.* **41:** 367–369.
Griffiths RA & Mylotte VJ (1988) Observations on the development of the secondary sexual characters of male newts, *Triturus vulgaris* and *T. helveticus. J. Herpetol.* **22:** 476–480.
Halliday TR (1974) Sexual behavior of the smooth newt, *Triturus vulgaris* (Urodela: Salamandridae). *J. Herpetol.* **8:** 277–292.

Halliday TR (1976) The libidinous newt. An analysis of variations in the sexual behaviour of the male smooth newt, *Triturus vulgaris*. *Anim. Behav.* **24:** 398–414.

Halliday TR (1977) The courtship of European newts: an evolutionary perspective. In *The Reproductive Biology of Amphibians*. DH Taylor & SI Guttman (eds), pp. 185–232. Plenum Press, New York.

Halliday TR (1983) The study of mate choice. In *Mate Choice*. P Bateson (ed.), pp. 3–32. Cambridge University Press, Cambridge.

Halliday TR (1987) Physiological constraints on sexual selection. In *Sexual Selection: Testing the Alternatives*. JW Bradbury & MB Andersson (eds), pp. 247–264. Wiley, Chichester.

Halliday TR (1990) The evolution of courtship behaviour in newts and salamanders. *Adv. Stud. Behav.* **19:** 137–169.

Halliday T (1996) Amphibians. In *Ecological Census Techniques: A Handbook*. WJ Sutherland (ed.), pp. 205–217. Cambridge University Press, Cambridge.

Halliday T & Arnold SJ (1987) Multiple mating by females: a perspective from quantitative genetics. *Anim. Behav.* **35:** 939–941.

Halliday T & Tejedo M (1995) Intrasexual selection and alternative mating behaviour. In *Amphibian Biology, Vol. 2. Social Behavior*. H Heatwole & B Sullivan (eds), pp. 419–468. Surrey Beatty and Sons, Chipping Norton, NSW.

Halliday TR & Verrell PA (1984) Sperm competition in amphibians. In *Sperm Competition and the Evolution of Animal Mating Systems*. RL Smith (ed.), pp. 487–508. Academic Press, New York.

Halliday TR & Verrell PA (1988) Body size and age in amphibians and reptiles. *J. Herpetol.* **22:** 253–265.

Harcourt AH, Harvey PH, Larson SG & Short RV (1981) Testis weight, body weight and breeding system in primates. *Nature* **293:** 55–57.

Hasumi M (1994) Reproductive behavior of the salamander *Hynobius nigrescens*: monopoly of egg sacs during scramble competition. *J. Herpetol.* **28:** 264–267.

Hasumi M & Iwasawa H (1990) Seasonal changes in body shape and mass in the salamander, *Hynobius nigrescens*. *J. Herpetol.* **24:** 113–118.

Hasumi M, Hasegawa Y & Iwasawa H (1993) Long-term maintenance of egg-fertilizability in water of the salamander *Hynobius nigrescens*. *Jpn. J. Herpetol.* **15:** 71–73.

Hedlund L (1990) Courtship display in a natural population of crested newts, *Triturus cristatus*. *Ethology* **85:** 279–288.

Himstedt W (1965) Beobachtungen zum Paarungsverhalten des Feuersalamanders. *Zool. Anz.* **175:** 295–300.

Hosie CA (1992) Female choice in newts. Ph.D. thesis, The Open University, Milton Keynes, UK.

Houck LD (1977) Life history patterns and reproductive biology of neotropical salamanders. In *The Reproductive Biology of Amphibians*. DH Taylor & SI Guttman (eds), pp. 43–71. Plenum Press, New York.

Houck LD (1980) Courtship behaviour in the plethodontid salamander *Desmognathus ochrophaeus*. *Am. Zool.* **20:** 825.

Houck LD (1988) The effect of body size on male courtship success in a plethodontid salamander. *Anim. Behav.* **36:** 837–842.

Houck LD & Schwenk K (1984) The potential for long-term sperm competition in a plethodontid salamander. *Herpetologica* **40:** 410–415.

Houck LD & Verrell PA (1993) Studies of courtship behavior in plethodontid salamanders: a review. *Herpetologica* **49:** 175–184.

Houck LD, Arnold SJ & Thisted RA (1985a) A statistical study of mate choice: sexual selection in a plethodontid salamander (*Desmognathus ochrophaeus*). *Evolution* **39:** 370–386.

Houck LD, Tilley SG & Arnold SJ (1985b) Sperm competition in a plethodontid salamander: preliminary results. *J. Herpetol.* **19:** 420–423.

Howard RD (1978a) The evolution of mating strategies in bullfrogs, *Rana catesbeiana*. *Evolution* **32:** 850–871.

Howard RD (1978b) The influence of male-defended oviposition sites on early embryo mortality in bullfrogs. *Ecology* **59:** 789–798.

Humphrey RR (1977) Factors influencing ovulation in the Mexican axolotl as revealed by induced spawning. *J. Exp. Zool.* **199:** 209–214.

Iwao Y (1989) An electrically mediated block to polyspermy in the primitive urodele *Hynobius nebulosus* and phylogenetic comparison with other amphibians. *Develop. Biol.* **134:** 438–445.

Jamieson BGM, Lee MSY & Long K (1993) Ultrastructure of the spermatozoon of the internally fertilizing frog *Ascaphus truei* (Ascaphidae: Anura: Amphibia) with phylogenetic considerations. *Herpetologica* **49:** 52–65.

Janzen FJ & Brodie ED (1989) Tall tails and sexy males: sexual behavior of rough-skinned newts (*Taricha granulosa)* in a natural breeding pond. *Copeia* **1989:** 1068–1071.

Jennions MD & Passmore NI (1993) Sperm competition in frogs: testis size and a sterile male experiment on *Chiromantis xerampelina* (Rhacophoridae). *Biol. J. Linn. Soc.* **50:** 211–220.

Jennions MD, Backwell PRY & Passmore NI (1992) Breeding behaviour of the African frog, *Chiromantis xerampelina*: multiple spawning and polyandry. *Anim. Behav.* **44:** 1091–1100.

Joly J (1966) Sur l'ethologie sexuelle de *Salamandra salamandra* L. *Z. Tierpsychol.* **23:** 8–27.

Jørgensen CB (1992) Growth and reproduction. In *Environmental Physiology of the Amphibians*. ME Feder & WW Burggren (eds), pp. 439–466. Chicago University Press, Chicago.

Kingsbury BF (1895) The spermatheca and methods of fertilization in some American newts and salamanders. *Proc. Am. Microsc. Soc.* **17:** 261–305.

Kluge AG (1981) The life history, social organization, and parental behavior of *Hyla rosenbergi* Boulenger, a nest-building gladiator frog. *Misc. Publ. Mus. Zool. Univ. Mich.* **160 vi:** 1–170.

Krenz JD & Scott DE (1994) Terrestrial courtship affects mating locations in *Ambystoma opacum*. *Herpetologica* **50:** 46–50.

Krenz JD & Sever DM (1995) Mating and oviposition in paedomorphic *Ambystoma talpoideum* precedes the arrival of terrestrial males. *Herpetologica* **51:** 387–393.

Kruse KC & Mounce M (1982) The effects of multiple matings on fertilization capability in male American toads (*Bufo americanus*). *J. Herpetol.* **16:** 410–412.

Kusano T (1980) Breeding and egg survival of a population of a salamander, *Hynobius nebulosus tokyoensis* Tago. *Res. Pop. Ecol.* **21:** 181–196.

Kusano T, Toda M & Fukuyama K (1991) Testes size and breeding systems in Japanese anurans with special reference to large testes in the treefrog, *Rhacophorus arboreus* (Amphibia: Rhacophoridae). *Behav. Ecol. Sociobiol.* **29:** 27–31.

Kuurabora K, Susuki N, Wakabayashi F, Ashikaga H, Inoue T & Kobara J (1989) Breeding the Japanese giant salamander. *Int. Zoo Year.* **28:** 22–31.

Labanick GM (1983) Inheritance of the red-leg and red-cheek traits in the salamander *Desmognathus ochrophaeus*. *Herpetologica* **39:** 114–120.

Lamb T (1984) Amplexus displacement in the southern toad, *Bufo terrestris*. *Copeia* **1984:** 1023–1025.

Maeda N & Matsui M (1990) *Frogs and Toads of Japan*, 2nd. edn. Bun-Ichi Sogo Shuppan, Tokyo.

Marynick SP (1971) Long term storage of sperm in *Desmognathus fuscus* from Louisiana. *Copeia* **1971:** 345–347.

Massey A (1988) Sexual interactions in red-spotted newt populations. *Anim. Behav.* **36:** 205–210.

Massey A (1990) Notes on the reproductive ecology of red-spotted newts (*Notophthalmus virodescens*). *J. Herpetol.* **24:** 106–107.

Mathis A (1991) Large male advantage for access to females: evidence of male–male competition and female discrimination in a territorial salamander. *Behav. Ecol. Sociobiol.* **29:** 133–138.

Mathis A, Jaeger RG, Keen WH, Ducey PK, Walls SC & Buchanan BW (1995) Aggression and territoriality by salamanders and a comparison with the territorial behavior of frogs. In *Amphibian Biology, Vol. 2. Social Behavior*. H Heatwole & BK Sullivan (eds), pp. 633–676. Surrey Beatty and Sons, Chipping Norton, NSW.

Matsuda M & Onitake K (1984) Fertilization of the eggs of *Cynops pyrrhogaster* (Japanese newt) after immersion in water. *Wilh. Roux. Arch. Develop. Biol.* **193:** 61–63.

McLaughlin EW & Humphries AA (1978) The jelly envelopes and fertilization of eggs of the newt, *Notophthalmus viridescens*. *J. Morphol.* **158:** 73–90.

McWilliams SR (1992) Courtship behavior of the small-mouthed salamander (*Ambystoma texanum*): the effects of conspecific males on male mating tactics. *Behaviour* **121:** 1–19.

Møller AP & Briskie JV (1995) Extra-pair paternity, sperm competition and the evolution of testis size in birds. *Behav. Ecol. Sociobiol.* **36:** 357–365.

Moore FL (1994) Complexity in the control of amphibian reproduction: reproductive behaviour and physiology of a salamander. In *Captive Management and Conservation of Amphibians and Reptiles*. JB Murphy, K Adler & JT Collins (eds), pp. 125–131. Society for the Study of Amphibians and Reptiles, Ithaca, NY.

Noble GK (1931) *The Biology of the Amphibia*. McGraw-Hill, New York.

Nussbaum RA (1985) The evolution of parental care in salamanders. *Misc. Publ. Mus. Zool. Univ. Mich.* **169:** 1–50.

d'Orgeix CA (1996) Multiple paternity and the breeding biology of the red-eyed treefrog, *Agalychnis callidryas*. PhD thesis, Virginia Tech., Blacksburg, Virginia, USA.

d'Orgeix CA & Turner BJ (1995) Multiple paternity in the red-eyed treefrog *Agalychnis callidryas* (Cope). *Molec. Ecol.* **4:** 505–508.

Park S-R, Park D-S & Yang SY (1996) Courtship, fighting behaviors and sexual dimorphism of the salamander, *Hynobius leechii*. *Kor. J. Zool.* **39:** 437–446.

Parker GA (1970) Sperm competition and its evolutionary consequences in the insects. *Biol. Rev.* **45:** 525–568.

Parker GA (1974) Courtship persistence and female guarding as male time investment strategies. *Behaviour* **48:** 157–184.

Parker GA (1984) Sperm competition and the evolution of animal mating strategies. In *Sperm Competition and the Evolution of Animal Mating Systems*. RL Smith (ed.), pp. 1–60. Academic Press, New York.

Passmore NI & Carruthers VC (1995) *South African Frogs*, 2nd edn. Southern Book Publishers and Witwarersrand University Press, Johannesburg.

Pecio A (1992) Insemination and egg laying dynamics in the smooth newt, *Triturus vulgaris*, in the laboratory. *Herpetol. J.* **2:** 5–7.

Pimental RA (1960) Inter- and intrahabitat movements of the rough-skinned newt, *Taricha torosa granulosa* (Skilton). *Am. Midl. Nat.* **63:** 470–496.

Pool TB & Hoage TR (1973) The ultrastructure of secretion in the spermatheca of the salamander *Manculus quadridigitatus* (Holbrook). *Tiss. Cell* **5:** 303–313.

Propper CR (1991) Courtship in the rough-skinned newt *Taricha granulosa*. *Anim. Behav.* **41:** 547–554.

Pyburn WF (1970) Breeding behavior of the leaf-frogs *Phyllomedusa callidryas* and *Phyllomedusa dacnicolor* in Mexico. *Copeia* **1970:** 209–218.

Rafinski JN (1981) Multiple paternity in a natural population of the alpine newt, *Triturus alpestris* (Laur). *Amphibia–Reptilia* **2:** 282.

Raxworthy CJ (1989) Courtship, fighting and sexual dimorphism of the banded newt, *Triturus vittatus ophryticus*. *Ethology* **81:** 148–170.

Roberts WE (1994) Explosive breeding aggregations and parachuting in a neotropical frog, *Agalychnis saltator* (Hylidae). *J. Herpetol.* **28:** 193–199.

Salthe SN (1967) Courtship patterns and phylogeny of the urodeles. *Copeia* **1967:** 100–117.

Sato T (1992) Reproductive behavior in the Japanese salamander *Hynobius retardatus*. *Jpn. J. Herpetol.* **14:** 184–190.

Sever DM (1991a) Comparative anatomy and phylogeny of the cloacae of salamanders (Amphibia: Caudata) I. Evolution at the family level. *Herpetologica* **47:** 165–193.

Sever DM (1991b) Sperm storage and degradation in the spermathecae of the salamander *Eurycea cirrigera*. *J. Morphol.* **210:** 71–84.

Sever DM (1991c) Comparative anatomy and phylogeny of the cloacae of salamanders (Amphibia: Caudata) II. Cryptobranchidae, Hynobiidae, and Sirenidae. *J. Morphol.* **207:** 283–301.

Sever DM (1992a) Comparative anatomy and phylogeny of the cloacae of salamanders (Amphibia: Caudata) III. Amphiumidae. *J. Morphol.* **211:** 63–72.

Sever DM (1992b) Comparative anatomy and phylogeny of the cloacae of salamanders (Amphibia: Caudata) IV. Salamandridae. *Anat. Rec.* **232:** 229–244.

Sever DM (1992c) Comparative anatomy and phylogeny of the cloacae of salamanders (Amphibia: Caudata) V. Proteidae. *Herpetologica:* **48:** 318–329.

Sever DM (1992d) Comparative anatomy and phylogeny of the cloacae of salamanders (Amphibia: Caudata) VI. Ambystomatidae and Dicamptodontidae. *J. Morphol.* **212:** 305–322.

Sever DM (1992e) Comparative anatomy and phylogeny of the cloacae of salamanders (Amphibia: Caudata) VII. Plethodontidae. *Herpet. Monogr.* **8:** 276–337.

Sever DM (1992f) Spermiophagy by the spermathecal epithelium of the salamander *Eurycea cirrigera*. *J. Morphol.* **212:** 281–290.

Sever DM (1993) Regionalization of eccrine and spermiophagic activity in spermathecae of the salamander *Eurycea cirrigera* (Amphibia: Plethodontidae). *J. Morphol.* **217:** 161–170.

Sever DM (1994) Observations on regionalization of secretory activity in the spermathecae of salamanders and comments on phylogeny of sperm storage in female amphibians. *Herpetologica* **50:** 383–397.

Sever DM (1995) Spermathecae of *Ambystoma tigrinum* (Amphibia: Caudata): development and a role for the secretions. *J. Herpetol.* **29:** 243–255.

Sever DM & Brunette SM (1993) Regionalization of eccrine and spermiophagic activity in spermathecae of the salamander *Eurycea cirrigera* (Amphibia: Plethodontidae). *J. Morphol.* **217:** 161–170.

Sever DM & Kloepfer NM (1993) Spermathecal cytology of *Ambystoma opacum* (Amphibia: Ambystomatidae) and the phylogeny of sperm storage organs in female salamanders. *J. Morphol.* **217:** 115–127.

Sever DM, Verrell PA, Halliday TR, Griffiths M & Waights V (1990) The cloaca and cloacal glands of the male smooth newt, *Triturus vulgaris vulgaris* (Linnaeus), with especial emphasis on the dorsal gland. *Herpetologica* **46:** 160–168.

Sever DM, Krenz JD, Johnson KM & Rania LC (1995) Morphology and evolutionary implications of the annual cycle of secretion and sperm storage in spermathecae of the salamander *Ambystoma opacum* (Amphibia: Ambystomatidae). *J. Morphol.* **223:** 35–46.

Sever DM, Rania LC & Krenz JD (1996a) Reproduction of the salamander *Siren intermedia* Le Conte with especial reference to oviductal anatomy and mode of fertilization. *J. Morphol.* **227:** 335–348.

Sever DM, Rania LC & Krenz JD (1996b) Annual cycle of sperm storage in spermathecae of the red-spotted newt, *Notophthalmus viridescens* (Amphibia: Salamandridae). *J. Morphol.* **227:** 155–170.

Shillington C & Verrell P (1996) Multiple mating by females is not dependent on body size in the salamander *Desmognathus ochrophaeus. Amphibia–Reptilia* **17:** 33–38.

Slater JR (1931) The mating behavior of *Ascaphus truei* Stijneger. *Copeia* **1931:** 62–63.

Smith-Gill S & Berven KA (1980) *In vitro* fertilization and assessment of male reproductive potential using mammalian gonadotrophin-releasing hormone to induce spermiation in *Rana sylvatica. Copeia* **1980:** 723–728.

Sparreboom M (1994) On the sexual behaviour of the sword-tailed newt, *Cynops ensicauda* (Hallowell, 1860). *Abh. Ber. Natur. Magdeburg* **17:** 151–161.

Sparreboom M (1996) Sexual interference in the sword-tailed newt, *Cynops ensicauda popei* (Amphibia: Salamandridae). *Ethology* **102:** 672–685.

Sparreboom M & Ota H (1995) Notes on the life-history and reproductive behaviour of *Cynops ensicauda popei* (Amphibia: Salamandridae). *Herpetol. J.* **5:** 310–315.

Specker JL & Moore FL (1980) Annual cycle of plasma androgens and testicular composition in the rough-skinned newt, *Taricha granulosa. Gen. Comp. Endocrinol.* **42:** 297–303.

Sullivan BK, Ryan MJ & Verrell PA (1995) Female choice and mating system structure. In *Amphibian Biology, Vol. 2. Social Behavior*. H Heatwole & BK Sullivan (eds), pp. 469–517. Surrey Beatty and Sons, Chipping Norton, NSW.

Tan A-M (1994) Chromosomal variation in the northwestern American newts of the genus *Taricha* (Caudata: Salamandridae). *Chrom. Res.* **2:** 281–292.

Tan A-M & Wake DB (1995) MtDNA phylogeography of the California newt, *Taricha torosa* (Caudata, Salamandridae). *Mol. Phylogen. Evol.* **4:** 383–394.

Tanaka K (1987) Body size and territorial behaviour of male *Hynobius takedai* in breeding season (Amphibia: Hynobiidae). *Jpn. J. Herpetol.* **12:** 45–49.

Tejedo M (1988) Fighting for females in the toad *Bufo calamita* is affected by the operational sex ratio. *Anim. Behav.* **36:** 1765–1769.

Thiesmeier B & Hornberg C (1990) Zur Fortpflanzung sowie zum Paarungsverhalten der Gebirgsmolche, Gattung *Euproctus* (Gené), im terrarium, unter besonderer Berücksichtigung von *Euproctus asper* (Dugès, 1852). *Salamandra* **26:** 63–82.

Thomas EO, Tsang L & Licht P (1993) Comparative histochemistry of the sexually dimorphic skin glands of anuran amphibians. *Copeia* **1993:** 133–143.

Thomas JS (1989) Courtship, male–male competition, and male aggressive behavior of the salamander *Eurycea bislineata*. MSc thesis, University of Southwestern Louisiana, Lafayette, LA.

Tilley SG & Hausman JS (1976) Allozymic variation and occurrence of multiple inseminations in populations of the salamander *Desmognathus ochrophaeus*. *Copeia* **1976:** 734–741.

Townsend DS, Stewart MM, Pough FH & Brussard PF (1981) Internal fertilization in an oviparous frog. *Science* **212:** 469–471.

Trauth SE, Sever DM & Semlitsch RD (1994) Cloacal anatomy of paedomorphic female *Ambystoma talpoideum* (Caudata: Ambystomatidae), with comments on intermorph mating and sperm storage. *Can. J. Zool.* **72:** 2147–2157.

Tsutsui Y (1931) Notes on the behavior of the common Japanese newt, *Diemyctylus pyrrhogaster* Boie. I. Breeding habit. *Mem. Fac. Sci. Kyoto Univ. Ser. Biol.* **7:** 159–179.

Twitty VC (1942) The species of California *Triturus*. *Copeia* **1942:** 65–76.

Twitty VC (1966) *Of Scientists and Salamanders*. Freeman, San Francisco.

Usuda H (1993) Reproductive behavior of *Hynobius nigrescens*, with special reference to male midwife behavior. *Jpn. J. Herpetol.* **15:** 64–70.

Verrell PA (1982) The sexual behaviour of the red-spotted newt, *Notophthalmus viridescens* (Amphibia: Urodela: Salamandridae). *Anim. Behav.* **30:** 1224–1236.

Verrell PA (1983) The influence of the ambient sex ratio and intermale competition on the sexual behavior of the red-spotted newt, *Notophthalmus viridescens* (Amphibia: Urodela: Salamandridae). *Behav. Ecol. Sociobiol.* **13:** 307–313.

Verrell PA (1984) Sexual interference and sexual defense in the smooth newt, *Triturus vulgaris* (Amphibia, Urodela, Salamandridae). *Z. Tierpsychol.* **66:** 242–254.

Verrell PA (1987) Limited male mating capacity in the smooth newt, *Triturus vulgaris* (Amphibia). *J. Comp. Psychol.* **100:** 291–295.

Verrell PA (1989) The sexual strategies of natural populations of newts and salamanders. *Herpetologica* **45:** 265–282.

Verrell P & Halliday T (1985a) The population dynamics of the crested newt *Triturus cristatus* at a pond in southern England. *Holarc. Ecol.* **8:** 151–156.

Verrell P & Halliday T (1985b) Reproductive dynamics of a population of smooth newts, *Triturus vulgaris*, in southern England. *Herpetologica* **41:** 386–395.

Verrell PA & McCabe N (1988) Field observations on the sexual behaviour of the smooth newt, *Triturus vulgaris vulgaris* (Amphibia: Salamandridae). *J. Zool. Lond.* **214:** 533–545.

Verrell P & Pelton J (1996) The sexual strategy of the central long-toed salamander, *Ambystoma macrodactylum columbianum*, in southeastern Washington. *J. Zool. Lond.* **240:** 37–50.

Verrell PA & Sever DM (1988) The cloaca and spermatheca of the female smooth newt, *Triturus vulgaris* L (Amphibia: Urodela: Salamandridae). *Acta Zool. (Stockh.)* **69:** 65–70.

Verrell PA, Halliday TR & Griffiths ML (1986) The annual reproductive cycle of the smooth newt (*Triturus vulgaris*) in England. *J. Zool. Lond. (A)* **210:** 101–119.

Waights V (1996) Female sexual interference in the smooth newt *Triturus vulgaris vulgaris*. *Ethology* **102:** 736–747.

Wake MH (1968) Evolutionary morphology of the caecilian urogenital system. I. The gonads and the fat bodies. *J. Morphol.* **126:** 291–332.

Wake MH (1970a) Evolutionary morphology of the caecilian urogenital system. II. The kidneys and urogenital ducts. *Acta Anat.* **75:** 321–358.

Wake MH (1970b) Evolutionary morphology of the caecilian urogenital system. III. The bladder. *Herpetologica* **26:** 120–128.

Wake MH (1972) Evolutionary morphology of the caecilian urogenital system. IV. The cloaca. *J. Morphol.* **136:** 353–356.

Wake MH (1977) The reproductive biology of caecilians. In *Reproductive Biology of Amphibians*. DH Taylor & SI Guttman (eds), pp. 73–101. Plenum, New York.

Wake MH (1978) The reproductive biology of *Eleutherodactylus jasperi* (Amphibia, Anura, Leptodactylidae), with comments on the evolution of live-bearing systems. *J. Herpetol.* **12:** 121–133.

Wake MH (1980) The reproductive biology of *Nectophrynoides malcolmi* (Amphibia, Bufonidae), with comments on the evolution of reproductive modes in the genus *Nectophrynoides*. *Copeia* **1980:** 193–209.

Wake MH (1981) Structure and function of the male Mullerian gland in caecilians, with comments on its evolutionary significance. *J. Herpetol.* **15:** 17–22.

Wake MH (1982) Diversity within a framework of constraints. Amphibian reproductive modes. In *Environmental Adaptation and Evolution*. D Mossakowski & G Roth (eds), pp. 87–106. Gustav Fischer, Stuttgart.

Wake MH (1992) Reproduction in caecilians. In *Reproductive Biology of South American Vertebrates*. WC Hamlett (ed.), pp. 112–120. Springer-Verlag, New York.

Wake MH (1993) Evolution of oviductal gestation in amphibians. *J. Exp. Zool.* **266:** 394–413.

Wells KD (1977) The social behaviour of anuran amphibians. *Anim. Behav.* **15:** 666–693.

Wernz JG (1969) Spring mating in *Ascaphus*. *J. Herpetol.* **3:** 167–169.

Zug GR (1993) *Herpetology*. Academic Press, San Diego.

Zuiderwijk A & Sparreboom M (1986) Territorial behaviour in crested newt *Triturus cristatus* and marbled newt *T. marmoratus* (Amphibia, Urodela). *Bijdr. Dierk.* **56:** 205–213.

13 Sexual Selection and Sperm Competition in Reptiles

M. Olsson and T. Madsen

The University of Sydney, School of Biological Sciences, Zoology Building AO8, N.S.W. 2006, Australia

I. INTRODUCTION

Our ambition with this chapter is (1), to review traits which determine male reproductive success in reptiles, and ongoing sexual selection on such traits arising from variance in mating success and (2), reflect on, for which of these traits and selective events that reptiles ought to serve as particularly suitable research models. Because of space constraints all aspects of sexual selection in this group cannot be covered in detail. However, we have tried also to provide key references in areas that we treat more superficially. Our imaginary axis through this chapter is the chronology of events during, and sometimes before, a reptile's breeding period. In Section II, we exemplify what traits that favour male access to females, such as male control of limited resources, factors that determine the outcome of contests between males over sexual partners, and how females may influence the outcome of male reproductive success by mate choice. In Section III we take a critical look at what factors that may influence testis size, whether testis size may be used as an index of capacity of sperm production, and to what extent testis size seems to vary depending on limiting resources, pathogens and, on a larger scale, phylogeny and climate. We also briefly exemplify how the morphology of

Sperm Competition and Sexual Selection
ISBN 0-12-100543-7

copulatory organs, time in coitus, and testis size may covary within, and differ between, reptilian groups. Of major importance for a male's probability of paternity is also how well his spermatozoa survive during prolonged storage in the female reproductive tract, and to what extent his seminal products and the chemical compounds produced by the female influences the motility of his spermatozoa within the female reproductive tract. These aspects form an interface to Section IV, which we begin by reviewing to what extent females mate multiply in different species of reptiles. We then go on to describe evidence of sperm competition, such as within-clutch multiple paternity, and determinants of probability of paternity in the few reptilian species for which data are available. Within this context, we also briefly address female benefits of multiple matings in two reptilian species (see Chapters 2 and 3).

We are deliberately speculative in some areas where data are few, but we hope that this approach will generate empirical tests of ideas that may emerge.

II. MATE ACQUISITION AND SEXUAL SELECTION

'Competition for mates is the defining aspect of all forms of sexual selection' (Andersson 1994, p. 12), and may be initiated by, for example, scrambles, contests, or competition based on endurance and is then, sometimes, continued by mate choice (Andersson 1994; Arnold and Duvall 1994; Møller 1994a; Andersson and Iwasa 1996). A male's success in any of these processes is determined by a plethora of factors of which we attempt to identify the most important.

Continued growth throughout life is characteristic for reptiles, although in some species it may become considerably retarded or completely arrested at maturation (Andrews 1982). When traits change with adult age, analyses of ongoing selection on these traits are confounded unless, for example, cohorts are analysed separately, or regression coefficients from individual-specific growth-equations of the traits are submitted to the statistical analyses (Lande 1982; Lande and Arnold 1983; Arnold and Wade 1984a,b). Ideal reptilian species for such selection analyses would be annual lizards such as *Urosaurus ornatus*, where less than 20% of males survive to breed in a second year (Thompson *et al.* 1993). Because virtually all males belong to the same cohort, the analyses are minimally confounded by age effects, and a single season of field work would return life-time reproductive success.

Body size is most often considered in studies of variance in reptilian male mating success, but age is not, probably because of the often considerable problems associated with age estimation (Halliday and Verrell 1988). Thus, although age-specific body size may vary considerably between individuals, focusing on studies which 'control' for body size is probably our best insurance that 'independent' determinants of mating success are not simply correlates of body size or age. However, we sometimes compromise this criteria in order to include valuable information from studies without explicit body size data.

Male–male contests for females occur in every reptilian order (Carpenter and Ferguson 1977). Since larger males win contests more often than smaller ones, it is not surprising that body size is one of the primary determinants of male mating success, which has been demonstrated in a number of species (Table 13.1). With the effects of body size removed, what other factors determine a reptilian male's mating success in processes of mate acquisition? We suggest (1) territory/home range attributes, (2) 'contest attributes', (3) alternative mating tactics, (4) some 'other factors' (Table 13.2), and (5) female choice (see Table 13.4).

Table 13.1. Species in which body size determines male success in mate acquisition in natural populations.

Species	Authority
Lizards	
Amblyrhynchus cristatus	Trillmich (1983), Rauch (1985), Wikelski *et al.* (1997)
Iguana iguana	Dugan (1982), Rodda (1992)
Conolophus subcristatus	Werner (1982)
Sauromalus obesus	Berry (1974), Ryan (1982)
Anolis garmani	Trivers (1976)
Anolis valencienni	Hicks and Trivers (1983)
Anolis carolinensis	Ruby (1984)
Sceloporus jarrovi	Ruby (1981)
Sceloporus virgatus	Smith (1985)
Uta palmeri	Hews (1990)
Ctenophorus maculosus	Mitchell (1973), Olsson (1995b)
Chamaeleo chamaeleon	Cuadro and Loman (1997)
Eumeces laticeps	Vitt and Cooper (1985a)
Podarcis muralis	Edsman (1989)
Lacerta agilis	Olsson (1992b)
Psammodromus algirus	Díaz (1993)
Ameiva plei	Censky (1995)
Snakes	
Vipera berus	Andrén (1986), Madsen *et al.* (1993), Madsen and Shine (1994a), Luiselli (1993)
Natrix natrix	Madsen and Shine (1993b)
Nerodia sipedon	Weatherhead *et al.* (1995)
Turtles	
Clemmys insculpta	Kaufmann (1992)
Crocodilians	
Crocodylus niloticus	Kofron (1990, 1993)

Table 13.2. Male traits under sexual selection.

Species	Examined trait	Authority
Territory quality		
Uta palmeri	Contested food resources	Hews (1990, 1993)
Ctenophorus maculosus	Territory size	Olsson (1995b)
Amblyrhynchus cristatus	Sites for thermoregulation	Trillmich (1983), Rauch (1985)
Contest attributes		
Morphology		
Euemeces laticeps	Head size	Cooper and Vitt (1985)
Uta palmer	Head size	Hews (1990)
Tiliqua rugosa	Head size	Bull and Pamula (unpublished data)
Lacerta agilis	Head size	Olsson (1992b)
Physiology		
Anolis carolinensis	Testosterone (aggression)	Tokarz (1985)
Sceloporus jarrovi	Testosterone (aggression)	Marler and Moore (1989, 1990)
Podarcis muralis	Thermoregulation (body temperature)	Edsman (1989)
Urosaurus ornatus	Innate aggressiveness	Hover (1985), Thompson and Moore (1991)
Behavioral asymmetries		
Anolis carolinensis	Residency, presence of partner	McMann (1993), Leuck (1995)
Lacerta agilis	Presence of mated partner	Olsson (1993a)
Cues to male fighting ability (status signals)		
Urosaurus ornatus	Dewlap coloration	Hover (1985), Thompson and Moore (1991), Thompson *et al.* (1993), Carpenter (1995a,b)
	Dorsal coloration	Zucker (1994)
Psammodromus algirus	Head coloration	Díaz (1993)
Lacerta agilis	Body coloration	Olsson (1994a)
Agama agama	Body coloration (labile, dominance related)	Harris (1964), Inoué and Inoué (1977), Madsen and Loman (1987)
Anolis garmani	Body coloration (labile, dominance related)	Trivers (1976)
Alternative mating tactics		
Eumeces laticeps	'Sneaking' (deferred agonistic encounters)	Cooper and Vitt (1987)
Lacerta agilis	'Sneaking' (deferred agonistic encounters)	Olsson (1994a–c)
Vipera berus	'Sneaking' (deferred agonistic encounters)	Madsen *et al.* (1993)
Podarcis muralis	Nomadism	Edsman (1989)

Table 13.2. Continued

Species	Examined trait	Authority
Conolophus subcristatus	Nomadism	Werner (1982)
Sauromalus obesus	Nomadism	Berry (1974), Ryan (1982)
Iguana iguana	Female mimicry, forced copulations	Rodda (1992)
Amlyrhynchus cristatus	Female mimicry, forced copulations	Wikelski *et al.* (unpublished data)
Ctenophorus maculosus	Forced copulations	Olsson (1995a)
Sauria (review)	Forced copulations	Rodda (1992)
Kinosternidae (turtles)	Forced copulations (bottom-walking sp.)	Berry and Shine (1980)
Other factors		
Iguana iguana	Display rate	Dugan (1982)
Sceloporus jarrovi	Activity	Ruby (1981)
Anolis carolinensis	Activity	Ruby (1984)
Scelporus virgatus	Activity	Smith (1985)
Vipera berus	Scent-trailing	Madsen *et al.* (1983)
Crotalus v. viridis	Search strategies, scent-trailing	Duvall and Schuett (unpublished data)

References are provided in the tables and appendices and occasionally in the text when the context so requires.

A. Territory/home range attributes

A male's territory could provide resources for females and, hence, the male may trade his resources for a high probability of paternity. Most territorial males in such species are resource defence polygynous (*sensu* Emlen and Oring 1977). This appears to be the case in *Uta palmeri*, in which females are more numerous on male territories containing relatively more seabird nests, around which invertebrates accumulate. That females indeed distribute themselves in relation to territory quality rather than male traits was experimentally confirmed (Hews 1993). In reptiles, and perhaps other ectotherms, territories which allow efficient thermoregulation may be limited and in *Amblyrhynchus cristatus*, males possessing such territories (large males) may also achieve more matings (Trillmich 1983).

Territory size alone could also determine male mating success, independent of any resources that the territory contains (i.e. 'Super-territoriality', *sensu* Stamps 1983). The lizard *Ctenophorus maculosus* inhabits dry salt pans in the Australian interior. Its diet consists of wind-blown insects and, hence, there are no predictable resources to defend. However, greater relative abundance of look-out sites (that were not used by females) allowed males to enlarge their territories and incorporate more partners.

The functional significance of male territoriality has been analysed by removing the female(s) sharing a male's territory. In the annual *Urosaurus ornatus* removal of females led to reduced male territorial defence. In the longer-lived *Sceloporus graciosus*, males kept defending territories in the absence of females, which suggests that there may be long-term benefits of territoriality to these males (M'Closkey *et al.* 1987; Deslippe and M'Closkey 1991). In the nonterritorial *Lacerta agilis*, home range size did not predict male mating success (Olsson 1992b).

B. 'Contest attributes' (which directly influence the outcome of escalated interactions)

Game theory, and perhaps in particular the sequential assessment game (Parker 1974; Enquist and Leimar 1983, 1987), is an attractive theoretical framework for making testable predictions about how animal conflicts will progress and what traits will determine how conflicts are resolved. The models posit that each behaviour in a repertoire (or trait contributing to fighting ability) has a corresponding asymmetry that can be estimated with some uncertainty, and that the asymmetries of these behaviours are intercorrelated. A straightforward prediction is, for example, that contests between similarly sized animals will be prolonged, since determining the asymmetry in fighting ability requires additional sampling by the contestants. This prediction, and several others, from the sequential assessment game was confirmed in *Lacerta agilis* (Olsson 1992a).

1. Morphology

Many lizard males use their jaws in contests over females and for holding on to the female during copulation (e.g. Stamps 1983) – male *Eumeces laticeps* even increase their head (jaw) size during the mating season. Large heads may also compensate for small body size; in *Tiliqua rugosa* small males still paired with females if they had larger than average heads, and similar results have been demonstrated for other species (Table 13.2). The male-biased dimorphic heads and jaws can be powerful weapons and male contests often result in injuries in, for example, *Anguis fragilis*, *Clammydosaurus kingii* and *Lacerta agilis* (Smith 1951; Shine 1990a; Olsson 1994a), and even death in *Eumeces laticeps*, *Eumeces okade* and *Varanus panoptes* (Vitt and Cooper 1985a; Hasegawa 1994, pers. comm.; T. Madsen, pers. obs.).

Male combat also occurs in snakes, and in these species large body size is favoured by selection (Shine 1994, and references therein). However, in some snake species males do not fight but aggregate around receptive females in 'mating balls', a situation in which long tails have been suggested to contribute to male mating success (Semlitsch and Gibbons 1982). However, there is still little firm evidence of this (King 1989; Madsen and Shine 1993a; Weatherhead *et al.* 1995).

2. Physiology

Several aspects of a male's physiology could be under sexual selection, such as plasma concentrations of hormones, which may determine level of aggression, and metabolic efficiency, which may limit male endurance and mate acquisition during prolonged breeding periods. At least two hormones have been identified which, in an antagonistic manner, determine male aggression and territory size and, hence, may have direct bearings on male mating success. In *Anolis carolinensis*, males with experimentally elevated testosterone levels acquired better territories (Tokarz 1985), and male *Sceloporus jarrovi* which received testosterone implants defended territories more actively, but suffered higher mortality due to energetic costs than control males (Marler and Moore 1988, 1989, 1991). In *Anolis carolinensis* (Greenberg *et al.* 1984), *Anolis sagrei* (Tokarz 1987a) and *Uta stansburiana* (DeNardo and Licht 1993; DeNardo and Sinervo 1994) corticosteroids have been demonstrated to counter the effects of androgens and may, hence, reduce aggression and territorial defence. Thus, males with above average androgen levels may increase their seasonal mating success, but not necessarily life-time reproductive success and, hence, sexual selection would be expected to drive the levels of corticosteroid and androgens (or the cells' responses to these hormones) towards optimal levels (see also Orr and Mann 1992).

In *Amblyrhynchus cristatus*, *Podarcis muralis* and *Liasis fuscus*, the reduction in male body condition during the mating season may be substantial (Trillmich 1983; Edsman 1989: Madsen and Shine, unpublished data), which may restrict mate acquisition and select for male endurance. A test of how endurance (running time on a treadmill) relates to outcome in contests rejected such a relationship in *Sceloporus occidentalis* (Garland *et al.* 1990); however, in the same study a male's maximum running speed was positively correlated with his success in contests, although the underlying mechanism for this relationship (e.g. muscle mass) is unclear.

An important modifier of male contest behaviour, perhaps unique to ectotherms, is body temperature. In *Podarcis muralis*, males fight over females and smaller males increased their possibility of winning contests over larger males by thermoregulating to a higher temperature (Edsman 1989). This important aspect of reptilian contest behaviour should be considered in experimental studies of contest traits since access to heat sources may cause asymmetries that can override and confound the effects of the traits which the study was intended to investigate.

3. Behavioural asymmetries

Factors such as residency, ownership, previous experience of mating (Barlow *et al.* 1986) and fighting (Jackson 1991) may determine the outcome of contests for resources such as females, in particular when males are closely matched for body size. In *Anolis carolinensis*, prior residency and presence of a female predicted the winner of contests (McMann 1993; Leuck 1995) and in *Lacerta agilis*, mated males which fought over their partners won all interactions with intruding males of approximately equal size (Olsson 1993a).

The importance of behavioural asymmetries have typically been evaluated by manipulating size-matched males in the laboratory (McMann 1993; Olsson 1994a; Leuck 1995). To what degree do these experiments reflect selection events in the wild? Contests between males with no measurable difference in body size in natural populations have been observed in *Urosaurus ornatus* (29% of the contests; Hover 1985), *Vipera berus* (49% of the contests; Madsen 1987) and *Lacerta agilis* (38% of the contests; Olsson 1992a), and hence seem to make up a significant part of contests for females, as predicted by game theory models (Parker 1974a; Maynard Smith 1982; Enquist and Leimar 1983, 1987).

4. Cues to male fighting ability ('status signals', *sensu* Rohwer 1975, 1982)

In some populations of *Urosaurus ornatus*, males occur predominantly in two different colour morphs and, in enclosure experiments, 'orange-blue' dominated 'orange' size-matched males, and contests never escalated between two orange males (Hover 1985; colour definitions following Thompson and Moore's 1991). By age 5 months (indoors), males developed throat patches which did not change throughout life and males with experimentally elevated androgen levels during this period were more likely to become orange-blue and aggressive as adults (Thompson *et al.* 1993; Hews and Moore 1996). In adults, steroid levels did not differ between the orange-blue and orange morphs, suggesting that adult strategies are determined by androgens early in life (Thompson *et al.* 1993; Hews *et al.* 1994; see also Stamps 1994). However, in other studies of *U. ornatus* throat patch variation exhibited a strong ontogenetic component: in some males throat patches faded when they were socially dominated (Carpenter 1995a,b), and in long-term enclosure experiments, throat patch colour did not predict dominance status, but dorsal darkening did (Zucker 1994). In the wild, there were also strong differences between populations in male mating tactics. Larger males established territories and had greater mating success in one population (M'Closkey *et al.*

1990), while only smaller (but more aggressive) males were territorial in a different population (Thompson and Moore 1991). At present there is insufficient evidence to explain these interpopulation differences in morphology and behaviour, and how these are linked to habitat-related opportunity for polygyny. Considering that these lizards are predominantly annual, a synchronized effort to obtain life-time reproductive success data (available in a single year) from disparate populations could yield some exciting results with the potential of demonstrating ongoing evolution and interpopulation divergence of the relevant male traits.

In a recent study of the lizard *Uta stansburiana*, Sinervo and Lively (1996; see also Maynard Smith 1996) demonstrated that colour polymorphism may be linked to three male mating strategies: 'sneaking', defence of a small territory with one female, or a large territory with many females. None of these strategies were evolutionary stable. Instead, the frequencies of the three morphs cycled. The fitness of the most aggressive morph (orange) declined when the frequency of 'sneakers' (yellow) increased, whose fitness in turn declined when the frequency of 'single-female' territorial males (blue) increased. This system is the first example of the 'Rock–Paper–Scissors' game in a natural population (Maynard Smith 1996).

In other species male coloration is labile and colours remain stable only during the breeding period (Rand 1988, 1990, 1992), or an episode of unaltered dominance status. In *Eumeces laticeps*, females painted like males elicited agonistic behaviours in males, and thus head coloration may be used in male–male communication (Cooper and Vitt 1988). In some species, male head and body coloration have been directly linked with male mating success (Table 13.2). Cues to fighting ability could perhaps also be moulded by sensory biases, and an example of this could be the lateral green breeding coloration in male *Lacerta agilis;* the wavelengths of the coloration correspond with the range in which the lizard is most perceptive (Swiezawska 1949).

Contests over mates may also occur in females, which was suggested by female–female aggression during the mating season in *Conolophus subcristatus* (Werner 1982) and *Carlia rostralis* (Whittier 1993; Whittier and Martin 1992). In *Carlia rostralis*, females have even evolved head-bobbing behaviour for consexual agonistic signalling (Whittier 1993). If females fought over food, aggression should be strongest when most resources for the eggs are accumulated – i.e. usually outside the mating season. Instead, Dugan and Wiewandt (1982) suggested that sperm could be a source of intrafemale conflict; in *Cyclura stejnegeri*, a male mated 11 times in 13 days with a concomitant reduction in duration of copula, possibly explained by declining sperm supplies. However, sources of conflict other than sperm limitation, such as oviposition sites, are possible during the mating season. In green sea turtles (*Chelonia mydas*), with a much larger clutch size than lizards, sperm depletion may explain why accumulated time in coitus is positively correlated with hatching success (Wood and Wood 1980).

C. Alternative mating tactics

When there is considerable variation in adult male body size, small males may (rarely) acquire matings through combat and alternative mating tactics may evolve. In such cases, young males may defer agonistic encounters over females until they become older and larger, which in some species coincides with male nomadism in early adult life (Table 13.2). In other species predominantly younger, smaller males copulate with females by force, sometimes preceded by mimicking female behaviours which decreases the attention of larger, dominant males. In *Thamnophis sirtalis*, male mimicry of receptive females is pheromonal, the female-mimicking by 'she-males' results in attraction of competitors (Mason and Crews 1985), but to what extent this increases 'she-male's' mating success is unknown.

D. Other male traits under sexual selection

In several species male activity, defined by number or proportion of active days, have been suggested to increase a male's mating success (Table 13.2). However, such activity indices may also reflect dominance relationships and be poor estimates of innate activity levels.

Many reptiles rely primarily on odour cues such as pheromones for mate location (Cooper and Vitt 1984, 1986; Mason 1992), and, hence, such traits could be under sexual selection. This is indeed the case, for example, in some snakes (Table 13.2).

E. Female choice

1. Species and sex recognition

One reason for female choice of mate(s) is that females usually make a larger parental investment, compared with males and therefore should benefit more from mating selectively with respect to a partner's available resources, genetic complementarity and viability (Darwin 1871; Parker 1983; Bradbury and Andersson 1987; Andersson 1994). Mate choice could arise within a lineage, for example, owing to selection against interspecific hybridization, i.e. targeting a partner's species identity. Heterospecific mating in the wild has not been reported, but hybridization between two lizard species, at least, suggest that it does occur occasionally (Arnold and Hodges 1995; see also Arnold 1986, and references therein). In experimental situations, female *Anolis carolinensis* preferred males with the conspecific red dewlaps rather than ones with dewlaps painted blue (Sigmund 1983; but see Cooper and Greenberg 1992; Tokarz 1995). Males of *Anolis marconi* become less aggressive towards conspecific males when these are painted as males of the sibling species *Anolis cybotes* (Losos 1985); thus, both sexes of two *Anolis* species use the dewlaps for species identification. In *Anolis nebulosus*, females discriminated against males with unfamiliar display sequences, replayed to them on film (Jenssen 1970), and character displacement in, for example, dewlap coloration may have evolved in some anoles (Rand and Williams 1970; Preston-Webster and Burns 1973).

Species and sex recognition may be facilitated by odours (Noble 1937; Cooper and Vitt 1984, 1986; Mason 1992) and colours (Losos 1985; Cooper and Vitt 1988), perhaps also in the ultraviolet range (Fleishmann *et al.* 1993). However, male homosexual matings and mating attempts have occasionally been recorded in natural populations of *Anolis garmani* (Trivers 1976), *Iguana iguana* (Rodda 1992), and *Amblyrhynchus cristatus* (Wikelski *et al.* 1996).

2. Female breeding coloration

Females of some lizard species develop extremely bright colours associated with the development of aggressive behaviours during the breeding season. This could, superficially, indicate sex-role reversal and these colours have received considerable interest (Table 13.3). Early reports showed that the coloration changes from 'dull' early in the ovarian cycle to 'bright' later in the cycle (Table 13.3). The same hormone, progesterone, strongly induces 'bright' colours (Cooper and Ferguson 1972b; Cooper and Crews 1987) and female rejection behaviour (Cooper and Crews 1987), and is generally associated with the post-ovulatory luteal phase in reptiles (Licht 1984; Whittier and Tokarz 1992).

The exact timing of 'brightness' and ovulation has been investigated because this information was considered important for inferring the function and evolution of female 'brightness'. However, female reptiles usually mate some time before ovulation and it is unknown when the receptivity period ends in relation to ovulation. What appears to be consistent between species is that the bright colours are stable once they become developed and that females reject males during most of the time that these colours are present (but see *Tropidurus delanonis*, Table 13.3).

Several adaptive hypotheses have been suggested for explaining the function and evolution of 'bright' female colours, of which the following two have received most attention:

(1) *Sex recognition*. Undoubtedly, males identify 'bright' females – as females – and in some species males painted as 'bright' females are courted by males (Vinegar 1972; Cooper 1984). However, males also copulate with 'dull' receptive females. Because males recognize females both when they are 'dull' and

Table 13.3. Coincidence of reproductive behaviour and female breeding coloration. B, A = denotes development of colours 'before' and 'after' a given reproductive event. If peak coloration coincides with mate rejection this is denoted (+), and (−) when it does not. Included also are studies that deal with the hormonal induction of female coloration.

Species	Coincidence with		Mate	
	Ovulation	Copulation	rejection	Authority
Tropidurus delanonis	B	B	–	Werner (1978)
Urosaurus ornatus	B (facultative)	?	?	Zucker and Boecklen (1990)
Sceloporus virgatus	A[a]	?	+	Vinegar (1972)
Crotaphytis collaris	A[a], B[b]	A[d]	+	Fitch (1956)[a], Mayhew (1968)[d], Cooper and Ferguson (1972a,b, 1973), Ferguson (1976)
Cophosaurus texanus	A[a]	?	+	Clarke (1965)
Holbrookia propinqua	A[a], B	A	+	Clarke (1965), Cooper (1984, 1986), Cooper and Clarke (1982), Cooper *et al.* (1983)
Holbrookia maculata	A[a]	?	+	Clarke (1965)
Gambelia wislizenii	A[a], B	A[c], A, B[d]	+	Turner *et al.* (1969), Medica *et al.* (1973), Brattstrom (1968) and Fitch (1968) cited in Milstead (1968[c], p. 75)
Callisaurus draconoides	A[a]	?	+	Clarke (1965)
Ctenophorus maculosus	A	A	+	Mitchell (1973), Olsson (1995a)
Uta palmeri	A	A	+	D. Hews, personal communication

[a] Coloration only stated as that of 'gravid female'.
[b] See body of text.
[c] [d] Reported in synonymously marked reference.

'bright', there appears to be no fitness benefit to the female from developing 'brightness' under a sex recognition hypothesis. Cooper (1984) rightly suggested that sex recognition cannot be the (only) agent selecting for female 'brightness'.

(2) *Mate rejection.* An unreceptive female conveying to males that she carries some other male's sperm could reduce costs associated with male associations (e.g. costs of predation, matings with nonpreferred males) by directing males to receptive females. The use of the bright colours are consistent with the mate rejection hypothesis in *Holbrookia maculata* and *Holbrookia propinqua*, in which the female turns up her tail and directly

> exposes the red coloration to pursuing males (Clarke 1965; Carpenter, p. 75, in Milstead 1968; Cooper 1984). In *Ctenophorus maculosus* the female even flips over onto her back thereby exposing her red belly patch and such flip-over behaviour has recently been observed in another agamid species, *Phrynocephalus persicus* (D. Modry, pers. comm.). Some support for reduced male pursuit of 'bright' females has been obtained (Clarke 1965; Cooper 1986). Furthermore, even if males attempt matings with 'bright', most often unreceptive, females (Mayhew 1968; Mitchell 1973; Olsson 1995a), males should prefer to mate with 'dull', receptive, females. It is consistent with this scenario that predominantly young males, inferior in competition with older, larger males, attempt copulations with 'bright' females (Fitch 1956).

In *Ctenophorus maculosus*, a mate rejection event is not significantly shorter than a mating by a receptive female (Olsson 1995a). However, and in conclusion, the selective advantage of 'brightness' may lie in the number or proportion of unwanted male associations that are successfully avoided and, hence, quantitative field data on mate rejections are needed for analyses of the female fitness benefit of 'bright' colours. The most parsimonious explanation does not support 'bright' colours as a response to sexual selection, i.e. used by females to deter some, but not all, males.

3. Female choice on male quantitative traits

For reptiles, Darwin's theory of sexual selection by female choice was met with scepticism by early herpetologists: 'Some naturalists would have us believe that . . . (the brilliance of colours) . . . has been evolved by natural selection, the female selecting the male as her mate who has the most brilliant display at his command. . . . This beautiful theory is pure nonsense in the case of lizards' (Wall 1922). What evidence of female mate choice in reptiles have accumulated since then? In fact, there is only one reptilian species, *Eumeces laticeps*, for which female choice has been experimentally demonstrated (Table 13.4); females preferred to mate with large rather than small males in staged mating experiments. Female spatial organization during the mating season also appears to be influenced by male body size in some species; in some other species females have been observed to reject copulation attempts by small males (Table 13.4).

Other observations in natural populations may suggest ongoing selection due to female choice but, because of the potentially confounding effects of territory quality, it is difficult to distinguish between female preferences for male and for territory traits. In some species the latter seems to be the case (Table 13.4). However, in the Iguania group, female

Table 13.4. Female reproductive behaviours in reptiles.

Species	Trait(s) examined	Authority
Confirmed female choice on a male quantitative trait		
Eumeces laticeps	Body size	Cooper and Vitt (1993)
Amblyrhynchus cristatus	Body size	Rauch 1985, Wikelski *et al.* (1996)
Rejection of some males within a female's receptive period		
Podarcis muralis	Body size (small), nonresident	Edsman (1989)
Iguana iguana	Body size (small), nonresident	Dugan (1982), Rodda (1992)
Confirmed resource defence polygyny		
Amblyrhynchus cristatus	Thermal properties	Trillmich (1983), Rauch (1985)
Uta palmeri	Food abundance	Hews (1990, 1993)
Female choice on male quantitative trait inconsistent or refuted		
Anolis carolinensis	Body size	Andrews (1985)
Uta palmeri	Morphology (several traits)	Hews (1990, 1993)
Lacerta agilis	Coloration, body size	Olsson and Madsen (1995)
Amblyrhynchus cristatus (high density)	Body size	Trillmich (1983)
Gopherus agassizii	Body size	Niblick *et al.* (1994)
Suggested visitation/sampling of several males and their resources		
Sauromalus obesus	?	Berry (1974)
Iguana iguana	Body size, display rate	Dugan (1982)
Conolophus stejnegeri	Territory quality	Wiewandt (1977)
Conolophus subcristatus	Territory quality (food)	Werner (1982)
Agama agama	?	Harris (1964)
Clemmys insculpta	Body size	Kaufmann (1992)
Kinosternid turtles	?	Berry and Shine (1980)
Suggested female solicitation of matings (no preference of a male trait identified unless stated)		
Sceloporus magister	?	D. Hews, personal communication
Anolis lineatopus	?	Rand (1967)
Anolis garmani	?	Trivers (1976)
Tropidurus delanonis	?	Werner (1978)
Iguana iguana	Body size (?)	Rodda (1992)
Amblyrhynchus cristatus	Body size	Wikelski *et al.* (1996)
Crocodilia	?	Lang (1987)
Clemmys insculpta	Body size	Kaufmann (1992)
Chelonia mydas	?	Crowell *et al.* (1990)

solicitation of matings in the wild have been observed in several species (Table 13.4), and in *Iguana iguana* females have larger home ranges than males (Dugan 1982), which is consistent with female visitation of several males. In *Amblyrhynchus cristatus*, female choice may be facultative and depend on population density. At high density, females clustered on the male territories with the best sites for thermoregulation, and males controlling these areas obtained most matings (Trillmich 1983; Rauch 1985). At lower density, and with equal basking opportunities on male territories, male size, condition and territory orography independently explained significant variance in female spatial distribution and concomitant male mating success (Wikelski *et al.* 1996).

In *Tiliqua rugosa* males and females remain in pairs (i.e. virtually side by side) for up to 8 weeks during the mating season and sometimes also between years (Bull 1988), but to what extent pairing is preceded by mate choice is not known. Such 'pairing' occurs in some sense also in *Anolis aeneus*, in which the number of days that partners' home ranges overlapped determined how quickly the females accepted copulation (J. A. Stamps, pers. comm.). However, the selective benefit to the female of 'partner familiarity' remains unknown.

Several researchers have looked for consistent female choice without finding any evidence for it, (e.g. *Anolis carolinensis*; Andrews 1985). Trivers (1976) concluded that female choice influenced male mating success in *Anolis garmani*, and that females may openly perform displays which convey that they are receptive. However, it is unclear if females displayed receptivity to all males, since, when given the opportunity, females sometimes also mated with small, nonterritorial males. Two experimental approaches have been adopted to distinguish between female choice of territorial males or the resources that the males control. In *Uta palmeri* females redistributed themselves around supplied food resources during the mating season, and consequently compromised any previously made mate choices (Hews 1993). M'Closkey *et al.* (1987) removed male *Urosaurus ornatus* but could not detect a concomitant redistribution of females. Thus, in this species also resources other than partners seemed to determine female spatial distribution.

In *Lacerta agilis*, long-lived males sired offspring with higher embryonic survival, which thus satisfies a fundamental premise of a 'good genes' model of female choice, but females did not prefer older or larger males (Olsson and Madsen 1995). As in most adult reptiles, age and size are poorly correlated in adult *L. agilis*, size would therefore be a poor indicator of male longevity (Halliday and Verrell 1988; Olsson and Shine 1996), which may explain why females mate indiscriminately (see also Hansen and Price 1995).

In other reptilian taxa, data on female choice are very limited and, at best, suggestive of such behaviours, e.g. in the snake *Agkistrodon contortrix* (Schuett and Duvall 1996). Owing to the formidable difficulties involved in detailed field work on most crocodilians, little is known about mate acquisition and ongoing sexual selection in these species, the only reptiles which are known to show paternal care (e.g. *Crocodylus palustris*; J. Lang unpublished data, reviewed by Webb and Manolis 1989). However, some crocodiles aggregate on what resemble leks (W. Magnusson, pers. comm.), and male *Crocodylus niloticus* have two types of calls which have been suggested to be used towards males and females, respectively (Cott 1961). In *Alligator mississipiensis* and *Alligator sinensis*, roaring bouts may be initiated by females and used for individual identification (Garrick 1975), and perhaps discrimination among potential partners.

4. Parasite resistance and showy males (*sensu* Hamilton and Zuk 1982)

Using the assumption that pathogens reduce colour brightness, Hamilton and Zuk (1982) suggested that females should prefer brightly coloured males with few parasites or high parasite resistance, in particular in highly parasitized species. Thus, species with high parasite abundance should have brighter colours. Data on coloration and prevalence of a broad range of endoparasites in 26 species of lizards from 14 genera and eight families did not support the Hamilton and Zuk hypothesis (Lefcort and Blaustein 1991). In *Sceloporus occidentalis*, malaria had strong detrimental effects in both sexes, and surprisingly intensified ventral coloration in males (contrary to Hamilton and Zuk's hypothesis; Ressel and Schall 1989), but female choice has never been demonstrated in this species (Schall 1983a,b; Schall and Sarni 1987; Dunlap and Schall 1995). However, malaria infestation reduces male contest success and probably also reproductive success (Schall and Dearing 1987). Ectoparasite load was not correlated with male colour traits in *Lacerta agilis* and females did not exhibit any preference for male models with more nuptial coloration (Olsson and Madsen 1995).

5. 'Sensory drive'

When there are pre-existing biases in a female's (or sometimes male's) sensory systems that yield concomitant biases in perception, individuals with traits in the 'optimal' range may be favoured by selection (*sensu* e.g. West-Eberhard 1984; Endler and McLellan 1988: Ryan 1990; reviewed by Andersson 1994). Few reptile studies deal exclusively with sensory drive. However, head bobbing in *Anolis* have been suggested to have evolved in response to selection arising from sensory drive (Fleishman 1992; see also Martins 1993). The red colour to which female *Tropidurus delanonis* shift at receptivity elicit strong responses in males (and females), whereas blue and green, for example, do not (Werner 1978), perhaps suggesting opportunity for 'sensory drive' in female attraction of males.

6. Fluctuating asymmetries

Small random deviations from bilateral symmetry (usually less than 1%) may convey phenotypic (such as parasites) or genetic stressors on developmental homeostasis, which is thought to become increasingly hard to maintain with increasing costs of a trait's development. This would apply to extreme ornaments for sexual displays (van Valen 1962; Palmer and Strobeck 1986; Møller 1991, 1992; Palmer 1994), and thus fluctuating asymmetries (FA) could function as a 'health certificate' which may influence female choice (Møller 1992) and, perhaps, male fighting tactics. FA has been recorded in lizards (Soulé 1967; Fox 1975; Sarre and Dearn 1991), and may be easy to monitor on several meristic characters (e.g. scales) in reptiles, but to which extent FA plays a part in sexual selection in this taxon is not known.

7. Costs of sexually selected traits

We have already mentioned that increased aggression and territorial defence may carry costs in terms of loss of body condition (Trillmich 1983; Edsman 1989) and reduced survival (Marler and Moore 1989, 1990, 1991) and, hence, behavioural traits favoured by sexual selection may be countered by natural selection in accordance with sexual selection theory (Fisher 1930; Andersson 1982). In *Psammodromus algirus*, males with experimentally elevated testosterone levels, yielding more pronounced nuptial coloration, also subsequently suffered a greater parasite load than control males (Salvador *et al.* 1995).

Morphological traits such as skin colours would also be expected to increase predation, however, brightly coloured model male *Lacerta agilis* were not attacked more often by predators than camouflaged female models (Olsson 1993b).

E. Conclusions and gaps in knowledge – sexual selection

Fights over mates have been more widely recognized as an important selective process than mate choice in reptiles. Interestingly, the low incidence of female choice, and lekking recorded so far in this group coincides with a dominating absence of paternal care. This could suggest that when direct benefits from paternal care are lacking, female choice may be less likely to evolve. However, subtle female choice, such as solicitation of matings from preferred males and rejection of low-ranking males, is likely to have gone undetected in many field studies. The significance of sex and species recognition as selective agents could be further clarified by experimental studies in which 'oestrus' females were allowed to reject or accept males painted as homosexual and heterospecific partners. Female rejection behaviours are easy to identify in most lizards, and male colours are easily manipulated by painting (but note the potential confounding effects of wavelengths in for human nonvisible spectra, such as ultraviolet; Fleishman 1992; Bennett *et al.* 1996). These characteristics could be exploited, for example, for teasing apart the relative importance of heterospecific mate rejection and choice on conspecific male's quantitative traits for the evolution of interspecific mating barriers and risk (or chance) of hybridization. In summary, more experiments targeting intra- and inter-specific female choice are needed, particularly in snakes (a taxon in which females apparently cannot be copulated by force).

Individual recognition has been demonstrated in several reptilian species (Glinski and Krekorian 1985; Olsson 1994c). Thus, some of the extensive variation often found in sexually dimorphic traits may convey more information to conspecifics than mere sex or species identification, and perhaps serve as indicators of phenotypic and genotypic quality. When intraspecific variation in male coloration has been linked to mating success, strong effects have been established, whereas no such relationship has yet been reported for odours.

Because of space constraints, we have not reviewed the role of sexual selection in the evolution of sexual size dimorphism (SSD). SSD has been linked to (1) mating system variation (Stamps 1983; Cooper and Vitt 1989; Andersson and Vitt 1990), (2) male–male aggression and combat (Carothers 1984; Shine 1978, 1994), (3) ecological determinants (Schoener 1967; Shine 1989, 1990b, 1991; Preest 1994), (4) sex-specific growth rates and age at maturation (Stamps 1993), (5) constraints on body size set by sex-specific reproductive costs (Madsen and Shine 1994b), (6) organizational effects of hormones (Hews and Moore 1995; Shine and Crews 1988), and (7) temperature effects during embryogenesis (Shine *et al.* 1995). Considering the wide range of nongenetic factors that may determine SSD in this taxon, reptiles should make challenging evolutionary models for future SSD research.

The intensity of sexual selection may covary with operational sex ratio in *Vipera berus* and directional sexual selection may rapidly shift a population's mean body size (Madsen and Shine 1992, 1993a). However, our knowledge is limited about how the intensity of sexual selection may vary between breeding periods and how such variation influences long-term changes in dimorphic traits. In *Lacerta agilis* (and perhaps most reptiles), correlations between morphological traits within cohorts become weaker with age (Olsson, unpublished data). Males in young cohorts have greater similarity in head size–body size relationships than males in older cohorts. This implies that shifts in age-distribution may cause between-year variation in (1) the variance in e.g. male body size, (2) the relative influence of morphological traits and behavioural asymmetries, such as 'ownership', on mating success, and (3) the degree to which multi-colinearity poses a problem in analyses of selection coefficients.

III. PRODUCTION, TRANSFER AND STORAGE OF SPERMATOZOA AND SEMEN

A. Determinants of testis size

The intensity of sperm competition may be predicted to covary with male primary sex traits such as testis size, ability to store sperm in the epididymis, and to transfer spermatozoa in the seminal plasma produced by the renal sex segment (RSS; Bishop 1959), the epididymis (Prasad and Reddy 1972; Fox 1977), and the ampulla of the ductus deferens (Akbarsha and Meeran 1995). There is considerable variation in testis size between species and populations (e.g. more than a fivefold difference between species in the lizard genus *Sceleporus*, Appendix B; see also Licht 1984), and larger conspecific adults have larger testis (e.g. Abts 1988; Mitchell and Zug 1984), and larger seminiferous tubules (Bauman and Metter 1977; Brackin 1978).

What factors may confound peak relative testis size (in relation to body mass) as an estimate of variation in rate of sperm production? Much of the variation in testis size (and associated structures) is hormonal (Licht 1984; Moore and Lindzey 1992), and antiandrogens may reduce the mass of the testis and the RSS (Tokarz 1987b). To what extent testis size covary with the degree of hypertrophy of androgen-secreting tissues such as interstitial cells is not entirely clear (Fox 1977; Licht 1984). However, in *Psammodromus algirus* (Díaz *et al.* 1994) and *Niveoscincus metallicus* (Swain and Jones 1994) testosterone peaks while the testes are still increasing in size. Furthermore, in male reptiles the production, and the transfer of spermatozoa, from the testis to the epididymis and from the epididymis to the female, may be separated by a shorter or longer period of sperm storage (i.e. 'associate' and 'dissociate' breeders; *sensu* Moore and Lindzey 1992; see also Licht 1984). In dissociate breeders, sperm production and copulation is typically separated by a longer storage of sperm in the epididymis. In dissociate breeders testis size peaks at maximal production of spermatozoa, not at peak level of plasma testosterone (usually at mating) (Licht 1984). Thus, it seems likely that most of the variation in testis size reflects the testis capacity to produce spermatozoa, as was suggested by Licht (1984; and references therein).

Can testis size (and tissue producing seminal constituents) be constrained by energetic costs? If so, we would predict (1) negative covariation between testis size and stored lipids, (2) positive covariation between testis size and food intake, and (3) that testis size/semen production is depressed by costly 'stress', such as pathogens.

1. Lipid content

The size of the testis and the fat bodies are strongly negatively correlated (Flemming and Hooker 1975; Fig. 13.1; Przystalski 1983; Licht 1984;

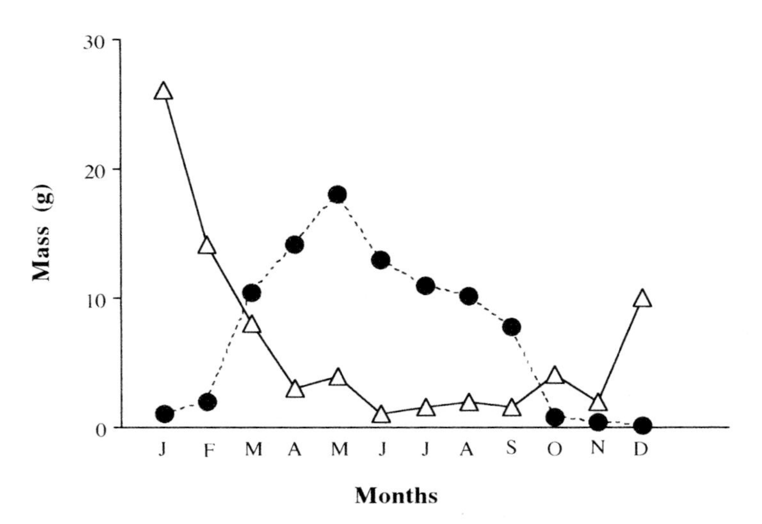

Fig. 13.1. *Annual reproductive and fat body cycles of* Anolis cupreus *males larger than 35 mm snout-vent length (SVL) in 1971, with additional data from 1972 and 1973. Triangles indicate mean fat body mass; filled circles indicate mean testis mass. Reproduced from Flemming and Hooker (1975) with permission of The Ecological Society of America.*

Seigel and Ford 1987; Moore and Lindzey 1992). Guillette and Casas-Andreu (1981) stated that fat bodies are used solely for costly mating behaviours and not 'testicular recrudescence'. However, according to Derickson (1976; and references therein) and Trauth (1979), the reduction in fat body size is due to spermatogenesis. In many lizards, fat-body size drops drastically at the onset of the mating season, in synchrony with an increase in testis size, and not gradually during the mating season as would be congruent with energy expenditure on reproductive behaviour (Mayhew 1966b; Ballinger 1974; Brackin 1979; Trauth 1979; Vitt and Goldberg 1983; but see Méndez de la Cruz *et al.* 1988). Cuellar (1973) removed fat bodies, manipulated food intake and follicle-stimulating hormone (FSH) levels on *Anolis carolinensis*, and concluded that fat-body removal did not affect testis recrudescence. However, Cuellar's experiment was carried out during late winter, and therefore it may be argued that males were not allocating resources to testis recrudescence as they would under natural circumstances. However, Cuellar also mimicked the hormone profile of reproductive males in two series of FSH-treated lizards (fat-body-ectomized and controls). In the FSH-treated males, males with the fat-bodies removed had consistently lower testis mass than controls, both in absolute numbers and in mean percentage of body mass over all treatments (2.0% vs. 1.8%, Table 13.4; Cuellar 1973, p. 68).

2. Food intake

Cuellar's experiment (1973) also gave strong support for positive feeding rate effects on testis size in *Anolis carolinensis*, both in FSH-treated males and controls. In a natural population of *Anolis acutus*, supplemental feeding resulted in both elevated testis mass and fat-body mass compared with controls (Rose 1982). Four supplementally fed *Anolis cristatellus* in a natural population did not deviate in testis size from a control group ($n = 11$; Licht 1974). However, in the males that were supplementally fed there was a perfect correlation between gain in body mass and testis mass (Licht 1974, Table II, p. 218, $r_s = 1.0$, $p = 0.0$, $N = 4$, unrelated to growth). Thus, males that put on relatively more weight also allocated relatively more energy in absolute terms into testis development, suggesting that testis recrudescence is energetically constrained.

Rainfall can have a strong effect on insect abundance (Dingle and Khamala 1972; Denlinger 1980), and on St Croix island in the Caribbean the rainfall is twice as high on the western side as on the eastern side. In the insectivorous *Anolis acutus*, males from the western side of the island had significantly larger testes than males from the eastern side (Rose 1982). Similarly, following good winter rains (1960), testis volume in adult *Uma scoparia* and *Uma notata* increased to about 170% and 125% of the testis volumes in years with low rainfall (Mayhew 1966a,b; Fig. 13.2). In *Lacerta vivipara*, testes mass was a maximal 1.22% of body mass in one year, and 2.50% in another year (Courty

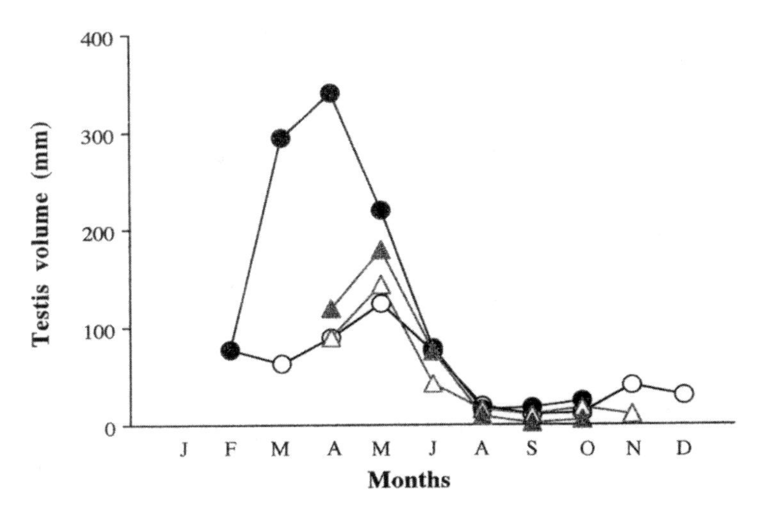

Fig. 13.2. *Changes in testis volumes (left testis) of adult male* Uma scoparia, *by year, from 1959 through 1962. Filled circles represent 1959; unfilled circles, 1960; unfilled triangles, 1961; filled triangles, 1962. Reproduced from Mayhew (1966a) with permission of The American Society of Ichthyologists and Herpetologists.*

and Dufaure 1979, 1982). In *Thamnophis radix* the ratio between yearly maximum and minimum testis size was 5.5 in one year, and 9.9 in the following year, suggesting much larger testis in the second year (Cieslak 1945). However, no reference to food supply was made in these studies.

3. Pathogens and other stresses

The effects of pathogens are relatively unstudied with one exception. Male *Sceloporus occidentalis* with malaria had smaller testis than nonparasitized males (Schall 1983a,b, 1996; Dunlap and Schall 1995), but the underlying mechanism for this is unclear.

Daily handling in captivity reduced testis size in the lizard *Anolis carolinensis* (Meier *et al.* 1973), which suggests that minimal, and equal, disturbance of experimental animals is desirable.

In summary, testis size and thus the capacity to produce spermatozoa appear to be significantly affected (to an extent that we have not previously appreciated) by a male's well-being and nutritional status.

B. *Testis size variation on a larger scale*

If testis size is constrained by energetics, one may predict that

(1) species with long mating seasons (e.g. 'all-year' reproducers, Appendix B) have low peak testis mass in relation to body mass (i.e. gonado-somatic index (GSI), percentage testes mass of body mass), and low relative annual GSI variation (GSI Max/GSI Min).

(2) Species with short mating seasons (largely temperate) have higher GSI and more annual GSI variation.

(3) If territoriality during the breeding season also make not only sexual partners but also food more available to males, territorial species could have larger GSI than nonterritorial species (see the example of maintained territoriality in the absence of females in Deslippe and M'Closkey 1991).

To test these predictions we gathered published data on male reproductive cycles (Appendix B). When GSI was not given explicitly in the cited papers, we calculated two GSI indices from figures and tables, depending on the published information; index 1 is testis mass (g) divided by body mass (g), multiplied by 100, and index 2 is testis mass (mg) divided by snout-vent length (mm). When testis size was given as a volume, we converted volume to mass using the volume–mass relationships of ellipsoid testis in Swain and Jones (1994) and, depending on the body size measurements given, classified these as index 1 or 2. When testis mass was given for one testis, we multiplied this figure by two. Thus, we ignored the possibility that the left and right testis may differ in size (Fox 1977), perhaps because maintaining two testes is associated with costs, such as immunosuppression by high androgen levels (Møller 1994b). Of course, our extrapolations do not yield exact GSIs, but errors should be random.

Data on territoriality were obtained from the literature. For species marked (*) in Appendix B, we have assumed presence or absence of territoriality based on general conclusions in Carpenter and Ferguson (1977), Stamps (1977, 1983), and Martins (1994), and the fact that teiids and New World skinks appear to mate guard and not be territorial (L. Vitt, pers. comm.). The phylogeny in the appendices adhere to standard works such as Etheridge and de Quiroz (1988), Estes *et al.* (1988), and Frost and Etheridge (1989).

A comparison across taxa is potentially confounded by phylogeny and therefore we restrict ourselves at this stage to giving descriptive results and leave out test statistics. However, we believe that our data are robust and invite in-depth analyses following standard comparative methods protocol. Indices 1 and 2 did not differ significantly in mean or variance, and we therefore pooled the data for the two indices. We restricted the descriptive comparison to lizards (but see data for turtles and snakes in Fig. 13.4) since available information were meagre for other taxonomic groups, and we excluded lizard families for which we had less than three observations. This also had the effect that the large varanids became excluded from the analyses, which was desired since GSIs characteristically decline in large species: in the remaining species body size is less than *c.* 100 g and strongly overlapping between all groups compared.

The results for the three trait categories were as follows:

(A) *Climate.* The mean ($\pm$SD) GSI and GSI ratio were, respectively, 1.0 ± 0.21, and 1.4 ± 0.31 for all-year reproducers, 1.31 ± 1.15 and 7.7 ± 9.69 for tropical species, and 2.1 ± 2.00 and 14.5 ± 11.50 for temperate species. Thus, in accordance with our predictions, the trends in the data appear to suggest that temperate zone species have larger testis at peak sperm production, and more variation in testis size over the year than species from other climate zones (Fig. 13.3).

(B) *Territoriality.* The mean GSI for territorial species was 2.1 ± 1.47 and 0.88 ± 0.60 for nonterritorial species; the corresponding mean GSI ratios were 11.5 ± 11.65 for territorial species and 10.12 ± 9.60 for nonterritorial species (Fig. 13.4).

(C) *Family.* There were strong effects of phylogeny on both GSI (Fig. 13.4) and GSI ratio. Mean ($\pm$SD) GSI varied from 0.73 ± 0.47 (Gekkonidae) to 4.73 ± 4.9 (Chamaeleonidae), and mean GSI ratio varied from 5.6 ± 5.57 (Polychridae) to 17.9 ± 13.0 (Phrynosomatidae) (Fig. 13.4).

We caution the reader that these analyses are highly preliminary, and

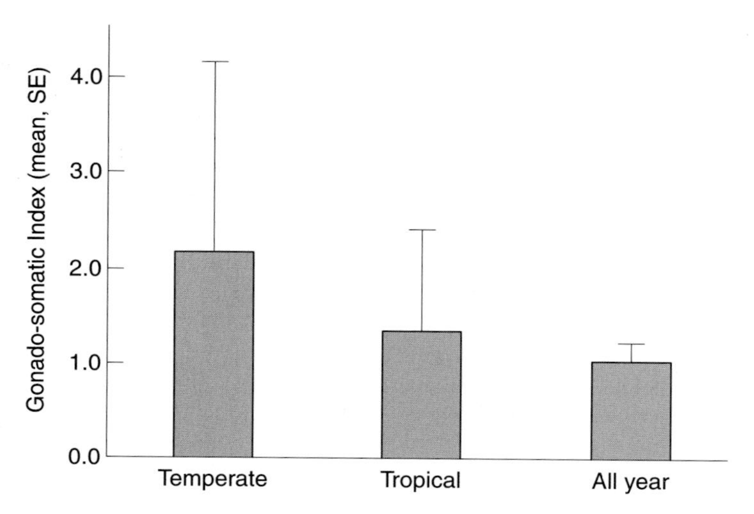

Fig. 13.3. *Mean gonado-somatic index (±SE) of temperate, tropical, and all-year reproducing lizards (data from Appendix 2).*

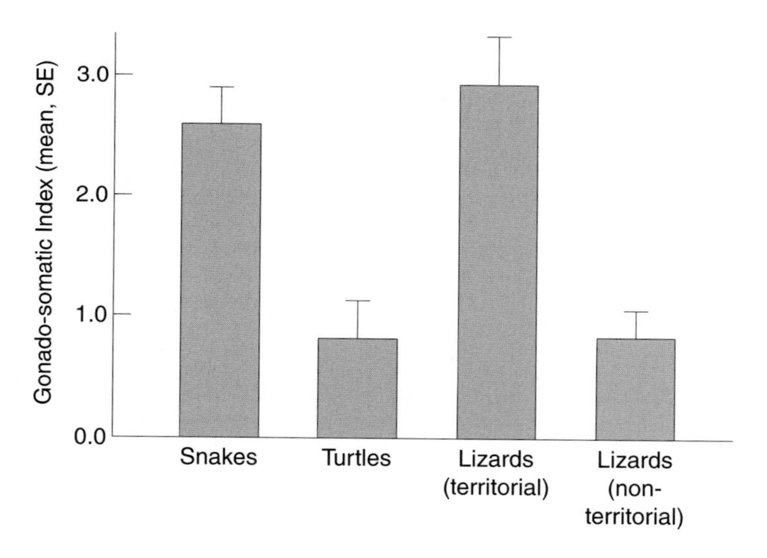

Fig. 13.4. *Mean gonado-somatic index (±SE) (see text for details) for snakes, turtles, and the territorial and nonterritorial lizard families for which we had the largest samples (Phrynosomatidae and Scincidae; Appendix B).*

that phylogeny has not yet been taken into account when presenting these descriptive results. Undoubtedly, phylogeny will explain significant variance in GSI and GSI ratio, and some of the differences in these traits between territorial and nonterritorial species, and between taxa from the different climate zones.

C. Production of semen and spermatozoa

Reptiles are ectotherms and a preferred body temperature (PBT), or window of temperatures, is maintained by thermoregulation. At temperatures close to PBT bodily functions are 'optimal', which may also apply to the rate of testis recrudescence, sperm production (Aldridge 1975; Weil and Aldridge 1979; Licht 1965) and possibly transport of sperm in the female reproductive tract (Ludwig and Rahn 1943). Temperatures maintained only 1–2°C above PBT can cause marked spermatogenic damage (Licht 1965).

All reptilian spermatozoa appear to be functional gametes, and vary in size from 20 to 30 μm (Crocodilians; Ferguson 1985), to 140–144 μm (Tuatara; Healy and Jamieson 1992). Spermatozoa mature in the male genital tract (Esponda and Bedford 1987), and in particular in the epididymis (Carcupino *et al.* 1989; Gist *et al.* 1992; Newton and Trauth 1992), a process which is finalized with maturation (= capacitation?) in the oviduct (Newton and Trauth 1992; Murphy-Walker and Haley, unpublished data). In *Nerodia sipedon* RSS secretions may contribute to sperm 'capacitation' (Weil 1984). Spermatozoa can be stored for considerable time in the epididymis, in dissociate breeders in particular (Licht 1984), and in crocodilians, in which the epididymis is virtually absent (as in snakes), spermatozoa is stored in the penial groove (Ferguson 1985).

D. Metabolism and motility of spermatozoa

In other vertebrates seminal carbohydrates, proteins, lipids, and ions, etc., can be metabolized by the spermatozoa (Cohen 1971; Mann and Lutwak-Mann 1981) and perhaps be used by the female for selecting spermatozoa for fertilization, or, by the male, for inducing 'female choice' of his spermatozoa (Eberhard and Cordero 1995). What is the evidence for such effects in reptiles? Amino acids and proteins are produced in the ductuli efferentes (Meeran 1994) and in the epididymis (Depeiges *et al.* 1987; Manimekalai and Akbarsha 1992; Averal *et al.* 1992), and these substances increase sperm motility (Hamid and Akbarsha 1989; Velmurugan 1992; Manimekalai 1993). Sperm motility also peaks in synchrony with plasma testosterone in *Lacerta vivipara* (Depeiges and Dacheux 1985) and is increased by RSS homogenate in *Anolis carolinensis* (Cuellar *et al.* 1972). The latter effect was also established in *Trachemys scripta* (Gartska and Gross 1990), and in this turtle RSS products also inhibited sperm motility outside the breeding period, which may prolong sperm survival. In several dissociate breeders the tight relationship between RSS secretions and mating, rather than spermatogenesis, further indicates the importance of RSS products for facilitation of sperm transfer (Aldridge *et al.* 1990).

E. Hemipenis morphology and copulation

All reptiles have internal fertilization. The sperm are transferred via cloacal apposition in the tuatara (Norris 1987), which lacks a penis. Turtles and crocodilians insert their single penis (Norris 1987), and squamates use one (of two) hemipenises during copulation. Hemipenises show extraordinary variation in size and shape, and the potential roles of sexual selection and sperm competition in explaining some of this variation are as obvious as the lack of studies attempting to do so (but see Eberhard 1985). In some species the hemipenis is bifurcated, as is the 'vagina' with its oviductal openings at the two anterior ends (Conner and Crews 1980); thus, the bifurcation may partition spermatozoa equally between the two oviducts. Cope (1896, p. 461) remarked that the lizard hemipenis is 'rarely spinous, as is so generally the case in the Ophidia'. The basal spines anchor the hemipenis which is then 'adjusted' (Pope 1941; Edgren 1953; Pisani 1976). Females of some species with spinous hemipenises have a thicker cloacal mucosa, but may still bleed after copulation (Noble 1937; Carpenter 1947).

We surveyed 14 major studies of hemipenis morphology of snakes and lizards and used Cope's (1896) initial observation of presence and absence of spines to categorize species into two groups (defined as 'present' when a spine was longer than its width at the base, and 'absent' when not) (Dowling and Savage 1960; Fleischmann 1902; Rosenberg 1967; Underwood 1967; Bellairs 1969; Presch 1978; Murphy and Barker 1980; Branch 1981; van Tienhoven 1983; Mao *et al.* 1984; Klaver and Böhme 1986; Böhme 1988; McCranie 1988). Our sample consisted of 112 species representing 10 families of lizards and four families of snakes. Spines were present in 35 of 41 species of snakes (85%), and in 14 out of 71 species of lizards (20%, one species was the legless *Anguis fragilis* and 10 species were chameleons).

We refrain from statistical analyses since these data are phylogenetically complex, but the general trend seems clear. Cope was indeed correct, snakes appear to have 'spinier' penises than lizards, which may facilitate, and select for, prolonged copulations in snakes. Snake copulations last 3 min to 28 h, whereas lizard copulations last 0.05 min to 1 h 6 min, but the bias towards longer copulations in snakes is much stronger than these ranges suggest (Fig. 13.5; Appendix A). An 'outlier' lizard, the legless *Anguis fragilis*, with spined hemipenises may copulate for 20 h. Interestingly, snakes which mate for much longer than lizards seem to have relatively larger testes than lizards, which would be consistent with an idea of prolonged sperm transfer (Wilcoxon paired signed-rank test, $Z = 2.39$, $P = 0.019$, Appendix B; note, this comparison requires appropriate control for phylogeny (from a phylogenetic perspective N equals one for both taxa)). However, prolonged copulations could be functionally related to sperm competition in more than one way: (A) more spermatozoa could be transferred during a prolonged ejacula-

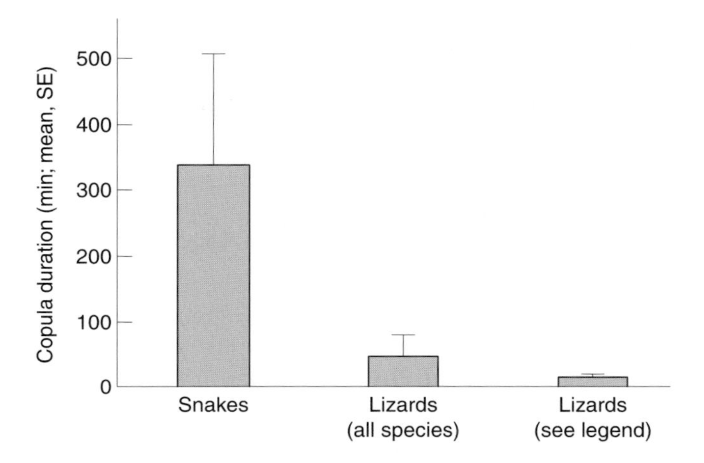

Fig. 13.5. *Copula durations in snakes, all lizard species, and lizards with* Anguis fragilis *excluded (data from Appendix A).*

tion, (B) more RSS secretions could be transferred, and (C) the male could physically block the cloaca.

'Ejaculatory thrusts' occur throughout most of the mating in squamates, in particular the first 75% of its duration (Ferguson 1970; Rodda 1992); in different populations of *Uta stansburiana* the thrusts varied between 12 and 160 per mating (correlated with coitus duration, Ferguson 1970). If semen was transferred at the end of the copulation only, semen should not 'leak' during the mating, which has been noted in *Coluber c. constrictor* (Noble 1937), and *Lampropeltis getulus* (Lewke 1979). In *Iguana iguana* (Rodda 1992) and *Natrix natrix* (T. Madsen, pers. obs.) natural or experimental interruption of copulations resulted in white viscous fluid oozing from the hemipenis.

It has been suggested that most of time spent in copula is used for transfer of substances other than spermatozoa (Blanchard and Blanchard 1942; Devine 1975), such as RSS secretions for a 'copulatory plug' (Devine 1975; see below). However, the longest copulations occur in rattlesnakes, in which copulatory plugs have never been reported, and some of the shortest matings occur in *Thamnophis* which have copulatory plugs (Appendix A). In *Elaphe o. quadrivittata*, considerable leakage of a whitish, clear fluid from the cloaca after copulation was microscopically confirmed to contain viable sperm (T. Helin, pers. obs., in captivity).

In *Amblyrhynchus cristatus*, semen appeared at the tip of the hemipenis in interrupted copulations after 2.8–3.1 min, and uninterrupted copulations by dominant males lasted for 9 min (Wikelski and Bäurle 1996). Dominant *Iguana iguana* males also copulated for longer than subdominant males (Wikelski and Bäurle 1996; Rodda 1992); in *Ctenophorus maculosus* relatively larger males in relation to females copulated longer by force (Olsson 1995a).

In conclusion, prolonged copulations may yield multiple fitness benefits such as transfer of relatively more spermatozoa and semen (such as RSS secretions), and may prevent sperm leakage and rival matings. At present, we lack information about the possible costs that may be associated with prolonged copulations.

F. Intrauterine transport and storage of spermatozoa

The transport of sperm up the female reproductive tract results partly from the sperms own swimming ability (motility), and could be facilitated by, for example, cilia in the oviduct (Parker 1931; Saint-Girons 1973; Conner and Crews 1980; Halpert *et al.* 1982). However, during the reproductive season nonciliated, secretory cells may replace the ciliated cells that have a potential transport function (Hoffman and Wimsatt 1972). In *Thamnopis sirtalis*, sperm may be stored in vaginal furrows, and the epithelium in this region is sloughed and forms 'carrier matrices' which, mixed with sperm, migrate to the infundibulum region, possibly under the influence of smooth muscle contractions induced by an oestrogen surge at mating (Halpert *et al.* 1982). In some mammals, semen contains gonadotrophins and other hormones which cause constriction of oviduct smooth muscles, facilitating sperm transport to the fallopian tube (Cohen 1971; Eberhard and Cordero 1995). If females promote sperm competition (Bellis and Baker 1990; Madsen *et al.* 1992) in the 'harsh' environment of the female reproductive tract (Birkhead *et al.* 1993; see also Keller and Reeves 1995), it is not clear why females would actively transport ejaculates (or provide carbohydrates that may be metabolized by the spermatozoa; Hoffman and Wimsatt 1972).

The movement of spermatozoa in the reproductive tract appears to be slower in reptiles than in mammals. Sperm may still be present in the female cloaca 2–3 days after copulation in *Natrix maura* (Hailey and Davies 1987); in *Anolis carolinensis*, spermatozoa reach the sperm storage tubules 2–6 h following copulation and in 6–24 h the infundibulum of the oviduct (Conner and Crews 1980). In mammals this process may take minutes, although the bulk of the semen take longer (Restall 1967; Mann and Lutwak-Mann 1981).

The capacity for prolonged sperm storage in the female reproductive tract is highly developed in reptiles (several years in some species, Birkhead and Møller 1993). Histological adaptations (or pre-adaptations) for sperm storage (sperm receptacles) have been reviewed by Fox (1956, 1977), Fox and Dessauer (1962), Cuellar (1966b), and Gist and Jones (1987). Since then, several case studies have followed: crocodilians (Davenport 1995), turtles (Palmer and Guillette Jr 1988; Gist and Jones 1989; Ruth Shantha Kumari *et al.* 1990; Gist and Fischer 1993), lizards (Picariello *et al.* 1989; Ruth Shantha Kumari *et al.* 1990; Manimekalai 1993; Whittier *et al.* 1994; Srinivas *et al.* 1995; Murphy-Walker and Haley (*Hemidactylus frenatus*), unpublished data; N. Zucker (in *Urosaurus ornatus*), pers. comm.), and snakes (Aldridge 1992; Bull *et al.* (absence of sperm receptacles in *Boiga irregularis*), unpublished data).

Virtually all female reptiles may store spermatozoa for at least some weeks between mating and ovulation. However, in several species spermatozoa cannot be stored between successive ovulations, e.g. in *Podarcis muralis* (Cooper 1965), *Lacerta agilis* (Olsson *et al.* 1994a), and *Cnemidophorus sexlineatus* (Newton and Trauth 1992). In *L. agilis*, sperm receptacles could not be detected in a preliminary study (Birkhead and Olsson, unpublished data). Although sperm receptacles increase both the motility and survival of spermatozoa (Manimekali 1993), the decreased fertility in subsequent clutches may be substantial. In *Uta stansburiana*, only 53% of the eggs in second clutches were fertile, and 0% of the eggs in third clutches (Cuellar 1966a), the corresponding figures for five subsequent clutches in *Chamaeleo ellioti* were 12–58% (Leptien 1989), and for *Chamaeleo hoehnelii*, 84–62% (Jun-Yi 1982). It is unclear if the reduction in fertility is due to sperm senility or mortality, passive loss of spermatozoa, or some effect on the female from being kept in captivity (although there is no reason to suspect this). In crocodilians, few surviving spermatozoa with low motility remain in the penial groove at the end of the mating season, which could be the result of sperm ageing (Kofron 1990).

Sperm receptacles may have evolved because they prevent sperm from being forced out of the oviduct by the first egg in a clutch of several (Fox 1956). However, sperm receptacles seem to be absent in several species producing large broods, and present in species such as *Anolis carolinensis* which lay a single egg.

Sperm storage has also been suggested to make females less dependent on multiple matings and ensure multiple fertilizations of subsequent clutches in the absence of males (Conner and Crews 1980; Gist and Jones 1987). However, sperm receptacles have evolved in species which occur in high densities (e.g. *Anolis carolinensis* and *Calotes versocolour*) and sperm storage occurs in multiply mating species (e.g. *Conolophus subcristatus*, Werner 1982). Sperm receptacles also occur in some *Anolis* species (Conner and Crews 1980). In *Anolis acutus*, females appear to have induced ovulation (Stamps 1975); thus, fertilization by stored sperm cannot occur in the absence of males.

The selective advantage of sperm storage does not appear to be long-term production of multiple clutches from a single mating, but rather the decoupling of copulation and ovulation, facilitating female choice and sperm competition (Birkhead and Møller 1993; see also Schuett 1992).

IV. SPERM COMPETITION

A major determinant of the opportunity for sexual selection and sperm competition is the mating system (Andersson 1994; Møller 1994a). Most reptiles appear to be polygynandrous (*sensu* Andersson 1994; Appendix A; see also Duvall *et al.* 1992; Arnold and Duvall 1994). Thus, there is ample opportunity for sperm competition arising from female matings with multiple partners. The data on number of a female's partners in Appendix A is strongly influenced by the interest of the male. In captivity, *Lacerta agilis* females may mate more than 15 times with different males in a few days (Olsson, pers. obs.), while in the wild they rarely get to mate more than *c.* 5–6 times during their receptive period (Olsson *et al.* 1996a).

A. Multiple paternity

The existence of ongoing sperm competition is reflected by multiple paternity within the same clutch. Somewhat paradoxically, the first observation of mixed paternity in a reptile was made on captive *Thamnophis* (Blanchard and Blanchard 1942), a genus that, for a long time, came to epitomize single-mating species (Ross and Crews 1977, 1978; Whittier and Tokarz 1992). Blanchard and Blanchard's results were confirmed for natural populations by Gibson and Falls (1975) and by Schwartz *et al.* (1989), who demonstrated mixed paternity in 50–72% of clutches. The corresponding figure for *Nerodia sipedon* was at least 86% (Barry *et al.* 1992), and for *Vipera berus* 17% and 80% in two different populations (Höggren 1995). Multiple paternity has been demonstrated in staged experiments in *Vipera berus* (Stille *et al.* 1986; Tegelström and Höggren 1994; Höggren and Tegelström 1995), *Lampropeltis getulus* (Zweifel and Dessauer 1983) and *Agkistrodon contortrix* (Schuett and Gillingham 1986). In turtles, multiple paternity has been demonstrated in *Caretta caretta* (Harry and Briscoe 1988), *Clemmys insculpta* (Kaufmann 1992) and *Chelydra serpentina* (Galbraith *et al.* 1993; see also Galbraith 1993). In lizards, *Lacerta agilis* exhibits mixed paternity both in the laboratory, and in around 80% of the clutches in the wild (Olsson *et al.* 1994a, 1996a).

B. The male perspective

When females copulate with more than one male, males that transfer more spermatozoa are predicted to be more successful. In *Anolis carolinensis*, males alternated between the two hemipenises in successive copulations, although individual males were biased towards either side (Crews 1978). This could increase the number of spermatozoa transferred per copulation, which Tokarz and coworkers confirmed in *Anolis sagrei.* Males used their hemipenises in an alternating fashion if the intercopulatory interval was less than approximately 73 h; after that time the sperm supplies were replenished (Tokarz 1988, 1989; Tokarz and Slowinski 1990; Tokarz and Kirkpatrick 1991). Observations in an enclosure suggested that *A. sagrei* mate frequently enough to make the use of alternating hemipenis an important part of male reproductive tactics in the wild (Tokarz and Kirkpatrick 1991). In *Lampropeltis getulus*, males that remated within 3 days alternated between hemipenises in 43 of 46 copulations, but when the interval was 6–28 days, only 10 of 21 shifted side (R. Zweifel, pers. comm.), which is in close agreement with Tokarz' results. In *Conolophus subcristatus*, males also used alternating hemipenises in copulations in a natural population (1982). Interestingly, in *Cnemidophorus lemniscatus*, copulations are left-biased (Müller 1971); reptiles may have testes of different size (Fox 1977; see also Møller 1994b) and males perhaps adjust copulation frequency in favour of the largest testis and/or epididymis.

When mating attempts by smaller males are frequently interrupted by larger males, rapid sperm transfer would be a selective advantage. Wikelski and Bäurle (1996) suggested that smaller male *Amblyrhynchus cristatus* ejaculated into the pouch of the hemipenis, after which the hemipenis could be inserted (often by force) resulting in almost instant sperm transfer (Wikelski and Bäurle 1996). However, ejaculation into the pouch could be adaptive for all males, especially if large males can transfer a second mating-ejaculation. Perhaps in support of this scenario, dried semen was found in the hemipenis pouches of every snake from which Mengden *et al.* (1980) sampled sperm. Ejaculation without cloacal contact has also been demonstrated in *Coleonyx variegatis* (Greenberg 1943) and *Chelydra serpentina* (Legler 1955).

In *Cnemidophorus*, mating is sometimes preceded by 'cloacal rubbing' (Carpenter 1962), which has been suggested to be a scent-marking behaviour (Mason 1992). An alternative hypothesis is perhaps that males rub the genital papilla situated between the two hemipenis openings (Mengden *et al.* 1980) and masturbate to achieve more rapid sperm transfer.

Males may also rub everted hemipenises, which may clean the sulcus spermaticus of dried semen (see also Mason 1992). Seminal plugs expelled this way (not to be confused with 'copulatory plugs') are monitored by reptile breeders because they announce an oncoming breeding period (T. Helin, pers. comm.). Similarly, *Iguana iguana* males may evert a hemipenis and exude a white, viscous liquid (presumably semen) on days when they do not copulate (Rodda 1992). Whether this is mere 'libidinousness', serves to cleanse the hemipenis or perhaps rejuvenate spermatozoa is unknown.

Perhaps the most well-known structure in studies of reptilian sperm competition is the copulatory plug, formed in the female cloaca predominantly by RSS secretions. Its function(s) could be multiple (Mason 1992), however, its major adaptive significance has been suggested to be as 'chastity belt', since it prevents copulations until it is expelled (Devine 1975). In *Thamnophis radix*, the plug has also been suggested to pheromonally make females unreceptive (Ross and Crews 1977, 1978); however, if mating plugs evolved first, it is unclear how selection could favour a male strategy which makes females unreceptive when females are incapable of mating. Expelling the plug takes *c.* 4 days at 5°C and 2 days at 21°C (Devine 1984). The preferred body temperature for the closely related *Thamnophis sirtalis* is 27.4°C (Gibson and Falls 1979), but it is not known for how long the plug is retained close to this temperature. In *T. sirtalis*, females become unreceptive immediately following copulation, but 2 days later more than 25% of mated females remate (Whittier and Tokarz 1992). According to Devine (1977), *Thamnophis* females stay at mating aggregations for long enough for multiple matings, and this is corroborated by the reports of mixed paternity reviewed above. Copulatory plugs were also recently demonstrated in a large number of lacertid lizards (in den Bosch 1994) such as *Lacerta agilis* (in den Bosch 1994; Olsson and Birkhead, unpublished).

In *Vipera berus*, mated females had constricted uteri, and an experiment was carried out to link this to RSS secretions (Nilson and Andrén 1982). A single female was administered with RSS secretions, mixed with urea and Bismarck Brown (stain), which resulted in harder uteri in the treatment female than in a female administered with 0.9% NaCl (without Bismarck Brown). No staining was found in the uteri, suggesting that the RSS secretions in the cloaca may induce a contraction of smooth muscle in the uterus. Nilson and Andrén (1982) stated that this functions as a paternity assurance mechanism for the first male.

In *Lacerta agilis*, we allowed vasectomized males to mate before a fertile male, but second males did not have reduced fertility (Olsson *et al.* 1994a). In unmanipulated males, there was no significant effect of mating order, whether copulations were separated by 1 h or 24 h, and second males sired, on average, 59% of the offspring (Olsson *et al.* 1996a). The factor that most strongly influenced male reproductive success was the time elapsed since the competing males' previous copulations, with longer intervals yielding greater reproductive success (Olsson *et al.* 1996a).

In *Vipera berus* the intervals between matings may affect male reproductive success. In staged experiments with three males copulating 1 day apart, the first male sired, on average, 75% of the offspring, the second 18%, and the third male 7% (n = 8, Höggren 1995). In a similar experiment using two males with 4 days between successive matings, the first male sired 62% and the second male 38% of the offspring (n = 5, Stille *et al.* 1986, 1987). This lends some support for Nilson and Andrén's (1982) paternity assurance hypothesis, although the underlying histochemical mechanism should be retested with appropriate sample sizes and controls. In natural populations of *V. berus* females mate approximately one month before ovulation (Nilson 1981). However, in species in which females accept copulations close to ovulation, a first-male advantage could perhaps arise simply from the timing of insemination in relation to ovulation; if the female ovulates while the ejaculates are still migrating towards the infundibulum (and before the ejaculates mix) a first-male advantage seems the probable outcome, independently of paternity assurance. Furthermore, in snakes the oviduct appears to be long compared with other vertebrates (Bracegirdle and Miles 1978); thus, to remove the potentially confounding 'timing effect', experiments designed to identify determinants of paternity and paternity assurance mechanisms must allow for adequate time for sperm transport to the infundibulum for all participating males. At present, we do not know to what extent this presents a problem for the interpretation of data.

Leuckart (1847, not seen but cited in Voss 1979) suggested that copulatory plugs may prevent sperm leakage, and Stille *et al.* (1986) suggested that this was probably the function of the uterus contraction in *Vipera berus*. Both 'sperm leakage' and 'paternity assurance' are consistent with sperm competition theory, although the selective agents differ. The 'sperm leakage' hypothesis has not been tested in a reptile. However, in rats, males with partially removed seminal vesicles are unable to produce fully formed vaginal plugs. Females mated by these males have a much reduced number of sperm in their uterus and, under the assumption that some transferred sperm must have disappeared from the female reproductive tract, sperm leakage seems plausible (Carballada and Esponda 1993). In *Lacerta agilis*, as in rats, the matrix of the copulatory plug consists in part of sperm (Olsson and Birkhead, unpublished data). In several lacertid lizards, the male bites the female over the anal shield following copulation (Verbeek 1972; Olsson, pers. obs.), thereby keeping the cloaca shut for up to 15 min (Verbeek 1972; pers. comm.). Whether this prevents sperm leakage is not known.

Postcopulatory mate guarding is another possible mechanism that can increase a male's probability of paternity. Male *Lacerta agilis* guarded their female for an average of 1.4 days after copulation in the wild (range 1–9 days). Larger males guarded significantly longer than smaller males, but female age and body size, clutch size, and time to ovulation (estimated by date of oviposition) did not affect guarding duration. The operational sex ratio also declined throughout the mating season but was unrelated to duration of mate guarding (Olsson *et al.* 1996a). In *Ameiva plei*, males guarded females for an average of 1.95 days (range 1–4 days) following copulation (Censky 1995); postcopulatory mate guarding has also been described in *Conolophus stejnegeri* (Wiewandt 1977), *Tropidurus delanonis* (Werner 1978), *Iguana iguana* (Dugan 1982), *Vipera berus* (Andrén 1986; Madsen *et al.* 1993), *Eumeces laticeps* (Cooper and Vitt, in prep.) and seems to occur in many teiids and New world skinks (L. Vitt, pers. comm.). However, co-occurrence of males and females may not necessarily be explained by mate-guarding behaviour. In the skink *Lampropholis guichenoti*, males and females bask on top of each other in nonsexual situations, and thus 'pairing' could be a thermoregulatory behaviour in this species (Torr and Shine 1993).

In a natural population of *Lacerta agilis*, larger males had more partners, but there was no correlation between male size and the proportion of the clutch that they sired (Olsson *et al.* 1996a). There was, however, considerable variation in number of partners at any given male body size, and number of partners was significantly correlated with the proportion of the clutch that a male sired. This suggests that suites of traits important for mate acquisition and sperm competition may be positively correlated. If so, trade-offs between, for example, number of partners and the proportion of sired offspring (*sensu* Parker 1974b) must be explored experimentally.

Provided that females mate equally often (but see Olsson 1993c), males could increase their fitness benefit per transferred ejaculate by mating with larger more fecund females. Males prefer large females in *Sceloporus jarrovi* and other *Sceloporus* species, *Lacerta agilis* and *Eumeces* (Fitzgerald 1982; Olsson 1993c; Cooper and Vitt, unpublished data). However, males also seem to prefer large females in *Anolis garmani* and *Anolis valencienni*, species in which females lay a single egg (Trivers 1976; Hicks and Trivers 1983; respectively). To what extent this exerts any selection on females, or renders males fitness benefits because larger females perhaps lay eggs more often, is unclear. In *Anolis sagrei*, males also preferred to mate with unfamiliar females (Tokarz 1992); avoiding familiar females could be an alternative strategy to mate guarding, thereby increasing the number of partners rather than the probability of paternity per brood (see also Stamps 1983; Ferguson 1969, 1972).

An untested hypothesis is that males in temperate regions time the copulation during the day so that it is followed by female basking, which could facilitate intrauterine sperm transport. In *Lacerta agilis*, 19 of 23 timed copulations on cloud-free days took place before 13.00 hours (Olsson, unpublished data); such temporal allocation of copulations does not appear to occur in tropical species (e.g. Rodda 1992; Censky 1995).

C. The female perspective

If females promote sperm competition, we may expect counter-adaptations against paternity assurance mechanisms to evolve in females. Devine (1984) suggested that *Thamnophis* females have evolved enzymes to rapidly dissolve and expel the copulatory plug. Better still would be if females could prevent transfer of the RSS secretions that make up the plug. In *Thamnophis marcianus* females mated normally for about 4 min, but then, in eight out of 12 copulations, rotated wildly resulting in interrupted copulations and no formation of copulatory plugs (Perry-Richardson *et al.* 1990).

In a laboratory experiment, *Lacerta agilis* females that were brought out of hibernation on the same day as the males they were allowed to mate with had 30% reduction in fertility compared with females that were brought out of hibernation 10 days after their mates. The reduction in fertility could possibly be due to incomplete production or maturation of spermatozoa (Olsson and Madsen 1996). In the wild, *L. agilis* females emerge, on average, *c.* 17 days after males and, hence, females may be selected to delay emergence to reduce the risk of mating with infertile males.

The underlying reason(s) for multiple matings in females could be several, and may vary between taxa (see Chapters 2 and 3). However, since our field studies of *Vipera berus* and *Lacerta agilis* have provided empirical support of genetic benefits of multiple matings in females, we will briefly address this issue. Our studies (Madsen *et al.* 1992; Olsson *et al.* 1994b,c) show that multiply sired broods have higher embryonic survival, fewer deformities and, in *Lacerta agilis*, the offspring are heavier and survive better during their first year of life. Furthermore, normal-looking young with malformed siblings (or half-siblings) in this natural population survive less well than young with no deformed siblings (Olsson *et al.* 1996b).

What can explain this in a species in which the ejaculate does not provide nutrients for the developing young? Bellis and Baker (1990) suggested that females may promote competition between sperm of different quality from different males (or from the same male; Cohen 1971). Madsen *et al.* (1992) proposed that this explains the increased offspring viability in *Vipera berus*. Parker (1992) cautioned against this interpretation because the haploid genome of the sperm was, at that time, believed to be unlikely to determine its performance. This was later clarified: morphological and functional traits of the spermatozoa develop primarily under the influence of the haploid genome (at least in mammals) and the sperm is now considered a model system for distinguishing between diploid and haploid expressed genes (Nayernia *et al.* 1996).

Both our study populations exhibit low genetic variation (Olsson *et al.* 1994b; Madsen *et al.* 1995) and thus we assumed that more detrimental recessive alleles could occur in the haploid sperm genome from a more inbred male, than from a less inbred one. Poor 'performance' of spermatozoa, e.g. reduced motility, is a common effect of inbreeding (Wildt *et al.* 1987). Thus, covariation between a male's sperm performance and the viability of his offspring could arise if the genome of the male dictates both the proportion of detrimental recessives in his sperm genome and the performance of his spermatozoa, independent of which genome (diploid or haploid) determine(s) sperm performance. In this scenario, the genome of a better-performing spermatozoa would have a higher probability of resulting in a zygote with increased viability. Our assumption(s) relies on the fact that there is enough genetic variation in the inbred population to express phenotypic variation in sperm performance and offspring viability (L. Keller, pers. comm.). We have not tested this assumption for sperm performance. However, staged sib-matings in *Lacerta agilis* resulted in increased proportions of deformed young (Olsson *et al.* 1996b), suggesting that our population was not 'purified' of recessives resulting from inbreeding (Hedrick 1994). Thus, although the genetic variation is low in this population, detrimental recessives still occur at frequencies high enough to make close inbreeding result in offspring with 'poor phenotypes'.

If low genetic variation explains our results, are they representative of other reptiles in particular, and organisms in general? We suggest that the comparatively high level of inbreeding in our study populations merely facilitated disclosure of fitness-enhancing effects of multiple matings. In support of this is an ongoing study of a large outbred population of *Liasis fuscus*, in which siblings show neonatal survival patterns similar to those reported for *Lacerta agilis* (T. Madsen and R. Shine, unpublished). Variation in offspring viability, due to genetically based determinants of metabolic physiology and immunological responses to infections, may be subtle and go undetected in many studies of natural populations. Therefore, we would not predict fitness-enhancing effects, apparent to the naked eye, in offspring in every staged multiple mating experiment. The probability of detecting such effects would depend on the genetic constitution of the population from which the participating animals were obtained. However, if sperm performance covaries with the genotypic quality of the male on the grounds stated above, multiple matings in females may be adaptive for this reason not only in inbred populations.

An alternative, or complementary, mechanism that may contribute to increased offspring viability resulting from multiple matings is that females (or eggs) can discriminate between sperm, and only allow fertilization by sperm from, for example, distantly related males. This appears to be the case in *Lacerta agilis* (Olsson *et al.* 1996d), where, under sperm competition, both in a natural population and in a laboratory experiment, more distantly related males sired a larger proportion offspring.

D. Summary and gaps in knowledge – sperm competition

Reptiles show traits, at virtually every level of reproductive biology, that potentially are adaptations evolved in response to selection arising from sexual selection and sperm competition. However, several of the relationships that we have discussed may be epiphenomena with respect to sperm competition and only careful future studies can test their validity. However, it is beyond doubt that sperm competition is an important selective agent in this taxon. In most species, females mate with more than one male, and mixed paternity occurs in 17–87% of the broods. However, we encourage colleagues to submit also single paternity results for publication to avoid a possible overestimation of the prevalence of multiple paternity in natural populations.

Most reptiles yolk their follicles and ovulate the entire clutch synchronously in contrast to, for example, birds, which forage between ovipositions and this may affect the nutritional status of the egg. Thus, in reptiles it is possible to reveal paternal (strictly genetic) viability-enhancing effects in mixed broods, with minimal maternal confounding effects (Olsson *et al.* 1996c). Such estimates of a male's ability to sire viable young could be correlated with, for example, estimates of his success in sperm competition and sexual characters.

Production and transfer of spermatozoa is separated by shorter or longer periods of time in associate and dissociate breeders, which could be exploited for investigation of behavioural adaptations in contexts of, for example, sperm depletion. Furthermore, squamate reptiles are unique in their separation of two functional units for production, storage and transfer of spermatozoa and other seminal products. This holds great promise for future experimental studies since, in many species, each separate unit can be surgically manipulated.

In mammals, the intrauterine transport of semen may be rapid (minutes) and the ovulation occurs within hours or days, whereas in reptiles this transport is relatively slow and ovulation may not follow until weeks later. This suggests that if seminal constituents influence the process of intrauterine sperm transport, or may influence female 'sperm choice', such compounds could be under selection arising from sperm competition. While squamate spermatozoa vary little in morphology, the intrauterine distances that they are transported vary considerably in this group, represented by 20 mm geckos and 8-m pythons, making them suitable candidates for research in this exciting unexplored field of sperm competition.

ACKNOWLEDGEMENTS

Many colleagues provided manuscripts, in press and other published and unpublished information. Our most sincere thanks to M. Akbarsha, F. Angelini, S. Arnold, H. in den Bosch, C. Carbone, M. Bull, E. Censky, J. Clobert, J. Congdon, B. Cooper, D. Crews, D. Duvall, M. Ferguson, W. Gibbons, S. Haider, P. Harlow, M. Hasegawa, J. Healey, T. Helin, B. Heulin, D. Hews, L. Hoffman, M. Höggren, S. Keogh, V. Lance, C. Limpus, L. Luiselli, B. Magnusson, M. Massot, D. Modry, S. Murphy-Walker, E. Panov, M. Paulissen, M. Plummer, N. Pratt, G. Rodda, H. Saint-Girons, A. Salvador, S. Sarre, R. Shine, L. Schwarzkopf, T. Shivanandappa, J. Stamps, H. Tegelström, R. Tokarz, S. Trauth, F. Trillmich, M. C. Tu, B. Verbeek, A. Walters, P. Weatherhead, G. Webb, J. Whittier, M. Wikelski, N. Zucker, and D. Zweifel. We are also most grateful to the personnel at the Biomedical Library, University of Göteborg, for their excellent help, and, in particular, to Malte Andersson, Lollo Bák-Olsson, Tim Birkhead, Scott Keogh, Anders Pape Møller, Rick Shine and Barry Sinervo for comments on an earlier draft of this manuscript. However, any erroneous interpretations of data or other mistakes are entirely our own. Mats Olsson thanks The Swedish Natural Science

Research Council and the Australian Research Council for financial support, and the Universities of Göteborg and Sydney for logistic support. Thomas Madsen thanks the Australian Research Council for financial support. Finally, we thank our loved ones for their constant help and encouragement throughout this work.

REFERENCES

Abts ML (1988) Reproduction in the saxicolous lizards, *Sauromalus obesus*: the male reproductive cycle. *Herpetologica* **44**(4): 404–415.

Akbarsha MA & Meeran MM (1995) Occurrence of ampulla in the ductus deferens of the Indian garden lizard *Calotes versicolor* Daudin. *J. Morphol.* **225:** 261–268.

Akester J (1989) Captive breeding of the rhinoceros-horned *Bitis nasicornis*. *Br. Herpetol. Soc. Bull.* **28:** 31–36.

Aldridge RD (1975) Environmental control of spermatogenesis in the rattlesnake *Crotalus viridis*. *Copeia* **3:** 493–496.

Aldridge RD (1992) Oviductal anatomy and seasonal sperm storage in the southern crowned snake (*Tantilla coronata*). *Copeia* **4:** 1103–1106.

Aldridge RD, Greenshaw JJ & Plummer MV (1990) The male reproductive cycle of the rough green snake (*Opheodrys aestivus*). *Amphibia–Reptilia* **11:** 165–172.

Altland P (1951) Observations of the structure of the reproductive organs of the box turtle. *J. Morphol.* **89:** 599–621.

Andersson M (1982) Sexual selection, natural selection and quality advertisement. *Biol. J. Linn. Soc.* **17:** 375–393.

Andersson M (1994) *Sexual Selection*. Princeton University Press, Princeton.

Andersson M & Iwasa Y (1996) Sexual selection. *Trends Ecol. Evol.* **11** (2): 53–58.

Anderson RA & Vitt LJ (1990) Sexual selection versus alternative causes of sexual dimorphism in Teiid lizards. *Oecologia* **84:** 145–157.

Andrén C (1986) Courtship, mating and agonistic behaviour in a free-living population of adders, *Vipera berus*. *Amphibia–Reptilia* **7:** 353–383.

Andrews RM (1982) Patterns of growth in reptiles. In *Biology of the Reptilia*, Vol. 13. C Gans & FH Pough (eds), pp. 273–320. Academic Press, New York.

Andrews RM (1985) Mate choice by females of the lizard, *Anolis carolinensis*. *J. Herpetol.* **19** (2): 284–289.

Angelini F, Brizzi R & Barone C (1979) The annual spermatogenetic cycle of *Podarcis sicula campestris* de Betta (Reptilia Lacertidae) 1. The spermatogenetic cycle. *Monit. Zool. Ital. (NS)* **13**: 279–301.

Antonioi FB (1980) Mating behavior and reproduction of the eyelash viper (*Bothrops schlegeli*) in captivity. *Herpetologica* **36** (3): 231–233.

Arnold EN (1986) Why copulatory organs provide so many useful taxonomic characters: the origin and maintenance of hemipenial differences in lacertid lizards (Reptilia: Lacertidae). *Biol. J. Linn. Soc.* **29:** 263–281.

Arnold ML & Hodges SA (1995) Are natural hybrids fit or unfit relative to their parents? *Trends Ecol. Evol.* **10** (2): 67–71.

Arnold SJ & Duvall D (1994) Animal mating systems: a synthesis based on selection theory. *Am. Nat.* **143** (2): 317–348.

Arnold SJ & Wade. MJ (1984a) On the measurement of natural and sexual selection: Theory. *Evolution* **38:** 709–719.

Arnold SJ & Wade MJ (1982b) On the measurement of natural and sexual selection: applications. *Evolution* **38:** 720–734.

Arslan M, Lobo J, Zaidi AA, Jalali S & Qazi MH (1978) Annual androgen rhythm in the spiny-tailed lizard, *Uromastix hardwicki*. *Gen. Comp. Endocrinol.* **36:** 16–22.

Auffenberg W (1988) *Gray's Monitor Lizard*. University of Florida Press, Gainesville.

Auffenberg W (1994) *The Bengal Monitor*. University of Florida Press, Gainesville.

Averal HI, Manimekalai M & Akbarsha MA (1992) Differentiation along the ductus epididymis of the Indian garden lizard *Calotes versicolor* (Daudin). *Bio. Struct. Morphogen.* **4** (2): 53–57.

Badir N (1958) Seasonal variation of the male urogenital organs of *Scincus scincus* L. and *Chalcides ocellatus*. *Forsk. Z. Wissen. Zool.* **160:** 290–351.

Baker RE & Gillingham JC (1983) An analysis of courtship behavior in Blanding's turtle, *Emydoidea blandingii*. *Herpetologica* **39** (2): 166–173.

Ballinger RE (1974) Reproduction of the Texas horned lizard, *Phrynosoma cornutum*. *Herpetologica* **30** (4): 321–327.

Ballinger RE & Ketels DJ (1983) Male reproductive cycle of the lizard *Sceloporus virgatus*. *J. Herpetol.* **17** (1): 99–102.

Barlow GW, Rogers W & Garley N (1986) Do Midas cichlids win through prowess or daring; it depends. *Behav. Ecol. Sociobiol.* **19:** 1–8.

Barry FE, Weatherhead PJ & Philipp DP (1992) Multiple paternity in a wild population of northern water snakes, *Nerodia sipedon*. *Behav. Ecol. Sociobiol.* **30:** 193–199.

Barwick RE (1959) The life history of the common New Zealand skink *Leiolepisma zelandica*. *Trans. Roy. Soc. NZ* **86** (3–4): 331–380.

Bauman MA & Metter DE (1977) Reproductive cycle of the northern water snake *Natrix s. sipedon* (Reptilia, Serpentes, Colubridae). *J. Herpetol.* **11**(1): 51–59.

Bauwens D & Verheyen RF (1985) The timing of reproduction in the lizard *Lacerta vivipara*: differences between individual females. *J. Herpetol.* **19** (3): 353–364.

Bellairs d'AA (1969) *The Life of Reptiles*, Vol 2, pp. 411–418. Weidenfeld and Nicholson, London.

Bellis MA & Baker RR (1990) Do females promote sperm competition? Data for humans. *Anim. Behav.* **40:** 997–999.

Bels V (1987) Observations of the courtship and the mating behaviour in the snake *Hydrodynastes gigas*. *J. Herpetol.* **21:** 350–352.

Bennett A, Cuthill IC, Partridge JC & Maier EJ (1996) Ultraviolet vision and mate choice in zebra finches. *Nature* **380:** 433–435.

Berry JF & Shine R (1980) Sexual size dimorphism and sexual selection in turtles (order Testudines). *Oecologia* **44:** 185–191.

Berry KH (1974) The ecology and social behavior of the chuckwalla, *Sauromalus obesus* Bairs. *Univ. Calif. Publ. Zool.* **101:** 1–60.

Birkhead TR & Møller AP (1993) Sexual selection and the temporal separation of reproductive events: sperm storage data from reptiles, birds and mammals. *Biol. J. Linn. Soc.* **50:** 295–311.

Birkhead TR, Møller AP & Sutherland WJ (1993) Why do females make it so difficult for males to fertilize their eggs? *J. Theor. Biol.* **161:** 51–60.

Bishop JE (1959) A histological and histochemical study of the kidney tubule of

by progesterone in the collared lizard (*Crotaphytis collaris*). *Herpetologica* **29:** 107–110.

Cooper WE, Jr & Greenberg N (1992) Reptilian coloration and behavior. In *Biology of the Reptilia, Vol. 18: Physiology E – Hormones, Brain and Behavior.* C Gans & D Crews (eds), pp. 299–400. The University of Chicago Press, Chicago.

Cooper WE, Jr & Vitt LJ (1984) Conspecific odour detection by the male broad-headed skink, *Eumeces laticeps*: effects of sex and site of odour source and of male reproductive condition. *J. Exp. Zool.* **230:** 199–209.

Cooper WE, Jr & Vitt LJ (1986) Tracking of female conspecific odor trails by male broad-headed skinks (*Eumeces laticeps*). **71:** 242–248.

Cooper WE, Jr & Vitt LJ (1987) Deferred agonistic behavior in a long-lived scincid lizard *Eumeces laticeps*. Field and laboratory data on the roles of body size and residency in agonistic strategy. *Oecologia* **72:** 321–326.

Cooper WE, Jr & Vitt LJ (1988) Orange head coloration of the male broad-headed skink (*Eumeces laticeps*, a sexually selected social cue. *Copeia* **1:** 1–6.

Cooper WE, Jr & Vitt LJ (1989) Sexual dimorphism of head and body size in an iguanid lizard: paradoxical results. *Am. Nat.* **133** (5): 729–735.

Cooper WE, Jr & Vitt LJ (1993) Female mate choice of large broad-headed skinks. *Anim. Behav.* **45:** 683–693.

Cooper WE, Jr, Adams CS & Dobie JL (1983) Female color change in the keeled earless lizard, *Holbrookia propinqua*: relationship to the reproductive season. *Southwest. Nat.* **28** (3): 275–280.

Cope ED (1896) On the hemipenis of the Sauria. *Proc. Nat. Sci. Acad. Philadel.* **48:** 461–467.

Cott HB (1961) Scientific results of an inquiry into the ecology and economic status of the Nile crocodile (*Crocodylus niloticus*) in Uganda and Northern Rhodesia. *Trans. Zool. Soc. Lond.* **29:** 211–356.

Courty Y & Dufaure JP (1979) Levels of testosterone in the plasma and testis of the viviparous lizard (*Lacerta vivipara* Jacquin) during the annual cycle. *Gen. Comp. Endocrinol.* **39:** 336–342.

Courty Y & Dufaure JP (1982) Circannual testosterone, dihydrotestosterone and androstenediols in plasma and testis of *Lacerta vivipara*, a seasonally breeding viviparous lizard. *Steroids* **39:** 517–529.

Crews D (1978) Hemipenile preference: stimulus control of male mounting behavior in the lizard *Anolis carolinensis*. *Science* **199:** 195–196.

Crowell Comuzzie DK & Owens, DW (1990) A quantitative analysis of courtship behavior in captive green sea turtles (*Chelonia mydas*). *Herpetologica* **46:** 195–202.

Cuellar HS (1973) Effect of adiposectomy and feeding level on FSH-induced testicular growth in male lizards *Anolis carolinensis* (Reptilia: Iguanidae). *J. Exp. Zool.* **185:** 65–72.

Cuellar HS, Roth JJ, Fawcett JD & Jones RE (1972) Evidence for sperm sustenance by secretions of the renal sexual segment of male lizards, *Anolis carolinensis*. *Herpetologica* **28:** 53–57.

Cuellar O (1966a) Delayed fertilization in the lizard *Uta stansburiana*. *Copeia* **3:** 549–552.

Cuellar O (1966b) Oviductal anatomy and sperm storage in lizards. *J. Morphol.* **19:** 7–20.

Daniel PM (1960) Growth and behavior in the west African lizard, *Agama agama africana*. *Copeia* **2:** 94–97.

Darwin C (1871) *The Descent of Man, and Selection in Relation to Sex*. Murray, London.

Davenport M (1995) Evidence of possible sperm storage in the caiman, *Paleosuchus palpebrosus*. *Herpetol. Rev.* **26** (1): 4–15.

Davis DD (1936) Courtship and mating behavior in snakes. *Zool. Ser. Field Mus. Nat. Hist* **XX**(22): 257–291.

DeNardo DF & Licht P (1993) Effects of corticosterone on social behavior of male lizards. *Horm. Behav.* **27:** 184–199.

DeNardo DF & Sinervo B (1994) Effects of steroid hormone interaction on activity and home-range size of male lizards. *Horm. Behav.* **28:** 273–287.

Denlinger DL (1980) Seasonal and annual variation of insect abundance in the Nairobi National Park, Kenya. *Biotropica* **12:** 100–106.

Deourcy KR & Jenssen TA (1994) Structure and use of male territorial headbob signals by the lizard *Anolis carolinensis*. *Anim. Behav.* **47:** 251–262.

Depeiges A & Dacheux JL (1985) Acquisition of sperm motility and its maintenance during storage in the lizard, *Lacerta vivipara*. *J. Reprod. Fertil.* **74:** 23–27.

Depeiges A, Force A & Dufaure J-P (1987) Production and glycolysation of sperm constitutive proteins in the lizard *Lacerta vivipara*. Evolution during the reproductive period. *Comp. Biochem. Physiol.* **86B** (2): 233–240.

Derickson WK (1976) Lipid storage and utilization in reptiles. *Am. Zool.* **16:** 711–723.

Deslippe RJ & M'Closkey RT (1991) An experimental test of mate defence in an iguanid lizard (*Sceloporus graciosus*). *Ecology* **72** (4): 1218–1224.

Devine MC (1975) Copulatory plugs in snakes: enforced chastity. *Science* **187:** 844–845.

Devine MC (1977) Copulatory plugs, restricted mating opportunities and reproductive competition among male garter snakes. *Nature* **267:** 345–346.

Devine MC (1984) Potential sperm competition in reptiles; behavioral and physiological consequences. In *Sperm Competition and the Evolution of Animal Mating Systems*. RL Smith (ed.), pp. 509–521. Academic Press, Orlando.

Dharmakumarsinhji KS (1946) Mating and the parental instinct of the marsh crocodile (*C. palustris* Lesson). *J. Bombay Nat. Hist. Soc.* **47:** 174–176.

Díaz JA (1993) Breeding coloration, mating opportunities, activity, and survival in the lacertid lizard *Psammodromus algirus*. *Can. J. Zool.* **71:** 1104–1110.

Díaz JA, Alonso-Gómez AL & Delgado MJ (1994) Seasonal variation of gonadal development, sexual steroids, and lipid reserves in a population of the lizard *Psammodromus algirus*. *J. Herpetol.* **8:** 199–205.

Dingle H & Khamala CPM (1972) Seasonal changes in insect abundance and biomass in an East African grassland with reference to breeding and migrating birds. *Ardea* **59:** 216–221.

Dowling HG & Savage JM (1960) A guide to the snake hemipenis: a survey of basic structures and systematic charactersistics. *Zoologica* **45:** 17–28.

Dugan B (1982) The mating behavior of the green iguana, *Iguana iguana*. In *Iguanas of the World – their Behavior, Ecology and Conservation*. GM Burghardt & AS Rand (eds), pp. 320–341. Noyes Publications, New Jersey.

Dugan B & Wiewandt TV (1982) Socio-ecological determinants of mating strategies in iguanine lizards. In *Iguanas of the World – their Behavior, Ecology and Conservation*. GM Burghardt & AS Rand (eds), pp. 303–319. Noyes Publications, New Jersey.

Dunlap KD & Schall JJ (1995) Hormonal alterations and reproductive inhibition in male fence lizards (*Sceloporus occidentalis*) infected with the malarial parasite *Plasmodium mexicanum*. *Physiol. Zool.* **68** (4): 608–621.

Duvall D & Schuett GW (1997) Straight-line movement and comparative mate searching in prairie rattlesnakes (*Crotalus viridis viridis*). *Animal Behaviour* **54**: 329–334.

Duvall D, Schuett GW & Arnold SJ (1987) Ecology and evolution of snake mating systems. In *Snakes: Ecology and Evolutionary Biology*. RA Seigel, JT Collins & SS Novak (eds), pp. 164–200. McGraw-Hill Publishing Company, New York.

Duvall D, Arnold SJ & Schuett GW (1992) Pitviper mating systems: ecological potential, sexual selection, and microevolution. In *The Biology of Pitvipers*. JA Campbell & ED Brodie (eds), pp. 165–200. Tyler, Selva, Texas.

Eberhard WG (1985) *Sexual Selection and Animal Genitalia*. Harvard University Press, Cambridge, MA.

Eberhard WG & Cordero C (1995) Sexual selection by cryptic female choice on male seminal products – a new bridge between sexual selection and reproductive physiology. *Trends Ecol. Evol.* **10** (12): 493–496.

Edgren RA (1953) Copulatory adjustment in snakes and its evolutionary implications. *Copeia* **3**: 162–164.

Edsman L (1989) Territoriality and competition in wall lizards. Ph.D. thesis, Department of Zoology, University of Stockholm, Sweden.

Emlen ST & Oring LW (1977) Ecology sexual selection, and the evolution of animal mating systems. *Science* **197**: 215–223.

Endler JA & McLellan T (1988) The processes of evolution: towards a newer synthesis. *Annot. Rev. Ecol. System.* **19**: 395–421.

Enquist M & Leimar O (1983) Evolution of fighting behaviour: decision rules and assessment of relative strength. *J. Theor. Biol.* **102**: 387–410.

Enquist M & Leimar O (1987) Evolution of fighting behaviour: the effect of variation in resource value. *J. Theor. Biol.* **127**: 187–205.

Ernst CH, Barbour RW & Lovich LE (1994) *Turtles of the United States and Canada*. Smithsonian Institution Press, Washington.

Esponda P & Bedford JM (1987) Post-testicular change in the reptile sperm surface with particular reference to the snake, *Natrix fasciata*. *J. Exp. Zool.* **241**: 123–132.

Estrada-Flores E, Villagran-Santa Cruz M, Mendez-de la Cruz FR & Casas-Andreau G (1990) Gonadal changes throughout the reproductive cycle of the viviparous lizard *Sceloporus mucronotus* (Sauria: Iguanidae). *Herpetologica* **46**: 43–50.

Estes R, de Queiroz K & Gauthier J (1988) Phylogenetic relationships within Squamata. In *Phylogenetic Relationships of the Lizard Families*. R Estes & G Pregill (eds), pp. 119–281. Stanford University Press, Stanford.

Etchberger CR & Stovall RH (1990) Seasonal variation in the testicular cycle of the loggerhead musk turtle, *Sternotherus minor minor*, from central Florida. *Can. J. Zool.* **68**: 1071–1074.

Etheridge R & de Queiroz K (1988) A phylogeny of Iguanidae. In *Phylogenetic Relationships of the Lizard Families*. R Estes & G Pregill (eds), pp. 283–367. Stanford University Press, Stanford.

Everett CT (1971) Courtship and mating of *Eumeces multivirgatus* (Scincidae). *J. Herpetol.* **5**: 3–4. 189–190.

Fearn S (1993) The tiger snake *Notechis scutatus* (Serpentes: Elapidae) in Tasmania. *Herpetofauna* **23**: 17–29.

Ferguson GW (1969) Interracial discrimination in male side blotched lizards, *Uta stansburiana*. *Copeia* **1**: 188–189.

Ferguson GW (1970) Mating behaviour of the side-blotched lizards of the genus *Uta* (Sauria: Iguanidae). *Anim. Behav.* **18:** 65–72.

Ferguson GW (1972) Species discrimination by male side-blotched lizards *Uta stansburiana* in Colorado. *Am. Midl. Nat.* **87** (2): 523–524.

Ferguson GW (1976) Color change and reproductive cycling in female collared lizards (*Crotaphytis collaris*). *Copeia* **3:** 491–494.

Ferguson MWJ (1985) Reproductive biology and embryology of the Crocodilians. In *Biology of the Reptilia, Vol. 14, Development A.* F Billett & PFA Maderson (eds), pp. 329–491. John Wiley and Sons, New York.

Fisher RA (1930) *The Genetical Theory of Natural Selection.* Clarendon Press, Oxford.

Fitch H (1955) Habits and adaptations of the great plains skink (*Eumeces obsoletus*). *Ecol. Monogr.* **25:** 59–83.

Fitch H (1956) An ecological study of the collared lizard (*Crotaphytis collaris*). *Univ. Kansas Mus. Nat. Hist. Publ.* **8:** 213–274.

Fitch H (1958) Natural history of the six-lined racerunner *Cnemidophorus sexlineatus. Univ. Kansas Mus. Nat. Hist. Publ.* **11:** 11–62.

Fitch H (1963) Natural history of the racer *Coluber constrictor. Univ. Kansas Mus. Nat. Hist. Publ.* **15:** 351–468.

Fitch H (1965) An ecological study of the garter snake *Thamnophis sirtalis. Univ. Kansas Mus. Nat. Hist. Publ.* **15**(10): 493–564.

Fitzgerald KT (1982) Mate selection as a function of body size and male choice in several lizard species. PhD. thesis, University of Colorado, Boulder, USA.

Fleischmann A (1902) Morphologische studien über Kloake und Phallus der Amnioten. *Morphol. Jahr.* **1902:** 539–581.

Fleishman LJ (1992) The influence of the sensory system and the environment on motion patterns in the visual displays of Anoline lizards and other vertebrates. *Am. Nat.* **139:** S36-S61.

Fleishmann LJ, Loew ER & Leal M (1993) Ultraviolet vision in lizards. *Nature* **365:** 397.

Flemming AF (1993) The male reproductive cycle of the lizard *Pseudocordylus m. melanotus* (Sauria: Cordylidae). *J. Herpetol.* **27** (4): 473–478.

Flemming AF (1994) Male and female reproductive cycles of the viviparous lizard, *Mabuya capensis* (Sauria: Scincidae) from South Africa. *J. Herpetol.* **28** (3): 334–341.

Flemming TH & Hooker RS (1975) *Anolis cupreus*: the response of a lizard to tropical seasonality. *Ecology* **56:** 1243–1261.

Flores-Villela OA & Zug GR (1995) Reproductive biology of the Chopontil, *Claudius angustatus* (Testudines: Kinosternidae) in Southern Veracruz, México. *Chelonian Conservation and Biology* **1** (3): 181–186.

Fox SF (1975) Natural selection on morphological phenotypes of the lizard *Uta stansburiana. Evolution* **29:** 95–107.

Fox W (1954) Genetic and environmental variation in the timing of the reproductive cycles of male garter snakes. *J. Morphol.* **95:** 415–450.

Fox W (1956) Seminal receptacles of snakes. *Anat. Res.* **124:** 519–533.

Fox W (1958) Sexual cycle of the male lizard, *Anolis carolinensis. Copeia* **1:** 22–29.

Fox W (1977) The urogenital system of the reptiles. In *Biology of the Reptilia, Vol. 6 – Morphology E.* C Gans & TS Parsons (eds), pp. 1–157. Academic Press, London.

Fox W & Dessauer H (1962) The single right oviduct and other urogenital structure of female *Typhlops* and *Leptotyphlops. Copeia* **3:** 590–597.

Frisch O, von (1958) Zur Biologie des Zwergchamäleons (*Microsaurus pumilus*). *Z. Tierpsychol.* **19:** 276–289.

Frost DR & Etheridge R (1989) A phylogenetic analysis and taxonomy of iguanian lizards (Reptilia: Squamata). *Univ. Kansas Mus. Nat. Hist. Misc. Publ.* **81:** 1–49.

Fyfe G & Munday B (1988) Captive breeding of desert adders (*Acanthophis pyrrhus*). *Herpetofauna* **18:** 21.

Galbraith DA (1993) Review: multiple paternity and sperm storage in turtles. *Herpetol. J.* **3:** 117–123.

Galbraith DA, White BN, Brooks RJ & Boag PT (1993) Multiple paternity in clutches of snapping turtles (*Chelydra serpentina*) detected using DNA fingerprints. *Can. J. Zool.* **71:** 318–324.

Garland T, Jr, Hankins E & Huey RB (1990) Locomotor capacity and social dominance in male lizards. *Funct. Ecol.* **4:** 243–250.

Garrick L (1975) Structure and pattern of the roars of Chinese alligators (*Alligator sinensis* Fauvel). *Herpetologica* **31** (1): 26–31.

Garstka WR & Gross M (1990) Activation and inhibition of sperm motility by kidney products in the turtle, *Trachemys scripta*. *Comp. Biochem. Physiol.* **95A:** 329–335.

Gibbons JRH & Watkins IF (1982) Behavior, ecology and conservation of south Pacific banded iguanas, *Brachylophus*, including a newly discovered species. In *Iguanas of the World – their Ecology, Behavior and Conservation*. GM Burghardt & AS Rand (eds), pp. 418–441. Noyes Publications, New Jersey.

Gibson AR & Falls JB (1975) Evidence for multiple insemination in the common garter snake, *Thamnophis sirtalis*. *Can. J. Zool.* **53:** 1362–1368.

Gibson AR & Falls JB (1979) Thermal biology of the common garter snake *Thamnophis sirtalis*. I. Temporal variation, environmental effects and sex differences. *Oecologia* **43:** 79–97.

Gillingham JC (1977) Further analysis of reproductive behavior in the western fox snake *Elaphe v. vulpina*. *Herpetologica* **33:** 349–353.

Gillingham JC & Chambers JA (1982) Courtship and pelvic spur use in the Burmese python, *Python molurus bivittatus*. *Copeia* **1:** 193–196.

Gillingham JC, Carpenter CC & Murphy JB (1983) Courtship, male combat and dominance in the western diamondback rattlesnake, *Crotalus atrox*. *J. Herpetol.* **17:** 265–270.

Gist DH & Fischer EN (1993) Fine structure of the sperm storage tubules in the box turtle oviduct. *J. Reprod. Fertil.* **97:** 463–468.

Gist DH & Jones JM (1987) Storage of sperm in the reptilian oviduct. *Scan. Microsc.* **1:** 1839–1849.

Gist DH & Jones JM (1989) Sperm storage within the oviduct of turtles. *J. Morphol.* **199:** 379–384.

Gist DH, Hess RA & Thurston RJ (1992) Cytoplasmic droplets of painted turtle spermatozoa. *J. Morphol.* **214:** 153–158.

Glinski TH & Krekorian O (1985) Individual recognition in free-living adult male desert iguanas, *Dipsosaurus dorsalis*. *J. Herpetol.* **19:** 544–546.

Gloyd HK (1947) Notes on the courtship and mating behavior of certain snakes. *Nat. Hist. Misc. Chicago Acad. Sci.* **12:** 1–4.

Goin OB (1957) An observation of mating in the broad-headed skink, *Eumeces laticeps*. *Herpetologica* **13:** 155–156.

Goldberg SR (1974) Reproduction in mountain and lowland populations of the lizard *Sceloporus occidentalis*. *Copeia* **1:** 176–182.

Goldberg SR (1976) Reproduction in a mountain population of the coastal whiptail lizard, *Cnemidophorus tigris multiscutatus*. *Copeia* **2:** 260–266.

Goldberg SR & Bezy RL (1974) Reproduction in the island night lizard *Xantusia riversiana*. *Herpetologica* **30** (4): 350–360.

Gorman GC, Licht P & McCollum F (1981) Annual reproductive patterns in three species of marine snakes from the central Philippines. *J. Herpetol.* **15** (3): 335–354.

Greenberg B (1943) Social behavior of the western banded gecko, *Coleonyx variegatis* Baird. *Physiol. Zool.* **XVI**(1): 110–122.

Greenberg B & Noble GK (1944) Social behavior of the American Chameleon. *Physiol. Zool.* **17:** 392–439.

Greenberg N, Chen T & Crews D (1984) Social status, gonadal status, and the adrenal stress response in the lizard, *Anolis carolinensis*. *Horm. Behav.* **18:** 1–11.

Guillette LJ, Jr (1983) Notes concerning reproduction of the montane skink *Eumeces copei*. *J. Herpetol.* **17** (2): 144–148.

Guillette LJ, Jr & Casas-Andreu G (1981) Seasonal variation in fat body weights of the Mexican high elevation lizard *Sceloporus grammicus microlepidotus*. *J. Herpetol.* **15:** 366–371.

Guillette LJ, Jr & Casas-Andreu G (1987) The reproductive biology of the high elevation Mexican lizard *Barisia imbricata*. *Herpetologica* **43** (1): 29–38.

Guillette LJ, Jr & Mendez-de la Cruz FR (1993) The reproductive cycle of the viviparous Mexican lizard *Sceloporus torquatus*. *J. Herpetol.* **27:** 168–174.

Guillette LJ, Jr & Sullivan WP (1985) The reproductive and fat body cycles of the lizard *Sceloporus formosus*. *J. Herpetol.* **19:** 474–480.

Haagner GV & Carpenter G (1988) Notes on the reproduction of captive forest cobras, *Naja melanoleuca* (Serpentes: Elapidae). *J. Herpetol. Assoc. Afr.* **34:** 35–37.

Hailey A & Davies PMC (1987) Maturity, mating and age-specific reproductive effort of the snake *Natrix maura*. *J. Zool.* **211:** 573–587.

Haldar C & Thapliyal JP (1977) Effects of pinealectomy on the annual testicular cycle of *Calotes versicolor*. *Gen. Comp. Endocrinol.* **32:** 395–399.

Halliday TR & Verrell PA (1988) Body size and age in amphibians and reptiles. *J. Herpetol.* **22:** 253–265.

Halpert AP, Gartska WR & Crews D (1982) Sperm transport and storage and its relation to the annual sexual cycle of the female red-sided garter snake, *Thamnophis sirtalis parietalis*. *J. Morphol.* **174:** 149–159.

Hamid KS & Akbarsha MA (1989) Utilization of seminal proteins by the house gecko, *Hemidactylus brooki* (Gray), sperm for motility. *Ind. J. Exp. Biol.* **27:** 930–933.

Hamilton WD & Zuk M (1982) Heritable true fitness and bright birds: a role for parasites. *Science* **218:** 384–387.

Hansen TF & Price DK (1995) Good genes and old age: do old mates provide superior genes. *J. Evol. Biol.* **8:** 759–778.

Harris V (1964) *The Life of the Rainbow Lizard*. Hutchinson and Co., London.

Harry JL & Briscoe DA (1988) Multiple paternity in the loggerhead turtle (*Caretta caretta*). *J. Hered.* **79:** 96–99.

Hasegawa M (1994) Demography, social structure and sexual dimorphism of the lizard *Eumeces okade*. In *Animal Societies – Individuals, Interactions and Organisation*. PJ Jarman & A Rossiter (eds), *Physiology and Ecology Japan* **29:** 248–263. Kyoto University Press, Kyoto.

Healy JM & Jamieson BGM (1992) Ultrastructure of the spermatozoon of the tuatara (*Sphenodon punctatus*) and its relevance to the relationships of the Sphenodontida. *Phil. Trans. Roy. Soc. Lond. Ser. B* **335:** 193–205.

Hedrick PW (1994) Purging inbreeding depression and the probability of extinction: full-sib mating. *Heredity* **73:** 363–372.

Heulin B (1988) Observations sur l'organisation de la reproduction et sur les comportements sexuels et agonistiques chez *Lacerta vivipara*. *Vie Milieu* **38** (2): 177–187.

Hews DK (1990) Examining hypotheses generated by field measures of sexual selection on male lizards, *Uta palmeri*. *Evolution* **44:** 1956–1966.

Hews DK (1993) Food resources affect female distribution and male mating opportunities in the iguanian lizard *Uta palmeri*. *Anim. Behav.* **46:** 279–291.

Hews DK & Moore MC (1995) Influence of androgens on differentiation of secondary sex characters in tree lizards, *Urosaurus ornatus*. *Gen. Comp. Endocrinol.* **97:** 86–102.

Hews D & Moore M (1996) A critical period for the organization of alternative male phenotypes of tree lizards by exogenous testosterone. *Physiol. Behav.* **60** (2): 425–429.

Hews D, Knapp R & Moore MC (1994) Early exposure to androgens affects adult expression of alternative male phenotypes in tree lizards. *Horm. Behav.* **28:** 96–115.

Hicks RA & Trivers RL (1983) The social behavior of *Anolis valencienni*. In *Advances in Herpetology and Evolutionary Biology: Essays in Honor of Ernest E. Williams*. AGJ Rhodin & K Miyata (eds), pp. 570–595. Museum of Comparative Zoology, Cambridge, MA.

Hoffman LH & Wimsatt WA (1972) Histochemical and electron microscopic observations on the sperm receptacles in the garter snake oviduct. *Am. J. Anat.* **134** (1): 71–96.

Höggren M (1995) Mating strategies and sperm competition in the adder (*Vipera berus*). Ph.D. thesis, Uppsala University, Sweden.

Höggren M & Tegelström H (1995) DNA fingerprinting shows within-season multiple paternity in the adder (*Vipera berus*). *Copeia* **2:** 271–277.

Hover EL (1985) Differences in aggressive behavior between two throat color morphs in a lizard, *Urosaurus ornatus*. *Copeia* **4:** 933–940.

Hughes B (1990) Captive breeding of the black-headed python (*Aspidites melanocephalus*). *Herpetofauna* **20:** 5–6.

Inger RF & Greenberg B (1966) Annual reproductive patterns of lizards from a Bornean rain forest. *Ecology* **47** (6): 1007–1021.

Inoué S & Inoué Z (1977) Colour changes induced by pairing and painting in the male rainbow lizard, *Agama agama agama*. *Experientia* **33:** 1443–1444.

Jackson JF & Telford SR, Jr (1974) Reproductive ecology of the Florida scrub lizard *Sceloporus woodi*. *Copeia* **3:** 689–694.

Jackson WM (1991) Why do winners keep winning? *Behav. Ecol. Sociobiol.* **28:** 271–276.

James CD (1991) Annual variation in reproductive cycles of Scincid lizards (*Ctenotus*) in central Australia. *Copeia* **3:** 744–760.

Jameson EW, Jr (1974) Fat and breeding cycles in a montane population of *Sceloporus graciocus*. *J. Herpetol.* **8:** 311–322.

Jameson EW, Jr & Alison A (1976) Fat and breeding cycles in two mountain populations of *Sceloporus occidentalis* (Reptilia, Lacertilia, Iguanidae). *J. Herpetol.* **10:** 211–220.

Jenssen TA (1970) Female response to filmed displays of *Anolis nebulosus*. *Anim. Behav.* **18:** 640–647.
Jenssen TA & Nunez SC (1994) Male and female reproductive cycles of the Jamaican lizard, *Anolis opalinus*. *Copeia* **3:** 767–780.
Johnson LF & Jacob JS (1984) Pituitary activity and reproductive cycle of male *Cenmidophorus sexlineatus* in west Tennessee. *J. Herpetol.* **18:** 396–405.
Jun-Yi, Lin (1982) Sperm retention in the lizard *Chamaeleo hoehnelii*. *Copeia* **2:** 488–489.
Kaufmann JH (1992) The social behavior of wood turtles, *Clemmys insculpta*, in central Pennsylvania. *Herpetol. Monogr.* **6:** 1–25.
Kay FR, Miller BW & Miller CL (1970) Food habits and reproduction of *Callisaurus draconoides* in death valley, California. *Herpetologica* **26:** 431–436.
Keller L & Reeve HK (1995) Why do females mate with multiple males? The sexually selected sperm hypothesis. *Adv. Stud. Behav.* **24:** 291–315.
Kennedy JL (1978) Field observations on courtship and copulation in the eastern king snake and the four-lined rat snake. *Herpetologica* **34** (1): 51–52.
Khaire A, Khaire N & Katdare M (1985) Observation on feeding, mating, egg laying and the seasonal moulting of the checkered keel back water snake (*Xenochrophis piscator*). *Snake* **17:** 25–30.
King R (1989) Sexual dimorphism in snake tail length: sexual selection, natural selection, or morphological constraint? *Biol. J. Linn. Soc.* **38:** 133–154.
Klauber LM (1956) *Rattlesnakes: their Habits, Life Histories, and Influence on Mankind*. University of California Press, Berkeley.
Klaver C & Böhme W (1986) Phylogeny and classification of the Chamaeleoniae (Sauria) with special reference to hemipenis morphology. *Bonn. Zool. Monogr.* **22:** 5–64.
Kofron CP (1990) The reproductive cycle of the Nile crocodile (*Crocodylus niloticus*). *J. Zool. Lond.* **221:** 477–488.
Kofron C (1993) Behavior of Nile crocodiles in a seasonal river in Zimbabwe. *Copeia* **2:** 463–469.
Krekorian CO (1976) Home-range size and overlap and their relationship to food abundance in the desert iguana, *Dipsosaurus dorsalis*. *Herpetologica* **32:** 405–412.
Krohmer RW & Aldridge RD (1985) Male reproductive cycle of the lined snake (*Tropidoclonion lineatum*). *Herpetologica* **41:** 33–38.
Kurfess JF (1967) Mating, gestation, and growth rate in *Lichanaura r. roseofusca*. *Copeia* **2:** 477–479.
Lance V (1987) Hormonal control of reproduction in Crocodilians. In *Wildlife Management: Crocodiles and Alligators*. GJW Webb, SC Manolis & PJ Whitehead (eds), pp. 409–41SW. Surrey Beatty and Sons Pty, Ltd., Chipping Norton.
Lande R (1982) A quantitative genetic theory of life-history evolution. *Ecology* **63:** 607–615.
Lande R & Arnold SJ (1983) The measurement of selection on correlated characters. *Evolution* **37:** 1210–1226.
Lang JW (1987) Crocodilian behaviour: implications for management. In *Wildlife Management: Crocodiles and Alligators*. GJW Webb, SC Manolis & PJ Whitehead (eds), pp. 273–293. Surrey Beatty and Sons Pty, Ltd., Chipping Norton.
Lardie RL (1975) Courtship and mating behavior in the yellow mud turtle, *Kinosternon flavescens flavescens*. *J. Herpetol.* **9** (2): 223–227.

Larsson H-O & Wihman J (1989) Breeding the Cuba crocodile, *Crocodylus rhombifer* at Skansen Aquarium. *Int. Zoo Year.* **28:** 110–113.

Lefcourt H & Blaustein AR (1991) Parasite load and brightness in lizards: an interspecific test of the Hamilton and Zuk hypothesis. *J. Zool.* **224:** 491–499.

Legge RE (1967) Mating behaviour of the American alligators, *Alligator mississippiensis. Int. Zoo Year.* **7:** 179–180.

Legler JM (1955) Observations on the sexual behavior of captive turtles. *Lloydia* **18:** 95–99.

Legler JM (1960) Natural history of the ornate box turtle, *Terrapene ornata ornata* Agassiz. *Univ. Kansas Publ. Mus. Nat. Hist.* **11** (10): 527–669.

Leptien R (1989) Zur Haltung eines Weibchens von *Chamaeleo ellioti* Günther, 1895, mit dem Nachweis von Amphigonia retardata. *Salamandra* **25** (1): 21–24.

Leuckart R (1847) *Zur Morphologie und Anatomie der Geschlectsorgane.* Gottingen.

Leviton AE (1961) Mating behavior of the panamint lizard, *Gerrhonotus panamintus* Stebbins. *Herpetologica* **17:** 204–206.

Leuck BE (1995) Territorial defence by male green anoles: an experimental test of the roles of residency and resource quality. *Herpetol. Monogr.* **9:** 63–74.

Lewke RE (1979) Neck-biting and other aspects of reproductive biology of the Yuma king snake (*Lampropeltis getulus*). *Herpetologica* **35** (2): 154–157.

Licht P (1965) The relation between preferred body temperatures and testicular heat sensitivity in lizards. *Copeia* **4:** 428–436.

Licht P (1974) Response of *Anolis* lizards to food supplementation in Nature. *Copeia* **1:** 215–221.

Licht P (1982) Endocrine patterns in the reproductive cycle of turtles. *Herpetologica* **38** (1): 51–61.

Licht P (1984) Reptiles. In *Marshall's Physiology of Reproduction – Vol. 1 – Reproductive Cycles of Vertebrates,* 4th edn. GE Lemming (ed.), pp. 206–282. Churchill Livingstone, Edinburgh.

Licht P & Gorman GC (1970) Reproductive and fat cycles in Caribbean *Anolis* lizards. *Univ. Calif. Publ. Zool.* **95:** 1–52.

Licht P & Gorman GC (1975) Altitudinal effects on the seasonal testis cycles of tropical *Anolis* lizards. *Copeia* **3:** 496–504.

Lofts B, Phillips JG & Tam WH (1966) Seasonal changes in the testis of the cobra, *Naja naja* (Linn.). *Gen. Comp. Endocrinol.* **6:** 466–475.

Losos JB (1985) An experimental demonstration of the species-recognition role of *Anolis* dewlap color. *Copeia* **4:** 905–910.

Lowe CH, Jr (1942) Notes on the mating of desert rattlesnakes. *Copeia* **4:** 261–262.

Ludwig M & Rahn H (1943) Sperm storage and copulatory adjustment in the prairie rattlesnake. *Copeia* **1:** 15–18.

Luiselli L (1993) Are sperm storage and within-season multiple mating important components of the adder reproductive biology? *Acta Oecol.* **14** (5): 705–710.

MacMahon JA (1957) Observations on mating in the corn snake, *Elaphe guttata guttata. Copeia* **3:** 232.

Madsen T (1987) Natural and sexual selection in grass snakes, *Natrix natrix*, and adders, *Vipera berus.* Ph.D. thesis, University of Lund, Sweden.

Madsen T & Loman J (1987) On the role of colour display in the social and spatial organization of male rainbow lizards (*Agama agama*). *Amphibia–Reptilia* **8:** 365–372.

Madsen T & Shine R (1992) A rapid, sexually selected shift in mean body size in a population of snakes. *Evolution* **46** (4): 1220–1224.

Madsen T & Shine R (1993a) Male mating success and body size in European grass snakes. *Copeia* **2:** 561–564.

Madsen T & Shine R (1993b) Temporal variability in sexual selection acting on reproductive tactics and body size in male snakes. *Am. Nat.* **141** (1): 167–171.

Madsen T & Shine R (1994a) Components of lifetime reproductive success in adders, *Vipera berus*. *J. Anim. Ecol.* **63:** 561–568.

Madsen T & Shine R (1994b) Costs of reproduction influence the evolution of sexual size dimorphism in snakes. *Evolution* **48** (4): 1389–1397.

Madsen T, Shine R, Loman J & Håkansson T (1992) Why do female adders copulate so frequently? *Nature* **355:** 440–441.

Madsen T, Shine R, Loman J & Håkansson T (1993) Determinants of mating success in male adders, *Vipera berus*. *Anim. Behav.* **45:** 491–499.

Madsen T, Stille B & Shine R (1995) Inbreeding depression in an isolated population of adders, *Vipera berus*. *Biol. Cons.* **75:** 113–118.

Mahmoud IY (1967) Courtship behavior and sexual maturity in four species of Kinosternid turtles. *Copeia* **2:** 314–319.

Mahmoud IY & Klicka J (1972) Seasonal gonadal changes in Kinosternid turtles. *J. Herpetol.* **6:** 3–4, 183–189.

Manimekalai M (1993) Investigation on the epididymis and its role in the secretion of seminal proteins in an agamid lizard and a fresh water turtle. Ph.D. thesis, Bharathidasan University, Tiruchirapalli, India.

Manimekalai M & Akbarsha MA (1992) Secretion of glycoprotein granules in the epididymis of the agamid lizard *Calotes versicolor* (Daudin) is region-specific. *Biol. Struct. Morphogen.* **4**(3): 96–101.

Mann T & Lutwak-Mann C (1981) *Male Reproductive Semen and Function.* Springer-Verlag, Berlin.

Mao SH, Yin FY & Guo YW (1984) The hemipenes of common Taiwanese venemous snakes. *Herpetologica* **40** (4): 406–410.

Marcelini DL (1978) The acoustic behavior of lizards. In *Behavior and Neurology of Lizards.* N Greenberg & PD MacLean (eds), pp. 287–300. National Institute of Mental Health, Rockville, Maryland.

Marion KR & Sexton OJ (1971) The reproductive cycle of the lizard *Sceloporus malachiticus*. *Copeia* **3:** 517–526.

Marler CA & Moore MC (1988) Evolutionary costs of aggression revealed by testosterone manipulations in free-living male lizards. *Behav. Ecol. Sociobiol.* **23:** 21–26.

Marler CA & Moore MC (1989) Time and energy costs of aggression in testosterone-implanted free-living, male mountain spiny lizards (*Sceloporus jarrovi*). *Physiol. Zool.* **62** (6): 1334–1350.

Marler CA & Moore MC (1991) Supplementary feeding compensates for testosterone-induced costs of aggression in male mountain spiny lizards, *Sceloporus jarrovi*. *Anim. Behav.* **42:** 209–219.

Martins E (1993) A comparative study of the evolution of *Sceloporus* push-up displays. *Am. Nat.* **142** (6): 994–1018.

Martins EP (1994) Phylogenetic perspectives on the evolution of lizard territoriality. In *Lizard Ecology – Historical and Experimental Perspectives.* LJ Vitt & ER Pianka (eds), pp. 117–144. Princeton University Press, Princeton.

Mason RT (1992) Reptilian pheromones. In *Biology of the Reptilia, Vol. 18, Physiology E: Hormones, Brain, and Behavior.* C Gans & D Crews (eds), pp. 114–228. The University of Chicago Press, Chicago.

Mason R & Crews D (1985) Female mimicry in garter snakes. *Nature* **316:** 59–60.

Mayhew WW (1963) Reproduction in the granite spiny lizard, *Sceloporus orcutti*. *Copeia* **1:** 144–152.
Mayhew WW (1966a) Reproduction in the psammophilus lizard, *Uma scoparia*. *Copeia* **1:** 114–122.
Mayhew WW (1966b) Reproduction in the arenicolous lizards, *Uma notata*. *Ecology* **47** (1): 9–18.
Mayhew WW (1968) Biology of desert amphibians and reptiles. In *Desert Biology*. GW Brown (ed.), p. 262. Academic Press, New York.
Mayhew WW (1971) Reproduction in the desert lizard, *Dipsosaurus dorsalis*. *Herpetologica* **27** (1): 57–77.
Maynard Smith J (1982) *Evolution and the Theory of Games*. Cambridge University Press, Cambridge.
Maynard Smith J (1996) The games lizards play. *Nature* **380:** 198–199.
M'Closkey RT, Baia KA & Russell RW (1987) Tree lizard (*Urosaurus ornatus*) territories: experimental perturbation of the sex ratio. *Ecology* **68** (6): 2059–2062.
M'Closkey RT, Deslippe RJ, Szpak CP & Baia KA (1990) Ecological correlates of the variable mating system of an iguanid lizard. *Oikos* **59** (1): 63–69.
McCranie JR (1988) Description of the hemipenis of *Sistrurus ravus* (Serpentes: Viperidae). *Herpetologica* **44:** 123–126.
McKinney RB & Marion KR (1985) Reproductive and fat body cycles in the male lizard, *Sceloporus undulatus*, from Alabama, with comparisons of geographic variation. *J. Herpetol.* **19** (2): 208–217.
McMann S (1993) Contextual signalling and the structure of dyadic encounters in *Anolis carolinensis*. *Anim. Behav.* **46:** 657–668.
McPherson RJ & Marion KR (1981) Seasonal testicular cycle of the stinkpot (*Sternotherus odoratus*) in central Alabama. *Herpetologica* **37:** 33–40.
Medica PA, Turner FB & Smith DD (1973) Hormonal induction of color change in female leopard lizards, *Crotaphytis wislizenii*. *Copeia* **4:** 658–661.
Meeran MM (1994) Lizard male reproductive tract: anatomy and role. Ph.D. thesis, Bharathidasan University, Tiruchirapalli, India.
Meier A, Trobec TN, Haymaker HG, MacGregor III R & Russo AC (1973) Daily variations in the effects of handling on fat storage and testicular weights in several vertebrates. *J. Exp. Zool.* **184:** 281–288.
Méndez de la Cruz F, Guillette LJ, Jr, Villagrán Santa Cruz M & Casas-Andreu G (1988) Reproductive and fat body cycles of the viviparous lizard, *Sceloporus mucronotus* (Sauria: Iguanidae). *J. Herpetol.* **22** (1): 1–12.
Mengden GA, Platz CG, Hubbard R & Quinn H (1980) Semen collection, freezing and artificial insemination in snakes. In *SSAR Contribution to Herpetology Number 1: Reproductive Biology and Diseases of Captive Reptiles*. JB Murphy & JT Collins (eds), pp. 71–78. Maseraull Printing Inc., Lawrence.
Miller MR (1948) The seasonal histological changes occurring in the ovary, corpus luteum and testis of the viviparous lizard *Xantusia vigilis*. *Univ. Calif. Publ. Zool.* **47:** 197–224.
Milstead WW (1968) *Lizard Ecology – a Symposium*. University of Missouri Press, Columbia.
Mishima S, Sawai Y, Yamasato S & Sawai K (1972) Studies on a natural monument, Sirohebi (albino *Elaphe climacophora*) on the Iwakuni in Japan. 3. Observations on copulation, egg-laying and hatching of the Shirohebi (1). *Snake* **9:** 14–26.
Mitchell CM & Zug GR (1984) Spermatogenic cycle of *Nerodia taxispilota* (Serpentes: Colubridae) in southern Virginia. *Herpetologica* **40:** 200–204.

Mitchell FJ (1973) Studies on the ecology of the agamid lizard *Amphibolurus maculosus* (Mitchell). *Trans. Roy. Soc. S. Aust.* **97:** 47–76.

Mitchell JC (1985) Variation in the male reproductive cycle in a population of painted turtles, *Chrysemys picta*, from Virginia. *Herpetologica* **41:** 45–51.

Moll EO & Legler JM (1971) The life history of a neotropical slider turtle, *Pseudemys scripta* (Schoepff) in Panama. *Bull. Los Ang. Co. Mus. Nat. Hist. Sci.* **11:** 1–102.

Møller AP (1991) Fluctuating asymmetry in male sexual ornaments may reliably reveal male quality. *Anim. Behav.* **40:** 1185–1187.

Møller AP (1992) Female swallow preference for symmetrical male sexual ornaments. *Nature* **357:** 238–240.

Møller AP (1994a) *The Barn Swallow and Sexual Selection.* Oxford University Press, Oxford.

Møller AP (1994b) Directional selection on directional asymmetry: testes size and secondary sexual characters in birds. *Proc. Roy. Soc. Lond. Ser. B* **258:** 147–151.

Montanucci RR (1982) Mating and courtship-related behaviours of the short-horned lizard, *Phrynosoma douglassi. Herpetologica* **4:** 971–974.

Moore MC & Lindzey J (1992) The physiological basis of sexual behavior in male reptiles. In *Biology of the Reptilia, Vol. 18, Physiology E – Hormones, Brain, and Behaviour.* C Gans & D Crews (eds), pp. 70–113. The University of Chicago Press, Chicago.

Mount RH (1963) The natural history of the red-tailed skink *Eumeces egregius* Baird. *Am. Midl. Nat.* **70:** 356–385.

Müller H (1971) Ecological and ethological studies on *Cnemidophorus llemniscatus* l (Reptilia: Tejidae) in Colombia. *Forma Et Functio* **4:** 189–224.

Murphy JB & Barker DG (1980) Courtship and copulation of the Ottoman viper (*Vipera xanthina*) with special reference to use of hemipenes. *Herpetologica* **36** (2): 165–170.

Murphy JB & Lamoreaux WE (1978) Mating behaviour in three Australian chelid turtles (Testudines: Pleurodira: Chelidae). *Herpetologica* **34:** 398–405.

Murphy JB, Tryon BW & Brecke BJ (1978) An inventory of reproduction and social behavior in captive grey-banded kingsnakes, *Lampropeltis mexicana alternata* (Brown). *Herpetologica* **34:** 84–93.

Nakamoto E & Toriba M (1983) Some observations on mating behavior of Burmese python, *Python molurus bivittatus* Kuhl. *Snake* **15:** 110–112.

Nayernia K, Adham I, Kremling H, Reim K, Schlicker M, Schlüter G & Engel W (1996) Stage and developmental specific gene expression during mammalian spermatogenesis. *Int. J. Develop. Biol.* **40:** 379–383.

Newlin ME (1976) Reproduction in the bunch grass lizard, *Sceloporus scalaris. Herpetologica* **32:** 171–184.

Newton WD & Trauth SE (1992) Ultrastructure of the spermatozoon of the lizard *Cnemidophorus sexlineatus* (Sauria: Teiidae). *Herpetologica* **48** (3): 330–343.

Niblick HA, Rostal DC & Classen T (1994) Role of male–male interactions and female choice in the mating system of the desert tortoise, *Gopherus agassizii. Herpetol. Monogr.* **8:** 124–132.

Nilson G (1981) Ovarian cycle and reproductive dynamics in the female adder, *Vipera berus* (Reptilia, Viperidae). *Amphibia–Reptilia* **2:** 63–82.

Nilson G & Andrén C (1982) Function of renal sex secretion and male hierarchy in the adder, *Vipera berus*, during reproduction. *Horm. Behav.* **16:** 404–413.

Noble GK (1937) The sense organs involved in the courtship of *Storeria, Thamnophis* and other snakes. *Bull. Am. Mus. Nat. Hist.* **73:** 673–725.
Noble GK & Bradley HT (1933) The mating behavior of lizards: its bearing on the theory of sexual selection. *Ann. NY Acad. Sci.* **XXXV:** 25–100.
Norris DO (1987) In *Hormones and Reproduction in Fishes, Amphibians, and Reptiles.* DO Norris & RE Jones (eds), pp. 339–340. Plenum Press, New York.
Oliver JA (1956) Reproduction in the king cobra, *Ophiophagus hannah* Cantor. *Zoologica* **41:** 145–152.
Olsson M (1992a) Contest success in relation to size and residency in male sand lizards, *Lacerta agilis. Anim. Behav.* **44:** 386–388.
Olsson M (1992b) Sexual selection and reproductive strategies in the sand lizard *(Lacerta agilis)*. Ph.D. thesis, University of Göteborg, Sweden.
Olsson M (1993a) Contest success and mate guarding in male sand lizards, *Lacerta agilis. Anim. Behav.* **46:** 408–409.
Olsson M (1993b) Nuptial coloration and predation risk in model sand lizards, *Lacerta agilis. Anim. Behav.* **46:** 410–412.
Olsson M (1993c) Male preference for large females and assortative mating for body size in the sand lizard (*Lacerta agilis*). *Behav. Ecol. Sociobiol.* **32:** 337–341.
Olsson M (1994a) Nuptial coloration in the sand lizard, *Lacerta agilis*: an intrasexually selected cue to fighting ability. *Anim. Behav.* **48:** 607–613.
Olsson M (1994b) Why are sand lizard males (*Lacerta agilis*) not equally green? *Behav. Ecol. Sociobiol.* **35:** 169–173.
Olsson M (1994c) Rival recognition affects male contest behavior in sand lizards (*Lacerta agilis*). *Behav. Ecol. Sociobiol.* **35:** 249–252.
Olsson M (1995a) Forced copulation and costly female resistance behavior in the Lake Eyre dragon, *Ctenophorus maculosus. Herpetologica* **51** (1): 19–24.
Olsson M (1995b) Territoriality in Lake Eyre dragons *Ctenophorus maculosus*: are males superterritorial? *Ethology* **101:** 222–227.
Olsson M & Madsen T (1995) Female choice on male quantitative traits in lizards – why is it so rare? *Behav. Ecol. Sociobiol.* **36:** 179–184.
Olsson M & Madsen T (1996) Costs of mating with infertile males selects for late emergence in female sand lizards (*Lacerta agilis* L.). *Copeia* **2:** 462–464.
Olsson M & Shine R (1996) How and why does reproductive success increase with age? A case study using sand lizards (*Lacerta agilis*). *Oecologia* **105:** 175–178.
Olsson M, Gullberg A & Tegelström H (1994a) Sperm competition in the sand lizard, *Lacerta agilis. Anim. Behav.* **48:** 193–200.
Olsson M, Gullberg A, Tegelström H, Madsen T & Shine R (1994b) Female 'promiscuity' enhances offspring fitness in a lizard. (Published under the Scientific correspondence 'Can adders multiply'.) *Nature* **368:** 528.
Olsson M, Madsen T, Shine R, Gullberg A & Tegelström H (1994c) Rewards of promiscuity. *Nature* **372:** 230.
Olsson M, Gullberg A & Tegelström H (1996a) Mate guarding in male sand lizards (*Lacerta agilis*). *Behaviour* **133:** 367–386.
Olsson M, Gullberg A & Tegelström H (1996b) Malformed offspring, sibling matings, and selection against inbreeding in the sand lizard (*Lacerta agilis*). *J. Evol. Biol.* **9:** 229–242.
Olsson M, Gullberg A, Shine R, Madsen T & Tegelström H (1996c) Paternal genotype influences incubation period, offspring size, and offspring shape in an oviparous reptile. *Evolution* **50:** 1328–1333.
Olsson M, Shine R, Gullberg A, Madsen T & Tegelström H (1996d) Female lizards

control the paternity of their offspring by selective use of sperm. *Nature* **383:** 585.

Orr TE & Mann DR (1992) Role of glucocorticoids in the stress-induced suppression of testicular steroidogenesis in adult male rats. *Horm. Behav.* **26:** 350–363.

Ortega A (1985) Fat body cycles in a montane population of *Sceloporus grammicus*. *J. Herpetol.* **20** (1): 109–111.

Palmer BD & Guillette LJ, Jr (1988) Histology and functional morphology of the female reproductive tract of the tortoise *Gopherus polyphemus*. *Am. J. Anat.* **183:** 200–211.

Palmer MG (1937) Notes on the breeding habits of the slow-worm, *Anguis fragilis*. *Naturalist*: 222.

Palmer AR (1994) Fluctuating asymmetry analyses: a primer. In *Developmental Instability: its Origins and Evolutionary Implications*. TA Markow (ed.), pp. 1–29. Kluwer Academic Publishers, Dordrecht.

Palmer AR & Strobeck C (1986) Fluctuating asymmetry: measurement, analyses, patterns. *Annot. Rev. Ecol. System.* **17:** 391–421.

Parker GA (1974a) Assessment strategy and the evolution of fighting behaviour. *J. Theor. Biol.* **47:** 223–243.

Parker GA (1974b) The reproductive behavior and the Nature of sexual selection in *Scatophaga stercoraria* L. IX. Spatial distribution of fertilization rates and evolution of male search strategy within the reproductive area. *Evolution* **28:** 93–108.

Parker GA (1983) Mate quality and mating decisions. In *Mate Choice*. P Bateson (ed.), pp. 141–166. Cambridge University Press, Cambridge.

Parker GA (1992) Snakes and female sexuality. *Nature* **355:** 395–396.

Parker GH (1931) The passage of sperms and eggs through the oviducts in terrestrial vertebrates. *Phil. Trans.* **219:** 381–419.

Perry-Richardson JJ, Wilson Schofield C & Ford NB (1990) Courtship of the garter snake, *Thamnophis marcianus*, with a description of a female behavior for coitus interruption. *J. Herpetol.* **24:** 76–78.

Picariello O, Ciarcia G & Angelino F (1989) The annual cycle of oviduct in *Tarentola m. mauretanica* L. (Reptilia, Gekkonidae). *Amphibia–Reptilia* **10:** 371–386.

Pisani GR (1976) Comments on the courtship and mating mechanisms of *Thamnophis* (Reptilia, Serpentes, Colubridae). *J. Herpetol.* **10** (2): 139–142.

Plummer MV (1977) Reproduction and growth in the turtle *Trionyx muticus*. *Copeia* **3:** 440–447.

Plummer MV & Mills NE (1996) Observations on trailing and mating behaviors in hognose snake (*Heterodon platirhinos*). *J. Herpetol.* **30:** 1, 80–82.

Pope CH (1941) Copulatory adjustment in snakes. *Zool. Ser. Field Mus. Nat. Hist.* **24:** 249–252.

Prasad MRN & Reddy PRK (1972) Physiology of the sexual segment of the kidney in reptiles. *Gen. Comp. Endocrinol. Suppl.* **3:** 649–662.

Preest MR (1994) Sexual size dimorphism and feeding energetics in *Anolis carolinensis*: why do females take smaller prey than males? *J. Herpetol.* **28** (3): 292–298.

Presch W (1978) Descriptions of the hemipenial morphology in eight species of microteiid lizards (Family Teiidae, Subfamily Gymnophtalminae). *Herpetologica* **34** (1): 108–112.

Preston Webster T & Burns JM (1973) Dewlap variation and electrophoretically detected sibling species in a Haitian lizard, *Anolis brevirostris*. *Evolution* **27:** 368–377.

Przystalski A (1983) Seasonal changes in the fat content in the sand lizard (*Lacerta agilis*). *Zool. Polon.* **30:** 1–4, 115–123.

Quinn HR (1979) Reproduction and growth of the Texas coral snake (*Micrurus fulvius tenere*). *Copeia* **3:** 453–463.

Ramírez-Bautista A, Uribe-Peña Z & Guillette LJ, Jr (1995) Reproductive biology of the lizard *Urosaurus bicarinatus bicarinatus* (Reptilia: Phrynosomatidae) from Rio Balsas basin, Mexico. *Herpetologica* **51** (1): 24–33.

Ramirez Pinilla MP (1994) Reproductive and fat body cycles of the oviparous lizard *Liolaemus scapularis*. *J. Herpetol.* **28** (4): 521–524.

Rand AS (1967) Ecology and social organization in the iguanid lizard *Anolis lineatopus*. *Proc. US Natl Mus Smithson. Inst.* **122:** 3595, 1–77.

Rand AS & Williams EE (1970) An estimation of redundancy and information content of anole dewlaps. *Am. Nat.* **104:** 99–103.

Rand MS (1988) Courtship and aggressive behavior in male lizards exhibiting two different sexual colorations. *Am. Zool.* **28:** 153A.

Rand MS (1990) Polymorphic sexual coloration in the lizard *Sceloporus undulatus erythrocheilus*. *Am. Midl. Nat.* **124:** 352–359.

Rand MS (1992) Hormonal control of polymorphic and sexually dimorphic coloration in the lizard *Sceloporus undulatus erythrocheilus*. *Gen. Comp. Endocrinol.* **88:** 461–468.

Rao RJ & Shaad FU (1985) Sexual cycle of the male freshwater turtle *Trionyx gangeticus* (Cuvier). *Herpetologica* **41** (4): 433–437.

Rauch N (1985) Female habitat choice as a determinant of the reproductive success on the territorial male marine iguana (*Amblyrhynchus cristatus*). *Behav. Ecol. Sociobiol.* **16:** 125–134.

Ressel S & Schall JJ (1989) Parasites and showy males: malarial infection and color variation in fence lizards. *Oecologia* **78:** 158–164.

Restall BJ (1967) The biochemical and the physiological relationships between the gametes and the female reproductive tract. In *Advances in Reproductive Physiology – Vol. 2*. A McLaren (ed.), pp. 181–212. Academic Press, New York.

Reynolds AE (1943) The normal seasonal reproductive cycle in the male *Eumeces fasciatus* together with some observations on the effects of castration and hormone administration. *J. Morphol. Philadel.* **72:** 331–373.

Robinson KM & Murphy GG (1978) The reproductive cycle of the eastern spiny softshell turtle (*Trionyx spiniferus spiniferus*). *Herpetologica* **34:** 137–140.

Rodda GH (1992) The mating behavior of *Iguana iguana*. *Smithsonian Contributions to Zoology*, **534**. Smithsonian Institution Press, Washington, DC.

Rohwer S (1975) The social significance of avian winter plumage variability. *Evolution* **46:** 226–234.

Rohwer S (1982) The evolution of reliable and unreliable badges of fighting ability. *Am. Zool.* **22:** 531–546.

Rose B (1982) Food intake and reproduction in *Anolis acutus*. *Copeia* **2:** 322–330.

Rosenberg HI (1967) Hemipenial morphology of some Amphisbaenids (Amphisbaenia: Reptilia). *Copeia* **2:** 349–361.

Ross P & Crews D (1977) Influence of the seminal plug on mating behaviour in the garter snake. *Nature* **267:** 344–345.

Ross P & Crews D (1978) Stimuli influencing mating behavior in the garter snake, *Thamnophis radix*. *Behav. Ecol. Sociobiol.* **4:** 133–142.

Ruby DE (1981) Phenotypic correlates of male reproductive success in the lizard, *Sceloporus jarrovi*. In *Natural Selection and Social Behaviour: Recent Research and New Theory*. R Alexander & D Tinkle (eds), pp. 96–107. Chiron Press, New York.

Ruby DE (1984) Male breeding success and differential access to females in *Anolis carolinensis*. *Herpetologica* **40:** 272–280.

Ruibal R, Philibosian R & Adkins JL (1972) Reproductive cycle and growth in the lizard *Anolis acutus*. *Copeia* **3:** 509–518.

Ruth Shantha Kumari T, Devaraj Sarkar HB & Shivanandappa T (1990) Histology and histochemistry of the oviductal sperm storage pockets of the agamid lizard *Calotes versicolor*. *J. Morphol.* **203:** 97–106.

Ryan MJ (1982) Variation in iguanine social organisation: mating systems in Chuckwallas. In *Iguanas of the World – their Behavior, Ecology and Conservation*. GM Burghardt & AS Rand (eds), pp. 380–390. Noyes Publications, New Jersey.

Ryan MJ (1990) Signals, species and sexual selection. *Oxf. Surv. Evol. Biol.* **7:** 157–195.

Saint-Girons H (1957) Le cycle sexuels chez *Vipera aspis* (L) dans l'quest de la France. *Bull. Biol.* **91:** 1–67.

Saint-Girons H (1973) Sperm survival and transport in the female genital tract of reptiles. In *The Biology of Spermatozoa*. ESE Hafez & CG Thibault (eds), pp. 105–113. S. Karger AG, Basel.

Salvador A, Veiga JP, Martin J, Lopez P, Abelenda M & Puerta M (1995) The cost of producing a sexual signal: testosterone increases the susceptability of male lizards to ectoparasitic infestation. *Behav. Ecol.* **6:** 382–387.

Sanyal MK & Prasad MRN (1967) Reproductive cycle of the Indian house lizard, *Hemidactylus flaviviridis* Rüppell. *Copeia* **3:** 627–633.

Sarre S & Dearn JM (1991) Morphological variation and fluctuating asymmetry among insular populations of the sleepy lizard, *Trachydosaurus rugosus* Gray (Squamata: Scincidae). *Aust. J. Zool.* **39:** 91–104.

Saxon JG (1968) Sexual behavior of a male checkered whiptail lizard, *Cnemidophorus tesselatus*. *Southwest. Nat.* **13** (4): 454–455.

Saylor Done B & Heatwole H (1977) Social behavior of some Australian skinks. *Copeia* **3:** 419–430.

Schall JJ (1983a) Lizard malaria: cost to vertebrate host's reproductive success. *Parasitology* **87:** 1–6.

Schall JJ (1983b) Lizard malaria: parasite-host ecology. In *Lizard Ecology – Studies of a Model Organism*. RB Huey, ER Pianka and TW Schoener (eds), pp. 84–100. Harvard University Press, Cambridge, MA.

Schall JJ (1996) Malarial parasites of lizards: diversity and ecology. In *Advances in Parasitology*, Vol. 37. JR Baker, R Muller & D Rollinson (eds), pp. 256–327. Academic Press, London.

Schall JJ & Dearing MD (1987) Malarial parasitism and male competition for mates in the western fence lizard, *Sceloporus occidentalis*. *Oecologia* **73:** 389–392.

Schall JJ & Sarni GA (1987) Malarial parasitism and the behavior of the lizard, *Sceloporus occidentalis*. *Copeia* **1:** 84–93.

Schoener TW (1967) The ecological significance of sexual size dimorphism in the lizard *Anolis conspersus*. *Science* **155:** 474–477.

Schoener TW & Schoener A (1980) Densities, sex ratios, and population structure in four species of Bahamian *Anolis* lizards. *J. Anim. Ecol.* **49:** 19–53.

Schrank GD & Ballinger RE (1973) Male reproductive cycles in two species of lizards (*Cophosaurus texanus* and *Cnemidophorus gularis*). *Herpetologica* **29:** 289–293.

Schuett G (1992) Is long-term sperm storage an important component of the reproductive biology of temperate pitvipers? In *The Biology of Pitvipers*. JA Campbell & ED Brodie (eds), pp. 169–184. Tyler, Selva, Texas.

Schuett GW & Duvall D (1996) Head lifting by female copperheads, *Agkistrodon contortrix*, during courtship: potential mate choice. *Anim. Behav.* **51:** 367–373.

Schuett GW & Gillingham JC (1986) Sperm storage and multiple paternity in the copperhead, *Agkistrodon contortrix*. *Copeia* **3:** 807–811.

Schuett GW & Gillingham JC (1988) Courtship and mating of the copperhead, *Agkistrodon contortrix*. *Copeia* **2:** 374–381.

Schwartz JM, McCracken GF & Burghardt GM (1989) Multiple paternity in wild populations of the garter snake, *Thamnophis sirtalis*. *Behav. Ecol. Sociobiol.* **25:** 269–273.

Secor SM (1987) Courtship and mating behavior of the speckled kingsnake, *Lampropetis getulus holbrooki*. *Herpetologica* **43:** 15–28.

Seigel RA & Ford NB (1987) Reproductive ecology. In *Snakes: Ecology and Evolutionary Biology*. RA Seigel, JT Collins & SS Novak (eds), pp. 210–225. McGraw-Hill Publishing Company, New York.

Semlitsch RD & Gibbons JW (1982) Body size dimorphism and sexual selection in two species of water snakes. *Copeia* **4:** 974–976.

Sexton OJ & Turner O (1971) The reproductive cycle of a neotropical lizard. *Ecology* **52:** 159–164.

Sexton OJ, Ortleb EP, Hathaway LM, Ballinger RE & Licht P (1971) Reproductive cycles of three species of Anoline lizards from the isthmus of Panama. *Ecology* **52:** 201–215.

Shine R (1977) Reproduction in Australian elapid snakes I. Testicular cycles and mating seasons. *Aust. J. Zool.* **25:** 647–653.

Shine R (1978) Sexual size dimorphism and male combat in snakes. *Oecologia* **33:** 269–277.

Shine R (1989) Ecological causes for the evolution of sexual dimorphism: a review of the evidence. *Quart. Rev. Biol.* **64** (4): 419–461.

Shine R (1990a) Function and evolution of the frill of the frillneck lizard, *Chlamydosaurus kingii* (Sauria: Agamidae). *Biol. J. Linn. Soc.* **40:** 11–20.

Shine R (1990b) Proximate determinants of sexual differences in adult body size. *Am. Nat.* **135:** 278–283.

Shine R (1991) Intersexual dietary divergence and the evolution of sexual dimorphism in snakes. *Am. Nat.* **138:** 103–122.

Shine R (1994) Sexual size dimorphism in snakes revisited. *Copeia* **2:** 326–346.

Shine R & Crews D (1988) Why male garter snakes have small heads: the evolution and endocrine control of sexual dimorphism. *Evolution* **42:** 1105–1110.

Shine R, Elphick M & Harlow P (1995) Sisters like it hot. *Nature* **378:** 451–452.

Shrivastava PC & Thapliyal JP (1965) The male sexual cycle of the chequered water snake, *Natrix piscator*. *Copeia* **4:** 410–415.

Sigmund WR (1983) Female preference for *Anolis carolinensis* males as a function of dewlap color and background coloration. *J. Herpetol.* **17:** 137–143.

Simonson WE (1951) Courtship mating of the fox snake, *Elaphe vulpina vulpina*. *Copeia* **4:** 309.

Sinervo B & Lively CM (1996) The rock–paper–scissors game and the evolution of alternative male strategies. *Nature* **380:** 240–243.

Singh DP (1977) Annual sexual rhythm in relation to environmental factors in a tropical pond turtle, *Lissemys punctata granosa*. *Herpetologica* **33** (2): 190–194.

Smith DC (1985) Home range and territory in the striped plateau lizard (*Sceloporus virgatus*). *Anim. Behav.* **33:** 417–427.

Smith M (1951) *The British Amphibians and Reptiles*, pp. 171–182. Collins, London.

Soulé M (1967) Phenetics of natural populations. II. Asymmetry and evolution in a lizard. *Am. Nat.* **101:** 918, 141–160.

Srinivas SR, Shivanandappa T, Hedge SN & Sarkar HBD (1995) Sperm storage in the oviduct of the tropical rock lizard, *Psammophilus dorsalis*. *J. Morphol.* **224:** 293–301.

Stamps JA (1975) Courtship patterns, estrus periods and reproductive condition in a lizard, *Anolis aeneus*. *Physiol. Behav.* **14:** 531–535.

Stamps JA (1977) Social behavior and spacing patterns in lizards. In *Biology of the Reptilia, Vol. 7 – Ecology and Behavior A*. C Gans & DW Tinkle (eds), pp. 265–334. Academic Press, London.

Stamps JA (1983) Sexual selection, sexual dimorphism, and territoriality. In *Lizard Ecology – Studies of a Model Organism*. RB Huey, ER Pianka and TW Schoener (eds), pp. 169–204. Harvard University Press, Cambridge, MA.

Stamps JA (1993) Sexual size dimorphism in species with asymptotic growth after maturity. *Biol. J. Linn. Soc.* **50:** 123–145.

Stamps J (1994) Early hormones and the development of phenotypic variation in tree lizards. *Trend Ecol. Evol.* **9:** 311–312.

Stebbins RC (1944) Field notes on a lizard, the mountain swift, with special reference to territorial behavior. *Ecology* **25:** 233–245.

Stille B, Madsen T & Niklasson M (1986) Multiple paternity in the adder, *Vipera berus*. *Oikos* **47:** 173–175.

Stille B, Niklasson M & Madsen T (1987) Within season multiple paternity in the adder, *Vipera berus*. *Oikos* **49:** 232–233.

Subba Rao MV & Rajabi BS (1972) Reproduction in the ground lizard, *Sitana pontiseriana* and the garden lizard, *Calotes nemoricola*. *Br. J. Herpetol.* 245–252.

Swain R & Jones S (1994) Annual cycle of plasma testosterone and other reproductive parameters in the Tasmanian skink, *Niveoscincus metallicus*. *Herpetologica* **50:** 502–509.

Swiezawska K (1949) Color-discrimination of the sand lizard, *Lacerta agilis* L. *Bull. Int. Acad. Pol. Sci. Lett. Ser. B Sci. Nat.* 1–20.

Swingland IR & Coe M (1978) The natural regulation of giant tortoise populations on the Aldabra Atoll. Reproduction. *J. Zool.* **186:** 285–309.

Tegelström H & Höggren M (1994) Paternity determination in the adder (*Vipera berus*) – DNA fingerprinting or random amplified polymorphic DNA? *Biochem. Genet.* **32**(7/8): 249–256.

Thorbjarnson JB (1994) Reproductive ecology of the spectacled caiman (*Caiman crocodilus*) in the Venezuelan Llanos. *Copeia* **4:** 919–970.

Thompson CW & Moore MC (1991) Throat colour reliably signals status in male tree lizards, *Urosaurus ornatus*. *Anim. Behav.* **42:** 745–753.

Thompson CW, Moore IT & Moore MC (1993) Social, environmental and genetic factors in the ontogeny of phenotypic differentiation in a lizard with alternative male reproductive strategies. *Behav. Ecol. Sociobiol.* **33:** 137–146.

Tienhoven A, van (1983) *Reproductive Physiology of Vertebrates*, pp. 108–109. Cornell University Press, Ithaca.

Tinkle DW (1976) Comparative data on the population ecology of the desert spiny lizard, *Sceloporus magister*. *Herpetologica* **32:** 1–6.

Tokarz RR (1985) Importance of androgens in male territorial acquisition in the lizard *Anolis sagrei*: an experimental test. *Anim. Behav.* **49:** 661–669.

Tokarz RR (1987a) Effects of corticosterone treatment on male aggressive behavior in a lizard (*Anolis sagrei*). *Horm. Behav.* **21:** 358–370.

Tokarz RR (1987b) Effects of antiandrogens cyproterone acetate and flutamide on male reproductive behavior in a lizard (*Anolis sagrei*). *Horm. Behav.* **21:** 1–16.

Tokarz RR (1988) Copulatory behaviour of the lizard *Anolis sagrei*: alternation of hemipenis use. *Anim. Behav.* **36:** 1518–1524.

Tokarz RR (1989) Pattern of hemipenis use in the male lizard *Anolis sagrei* after unilateral castration. *J. Exp. Zool.* **250:** 93–99.

Tokarz RR (1992) Male mating preference for unfamiliar females in the lizard, *Anolis sagrei*. *Anim. Behav.* **44:** 843–849.

Tokarz RR (1995) Mate choice in lizards: a review. *Herpetol. Monogr.* **9:** 17–40.

Tokarz RR & Kirkpatrick SJ (1991) Copulation frequency and pattern of hemipenis use in males of the lizard *Anolis sagrei* in a semi-natural enclosure. *Anim. Behav.* **41:** 1039–1044.

Tokarz RR & Slowinski JB (1990) Alternation of hemipenis use as a behavioural means of increasing sperm transfer in the lizard *Anolis sagrei*. *Anim. Behav.* **40:** 374–379.

Tollestrup K (1981) The social behavior and displays of two species of horned lizards, *Phrynosoma platyrhinos* and *Phrynosoma coronatum*. *Herpetologica* **37:** 130–141.

Tomko DS (1972) Autumn breeding of the desert tortoise. *Copeia* **1972:** 895.

Torr G & Shine R (1993) Experimental analysis of thermally dependent behaviour patterns in the scincid lizard *Lampropholis guichenoti*. *Copeia* **3:** 850–854.

Trauth SE (1979) Testicular cycle and timing of reproduction in the collared lizard (*Crotaphytis collaris*) in Arkansas. *Herpetologica* **35:** 184–192.

Trillmich KGK (1979) Feeding behaviour and social behaviour of the marine iguana. *Noticias Galápagos* **29:** 19–20.

Trillmich KGK (1983) The mating system of the marine iguana (*Amblyrhynchus cristatus*). *Z. Tierpsychol.* **63:** 141–172.

Trivers RL (1976) Sexual selection and resource-accruing abilities in *Anolis garmani*. *Evolution* **30:** 253–269.

Tryon BW (1979) Reproduction in captive forest cobras, *Naja melanoleuca* (Serpentes: Elapidae). *J. Herpetol.* **13:** 499–504.

Turner FB, Lannom JR, Jr, Medica PA & Hoddenbach GA (1969) Density and composition of fenced populations of leopard lizards (*Crotaphytis wislizenii*) in southern Nevada. *Herpetologica* **25:** 247–257.

Turner G (1984) Captive breeding of *Unechis flagellum*. *Herpetofauna* **16:** 53.

Underwood G (1967) A contribution to the classification of snakes. *Br. Mus. Nat. Hist. Publ* **653:** 1–179.

Velmurugan P (1992) Albumins enhance motility of sperm of spotted Indian house gecko *Hemidactylus brooki* (Gray). *Indian J. Exp. Biol.* **30:** 530–532.

Verbeek B (1972) Ethologische untersuchungen an einigen europäischen Eidechsen. *Bonn. Zool. Beit.* **2:** 122–151.

Vinegar MB (1972) The function of breeding coloration in the lizard, *Sceloporus virgatus*. *Copeia* **4:** 660–664.

Vitt L (1983) Reproduction and sexual dimorphism in the tropical Teiid lizard *Cnemidophorus ocellifer*. *Copeia* **2:** 359–366.

Vitt LJ & Cooper WE, Jr (1985a) The evolution of sexual dimorphism in the skink *Eumeces laticeps*: an example of sexual selection. *Can. J. Zool.* **63:** 995–1002.

Vitt LJ & Cooper WE, Jr (1985b) The relationship between reproduction and lipid cycling in the skink *Eumeces laticeps* with comments on brooding biology. *Herpetologica* **41:** 419–432.

Vitt LJ & Goldberg SR (1983) Reproductive ecology of two tropical iguanid lizards: *Tropidurus torquatus* and *Platynotus semitaeniatus*. *Copeia* **1:** 131–141.

Vitt LJ & Ohmart RD (1974) Reproduction and ecology of a Colorado river population of *Sceloporus magister* (Sauria: Iguanidae). *Herpetologica* **30:** 410–417.

Vitt LJ & Ohmart RD (1977a) Ecology and reproduction of lower Colorado river lizards: *Callisaurus draconoides* (Iguanidae). *Herpetologica* **33:** 214–222.

Vitt LJ & Ohmart RD (1977b) Ecology and reproduction of lower Colorado river lizards II. *Cnemidophorus tigris*, Teiidae (eds), with comparisons. *Herpetologica* **33:** 223–224.

Voss R (1979) Male accessory glands and the evolution of copulatory plugs in rodents. *Occas. Pap. Mus. Zool. Univ. Mich.* **689:** 1–27.

Wall F (1922) Notes and some lizards, frogs, and human beings in the Nilgiri Hills. *J. Bombay Nat. Hist.* Soc. **XXVIII:** 687–689.

Weatherhead PJ, Berry FE, Brown GP & Forbes MRL (1995) Sex ratios, mating behavior and sexual dimorphism of the northern water snake, *Nerodia sipedon*. *Behav. Ecol. Sociobiol.* **36:** 301–311.

Webb G & Manolis C (1989) *Crocodiles of Australia*. Reed Books Pty, Ltd., Frenches Forest.

Weil MR (1984) Seasonal histochemistry of the renal sexual segment in male common water snakes, *Nerodia sipedon*. *Can. J. Zool.* **62:** 1737–1740.

Weil MR & Aldridge RD (1979) The effect of temperature on the male reproductive system of the common water snake (*Nerodia sipedon*). *Exp. Zool.* **210:** 327–332.

Werner DI (1978) On the biology of *Tropidurus delanonis*, Baur (Iguanidae). *Z. Tierpsychol.* **47:** 337–395.

Werner DI (1982) Social organization and ecology of land iguanas *Conolophus subcristatus*, on Isla Fernandia, Galápagos. In *Iguanas of the World – their Behavior, Ecology and Conservation*. GM Burghardt & AS Rand (eds), pp. 342–365. Noyes Publications, New Jersey.

West-Eberhard MJ (1984) Sexual selection, competitive communication and species-specific signals in insects. In *Insect Communication*. T Lewis (ed.), pp. 283–324. Academic Press, New York.

White DR, Mitchell JC & Woodcoot WS (1982) Reproductive cycle and embryonic development of *Nerodia taxispilota* (Serpentes: Colubridae) at the north eastern edge of its range. *Copeia* **1982:** 646–652.

Whittier JM (1993) Behavioural repertoire of *Carlia rostralis* (Scincidae) in the wet tropics of Queensland, Australia. In *Herpetology in Australia*. D Lunney & D Ayers (eds), pp. 305–310. Royal Zoological Society of New South Wales, Chipping Norton.

Whittier JM & Martin J (1992) Aspects of social behaviour and dominance in male rainbow skinks, *Carlia rostralis*. *Aust. J. Zool.* **40:** 73–79.

Whittier JM & Tokarz RR (1992) Physiological regulation of sexual behavior in female reptiles. In *Biology of the Reptilia Vol. 18: Physiology E – Hormones, Brain, and Behavior*. C Gans and D Crews (eds), pp. 24–69. The University of Chicago Press, Chicago.

Whittier JM, Mason RT & Crews D (1985) Mating in the red-sided garter snake,

Thamnophis sirtalis parietalis: differential effects on male and female sexual behaviour. *Behav. Ecol. Sociobiol.* **16:** 257–261.

Whittier JM, Stewart D & Tolley L (1994) Ovarian and oviductal morphology of sexual and parthenogenetic geckos of the *Heteronotia bionei* complex. *Copeia* **2:** 484–491.

Wiewandt T (1977) Ecology behavior, and management of the Mona island ground iguana, *Cyclura stejnegeri*. Ph.D. thesis, Cornell University, Ithaca, NY, USA.

Wiewandt T (1979) La Gran Iguana de Mona. *Nat. Hist.* **88:** 56–65.

Wikelski M, Carbone C & Trillmich F (1996) Lekking in marine iguanas: female grouping and male strategies. *Anim. Behav.* **52:** 581–596.

Wikramanayake ED & Dryden GL (1988) The reproductive ecology of *Varanus indicus* on Guam. *Herpetologica* **44:** 338–344.

Wildt DE, Bush M, Goodrowe KL, Packer C, Pusey AE, Brown JL, Joslin P & O'Brien SJ (1987) Reproductive and genetic consequences of founding isolated lion populations. *Nature* **329:** 328–331.

Wilhoft DC (1963) Reproduction in the tropical Australian skink, *Leiolepisma rhomboidalis*. *Am. Midl. Nat.* **70** (2): 442–461.

Wilhoft DC & Reiter EO (1965) Sexual cycle of the lizard, *Leiolepisma fuscus*, a tropical Australian skink. *J. Morphol.* **116:** 379–388.

Wood JR & Wood FE (1980) Reproductive biology of captive green sea turtles *Chelonia mydas*. *Am. Zool.* **20:** 499–505.

Woodbury M & Woodbury AM (1945) Life-history studies of the sagebrush lizard *Sceloporus g. graciosus* with special reference to cycles in reproduction. *Herpetologica* **2:** 175–196.

Wyk JH, van (1990) Seasonal testicular activity and morphometric variation in the femoral glands of the lizard *Cordylus polyzonus polyzonus* (Sauria: Cordylidae). *J. Herpetol.* **24** (4): 405–409.

Zucker N & Boecklen W (1990) Variation in female throat coloration in the tree lizard (*Urosaurus ornatus*): relation to reproductive cycle and fecundity. *Herpetologica* **46** (4): 387–394.

Zucker N (1994) A dual status-signalling system: a matter of redundancy or differing roles? *Anim. Behav.* **47:** 15–22.

Zweifel RG & Dessauer HC (1983) Multiple insemination demonstrated experimentally in the kingsnake (*Lampropeltis getulus*). *Experientia* **39:** 317–319.

APPENDIX A

Mating data

Species	No. of matings	No. of partners	Copula duration	Captive/ Field	Authority
Lizards					
Conolophus stejnegeri	1–3	1–3	?	f	Wiewandt (1977, 1979)
Conolophus subcristatus	> 1	1–2	0.5–4 min	f	Dugan (1982)
Brachylophus fasciatus	> 5	?	15–20 min	c	Gibbons and Watkins (1982)
Iguana iguana	1–5	$^{3}1$	7.4 min	f	Dugan (1982)
Iguana iguana	3–4.8 (mean)	1–several	9.0 min (mean)	f	Rodda (1992)
Amblyrhynchus cristatus	1	1	2.8–3.1 min	f	Trillmich (1979), Wikelski *et al.* (1997)
Anolis lineatopus	> 1	> 1	2–3 min	f	Rand (1967)
Anolis carolinensis	1	1	> 3 min	f	Ruby (1984)
Anolis carolinensis	?	?	9.42, 0.7 (SD)	c	Greenberg and Noble (1944)
Anolis valencienni	> 2	> 2	?	f	Hicks and Trivers (1983)
Anolis sagrei	> 3	?	?	c	Tokarz (1992)
Anolis garmani	1	1	10–25 min	f	Trivers (1976)
Anolis aeneus	1	1	1–60 min	c, f	Stamps (1975)
Tropidurus delanonis	?	?	0.5–1.5 min	f	Werner (1978)
Sceloporus jarrovi	1	1	?	f	Ruby (1981)
Sceloporus virgatus	1–5	1–2	–5 min	c, f	Carpenter (1962), Smith (1985)
Phrynosoma douglassi	> 1	?	3.5–4 min	f	Montanucci (1982)
Phrynosoma platyrhinos	?	?	7 min	f	Tollestrup (1981)
Uta stansburiana	?	?	0.2–2.7 min*	f, c	Ferguson (1970)
Gerrhonotus panamintus	> 1	?	1 h 6 min	c	Leviton (1961)
Ctenophorus maculosus	?	?	0.4 min	f	Mitchell (1973), Olsson (1995a)
Sitana ponticeriana	?	?	2–6 min	f	Chopra (1967), Subba Rao and Rajabi (1972)

Continued

Mating data—Continued

Species	No. of matings	No. of partners	Copula duration	Captive/ Field	Authority
Trapelus sanguinolenta	> 1	1	?	f	Panov, personal communication
Chamaeleo chamaeleon	1–5	> 1	?	f	Cuadro and Loman (1997)
Ameiva plei	2–3	1	?	f	Censky (1995)
Ameiva chrysolaema	?	?	< 1 min	c	Noble and Bradley (1933)
Cnemidophorus sexlineatus	?	?	1–15 min	c, f	Noble and Bradley (1933), Carpenter (1962)
Cnemidophorus lemniscatus	?	?	0.03–0.1 min	f	Müller (1971)
Cnemidophorus tesselatus	?	?	5–10 min	c	Saxon (1968)
Cnemidophorus tigris	?	?	7 min	f	Vitt and Ohmart (1977b)
Tiliqua rugosa	1	1	?	f	Bull (1988), personal communication
Eumeces okadae	3.5 s, 0.4 (SE)	?	3.7, 0.4 (SE)	f	Hasegawa (1994)
Eumeces laticeps	1–2	> 1	?	c	Goin (1957), Cooper and Vitt (1993)
Eumeces obsoletus	> 1	> 1	*c*. 3 min	f	Fitch (1955)
Eumeces egregius	1.2 s (mean)	?	15–30 min	f	Mount (1963)
Eumeces multivarigatus	> 1	?	5 min	c	Everett (1971)
Sphenomorphus kosciuskoi	?	?	*c*. 0.05 min	c	Saylor Done and Heatwole (1977)
Lacerta sicula	?	?	0.4–0.9 min	c	Verbeek (1972)
Lacerta melisillensis	?	?	7–9 min	c	Verbeek (1972)
Lacerta hispanica	?	?	5.5–59 min	c	Verbeek (1972)
Lacerta muralis	?	?	0.5 min	c	Verbeek (1972)
Lacerta vivipara	2–3	> 1	35–53 min	c, f	Verbeek (1972), Bauwens and Verheyen (1985), Heulin (1988)
Lacerta agilis	3.2	1.7	2–4 min	f	Olsson (1992b), Olsson *et al.* (1994a)
Lacerta parva	?	?	11.1 s, 3.2 SD	c	in den Bosch (1990)
Algyroides moreoticus	> 1	?	?	c	in den Bosch (1983)
Psammodromus hispanicus	?	?	0.7 min	c	in den Bosch (1986)

Anguis fragilis	> 1	> 1	> 20 h	c	Palmer (1937)
Brookesia pumilis (as determined by mating scars)	?	?	11 min	c	von Frisch (1958)
Snakes					
Naja melanoleuca	?	?	1–6 h	c	Tryon (1979), Haagner and Carpenter (1988)
Naja naja	> 2	?	1 h	c	Campbell and Quinn (1975)
Acantophis pyrrhus	?	?	c. 2 h	c	Fyfe and Munday (1988)
Unechis flagellum	?	?	4–6 h	c	Turner (1984)
Pseudonaja nuchalis	?	?	2 h 55 min	c	Bush (1989)
Notechis scutatus	?	?	–7 h	?	Fearn (1993)
Ophiophagus hannah	?	?	57 min	c	Oliver (1956)
Crotalus spp.	?	?	–28 h	c, f	Klauber (1956), Gillingham *et al.* (1983)
Crotalus cerastes	?	?	2 h	c	Lowe (1942)
Crotalus v. viridis	> 1	> 1	?	f	Duvall, personal communication
Crotalus atrox	> 1	> 1	?	f	Duvall, personal communication
Bothrops schlegeli	?	?	2 h 53 min	c	Antonioi (1980)
Agkistrodon contortrix	> 1	> 1	6.1 h	c	Schuett and Gillingham (1986, 1988)
Bitis nasicornis	?	?	3 h	c	Akester (1989)
Vipera aspis	> 1	> 1	?	f	Luiselli, personal communication; Saint Girons, personal communication
Vipera berus	3.7	> 3	?	f	Madsen *et al.* (1992) personal observation
Vipera berus	1 or more	1 or more	?	c	Nilson and Andrén (1982), Luiselli (1993), Capula and Luiselli (1994)
Vipera berus	> 2	> 2	?	c	Höggren and Tegelström (1995)
Aspidites melanocephalus	?	?	–8 h	c	Hughes (1990)

Continued

Mating data—Continued

Species	No. of matings	No. of partners	Copula duration	Captive/ Field	Authority
Python mulurus	?	?	30 min–3 h	?	Nakomoto and Toriba (1983)
Python m. bivittatus	?	?	6 h 45 min	c	Gillingham and Chambers (1982)
Lichanaura roseofusca	?	?	30 min	c	Kurfess (1967)
Hydrodynastes gigas	?	?	37.4 min	c	Bels (1987)
Coluber constrictor	?	?	23 min	?	Fitch (1963)
Elaphe climacophora	> 1	?	2–5 h	?	Mishima *et al.* (1972)
Elaphe vulpina	?	?	3-25 min	c, f	Carpenter (1947), Simonson (1951), Gillingham (1977), Kennedy (1978)
Elaphe g. guttata	> 1	?	?	c	MacMahon (1957)
Elaphe longissima	> 1	> 1	?	c, f	Luiselli, personal communication
Elaphe quatorlineata	> 1 (4)	> 1	?	c, f	Luiselli, personal communication
Lampropeltis getulus	2	2	1.4–5 h	c	Lewke (1979), Secor (1987), Zweifel and Dessauer (1983)
Lampropeltis mexicana	?	?	4–15 min	c	Murphy *et al.* (1978)
Storeria dekayi	?	?	24 min	c	Noble (1937)
Pituophis sayi affinis	> 1	?	c. 20 min	c	Gloyd (1947)
Thamnophis marcianus	?	?	6 min	c	Perry-Richardson *et al.* (1990)
Thamnophis butleri	?	?	40 min	c	Noble (1937)
Thamnophis sirtalis	> 1	> 1	15–20 min	c, f	Blanchard and Blanchard (1942), Gibson and Falls (1975), Fitch (1965), Whittier *et al.* (1985), Schwartz *et al.* (1989)
Thamnophis radix	> 1	> 1	?	f	Davis (1936)
Natric maura	> 1	?	17–60 min	f	Hailey and Davies (1987)
Natrix natrix	?	?	c. 1 h	f	T. Madsen, personal observation
Natrix sipedon	> 1	> 1	?	f	Barry *et al.* (1992)
Heterodon platyrhinos	> 1	> 1	?	f	Plummer and Mills (1996)

Xenochrophis piscator	?	?	3.5 h	?	Khaire *et al.* (1985)
Turtles and Tortoises					
Chelydra serpentina	> 2	> 2	?	f	Galbraith *et al.* (1993)
Caretta caretta	> 2	> 2	?	f	Harry and Briscoe (1988)
Chelonia mydas	> 2	> 2	16.2 min	f	Ernst *et al.* (1994)
Clemmys insculpta	> 2	> 2	1–2 h	f	Kaufmann (1992), Galbraith (1993)
Clemmys guttata	?	?	1 h	?	Ernst *et al.* (1994)
Terrapene ornata	> 1	> 1	?	f	Ernst *et al.* (1994)
Terrapene c. carolina	?	?	1 h 45 min	c	Cahn and Conder (1932)
Clemmys muhlenbergii	> 1	> 1	5–20 min	f	Ernst *et al.* (1994)
Sternotherus odoratus	> 2	?	?	f	Ernst *et al.* (1994)
Sternotherus minor	?	?	67 min	?	Ernst *et al.* (1994)
Kinosternon flavsescens	?	?	5 min–2.5 h	c, f	Mahmoud (1967), Lardie (1975), Ernst *et al.* (1994)
Gopherus berlandieri	?	?	< 10 min	?	Ernst *et al.* (1994)
Malaclemys terrapin	?	?	1–2 min	?	Ernst *et al.* (1994)
Graptemys pseudographica	?	?	0.2 min–4 h	?	Ernst *et al.* (1994)
Pseudemys nelsoni	?	?	4 min	?	Ernst *et al.* (1994)
Trachemys scripta	?	?	15 min	?	Ernst *et al.* (1994)
Emydoidea blandingii	?	?	16.5–29.3 min	?	Ernst *et al.* (1994)
Emydoidea blandingii	?	?	23 min	c	Baker and Gillingham (1983)
Chelodina longicollis	?	?	3 min	c	Murphy and Lamoreaux (1978)
Gopherus agassizi	?	?	> 70 min	f	Tomko (1972)
Testudo horsfieldii	1	1	?	f	Panov, personal communication
Crocodilians					
Crocodylus rhombifer	?	?	5 min	c	Larsson and Wihman (1989)
Crocodylus palustris	?	?	2–3 min	f	Dharmakumarsinhji (1946)
Crocodylus niloticus	> 1	> 1	?	f	Lance, personal communication
Alligator mississippiensis	?	?	2–3 min	c	Legge (1967)
Crocodilia spp.	> 1	> 1	2–15 min	f	Lang (1987), Webb and Manolis (1989)

APPENDIX B

Reproduction data

The columns read from left to right, Species, climatic region (TeL, TeH = temperate, low and high altitude, respectively; TrL and TrH, tropical region denoted for altitude, All = all year reproducer), GSI (sample size), Breeding pattern (Associate, Dissociate, Mixed, i.e. testis size peaking in, or out of, synchrony with mating), GSI ratio (max GSI/min GSI), Male combat in snakes (±), or Territoriality in Lizards followed by Authority within parentheses, and finally Authority for the GSI data.

	GSI(*n*)	**Breeding pattern**	**GSI ratio**	**Territoriality (lizards)/ Male combat (snakes)**	**References**
GSI = Testis mass (g)/Body mass (g) × *100*					
Lizards					
Polychridae					
Anolis opalinus, All	1.15 (11)	A	1.9	+	Jenssen and Nunez (1994)
Phrynosomatidae					
Sceloporus graciosus, TeL	—	A	8.1e, v	+(2)	Woodbury and Woodbury (1945)
S. graciosus, TeH	4.71 (18)	A	12.4	+(2)	Jameson (1974)
S. grammicus, TeH	4.90 (^{3}9)	A	16.3	+	Ortega (1985)
S. grammicus, TeH	1.52 (60c)	A	7.6	+	Guillette and Casa-Andreu (1981)
S. magister, TeL	—	A	> 40.0e	+(*1*)	Vitt and Ohmart (1974)
S. mucronotus, TeH	1.2 v(^{3}10)	A	6.0 v	+(*)	Mendez-de la Cruz *et al.* (1988)
S. mucronotus, TeH	—	A	5.7e, v	+(*)	Estrada-Flores *et al.* (1990)
S. occidentalis, TeH	3.80 (9)	A	19.0	+	Goldberg (1974)
S. occidentalis, TeL	3.40 (7)	A	13.6	+	Goldberg (1974)
S. scalaris, TeH	3.65 (6)	A	45.6	– (*)	Newlin (1976)
S. torquatus, TeH	1.48 (124c)	A	14.8	+(*)	Guillette and Mendez-de la Cruz (1993)
S. undulatus, TeL	2.34 (5)	A	7.3	+(*15*)	McKinney and Marion (1985)

Phrynosoma cornutum, TeL	2.80 (9)	A	9.3	—	Ballinger (1974)
Cophosaurus (Holbrookia) *texanus*, TeL	—	A	13.0e	+(*3*)	Schrank and Ballinger (1973)
Tropiduridae					
Tropidurus torquatus, TrL	0.43 (18)	A	3.6	—	Vitt and Goldberg (1983)
Liolaemus bitaeniatus, TeH	—	A	5.8	—	Ramirez Pinilla (1994)
Platynotus semitaniatus, TrL	0.49 (10)	A	4.9	—	Vitt and Goldberg (1983)
Chamaeleonidae					
Calotes versicolor, TrL	1.44 (5)	A	36.0	+(*11*)	Haldar and Thapliyal (1977)
Lacertidae					
Lacerta vivipara, TeL	1.22 (14)	A	4.1	–	Courty and Defaure (1979)
L. vivipara, TeL	2.50 (6)	A	4.4	–	Courty and Defaure (1982)
L. agilis, TeL	2.25 (>6)	A	4.5	–	Przystalski (1983)
Podarcis sicula, TeL	0.52 (3–4)	A	13.0	–	Angelini *et al*. (1979)
Teiidae					
Cnemidophorus ocellifer, All	0.85 (43)	A	1.3	– (*)	Vitt (1983)
C. sexlineatus, TeL	1.20 (24)	A	5.0	–	Brackin (1979)
C. tigris multiiscutatus, TeH	0.97 (15)	A	8.1	–	Goldberg (1976)
C. gularis, TeL	—	A	7.8e	– (*)	Schrank and Ballinger (1973)
Gekkonidae					
Hemidactylus flaviviridis, TrL	1.02 (12)	A	34.0	+(*12*)	Sanyal and Prasad (1967)
Scincidae					
Niveoscincus microlepidotum, TeH	1.80 (5)	A	1.4	–	Olsson, Birkhead and Shine, unpublished data
Tiliqua rugosa, TeL	0.26 (11)	A	5.4	–	Bourne *et al*. (1986)
Xantusidae					
Xantusia riversiana, TeL	0.97 (8)	A	8.1	—	Goldberg and Bezy (1974)
X. vigilis, TeL	—	—	7.3e	—	Miller (1948)

Continued

Reproduction data—Continued

	GSI(*n*)	Breeding pattern	GSI ratio	Territoriality (lizards)/ Male combat (snakes)	References
Varanidae					
Varanus indicus, TrL	0.22 (2)	A	3.4	– (*)	Wikramanayake and Dryden (1988)
V. grayi, TrL	0.14 (—)	A	3.7	– (*)	Auffenberg (1988)
GSI = Testis mass (mg)/SVL (mm)					
Polycridae					
Anolis acutus, TrL	0.38v (21)	—	1.90v	+ (7)	Ruibal *et al.* (1972)
A. auratus, TrL	0.96 (69)	A	9.6	+	Sexton *et al.* (1971)
A. carolinensis, TrL	1.12 (7)	A	18.7	+ (8)	Fox (1958)
A. cristatellus, TrL	0.96 (> 10)	—	2.4	+ (*13*)	Licht and Gorman (1970)
A. cybotes, TrL	1.95 (12)	—	3.4	+ (*13*)	Licht and Gorman (1970)
A. grahami, Jamaica, TrL	1.31 (7)	—	3.2	+ (*13*)	Licht and Gorman (1970)
A. grahami, Bermuda, TrL	1.63 (15)	—	8.6	+ (*13*)	Licht and Gorman (1970)
A. limifrons, TrL	0.88 (49)	A	1.7	+	Sexton *et al.* (1971)
A. lineatopus, TrL	1.34 (11)	—	3.2	+ (5)	Licht and Gorman (1970)
A. pulchellus, TrL	0.37 (> 10)	—	2.0	+ (*13*)	Licht and Gorman (1975)
A. richardi, TrL	2.00 (7)	—	1.6	+ (*13*)	Licht and Gorman (1970)
A. sagrei, TrL	1.89 (16)	—	2.7	+ (6)	Licht and Gorman (1970)
A. stratulus, TrH	1.48 (> 10)	—	18.5	+ (*13*)	Licht and Gorman (1975)
A. stratulus, TrL	1.25 (> 10)	—	3.3	+ (*13*)	Licht and Gorman (1975)
A. trinitatis, TrL	0.70 (9)	—	3.1	+ (*13*)	Licht and Gorman (1970)
A. tropidogaster, TrL	0.96 (67)	A	9.6	+	Sexton *et al.* (1971)
Crotaphytidae					
Crotaphytis collaris, TeL	—	A	12.6e, v	—	Trauth (1979)

Iguanidae					
Dipsosaurus dorsalis, TeL	3.23v (89)	A	13.4v	+(*4*)	Mayhew (1971)
Phrynosomatidae					
Callisaurus draconoides, TeL	2.10 (5)	A	21.0	—	Vitt and Ohmart (1977a)
C. draconoides, TeL	—	—	48.3e, v	—	Kay *et al.* (1970)
Sceloporus formosus, TeH	1.36 (5)	—	15.1v	—	Guillette and Sullivan (1985)
S. malachiticus, TrH	6.15 (10)	A	10.4	+(*)	Marion and Sexton (1971)
S. occidentalis, TeH	—	A	20.8c	+	Jameson and Allison (1976)
S. orcutti, TeL	3.03v (17)	A	7.2v	+(*)	Mayhew (1963)
S. virgatus, TeL	5.48 (14)	A	—	+	Ballinger and Ketels (1983)
S. woodi, TrL	1.58v (10)	A	1.5v	+(*)	Jackson and Telford (1974)
Uma scoparia, TeL	3.10v (51)	A	20.7v	+(*)	Mayhew (1966a)
U. notata, TeL	1.82v (45)	A	18.2v	+(*)	Mayhew (1966b)
Urosaurus b. bicarinatus, TrH	1.23v (155c)	A	308.0v, d	+(*)	Ramírez-Bautista *et al.* (1995)
U. b. bicarinatus, TrL	1.49v (83c)	A	395.0v, d	+(*)	Ramírez-Bautista *et al.* (1995)
Tropiduridae					
Liolaemus scapularis, TeH	1.30v (> 3)	A	32.5v	—	Ramirez Pinilla (1994)
Chamaeleonidae					
Agama a. africana, TrL	—	A	1.1g	+(*14*)	Daniel (1960)
Draco melanopogon, All	—	A	1.1g (365c)	—	Inger and Greenberg (1966)
D. quinquefasciatus, All	—	A	1.1g (266c)	—	Inger and Greenberg (1966)
Japalura swinhonis formosensis, TeL	2.36 (2)	A	11.2	+	Cheng and Lin (1977)
Uromastix hardwicki, TeL	10.40 (5)	A	21.2	—	Arslan *et al.* (1978)
Teiidae					
Cnemidophorus tigris, TeL	—	A	> 6h	−(*)	Vitt and Ohmart (1977b)
C. sexlineatus, TeL	—	A	5.0c (13, 8)	−(*16*)	Johnson and Jacob (1984)
Lacertidae					
Takydromus septentrionalis, TeL	0.75 (13)	M	4.7	−(*)	Cheng and Lin (1977)

Continued

Reproduction data—Continued

	GSI(*n*)	Breeding pattern	GSI ratio	Territoriality (lizards)/ Male combat (snakes)	References
Scincidae					
Chalcides ocellatus, TeL	—	—	14e	– (*)	Badir (1958)
Ctenotus brooksi, TeL	0.22 (35)	A	22.0, v	– (*)	James (1991)
C. helenae, TeL	0.49 (4)	A	8.8, v	– (*)	James (1991)
C. leonhardii, TeL	0.21 (29)	A	10.5, v	– (*)	James (1991)
C. pantherinus, TeL	0.83 (3)	A	42.0, v	– (*)	James (1991)
C. schomburgkii, TeL	0.14 (44)	A	28.0, v	– (*)	James (1991)
Eumeces copei, TeH	—	A	22.2v, e	—	Guillette (1983)
E. fasciatus, TeL	—	—	5.0e	– (*)	Reynolds (1943)
E. laticeps, TeL	—	A	12.5	– (9)	Vitt and Cooper (1985b)
Leiolepisma rhomboidalis, TrL	1.05 (24)	M	1.6	– (*)	Wilhoft (1963)
L. fuscum, TrL	1.45 (9)	D	14.5	– (*)	Wilhoft and Reiter (1965)
L. zelandica, TeL	—	A	4.8e	—	Barwick (1959)
Mabuya capensis, TeL	0.97 (6)	—	12.1v	—	Flemming (1994)
Niveoscincus metallicum, TeL	1.35 (7)	A	4.1	–	Swain and Jones (1994)
Scincus scincus, TeL	—	—	22e	– (*)	Badir (1958)
Gekkonidae					
Cyrtodactylus malayanus, All	—	A	1.3g (131c)	—	Inger and Greenberg (1966)
C. pubisulcus, All	—	A	1.6g (56c)	—	Inger and Greenberg (1966)
Gonatodes albogularis, TrL	0.19 (89c)	A	2.4	– (*)	Sexton and Turner (1971)
Hemidactylus frenatus, TeL	0.99 (11)	A	2.5	+ (*12*)	Cheng and Lin (1977)
Anguidae					
Barisia i. imbricata, TeH	0.19v (125c)	—	19.0v	– (*)	Guillette and Casas-Andreu (1987)
Cordylidae					
Cordylus p. polyzonus, TeL	—	A	46.0e, v	—	van Wyk (1990)

Pseudocordylus m. melanotus, TeL	1.48 (5)	—	8.7v	—	Flemming (1993)
Varanidae					
Varanus bengalensis, TrL	—	A	36.0c	—	Auffenberg (1994)
Snakes					
GSI = Testis mass (g)/Body mass (g) × 100					
Elapidae					
Micrurus fulvius tenere, TeL	1.82 (2)	A (?)	2.7	—	Quinn (1979)
Naja naja, TeL	3.56 (³7)	M	9.0	+ (*I*)	Lofts *et al*. (1966)
Notechis scutatus, TeL	5.10 (> 6)	M	2.0	+ (*I*)	Shine (1977)
P. nuchalis, TeL	4.60 (> 6)	A	4.6	− (*I*)	Shine (1977)
P. porphyriacus, TeL	4.00 (> 6)	A	6.7	+ (*I*)	Shine (1977)
P. textilis, TeL	2.90 (3)	A	—	+ (*I*)	Shine (1977)
Unechis gouldii, TeL	3.80 (3)	A	2.5	− (*I*)	Shine (1977)
Laticaudidae					
Laticauda colubrina, All	0.24b (10)	A	1.3	− (*I*)	Gorman *et al*. (1981)
Viperidae					
Vipera aspis, TeL	1.64 (3)	D	2.3	+ (*I*)	Saint-Girons (1957)
Colubridae					
Elaphe taeniura, TeL	1.30 (6)	D	13.0	—	Chiu and Wong (1974)
Thamnophis radix, TeL	—	—	9.9e	− (*I*)	Cieslak (1945)
T. radix, TeL	—	—	5.0e	− (*I*)	Cieslak (1945)
Achrochordidae					
Acrochordus granulatus, TrL	1.11b (19)	A	22.2	—	Gorman *et al*. (1981)
Homalopsinidae					
Cerberus rhynchops, TrL	1.89b (53)	A	2.0	—	Gorman *et al*. (1981)

Continued

Reproduction data—Continued

	GSI(*n*)	Breeding pattern	GSI ratio	Territoriality (lizards)/ Male combat (snakes)	References
GSI = Testis mass (mg)/SVL (mm)					
Elapidae					
Hemiaspis signata, TeL	3.36 (3)	A	2.8	– (*l*)	Shine (1977)
Austrelaps superbus, TeL	2.60 (> 6)	A	2.5	– (*l*)	Shine (1977)
Colubridae					
Thamnophis elegans terrestris, TeL	2.56 (2)	D	10.7	– (*)	Fox (1954)
T. sirtalis tetrataenia, TeL	2.00 (6)	A	7.7	– (*)	Fox (1954)
Natrix piscator, TrL	3.43b (2)	—	3.7	—	Shrivastava and Thapliyal (1965)
N. taxispilota, TeL	0.72 (10)	D	7.2	—	White *et al.* (1982)
Tropidoclonion lineatum, TeL	—	A	7.4e	—	Krohmer and Aldridge (1985)
Turtles and Tortoises					
GSI = Testis mass (g)/Body mass (g) × 100					
Chrysemys picta, TeL	—	—	4.2e (17, 19)	—	Mitchell (1985)
Geochelone gigantea, TrL	—	A	15.0e	—	Swingland and Coe (1978)
Lissemys punctata granosa, TrL	—	A	3.9	—	Singh (1977)
Pseudemys scripta, TrL	—	A	6.0e	—	Moll and Legler (1971)
Terrapene carrolina, TeL	0.42 (8)	—	4.8e	—	Altland (1951)
T. o. ornata, TeL	—	D	2.4e	—	Legler (1960)
Trionychidae					
Trionyx gangeticus, TrL	0.10 (10)	A	33.7	—	Rao and Shaad (1985)
T. muticus, TeL	—	—	4.8e, v	—	Plummer (1977)
T. sinensis, TeL	2.05 (^{3}4)	A	5.8	—	Licht (1982)
Trionyx s. spiniferus, TeL	0.83 (^{3}4)	—	8.3	—	Robinson and Murphy (1978)

Kinosternidae					
Claudius angustatus, TrL	—	—	3.0e	—	Flores-Villela and Zug (1995)
Kinosternon f. flavescens, TeL	—	A	6.1e	—	Mahmoud and Klicka (1972)
K. subrurumhippocrepis, TeL	—	A	3.1e	—	Mahmoud and Klicka (1972)
Sternotherus carinatus, TeL	—	A	5.8e	—	Mahmoud and Klicka (1972)
S. m. minor, TeL	0.50 (3)	A	16.7	—	Etchberger and Stovall (1990)
S. odoratum, TeL	—	A	9.0e	—	Mahmoud and Klicka (1972)
S. odoratum, TeL	0.95 (—)	A	11.9	—	McPherson and Marion (1981)
Crocodiilians					
Alligator mississipiensis, TrL	—	A	26.0e	—	Lance (1987)
Caiman crocodilus, TrL	—	A	6.1e	—	Thorbjarnsson (1994)

General notes

b Mean body mass calculated as means of given minimum and maximum values; c Sample for all year; d Deleted from comparative analyses because the minimum GSI estimate was derived from inactive animals (in winter); e Ratio of maximum and minimum testis mass only; g Indexed by taking the ratio between testis length and SVL. Testis length and mass are near perfectly correlated in *Sceloporus magister*, Vitt and Ohmart (1974); h Minimum testis size was calculated as the mean of the two lowest samples owing to low sample size; v Volumes converted to mass (see body of text).

References within tables (italicised number)

Snakes: 1. Shine (1978).

Lizards: 1. Tinkle (1976); 2. Stebbins (1944); 3. Clarke (1965) (semi-natural enclosures); 4. Krekorian (1976); 5. Rand (1967); 6. Schoener and Schoener (1980); 7. Stamps (1983); 8. McMann (1993), Deourcy and Jenssen (1994), Leuck (1995); 9. Vitt and Cooper (1985a); 10. Méndez *et al.* (1988); 11. T. Shivanandappa, personal communication; 12. Marcelini (1978); 13. Tokarz (1995); 14. Harris (1964), Madsen and Loman (1987); 15. Tinkle (1976); 16. Finch (1958); 17. Martins (1994).

14 Sperm Competition in Birds: Mechanisms and Function

T. R. Birkhead

Department of Animal and Plant Sciences, The University, Sheffield S10 2TN, UK

I. INTRODUCTION

The study of sexual selection has traditionally been concerned with the acquisition of partners through male–male competition or female choice – that is, the events occurring before copulation (Darwin 1871; Andersson 1994). Parker (1970, 1984) extended this view by showing that post-mating male–male competition, in the form of sperm competition is also an important component of sexual selection. Subsequently, it has become clear that the effects of sexual selection extend even further. Møller (1994, p. 89; see also Chapters 2 and 17) has shown that sexual selection comprises a succession of selection episodes throughout all stages of the breeding cycle with an individual's fitness being determined by the factors that affect its differential mating success and the differential fecundity of its mate(s). Sperm competition (comprising both male and female perspectives: see Chapter 17) represents just one of these

Sperm Competition and Sexual Selection
ISBN 0-12-100543-7

selection episodes but, as we shall see, its effects are far reaching and impinge on many of the other factors affecting fitness.

The potential for sperm competition was known to Aristotle in 300 BC, who also provided a clear statement about last-male sperm precedence in birds (Peck 1943). Darwin was also aware of sperm competition, having observed extra-pair behaviour in his pigeons and been told about a case of multiple paternity in geese (Darwin 1868, 1871; Birkhead 1997). Despite these illustrious predecessors, the study of sperm competition in birds (and most other taxa) lay dormant until the publication of the seminal papers by Parker (1970) and Trivers (1972). Prior to this, and indeed for some time afterwards, it was assumed that because most birds were socially monogamous, with partners cooperating to rear offspring (Lack 1968), they were also genetically monogamous (Wittenberger and Tilson 1980). However, following the lead provided by Parker and Trivers, studies of avian sperm competition started to appear in the latter part of the 1970s (Hoogland and Sherman 1976; Birkhead 1978, 1979; Beecher and Beecher 1979; Gladstone 1979) and McKinney *et al.*'s (1984) paper in Smith's volume was the first attempt to review the field. The information available at that time was limited and the review comprised mainly an annotated list of species in which extra-pair copulations had been recorded, together with a description of the small number of experimental studies of sperm competition in poultry. In the following few years, however, the field rapidly expanded, with a large number of studies covering a range of topics (reviewed by Birkhead and Møller 1992a). Since 1992 the field has continued to expand, and the aim of this chapter is to provide a broad overview, focusing on recent developments.

At the time of McKinney *et al.*'s (1984) review of sperm competition in birds the emphasis was on male behaviours and it was assumed that most extra-pair copulations were forced on reluctant or acquiescent females. Since then it has become clear that far from being forced on females, in some cases extra-pair copulations are initiated by females. As well as recognizing the important role of females, there have been other changes, including an explosion of paternity studies using DNA fingerprinting, and a shift, across the whole of behavioural ecology, away from purely functional questions towards an integration of functional and mechanistic approaches (Krebs and Davies 1997).

Although the study of sperm competition in birds initially lagged behind that of insects, birds have subsequently proved to be an ideal group in which to examine sperm competition, for a number of reasons. (1) Most species can be caught, individually marked and their natural behaviour easily observed. (2) They rear their offspring in nests which are relatively easy to find, so that true parentage (using molecular or other techniques) can be measured. (3) The extensive parental care in socially monogamous species has allowed biologists to address the fundamental question of the relationship between paternity and paternal investment (Westneat and Sargent 1996; see Chapter 4). Finally (iv) the variation in mating systems provides a range of opportunities for testing sexual selection hypotheses.

A number of unique features affect the potential for sperm competition in birds. Flight provides tremendous mobility and gives birds behavioural flexibility and the ability to rapidly exploit sperm competition opportunities. Conversely, flight also has important constraints on avian anatomy and physiology. These include weight-saving devices (such as (1) the sequential ovulation (and fertilization) of successive large yolky ova, which are incubated outside the body (and usually requiring extensive parental care) and (2) the regression of the reproductive system of both sexes outside the breeding season), which influence the mechanics of gamete production, fertilization and oviposition.

The two most fundamental questions in the field of sperm competition as a whole are: (1) why do females copulate with multiple males? and (2) what are the mechanisms that determine the outcome of sperm competition? In other words how do copulations translate into fertilizations? For birds and other taxa the first question has not been satisfactorily answered. The second question has started to be answered and we now have a reasonable understanding of the mechanism of sperm competition in birds. This, in turn, allows us to make better predictions about the optimal behavioural strategies for each individual involved in sperm competition.

II. MALE MORPHOLOGY AND SPERM COMPETITION

The male reproductive system comprises the paired testes and ductus deferens, and in a very few species, a phallus (King 1981; Lake 1981). Passerines and nonpasserines differ in that the former possess a highly convoluted distal region of ductus deferens, the seminal glomera, in which sperm mature and are stored at temperatures lower than that of the body; they are functionally similar to the distal epididymi in scrotal mammals (Wolfson 1954; Bedford 1990). The avian reproductive system shows seasonal variation in size, and is largest during the breeding season, after which it regresses (Lake 1981). The relative size of the testes and seminal glomera are larger in species in which sperm competition is intense (see below). It appears likely that the possession of a phallus is also linked to sperm competition in some way, but comparative studies have not been able to demonstrate such an effect (Briskie and Montgomerie 1997).

Breeding season testes mass (both testes combined) scales allometrically with body mass, with an exponent of 0.67, and the relationships for passerines and nonpasserines do not differ (Møller 1991a). As predicted, relative testes mass is smaller in species which use mate guarding as their main paternity guard (see below), and larger in species in which frequent copulation occurs (Møller 1991a). Subsequently, Møller and Briskie (1995) showed that relative testes size was also positively correlated with the intensity of sperm competition, as reflected by the proportion of extra-pair offspring, and proposed two explanations for this relationship. Relatively large testes are required to engage in frequent extra-pair copulations, or, more likely as a way for pair males to maintain a high sperm input to their partner to minimize the risk of being cuckolded.

Across passerine species the mass of the seminal glomera is also positively correlated with body mass, and with relative testes mass (Birkhead *et al.* unpublished data). The larger the seminal glomera the greater number of sperm they can store and are available for copulation. Comparisons of the number of sperm in the seminal glomera across species is confounded by the time since the last ejaculation (Birkhead *et al.* 1995a), but despite this, seminal glomera sperm numbers (range: $1–1500 \times 10^6$) are positively correlated with both copulation frequency and the intensity of sperm competition, as measured by the proportion of extra-pair offspring (Birkhead *et al.* 1993a; Birkhead *et al.* unpublished data). The relative size of testes and seminal glomera reach their maximum in polygynandrous or promiscuous passerines in which sperm competition is particularly intense (e.g. the dunnock *Prunella modularis* (Birkhead *et al.* 1991; Davies 1992), alpine accentor *Prunella collaris* (Nakamura 1990; T. R. Birkhead, unpublished data), Smith's longspur *Calcarius pictus* (Briskie 1993), the aquatic warbler *Acrocephalus paludicola* (Schulze-Hagen *et al.* 1995), stitchbird *Notiomystis cincta* (Castro *et al.* 1996) and sharp-tailed sparrow *Ammodramus caudacutus* (Rising 1996)).

Relatively large seminal glomera provide birds with the option of either frequent copulation, or relatively large ejaculates. Unfortunately, we know very little about natural ejaculate size in birds (but see Pellatt and Birkhead 1994). In a pioneering comparative study Møller (1988a) used data on the number of sperm manually extruded during artificial insemination studies to estimate ejaculate size in both passerines and nonpasserines and showed that the number of sperm per 'ejaculate' was positively correlated with testes mass, but unrelated to body mass. These results must be considered with caution because (1) we know nothing about the relationship between true ejaculate size and the number of sperm manually extruded, and (2) the size of natural ejaculates varies considerably, even within the same individuals, due in part to the time since the last ejaculation (Birkhead *et al.* 1995a). In the zebra finch the mean ejaculate size of a rested male (i.e. no copulation in previous 7 days) is about $6–10 \times 10^6$ sperm, but a male that is regularly copulating through a breeding cycle produces about 1×10^6 sperm per ejaculate (Birkhead *et al.* 1995a). Male zebra finches have relatively small seminal glomera and in rested males these contain about 10×10^6 sperm. However, if they copulate three times at hourly intervals, their sperm supplies are completely depleted and since sperm production rates are low ($1.885 \times 10^6 \,\mathrm{day}^{-1}$, or $35 \times 10^6 \,\mathrm{day}^{-1}\,\mathrm{g}^{-1}$ testis tissue), it takes

5 days for the seminal glomera to be fully replenished. At the other extreme, the seminal glomera of the polygynandrous or promiscuous species listed above contain between 200 and 1000 $\times$ 10^6 sperm, and in the splendid fairy wren *Malurus splendens and* white-winged fairy wren *Malurus leucopterus*, the seminal glomera contain 2756 $\times$ 10^6 and 3498 $\times$ 10^6 sperm, respectively (Tuttle *et al.* 1996). Ejaculate size in these species is unknown, but variation in their copulation behaviour indicates that sperm allocation differs between them. The dunnock, alpine accentor and Smith's longspur copulate several hundred times per clutch, but fairy wrens and the aquatic warbler appear to copulate infrequently (Mulder and Cockburn 1993; Schulze-Hagen *et al.* 1995), suggesting that in these species ejaculate size is likely to be large (see also Tuttle *et al.* 1996).

In contrast to mammals, avian sperm does not need to undergo capacitation prior to fertilization (Howarth 1970, 1995). Sperm morphology in birds varies considerably, both between and, to a lesser extent, within species. Nonpasserine sperm are of the basic reptilian type, whereas passerine sperm are typically spiral-shaped (McFarlane 1963). Sperm length varies between species (passerines: 50–300 μm; McFarlane 1963; Briskie and Montgomerie 1992) and are generally shorter in nonpasserines (Lake 1981). Even within a single species sperm length can vary significantly between individuals (McFarlane 1963; T. R. Birkhead unpublished data). In comparative studies of passerines, sperm length is positively correlated with extra-pair paternity (intensity of sperm competition) (Briskie *et al.* 1997). The functional significance of this relationship is unclear. In mammals it has been suggested that longer sperm swim faster than shorter ones (Roldan *et al.* 1992), However, empirical data do not support this: no relationship was found between *in vitro* sperm velocity and length in passerines (Birkhead *et al.* in prep.). Indeed, theoreticians do not predict such a relationship because any increase in power generated by a longer flagellum is offset by the increased drag (R. M. Alexander, personal communication: Wu *et al.* 1975; Wu 1977). The relationship between sperm length and sperm competition therefore requires some other explanation and further investigation.

III. FEMALE MORPHOLOGY AND SPERM COMPETITION

Whereas male morphology is often blatantly correlated with the intensity of sperm competition, female morphology is much more conservative. In birds the female reproductive system comprises a single ovary (in most species, and usually the left one), and its oviduct. Successive ova develop out of phase with each other, and are ovulated and fertilized sequentially at intervals of 24 h or longer. In a typical passerine the rapid growth phase on an ovum lasts 4–6 days, after which ovulation occurs (usually in the morning). Fertilization takes place within the infundibulum at the top of the oviduct within 1 h of ovulation. The ovum then passes fairly rapidly to the uterus or shell gland, where it spends about 20 h; oviposition of the fully formed egg occurs about 24 h after the ovum was ovulated. The next ovum is then ovulated and the process repeats itself until the clutch is complete (Johnson 1986). Because each egg is fertilized separately, the storage of sperm in the female tract may have initially evolved to avoid the necessity of copulating prior to the ovulation of each egg. Sperm are stored in the sperm storage tubules (SSTs) located at the uterovaginal junction. The number of SSTs varies between 500 and 20 000 in different species and larger birds tend to have more SSTs, presumably to avoid the dilution effect of a larger reproductive tract (Birkhead and Møller 1992b; Briskie and Montgomerie 1993). SST size also varies (range: 70–700 μm); shorter SSTs tend to occur in species with large numbers of SSTs (Briskie and Montgomerie 1993), in part because space in the uterovaginal region is limited. The SSTs are blind ending; in some species they are branched, but in most they comprise a single, sausage-shaped tube. On average, SSTs are about three times as long as the sperm they store (Briskie and Montgomerie 1993). The positive correlation between sperm length and SST length suggests coevolution of male and female traits, and the fact that SST length is also positively correlated with sperm length is consistent with Briskie *et al.*'s (1997) suggestion that in some way this may allow females to retain control over fertilization.

Sperm transfer occurs when male and female cloacae make contact during copulation. In most small birds mounting and cloacal contact is brief, lasting only 1 or 2 s; in larger birds mounting can be more protracted but cloacal contact is still relatively brief (Birkhead *et al.* 1987). In the domestic fowl, the female everts her oviduct during copulation such that the region with the vaginal opening, the vaginal mound, protrudes into the cloacal opening. In this way the male deposits semen directly onto the opening of the vagina (Lake 1981). It seems plausible that copulation is similar in other species. However, during detailed observations of copulation in Adelie penguins *Pygoscelis adeliae* and pigeons *Columba livia*, semen was often seen to be deposited on the outside of the cloaca and then drawn into the cloaca by rhythmic pumping movements (Lovell-Mansbridge 1995; Hunter *et al.* 1996).

Within minutes of copulation and insemination a large number of the inseminated sperm may be ejected with faeces by the female (Howarth 1971). Only a small proportion of those retained survive the vagina, which is particularly hostile to sperm (Steele and Wishart 1992; Birkhead 1996), so that less than 2% of the sperm originally inseminated enter the SSTs. Following insemination, some sperm (an unknown proportion) bypass the uterovaginal junction and are transported directly to the infundibulum. If no ovulation occurs, these sperm may remain there for several days before being lost into the body cavity (Brillard and

Antoine 1990; Brillard and Bakst 1990). Sperm in the SSTs usually reside with their heads facing the distal end: there is no convincing evidence that sperm from successive ejaculates remain stratified within the SSTs (Birkhead *et al.* 1990; see below). Sperm subsequently leak out of the SSTs at a constant rate over a matter of days, or weeks and are transported to the infundibulum, where fertilization takes place. In different species the median duration of sperm storage varies between 6 and 45 days; in most species it is about 10 days (Birkhead and Møller 1992b). Some of the variation between species in sperm storage duration can be explained by the spread of laying (clutch size × laying interval) or by the ecology of particular species (Birkhead and Møller 1992b). The duration of sperm storage is a heritable trait (Beaumont *et al.* 1992). The rate at which sperm are lost from the SSTs can be estimated noninvasively by counting the numbers of sperm on the outer perivitelline layers of laid eggs (Wishart 1987; Birkhead *et al.* 1994): rates of loss vary between species, and the instantaneous per capita rate of loss (see Lessells and Birkhead (1990) for a definition of this term) in most passerines appears to be 0.02–0.03 sperm h^{-1} (Birkhead *et al.* 1993b, 1994), while in some nonpasserines it is lower (0.013 sperm h^{-1} in the domestic fowl; 0.003 sperm h^{-1} in the turkey; Wishart 1988). The rate of loss, which may be species specific, combined with the numbers of sperm inseminated, determines the duration of sperm storage and hence the fertile period (Wishart 1987; Birkhead and Fletcher 1993; J. P. Brillard, pers. comm.). The rate at which sperm are lost from the female reproductive tract is a key parameter in modelling the mechanism of sperm competition (see below). Moreover, because sperm storage is protracted (relative to mammals: see Chapter 16) and each ovum is fertilized separately, each egg can potentially be fertilized by a different male, as is demonstrated by the aquatic warbler (Schulze-Hagen *et al.* 1993).

IV. MULTIPLE MATING BY FEMALES

The extent to which females copulate with more than one male and hence the intensity of sperm competition and the extent of multiple paternity vary between different mating systems. In simultaneous polyandrous species such as the dunnock, alpine accentor, and Smith's longspur (Davies 1992; Briskie 1993; Davies *et al.* 1996), sperm competition is almost inevitable. It is less obvious that sperm competition will occur in sequentially polyandrous species, such as the spotted sandpiper *Actitis macularia*, but Oring *et al.* (1992) found that 10% of eggs were fertilized by a male other than the female's current partner – usually a male from earlier in the season (see below).

Among communal breeders the extent of sperm competition varies considerably. In those species which helping behaviour depends on shared paternity (cooperative polyandry) sperm competition occurs routinely between male group members. In the pukeko *Porphyrio porphyrio* and Galapagos hawk *Buteo galapagoensis*, for example, male group members show no mate guarding or overt competition over copulations, and paternity appears to be shared equally between group members (Jamieson *et al.* 1994; Faaborg *et al.* 1995). In other communal systems in which helping depends upon collateral kinship with the brood, helpers are usually close relatives and only rarely copulate with the breeding female. Examples include the Florida scrub jay *Aphelocoma coerulescens* (G. Woolfenden and J. S. Quinn, pers. comm. cited in Birkhead and Møller 1992a) and the Siberian jay *Perisoreus infaustus* (J. Ekman, pers. comm.), in which multiple paternity is zero, and the European bee-eater *Merops apiaster* (Jones *et al.* 1991), bicolored wren *Campylorhynchus griseus*, stripe-backed wren *Campylorhynchus nuchalis* (Haydock *et al.* 1996; Rabenold *et al.* 1990) and fairy wrens *Malurus* spp. (Brooker *et al.* 1990; Mulder *et al.* 1994) in which within-group extra-pair paternity is low. However, in the fairy wrens extra-pair paternity is extremely high (60–70%) as a result of extra-pair copulations by males from other groups (Brooker *et al.* 1990; Mulder *et al.* 1994) and this is reflected in the morphological attributes of males (Mulder and Cockburn 1993; Tuttle *et al.* 1996).

In many socially polygynous species, such as the red-winged blackbird *Agelaius phoeniceus*, extra-pair paternity is common (Gibbs *et al.* 1990; Westneat 1993; Gray 1996), although no more so than in socially monogamous species: Møller and Birkhead 1994). Polygyny increases the reproductive success of successful males through the acquisition of additional females. It occurs in situations in which either territory quality or male quality varies sufficiently to make it worthwhile for females to join an already-mated male. Overall, the opportunities for males to become polygynous are limited and, as a result, the majority of birds are socially monogamous (Lack 1968; Wittenberger and Tilson 1980; Davies 1992). Davies (1992) has suggested that all mating systems lie on a continuum, from polyandry, where females achieve their preferred optimum, through monogamy (where neither sex does), to polygyny, where males achieve their preferred optimum. Extra-pair copulation is part of this continuum and constitutes a way in which either sex can attempt to modify its reproductive success within a particular mating system (Møller 1992a). For example, in the polygynous red-winged blackbird both sexes may benefit (albeit in different ways) from extra-pair copulations. An important aspect of polygyny is that in the majority of cases females rear offspring with very little male assistance: male care is not essential. This fact may explain why extra-pair paternity is frequent in polygynous species. Birkhead and Møller (1996) have argued that in situations where male care is essential, as in many socially monogamous non-passerines (e.g. seabirds and raptors), either the costs of extra-pair copulation for females are too great, or the benefits too small to make it worthwhile, hence their low levels of extra-pair paternity. The most important potential cost of extra-pair behaviour for females of these species is the loss of male care. In polygynous species however, this cost is negligible, hence the regular occurrence of extra-pair paternity.

Lekking species comprise a special subset of polygyny: females obtain no parental assistance from males and are less constrained in their choice of copulation partner than in other mating systems (Møller 1992a). As a consequence, females tend to copulate infrequently and usually with a single male so that multiple paternity is relatively infrequent (Höglund and Alatalo 1995).

Although monogamy is the most frequent mating system in birds, parentage studies have revealed high levels of extra-pair paternity, particularly among passerines, so we can refer to birds only as *socially* monogamous. This has caused problems in terms of how mating systems are defined, and the situation where a male and a female rear young together is now referred to as social monogamy, to allow for the fact that extra-pair paternity may occur.

Parker (1984) suggested that males stood to gain more than females from engaging in multiple matings and, as a consequence, selection operated more intensely on males than females. Parker's idea was that multiple mating increased the number offspring a male fathered, whereas for a female it increased only the quality of her offspring. One reason why the initial emphasis was on males is that their behaviour is often much more obvious than that of females. Among socially monogamous birds, the males of many ducks and colonial birds for example, attempt to mount females forcibly, and sometimes in groups (Birkhead and Møller 1992a). Even in species where forced extra-pair copulations are absent or rare it is also males that initiate most extra-pair incidents, leaving their own territory to visit other females (e.g. magpie *Pica pica*; Buitron 1983). Female behaviour is much more covert, even in those cases where females leave their own territory in search of extra-pair males (e.g. tits; Kempenaers *et al.* 1992; Otter *et al.* 1994; Gray 1996). Moreover, in almost all species, females engage in extra-pair copulations when their partner is absent, suggesting that detection of infidelity carries costs for females. Because female behaviour is often furtive, it is difficult to obtain unbiased estimates of extra-pair copulation frequency across species (Dunn and Lifjeld 1994; Birkhead and Møller 1995).

V. PATERNITY GUARDS

The two main paternity guards in birds comprise: (1) mate guarding by close following and (2) frequent copulation (Birkhead *et al.* 1987; Birkhead and Møller 1992a). A number of other paternity guards have also been proposed: (3) territoriality, (4) song, (5) aggression, and (6) the deceptive use of alarm calls (Birkhead and Møller 1992a).

A. Mate guarding

As an undergraduate I found David Lack's book *Ecological Adaptations for Breeding Birds* a revelation and a mine of information. In a section headed 'Pair Formation' Lack (1968, p. 159) refers to the Reverend J. M. McWilliam, Scotland's 'minister of ornithology', who apparently said that if he took his wife to Edinburgh and did not want to lose her, he had two alternatives: (1) to have a fixed address to which they could both return, and (2) to follow her about continuously. Lack used this story to illustrate the fact that birds use the same two strategies in pair formation. However, in the light of current knowledge this analogy is misleading on one count and incorrect on the other. First, the Reverend McWilliam's following behaviour was much more consistent with mate guarding than pair formation – he was already married – and second, we now know that the following behaviour seen among birds in the spring is usually between already-paired birds, and not between birds attempting to form a bond.

Mate guarding in most species comprises a male remaining close to his fertile partner and following her when she moves. By remaining in close proximity a male is able to deter any extra-pair males which might approach the female, and may also be able to deter his female from initiating and engaging in extra-pair copulations. In many species guarding males remain within 5 or 10 m of their partner for much of their fertile period (e.g. Davies 1992), but there is considerable interspecific variation in guarding intensity. In the yellowhammer *Emberiza citrinella*, indigo bunting *Passerina cyanea* and red-winged blackbird for example, mate guarding is relatively lax (Sundberg 1992; Westneat 1987a, 1993). At the other extreme, magpies guard their females assiduously (Birkhead 1979), and in the feral pigeon the mean distance between pair members over the entire 10-day fertile period is just 61 cm (Lovell-Mansbridge 1995). In two other species guarding may be even more extreme, with males remaining in direct physical contact with their partner. The male aquatic warbler clings on to the female's back for up to 50 min at a time, inseminating her every few minutes (Schulze-Hagen *et al.* 1995), and in the vasa parrot *Caracopsis vasa*, pairs form a copulatory tie (as in canids) lasting up to 100 min (Wilkinson and Birkhead 1995).

Males guard their partner when she is fertile. The fertile period varies between species, depending on the duration of sperm storage. In most species the fertile period starts about 10 days before the first egg is fertilized and ends when the final egg of the clutch is fertilized (Birkhead and Møller 1992b). Although data on the duration of sperm storage and hence the onset of the fertile period are lacking for most wild birds, field observations show that mate guarding occurs mainly in the middle portion of the female's fertile period, starting about 5 days or less before the first egg is laid and usually terminating before the last ovum is fertilized (Birkhead 1982; Møller 1987a). The fact that mate guarding does not cover the entire time females are fertile suggests that it is costly for males and that they concentrate their efforts during the time when the risks of extra-pair fertilization are greatest.

Mate guarding intensity also varies intraspecifically and is usually more intense the greater the risk of the female engaging in an extra-pair copulation (Dickinson and Leonard 1997). In the barn swallow *Hirundo rustica*, females in colonies, where the likelihood of extra-pair copulation is greater, tended to be guarded more closely than females of solitary pairs (Møller 1987a); within colonies, guarding was more intense when the operational sex ratio was low, and the number of potential extra-pair males high (Møller 1987a). In the dunnock, males in polyandrous trios started guarding earlier and guarded more intensively than those in monogamous pairs (Davies 1985). In the blue tit *Parus caeruleus* poor-quality males guarded more intensively than good-quality birds and appeared to be making the best of a bad job, since females paired to poor-quality males actively sought extra-pair copulations from good-quality males (Kempenaers *et al.* 1992, 1995). Similar patterns occur in the bluethroat *Luscinia svecica* (Johnsen and Lifjeld 1995) and purple martin *Progne subis* (Wagner *et al.* 1996).

The function of mate guarding is to prevent males being cuckolded and this occurs either as a result of other males approaching fertile females, or fertile females seeking out extra-pair males. The effectiveness of mate guarding has been tested by the experimental removal of males during their partner's fertile period. In all studies published to date, male removal resulted in increased extra-pair courtship and copulation attempts on the female (pied flycatcher *Ficedula hypoleuca*, Björklund and Westman 1983; swallow, Møller 1988b; zebra finch, Birkhead *et al.* 1989; great tit, Björklund *et al.* 1991; blue tit, Kempenaers *et al.* 1995; red-winged blackbird, Westneat 1994; western bluebird *Sialia mexicana*, Dickinson 1997; yellowhammer, Sundberg 1992; feral pigeon, Lovell-Mansbridge 1995). The propensity of either sex to leave the territory in search of extra-pair opportunities apparently differs between species; for example in the red-winged blackbirds studied by Westneat (1995), and in magpies (Birkhead 1982; Buitron 1983; Parrott 1995), males leave to seek extra-pair copulations. In other species such as the black-capped chickadee *Parus atricapillus* (Smith 1988; Otter *et al.* 1994) and blue tit (Kempenaers *et al.* 1992), females actively seek extra-pair copulations outside their own territory.

Although the paternity assurance hypothesis for mate guarding is intrinsically difficult to test (Dickinson and Leonard 1997), overall, the evidence is most consistent with this hypothesis. There is very little evidence for any of the alternative hypotheses for the proximity between partners (e.g. to gain access for copulation, to strengthen the pair bond, to reduce predation risks on the female, to facilitate courtship feeding, as a benefit to females to minimize harassment from males, and as a passive consequence of a reduced foraging range by females; Lumpkin *et al.*

1982; Gowaty and Pilssner 1987; reviewed by Birkhead and Møller 1992a; see also Dickinson and Leonard 1997).

The intra- and inter-specific variation in mate guarding may result from the relative benefits and costs it entails. The benefits are a reduced probability of being cuckolded (above). The costs of mate guarding include increased energetic demands on males and reduced foraging efficiency (Davies 1985; Lamprecht 1989; Møller 1987a; Westneat 1994), and possibly increased vulnerability to predators. Mate guarding may also be costly in terms of reduced opportunities for extra-pair copulation (Grafen 1980): however, in most species guarding has priority over seeking extra-pair copulations and the two activities are often temporally separated (Birkhead and Møller 1992a).

B. Frequent copulation

In reviewing the copulation behaviour of birds Birkhead *et al.* (1987) showed that copulation rates tend to be relatively high in two categories of birds: (1) polyandrous species, such as the dunnock and Smith's longspur in which sperm competition is intense (Davies 1992; Briskie 1992), and (2) socially monogamous species in which mate guarding by close following does not occur. In both of these categories females copulate up to several hundred times per clutch, even though it is known that one or a few inseminations can fertilize an entire clutch (Lake 1975; Birkhead 1988; Birkhead *et al.* 1988a, 1989; Adkins-Regan 1995). Frequent copulation is used as a paternity guard in birds of prey (including owls), and many colonial species, such as seabirds, since in neither of these groups is mate guarding practical. In raptors, the female usually remains at the nest while the male forages and brings food back to her, and in colonial species each member of the pair take turns of remaining at and defending the nest site while the other forages away from the colony. The majority of bird species use either frequent copulation or mate guarding as the main paternity guard (Møller and Birkhead 1991), although there are some species in which both frequent copulation and mate guarding occur (e.g. swallow, Møller 1985; chaffinch *Fringilla coelebs*, Sheldon and Burke 1994; western bluebird, Dickinson and Leonard 1997), and two species in which mate guarding is negligible and pair copulations rare (a Norwegian population of pied flycatchers (Chek *et al.* 1996) and the Capricorn silvereye *Zosterops lateralis* (Robertson 1996)).

An assumption of the hypothesis that frequent copulation acts as a paternity guard is that it results in the transfer of more sperm (see Parker 1984). This seems likely to be true since species with high copulation rates tend to have relatively large testes, and larger testes are known to have higher rates of sperm production (Møller 1988a 1991a; see also Birkhead *et al.* 1995a). In addition, Møller and Briskie (1995) have shown that the intensity of sperm competition is positively associated with relative testes size: they propose that this has resulted from selection pressure on pair males to protect paternity, rather than on males to secure extra-pair fertilizations. This is because the costs of lost paternity and misdirected parental care are high and in a situation where paternity is largely under female control, selection will favour males that can inseminate large numbers of sperm into their partner. Moreover, experimental studies show that the numbers of sperm inseminated by different males play an important part in determining the outcome of sperm competition (see below). Frequent copulation therefore clearly has the potential to serve as an effective paternity guard. Among red-winged blackbirds in the wild, Westneat (1995) showed that pairs which copulated most had fewer extra-pair offspring. Previously, there has been speculation about whether frequent copulation is more effective than mate guarding (Birkhead and Møller 1992a; Møller and Birkhead 1993a). If all other things were equal, then frequent copulation should be the more effective paternity guard, but the conflicting interests of males and female partners over extra-pair copulations mean that other factors are rarely equal. Male partners stand to gain from frequent pair copulations because it increases their likelihood of fertilization. However, it is not clear why females often comply with males and copulate as often as they do. After all, females control the frequency and success of copulations and require only a few inseminations to ensure the fertilization of their eggs. There are numerous hypotheses for why females copulate so frequently (Petrie 1992; Hunter *et al.* 1993; see Chapter 3), but perhaps the single most important one is that females need to 'convince' their partner of his paternity in order to ensure his continued investment in the offspring (Davies 1985; see also Chapter 4). However, females also need to retain the option of engaging in an extra-pair copulation, should an appropriate male be available. This may explain why, in most species, females cease copulating before the end of their fertile period: by doing so they retain the option of securing extra-pair fertilizations (Birkhead and Møller 1993a). Mate guarding appears to be effective because it constrains female extra-pair activity: females rarely engage in extra-pair copulations with their partner present – presumably because they are invariably disrupted and are potentially costly. However, males are not infallible and females are clearly sometimes able to elude their guarding partner and obtain an extra-pair copulation (Kempenaers *et al.* 1992, 1995).

Frequent copulation may also be costly for males. The relatively large testes required to sustain a high output of sperm may be energetically expensive to maintain and will increase the energetic costs of flight. The cost of sperm production itself may also add to the male's energy budget. However, the physiological cost of sperm production is not known for any species, but it may be relatively low in birds since production appears to be continuous during the breeding season, with noncopulating individuals shedding large numbers of sperm with the faeces (Birkhead 1996).

C. Other paternity guards

1. Territoriality

In many species, male territorial behaviour peaks during their female's fertile period indicating that territorial aggression may serve as a paternity guard (Birkhead 1979; Møller 1987b; Langmore 1996). Experiments in which male red-winged blackbirds were removed from their territory demonstrated that their presence on the territory reduces the incidence of extra-pair copulation attempts (Westneat 1994). Møller (1990a) found that territory size in swallows and yellowhammers was greatest during the fertile period, and suggested that this variation in territory size comprised part of a suite of paternity guards (but see Dunn 1992; Møller 1992b). In the dunnock Langmore (1996) showed that breeding territoriality functions exclusively for mate defence.

2. Song

Males of many species use song to deter territorial intruders and since in some species at least, song is most frequent when females are (1) close to egg laying, and (2) early in the morning – both times when females are most fertile (see Birkhead *et al.* 1996) – Møller (1991b) suggested that male song might serve as a paternity guard. Specifically, Møller suggested that because singing is costly, males advertise their overall quality through song to other males. If song is a 'keep-out' signal it will pay to sing frequently to minimize intrusions. Subsequent studies have found rather little evidence for this hypothesis and in a number of species song rate actually declines during the female's fertile period (Hanski and Laurila 1993; Westneat 1993; Sheldon 1994; Rodrigues 1996; Evensen *et al.* 1997). Conversely, Langmore (1996) showed convincingly that in the polygynandrous dunnock and alpine accentor, song serves as a paternity guard.

3. Alarm calls

Møller (1990b) also suggested that males might use alarm calls deceptively in paternity defence. When male swallows uttered alarm calls, conspecifics took flight and extra-pair copulation attempts were disrupted. Male zebra finches also utter alarm calls when separated from their fertile female partner (Birkhead *et al.* 1989), but Fletcher (1996) found no evidence that alarm calls altered the behaviour of either the extra-pair male or the pair female.

VI. MEASURING EXTRA-PAIR PATERNITY

Four main techniques have been used to detect the occurrence of sperm competition in birds.

(a) Sex-specific heritability of morphological characters
This method, first used by Alatalo *et al.* (1984), is based on the fact that offspring resemble their biological parents. If the similarity between offspring and the putative mother is greater than that between the offspring and the putative father then extra-pair paternity is likely. This method assumes that: (1) there are no environmental correlations between offspring and putative parents; (2) the putative mother is the true mother, i.e. that intraspecific brood parasitism is very rare; and (3) there is no sexual dimorphism in the character used for comparison (e.g. tarsus length). The degree of extra-pair paternity can be estimated from the difference in the 'heritability', that is, the slopes of the male and female regressions (Alatalo *et al.* 1984, 1989).

In a number of studies in which both the heritability method and another method have been used to estimate the extent of extra-pair paternity, the two estimates were similar (Møller and Birkhead 1992). Hasselquist *et al.* (1995) showed in the great reed warbler *Acrocephalus arundinaceus* that the heritability method grossly overestimated the true incidence of extra-pair paternity determined using multilocus DNA fingerprinting. Their study demonstrates that, in the great reed warbler at least, the fundamental assumption that the mother and father contribute equally to offspring traits is incorrect. For this reason, and because (1) the standard errors around parent–offspring regressions are large (Lifjeld and Slagsvold 1989) and (2) molecular techniques are so much better, the heritability method is an unreliable way to estimate the incidence of extra-pair paternity.

(b) Genetic markers
Laboratory studies of sperm competition in a range of taxa have relied on genetic markers to assign parentage. In birds, these are usually plumage markers. In the zebra finch, for example, wild-type (grey) plumage is dominant over fawn (a sex-linked recessive). Fawn pairs produce only fawn offspring, but homozygous grey males paired with fawn females produce only grey offspring. When a fawn female copulates with both types of male the paternity of the offspring can be established at hatching (Birkhead *et al.* 1988a). A common problem in using genetic markers is that one genotype usually exhibits differential fertilizing capacity (Lanier *et al.* 1979), even when sperm numbers are controlled, as in studies using artificial insemination. However, analytical techniques are now available which take differential fertilizing capacity into account when analysing sperm competition data (Birkhead *et al.* 1995b).

(c) Multiple allozymes
Initially this was the main technique for examining parentage in various taxa, with the notable exception of birds. It relies on detecting enzyme polymorphisms from blood or other tissue. However, in birds the genetic variability detectable by electrophoresis is so low that this method is generally impractical. In the few cases where sufficient variation exists (e.g. Westneat 1987b), the technique still provides only an indirect estimate of the degree of extra-pair paternity within a population, and can only rarely assign paternity to a particular individual (Westneat and Webster 1994).

(d) DNA techniques
To date, the most successful and widely used method for measuring parentage directly in wild birds is multilocus DNA fingerprinting (Burke and Bruford 1987; Wetton *et al.* 1987). This technique has the important advantage that it can assign parentage to individuals, provided that the number of potential candidates is fairly limited (e.g. when extra-pair males are territorial neighbours). Where the number of potential extra-pair fathers is large single-locus fingerprints from cloned minisatellites provide a solution (Dixon *et al.* 1994; Wetton *et al.* 1995). DNA techniques are now the standard for paternity assignment. However, there have been no published studies of kinship in birds based on rapid amplified polymorphic DNA (RAPD) techniques (Westneat and Webster 1994).

VII. PATTERNS OF EXTRA-PAIR PATERNITY

The number of extra-pair recorded in different species offspring varies from zero to over 60% (Westneat and Webster 1994; Birkhead and Møller 1995). Data are currently available for over 100 species and the list continues to grow. The patterns of extra-pair paternity across species have helped to identify adaptations to sperm competition and may also help reveal the adaptive significance of extra-pair fertilizations. One clear pattern to emerge concerns the anatomical correlates of sperm competition. Species in which sperm competition is intense (as measured by the proportion of extra-pair offspring) have relatively larger testes, seminal glomera and cloacal protuberances (Møller and Briskie 1995; Birkhead *et al.* 1993a). Similar patterns occur in other taxa (see Chapters 10–16).

Other correlates of sperm competition intensity include: (1) sexual plumage dimorphism, which is positively correlated with the intensity of sperm competition (measured by the extent of extra-pair paternity), suggesting that female choice of extra-pair partners has resulted in intense sexual selection (Møller and Birkhead 1994); (2) male parental care, which is negatively correlated with sperm competition intensity (Møller and Birkhead 1993b), as some models predict (see Chapter 4); and (3) the duration of the pair bond, which is also negatively correlated with sperm competition intensity (Cezilly and Nager 1995). A fourth possible correlate is breeding dispersion. From the time that behavioural ecologists first started to look at extra-pair behaviour it was suggested that where individuals bred in close proximity, such as colonies, the opportunities for extra-pair behaviour would be increased (Birkhead 1979; Gladstone 1979; Møller and Birkhead 1993a). Although in some species density and extra-pair behaviour are positively correlated, across species there is no simple relationship between proximity and extra-pair paternity (Birkhead and Møller 1996; Westneat and Sherman 1997). Many colonial species breed highly synchronously (Gochfeld 1980), and Stutchbury and Morton (1995) obtained some evidence that extra-pair paternity covaried positively with breeding synchrony. However, they did not control for phylogeny or other potentially confounding variables, such as breeding dispersion.

Because extra-pair copulations are at least sometimes initiated by females, they must obtain some benefit from behaving in this way. Female benefits from extra-pair copulations can be either direct or indirect (genetic) (Westneat *et al.* 1990; Birkhead and Møller 1992a). It was originally assumed that genetic variation in fitness was low, and that females were therefore unlikely to gain any genetic benefit from extra-pair copulations, and so any benefit that they might obtain would be direct. Of the direct benefits, the most frequently proposed one was fertility insurance. That is, by copulating with more than one male females increase their chances of obtaining sufficient sperm to fertilize their eggs. However, as Table 14.1 shows there is little evidence that females of socially monogamous species obtain any direct fitness benefits from extra-pair copulations. Conversely, genetic variance in fitness traits may not be as low as was once thought (Burt 1995; Pomiankowski and Møller 1995) and it is therefore feasible that females could obtain genetic benefits from extra-pair fertilizations.

One frequently cited potential benefit of multiple mating is an increase in genetic diversity offspring (Williams 1975). However, two factors are thought to work against this hypothesis: (1) mating with more than one male generates a relatively small increase in genetic diversity, and (2) any benefits of increased genetic diversity are offset by increased sib competition (Birkhead and Møller 1992a). Another category of genetic benefit, suggested by Zeh and Zeh (1996, 1997) is that females benefit from multiple mating because it increases the likelihood of finding a genetically compatible partner. Their genetic incompatibility hypothesis for polyandry predicts an increase in the number (rather than the quality) of surviving offspring by females copulating with multiple males. Polyandry, as a mechanism to reduce the costs of inbreeding in birds, has been proposed (e.g. Brooker *et al.* 1990) and is a special case of the genetic incompatibility hypothesis. One component of Zeh and Zeh's (1996, 1997) hypothesis is that the fitness consequences of genetic compatibility depend upon an interaction between parental genomes, and are

Table 14.1. Examples of direct benefits to females from copulating with more than one male. Note, however, that none of these direct benefits provides a general explanation for multiple mating by females in socially monogamous species.

Benefit	Species	References
Fertility insurance	House Sparrow *Passer domesticus*	Wetton and Parkin (1991) claimed this effect, but see Birkhead and Møller (1992a), Lifjeld (1994), Birkhead and Fletcher (1995), Birkhead *et al.* (1995c, 1997)
	Red-winged blackbird *Agelaius phoeniceus*	Gray (1997a)
Acquisition of nutrients	Red-billed gull *Larus novaehollandiae*	Mills (1994); see also Wolf (1975), Cronin and Sherman (1976), Gray (1997b)
Paternal care	Cooperatively polygynandrous species:	
	Dunnock *Prunella modularis*	Davies *et al.* (1996)
	Alpine accentor *P. collaris*	Davies *et al.* (1996)
	Galapagos hawk *Buteo galapagoensis*	Faabourg *et al.* (1995)
	Pukeko *Porphyrio porphyrio*	Jamieson *et al.* (1994)
Change in partner	Oystercatcher *Haematopus ostralegus*	Ens *et al.* (1993), Heg *et al.* (1993): in nonpasserines extra-pair copulations may facilitate mate change
	Spotted sandpiper *Actitis macularia*	Colwell and Oring (1989)

therefore nonadditive (in contrast to a good genes hypotheses: see below).

One pattern to emerge from the extra-pair data is that in many species females tend to perform extra-pair copulations with males with particular attributes. These attributes differ between species, but most studies suggest that females prefer better 'quality' males as extra-pair copulation partners. Since there is no evidence for direct benefits (Table 14.1), this indicates that females obtain indirect benefits, that is, genes for either viability or attractiveness from extra-pair copulations (Table 14.2; see Chapter 2).

There is some evidence, albeit rather indirect so far, for the viability genes hypothesis, which predicts that the viability of offspring fathered by extra-pair males is greater than that of the pair male. Sheldon *et al.* (1997) have shown in the collared flycatcher *Ficedula albicollis* that extra-pair offspring generally fledge at a greater mass than non-extra-pair offspring, and that females appear to prefer males with larger fore-

Table 14.2. Studies in which females apparently do and do not prefer better quality males as EPC partners.

Species	Effect	Male trait examined	Reference
Purple martin *Progne subis*	Yes	Age	Wagner *et al.* (1996)
Swallow *Hirundo rustica*	Yes	Tail length	Møller (1988d), Saino *et al.* (1997)
Black-capped chickadee *Parus atricapillus*	Yes	Dominance	Smith (1988), Otter *et al.* (1994)
Zebra finch *Taeniopygia guttata*	Yes	Bill colour	Burley *et al.* (1994)
Great reed warbler *Acrocephalus arundinaceus*	Yes	Song complexity	Hasselquist *et al.* (1996)
Blue tit *Parus caeruleus*	Yes	Survival Tarsus length	Kempenaers *et al.* (1992)
Indigo bunting *Passerina cyanea*	Yes	Age	Westneat (1990)
Red-billed gull *Larus novahollandiae*	Yes	Courtship feeding	Mills (1994)
Shag *Phalacrocorax aristotelis*	Yes	? nest site?	Graves *et al.* (1993)
Superb fairy wren *Malurus cyaneus*	Yes	Genotype	Mulder *et al.* (1994)
House sparrow *Passer domesticus*	Yes	Plumage colour Age	Møller (1988c) Wetton *et al.* (1995)
Savannah sparrow *Passerculus sandwichensis*	Yes	Feeding rate	Freeman-Gallant (1996)
Bearded tit *Panurus biarmicus*	Yes	Chasing ability	Hoi *et al.* (1997)
Collared flycatcher *Ficedula albicollis*	Yes	Plumage	Sheldon *et al.* (1997)
Yellowhammer *Emberiza citrinella*	Yes	Age/plumage	Sundberg and Dixon (1996)
Reed bunting *Emberiza schoeniculus*	No	Various	Dixon (1993)
Red-winged blackbird *Agelaius phoeniceus*	No	Age	Weatherhead and Boag (1995)
	Yes	Body size	
	No	Territory quality	Gray (1997b)
	Yes	Harem size	Gray (1997a)
House finch *Carpodacus mexicanus*	No	Plumage colour	Hill *et al.* (1994)
Pied flycatcher *Ficedula hypoleuca*	No	Plumage colour	Ratti *et al.* (1995)
Hooded warbler *Wilsonia citrina*	No	Size/plumage	Stutchbury *et al.* (1997)

head patch; the greater the difference in this trait between the pair male and the extra-pair male, the greater the difference in the fledging mass of extra-pair vs. non-extra-pair chicks. Studies of other species have failed to find similar effects.

This still leaves unresolved the question of what component of viability is important. A scenario that identifies this component and is consistent with the observed patterns of extra-pair behaviour is that certain types of habitats allow plumage dimorphism to evolve (e.g. few predators) and females preferentially copulate with the most extravagant males. This increases the frequency of horizontal parasite transmission, which selects for increased virulence among the parasites. This in turn increases the need for more sexual selection. In other words, by engaging in extra-pair copulations with relatively attractive males, females are seeking parasite-resistance genes (see Chapter 2).

VIII. MECHANISMS OF SPERM COMPETITION

A. *Last-male sperm precedence: what is the mechanism?*

Most experimental studies of sperm competition show that sperm from the second of two inseminations or the last of a series of inseminations fertilize the majority (0.6–0.8) of eggs (reviewed by Birkhead and Møller 1992a). Last-male sperm precedence occurs regardless of whether the study comprises natural matings or artificial insemination.

Two early studies of sperm competition in the domestic fowl *Gallus domesticus* shaped our initial thoughts about the mechanism responsible for last-male precedence (see McKinney *et al.* 1984). First, Martin *et al.* (1974) inseminated females just once with a mixture of different ratios of sperm from two genotypes and found that paternity was proportional to the relative number of sperm from each genotype. Second, Compton *et al.* (1978) inseminated females twice with equal numbers of sperm from two different genotypes and found that, regardless of the order in which each genotype's sperm was inseminated, the second insemination fertilized the majority (average 77%) of eggs. Cheng *et al.* (1983) subsequently conducted similar studies on domestic mallard ducks *Anas platyrhynchos* and found that when two inseminations were made just 1 h apart there was no last-male effect, but with a 3 h interval, and particularly after 6 h, the last-male effect was pronounced (0.79). On the basis of these studies, Cheng *et al.* (1983) and McKinney *et al.* (1984) suggested that the timing of successive inseminations determined the degree of last-male sperm precedence. Moreover, they accepted Compton *et al.*'s (1978) explanation for the mechanism: when inseminations were close together sperm mix before going into the sperm-storage tubules, but with an interval of 4 h or more successive ejaculates remain stratified within the SSTs, and a last in–first out system operates.

Subsequent studies indicated that the stratification of ejaculates seemed an unlikely explanation for last-male precedence (Birkhead *et al.* 1990). To explore this further Lessells and Birkhead (1990) constructed mathematical models of sperm competition in an attempt to identify the most plausible mechanism which would account for the level of precedence observed in Compton *et al.*'s (1978) study; this study was chosen because at that time there was more information available on the relevant reproductive parameters in the domestic fowl than for any other species. Three broad categories of models were constructed.

1. Passive sperm loss: here second male precedence occurs simply because by the time the second insemination has occurred, sperm from the first insemination have been lost from the female tract (Fig. 14.1a). However, to obtain 0.77 precedence, as Compton *et al.* did, the rate at which sperm are lost from the female tract would have to have been over 20 times greater than that measured empirically (Wishart 1987). The passive sperm loss model could not therefore account for Compton *et al.*'s results.
2. Stratification: this model tests Compton *et al.*'s idea of a last in–first out mechanism. It has not been possible to determine empirically whether ejaculates from different males do or do not remain stratified in the female's sperm storage tubules. However, this model does make a clear prediction, which is that as sperm from the second insemination are used up over time and the first male's sperm are 'uncovered', the proportion of offspring fathered by the first insemination should increase. Since Compton *et al.* (1978) demonstrated unequivocally that the ratio offspring did not change over time, the stratification model seemed unlikely.
3. Displacement: the idea here is that space is limited in the female's sperm store tubules and that sperm from the second insemination displace stored sperm. Lessells and Birkhead (1990) showed that, depending on the proposed degree of displacement, this model could account for Compton *et al.*'s results. It has subsequently become clear that there is rather little empirical evidence that sperm storage space is limited.

Birkhead *et al.* (1995b) repeated Compton *et al.*'s (1978) experiment and found no evidence for last-male precedence in domestic fowl with a 4 h interval between inseminations. Instead, Birkhead *et al.*'s results were more consistent with those predicted by the passive sperm loss model. The two studies differed in one important way – the timing of inseminations relative to when females laid. In Birkhead *et al.*'s study the first insemination took place at 19.00 hours, about 7 h after the female had laid, and the second insemination 4 h later at 23.00 hours. In

Compton *et al.*'s experiments however, the first insemination took place soon after the first egg was laid. Compton *et al.* (1978) were aware that inseminations made within 90 min before oviposition had a reduced likelihood of fertilization, but at that time it was assumed that inseminations made soon after laying had a similar fertilization success to those made later. However, Leman (1975) showed that inseminations made soon after laying were much less likely to result in fertilization, and Brillard *et al.* (1987) demonstrated experimentally that this was because fewer sperm were taken into the female's sperm-storage tubules. The marked last-male effect found in Compton *et al.*'s experiments with a 4 h interval between inseminations was therefore a consequence of making the first inseminations soon after oviposition. Last-male precedence does occur in birds when equal numbers of sperm are inseminated but only when the interval between inseminations is great enough for some sperm from the first insemination to have been lost (see Fig. 14.1; note that the relative number of sperm from each male also affects the number offspring fathered by each male). Compton *et al.*'s result highlighted the fact that the timing of inseminations relative to oviposition has an important effect on the outcome of sperm competition.

The passive sperm loss model also accounts for last-male sperm precedence in the zebra finch. Birkhead *et al.* (1988a) conducted two experiments on domesticated zebra finches (Fig. 14.2), both of which were similar to situations that could occur in the wild. In the first, a female copulated with one male for several days before being replaced by

---

Fig. 14.1. *The passive sperm loss model. Once sperm enter the sperm-storage tubules they are released at a constant rate. This can be estimated empirically by counting the numbers of sperm on the outer perivitelline layer of successively laid eggs either after a single insemination or after copulation has ceased. The slope of the relationship between* $\log_N$ *sperm numbers and time (h) is the instantaneous per capita rate of loss (Wishart 1987; Lessells and Birkhead 1990). The timing of inseminations are indicated by tailless arrows, and the slopes represent the rate of loss of sperm from the female tract. Long arrows indicate the timing of ovulation and fertilization (here, just one egg for simplicity). (a). The simplest situation in which two inseminations containing the same number of sperm are made at different times. Here, the number of sperm from each male in the female's reproductive tract at any time is determined by the interval between the inseminations and there will be more sperm from the second insemination at the time of fertilization. (b) Two inseminations again, but here the first insemination contains many more sperm than the second, and hence there are more sperm from the first insemination at the time of fertilization. (c). A more complex situation in which the first male makes repeated inseminations and, as a result, his sperm accumulate in the female tract. There are then two equal-sized inseminations (x and y) by another male. Insemination x occurs relatively early and is swamped by sperm from the first male but, more importantly in this scenario, all the sperm from this insemination have been lost by the time of fertilization. Conversely, insemination y is made relatively late, after most of the sperm from the first male have been lost, and hence would have a high chance of fertilization.*

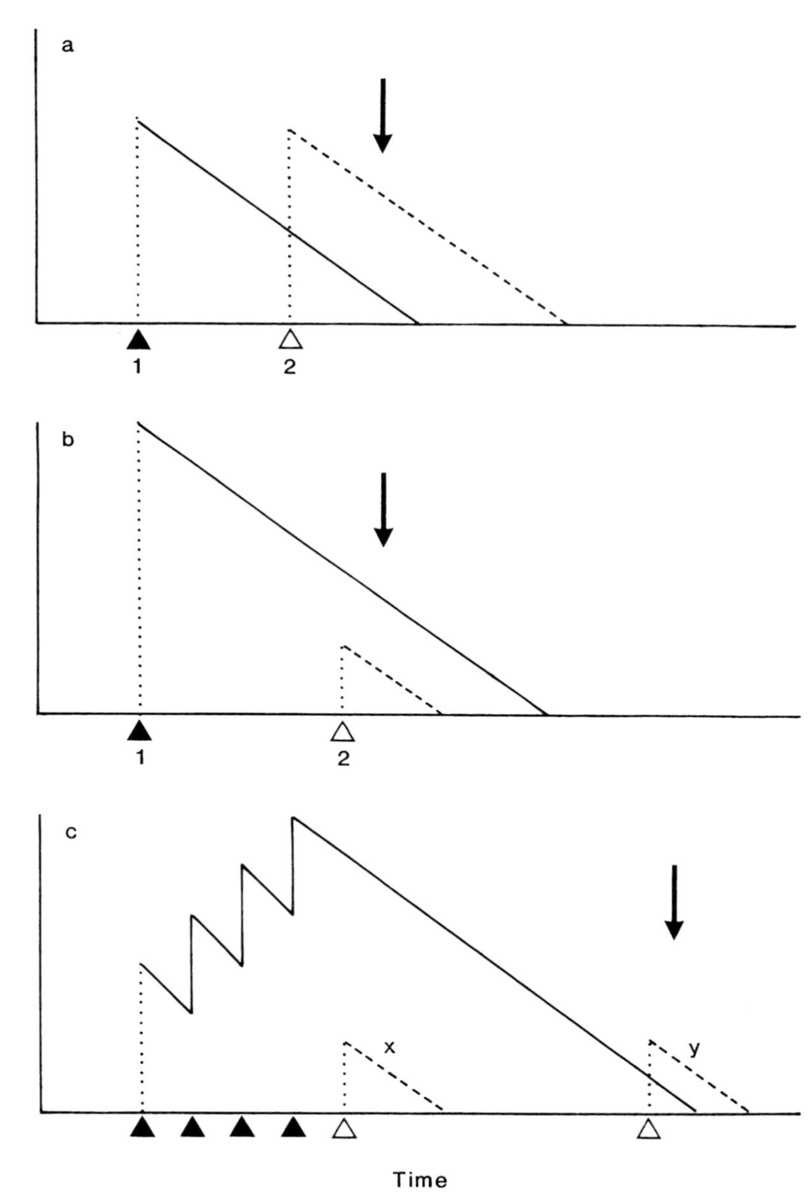
a
1
2
b
1
2
c
x
y
Number of sperm from each male in female reproductive tract
Time

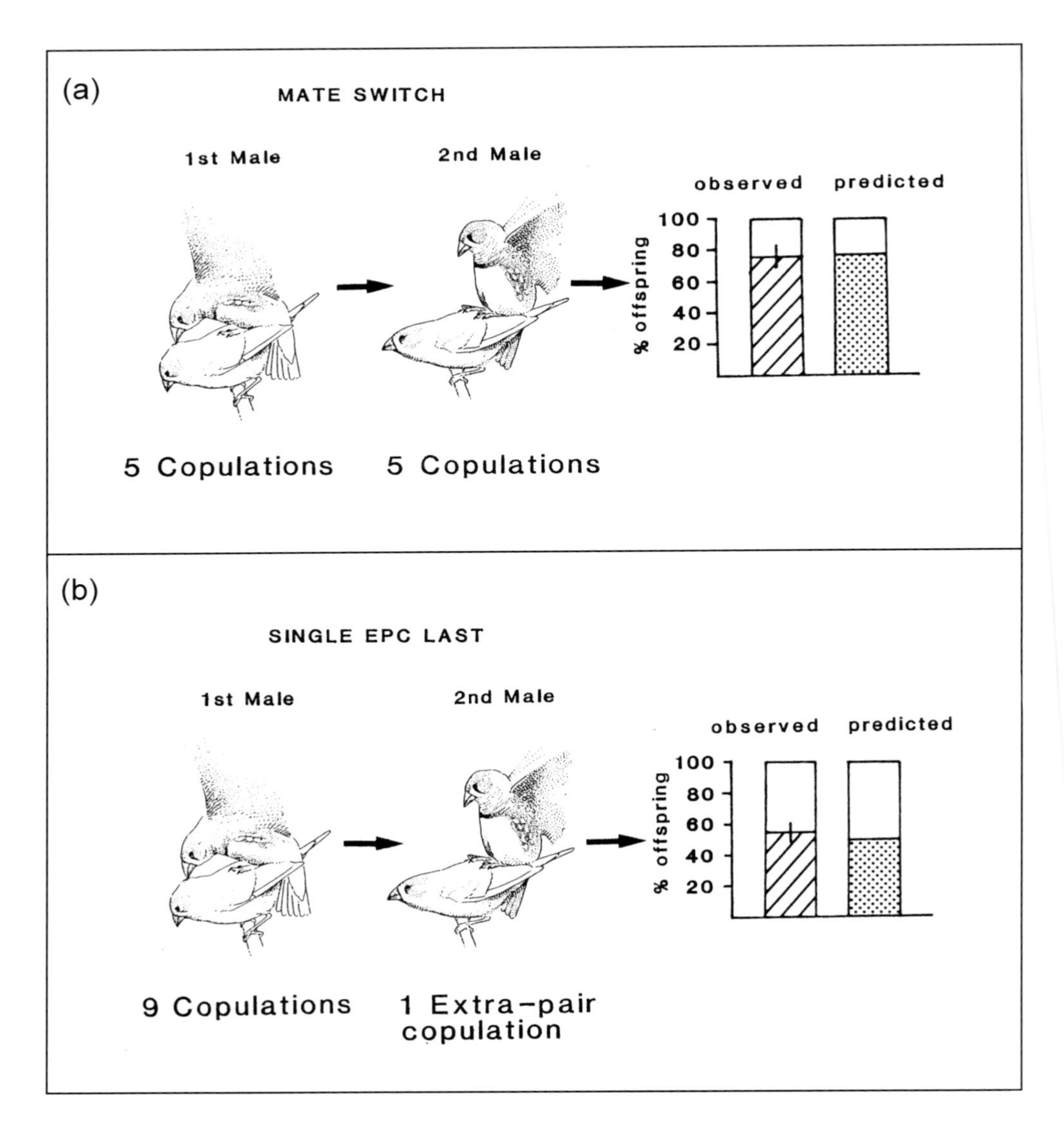

Fig. 14.2. *Sperm competition in the zebra finch. The figure shows the protocol and observed outcome of two experiments. (a) Mate switch, in which two males sequentially obtain similar numbers of copulations, and (b) single extra-pair copulation last (see text). Observed levels of second-male precedence are compared with levels predicted by the passive sperm loss model. The predicted level of second-male paternity was determined using information on the timing of all copulations, the instantaneous per capita rate of loss of sperm from the female zebra finch tract, and the estimated number of sperm inseminated by each male (from Birkhead* et al. *1988a; Colegrave* et al. *1995).*

another male. Despite the fact that the female copulated with each male a similar number of times, the second male fertilized most eggs. The second experiment was designed to test the efficacy of a single 'extra pair copulation' when it was the last copulation a female received before egg laying. Here, the first male obtained about nine copulations, but despite only a single copulation the second male still fertilized over half of all eggs. To test the passive sperm loss model three pieces of information were required: (1) the timing of inseminations, (2) the numbers of sperm inseminated, and (3) the rate at which sperm are lost from the female zebra finch reproductive tract. The first was obtained from continuous video recording. The second was determined in a separate study, by persuading males to copulate with a model female fitted with a false cloaca (Pellatt and Birkhead 1994). This showed that the time since the last ejaculation was the single most important factor affecting the number of sperm ejaculated (Birkhead *et al.* 1995a). The third was estimated by recording the numbers of sperm on the perivitelline layers of successive eggs after copulation had ceased (Birkhead *et al.* 1993b). Using these variables, the passive sperm loss model predicted levels of precedence that were remarkably close to those observed (Fig. 14.2).

Birkhead and Biggins (1998) have also shown that passive sperm loss convincingly accounts for the patterns of paternity in the domestic fowl and turkey *Meleagris gallopavo*. Moreover, they showed that two other models, stratification and displacement (originally described by Lessells and Birkhead 1990) could not explain the observed patterns of paternity. Since the passive loss of sperm from the female reproductive tract is a fixed feature of avian reproductive biology, Birkhead and Biggins modified the stratification and displacement models to incorporate this phenomenon. They then compared the predictions of all three models: (1) stratification (with passive sperm loss) (2) displacement (with passive sperm loss) and (3) passive sperm loss alone, with the observed patterns (see Fig. 14.3). In all the cases they analysed (three additional studies of the domestic fowl and one for the turkey) the observed patterns of paternity were most consistent with the passive sperm loss alone model.

In conclusion, all these studies suggest that the passive sperm loss model is the basic mechanism of last-male sperm precedence in birds. This means that the outcome of sperm competition depends upon the relative numbers of sperm from two or more males at the site of fertilization at the time of fertilization. This is determined by several factors: (1) the numbers of sperm inseminated by each male, (2) the interval between the inseminations from each male, and (3) the number of sperm from each male retained by the female. This, in turn, is affected by the timing of insemination relative to oviposition. Inseminations made close to the time of oviposition have a reduced likelihood of fertilization (see below). An additional factor may also be important: Brillard and Bakst (1990) showed that female turkeys inseminated during the pre-laying period had a significantly faster and greater uptake of sperm than those inseminated after egg laying. It is possible that this effect also occurs in other birds (Birkhead and Møller 1993a). It appears likely that other factors will also be found that modify uptake of sperm by the female and possible contenders include differences in the quality of a particular male's sperm, and female choice (see below). The main factors known to affect the outcome of sperm competition in birds are summarized in Fig. 14.4.

B. The insemination window – a misunderstood phenomenon

The fact that inseminations made close to the timing of oviposition are much less likely to result in fertilization appears to contradict a study by Cheng *et al.* (1983). In the domestic fowl an insemination made soon after oviposition can sometimes result in the fertilization of the next egg to be laid (Johnston and Parker 1970): Cheng *et al.* (1983) found the same effect in domestic mallards *Anas platyrhynchos*, and referred to the hour following oviposition as an 'insemination window', suggesting that it was 'an especially favourable period' for males to obtain extra-pair copulations. Unfortunately, Cheng *et al.*'s results have been misinterpreted by subsequent workers, who assumed that the insemination window represents a peak of female fertility, and as a consequence predicted a peak of copulation or mate guarding behaviour at this time. However, the hour following laying is a favourable time for insemination

Fig. 14.3. *Schematic representation of the three proposed mechanisms of sperm competition in birds. The left side of the figure shows the proposed events in terms of the timing of the two inseminations (1st and 2nd) separated by interval (T) and the distribution of sperm from the first (dashed) and second (dotted) male in a single (representative) sperm-storage tubule. The right side of the figure shows the predicted relationship between the time interval (T) between two inseminations and the logit proportion of offspring of one of the two males ($\ln(p/(1-p))$): see Birkhead and Biggins (1998). Negative and positive T-values refer to reciprocal experiments in which the sperm of one particular male are inseminated first or second (see Birkhead and Biggins 1998). The three models are as follows. (1) Stratification with passive sperm loss: this predicts first-male precedence (reflected here by the higher values in the –T region), and a 'broken stick' pattern, caused by the switch from the first to the second male's sperm. In this model (and in the other two models), the relationship between T and $\ln(p/(1-p))$ is positive because passive sperm loss always results in an advantage to the second insemination the longer the interval between inseminations. (2) Displacement with passive sperm loss: this predicts another broken stick pattern, which is caused by the immediate loss (displacement) of the first male's sperm as the second male's sperm are inseminated. (3) Passive sperm loss alone: this predicts a linear relationship between T and $\ln(p/(1-p))$, with a slope (μ) equal to the instantaneous rate of sperm loss. The relationship would lie through the origin if there was no differential fertilizing capacity (Birkhead* et al. *1995b; Birkhead and Biggins 1998). Data from all studies examined to date are consistent with passive sperm loss alone (see text).*

PROCESS MODEL PREDICTION

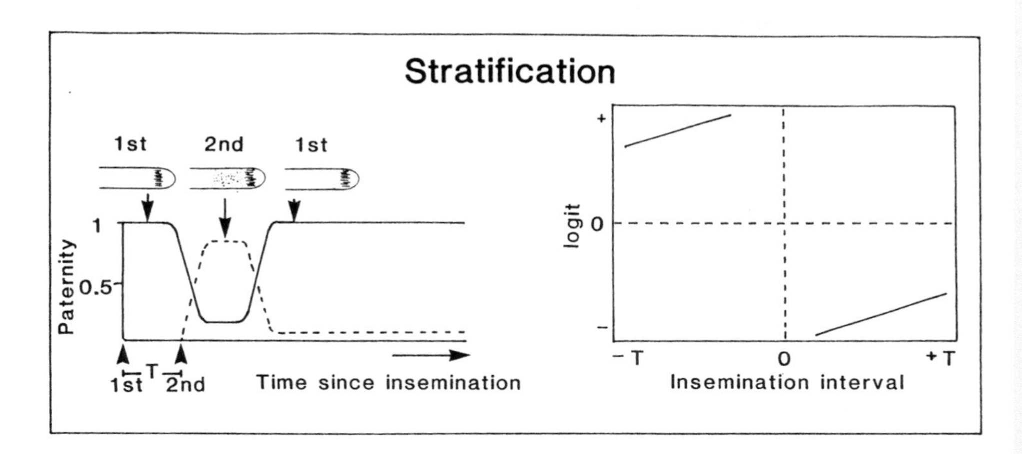

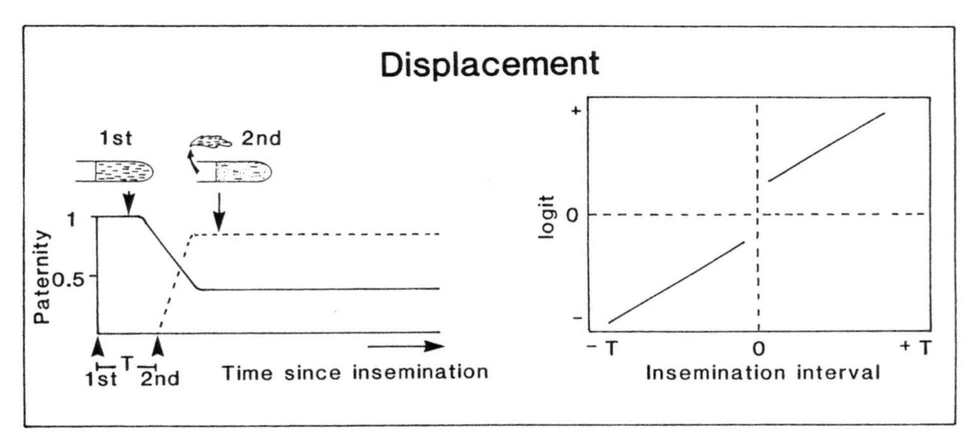

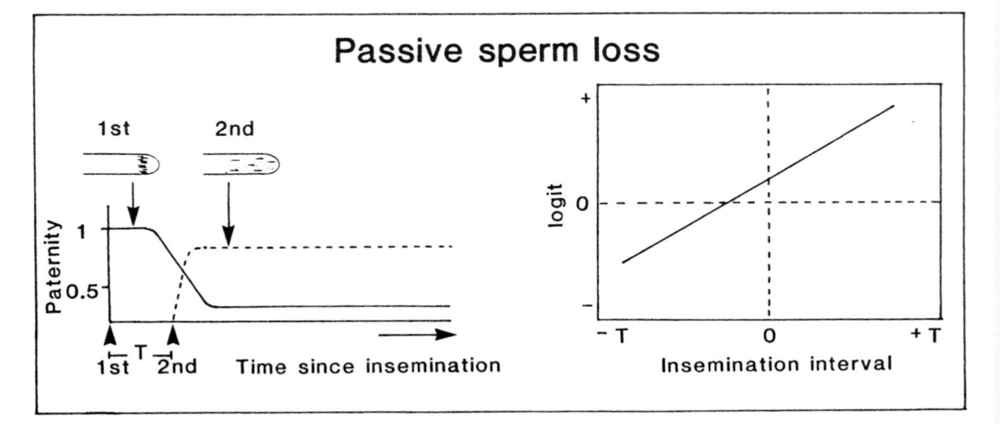

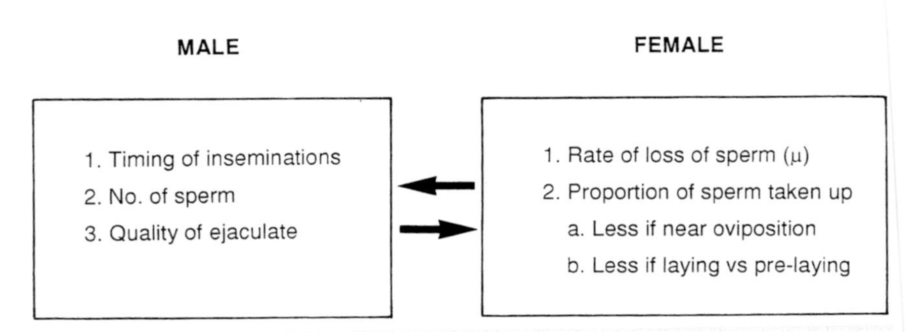

Fig. 14.4. *Summary of the main male and female factors which determine the outcome of sperm competition in birds.*

only for next ovum to be fertilized (i.e. the next egg to be laid). An insemination after this cannot fertilize the next egg to be laid. In fact, as Birkhead *et al.* (1996) have shown by further analysis of Cheng *et al.*'s (1983) data, and from other information in the literature, in general terms, the period spanning several hours before and several hours after oviposition is the worst time for an insemination (Johnston and Parker 1970). Consistent with this is the fact that very few birds copulate at this time and very few change their paternity guards during the insemination window (Birkhead *et al.* 1996).

C. Sperm competition in wild birds

The results obtained from field studies which comprise detailed measurements of copulation behaviour and paternity can be compared with those predicted by the passive sperm loss mechanism of last-male precedence. In the dunnock, Davies *et al.* (1992: Fig. 14.2) showed that a male's assumed share of copulations was positively and significantly correlated with his actual paternity, although the variation around this relationship was, as expected, considerable. Davies *et al.* (1992) also showed that following the removal of males, replacement males fertilized most eggs during the removal period. Both of these results are consistent with the passive sperm loss model and are presumably a consequence of the extraordinarily high copulation rates in this species (Davies 1992). Similarly, when territorial male red-winged blackbirds, European starlings and pied flycatchers were removed, the closer the removal occurred to the start of laying, the greater the loss of paternity from extra-pair males (Westneat 1994; Smith *et al.* 1996; Lifjeld *et al.* 1997). These results are also consistent with last-male precedence. However, I suspect that it will not be easy to critically test the passive sperm loss model with birds in the wild.

IX. MALE OR FEMALE CONTROL?

The gain in fitness that males obtain from extra-pair fertilizations is obvious, but the female benefits are less obvious. As Parker (1970, 1984) stated, males benefit in terms of the number offspring they father whereas females might benefit only in the quality of their offspring and, as a consequence, selection will always operate more intensively on males. Hence, the traditional male-orientated view of sperm competition. However, since it now seems clear that females can benefit from extra-pair fertilization it follows that they will benefit from being able control the paternity of their offspring. This in turn automatically creates conflicts of interest between males and females (Parker 1979, 1984; Stockley 1997), the outcome of which may often be a compromise, with neither sex having absolute control.

Females are potentially able to influence the paternity of their offspring in one of two ways, either behaviourally, or physiologically.

A. Behavioural control

It is clear that in the majority of bird species females can determine whether copulations (either pair or extra-pair) take place or not, and whether they result in cloacal contact and sperm transfer (Birkhead and Møller 1992a). For example, in the wild, female zebra finches were repeatedly courted by extra-pair males, but in only two out of 84 (2.3%) cases did females solicit and perform a copulation: in the rest they simply flew away from the extra-pair male (Birkhead *et al.* 1988b). A similar situation occurs in many colonial seabirds, where male extra-pair copulation attempts are frequent, but females usually reject such males (e.g. gulls, MacRoberts 1973). In an experimental study of the tree swallow *Tachycineta bicolor* in which pair males were removed, replacement males only rarely succeeded in fertilizing any eggs, even though there was ample time for this to occur. This suggests that either that females expected their partner to return or that they were very particular about which males fertilize their eggs and that they can exert this choice through their copulation behaviour (Lifjeld and Robertson 1992).

Other species, such as the blue tit (Kempenaers *et al.* 1992) and black-capped chickadee (Smith 1988; Otter *et al.* 1994) have taken this one stage further and some females leave their territory to actively seek extra-pair copulations.

B. Physiological control

The idea that females might be able to determine or influence the paternity of their offspring at a physiological level is referred to as cryptic female choice (Thornhill 1983; Birkhead and Møller 1993b; Eberhard 1996). There is some evidence for cryptic female choice in the ascidian *Diplosoma listerianum* (Bishop *et al.* 1996). Olsson *et al.* (1996) showed in the sand lizard *Lacerta agilis* that females typically copulate with several males, but apparently had no behavioural control over their copulation partners. When females copulated with two males, the sperm from males that were less closely related to females were more likely to father offspring, indicating that sperm from close relatives were avoided. Mixed paternity in the sequentially polyandrous spotted sandpiper has been interpreted as a case of cryptic female choice (Oring *et al.* 1992; Adkins-Regan 1995). In this species extra-pair fertilizations were least common in first clutches of the season. In subsequent clutches extra-pair paternity was usually attributable to the female's first male of the season and Oring *et al.* (1992) concluded that the prolonged storage (which they considered to be entirely under female control) of the first male's sperm enabled them to cuckold later males. They estimated that the mean minimum interval between the time that the initial male last associated with the female and the ovulation of the egg he fertilized was 18.6 days (range: 3–31 days). While we cannot rule out the possibility of cryptic female choice, studies of sperm competition in other birds suggest a more parsimonious explanation. A female spotted sandpiper's initial partner was usually older and returned to the breeding grounds before younger birds. Moreover, older birds had higher copulation rates (Oring *et al.* 1992), and may well have had larger ejaculates or better quality sperm than subsequent male partners, thus explaining their greater success. In other, experimental studies of sperm competition involving mate switching, the first male may continue to fertilize some eggs for several days or weeks after males have been switched (Birkhead *et al.* 1988a; Birkhead and Møller 1992a, p. 73), simply as a consequence of the way in which sperm are utilized by females (Birkhead *et al.* 1995b).

The unequivocal demonstration of cryptic female choice of sperm is technically difficult and requires a number of criteria to be fulfilled (see Chapter 17). Specifically, it requires clear evidence for a preference for the sperm of particular males, and an unequivocal demonstration that this preference is largely under female physiological control. This may not be as straightforward as it sounds. Given that females seek extra-pair copulations with relatively attractive males, a simple protocol might be to see whether females preferentially utilize the sperm of such males. Although there is now abundant evidence for female choice (Andersson 1994), the criteria that females use are not always clear. For example, in the zebra finch, which is one of the best studied species in this respect, it has variously been suggested that females prefer males with: (1) high song and display rates (Immelmann 1959; Collins *et al.* 1994), (2) red beaks (Burley and Price 1991), (3) red colour bands (Burley *et al.* 1982, 1992), (4) symmetrical chest plumage (Swaddle and Cuthill 1994), and (5) symmetric colour bands (Swaddle and Cuthill 1994). However, there is apparently relatively little consensus among researchers, or female zebra finches, about what constitutes an attractive male. Collins *et al.* (1994) showed that the effect of beak colour was confounded by song and display rate (see also Collins and ten Cate 1996). Birkhead and Fletcher (1995) found that song rate and chest plumage symmetry did not covary, and two studies failed to find any consistent effect of colour bands (Ratcliffe and Boag 1987; Sullivan 1994), or symmetry of colour bands (Jennions 1996). There are several possible explanations for these inconsistencies. Nevertheless, without unequivocal evidence for a female preference it is difficult to test the idea of cryptic female choice in this or any other species.

Even in cases where male behaviour and morphology are controlled, as in some external fertilizers and in artificial insemination studies, a clear difference in the paternity of two or more different males does not constitute evidence for cryptic female choice. Differences in paternity can arise for several reasons, independently of female choice. For example, the sperm of some males may be inherently more likely to fertilize ova, a phenomenon known as differential fertilizing capacity (DFC; Lanier *et al.* 1979). DFC is well known among animal breeders and comprises one male (or a particular genotype) fertilizing a disproportionate number of ova even after controlling for sperm numbers (Dziuk 1996). Moreover, since DFC can occur across numerous individual females, it is generally regarded as an attribute of the male (Dziuk 1996): attempts to explain any of the variance via female effects in birds and mammals have been unsuccessful (P. J. Dziuk, pers. comm.). Avian studies require methods that can distinguish between a male (sperm quality) effect, gamete compatibility (Zeh and Zeh 1996) and cryptic female choice (Simmons *et al.* 1996).

The most likely situation in which we would expect cryptic female choice in birds to occur is when females have rather little behavioural choice over their copulation partner, as in forced copulations. These occur in the stitchbird or hihi *Notiomystis cincta* (Castro *et al.* 1996) and in many waterfowl (McKinney *et al.* 1983). To date, however, there is no evidence that females of any bird species exhibit cryptic female choice.

What are the potential mechanisms by which cryptic female choice might occur? Eberhard (1996) lists all potential mechanisms and in birds, perhaps the most obvious mechanism is the proportion of sperm a female retains immediately after insemination. It has been shown in both the zebra finch and poultry (Howarth 1974; Birkhead *et al.* 1993b; Birkhead 1996) – and it probably occurs in other birds – that females eject a large proportion of sperm soon after insemination and only 1 or 2% of inseminated sperm enter the sperm storage tubules. The basis for this rejection is not known; it could be passive, such that any semen that have not been drawn from the cloaca into the vagina after a certain time are simply voided when the female next defecates (T. R. Birkhead and E. J. Pellatt unpublished data). It is possible however, that females control when they defecate and thus how many sperm they void. If females can do this and adjust the proportion of sperm they eject, they can bias paternity. The potential for this type of control is nicely illustrated by the sperm competition data from chickens: inseminations made close to oviposition result in fewer sperm getting into the sperm storage tubules, and hence entering the fertilizing set (Brillard *et al.* 1987; Birkhead *et al.* 1995b). Adkins-Regan (1995) reported this type of effect in Japanese quail *Coturnix japonica*: females that initially ran away from males but were subsequently mounted and inseminated were less likely to lay fertile eggs than those females that more readily accepted a copulation from a male.

A second potential mechanism in birds concerns the selection of sperm within the vagina. The vagina is a particularly hostile region of the female tract and it is known that only a small proportion of the sperm traverse this region of the female tract and reach the sperm storage tubules (Bakst *et al.* 1994). Steele and Wishart (1992) suggested that the surface characteristics of sperm which are accepted by the sperm storage tubules differ from those which do not and that the selection of sperm is based on an immunological response. In other words, only those sperm with certain characteristics are capable of getting through the vagina. This raises the question of why males produce such a small proportion of acceptable sperm. One answer is that sperm characteristics are sexually selected (Birkhead *et al.* 1993c; Keller and Reeve 1995).

A third potential mechanism would occur in the infundibulum at the time of fertilization. The female pronucleus is a relatively tiny target for sperm in the large yolky ovum. Not only do sperm have to locate the pronucleus they also have to penetrate the inner perivitelline layer surrounding the ovum to get to it (Howarth 1984; Birkhead *et al.* 1994). Not withstanding the potential for choice at this stage of the reproductive cycle, it seems likely that the greatest potential for cryptic choice occurs between ejaculation and the uptake of sperm into the sperm-storage tubules since this is where the greatest proportional reduction in sperm numbers occurs (see also Eberhard 1996).

If we know about the underlying mechanisms associated with sperm competition (e.g. the way that females utilize sperm, the size of a male's sperm store, rate at which they produce sperm, and the numbers they ejaculate), we can start to predict optimal behaviour patterns for each sex. Consider the zebra finch; in terms of the number of eggs the extra-pair male will fertilize, the optimum time to perform an extra-pair copulation is on day −1 (the day before the first egg is laid) of the female's cycle (Colegrave *et al.* 1995). The fact that extra-pair copulations do not peak on this day (even though they almost always occur during the female's fertile period) reflects the conflict of interests of the different parties involved (Birkhead 1996). As with other species, female zebra finches engage in extra-pair copulations only when their partner is out of sight (Birkhead *et al.* 1988b, 1989), so mate guarding by the male partner can severely constrain the optimum timing of copulations by other males, and the pair female herself. Because pair females actively solicit extra-pair copulations it seems reasonable to assume that they are seeking extra-pair fertilization and hence the optimal time of an extra-pair copulation for them would be the same as for the extra-pair male. However, this need not be true because a female may have to allow their partner a certain amount of paternity to persuade him to invest in her offspring, so she may not want to maximize the number of extra-pair offspring. Finally, pair males attempt to minimize the risk of being cuckolded through a combination of different paternity guards (above), although as we have seen, they are not always successful.

X. CONCLUSIONS

Birds have proved to be among the best organisms for examining sperm competition, from both a functional and causal perspective. This is reflected by the large numbers of workers studying this aspect of avian biology. What does the future hold for this field? The question of why females engage in extra-pair copulations still needs to be resolved. The available evidence suggests that, except for a rather small number of special cases (Table 14.1), females are unlikely to obtain direct benefits from copulating with more than one male. Although it was once thought unfeasible that females could obtain genetic benefits from multiple mating, resolutions to the so-called 'paradox of the lek' (Kirkpatrick and Ryan 1991) now exist (Pomiankowski and Møller 1995; Rowe and Houle 1996; but see Alatalo *et al.* 1984) and there is increasing circumstantial evidence for genetic benefits (Table 14.2). However, rigorous experimental evidence is still required before this hypothesis can be accepted. Even if genetic benefits drive extra-pair behaviour we still have to explain how this could account for the considerable interspecific variation in levels of extra-pair paternity.

In terms of the mechanisms of sperm competition, we now have a reasonable understanding of the processes and the main factors involved. Progress in this area was hampered because earlier workers were unable to disentangle the various factors that determine the outcome of sperm competition. The passive loss of sperm from the female tract, coupled with an oviposition effect and possibly a laying period effect, now provide a general framework for understanding what determines which male will fertilize most ova. Because the processes associated with fertilization, egg formation and oviposition appear to be similar in all birds and because we can estimate the rate at which sperm are lost from the female's sperm-storage tubules, birds provide an excellent system to establish empirically the events taking place in the female reproductive tract. Future studies in this area should focus on: (1) refining estimates of the various parameters (e.g. ejaculate size, the rate of sperm loss from the female tract), (2) checking whether passive sperm loss accounts for observed patterns of precedence in other species, (3) checking whether the laying-period effect observed in turkeys also applies to other species, and (4) testing the idea that cryptic female choice explains some of the variation in the patterns of sperm precedence. Finally, using our understanding of the mechanisms of sperm competition we should start to predict the outcome of extra-pair copulations and predict the optimal behaviour patterns of males and females in the field.

REFERENCES

Adkins-Regan E (1995) Predictors of fertilization in the Japanese quail, *Coturnix japonica*. *Anim. Behav.* **50:** 1405–1415.

Alatalo R, Gustafsson L & Lundberg A (1984) High frequency of cuckoldry in pied and collared flycatchers. *Oikos* **42:** 41–47.

Alatalo R, Gustafsson L & Lundberg A (1989) Extra-pair paternity and heritability estimates of tarsus length in pied and collared flycatchers. *Oikos* **56:** 54–58.

Alatalo RV, Mappes J & Elgar MA (1987) Heritabilities and paradigm shifts. *Nature* **385:** 402–403.

Andersson M (1994) *Sexual Selection.* Princeton University Press, Princeton.

Bakst MR, Wishart GJ & Brillard JP (1994) Oviducal sperm selection, transport, and storage in poultry. *Poultry Sci. Rev.* **5:** 117–143.

Beaumont C, Brillard JP, Millet N & de Reviers M (1992) Comparison of various characteristics of the duration of fertility in hens. *Br. Poultry Sci.* **33:** 639–652.

Bedford MJ (1990) Sperm dynamics in the epididymis. In *Gamete Physiology.* RH Asch, JP Balmaceda & I Johnson (eds), pp. 53–67. Serono Symposia, Norwell.

Beecher MD & Beecher IM (1979) Sociobiology of bank swallows: reproductive strategy of the male. *Science* **205:** 1282–1285.

Birkhead TR (1978) Behavioral adaptations to high density breeding in the common guillemot *Uria aalge*. *Anim. Behav.* **26:** 321–331.

Birkhead TR (1979) Mate guarding in the magpie *Pica pica*. *Anim. Behav.* **27:** 866–874.

Birkhead TR (1982) Timing and duration of mate guarding in magpies, *Pica pica*. *Anim. Behav.* **30:** 277–283.

Birkhead TR (1988) Behavioral aspects of sperm competition in birds. *Adv. Stud. Behav.* **18:** 35–72.

Birkhead TR (1996) Sperm competition: evolution and mechanisms. *Curr. Top. Develop. Biol.* **33:** 103–158.

Birkhead TR (1997) Darwin on sex. *The Biologist* **44:** 397–399.

Birkhead TR & Biggins JD (1998) Sperm competition mechanisms in birds: models and data. *Behav. Ecol.*, in press.

Birkhead TR & Fletcher F (1993) Sperm storage and release of sperm from the sperm storage tubules in Japanese Quail *Coturnix japonica*. *Ibis* **136:** 101–105.

Birkhead TR & Fletcher F (1995) Male phenotype and ejaculate quality in the zebra finch *Taeniopygia guttata*. *Proc. Roy. Soc. Lond. Ser. B* **262:** 329–334.

Birkhead TR & Møller AP (1992a) *Sperm Competition in Birds: Evolutionary Causes and Consequences.* Academic Press, London.

Birkhead TR & Møller AP (1992b) Numbers and size of sperm storage tubules and the duration of sperm storage in birds: a comparative study. *Biol. J. Linn. Soc.* **45:** 363–372.

Birkhead TR & Møller AP (1993a) Why do male birds stop copulating while their partners are still fertile? *Anim. Behav.* **45:** 105–118.

Birkhead TR & Møller AP (1993b) Female control of paternity. *Trends Ecol. Evol.* **8:** 100–104.

Birkhead TR & Møller AP (1995) Extra-pair copulation and extra-pair paternity in birds. *Anim. Behav.* **49:** 843–848.

Birkhead TR & Møller AP (1996) Monogamy and Sperm Competition in Birds. In *Partnerships in Birds: the Ecology of Monogamy.* JM Black (ed.). Oxford University Press, Oxford.

Birkhead TR, Atkin L & Møller AP (1987) Copulation behaviour of birds. *Behaviour* **101:** 101–138.

Birkhead TR, Pellatt JE & Hunter FM (1988a) Extra-pair copulation and sperm competition in the zebra finch. *Nature* **334:** 60–62.

Birkhead TR, Clarkson K & Zann R (1988b) Extra-pair courtship, copulation and mate guarding in wild zebra finches *Taeniopygia guttata. Anim. Behav.* **35:** 1853–1855.

Birkhead TR, Hunter FM & Pellatt JE (1989) Sperm competition in the zebra finch, *Taeniopygia guttata. Anim. Behav.* **38:** 935–950.

Birkhead TR, Hatchwell BJ & Davies NB (1991) Sperm competition and the reproductive organs of the male and female dunnock (*Prunella modularis*). *Ibis* **133:** 306–311.

Birkhead TR, Pellatt JE & Hunter FM (1990) Numbers and distribution of sperm in the uterovaginal sperm storage tubules of the zebra finch. *Condor* **92:** 508–516.

Birkhead TR, Briskie JV & Møller AP (1993a) Male sperm reserves and copulation frequency in birds. *Behav. Ecol. Sociobiol.* **32:** 85–93.

Birkhead TR, Pellatt EJ & Fletcher F (1993b) Selection and utilization of spermatozoa in the reproductive tract of the female zebra finch *Taeniopygia guttata. J. Reprod. Fertil.* **99:** 593–600.

Birkhead TR, Møller AP & Sutherland WJ (1993c) Why do females make it so difficult for males to fertilize their eggs? *J. Theor. Biol.* **161:** 51–60.

Birkhead TR, Sheldon BC & Fletcher F (1994) A comparative study of sperm–egg interactions in birds. *J. Reprod. Fertil.* **101:** 353–361.

Birkhead TR, Fletcher F, Pellatt EJ & Staples A (1995a) Ejaculate quality and the success of extra-pair copulations in the zebra finch. *Nature* **377:** 422–423.

Birkhead TR, Wishart GJ & Biggins JD (1995b) Sperm precedence in the domestic fowl. *Proc. Roy. Soc. Lond. Ser. B* **261:** 285–292.

Birkhead TR, Veiga JP & Fletcher F (1995c) Sperm competition and unhatched eggs in the house sparrow. *J. Avian Biol.* **26:** 343–345.

Birkhead TR, Cunningham EJA & Cheng KM (1996) The insemination window provides a distorted view of sperm competition in birds. *Proc. Roy. Soc. Lond. Ser. B* **263:** 1187–1192.

Birkhead TR, Buchanan KL, Devoogd T, Pellatt EJ, Szekely T & Catchpole CK (1997) Song, sperm quality and testes asymmetry in the sedge warbler *Acrocephalus schoenbaenus. Anim. Behav.* **53:** 965–971.

Bishop JDD, Jones CS & Noble LR (1996) Female control of paternity in the internally fertilizing compound ascidian *Diplosoma listerianum.* II. Investigation of male mating success using RAPD markers. *Proc. Roy. Soc. Lond. Ser. B* **263:** 401–407.

Björklund M & Westman B (1983) Extra-pair copulations in the pied flycatcher (*Ficedula hypoleuca*). *Behav. Ecol. Sociobiol.* **13:** 271–275.

Björklund M, Møller AP, Sundberg J & Westman B (1991) Female great tits *Parus major* avoid extra-pair copulation attempts. *Anim. Behav.* **43:** 691–693.

Brillard JP & Antoine H (1990) Storage of sperm in the uterovaginal junction and its incidence on the numbers of spermatozoa present in the perivitelline layer of hens' eggs. *Br. Poultry Sci.* **31:** 635–644.

Brillard JP & Bakst MR (1990) Quantification of spermatozoa in the sperm-storage tubules of turkey hens and its relation to sperm numbers in the perivitelline layer of eggs. *Biol. Reprod.* **43:** 271–275.

Brillard JP, Galut O & Nys Y (1987) Possible causes of subfertility in hens following insemination near the time of ovulation. *Br. Poultry Sci.* **28:** 307–318.

Briskie JV (1992) Copulation patterns and sperm competition in the polygynandrous Smith's longspur. *Auk* **109:** 563–575.

Briskie JV (1993) Anatomical adaptations to sperm competition in Smith's longspurs and other polygynandrous passerines. *Auk* **110:** 875–888.

Briskie JV & Montgomerie R (1992) Sperm size and sperm competition in birds. *Proc. Roy. Soc. Lond. Ser. B* **247:** 89–95.

Briskie JV & Montgomerie R (1993) Patterns of sperm storage in relation to sperm competition in passerine birds. *Condor* **95:** 442–454.

Briskie JV & Montgomerie R (1997) Sexual selection and the intromittent organ of birds. *J. Avian Biol.* **28:** 78–86.

Briskie JV, Montgomerie R & Birkhead TR (1997) The evolution of sperm size in birds. *Evolution* **51:** 937–945.

Brooker MG, Rowley I, Adams M & Baverstock PR (1990) Promiscuity: an inbreeding avoidance mechanism in a socially monogamous species? *Behav. Ecol. Sociobiol.* **26:** 191–199.

Buitron D (1983) Extra-pair courtship in black-billed magpies. *Anim. Behav.* **31:** 211–220.

Burke T & Bruford MW (1987) DNA fingerprinting in birds. *Nature* **327:** 149–152.

Burley N, Krantzberg G & Radman P (1982) Influence of color banding on the conspecific preferences of zebra finches. *Anim. Behav.* **30:** 444–455.

Burley N, Price DK & Zann RA (1992) Bill color, reproduction and condition effects in wild and domesticated zebra finches. *Auk* **109:** 13–23.

Burley NT & Price DK (1991) Extra-pair copulation and attractiveness in zebra finches. *Proc. Int. Ornithol. Congr.* **20:** 1367–1372.

Burley NT, Enstrom DA & Chitwood L (1994) Extra-pair relations in zebra finches: differential male success results from female tactics. *Anim. Behav.* **48:** 1031–1041.

Burt A (1995) The evolution of fitness. *Evolution* **49:** 1–8.

Castro I, Minot E, Fordham R & Birkhead TR (1996) Polygynandry, face-to-face copulation and sperm competition in the hihi *Notiomystis cincta* (Aves: Meliphagidae). *Ibis* **138:** 765–771.

Cezilly F & Nager RG (1995) Comparative evidence for a positive association between divorce and extra-pair paternity in birds. *Proc. Roy. Soc. Lond. Ser. B* **262:** 7–12.

Chek AA, Lifjeld JT & Robertson RJ (1996) Lack of mate-guarding in a territorial passerine bird with a low intensity of sperm competition, the pied flycatchers (*Ficedula hypoleuca*). *Ethology* **102:** 134–145.

Cheng KM, Burns JT & McKinney F (1983) Forced copulation in captive mallards. III. Sperm competition. *Auk* **100:** 302–310.

Colegrave N, Birkhead TR & Lessells CM (1995) Sperm precedence in zebra finches does not require special mechanisms of sperm competition. *Proc. Roy. Soc. Lond. Ser. B* **259:** 223–228.

Collins SA (1994) Male displays: cause or effect of female preference? *Anim. Behav.* **48:** 371–375.

Collins SA & ten Cate C (1996) Does beak colour affect female preference in zebra finches? *Anim. Behav.* **52:** 105–112.

Collins SA, Hubbard C & Houtman AM (1994) Female mate choice in the zebra finch – the effect of male beak colour and song. *Behav. Ecol. Sociobiol.* **35:** 21–25.

Colwell MA & Oring LW (1989) Extra-pair mating in the spotted sandpiper: a female mate acquisition tactic. *Anim. Behav.* **38:** 675–684.

Compton MM, van Krey HP & Siegel PB (1978) The filling and emptying of the uterovaginal sperm-host glands in the domestic hen. *Poultry Sci.* **57:** 1696–1700.

Cronin JWJ & Sherman PW (1976) A resource based mating system: the orange-rumped honeyguide. *Living Bird* **15:** 5–32.

Darwin C (1868) *The Variation of Animals and Plants Under Domestication.* (2 Vols). Murray, London.

Darwin C (1871) *The Descent of Man, and Selection in Relation to Sex.* John Murray, London.

Davies NB (1985) Cooperation and conflict among dunnocks, *Prunella modularis*, in a variable mating system. *Anim. Behav.* **33:** 628–648.

Davies NB (1992) *Dunnock Behaviour and Social Evolution.* Oxford University Press, Oxford.

Davies NB, Hatchwell BJ, Robson T & Burke T (1992) Paternity and parental effort in dunnocks *Prunella modularis*: how good are male chick-feeding rules? *Anim. Behav.* **43:** 729–746.

Davies NB, Hartley IR, Hatchwell BJ & Langmore NE (1996) Female control of copulations to maximize male help: a comparison of polygynandrous alpine accentors, *Prunella collaris*, and dunnocks *P. modularis*. *Anim. Behav.* **51:** 27–47.

Dickinson JL (1997) Male detention affects extra-pair copulation frequency and pair behavior in western blue birds. *Anim. Behav.* **53:** 561–571.

Dickinson JL & Leonard ML (1997) Mate-attendance and copulatory behaviour in western bluebirds: evidence of mate guarding. *Anim. Behav.* **52:** 981–992.

Dixon A (1993) Parental investment and reproductive success in the reed bunting (*Emberiza schoenicius*) investigated by DNA fingerprinting. Ph.D. thesis, University of Leicester, UK.

Dixon A, Ross D, O'Malley SLC & Burke T (1994) Paternal investment inversely related to degree of extra-pair paternity in the reed bunting (*Emberiza shoeniclus*). *Nature* **371:** 698–700.

Dunn PO (1992) Do male birds adjust territory size to the risk of cuckoldry? *Anim. Behav.* **43:** 857–859.

Dunn PO & Lifjeld JT (1994) Can extra-pair copulations be used to predict extra-pair paternity in birds? *Anim. Behav.* **47:** 983–985.

Dziuk PJ (1996) Factors that influence the proportion offspring sired by a male following heterospermic insemination. *Anim. Reprod. Sci.* **43:** 65–88.

Eberhard WG (1996) *Female Control: Sexual Selection by Cryptic Female Choice.* Princeton University Press, Princeton.

Ens BJ, Safriel UN & Harris MP (1993) Divorce in the long-lived and monopgamous oystercatcher, *Haemotopus ostralegus*: incompatability or choosing a better option? *Anim. Behav.* **45:** 1199–1217.

Faaborg J, Parker PG, Delay L, de Vries TJ, Bernardz JC, Maria Paz S, Naranjo J & Waite TA (1995) Confirmation of cooperative polyandry in the Galapagos hawk (*Buteo galapagoensis*). *Behav. Ecol. Sociobiol.* **36:** 83–90.

Fletcher FJC (1996) Male and female aspects of sperm competition in the zebra finch *Taeniopygia guttata*. Ph.D. thesis, University of Sheffield, UK.

Freeman-Gallant CR (1996) DNA fingerprinting reveals female preference for male parental care in Savannah Sparrows. *Proc. Roy. Soc. Lond. Ser. B* **263:** 157–160.

Gibbs HL, Weatherhead PJ, Boag PT, White BN, Tabak LM & Hoysak DJ (1990) Realized reproductive success of polygynous red-winged blackbirds revealed by DNA markers. *Science* **250:** 1394–1397.

Gladstone DE (1979) Promiscuity in monogamous colonial birds. *Am. Nat.* **114:** 545–557.

Gochfeld M (1980) Mechanisms and adaptive value of reproductive synchrony in colonial seabirds. In *Behaviour of Marine Animals; Current Perspectives in Research*. J Burger, BL Olla & HE Winn (eds), pp. 207–270. Plenum Press, New York and London.

Gowaty PA & Pilssner JH (1987) Association of male and female American robins (*Turdus migratorius*) during the breeding season: paternity assurance by sexual access or mate-guarding? *Wilson Bull.* **99:** 56–62.

Grafen A (1980) Opportunity cost, benefit and the degree of relatedness. *Anim. Behav.* **28:** 967–968.

Graves J, Ortega-Ruano J & Slater PJB (1993) Extra-pair copulations and paternity in shags: do females choose better males? *Proc. Roy. Soc. Lond. Ser. B* **253:** 3–7.

Gray EM (1996) Female control offspring paternity in a western population of red-winged blackbirds (*Agelaius phoeniceus*). *Behav. Ecol. Sociobiol.* **38:** 267–278.

Gray EM (1997a) Do female red-winged blackbirds benefit genetically from seeking extra-pair copulations? *Anim. Behav.* **53:** 605–623.

Gray EM (1997b) Female red-winged blackbirds accrue material benefits from copulating with extra-pair males. *Anim. Behav.* **53:** 625–639.

Hanski IK & Laurila A (1993) Variation in song rate during the breeding cycle of the chaffinch *Fringilla coelebs*. *Ethology* **93:** 161–169.

Hasselquist D, Bensch S & von Schantz T (1995) Estimating cuckoldry in birds: the heritability method and DNA fingerprinting give different results. *Oikos* **72:** 173–178.

Hasselquist D, Bensch S & von Schantz T (1996) Correlation between male song repertoire, extra-pair paternity and offspring survival in the great reed warbler. *Nature* **381:** 229–232.

Haydock J, Parker PG & Rabenold KN (1996) Extra-pair paternity in the cooperatively breeding bicolored wren. *Behav. Ecol. Sociobiol.* **38:** 1–16.

Heg D, Ens BJ, Burke T, Jenkins L & Kruijt JP (1993) Why does the typically monogamous oystercatcher (*Haemotopus ostralegus*) engage in extra-pair copulations? *Behaviour* **126:** 247–287.

Hill GE, Montgomerie R, Rieder C & Boag P (1994) Sexual selection and cuckoldry in a monogamous songbird: implications for sexual selection theory. *Behav. Ecol. Sociobiol.* **35:** 193–199.

Hoi H (1997) Assessment of the quality of copulation partners in the monogamous bearded tit. *Anim. Behav.* **53:** 277–286.

Höglund J & Alatalo RV (1995) *Leks*. Princeton University Press, Princeton.

Hoogland JL & Sherman PW (1976) Advantages and disadvantages of bank swallow (*Riparia riparia*) coloniality. *Ecol. Monogr.* **49:** 682–694.
Howarth B (1970) An examination for sperm capacitation in the fowl. *Biol. Reprod.* **3:** 338–341.
Howarth B (1971) Transport of spermatozoa in the reproductive tract of turkey hens. *Poultry Science* **50:** 84–89.
Howarth B (1974) Sperm storage as a function of the female reproductive tract. In *The Oviduct and its Functions.* AD Johnson & CE Foley (eds). Academic Press, New York.
Howarth B (1984) Maturation of spermatozoa and mechanism of fertilisation. In *Reproductive Biology of Poultry.* EJ Cunningham, PE Lake & D Hewitt (eds), pp. 161–174. Longman, Harlow.
Howarth B (1995) Physiology of reproduction: the male. In *Poultry Production. World Animal Science. Subseries C: Production System Approach*, Vol. 9. P. Hunton (ed.), pp. 243–270. Elsevier, Amsterdam.
Hunter FM, Davis LS & Miller GD (1996) Sperm transfer in the Adelie penguin. *Condor* **98:** 410–413.
Hunter FM, Petrie M, Otronen M, Birkhead TR & Møller AP (1993) Why do females copulate repeatedly with one male? *Trends Ecol. Evol.* **8:** 21–26.
Immelmann K (1959) Experimentelle untersuchungen uber die biologische bedeutung artsspezifischer merkmale beim zebrafinken (*Taeniopygia guttata* Gould). *Zool. Jahr. Abt. System. Okol. Geogr. Tiere* **86:** 437–592.
Jamieson IG, Quinn JS, Rose PA & White BN (1994) Shared paternity among non-relatives is a result of an egalitarian mating system in a communally breeding bird, the pukeko. *Proc. Roy. Soc. Lond. Ser. B* **257:** 271–277.
Jennions MD (1996) *Signalling and Sexual Selection in Animals and Plants.* D.Phil. thesis, University of Oxford, UK.
Johnsen A & Lifjeld JT (1995) Unattractive males guard their mates more closely: an experiment with bluethroats (Aves, Turdidae: *Luscinia s. svecica*). *Ethology* **101:** 200–212.
Johnson AL (1986) Reproduction in the male. In *Avian Physiology.* PD Sturkie (ed.), pp. 432–451. Springer-Verlag, New York.
Johnston NP & Parker JE (1970) The effect of time of oviposition in relation to insemination of chicken hens. *Poultry Sci.* **49:** 325–327.
Jones CS, Lessells CM & Krebs JR (1991) Helpers-at-the-nest in European bee-eaters (*Merops apiaster)*: a genetic analysis. In *DNA Fingerprinting: Approaches and Applications.* T Burke, G Dolf, AJ Jeffreys and R Wolff (eds). Birkhäuser, Basle.
Keller L & Reeve HK (1995) Why do females mate with multiple males? The sexually selected sperm hypothesis. *Adv. Stud. Behav.* **24:** 291–315.
Kempenaers B, Verheyen GR, Broeck MV de, Burke T, Broeckhoven CV & Dhondt AA (1992) Extra-pair paternity results from female preference for high-quality males in the blue tit. *Nature* **357:** 494–496.
Kempenaers B, Verheyen GR & Dhondt AA (1995) Mate guarding and copulation behavior in monogamous and polygynous blue tits: do males follow a best-of-a-bad-job strategy? *Behav. Ecol. Sociobiol.* **36:** 33–42.
King AS (1981) Phallus. In *Form and Function in Birds.* AS King & J McLelland (eds), pp. 107–147. Academic Press, London.
Kirkpatrick M & Ryan MJ (1991) The evolution of mating preferences and the paradox of the lek. *Nature* **350:** 33–38.
Krebs JR & Davies NB (1997) *Behavioural Ecology: An Evolutionary Approach*, 4th edn. Blackwell Scientific Publications, Oxford.

Krokene C, Anthonisen K, Lifjeld JT & Amundsen T (1996) Paternity and paternity assurance behaviour in the bluethroat, *Luscinia s. svecica*. *Anim. Behav.* **52:** 405–417.

Lack D (1968) *Ecological Adaptations for Breeding in Birds.* Chapman and Hall, London.

Lake PE (1975) Gamete production and the fertile period with particular reference to domesticated birds. *Symp. Zool. Soc. Lond.* **35:** 225–244.

Lake PE (1981) Male genital organs. In *Form and Function in Birds.* AS King & J McLelland (eds), pp. 1–61. Academic Press, London.

Lamprecht J (1989) Mate guarding in geese: awaiting female receptivity, protection of paternity or support of female feeding? In *The Sociobiology of Sexual and Reproductive Strategies.* AE Rasa, E Vogel & E Voland (eds). Chapman and Hall, London and New York.

Langmore N (1996) Territoriality and song as flexible paternity guards in dunnocks and alpine accentors. *Behav. Ecol.* **7:** 183–188.

Lanier DL, Estep DG & Dewsbury DA (1979) Role of prolonged copulatory behaviour in facilitating reproductive success in a competitive mating situation in laboratory rats. *J. Comp. Physiol. Psychol.* **93:** 781–792.

Leman AD (1975) *A study of factors that may influence heterospecific dominance.* Ph.D. thesis, University of Champaing-Urbana, USA.

Lessells CM & Birkhead TR (1990) Mechanisms of sperm competition in birds: mathematical models. *Behav. Ecol. Sociobiol.* **27:** 325–337.

Lifjeld JT (1994) Do female house sparrows copulate with extra-pair mates to enhance their fertility? *J. Avian Biol.* **25:** 75–76.

Lifjeld JT & Robertson RJ (1992) Female control of extra-pair fertilization in tree swallows. *Behav. Ecol. Sociobiol.* **31:** 89–96.

Lifjeld JT & Slagsvold T (1989) How frequent is cuckoldry in pied flycatchers *Ficedula hypoleuca*? Problems with the heritability estimates of tarsus length. *Oikos* **54:** 205–210.

Lifjeld JT, Slagsvold T & Ellegren H (1997) Experimental mate switching in pied flycatchers: male copulatory access and fertilization success. *Anim. Behav*, **53:** 1225–1232.

Lovell-Mansbridge C (1995) Sperm competition in the feral pigeon *Columba livia*. Ph.D. thesis, University of Sheffield, UK.

Lumpkin S, Kessel K, Zenone PG & Erickson CJ (1982) Proximity between the sexes in ring doves: social bonds or surveillance? *Anim. Behav.* **30:** 506–513.

MacRoberts MH (1973) Extramarital courting in lesser black-backed and herring gulls. *Z. Tierpsychol.* **32:** 62–74.

Martin PA, Reimers TJ, Lodge JR & Dziuk PJ (1974) The effect of ratios and numbers of spermatozoa mixed from two males on proportions offspring. *J. Reprod. Fertil.* **39:** 251–258.

McFarlane RW (1963) The taxonomic significance of avian sperm. *Proc. XIII Int. Ornithol. Congr.*, pp. 91–102.

McKinney F, Derrickson SR & Mineau P (1983) Forced copulation in waterfowl. *Behaviour* **86:** 250–294.

McKinney F, Cheng KM & Bruggers DJ (1984) Sperm competition in apparently monogamous birds. In *Sperm Competition and the Evolution of Animal Mating Systems.* RL Smith (ed.), pp. 523–545. Academic Press, Orlando.

Mills JA (1994) Extra-pair copulations in the red-billed gull: females with high quality, attentive males resist. *Behaviour* **128:** 41–64.

Møller AP (1985) Mixed reproductive strategy and mate guarding in a semi-colonial passerine, the swallow *Hirundo rustica*. *Behav. Ecol. Sociobiol.* **17:** 401–408.

Møller AP (1987a) Extent and duration of mate guarding in swallows *Hirundo rustica*. *Ornis Scandinavica* **18:** 95–100.

Møller AP (1987b) Intruders and defenders on avian breeding territories: the effect of sperm competition. *Oikos* **48:** 47–54.

Møller AP (1988a) Testis size, ejaculate quality, and sperm competition in birds. *Biol. J. Linn. Soc.* **33:** 273–383.

Møller AP (1988b) Paternity and parental care in the swallow *Hirundo rustica*. *Anim. Behav.* **36:** 996–1005.

Møller AP (1988c) Badge size in the house sparrow *Passer domesticus:* effects of intra- and intersexual selection. *Behav. Ecol. Sociobiol.* **22:** 373–378.

Møller AP (1988d) Female choice selects for male sexual tail ornaments in the monogamous swallow. *Nature* **332:** 640–642.

Møller AP (1990a) Changes in the size of avian breeding territories in relation to the nesting cycle. *Anim. Behav.* **40:** 1070–1079.

Møller AP (1990b) Deceptive use of alarm calls by male swallows *Hirundo rustica*: a new paternity guard. *Behav. Ecol.* **1:** 1–6.

Møller AP (1991a) Sperm competition, sperm depletion, paternal care and relative testis size in birds. *Am. Nat.* **137:** 882–906.

Møller AP (1991b) Why mated songbirds sing so much: mate guarding and male announcement of mate fertility status. *Am. Nat.* **138:** 994–1014.

Møller AP (1992a) Frequency of female copulations with multiple males and sexual selection. *Am. Nat.* **139:** 1089–1101.

Møller AP (1992b) Relative size of avian breeding territories and the risk of cuckoldry. *Anim. Behav.* **43:** 860–861.

Møller AP (1994) *Sexual Selection and the Barn Swallow*. Oxford University Press, Oxford.

Møller AP & Birkhead TR (1991) Frequent copulations and mate guarding as alternative paternity guards in birds: a comparative study. *Behaviour* **118:** 170–186.

Møller AP & Birkhead TR (1992) Validation of the heritability method to estimate extra-pair paternity in birds. *Oikos* **64:** 485–488.

Møller AP & Birkhead TR (1993a) Cuckoldry and sociality: a comparative study of birds. *Am. Nat.* **142:** 118–140.

Møller AP & Birkhead TR (1993b) Certainty of paternity covaries with paternal care in birds. *Behav. Ecol. Sociobiol.* **33:** 261–268.

Møller AP & Birkhead TR (1994) The evolution of plumage brightness in birds is related to extra-pair paternity. *Evolution* **48:** 1089–1100.

Møller AP & Briskie JV (1995) Extra-pair paternity, sperm competition and the evolution of testis size in birds. *Behav. Ecol. Sociobiol.* **36:** 357–365.

Mulder RA & Cockburn A (1993) Sperm competition and the reproductive anatomy of male superb fairy-wrens. *Auk* **110:** 588–593.

Mulder RA, Dunn PO, Cockburn A, Lazenby-Cohen KA, Howell MJ (1994) Helpers liberate female fairy-wrens from constraints on extra-pair mate choice. *Proc. Roy. Soc. Lond. Ser. B* **255:** 223–229.

Nakamura M (1990) Cloacal protuberance and copulatory behaviour of the Alpine accentor (*Prunella collaris*). *Auk* **107:** 284–295.

Olsson M, Shine R, Madsen T, Gullberg A & Tegelstrom H (1996) Sperm selection by females. *Nature* **383:** 585.

Oring LW, Fleischer RC, Reed JM & Marsden KE (1992) Cuckoldry through stored sperm in the sequentially polyandrous spotted sandpiper. *Nature* **359:** 631–633.

Otter K, Ratcliffe L, Boag PT (1994) Extra-pair paternity in the black-capped chickadee. *Condor* **96:** 218–222.

Parker GA (1970) Sperm competition and its evolutionary consequences in the insects. *Biol. Rev.* **45:** 525–567.

Parker GA (1979) Sexual selection and sexual conflict. In *Sexual Selection and Reproductive Competition in Insects.* MS Blum & NA Blum (eds), pp. 123–166. Academic Press, New York.

Parker GA (1984) Sperm competition and the evolution of animal mating strategies. In *Sperm Competition* and *the Evolution of Animal Mating Systems.* RL Smith (ed.), pp. 1–60. Academic Press, Orlando.

Parrott D (1995) Social organisation and extra-pair behaviour in the European black-billed magpie *Pica pica*. Ph.D. thesis, University of Sheffield, UK.

Peck AL (1943) *Aristotle's Generation of Animals.* Heinemann, London.

Pellatt EJ & Birkhead TR (1994) Ejaculate size in zebra finches *Taeniopygia guttata* and a method for obtaining ejaculates from passerine birds. *Ibis* **136:** 97–101.

Petrie M (1992) Copulation frequency in birds: Why do females copulate more than once with the same male? *Anim. Behav.* **44:** 790–792.

Pomiankowski A & Møller AP (1995) A resolution of the lek paradox. *Proc. Roy. Soc. Lond. Ser. B* **260:** 21–29.

Rabenold PP, Rabenold KN, Piper WH, Haydock J & Zack SW (1990) Shared paternity revealed by genetic analysis in cooperatively breeding tropical wrens. *Nature* **348:** 538–540.

Ratcliffe LM & Boag PT (1987) Effects of colour bands on male competition and sexual attractiveness in zebra finches (*Poephila guttata*). *Can. J. Zool.* **65:** 333–338.

Ratti O, Hovi M, Lundberg A, Tegelstron H & Alatalo RV (1995) Extra-pair paternity and male characteristics in the pied flycatcher. *Behav. Ecol. Sociobiol.* **37:** 419–425.

Rising JD (1996) Relationship between testis size and mating systems in American sparrows (Emberizidae). *Auk* **113:** 224–228.

Robertson BC (1996) The mating system of the Capricorn Silvereye, *Zosterops lateralis chlorocephala*: a genetic and behavioural assessment. Ph.D. thesis, University of Queensland, Australia.

Rodrigues M (1996) Song activity in the chiffchaff: territorial defence or mate guarding? *Anim. Behav.* **51:** 709–716.

Roldan ERS, Gomendio M & Vitullo AD (1992) The evolution of eutherian spermatozoa and underlying selective forces: female selection and sperm competition. *Biol. Rev.* **67:** 1–43.

Rowe L & Houle D (1996) The lek paradox and the capture of genetic variance by condition dependent traits. *Proc. Roy. Soc. Lond. Ser. B* **263:** 1415–1421.

Saino N, Primmer C, Ellegren H & Møller AP (1997) An experimental study of paternity and tail ornamentation in the barn swallow (*Hirundo rustica*). *Evolution* **51:** 562–570.

Schulze-Hagen K, Swatschek I, Dyrcz A & Wink M (1993) Multiple vaterschaften in bruten des seggenrohrsangers *Acrocephalus paludicola*: erste ergebnisse des DNA-fingerprintings. *J. Ornithol.* **134:** 145–154.

Schulze-Hagen K, Leisler B, Birkhead TR & Dyrcz A (1995) Prolonged copulation, sperm reserves and sperm competition in the aquatic warbler *Acrocephalus paludicola*. *Ibis* **137:** 85–91.

Sheldon BC (1994) Song rate and fertility in the chaffinch. *Anim. Behav.* **47:** 986–987.

Sheldon BC & Burke T (1994) Copulation behaviour and paternity in the chaffinch. *Behav. Ecol. Sociobiol.* **34:** 149–156.

Sheldon BC, Merila J, Qvarnstrom A, Gustafsson L & Ellegren H (1997) Paternal genetic contribution to offspring condition predicted by size of a male secondary sexual character. *Proc. Roy. Soc. Lond. Ser. B* **264:** 297–302.

Simmons LW, Stockley P, Jackson RL & Parker GA (1996) Sperm competition or sperm selection: no evidence for female influence over paternity in yellow dung flies *Scatophaga stercoraria*. *Behav. Ecol. Sociobiol.* **38:** 199–206.

Smith SM (1988) Extra-pair copulations in black-capped chickadees: the role of the female. *Behaviour* **107:** 15–23.

Smith HG, Wennerberg L & von Schantz T (1996) Sperm competition in the European starling (*Sturnus vulgaris*): an experimental study of mate switching. *Proc. Roy. Soc. Lond. Ser. B* **263:** 797–801.

Steele MG & Wishart GJ (1992) Characterisation of mechanism impeding sperm transport through the vagina of the chicken. *Proc. 12th Int. Congr. Anim. Reprod.* **3:** 474–476.

Stockley P (1997) Sexual conflict resulting from adaptations to sperm competition. *Trends Ecol. Evol.* **12:** 154–159.

Stutchbury BJ & Morton ES (1995) The effect of breeding synchrony on extra-pair mating systems in songbirds. *Behaviour* **132:** 675–690.

Stutchbury J, Piper WH, Neudorf DL, Tarof SA, Rhymer JM, Fuller G & Fleischer RC (1997) Correlates of extra-pair fertilizations in hooded warblers. *Behav. Ecol. Sociobiol.* **40:** 119–126.

Sullivan MS (1994) Discrimination among males by female zebra finches based on past as well as current phenotype. *Ethology* **96:** 97–104.

Sundberg J (1992) Absence of mate guarding in the Yellowhammer (*Emberiza citrinella*)? *Ethology* **92:** 242–256.

Sundberg J & Dixon A (1996) Old, colourful male yellowhammers, *Emberiza citrinella*, benefit from extra-pair copulations. *Anim. Behav.* **52:** 113–122.

Swaddle JP & Cuthill IC (1994) Female zebra finches prefer males with symmetric chest plumage. *Proc. Roy. Soc. Lond. Ser. B* **258:** 267–271.

Thornhill R (1983) Cryptic female choice and its implications in the scorpionfly *Harpobittacus nigriceps*. *Am. Nat.* **122:** 765–788.

Trivers RL (1972) Parental investment and sexual selection. In *Sexual Selection and the Descent of Man, 1871–1971*. B Campbell (ed.), pp. 136–179. Aldine-Atherton, Chicago.

Tuttle EM, Pruett-Jones S & Webster MS (1996) Cloacal protuberances and extreme sperm production in Australian fairy-wrens. *Proc. Roy. Soc. Lond. Ser. B* **263:** 1359–1364.

Wagner RH, Schug MD & Morton ES (1996) Condition-dependent control of paternity by female purple martins: implications for coloniality. *Behav. Ecol. Sociobiol.* **38:** 379–389.

Weatherhead PJ & Boag PT (1995) Pair and extra-pair mating success relative to male quality in red-winged blackbirds. *Behav. Ecol. Sociobiol.* **37:** 81–91.

Westneat DF (1987a) Extra-pair copulations in a predominantly monogamous bird: observations of behaviour. *Anim. Behav.* **35:** 865–876.

Westneat DF (1987b) Extra-pair fertilizations in a predominantly monogamous bird: genetic evidence. *Anim. Behav.* **35:** 877–886.

Westneat DF (1990) Genetic parentage in the indigo bunting: a study using DNA fingerprinting. *Behav. Ecol. Sociobiol.* **27:** 67–76.

Westneat DF (1993) Temporal patterns of within-pair copulations, male mate-guarding, and extra-pair events in eastern red-winged blackbirds (*Agelaius phoeniceus*). *Behaviour* **124:** 267–290.

Westneat DF (1994) To guard or go forage: conflicting demands affect the paternity of male red-winged blackbirds. *Am. Nat.* **144:** 343–354.

Westneat DF (1995) Paternity and paternal behaviour in the red-winged blackbird, *Agelaius phoeniceus. Anim. Behav.* **49:** 21–35.

Westneat DF, Sherman PW & Morton ML (1990) The ecology and evolution of extra-pair copulations in birds. *Cur. Ornithol.* **7:** 331–369.

Westneat DF & Sargent RC (1996) Sex and parenting: the effects of sexual conflict and parentage on parental strategies. *Trends Ecol. Evol.* **11:** 87–91.

Westneat DF & Webster MS (1994) Molecular analysis of kinship in birds: Interesting questions and useful techniques. In *Molecular Ecology and Evolution: Approaches and Applications.* B Schierwater, B Streit, GP Wagner & R DeSalle (eds). Birkhauser Verlag, Basel.

Wetton J, Burke T, Parkin DT & Cairns E (1995) Single-locus DNA fingerprinting reveals that male reproductive success increases with age through extra-pair paternity in the house sparrow (*Passer domesticus*). *Proc. Roy. Soc. Lond. Ser. B* **260:** 91–98.

Wetton JH & Parkin DT (1991) An association between fertility and cuckoldry in the house sparrow *Passer domesticus. Proc. Roy. Soc. Lond. Ser. B* **245:** 227–233.

Wetton JH, Carter RE, Parkin DT & Walters D (1987) Demographic study of a wild house sparrow population by DNA fingerprinting. *Nature* **327:** 147–149.

Wilkinson R & Birkhead TR (1995) Copulation behaviour in the vasa parrots *Coracopsis vasa* and *C. nigra. Ibis* **137:** 117–119.

Williams GC (1975) *Sex and Evolution.* Princeton University Press, Princeton.

Wishart GJ (1987) Regulation of the length of the fertile period in the domestic fowl by numbers of oviductal spermatozoa as reflected by those trapped in laid eggs. *J. Reprod. Fertil.* **80:** 493–498.

Wishart GJ (1988) Numbers of oviductal spermatozoa and the length of the fertile period in different avian species. *11th Int. Congr. Anim. Reprod. Artif. Insem.* **3:** 362–364.

Wittenberger JL & Tilson RL (1980) The evolution of monogamy: hypotheses and evidence. *Annu. Rev. Ecol. System.* **11:** 197–232.

Wolf LL (1975) 'Prostitution' behaviour in a tropical hummingbird. *Condor* **77:** 140–144.

Wolfson A (1954) Sperm storage at lower-than-body temperature outside the body cavity in some passerine birds. *Science* **120:** 68–71.

Wu TY (1977) Introduction to the scaling of aquatic animal locomotion. In *Scale Effects in Animal Locomotion.* TJ Pedley (ed.), pp. 203–232. Academic Press, London.

Wu YT, Brokaw CJ & Brennen C (1975) *Swimming and Flying in Nature.* Plenum Press, New York.

Zeh JA & Zeh DW (1996) The evolution of polyandry I: intragenomic conflict and genetic incompatability. *Proc. Roy. Soc. Lond. Ser. B* **263:** 1711–1717.

Zeh JA & Zeh DW (1997) The evolution of polyandry II: post-copulatory defences against genetic incompatibility. *Proc. Roy. Soc. Lond. Ser. B* **264:** 69–75.

15 Reproduction, Mating Strategies and Sperm Competition in Marsupials and Monotremes

D. A. Taggart[1], *W. G. Breed*[2], *P. D. Temple-Smith*[3], *A. Purvis*[4] *and G. Shimmin*[1]

[1] Department of Anatomy, Monash University, Clayton, Victoria, Australia; [2] Department of Anatomical Sciences, University of Adelaide, Adelaide, South Australia, Australia; [3] Zoological Park and Gardens Board of Victoria, P.O. Box 74, Parkville, Victoria, Australia 3052; [4] Imperial College, Silwood Park, Ascot, Berks SL5 7PY, UK

I. INTRODUCTION

The theory of resource competition and natural selection in vertebrates was originally proposed by Darwin in the 1800s and since that time has generated considerable interest among biologists, with much attention being focused on the significance of differential success in competition for resources (food, shelter and mates) and its effects on an animal's ability to contribute genes to the next generation. More recently however, the debate has grown to include discussions of the causative factors that determine the amount of investment a male or female makes in gamete production and its relationship to mating strategies and successful reproduction.

Sperm Competition and Sexual Selection
ISBN 0-12-100543-7

In eutherian mammals, as in many other vertebrates, there is an allometric relationship between adult testis mass and body mass (Kenagy and Trombulak 1986; Møller and Briskie 1995) and, for primates at least, the testis mass correlates with the number of sperm ejaculated (Møller 1989). Over and above this allometric relationship, there are, in primates, significant species differences in relative testis mass. A considerable body of evidence has now been amassed for this order of mammals, which indicates that this difference relates to the likelihood of sperm competition within the female reproductive tract. Thus, species that have relatively large testes generally occur in multimale breeding groups where a female is likely to mate with two or more males within the group at the one oestrus, with the consequence that sperm competition within the female reproductive tract is likely. Conversely those species that occur in monogamous pairs or as single male breeding groups tend to have relatively smaller testes because of the unlikely occurrence of sperm competition and lower copulatory frequencies (Short 1979; Harcourt *et al.* 1981; Harvey and Harcourt 1984). This interspecific variation in relative testis mass appears also to relate to the quality of the ejaculates (Møller 1988, 1989). Recently, it has been claimed that not only the relative numbers of sperm produced, but also their relative size, may relate to an animal's mating system (Gomendio and Roldan 1991), although this causal relationship has been questioned (Harcourt 1991).

Despite considerable information on the relationships between body size, testis mass, sperm numbers, and sperm length in eutherian mammals, there is no data available for the marsupials or monotremes, apart from a few data on individual species (Bedford *et al.* 1984; Tyndale-Biscoe and Renfree 1987). These groups were not included in the 1984 review on sperm competition and sexual selection in vertebrates (Smith 1984), hence, inclusion in this volume is appropriate and may serve to broaden our understanding of these associations. Here, we present both phylogenetic and nonphylogenetic comparative analyses of how testes mass, sperm numbers, and sperm length scale with body mass in these groups of mammals. From our analyses we make predictions about the likely breeding system of the species concerned and examine the available evidence to determine whether it supports our predictions.

II. THE MONOTREMES

The monotremes are found only in Australia, New Guinea, Irian Jaya and a few close Indonesian islands and are represented by three extant species – the platypus *Ornithorhynchus anatinus* (restricted to the waterways of Eastern Australia, Tasmania and Kangaroo Island), the short-beaked echidna *Tachyglossus aculeatus* (found right across mainland Australia and

Tasmania), and the long-beaked echidna *Zaglossus bruinjii* which occurs only in New Guinea (Flannery 1995; Grant 1995). Little is known of the reproductive biology of *Zaglossus* and there are large gaps in our knowledge of the reproduction in the other two species, although some basic information is now available (Griffith 1968, 1978, 1984; Temple-Smith 1973; Hughes and Carrick 1978). Monotremes may represent an early divergence from the stem groups of early mammals and they have retained many reproductive characteristics, such as sperm structure, egg laying habit and egg structure, which are found in extant reptiles and presumably occurred in the common ancestors of reptiles and mammals.

A. Relative testis size and mating strategies

The platypus and short-beaked echidna are both seasonal breeders (Griffith 1968, 1978, 1984; Temple-Smith 1973). In the echidna, breeding occurs mostly during the Australian winter months (July–August) when testicular size (Table 15.1) is maximal (Augee 1978; Griffith 1978). Peak testicular size in south-eastern Australian populations of echidnas and platypus occurs in July and August, respectively (Griffith 1978, 1984; Temple-Smith 1973) and, in the platypus, large numbers of spermatozoa are found in the caudal region of the epididymis about 1 month later, corresponding to peak epididymal weight (Temple-Smith 1973).

In all three species the testes and epididymides are positioned abdominally (Home 1802; Oudemans 1892; Rodger and Hughes 1973; Temple-Smith 1973; Griffith 1978) and the testes are large compared with those of most marsupials and eutherians, with testes mass of adult males during the breeding season being about 1.0–1.2% of body mass (Table 15.1).

Because the testes are relatively large one might predict that a polygynous or promiscuous mating strategy occurs. In studies on the short-

Table 15.1. Comparisons of paired testis/body mass relationships ($\bar{x} \pm 5.0$) in monotremes.

Species	Adult male body mass (kg)	Paired testicular mass (g)	Testis mass as % body mass
Ornithorhynchus anatinus[a] (n = 10)	1.5 ± 0.2	17 ± 4	1.15 ± 0.30
Tachyglossus aculeatus[b] (n = 1)	5	50	1.0
Zaglossus bruijnii[c] (n = 5)	6.9 ± 0.7	81 ± 15	1.22 ± 0.25

[a] Data from 10 adult males collected July–October 1970/71 (Temple-Smith 1973).
[b] Data from Griffiths (1984).
[c] Data from Griffiths (1978).

beaked echidna Rismiller (1992, 1993) found that several male echidnas associate with a female during the breeding season and that one group of males remained with a female for 44 days. Rismiller (1993) suggested that strong competition occurs between males for access to oestrus females. However, female echidnas appear not to mate more than once during a single period of oestrus. Most males leave the female directly after the completion of copulation. Unsuccessful males may return to a mated female for a day or two after she has mated but no further copulations were found to take place, suggesting the absence of sperm competition between males in this species.

In the platypus even less information is available on social and sexual interactions during the breeding season. Data from wild populations suggests that male platypus are territorial and during the breeding season occupy home ranges which overlap with those of several females (Serena 1994). The observations of Hawkins and Fanning (1992) on a pair of captive platypus indicate that the female is receptive for only a few days and that several matings take place during this time. No evidence is available which indicates whether a male will mate with several females during the course of a breeding season, although this has been suggested by several authors (Hawkins and Fanning 1992; Serena 1994).

B. *Sperm length*

Spermatozoa of the platypus and short-beaked echidna are similar in structure (Carrick and Hughes 1982) although platypus spermatozoa are about 20% shorter (102 μm) than those of the echidna (125 μm) (Griffith 1984). Monotreme spermatozoa are similar in total length to marsupial spermatozoa from the possum and macropod families (Cummins and Woodall 1985).

C. *Anatomy of the female urogenital tract*

All female monotremes lack a vagina and in the platypus only the left ovary is functional. The female reproductive tract of the echidna (Fig. 15.1) consists of paired ovaries enclosed in the thin infundibular processes of the oviducts, and long sparsely convoluted oviducts connected to paired uteri which open directly and independently into a central urogenital sinus (Griffith 1968, 1978; Hughes and Carrick 1978). The ureters enter the urogenital sinus at about the same level as the uteri adjacent to the bladder neck (Griffith 1968) and, unlike other mammals, are not incorporated into the neck of the bladder (Hughes and Carrick 1978).

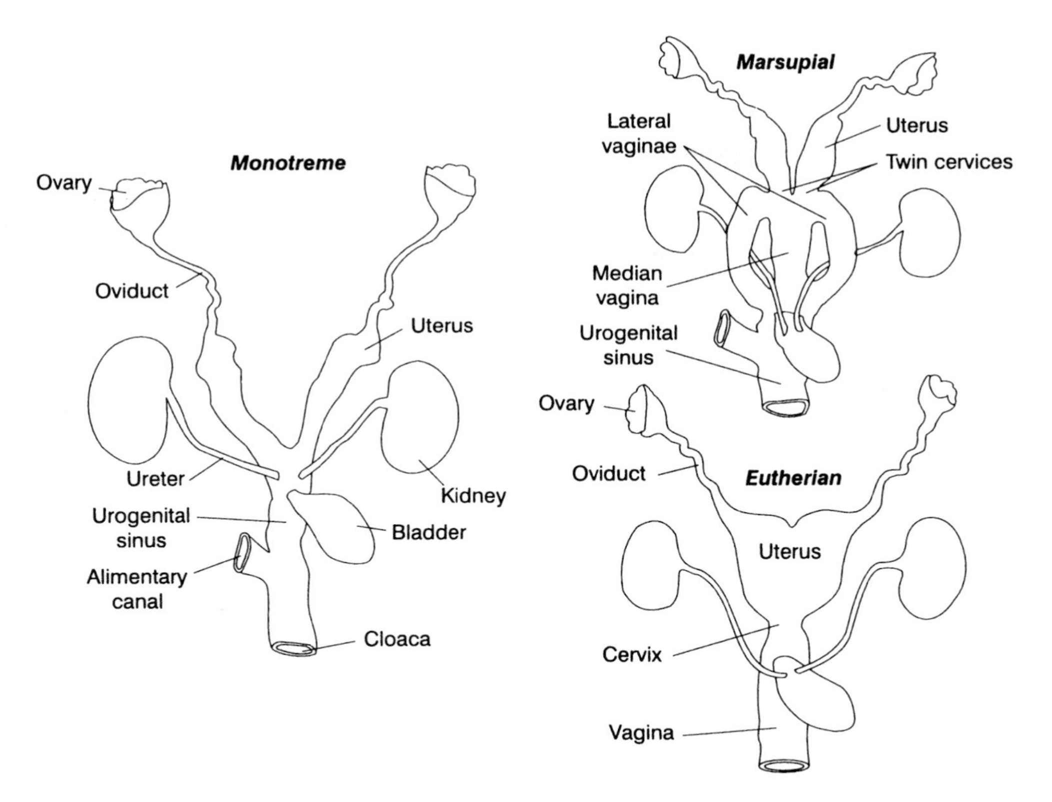

Fig. 15.1. *Female reproductive tract of a Monotreme (echidna), a marsupial and a eutherian mammal. In monotremes the separate uteri, bladder and ureters all open into a common urogenital sinus. Marsupials have paired, but separate uteri, twin cervices and a vaginal complex consisting of lateral vaginae for sperm transport and a median vagina which acts as a birth canal; in many species, it is patent only around the time of birth. In eutherian mammals there is a single vagina.*

D. Sperm transport and storage in the female reproductive tract

Although there are no data available on the dynamics of sperm transport in the female reproductive tract in monotremes, nor on the barriers to sperm passage, sperm storage has been suggested to occur in the oviducts of the short-beaked echidna (Griffith 1978) as an explanation for an unusually long gestation period in one individual. Hibernation-related sperm storage has also been hypothesized for the echidna by Geiser and Seymour (1989), but more recent observations from Beard *et al.* (1992) were more consistent with copulation occurring post-hibernation. Sperm storage in the female has also been suggested for the platypus by Griffith (1978) from observations by Flyn and Hill (1939) of spermatozoa in the uterus and uterine glands of a female with oocytes that were not yet developed.

E. Copulatory and breeding behaviour

Courtship behaviour and copulation have only been described for the platypus (Strahan and Thomas 1975; Fleay 1980; Hawkins and Fanning 1992) and short-beaked echidna (Broom 1895; Augee *et al.* 1975; Griffith 1978; Rismiller 1992). In these species reproductive activity is highly seasonal and individuals of both species are essentially solitary (Griffith 1978; Temple-Smith 1973). Copulation in the platypus has been described only between captive pairs (Strahan and Thomas 1975; Fleay 1980; Hawkins and Fanning 1992). Copulatory behaviour may be initiated by either the male or the female, and coupling lasts from 1 to 28 min and occurs in water (Hawkins and Fanning 1992).

In the short-beaked echidna courtship behaviour appears to differ depending on climatic conditions. In cold climates echidnas form pairs and mate almost immediately after arousal from hibernation; mating usually occurs within a sheltered retreat rather than above ground (Beard *et al.* 1992). In the warmer winter climate on Kangaroo Island in South Australia, male echidnas form mating trains of up to 11 males nose-to-tail behind an oestrus female (Augee *et al.* 1975; Griffith 1978; Rismiller 1992, 1993). The trains and courtship behaviour last for between 14 and 44 days and the trains have a hierarchical structure in which the largest male is at the head of the line behind the female and the smallest male, often a subadult, is last (Rismiller 1992). Large males use their body size to prevent other smaller males from gaining access to the female during competition between train members to mate with her. The duration of copulation in wild echidnas is from about 30 to 180 min (Rismiller 1993) and, although there is no evidence of any form of mate guarding after copulation, the formation of trains of male echidnas during the courtship period on Kangaroo Island is a form of premating mate guarding which presumably ensures that the most vigorous male in the vicinity mates with each oestrus female (Rismiller 1993). The absence of trains during courtship of cold climate echidnas suggests that perhaps two different forms of mate selection are operating in echidna populations.

F. Sexual dimorphism

Platypus show a marked sexual dimorphism in size with adult males being, on average, about 40% heavier and about 10% longer than adult females (Table 15.2). In contrast, echidnas show similar variations in body size and weight, pelage colour, and spine density and development between males and females.

In summary, the available reproductive data, together with observations on the copulatory and breeding behaviour of monotremes suggest that in the short-beaked echidna intermale sperm competition is unlikely. There is no relevant comparable data for the platypus and long-beaked echidna.

Table 15.2. Comparisons of male and female body mass and dimensions (mean ± SD) in platypus from southern New South Wales populations.

	Females (n = 110)	Males (n = 106)	t value
Mass (g)	1089 ± 208	1467 ± 310	10.496 (df = 214)***
Total length (cm)	46 ± 5	52 ± 3	10.540 (df = 214)***
Bill length (cm)	5.3 ± 0.3	5.9 ± 0.4	14.030 (df = 214)***

Data from Temple-Smith (1973).
*** $P < 0.001$.

III. THE MARSUPIALS

Marsupials, or pouched mammals are native to South America, Australia and the New Guinea Islands, and are represented by 17 families comprising approximately 270 species (Flannery 1995; Kirsch *et al.* 1997). Like the monotremes, little is known of the reproductive biology of many marsupial species, however, considerable data is available for 32 species, representing 6 families, and a lesser amount for a further 45 species (Tyndale-Biscoe and Renfree, 1987). As much of the diversity in reproductive processes in marsupials has occurred since their separation from eutherian mammals more than 100 million years ago, comparisons of reproductive structures and strategies within this group can help provide a greater understanding of sperm competition in general and its influence on testis size, sperm number and sperm length, particularly in the Class Mammalia.

A. Interpretation of the comparative analyses

The relationship between testis mass, sperm number, sperm tail length and body mass for marsupials are shown in Figures 15.3–15.6. Part (a) of each figure shows the species data. However, close relatives are often similar through shared inheritance rather than through independent adaptation, invalidating statistical tests that treat species values as independent points (Harvey and Pagel 1991; Felsenstein 1995). Thus, we have also analysed the data using the comparative analysis by independent contrasts (CAIC) package (Purvis and Rambaut 1995), which calculates phylogenetically independent contrasts (roughly, differences between sister taxa) and scales them to have common variance: contrasts are shown in part (b) of each figure. A contrast is computed for each branching point in the phylogeny shown in Fig. 15.2. Most of this phylogeny comes from Kirsch *et al.* (1997), whose DNA–DNA hybridization

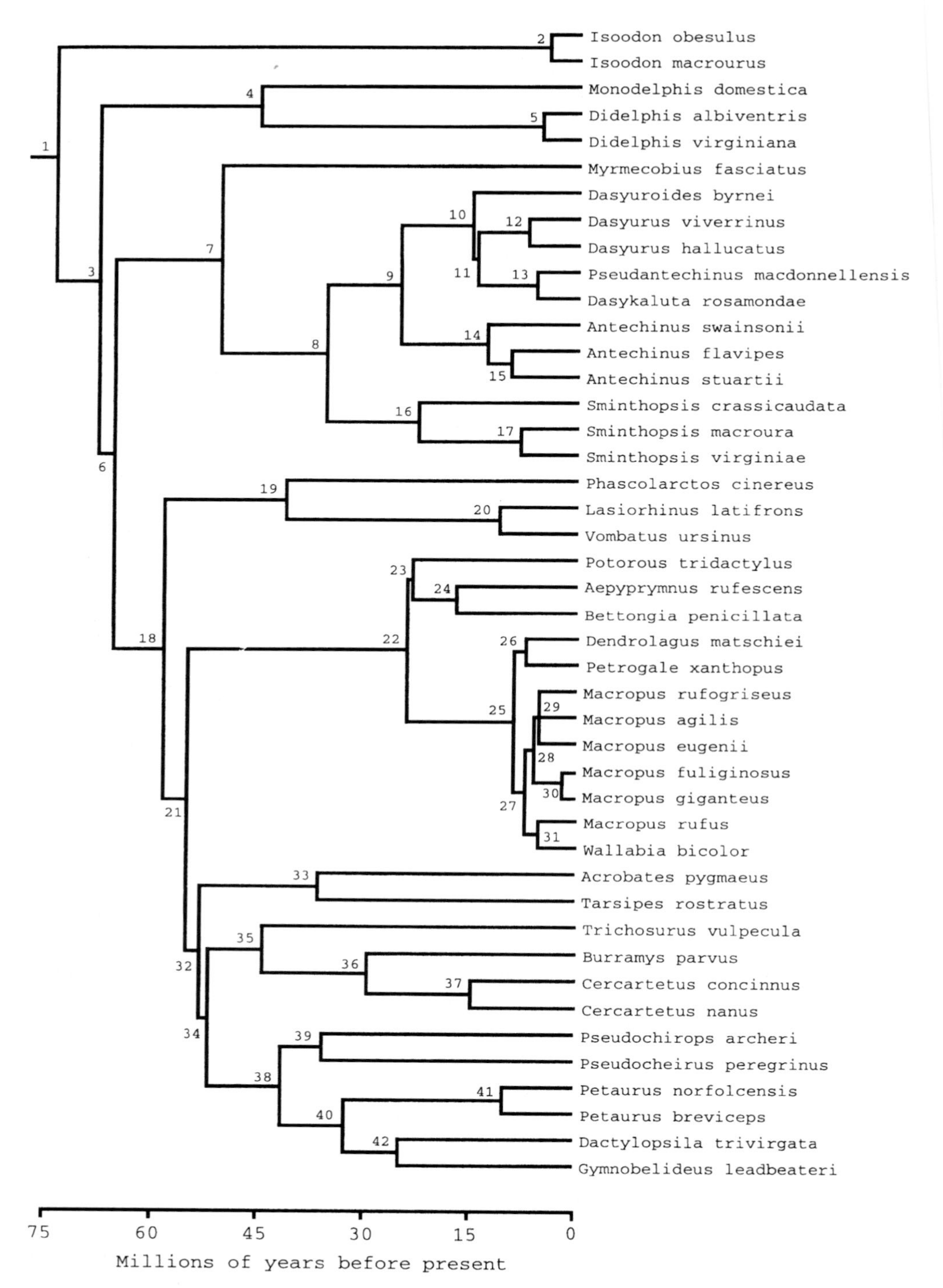
Isoodon obesulus
Isoodon macrourus
Monodelphis domestica
Didelphis albiventris
Didelphis virginiana
Myrmecobius fasciatus
Dasyuroides byrnei
Dasyurus viverrinus
Dasyurus hallucatus
Pseudantechinus macdonnellensis
Dasykaluta rosamondae
Antechinus swainsonii
Antechinus flavipes
Antechinus stuartii
Sminthopsis crassicaudata
Sminthopsis macroura
Sminthopsis virginiae
Phascolarctos cinereus
Lasiorhinus latifrons
Vombatus ursinus
Potorous tridactylus
Aepyprymnus rufescens
Bettongia penicillata
Dendrolagus matschiei
Petrogale xanthopus
Macropus rufogriseus
Macropus agilis
Macropus eugenii
Macropus fuliginosus
Macropus giganteus
Macropus rufus
Wallabia bicolor
Acrobates pygmaeus
Tarsipes rostratus
Trichosurus vulpecula
Burramys parvus
Cercartetus concinnus
Cercartetus nanus
Pseudochirops archeri
Pseudocheirus peregrinus
Petaurus norfolcensis
Petaurus breviceps
Dactylopsila trivirgata
Gymnobelideus leadbeateri
75
60
45
30
15
0
Millions of years before present

Fig. 15.2. *Phylogeny of the species used in the comparative analysis (Kirsch* et al. *1997).*

phylogeny of over a 100 marsupial species includes 31 of those in our data set. Kirsch *et al.* (1997) note that several groupings in their final tree were not found in all of their analyses so should be treated as provisional; e.g. the placement of *Isoodon* outside the other species in our study. We have followed Kirsch *et al.*'s final tree, even for such less robust and more controversial placements, because independent contrast analyses are little affected by the precise order of branching among higher taxa provided that the branches separating the nodes are short (Purvis and Garland 1993), as they are here. The remaining species were added to the backbone, provided by Kirsch *et al.* (1997), from Baverstock *et al.* (1982, 1990), Richardson and McDermid (1978), and Wilson and Reeder (1993). The resulting tree has 42 nodes, 33 of which were dated by Kirsch *et al.* (1997). Dates for five further nodes were obtained using information from the above references and the remaining dates estimated by splitting branches into segments of equal length.

All species data were logarithmically transformed (base 10) before the contrasts were calculated. Least-squares regression through the origin (Garland *et al.* 1992) showed that body mass was a significant predictor of testes mass ($P < 0.001$), sperm number in the cauda epididymides ($P = 0.001$), total sperm number ($P = 0.006$), and sperm tail length ($P = 0.009$); the first three associations were positive and the last negative. However, estimating the 'true' slopes of the allometric lines is problematic if the X-value contains sampling error (see Harvey and Pagel 1991 for a review), as will always be the case with comparative data. Sampling error matters most when closely related lineages are being compared: the real (evolved) difference in X between such lineages is probably small so the error in sampling could easily swamp it. The evolved differences among higher taxa are greater, so sampling error matters much less. In order to better estimate the true slopes, we have therefore split the contrasts into two equal-sized groups based on the ages of the corresponding nodes in the phylogeny, and used only the contrasts from older nodes – those less affected by sampling error – in the slope estimation (following Purvis and Harvey 1995). In justification we note that, although all regressions were highly significant whether or not younger nodes were excluded, they all became much tighter (higher r^2) when we excluded the younger nodes. Finally, we calculated residual values by fitting these regression lines through the species data.

Figure 15.3c shows how relative testes mass varies across the phylogeny. For ease of presentation, residual testes mass from the regression above was recoded as a three-state ordered character, with equal numbers of species in each state. The evolution of this character was then mapped onto the phylogeny using parsimony implemented by MacClade (Maddison and Maddison 1992). Within *Macropus*, two different reconstructions are equally parsimonious; the affected branches are labelled as 'equivocal'. (Figure 15.6c, below, was constructed in an analogous way and shows how relative sperm tail length varies across the phylogeny.)

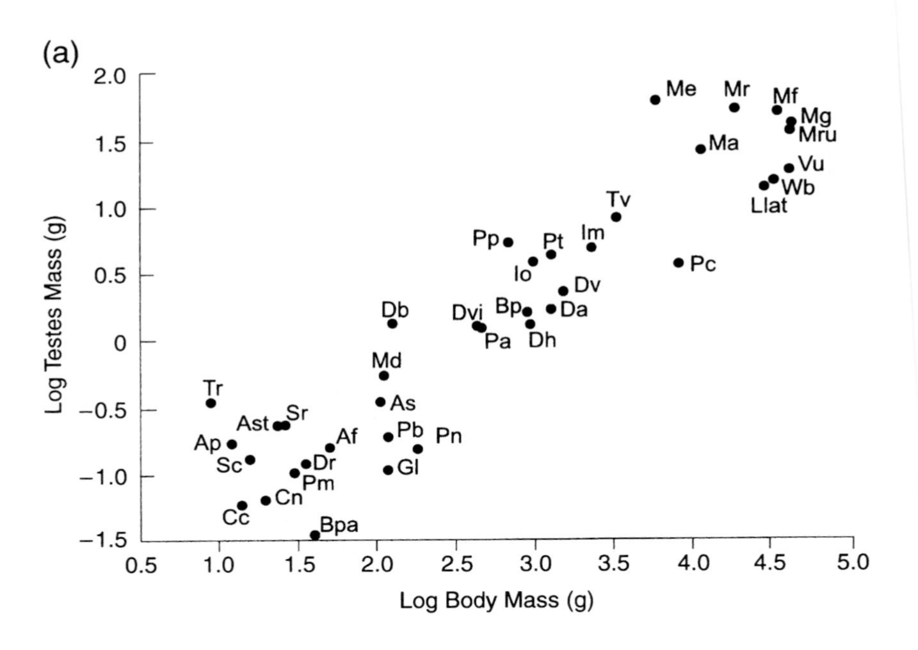
(a)
Log Testes Mass (g)
Log Body Mass (g)
Me
Mr
Mf
Mg
Mru
Ma
Vu
Wb
Llat
Tv
Im
Pp
Pt
Io
Pc
Dv
Db
Dvi
Bp
Da
Pa
Dh
Md
Tr
As
Sr
Ast
Ap
Af
Pb
Pn
Sc
Dr
Gl
Pm
Cn
Cc
Bpa

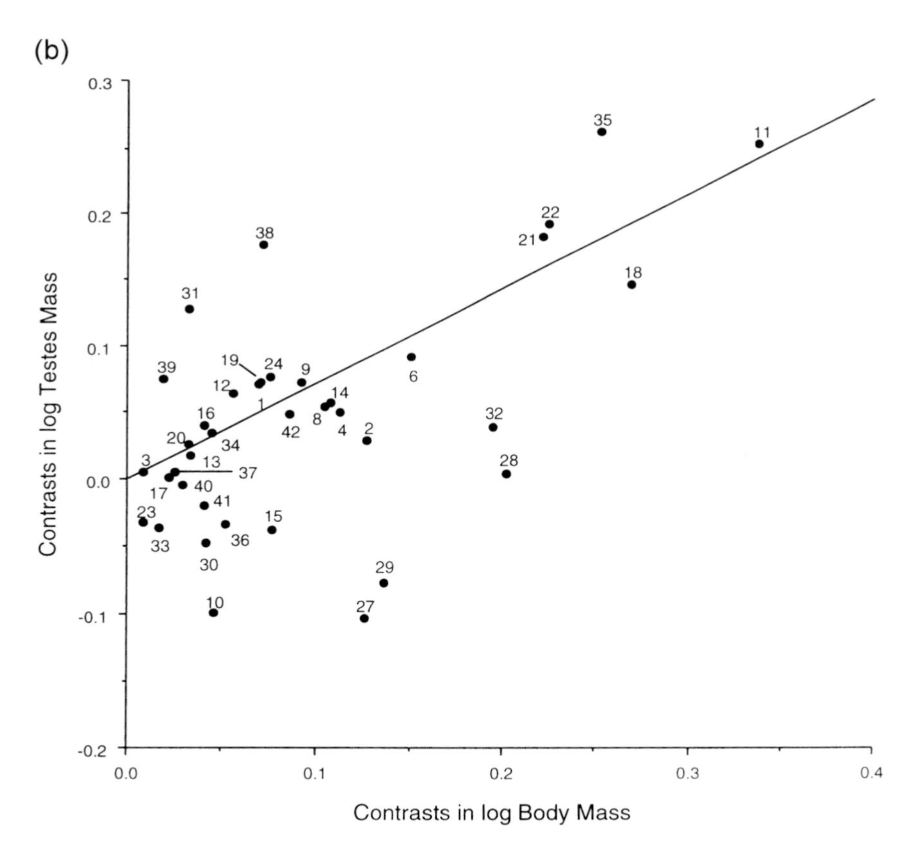
(b)
Contrasts in log Testes Mass
Contrasts in log Body Mass

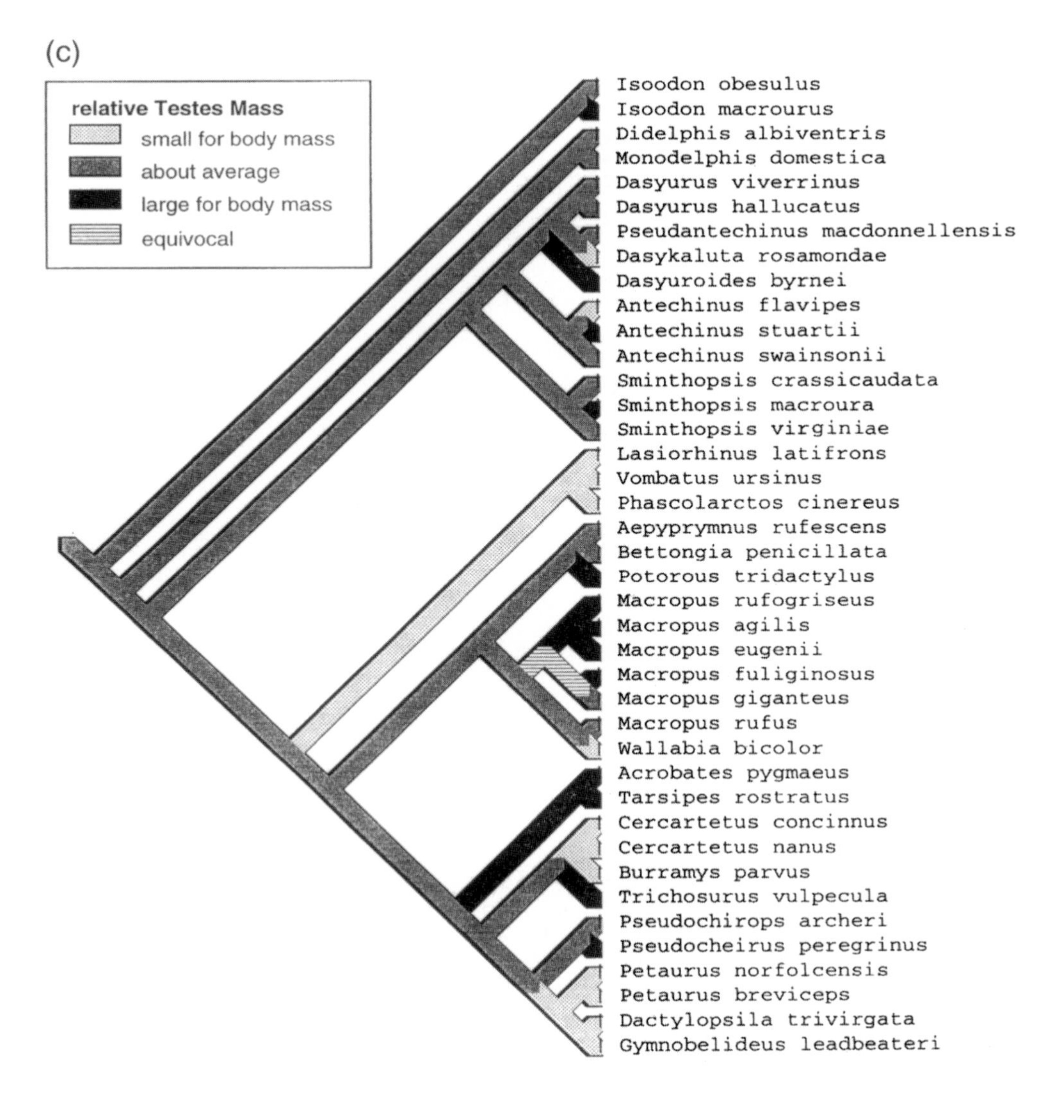

Fig. 15.3. *(a) Regression of testes mass against body mass for all available marsupial species. (For species codes refer to Table 15.3). (b) Independent contrasts testes mass and body size. Numbers refer to nodes on Fig. 15.2. The line is the least-squares regression through the origin, fitted through the older contrasts only; see text for explanation. (c) A possible phylogram of change in relative testes mass (see text for explanation).*

B. Comparison of relative testis size in marsupials

A positive allometric relationship between body and testes mass was observed across marsupial species (Fig. 15.3a; $y = 1.11 + 0.75x$, $r^2 = 0.87$). Inspection of the individual data (Fig. 15.3a,b; Table 15.3) show

Table 15.3. Body mass (kg), combined testes mass (mg), number of sperm in cauda epididymides and in total epididymides per animal, and sperm size (μm) as indicated by tail length of marsupial species. Sources of data: D. A. Taggart published and unpublished data (D.T.), W. G. Breed published and unpublished data (W.B.), Tyndale-Biscoe and Renfree (1987) (H.T./M.R.), Cummins and Woodall (1985) (J.C./P.W.).

Family	Genus	Species		*n*	Body mass, mean ± SD (g)	Testes mass, mean ± SD (mg)	Relative testis mass	Mean no. of sperm in cauda emidymides ($\times 10^6$)	Mean no. per animal ($\times 10^6$)	Mean sperm tail length (μm)	Source
Didelphidae	*Didelphis*	*albiventris*	Da	16	1291	1760					H.T./M.R.
	Monodelphis	*domestica*	Md	8	110 ± 7	570 ± 98	0.85	2.36	4.2	204.3	D.T.
	Didelphis	*virginiana*	Dvi	4	3700			26.2	37.2		H.T./M.R.
Dasyuridae	*Dasykaluta*	*rosamondae*	Dr	6	35.3 ± 8.7	126 ± 560	0.36				H.T./M.R.
	Dasyuroides	*byrnei*	Db	5	123.6 ± 11.72	1380.8 ± 161	0.56	0.9 ± 0.14	1.7 ± 1.13	242.1	W.B.
	Dasyurus	*hallucatus*	Dh	7	940 ± 84.1	1350 ± 210	0.3				H.T./M.R.
	Dasyurus	*viverrinus*	Dv	2	1525 ± 106.1	2357.5 ± 1899	0.15		4		W.B.
	Pseudantechinus	*macdonnellensis*	Pm	8	27.1 ± 5.4	110 ± 60	0.41				H.T./M.R.
	Antechinus	*flavipes*	Af	4	49.76 ± 2.0	164 ± 25.2	0.33	3.24	8.22		D.T.
	Antechinus	*stuartii*	Ast	4	22.98 ± 3.71	240 ± 6	1.04	0.64	7	259.7	D.T.
	Antechinus	*swainsonii*	As	3	104.7 ± 36.2	362.2 ± 80	0.17	5.6	7.2	260	D.T. & W.B.
	Sminthopsis	*crassicaudata*	Sc	8	15.5 ± 1.1	137.8 ± 26.4	0.89	0.54	1.22	252.2	D.T. & W.B.
	Sminthopsis	*macroura*	Sm	4	25.4 ± 0.45	246.6 ± 69.6	0.76	0.22	1.06		D.T. & W.B.
	Sminthopsis	*virginiae*	Sr	1	31	252	0.81				D.T.
Myrmecobiidae	*Myrmecobius*	*fasciatus*	Mf	2						127.4	D.T.
Peramelidae	*Isoodon*	*obesulus*	Io	3	978.3 ± 183	3880 ± 501	0.4	56	96	162	D.T. & H.T./M.R. & W.B.
	Isoodon	*macrourus*	Im	3	2300 ± 410	4720 ± 220	0.21	118	199	165.1	D.T.
Phascolarctidae	*Phascolarctos*	*cinereus*	Pc	2	8150 ± 1340	3720 ± 340	0.05			72.1	D.T.
Vombatidae	*Lasiorhinus*	*latifrons*	Llat	6	28290 ± 2780	13800 ± 330	0.049		146.4	72	D.T.
	Vombatus	*ursinus*	Vu	1	40100	18420	0.05			87.9	D.T.
Burramyidae	*Burramys*	*parvus*	Bpa	1	39	37.2	0.1				D.T.
	Cercartetus	*concinnus*	Cc	5	14.3 ± 2.4	62 ± 13	0.44				H.T./M.R.
	Cercartetus	*nanus*	Cn	1	19.5	67	0.34				D.T.
Petauridae	*Gymnobelideus*	*leadbeateri*	Gl	1	115.5	112	0.048				D.T.
	Petaurus	*breviceps*	Pb	1	119	200	0.17			107.1	W.B.
	Petaurus	*norfolcensis*	Pn	1	180	162	0.09				W.B.
	Dactylopsila	*trivirgata*	Dt	2	467	250	0.054				D.T.
Pseudocheiridae	*Pseudocheirus*	*peregrinus*	Pp	3	684.5 ± 46.32	5425 ± 130.81	0.8			112.8	W.B.
	Pseudochirops	*archeri*	Pa	2	477 ± 12.7	1250 ± 60	0.26				D.T.
Tarsipedoidea	*Tarsipes*	*rostratus*	Tr	24	8.9 ± 1.9	365 ± 64	4.12			356	H.T./M.R. & J.C./P.W.
Acrobatidae	*Acrobates*	*pygmaeus*	Ap	4	12.3 ± 2.2	178 ± 80	1.45				H.T./M.R.
Phalangeridae	*Trichosurus*	*vulpecula*	Tv	22	3350 ± 410	8260 ± 1130	0.25	261.4	350.6	94.2	H.T./M.R. & W.B.
Potoroidae	*Aepyprymnus*	*rufescens*	Ar	1	2400	4680	0.2	57.8	83.8	99.0 ± 5.5	D.T. & W.B.
	Bettongia	*penicillata*	Bp	2	872 ± 20	1667 ± 188	0.19			155	D.T. & W.B.
	Potorous	*tridactylus*	Pt	2	1280 ± 305	4380 ± 260	0.34	38.8	97.2	156	D.T.
Macropodidae	*Dendrolagus*	*matschiei*	Dm							97.2	D.T.
	Macropus	*agilis*	Ma	6	11400 ± 2400	25640 ± 6900	0.23			105.7	H.T./M.R.
	Macropus	*eugenii*	Me	17	5850 ± 640	62000 ± 8800	0.53			99	H.T./M.R. & D.T.
	Macropus	*fuliginosus*	Mf	10	34150 ± 944	51620 ± 10080	0.15	1266	2064		H.T./M.R.
	Macropus	*giganteus*	Mg	9	40720 ± 994	42020 ± 10620	0.1			111.6	H.T./M.R.
	Macropus	*rufogriseus*	Mr	9	18500 ± 3000	54570 ± 9470	0.3	1460	1600		H.T./M.R.
	Macropus	*rufus*	Mrufus	4	39825 ± 7283.0	38180 ± 6181.8	0.096	388 ± 212	522 ± 254	118.8	W.B.
	Petrogale	*xanthopus*	Mx							93.5	D.T.
	Wallabia	*bicolor*	Wb	2	31500	14751 ± 1770	0.05	305	354	101.7	D.T. & W.B.

that among species of small body mass the lineage represented by *Tarsipes rostratus* (testes = 4.12% body mass) and *Acrobates pygmaeus* (testes = 1.45% body mass) have large testes relative to body mass. Data for the semelparous dasyurids must be viewed with caution since total spermatogenic failure occurs prior to the onset of mating in some semelparous species (Lee *et al.* 1982; Kerr and Hedger 1983). Data are available for three species of semelparous dasyurid which fall into this group: *Antechinus stuartii, Antechinus swainsonii* and *Antechinus flavipes.* Although the data on the last two species was recorded at the height of the breeding season, this occurred well after spermatogenic failure (Taggart and Temple-Smith 1994). Only in *A. stuartii* was body mass recorded at the time of maximal testicular mass, which is approximately 1 month prior to the commencement of the mating season (Taggart and Temple-Smith 1990a). In *A. stuartii* testis mass was found to be large relative to body mass (testes = 1.04% body mass), whereas the burramyid lineage, containing *Burramys* and two *Cercartetus* spp., have small testes relative to body mass (testes = 0.34–0.44% body mass). Of the marsupials with a body mass of between 100 and 200 g, *Dasyuroides byrnei* (testes = 0.56% body mass) has relatively large testes whereas those of the four species from the petaurid lineage examined (e.g. sugar glider *Petaurus breviceps,* testes = 0.17% body mass; Leadbeater's possum *Gymnobelideus leadbeateri,* testes = 0.048% body mass) are small relative to body mass (Fig. 15.3a–c; Table 15.3).

Among the larger marsupials, most members of the macropod lineage such as *Macropus eugenii* and *Macropus rufogriseus* have a relatively large testis mass for body mass (testes = 0.53% and 0.3% body mass, respectively) (Fig. 15.3a,c) although in *Wallabia bicolour* it was small relative to body weight. All representatives of the Vombatiformes lineage examined (i.e. the koala and two species of wombat) have small testes relative to body mass (testes $\approx$ 0.05% body mass) (Fig. 15.3a,c; Table 15.3).

C. Comparisons of epididymal sperm number

Far fewer data were available for numbers of spermatozoa in the epididymis. For total number of sperm in the epididymides the allometric relationship across species is $y = -0.78 + 0.82x$, $r^2 = 0.67$ (Figs 15.4 and 15.5; Table 15.3). Sperm counts for the species with a smaller body mass all come from members of the dasyurid lineage where between 1 and 7×10^6 sperm per animal were found, regardless of body mass (range 15 g for *Sminthopsis crassicaudata* to approximately 1.5 kg for *Dasyurus viverrinus*). All have low to average numbers of sperm relative to body mass (Figs 15.4 and 15.5; Table 15.3). The two representatives of the didelphid lineage also had low numbers of epididymal sperm relative to body mass whereas those of the peramelid lineage were high relative to body mass (e.g. *Isoodon obesulus* and *Isoodon macrourus* have about $100–200 \times 10^6$

sperm per animal), as were *Trichosurus vulpecula* (350×10^6 sperm per animal) and *Potorous tridactylus* (Figs 15.4 and 15.5; Table 15.3).

Of the larger marsupials, the values for some of the representatives of the macropod lineage (e.g. the western grey kangaroo *Macropus fuliginosus* and red-necked wallaby *Macropus rufogriseus*) were high relative to body mass and contrasted with those of the swamp wallaby *Wallabia bicolour* and the wombat *Lasiorhinus latifrons*, which were low relative to body mass (Figs 15.4 and 15.5; Table 15.3).

D. Comparison of sperm tail length

In marked contrast to the data for testes mass and sperm number, a negative relationship was found between sperm tail length and body mass (regression of species data: $y = 2.58 - 0.14x$, $r^2 = 0.71$) (Fig. 15.6; Table 15.3). Upon examination of the data for species <1 kg (Fig. 15.6a; Table 15.3) it is evident that the value for sperm tail length for the honey possum *Tarsipes rostratus* (356 μm), the kowari *Dasyuroides byrnei* (242 μm) and the dusky marsupial mouse *Antechinus swainsonii* (260 μm), are all high relative to body mass and that in the lineage containing the two gliders it is short relative to body mass. Within the 1 kg–3.5 kg weight range the bandicoots had about average or longer than average spermatozoa relative to body mass ($\approx$160 μm), and the rufus bettong (*Aepyprymnus rufescens*, $\approx$99 μm) and the brush-tailed possum (*Trichosurus vulpecula*, $\approx$94 μm) also had relatively short spermatozoa relative to body weight. Of the larger marsupials the macropod lineage had about average or larger than average (e.g. eastern grey kangaroo *Macropus giganteus* $\approx$112 μm; red kangaroo *Macropus rufus* $\approx$119 μm; and swamp wallaby *Wallabia bicolour* $\approx$102 μm) sperm size for body mass, whereas the spermatozoa of species in the wombat/koala lineage were about average or small for body mass ($\approx$72–82 μm) (Fig. 15.6; Table 15.3).

E. Anatomy of the female urogenital tract

Testes size and epididymal sperm number are not the only factors that determine the number of sperm reaching the upper reaches of the female reproductive tract in mammals. Other reproductive features, such as the

---

Fig. 15.4. *(a) Regression of sperm number in cauda epididymides against body mass for all available marsupial species. For an explanation of species codes refer to Table 15.3. (b) Independent contrasts in sperm number in cauda epididymides and body size. Numbers refer to node numbers on Fig. 15.2. The line is the least-squares regression through the origin, fitted through the older contrasts only; see text for explanation.*

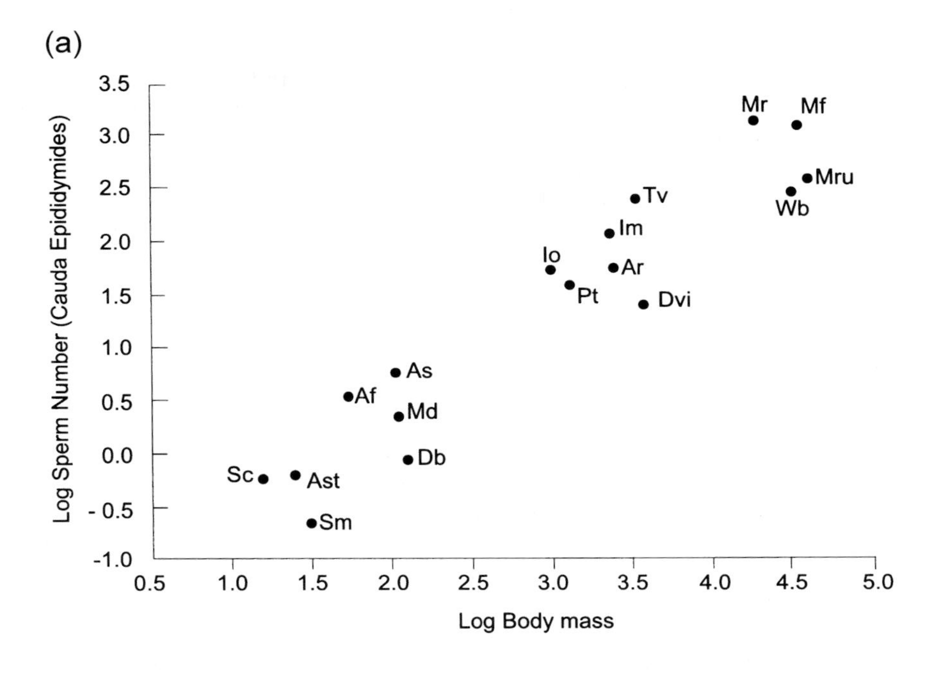
(a)
Log Sperm Number (Cauda Epididymides)
Log Body mass
3.5
3.0
2.5
2.0
1.5
1.0
0.5
0.0
- 0.5
-1.0
0.5
1.0
1.5
2.0
2.5
3.0
3.5
4.0
4.5
5.0
Mr
Mf
Mru
Wb
Tv
Im
Io
Ar
Pt
Dvi
As
Af
Md
Db
Sc
Ast
Sm

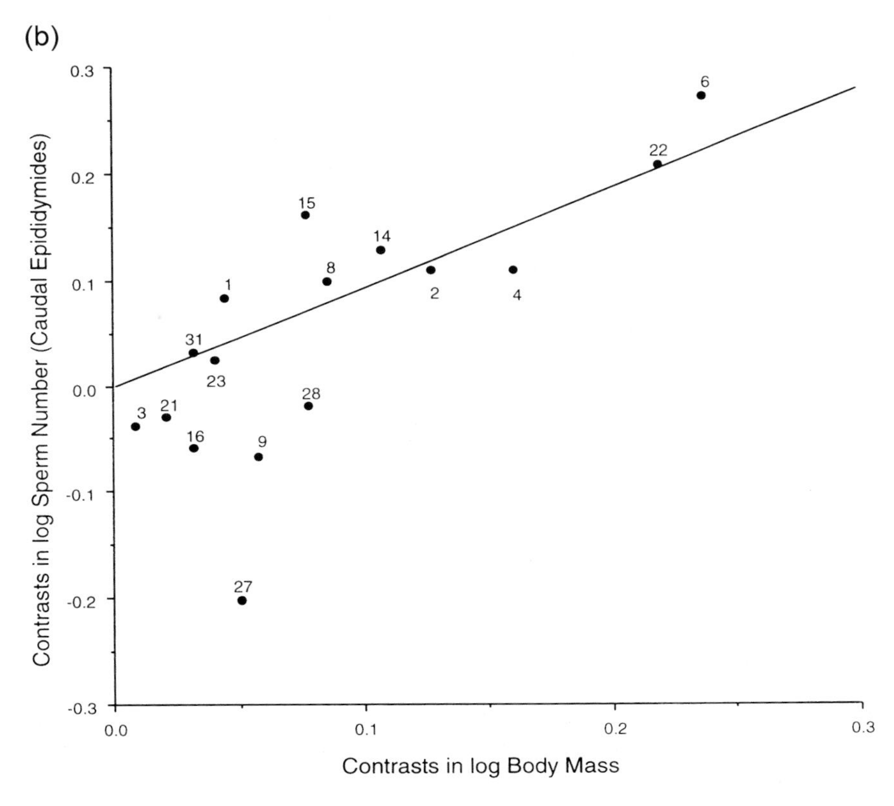
(b)
Contrasts in log Sperm Number (Caudal Epididymides)
Contrasts in log Body Mass
0.3
0.2
0.1
0.0
-0.1
-0.2
-0.3
0.0
0.1
0.2
0.3
6
22
15
14
8
2
4
1
31
23
3
21
28
16
9
27

morphology of the female reproductive tract, the presence or absence of sperm storage regions and factors that influence the viability of sperm in the tract all affect the chances of a particular male fertilizing the ovulated oocytes of the female.

The female reproductive tract of marsupials is highly divergent from that of monotremes and eutherians (Tyson 1698; Pearson 1945; Arnold and Shorey 1985; Tyndale-Biscoe and Renfree 1987), in having two lateral and one median vaginae, twin cervices, and paired, but separate, uteri (Tyndale-Biscoe and Renfree 1987) (Fig. 15.1). In most species the two lateral vaginae originate in the urogenital sinus and, following mating, become large and distended with ejaculatory and vaginal fluids. Each connects to one uterus to form the vaginal cul-de-sac. It is here that the twin uterine cervices open. The cervices are connected via a uterine neck, which varies in length depending upon the species, to the body of the uterus of the same side. The oviducts have an isthmic segment which is tightly coiled and wider than the upper ampulla segment (Pearson 1945; Tyndale-Biscoe and Renfree 1987; Arnold and Shorey 1985; Rodger 1991; Bedford and Breed 1994) (Fig. 15.1).

F. Barriers to sperm transport in marsupials

Upon ejaculation semen is deposited in the upper part of the urogenital sinus and sperm travel rapidly to the cervix (Hartman 1924; McCrady 1938; Hughes and Rodger 1971; Tyndale-Biscoe and Rodger 1978; Tyndale-Biscoe and Renfree 1987), which may act as a reservoir for spermatozoa and as a selective barrier to further transport (Thibault 1973; Tyndale-Biscoe and Rodger 1978; Hunter 1988; Taggart 1994). In eutherian mammals the uterotubal junction forms a secondary barrier to sperm migration but whether this occurs in marsupials is not known. In didelphid (Bedford *et al.* 1984) and dasyurid marsupials (Breed *et al.* 1989; Taggart and Temple-Smith 1990a,b, 1991) there appears to be extremely efficient transport of ejaculated spermatozoa to the isthmic region of the oviduct (Virginia opossum *Didelphis virginiana*, $\approx$1:20; fat-tailed dunnart *Sminthopsis crassicaudata*, $\approx$1:10; brown marsupial mouse *Antechinus stuartii*, $\approx$1:1 to 1:7) which contrasts dramatically with the small percentage of ejaculated spermatozoa reaching the oviduct in most eutherian mammals studied (e.g. $\approx$1:10000 in the rabbit; Overstreet and Cooper 1978, 1979).

Fig. 15.5. *(a) Regression of total sperm number in epididymides against body mass for all available marsupial species. For explanation of species codes refer to Table 15.3. (b) Independent contrasts in total sperm number in epididymides and body size. Numbers refer to nodes on Fig. 15.2. The line is the least-squares regression through the origin, fitted through the older contrasts only; see text for explanation.*

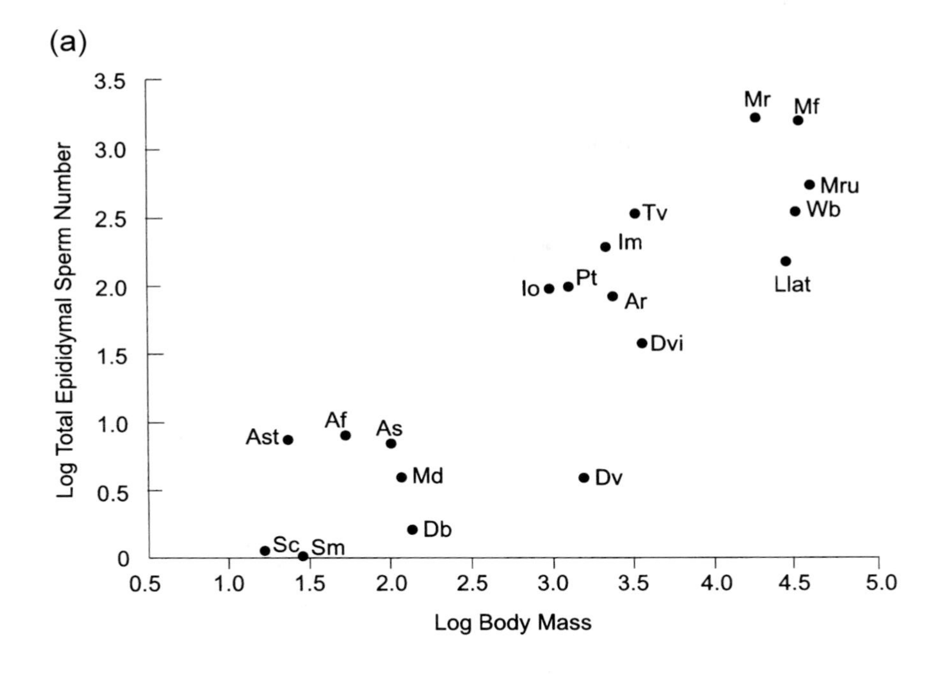
(a)
Log Total Epididymal Sperm Number
Log Body Mass
Mr
Mf
Mru
Tv
Wb
Im
Llat
Io
Pt
Ar
Dvi
Ast
Af
As
Md
Dv
Db
Sc
Sm
3.5
3.0
2.5
2.0
1.5
1.0
0.5
0
0.5
1.0
1.5
2.0
2.5
3.0
3.5
4.0
4.5
5.0

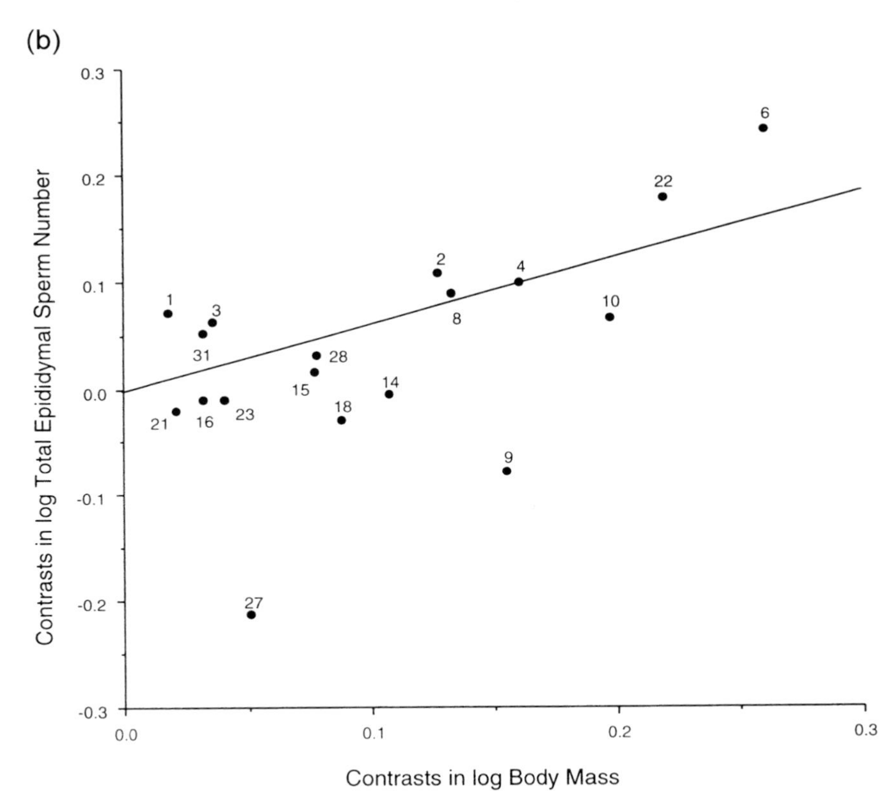
(b)
Contrasts in log Total Epididymal Sperm Number
Contrasts in log Body Mass
0.3
0.2
0.1
0.0
-0.1
-0.2
-0.3
0.0
0.1
0.2
0.3
6
22
2
4
8
10
1
3
31
28
15
14
18
21
16
23
9
27

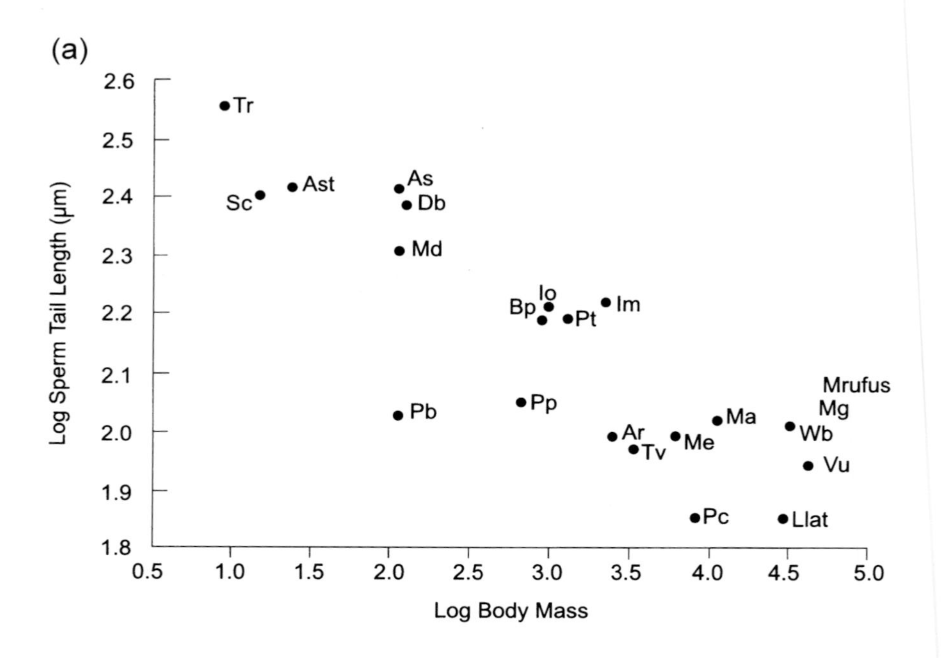
(a)
Log Sperm Tail Length (μm)
Log Body Mass
Tr
Sc
Ast
As
Db
Md
Bp
Io
Pt
Im
Pb
Pp
Ar
Tv
Me
Ma
Mrufus
Mg
Wb
Vu
Pc
Llat

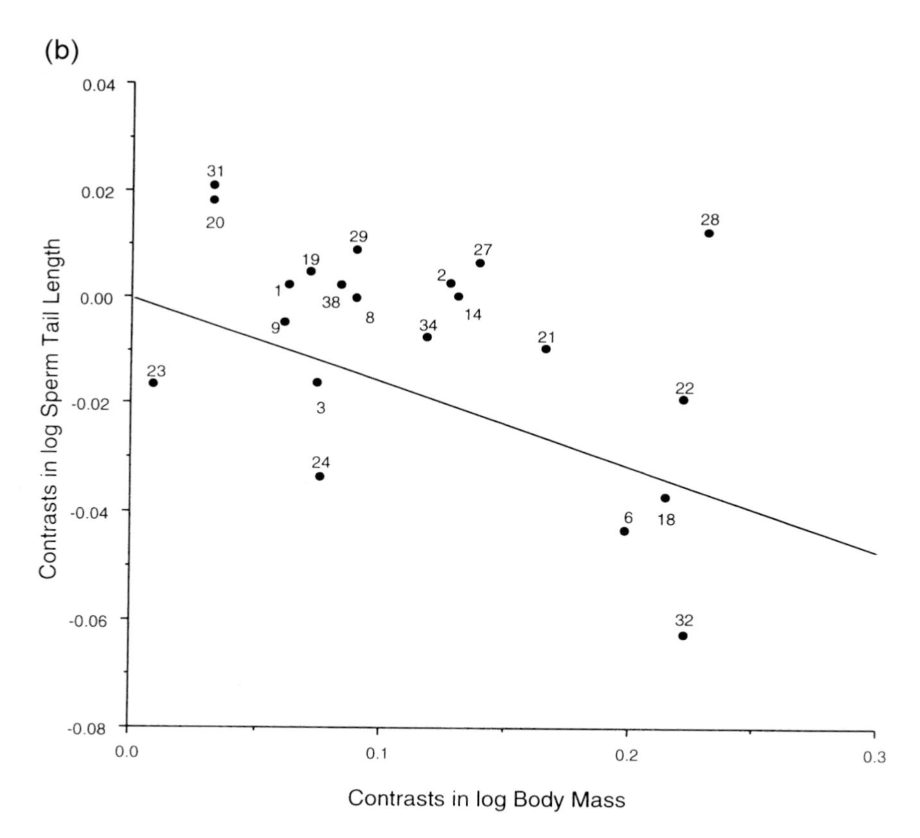
(b)
Contrasts in log Sperm Tail Length
Contrasts in log Body Mass

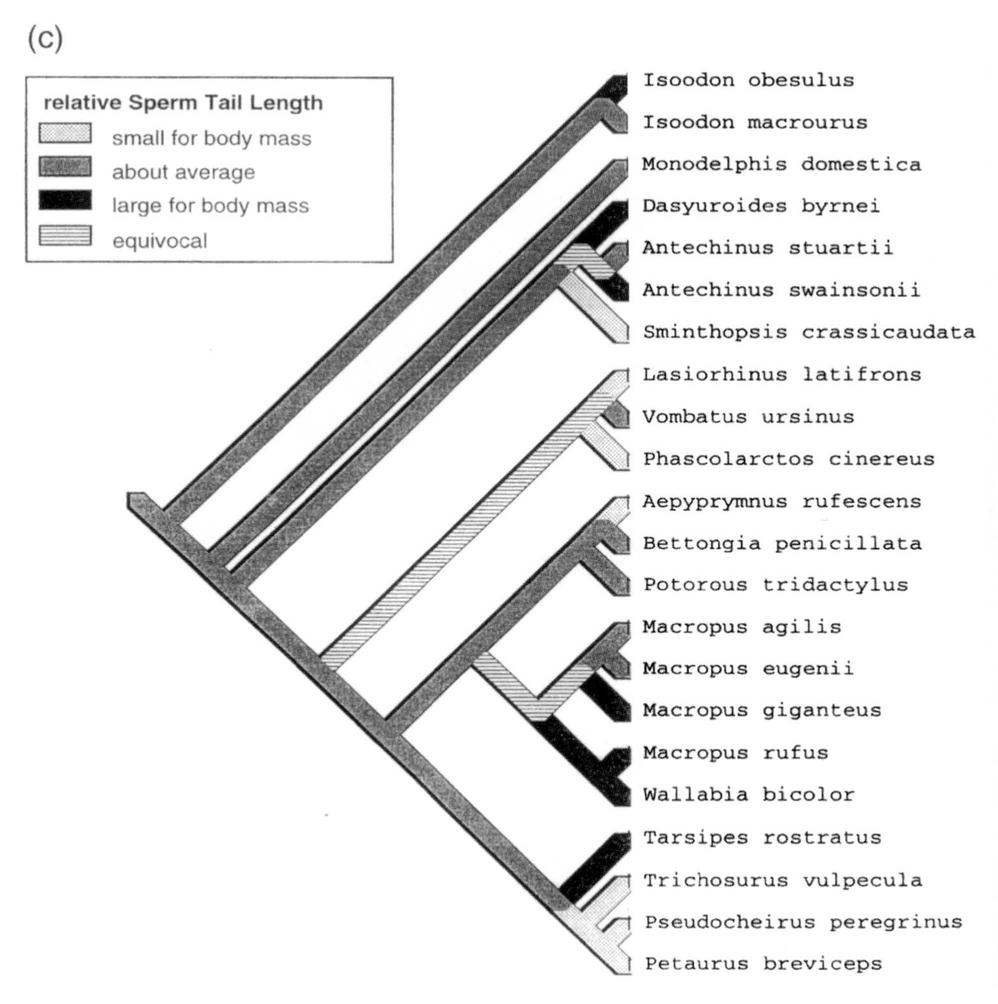

Fig. 15.6. *(a) Regression of sperm tail length (μm) against body mass for all available marsupial species (see Table 15.3. for species codes). (b) Independent contrasts in sperm tail length and body size. Numbers refer to nodes on Fig. 15.2. The line is the least-squares regression through the origin, fitted through the older contrasts only; see text for explanation. (c) The evolution of relative sperm tail length; see text for explanation.*

G. Sperm storage and release in the female reproductive tract

Sperm storage in the female reproductive tract is a relatively common phenomenon in insects, lower vertebrates, reptiles and birds (Walker 1980; Thomas and Zeh 1984; Gist and Jones 1987; Ridley 1989; Birkhead and Møller 1992, 1993; Ward 1993; Bakst *et al.* 1994). In eutherian mammals, because fertilization generally occurs within 24 h of

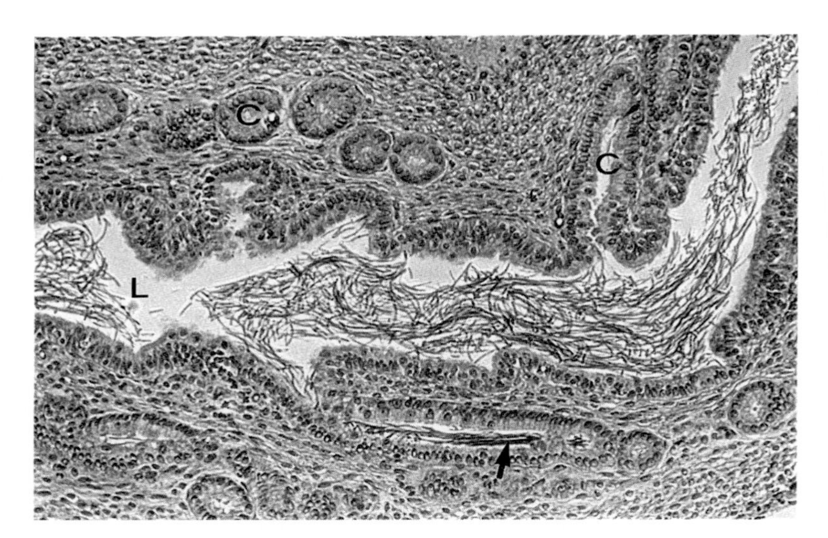

Fig. 15.7. *Longitudinal section (× 125) through the lower isthmus region of the oviduct of the brown marsupial mouse,* Antechinus stuartii, *showing the lumen (L) and spermatozoa (S) and, originating from the lumen, crypts (C) containing spermatozoa (arrow).*

mating, spermatozoa survive only for short periods in the female tract (Bishop 1970). Long-term sperm storage is therefore extremely rare, with insectivorous bats being the best known exception (Racey and Potts 1970; Racey 1979; Racey *et al.* 1987), although somewhat prolonged storage may also occur in the European hare and domestic dog (Birkhead and Møller 1993).

In marsupials extended periods of sperm storage in the female tract (up to 2–3 weeks) have been reported for three families. In the Dasyuridae (Selwood 1980; Selwood and McCallum 1987; Breed *et al.* 1989; Taggart and Temple-Smith 1991) and the Didelphidae (Rodger and Bedford 1982a,b; Bedford *et al.* 1984; Taggart and Moore unpublished observations) sperm storage occurs in specialized isthmic crypts (Fig. 15.7), whereas in peramelids (Lyne and Hollis 1977) it occurs in the vaginal caeca. Relatively low fertility levels were found in female *Antechinus* which had stored spermatozoa for less than 5 days prior to ovulation or for more than 13 days prior to ovulation (Selwood and McCallum 1987).

The release of spermatozoa from the isthmic storage crypts has been studied in the fat-tailed dunnart *Sminthopsis crassicaudata* using transmitted light (Bedford and Breed 1994). Surprisingly, the heads of all sperm stored in the isthmic crypts initially lay parallel to their tails (i.e. appeared spear-shaped) except for a small 'vanguard' population of approximately 200, located in crypts closest to the ovary, where the heads became T-shaped. Following ovulation T-shaped sperm migrated from the crypts to the site of fertilization higher up the oviduct (Bedford and Breed 1994).

H. Length of copulation and copulatory behaviour

With the exception of dasyurid and macropod marsupials few detailed observations are available on copulatory behaviour (dasyurids: Marlow 1961; Woolley 1971a,b, 1988, 1990a,b, 1991; Morton 1978b; Fanning 1982; Woolley and Ahern 1983; Read 1984; Dickman 1985, 1993; Shimmin and Temple-Smith 1996a; macropodids: Sharman *et al.* 1966; Kaufmann 1974; Croft 1981a,b; Jarman 1983; Watson *et al.* 1992; Rudd 1994; peramelids: Stodart 1966; vombatids: Gaughwin 1981; didelphids: Hunsaker 1977; Barnes and Barthold 1969; Trupin and Fadem 1982; Taggart and Moore unpublished observations) (Table 15.4).

In dasyurids the length of copulation varies from 2 to 12 h depending upon the species (see Dickman 1993 for summary) (Table 15.4). In general, the semelparous dasyurid species, such as *Antechinus stuartii* (Lee *et al.* 1982) have the longest copulation (≈7.7–18 h). Lengthy copulations have also been reported in some didelphids (McManus 1970). In contrast, the macropods mate for between 5 and 53 min (Sharman and Calaby 1964; Sharman *et al.* 1966; Tyndale-Biscoe and Rodger 1978; Rudd 1994), wombats for approximately 30 min (Gaughwin 1981), and other didelphids for 4–40 min (Barnes and Barthold 1969; Trupin and Fadem 1982; Taggart and Moore unpublished observations) (Table 15.4). The shortest copulation (less than 30 s), has been reported for bandicoots (Stodart 1966) (Table 15.4).

Table 15.4. Maximum length of copulation in various marsupial species.

Marsupial family	Species	Maximum duration of copulation
Dasyuridae	Brown antechinus (*Antechinus stuartii*)	18.0 h
(Semelparous)	Dusky antechinus (*Antechinus swainsonii*)	9.5 h
	Yellow footed antechinus (*Antechinus flavipes*)	11.0 h
(Iteroparous)	Stripe-faced dunnart (*Sminthopsis macroura*)	2.5 h
	Kowari (*Dasyuroides byrnei*)	3.0 h
	White-footed dunnart (*Sminthopsis leucopus*)	1.8 h
	Fat-tailed dunnart (*Sminthopsis crassicaudata*)	11 h
Peramelidae	Long-nosed bandicoot (*Perameles nasuta*)	< 30 s
Potoroidae	Long-nosed potoroo (*Potorous tridactylus*)	2 min
Macropodidae	Tammar wallaby (*Macropus eugenii*)	8 min
	Eastern grey kangaroo (*Macropus giganteus*)	50 min
	Red kangaroo (*Macropus rufus*)	15–20 min
	Red-necked wallaby (*Macropus rufogriseus*)	8 min
	Parma wallaby (*Macropus parma*)	5 min
Vombatidae	Southern hairy-nosed wombat (*Lasiorhinus latifrons*)	~30 min
Didelphidae	Grey short-tailed opossum (*Marmosa domestica*)	4–40 min
	Mouse opossum (*Marmosa robinsoni*)	> 6 h

Shimmin and Temple-Smith (1996a,b) have extensively examined the copulatory behaviour of male *Antechinus* under laboratory conditions. Although only a single male was permitted access to a female at any one time, changes in copulatory behaviour associated with order of mating, time after initial mating, and time relative to ovulation were examined in detail. It was found that the time at which males were given access to females within the oestrus period influenced the length of copulation (Shimmin and Temple-Smith 1996a), because significant differences were evident between males given the same length of access but at different times within the mating season (virgin male given second access after a 1-day delay or second access after a 3-day delay); those males mating closer to the time of ovulation consistently mated for less time than those mating earlier in oestrus (Fig. 15.8). It is not clear whether this reduction in copulation time is the result of satiation, males tiring earlier, females initiating the dismount, or other factors.

Bouts of thrusting activity that last between 10 and 100 seconds occur throughout copulation in *Antechinus stuartii* (Shimmin and Temple-Smith 1996a). It is not known, however, what proportion of these bouts results in sperm and/or other ejaculate component transfer, although, given the amount of thrusting activity over the total copulation time it is

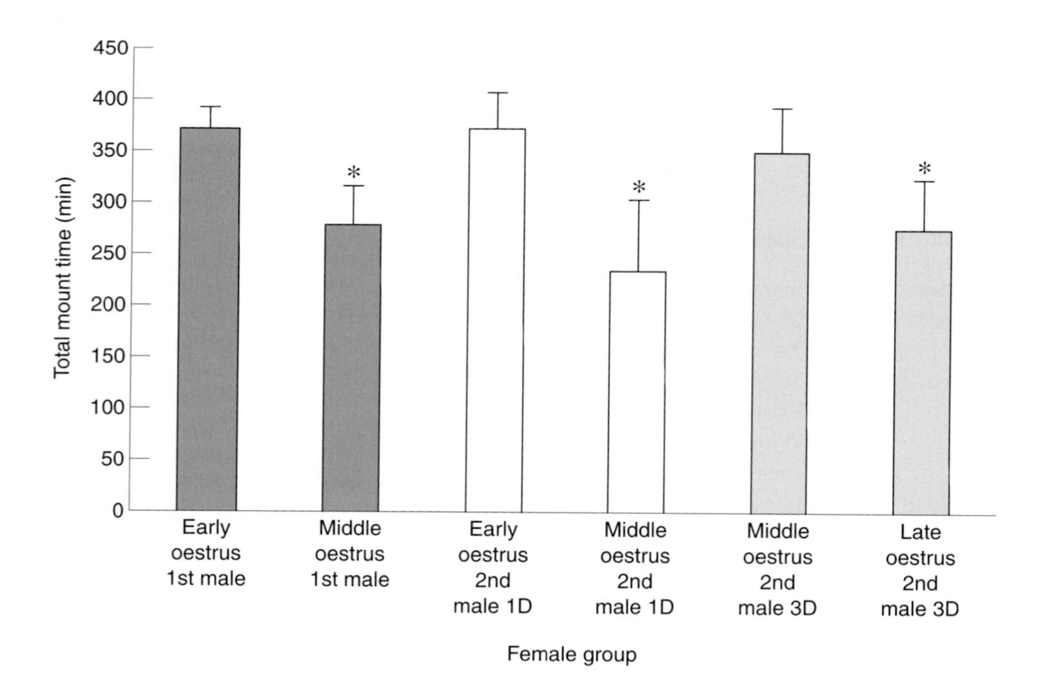

Fig. 15.8. *Effects of time of mating in relation to ovulation and mating order on total mount time in* Antechinus stuartii; *1st or 2nd male refers to the order of access to the oestrus female; 1D and 3D indicate whether there was 1 or 3 days delay between 1st and 2nd male mating;* * = $P < 0.001$.

possible that multiple ejaculations occur (Shimmin and Temple-Smith 1996a).

Among macropods the tammar wallaby *Macropus eugenii* has been the most extensively studied (Tyndale-Biscoe and Renfree 1987). Female tammars come into oestrus approximately 1.3 h after giving birth, whereas in the swamp wallaby *Wallabia bicolour* oestrus occurs 3 days post partum (Sharman *et al.* 1966) and in the red kangaroo 2 days post partum (Poole and Merchant 1987). In other macropod species it does not appear to be related to parturition (e.g. eastern grey kangaroo, western grey kangaroo, whip-tail wallaby *Macropus parryi* and Parma wallaby; Tyndale-Biscoe and Renfree 1987). In tammars (Rudd 1994), whip-tails (Kaufmann 1974) and red-necked wallabies *Macropus rufogriseus banksianus* (Johnson 1989) following birth and the initiation of oestrus, females are subsequently vigorously pursued by the males within the group (Kaufmann 1974; Rudd 1994). There is intense intermale aggression in these mating chases and, in the tammar at least (Rudd 1994), this results in delaying the time of the first successful ejaculation until 1.3 ± 0.8 h post partum (Rudd 1994). The first ejaculation is usually secured by the dominant (alpha) male, which is usually the largest male within the group (Kaufmann 1974; Rudd 1994). Bouts of thrusting activity similar to those described in *Antechinus* are also observed in the tammar; however, the tammar ejaculates after each bout, although the components of each ejaculate are not known (Rudd 1994).

I. Instance and duration of mate guarding in marsupials

No evidence of mate guarding has been found in laboratory studies of the brown marsupial mouse but it was evident that early in the mating the physical dominance of the male over the female determined the mating behaviour of the pair (Shimmin and Temple-Smith 1996a) although Shimmin and Temple-Smith (1996a,b) concluded that it was the female who had final control over when penile withdrawal occurred. Females appeared to be more receptive to the second male than the first, provided that there was a delay of more than 1 day between exposure to the first and second males (Shimmin and Temple-Smith 1996a).

In the tammar wallaby, the dominant male always copulates and ejaculates first (Jarman 1983), and subsequently guards the female from the advances of other males (by chasing, biting and kicking) for up to 8 h (Rudd 1994). In macropods, nonmating male tammars, red-necked wallabies and red kangaroos respond violently to mating by biting and kicking the copulating male until he releases his hold on the female (Sharman and Calaby 1964; Johnson 1989; Rudd 1994). After the dominant male has finished guarding the female some subordinate males may mate with her. In tammars and red-necked wallabies, the dominant male has considerable mating advantages over subordinate males in terms of timing, and is probably the most reproductively successful (Watson *et al.* 1992; Rudd 1994). As ovulation does not occur until 40 h post partum in the latter species it is likely that the copulatory plug deposited by the dominant male plays a significant role in ensuring a high rate of paternity success (Tyndale-Biscoe and Rodger 1978).

In the long-nosed bandicoot the male closely followed the female for several nights preceding copulation (Stodart 1966, 1977) and, although length of copulation was short (less than 30 s), the frequency was quite high with successive mounts occurring at intervals of several minutes. A peak in activity occurs about 2 h later when about 13 mounts with intromission follow in quick succession; this is followed by a steady waning in attraction. Whether multiple ejaculation occurs during this period has not been determined. A similar pattern of multiple, but brief, copulations has also been reported in members of the Potoroidae (Seebeck and Rose 1989).

In the grey short-tailed opossum, only a single intromission/ejaculation per male has been observed (Trupin and Fadem 1982; Taggart and Moore unpublished observations), whereas in *Marmosa robinsoni* and *Didelphis virginiana* there are multiple intromissions and/or ejaculations (Barnes and Barthold 1969; Dewsbury 1972; Hunsaker 1977; Trupin and Fadem 1982). Locking at the conclusion of mating immediately prior to dismount is also a feature of copulation in many didelphid opossums (Dewsbury 1972; Moore 1992; Moore and Taggart unpublished observations) and has also been observed in the yellow-footed (*Antechinus flavipes*) and brown marsupial mice (Marlow 1961; Woolley 1966; Shimmin unpublished observations).

J. Prevalence of mating plugs

Copulatory plugs have been observed in the urogenital sinus and/or lateral vaginae following mating and ejaculation in didelphid opossums (Hartman 1924; McCrady 1938; Taggart and Moore unpublished observations), macropods (Tyndale-Biscoe and Rodger 1978; Tyndale-Biscoe and Renfree 1987), phalangerids (Hughes and Rodger 1971), vombatids (Taggart *et al.* 1998), dasyurids (Breed 1994) and the numbat *Myrmecobius fasciatus* (Taggart and Friend unpublished observations). The copulatory plug is thought to result from the mixing of semen and vaginal secretions, however, coagulation can occur in the absence of female tract secretions in macropods. In *Macropus eugenii* the mating plug is devoid of spermatozoa soon after ejaculation (Hartman 1924; McCrady 1938; Tyndale-Biscoe and Rodger 1978). Typically, it appears as a pale creamy-coloured rubbery mass, and in macropods can often be seen protruding from the urogenital sinus for up to 24 h after mating (Hartman 1924; Tyndale-Biscoe and Rodger 1978; Rudd 1994). Copulatory plugs in marsupials may prevent leakage of spermatozoa, act to retain spermatozoa in the vaginae close to the cervical canal, ensuring maximal access for spermatozoa to the cervix, and/or perhaps act as a temporary physical barrier to subsequent matings by other males (Harper 1988; Taggart 1994). In the fat-tailed dunnart, the plug is thought to be largely, if not entirely, of prostatic origin and consists of an eosinophilic sperm-rich cranial portion, a middle portion almost completely devoid of spermatozoa except at its margins, and a caudal sperm-rich region that is periodic acid–Schiff reagent positive (Breed 1994).

K. *Sexual dimorphism in body weight*

For data on male and female body mass in various marsupial species see Appendix A of Lee and Cockburn (1985) and Lee *et al.* (1982). Within the Dasyuridae, the greatest male-biased sexual dimorphism in body mass occurs in the semelparous species (e.g. *Antechinus stuartii*, northern quoll; tiger quoll and Tasmanian devil) (Lee and Cockburn 1985; Lee *et al.* 1982). These species are either monoestrus, or if polyoestrus tend to reproduce synchronously and once yearly and so are facultative monoestrus (Lee and Cockburn 1985). Sexual dimorphism of body mass is minimal in the iteroparous dasyurid species which are polyoestrus and have either an extended seasonal breeding period or display reproduction throughout the year (e.g. fat-tailed dunnart; kultarr, *Antechinomys laniger*; *Antechinus melanurus*) (Lee and Cockburn 1985; Lee *et al.* 1982). Male body weight is also much greater than that of females in peramelids (e.g. northern brown bandicoot), phalangerids (e.g. common brushtail possum), acrobatids (feathertail glider *Acrobates pygmaeus;* Ward 1990a) and macropodids (e.g. tammar wallaby, red kangaroo). In contrast, a similarity in male and female body weight occurs in vombatids (e.g. northern and southern hairy-nosed wombats), phascolarctid (koala), burramyids (e.g. eastern and western pygmy possum; Ward 1990b), peturids (e.g. sugar glider, Leadbeater's possum) and pseudocheirids (e.g. lemuroid and green ringtail possums). Greater female body weight occurs in the honey possum *Tarsipes rostratus* (Renfree *et al.* 1984; Russell and Renfree 1989). Among the American opossums, greater male body weight occurs in the grey short-tailed opossum (Taggart and Moore unpublished observations), while this is not evident in *Philander opossum* and *Caluromys philander* (Atramentowicz 1982).

L. *Evidence of multiple paternity within litters*

The best evidence for sperm competition within marsupials comes from studies of captive colonies of two dasyurid species, the brown marsupial mouse (*Antechinus stuartii*; Shimmin and Temple-Smith 1996b) and the brush-tailed phascogale (*Phascogale tapoatafa*; Millis *et al.* 1995). Both studies examined paternity within litters associated with competitive mating trials between two males. These studies indicate that spermatozoa from more than one male can occupy the isthmic sperm storage crypts concurrently prior to ovulation, and also that multiple paternity occurs within the one litter. Fertility studies undertaken in the brown marsupial mouse (Selwood and McCallum 1987) suggested that spermatozoa from second and third inseminations can contribute spermatozoa for fertilization. In studies on the brown marsupial mouse (Shimmin and Temple-Smith 1996b), of the 61 young to which DNA paternity was assigned 72% were sired by the second-mating male when both matings occurred early in oestrus, 62% were sired by the second-mating male when one mating occurred early oestrus and one in mid-oestrus and 58% were sired by the second-mating male when both matings occurred in mid-oestrus. Overall, 64% of young were sired by the second mating male. Support for these findings also come from field studies of the brown marsupial mouse (Scott and Tan 1985) in which radionucleotide labels, individually recognizable by their spectral properties, were injected into males at the beginning of the breeding period. These labels passed to females during ejaculation were identified and counted following female capture to determine male mating success, and subsequently, demonstrated that males and females did indeed exhibit a promiscuous mating strategy in the wild.

M. Male dominance and paternity in macropodids

The relationship between male dominance and paternity has been examined in captive colonies of red-necked wallabies (*Macropus rufogriseus*; Watson *et al.* 1992) and tammar wallabies (*Macropus eugenii*; Ewen *et al.* 1993) using electrophoretic and DNA fingerprinting techniques. Within groups of red-necked wallabies the dominant male sired at least 70% of young surviving to the age of pouch emergence, with 30% or less surviving young being sired by subordinate males (Watson *et al.* 1992). In the tammar wallabies (Ewen *et al.* 1993), however, results from paternity analysis of a captive group showed that first mating and subsequent mate guarding by the dominant male did not significantly skew the outcome of paternity towards the dominant male. This result however, must be viewed with caution as only one group of animals was observed, females sometimes entered oestrus at the same time (Ewen *et al.* 1993), and mate-guarding ability/duration varies between males (Rudd 1994).

N. *Inferred mating systems from behavioural observations – do they fit the reproductive data?*

I. Dasyuridae

Comparisons of reproductive data and social organization for the semelparous dasyurids will be made only for *Antechinus stuartii* for the reasons given earlier (see text on relative testis mass in marsupials earlier in this section). In this species sexual maturity occurs at 11 months old and mating takes place within a short, highly synchronized, period each year (Lee and Cockburn 1985). As a consequence of spermatogenic failure prior to the mating season (Woolley 1966; Kerr and Hedger 1983), males must rely on stored epididymal spermatozoa for fertilizing females (Taggart and Temple-Smith 1989) and within a few weeks of mating, all males die (Lee *et al.* 1977; Bradley *et al.* 1980; Lee and Cockburn 1985). Although territoriality is weak, it is likely that the heaviest males occupy the most resource-rich habitat, mate first, and subsequently, are the first to die (Lee and Cockburn 1985). Monoestry and long periods of female receptivity (up to 2 weeks) have been reported for this species (Marlow 1961; Woolley 1966). Male and female brown antechinus are known to be promiscuous (Wittenberger 1979; Scott and Tan 1985; Lee and Cockburn 1985). Information presented on copulatory behaviour, sperm transport and isthmic sperm storage in females (Selwood 1980; Selwood and McCallum 1987; Taggart and Temple-Smith 1991) and paternity studies (Shimmin and Temple-Smith 1996b; Millis *et al.* 1995) support this conclusion. These data, together with the values presented here for relative testis mass and total sperm number, suggest that sperm competition between males is very likely to occur in this species and probably in other semelparous dasyurids.

Social organization in the iteroparous dasyurids, like *Sminthopsis crassicaudata*, contrasts with that of *Antechinus* in that males do not die after mating so that a proportion of males as well as females survive to breed in a second year. Populations are generally less dense than for the semelparous species and the social organization is loose by comparison with *Antechinus* with no long term bonds formed between individuals or between an individual and its range (Morton 1978a). Males do not defend a territory or home range (Morton 1978a). Females are polyoestrus, receptivity is relatively short (1–3 days), and sperm storage within the female tract, occurs for a significantly shorter time than for antechinus (Morton 1978b; Breed *et al.* 1989). As only a single adult male has been found with pro-oestrus or oestrus females, litters are probably sired by only a single male (Morton 1978a; Lee and Cockburn 1985). A mate defence polygyny has been described for *S. crassicaudata* whereby a male defends an oestrus female for several days during the period of receptivity (Wittenberger 1979; Morton 1978a; Lee and Cockburn 1985). Although, the absolute testis size does not vary markedly from that of the brown marsupial mouse the values for relative testes mass, epididymal sperm number and sperm tail length are generally lower than those reported for *Antechinus*. Sperm competition is therefore unlikely to occur in the iteroparous dasyurids.

2. Acrobatidae

Feathertail gliders *Acrobates pygmaeus* show a high degree of social tolerance especially during the breeding season. Although male–female pairs are found frequently, reproductively active females have also been found with several males during the one season and at the one time, even sharing nest boxes (Fleming and Frey 1984; Ward 1990a). Nest sites of feathertail gliders containing as many as 29 individuals have been reported during the breeding season (Fleming and Frey 1984; Ward 1990a). There is little evidence of prolonged associations between males and females (Ward 1990a). Females often produce two litters annually and both males and females usually survive for 2–3 years. On the basis of these life history characteristics and behavioural observations this species has been classified as promiscuous (Wittenberger 1979; Fleming and Frey 1984; Ward 1990a); this, together with data from the present study, which shows that *A. pygmaeus* has relatively large testes for body mass, suggests that sperm competition is likely to occur.

3. Burramyidae

In contrast to the Acrobatidae, the burramyids are, in general, solitary animals, especially during the breeding season, with only limited social interactions (Fleming and Frey 1984; Ward 1990b,c). Male eastern and western pygmy possums (*Cercartetus nanus* and *Cercartetus concinnus*, respectively) are reproductively active throughout the year (Clark 1967; Ward 1990b,c) with births occurring all year in some populations. Low levels of aggression have been reported between adult males, even in the presence of females (Ward 1990b,c). A broad overlap in home ranges has been reported for *C. nanus*. Females are behaviourally dominant in the mountain pygmy possum *Burramys parvus* and probably also in *C. nanus* (Kerle 1984; Mansergh 1984; Turner 1985). A promiscuous mating system has been suggested for species within this group since no prolonged association between the sexes is maintained (based on the definition of Wittenberger 1979) (Ward 1990b,c). Although only data on relative testes mass were available for three species within this group, all values were small for body mass. These data, coupled with an absence of sexual dimorphism in body mass and the relatively solitary nature of individuals, suggests that the likelihood of sperm competition between males in this group is low and that a type of female monogamy may occur.

4. Tarsipedidae

The honey possum *Tarsipes rostratus*, has very large spermatozoa (356 μm) and a relative testis size (testes = 4.12% body mass), which is one of the largest recorded for any mammal (Renfree *et al.* 1984; Cummins and Woodall 1985; Cummins *et al.* 1986; Russell and Renfree 1989). This species reproduces throughout the year; females are polyoestrus and have a post partum oestrus. Home ranges typically overlap and groups of females nesting together are common. A hierarchy is evident among females but overt aggression has not been observed, although a dominant female is very aggressive towards males (Russell and Renfree 1989). Little is known of the mating system but there is no evidence of a monogamous relationship (Russell and Renfree 1989) and a mating system in which several males compete for access to a receptive female is likely (Russell and Renfree 1989). Large relative testes mass and sperm tail length, together with sexual dimorphism in body mass and social structure would support this hypothesis and suggest that sperm competition between males probably occurs.

5. Petauridae

The social organization of most petaurids studied is very different from that of other marsupials (Smith 1980; Klettenheimer *et al.* 1996). Colonies of Leadbeater's possums *Gymnobelideus leadbeateri*, sugar gliders *Petaurus breviceps* and yellow-bellied gliders *Petaurus australis* are territorial, and individuals, within a colony, share a common nest (Smith 1980; Henry and Craig 1984; Russell 1984; Klettenheimer *et al.* 1996). Little, if any, overlap occurs in home ranges between adjacent groups (Henry and Craig 1984; Russell 1984; McKay 1989). DNA fingerprinting studies have recently indicated that, in the sugar glider, and probably Leadbeater's possum, dominant males within each group are closely related (Smith 1980; Klettenheimer *et al.* 1996). The basic group unit consists of a monogamous pair, plus one or more generations of offspring (Smith 1980; Klettenheimer *et al.* 1996) and, occasionally, additional unrelated adult males. Only one of the adult males in the colony is reproductively active (Smith 1980; Klettenheimer *et al.* 1996) with related males forming a coalition (Klettenheimer *et al.* 1996) and cooperating in territory defence, suppression of unrelated males (territorial, social and reproductive) and caring for young (Smith 1980; McKay 1989; Klettenheimer *et al.* 1996). On the basis of Wittenberger's (1979) classification of mating systems, the social organization of these animals suggests territorial monogamy (Lee and Cockburn 1985; McKay 1989). Values for relative testes mass and relative sperm tail length for *G. leadbeateri* and *P. breviceps* are small for body mass and suggest that intermale sperm competition is unlikely, which supports the conclusion derived from behavioural observations. In the striped possum *Dactylopsila trivirgata*, relative testes mass is also low for body mass suggesting that, like the other members of this group, intermale sperm competition is also probably minimal.

6. Peramelidae

The social organization of bandicoots is poorly known (Gordon and Hulbert 1989). Females are polyoestrus, fecundity is high (up to four litters being produced each year) and length of gestation and lactation are very short (12.5 days and 60 days, respectively). Home range overlap is common, with males occupying home ranges up to 10 times the size of those occupied by females (Gordon and Hulbert 1989). Most bandicoots are solitary and only come together to mate (Heinsohn 1966; Lee and Cockburn 1985). Older, larger male northern and southern brown bandicoots (*Isoodon obesulus*) dominate optimal habitat (Gordon 1974; Stoddart and Braithwaite 1979). Although the evidence is very patchy it has been claimed that the mating system of these species is probably promiscuous or polygynous (Lee and Cockburn 1985). Reproductive evidence, particularly the presence of sperm storage regions associated with the female reproductive tract and a strong male-biased sexual dimorphism in body mass, supports this claim. Data on relative testis mass indicates that *Isoodon macroura* has relatively large testes for body mass, which also supports this assumption and suggests that sperm competition may occur in this species.

7. Phascolarctidae

Adult male koalas (*Phascolarctos cinereus*) have larger home ranges than females. (Mitchell 1990a). Although home ranges of both sexes and of all ages overlap extensively, koalas are essentially solitary animals. Males occupying an area develop stable dominance hierarchies, with the dominant male moving most frequently and widely and having the highest frequencies of associations with conspecifics (Mitchell 1990a,b). Dominant males are particularly aggressive toward subordinate males in the presence of females and may limit the mating success of young males by attacking any subordinate male that attempts to mate (Mitchell 1990b). Young males disperse to avoid competition with the dominant male. A mating system comprising territorial polygyny has been proposed for this species where several females are paired with at least some of the territorial males (Emlen and Oring 1977; Wittenberger 1979; Lee and Cockburn 1985; Mitchell 1990b). Data on the social organization suggest that single-male breeding units occur, with the dominant male monopolizing matings within his home range. This is consistent with the relatively small testes and short sperm in this species, suggesting little, if any, inter-male sperm competition.

8. Vombatidae

The focus of the social organization of the hairy-nosed wombats *Lasiorhinus latifrons* (Wells 1973, 1978; Gaughwin 1981; Johnson and Crossman 1991) is the warren and for the common wombat *Vombatus ursinus* is the burrow (McIlroy 1973). A large warren may have 10 or more burrows and be inhabited by up to 10 wombats (Wells 1973; Gaughwin 1981; Johnson and Crossman 1991). Female southern hairy-nosed wombats show greater burrow preference than males. A dominance hierarchy exists among males within a warren (Gaughwin 1981). Inter-warren territoriality is strong between male *L. latifrons* (Gaughwin 1981) and has also been implicated in the social organization of *V. ursinus* (McIlroy 1973). Warrens are connected by a network of trails and territoriality is maintained by olfactory cues, fighting and chasing (Wells 1973; McIlroy 1973; Gaughwin 1981; Johnson and Crossman 1991). Female southern hairy-nosed wombats are polyoestrus and breeding appears to relate closely to rainfall and growth of pasture (Wells 1989). The mating system employed by species within this group has not been alluded to previously. Relative testes mass, sperm number and sperm tail length for body mass are low and these observations, together with the limited behavioural observations and lack of sexual dimorphism in body mass suggest that sperm competition between males is unlikely. Perhaps in the three wombat species males are polygynous, whereas females only mate with one male at any one oestrus.

9. Phalangeridae

The social organization of the phalangerids is probably best known for the common and mountain brushtail possums (*Trichosurus vulpecula* and *Trichosurus caninus*, respectively; Dunnet 1964; Winter 1977; How 1978, 1981). Home ranges of male common brushtail possums are virtually exclusive of one another but each overlaps with the home ranges of several females which, in turn, overlap extensively (Dunnet 1964). Social status of males increases with age and size, with dominant males occupying stable home ranges with core areas from which all other established males are excluded (Winter 1977). Dominant males repel advances by other mature males to resident females within their home ranges (Winter 1977). However, some females have no particularly strong association with any male, will mate with several males and are defended by none of the mating males (Winter 1977). It has been suggested that the predominant mating system for the common brushtail possum is serial polygyny, where males mate with a number of females within each season (Winter 1977; Wittenberger 1979; Lee and Cockburn 1985). Relative testes mass and relative epididymal sperm number are large for the body mass of this species, whereas relative sperm tail length is small. Sexual dimorphism in body mass strongly favours males in this species. Overall, the data suggest that intermale sperm competition is likely, which seems particularly relevant for the cohort of females that are not actively defended by any male following mating.

10. Macropodidae

Within the macropods social organization varies widely. The swamp wallaby *Wallabia bicolour*, yellow-footed rock wallaby *Petrogale xanthopus* and some of the tree kangaroo species (*Dendrolagus lumholtzi, Dendrolagus bennettianus*) have either a solitary existence (except at mating) or they tend toward unimale groups. Individuals use closed shelter, are nocturnal and are strongly territorial (Edwards and Ealey 1974; Crebbin 1982; Proctor-Gray and Ganslosser 1986; Croft 1989; Martin 1995). Other species, such as the tammar *Macropus eugenii*, red-necked wallaby *Macropus rufogriseus* and red kangaroo *Macropus rufus* are either sometimes solitary but aggregate on favoured resource patches, or are gregarious (Taylor 1983; Johnson 1983, 1987; Croft 1989). These species are also usually partially diurnally active and often occupy more open habitats, feeding predominantly on grass; home ranges of both sexes overlap and territorial behaviour is absent (Jarman 1983; Croft 1989). In general, male macropods have either a similar size or, more usually, a larger home range than that of females (Croft 1989). In species such as the tammar, red-necked wallaby and red kangaroo the size of male home range tends to increase with an increase in male size rank, with high-ranked males centring their range on the densest populations of females, and having high levels of mating success (Batchelor 1980; Croft 1981a, 1989; Russell 1984; Johnson 1985). Competition among male macropods for oestrus females is greatest in species that breed seasonally (e.g. tammar and red-necked wallaby), suggesting that for continuous breeders, such as the red kangaroo, oestrus females are difficult to defend for any extended period. The mating system in both these groups of macropods is therefore likely to be promiscuous (Croft 1981a, 1989; Russell 1984).

The relative testis size of the several *Macropus* species studied are either average or large for body mass and the relative number of spermatozoa in the epididymis are large for the body mass of the two grey kangaroo species. In contrast, the relative testes mass and epididymal sperm number for the swamp wallaby are low for body mass. The data from the *Macropus* species therefore suggest a multimale breeding system with the possibility of sperm competition, the likelihood of which appears greatest for the seasonally breeding species. Behavioural observations support this conclusion (Croft 1981a, 1989; Russell 1984; Johnson 1989; Watson *et al.* 1992; Rudd 1994). In the swamp wallaby (and probably also some of the tree kangaroos; T. Flannery and R. Martin, pers. comm.), data on social organization suggest that the likelihood of sperm competition is low. Data for relative testes mass and epididymal sperm number in general support the inferred mating strategies from behavioural observation of species within this group.

11. Didelphidae

There is a general paucity of data available on the social structure of species in the Didelphidae. The species for which most information is available are the Virginia and grey short-tailed opossums (*Didelphis virginiana* and *Monodelphis domestica*, respectively). Most species studied are seasonal breeders, polyoestrus and polytocous although, the grey short-tailed opossum may breed for most of the year in Brazil (Streilein 1982; Fadem and Rayve 1985; Baggott *et al.* 1987; Perret and M'Barek 1991). In captivity, at least, female grey short-tailed opossums will mate with 2–3 males during the one oestrus (Taggart and Moore unpublished observations). Relative testes mass and sperm tail length of the two didelphid species examined are average for body mass, as is relative epididymal sperm number in *Monodelphis*. Prediction of whether intermale sperm competition occurs in these species is made difficult owing to a number of confounding factors, in addition to the lack of data on social organization. For example, epididymal sperm pairing, a characteristic of all didelphids, directly affects sperm motility (Biggers and Creed 1962; Biggers and De Lamater 1965; Phillips 1972; Temple-Smith and Bedford 1980; Taggart *et al.* 1993a,b; Moore and Taggart 1995), sperm transport in the female tract is highly efficient (Bedford *et al.* 1984), and, as indicated earlier, sperm storage occurs in the lower oviduct (Bedford *et al.* 1984). In *Monodelphis*, when all these factors are taken into account, in combination with copulatory behaviour and a strongly male-orientated sexual dimorphism in body mass, the occurrence of intermale sperm competition appears to be likely.

IV. CONCLUDING COMMENTS

The present analyses have shown that, as with eutherian mammals (Harcourt *et al.* 1981; Kenagy and Trombulak 1986) there is, in marsupials, a clear positive correlation between body mass and relative testis mass and in the number of sperm in the epididymides. Also, as in eutherian mammals (Cummins and Woodall 1985), a negative correlation between body mass and sperm size, as indicated by the length of the sperm tail, occurs. When data for individual species are examined it is clear that for species within the same genus and often family the data shows similar trends suggesting a phylogenetic influence, and that, for some groups there is a marked deviation from the average for body mass. Data on social organization and information on breeding systems, copulatory behaviour and paternity guards are relatively sparse but, in some cases, the occurrence of a multimale breeding system is evident, with the potential for intermale sperm competition. These species tend to have large relative testis size and high sperm numbers, thus the presence or absence of intermale sperm competition is likely to be one of the factors responsible for the variation in relative mass of the testis in marsupial species. We present these data as a challenge to test the hypothesis that sperm competition is one of the driving forces that influence testis size and sperm numbers in this group of animals. Whether this is also one of the factors that has brought about the differences in sperm size in marsupials is less convincing. Clearly, other factors need also to be considered.

ACKNOWLEDGEMENTS

The authors thanks Ms Sue Simpson for preparation of diagrams. The regression analyses of species data in this review were funded on ARC grant no. 9531986 to W.G.B and carried out by Matthew Breed. D.A.T. was employed on ARC grant no. A09330847 to Dr Peter Temple-Smith. A.P. is funded by the Royal Society. We also thank Professor R. V. Short for constructive criticism of this chapter.

REFERENCES

Arnold R & Shorey C (1985) Structure of the oviducal epithelium of the brush-tailed possum (*Trichosurus vulpecula*). *J. Reprod. Fertil.* **73:** 9–19.

Atramentowicz M (1982) Influence du milieu sur l'activite' locomotrice et la reproduction de *Caluromys philander* (L). *Rev. Ecol.* (*Terre et Vie*) **36:** 373–395.

Augee ML (1978) Monotremes and the evolution of homeothermy. *Aust. Zool.* **20:** 111–119.

Augee ML, Ealey EH & Price IP (1975) Movements of echidnas, *Tachyglossus aculeatus*, determined by marking recapture and radio-tracking. *Aust. Wildl. Res.* **2:** 93–101.

Baggott L, Davis-Butler S and Moore HDM (1987) Characterization of oestrus and timed collection of oocytes in the grey short-tailed opossum, *Monodelphis domestica*. *J. Reprod. Fertil.* **79:** 105–114.

Bakst MR, Wishart GJ & Brillard JP (1994) Oviductal sperm selection, transport and storage in poultry. *Poultry Sci. Rev.* **5:** 117–143.

Barnes RD & Barthold SW (1969) Reproduction and breeding behaviour in an experimental colony of *Marmosa mitis bangs* (Didelphidae). *J. Reprod. Fertil. Suppl.* **6:** 477–482.

Batchelor TA (1980) The social organization of the brush-tailed rock wallaby (*Petrogale penicillata penicillata*) on Motutapu Island. M.Sc. thesis, University of Auckland, New Zealand.

Baverstock PR, Archer M, Adams M & Richardson BJ (1982) Genetic relationships among 32 species of Australian dasyurid marsupials. In *Carnivorous Marsupials.* M Archer (ed.), pp. 641–650. Royal Zoological Society of New South Wales, Sydney.

Baverstock PR, Krieg M & Birrell J (1990) Evolutionary relationships of Australian marsupials as assessed by albumin immunology. *Aust. J. Zool.* **37:** 273–287.

Beard LA, Grigg GC & Augee ML (1992) Reproduction by echidnas in a cold climate. In *Platypus and Echidnas.* ML Augee (ed.), pp. 93–100. Royal Society of New South Wales, Mosman.

Bedford JM, Rodger JC & Breed WG (1984) Why so many mammalian spermatozoa – a clue from marsupials? *Proc. Roy. Soc. Lond.* **221:** 221–233.

Bedford JM & Breed WG (1994) Regulated storage and subsequent transformation of spermatozoa in the fallopian tubes of an Australian marsupial, *Sminthopsis crassicaudata. Biol. Reprod.* **50:** 845–854.

Biggers JD & Creed RF (1962) Conjugate spermatozoa of the North American opossum. *Nature* **196:** 1112–1113.

Biggers JD & De Lamater ED (1965) Marsupial spermatozoa pairing in the epididymis of American forms. *Nature* **208:** 402–404.

Birkhead TR & Møller AP (1992) Numbers and size of sperm storage tubules and the duration of sperm storage in birds: a comparative study. *Biol. J. Linn. Soc.* **45:** 363–372.

Birkhead TR & Møller AP (1993) Sexual selection and the temporal separation of reproductive events: sperm storage data from reptiles, birds and mammals. *Biol. J. Linn. Soc.* **50:** 295–311.

Bishop MWH (1970) Aging and reproduction in the male. *J. Reprod. Fertil. Suppl.* **12:** 65–88.

Bradley, AJ, McDonald IR & Lee AK (1980) Stress and mortality in a small marsupial (*Antechinus stuartii* Macleay). *Gen. Comp. Endocrinol.* **40:** 188–200.

Breed WG (1994) How does sperm meet egg? – In a marsupial. *Reprod. Fertil. Devel.* **6:** 485–506.

Breed WG, Leigh CM, and Bennett JH (1989) Sperm morphology and storage in the female reproductive tract of the fat-tailed dunnart, *Sminthopsis crassicaudata* (Marsupialia: Dasyuridae). *Gamete Res.* **23:** 61–75.

Broom R (1895) Notes on the period of gestation in echidna. *Proc. Linn. Soc. NSW* **10:** 576–577.

Carrick FN & Hughes RL (1982) Aspects of the structure and development of monotreme spermatozoa and their relevance to the evolution of mammalian sperm morphology. *Cell Tiss. Res.* **222:** 127–141.

Clark MJ (1967) Pregnancy in the lactating pigmy possum *Cercartetus concinnus. Aust. J. Zool.* **15:** 673–687.

Crebbin A (1982) Social organization and behaviour of the swamp wallaby, *Wallabia bicolor* (Desmarest) (Marsupialia: Macropodidae) in captivity. B.Sc. Honours thesis, University of New South Wales, Sydney, Australia.

Croft DB (1981a) Behaviour of red kangaroos, *Macropus rufus* (Desmarest, 1822), in north western New South Wales. *Aust. Mammal.* **4:** 5–58.

Croft DB (1981b) Society behaviour of the euro, *Macropus robustus* (Gould), in the Australian arid zone. *Aust. Wildl. Res.* **8:** 13–49.

Croft DB (1989) Social organization of the Macropodoidea. In *Kangaroos, Wallabies and Rat Kangaroos.* G Grigg, P Jarman & ID Hume (eds), pp. 505–525. Surrey-Beatty and Sons Ltd, Sydney.

Cummins JM & Woodall PF (1985) On mammalian sperm dimensions. *J. Reprod. Fertil.* **75:** 153–175.

Cummins JM, Temple-Smith PD & Renfree MB (1986) Reproduction in the male honey possum (*Tarsipes rostratus*: Marsupialia): the epididymis. *Am. J. Anat.* **177:** 385–401.

Dewsbury DA (1972) Patterns of copulatory behaviour in mammals. *Quart. Rev. Biol.* **47:** 1–33.

Dickman CR (1985) Effects of photoperiod and endogenous control on timing of reproduction in the marsupial genus *Antechinus. J. Zool.* **206A**, 509–524.

Dickman CR (1993) Evolution of semelparity in male dasyurid marsupials: a critique and an hypothesis of sperm competition. In *The Biology and Management of Australasian Carnivorous Marsupials.* M Roberts, J Carnio, G Crawshaw & M Hutchins (eds), pp. 25–38. American Association of Zoological Parks and Aquariums Publishers, Toronto.

Dunnet GM (1964) A field study of local populations of the brushtail possum, *Trichosurus vulpecula* in eastern Australia. *Proc. Zool. Soc. Lond.* **142:** 665–695.

Edwards GP & Ealey EMH (1974) Aspects of the ecology of the swamp wallaby, *Wallabia bicolor* (Marsupialia: Macropodidae). *Aust. Mammal.* **1:** 307–317.

Emlen ST & Oring LW (1977) Ecology, sexual selection and evolution of mating systems. *Science* **197:** 215–223.

Ewen KR, Temple-Smith PD, Bowden DK, Marinopoulos J, Renfree MB & Yan H (1993) DNA fingerprinting in relation to male dominance and paternity in a captive colony of tammar wallabies (*Macropus eugenii*). *J. Reprod. Fertil.* **99:** 33–37.

Fadem BH & Rayve R (1985) Characteristics of the oestrus cycle and influence of social factors in grey short-tailed opossums (*Monodelphis domestica*). *J. Reprod. Fertil.* **73:** 337–342.

Fanning FD (1982) Reproduction, growth and development in *Ningaui* sp. (Dasyuridae, Marsupialia) from the Northern Territory. In *Carnivorous marsupials.* M Archer (ed.), pp. 23–37. Royal Zoological Society of NSW, Sydney.

Felsenstein J (1985) Phylogenies and the comparative method. *Am. Nat.* **125:** 1–15.

Flannery T (1995) Mammals of New Guinea. Australian Museum Reed Books, Sydney. pp. 66–73.

Fleay D (1980) *Paradoxical Platypus: Hobnobbing with Duckbills.* Jacaranda Press, Milton.

Fleming MR & Frey H (1984) Aspects of the natural history of feathertail gliders (*Acrobates pygmaecus*) in Victoria. In *Possums and Gliders.* AP Smith & ID Hume (eds), pp. 403–408. Australian Mammal Society, Sydney.

Flyn TT & Hill JP (1939) The development of the Monotremata. Part 4: Growth of the ovarian ovum, maturation, fertilization and early cleavage. *Trans. Roy. Soc. Lond.* **24:** 445–622.

Garland T, Harvey PH & Ives AR (1992) Procedures for the analysis of comparative data using phylogenetically independent contrasts. *Syst. Biol.* **41:** 18–32.

Gaughwin MD (1981) Socio-ecology of the hairy-nosed wombat (*Lasiorhinus latifrons*) in the Blanche Town Region of South Australia. Ph.D. thesis, University of Adelaide, Australia.

Geiser F & Seymour RS (1989) Torpor in a pregnant echidna, *Tachyglossus aculeatus* (onotremata: Tachyglossidae). *Aust. Mammal.* **12:** 81–82.

Gist DH & Jones JM (1987) Storage of sperm in the reptilian oviduct. *Scann. Microsc.* **1:** 1839–1849.

Gomendio M & Roldan ER (1991) Sperm competition influences sperm size in mammals. *Proc. Roy. Soc. Lond.* **243:** 181–185.

Gordon G (1974) Movements and activity of the short-nosed bandicoot *Isoodon macrourus* Gould (Marsupialia). *Mammalia.* **38:** 40–431.

Gordon G & Hulbert AJ (1989) Peramelidae. In *Fauna of Australia.* DW Walton & BJ Richardson (eds), pp. 603–624. Australian Government Publ. Service, Canberra.

Grant T (1995) *Platypus: A Unique Mammal.* New South Wales University Press, Sydney.

Griffith M (1968) *Echidnas.* Pergamon Press, Oxford.

Griffith M (1978) *The Biology of Monotremes.* Academic Press, New York.

Griffith M (1984) Mammals: Monotremes. In *Marshalls Physiology of Reproduction,* Vol. 1. GE Lamming (ed.), pp. 351–385. Churchhill Livingstone, Edinburgh.

Harcourt AH (1991) Sperm competition and the evolution of nonfertilizing sperm in mammals. *Evolution* **45:** 314–328.

Harcourt AH, Harvey PH, Larson SG & Short RV (1981) Testis weight, body weight and breeding system in primates. *Nature* **293:** 55–57.

Harper MJK (1988) Gamete and zygote transport. In *The Physiology of Reproduction.* E Knobil & J Neill (eds), Ch 4, pp. 103–134. Raven Press, New York.

Hartman CG (1924) Observations on the motility of the opossum genital tract and the vaginal plug. *Anat. Rec.* **27:** 293–303.

Harvey PH & Harcourt AH (1984) Sperm competition, testes size and breeding systems in primates. In *Sperm Competition and the Evolution of Animal Mating Systems.* RL Smith (ed.), pp. 589–600. Academic Press, Orlando.

Harvey PH & Pagel MD (1991) *The Comparative Method of Evolutionary Biology.* Oxford University Press, Oxford.

Hawkins M & Fanning D (1992) Courtship and mating behaviour of captive platypuses at Taronga Zoo. In *Platypus and Echidnas.* ML Augee (ed.), pp. 106–114. Royal Society of New South Wales, Mosman.

Heinsohn GE (1966) Ecology and reproduction of the Tasmanian bandicoots (*Perameles gunnii* and *Isoodon obesulus*). *Univ. Calif. Publ. Zool.* **80:** 1–95.

Henry SR & Craig SA (1984) Diet, ranging behaviour and social organization of the yellow bellied glider (*Petaurus australis* Shaw) in Victoria. In *Possums and Gliders.* AP Smith & ID Hume (eds), pp. 331–341. Surrey Beatty and Sons with the Australian Mammal Society, Sydney.

Home E (1802) A description of the anatomy of the *Ornithorhynchus paradoxus. Phil. Trans. Roy. Soc. Lond. 1802*, pp. 67–84.

How RA (1978) Population strategies of four species of Australian possum. In *The Ecology of Arboreal Folivores.* GG Montgomery (ed.), pp. 305–313. Smithsonian Institute Press, Washington, DC.

How RA (1981) Population parameters of two congeneric possums, *Trichosurus* spp. in north-eastern New South Wales. *Aust. J. Zool.* **29:** 205–215.

Hughes RL & Rodger JC (1971) Studies on the vaginal mucus of the marsupial, *Trichosurus vulpecula. Aust. J. Zool.* **19:** 19–33.

Hughes RL & Carrick FN (1978) Reproduction in female monotremes. In *Monotreme Biology*. ML Augee (ed.), pp. 233–253. Royal Zoological Society of New South Wales, Sydney.

Hunsaker D (1977) Ecology of New World marsupials. In *The Biology of Marsupials*. Academic Press, New York.

Hunter RHF (1988) Transport of gametes, selection of spermatozoa and gamete lifespans in the female tract. In *The Fallopian Tubes: Their Role in Fertility and Infertility*. RH Hunter (ed.), pp. 53–74. Springer-Verlag, London.

Jarman P (1983) Mating system and sexual dimorphism in large terrestrial, mammalian herbivores. *Biol. Rev.* **58:** 485–520.

Johnson CN (1983) Variation in group size and composition in red and western grey kangaroos, *Macropus rufus* (Desmarest) and *Mfuliginosus* (Desmarest). *Aust. Wildl. Res.* **10:** 25–31.

Johnson CN (1985) Ecology, social behaviour and reproductive success in a population of red necked wallabies. Ph.D. thesis, University of New England, Armidale, Australia.

Johnson CN (1987) Macropod studies at Wallaby Creek. IV. Home range and movements of the red necked wallaby. *Aust. Wildl. Res.* **14:** 125–132.

Johnson CN & Crossman DG (1991) Dispersal and social organization of the northern hairy-nosed wombat *Lasiorhinus krefftii*. *J. Zool.* **225:** 605–613.

Johnson KA (1989) Thylacomyidae. In *Fauna of Australia, Vol. 1b. Mammalia*. DW Walton & BJ Richardson (eds), pp. 625–635. Australian Government Publishing Service, Canberra.

Kaufmann JH (1974) Social ethology of the whiptail wallaby, *Macropus parryi*, in north eastern NSW. *Anim. Behav.* **22:** 281–369.

Kenagy GJ & Trombulak SC (1986) Size and function of mammalian testes in relation to body size. *J. Mammal.* **67:** 1–22.

Kerle JA (1984) Growth and development of *Burramys parvus* in captivity. In *Possums and Gliders*. AP Smith & ID Hume (eds), pp. 409–412. Australian Mammal Society, Sydney.

Kerr JB & Hedger MP (1983) Spontaneous spermatogenic failure in the marsupial mouse *Antechinus stuartii* MacLeay (Dasyuridae: Marsupialia). *Aust. J. Zool.* **31:** 445–466.

Kirsch JA, Lapointe FJ & Springer MS (1997) DNA hybridization studies of marsupials and their implications for metatherian classification. *Aust. J. Zool.* **45:** 211–280.

Klettenheimer BS, Temple-Smith PD & Sofronidis G (1996) Father and son sugar gliders: more than a genetic coalition. *J. Zool.* **242:** 741–750.

Lee AK & Cockburn A (1985) *Evolutionary Ecology of Marsupials*. Cambridge University Press, Cambridge.

Lee AK, Bradley AJ & Braithwaite RW (1977) Corticosteroid levels and male mortality in *Antechinus stuartii*. In *The Biology of Marsupials*. B Stonehouse & D Gilmore (eds), pp. 209–220. MacMillan, London.

Lee AK, Woolley P & Braithwaite RW (1982) Life history strategies of dasyurid marsupials. In *Carnivorous Marsupials*. M Archer (ed.), pp. 1–11. Royal Zoological Society of NSW, Sydney.

Lyne AG & Hollis DE (1977) The early development of marsupials with special reference to bandicoots. In *Reproduction and Evolution*. JH Calaby & CH Tyndale-Biscoe (eds), pp. 293–302. Australian Academy of Science, Canberra.

Maddison WP & Maddison DR (1992) *MacClade: Analysis of Phylogeny and Character Evolution. Version 3.01*. Sinauer, Sunderland, MA.

Mansergh IM (1984) *Burramys parvus* (Broom), a short review of the current state of knowledge concerning the species. In *Possums and Gliders.* AP Smith & ID Hume (eds), pp. 413–416. Australian Mammal Society, Sydney.

Marlow BJ (1961) Reproductive behaviour of the marsupial mouse, *Antechinus flavipes* (Waterhouse) (Marsupialia) and the development of pouch young. *Aust. J. Zool.* **9:** 203–218.

Martin R (1995) Ecological studies of Bennett's tree-kangaroo (*Dendrolagus bennettianus*) with emphasis on processes affecting distribution and abundance. Annual General Meeting James Cook University, Townsville. *Proc. Aust. Mamm. Soc.*, pp. 45.

McCrady E (1938) The embryology of the opossum. *Am. Anat. Mem.* **16:** 1–233.

McIlroy JC (1973) Aspects of the ecology of the common wombat, *Vombatus ursinus* (Shaw, 1800). Ph.D. thesis, Australian National University, Canberra, Australia.

McKay GM (1989) Family Peturidae. In *Fauna of Australia, Vol. 1b Mammalia.* DW Walton & BJ Richardson (eds), pp. 665–678. Australian Government Publishing Service, Canberra.

McManus JJ (1970) Behaviour of captive opossums, *Didelphis marsupialis virginiana. Am. Midl. Natur.* **84:** 144–169.

Millis A, Taggart DA, Phelan J & Temple-Smith PD (1995) Investigation of multiple paternity in two species of dasyurid marsupial, *Phascogale tapoatafa* and *Dasyurus hallucatus. Proc. Australian Mammal Soc. Meeting 1995*, Townsville, Queensland, pp. 1.

Mitchell P (1990a) The home ranges and social activity of koalas – a quantitative analysis. In *Biology of the Koala.* AK Lee, KA Handasyde & GD Sanson (eds), pp. 171–187. Surrey Beatty and Sons, Sydney.

Mitchell P (1990b) Social behaviour and communication of koalas. In *Biology of the Koala.* AK Lee, KA Handasyde & GD Sanson (eds), pp. 152–170. Surrey Beatty and Sons, Sydney.

Møller AP (1988) Ejaculate quality, testis size and sperm competition in primates. *J. Hum. Evol.* **17:** 479–488.

Møller AP (1989) Ejaculate quality, testes size and sperm production in mammals. *Funct. Ecol.* **3:** 91–96.

Møller AP & Briskie JV (1995) Extra-pair paternity, sperm competition and the evolution of testis size in birds. *Behav. Ecol. Sociobiol.* **36:** 357–365.

Moore HDM (1992) Reproduction in the grey short-tailed opossum, *Monodelphis domestica.* In *Reproductive Biology of South American Vertebrates.* WC Hamlett (ed.), pp. 229–241. Springer-Verlag, New York.

Moore HDM & Taggart DA (1995) Sperm pairing in the opossum increases the efficiency of sperm movement in a viscous environment. *Biol. Reprod.* **52:** 947–953.

Morton SR (1978a) An ecological study of *Sminthopsis crassicaudata* (Marsupialia: Dasyuridae) II. Behaviour and social organization. *Aust. Wildl. Res.* **5:** 163–182.

Morton SR (1978b) An ecological study of *Sminthopsis crassicaudata* (Marsupialia: Dasyuridae) III. Reproduction and life-history. *Aust. Wildl. Res.* **5:** 183–211.

Oudemans JT (1892) Die accessorishen Geschlechtsdrusen der Saugetthiere – Monotremata. *Nat. Verh. vd. Holl. Maatsh. dWetensch. Haarlem 3 Verz. Deel* **5**(2): 11–14.

Overstreet JW & Cooper GW (1978) Sperm transport in the reproductive tract of the female rabbit 1. The rapid transit phase of transport. *Biol. Reprod.* **19:** 101–114.

Overstreet JW & Cooper GW (1979) Effect of ovulation and sperm motility on the migration of rabbit spermatozoa to the site of fertilization. *J. Reprod. Fertil.* **55:** 53–59.

Pearson J (1945) The female urogenital system of the Marsupialia with special reference to the vaginal complex. *Pap. Proc. Roy. Soc. Tasm.* **1944:** 71–98.

Perret M & M'Barek SB (1991) Male influence on oestrus cycles in female woolly opossum (*Caluromys philander*). *J. Reprod. Fertil.* **91:** 557–566.

Phillips DM (1972) Comparative analysis of mammalian sperm motility. *J. Cell Biol.* **53:** 561–573.

Poole WE & Merchant JC (1987) Reproduction in captive wallaroos: the eastern wallaroo, *Macropus robustus robustus*, the euro, *M. robustus erubescens* and the antilopine wallaroo, *M. antilopinus*. *Aust. Wildl. Res.* **14:** 225–242.

Proctor-Gray E & Ganslosser U (1986) The individual behaviour of Lumholtz's tree kangaroo: repertoire and taxonomic implications. *J. Mammal.* **67:** 343–352.

Purvis A & Garland T (1993) Polytomies in comparative analyses of continuous characters. *System. Biol.* **42:** 569–575.

Purvis A & Harvey PH (1995) Mammal life history: a comparative test of Charnov's model. *J. Zool.* **237:** 259–283.

Purvis A & Rambaut A (1995) Comparative analysis by independent contrasts (CAIC): an apple Macintosh application for analysing comparative data. *Comp. Appl. BioSci.* **11:** 247–251.

Racey PA (1979) The prolonged storage and survival of spermatozoa in Chiroptera. *J. Reprod. Fertil.* **56:** 391–402.

Racey PA & Potts DM (1970) Relationship between stored spermatozoa and the uterine epithelium in the pipestrelle bat (*Pipistrellus pipistrellus*). *J. Reprod. Fertil.* **22:** 57–63.

Racey PA, Uchida TA, Mori T, Avery MI & Fenton MB (1987) Sperm–epithelium relationships in relation to time of insemination in little brown bats (*Myotis lucifugus*). *J. Reprod. Fertil.* **80:** 445–454.

Read DG (1984) Reproduction and breeding season of *Planigale gilesi* and *P. tenuirostris* (Marsupialia: Dasyuridae). *Aust. Mammal.* **7:** 161–173.

Renfree MB, Russell EM & Wooller RD (1984) Reproduction and life history of the honey possum, *Tarsipes rostratus*. In *Possums and Gliders*. AP Smith & ID Hume (eds), pp. 427–437. Australian Mammal Society, Sydney.

Richardson BJ & McDermid EM (1978) A comparison of genetic relationships within the Macropodidae as determined from allozyme, cytological and immunological data. *Aust. Mammal.* **2:** 43–51.

Ridley M (1989) The incidence of sperm displacement in insects: four conjectures, one corroboration. *Biol. J. Linn. Soc.* **38:** 349–367.

Rismiller PD (1992) Field observations on Kangaroo Island echidnas (*Tachyglossus aculeatus multiaculeatus*) during the breeding season. In *Platypus & echidnas*. ML Augee (ed.), pp. 101–105. Royal Society of New South Wales, Mosman.

Rismiller PD (1993) Overcoming a prickly problem. *Aust. Nat. Hist.* **24:** 22–29.

Rodger JC (1991) Fertilization in marsupials. In *A Comparative Overview of Mammalian Fertilization*. BS Dunbar & MG O'Rand (eds), pp. 117–135. Plenum Press, New York.

Rodger JC & Bedford JM (1982a) Induction of oestrus, recovery of gametes and timing of fertilization events in the opossum, *Didelphis virginiana*. *J. Reprod. Fertil.* **64:** 159–169.

Rodger JC & Bedford JM (1982b) Separation of sperm pairs and sperm–egg interaction in the opossum, *Didelphis virginiana*. *J. Reprod. Fertil.* **64:** 171–179.

Rodger JC & Hughes RL (1973) Studies of accessory glands of male marsupials. *Aust. J. Zool.* **21:** 303–320.

Rudd CD (1994) Sexual behaviour of male and female tammar wallabies (*Macropus eugenii*) at post-partum oestrus. *J. Zool.* **232:** 151–162.

Russell R (1984) Social behaviour of the yellow bellied glider, *Petaurus australis reginae* in north Queensland. In *Possums* and *Gliders*. AP Smith & ID Hume (eds), pp. 343–353. Surrey Beatty and Sons/Australian Mammal Society, Sydney.

Russell E & Renfree MB (1989) Tarsipidae. In *Fauna of Australia, Vol. 1b, Mammalia*. DW Walton & BJ Richardson (eds), pp. 769–782. Australian Government Publishing Service, Canberra.

Scott MP & Tan TN (1985) A radiotracer technique for the determination of male mating success in natural populations. *Behav. Ecol. Sociobiol.* **17:** 29–33.

Seebeck JH & Rose RW (1989) Potoroidae. In *Fauna of Australia, Vol. 1b Mammalia*. DW Walton & BJ Richardson (eds), pp. 716–739. Australian Government Publishing Service, Canberra.

Selwood L (1980) A timetable for embryonic development of the dasyurid marsupial, *Antechinus stuartii* (Macleay). *Aust. J. Zool.* **28:** 649–668.

Selwood L & McCallum F (1987) Relationship between longevity of spermatozoa after insemination and the percentage of normal embryos in brown marsupial mice (*Antechinus stuartii*). *J. Reprod. Fertil.* **79:** 495–503.

Serena M (1994) Use of time and space by platypus (*Ornithorhynchus anatinus: Monotremata*) along a Victorian stream. *J. Zool.* **232:** 117–131.

Sharman GB & Calaby JH (1964) Reproductive behaviour in the red kangaroo, *Megaleia rufa*, in captivity. *CSIRO Wildl. Res.* **9:** 58–85.

Sharman GB, Calaby JH & Poole WE (1966) Patterns of reproduction in female diprotodont marsupials. *Symp. Zool. Soc. Lond.* **15:** 205–232.

Shimmin GA & Temple-Smith PD (1996a) Mating behaviour in the brown marsupial mice (*Antechinus stuartii*). *42nd Ann. Conf. Australian Mammal Society*. Melbourne, Australia. pp. 39.

Shimmin, GA & Temple-Smith PD (1996b) DNA fingerprinting to determine sperm competition winners in the brown marsupial mouse (*Antechinus stuartii*). *4th Int. DNA Fingerprinting Conf.* Melbourne, Australia. pp. 117.

Short RV (1979) Sexual selection and its component parts, somatic and genital selection as illustrated by man and the great apes. *Adv. Stud. Behaviour.* **9:** 131–158.

Smith AP (1980) The diet and ecology of leadbeaters possum and the sugar glider. Ph.D. thesis, Monash University, Clayton, Victoria, Australia.

Smith RL (1984) *Sperm Competition and the Evolution of Animal Mating Systems*. Academic Press, Orlando.

Stodart E (1966) Management and behaviour of breeding groups of the marsupial *Perameles nasuta* Geoffroy in captivity. *Aust. J. Zool.* **14:** 611–623.

Stodart E (1977) Breeding and behaviour of Australian bandicoots. In *The Biology of Marsupials*. B Stonehouse & D Gilmore (eds), pp. 179–191. MacMillian Press, London.

Stoddart DM & Braithwaite RW (1979) A strategy for utilization of regenerating heathland habitat by the brown bandicoot (*Isoodon obesulus*; Marsupialia: Peramelidae). *J. Anim. Ecol.* **48:** 165–179.

Strahan R & Thomas DE (1975) Courtship of the platypus, *Ornithorhynchus anatinus. Aust. Zool.* **18:** 165–178.

Streilein KE (1982) Behaviour, ecology and distribution of the South American marsupials. In *Mammalian Biology of South America*. MA Mares & HH Genoways (eds), pp. 231–250. University of Pittsburgh Press, Pittsburg.

Taggart DA (1994) A comparison of sperm and embryo transport in the female reproductive tract of marsupial and eutherian mammals. *Reprod. Fertil. Devel.* **6:** 1–22.

Taggart DA & Temple-Smith PD (1989) Structural features of the epididymis in a dasyurid marsupial (*Antechinus stuartii*). *Cell Tiss. Res.* **258:** 203–210.

Taggart DA & Temple-Smith PD (1990a) The effects of breeding season and mating on total number and relative distribution of spermatozoa in the epididymis of the brown marsupial mouse, *Antechinus stuartii. J. Reprod. Fertil.* **88:** 81–91.

Taggart DA & Temple-Smith PD (1990b) An unusual mode of progressive motility in spermatozoa from the dasyurid marsupial, *Antechinus stuartii. Reprod. Fertil. Devel.* **2:** 107–114.

Taggart DA & Temple-Smith PD (1991) Transport and storage of spermatozoa in the female reproductive tract of the brown marsupial mouse, *Antechinus stuartii* (Dasyuridae). *J. Reprod. Fertil.* **93:** 97–110.

Taggart DA & Temple-Smith PD (1994) Comparative epididymal morphology and sperm distribution studies in dasyurid marsupials. *J. Zool.* **232:** 365–381.

Taggart DA, O'Brien H & Moore HDM (1993a) Ultrastructural characteristics of *in vivo* and *in vitro* fertilization in the grey short-tailed opossum, *Monodelphis domestica. Anat. Rec.* **237**(1): 21–37.

Taggart DA, Johnson JL, O'Brien HP & Moore HDM (1993b) Why do spermatozoa of American marsupials form pairs? A clue from the analysis of sperm pairing in the grey short-tailed opossum, *Monodelphis domestica. Anat. Rec.* **236:** 465–478.

Taggart DA, Steele VR, Schultz D, Dibben R, Dibben J & Temple-Smith PD (1997) Semen collection and cryopreservation in the southern hairy-nosed wombat, *Lasiorhinus latifrons*; implications for conservation of the northern hairy-nosed wombat, *Lasiorhinus krefftii*. In *Wombats*. R Wells & P Pridmore (eds), Chapter 17, pp. 195–206. Surrey Beatty & Sons, Chipping Norton, Australia.

Taylor RJ (1983) Association of social classes of the wallaroo, *Macropus robustus* (Marsupialia: Macropodidae). *Aust. Wildl. Res.* **10:** 229–237.

Temple-Smith PD (1973) Seasonal breeding biology of the platypus, *Ornithorhynchus anatinus* (Shaw 1799), with special reference to the male. Ph.D. thesis, Australian National University, Canberra.

Temple-Smith PD & Bedford JM (1980) Sperm maturation and formation of sperm pairs in the epididymis of the opossum, *Didelphis virginiana. J. Exp. Zool.* **214:** 161–171.

Thibault C (1973) Sperm transport and storage in vertebrates. *J. Reprod. Fertil.* Suppl. **18:** 39–53.

Thomas RH & Zeh DW (1984) Sperm transfer in arachnids: ecological and morphological constraints. In *Sperm Competition and the Evolution of Animal Mating Systems*. RL Smith (ed.), pp. 179–221. Academic Press, London.

Trupin GL & Fadem BH (1982) Sexual behaviour of the grey short-tailed opossum *Monodelphis domestica. J. Mammal.* **63:** 409–414.

Turner V (1985) The ecology of the eastern pygmy possum, *Cercartetus nanus*, and its association with banksia. Ph.D. thesis, Monash University, Melbourne, Australia.

Tyndale-Biscoe CH & Rodger JC (1978) Differential transport of spermatozoa into the two sides of the genital tract of a monovular marsupial, the tammar wallaby (*Macropus eugenii*). *J. Reprod. Fertil.* **52:** 37–43.

Tyndale-Biscoe CH & Renfree MB (1987) *Reproductive Physiology of Marsupials*. Cambridge University Press, Cambridge.

Tyson E (1698) Carigueya, seu Marsupiale Americanum or the anatomy of an opossum dissected at Gresham college. *Phil. Trans. Roy. Soc.* **20:** 105–164.

Walker WF (1980) Sperm utilization strategies in nonsocial insects. *Am. Nat.* **115:** 780–799.

Ward PI (1993) Females influence sperm storage and use in the yellow dung fly *Scathophaga stercoraria* (L). *Behav. Ecol. Sociobiol.* **32:** 313–319.

Ward SJ (1990a) Life history of the feathertail glider, *Acrobates pygmaeus* (Acrobatidae: Marsupialia) in south eastern Australia. *Aust. J. Zool.* **38:** 503–517.

Ward SJ (1990b) Life history of the eastern pygmy possum, *Cercartetus nanus* (Burramyidae: Marsupialia), in south eastern Australia. *Aust. J. Zool.* **38:** 287–304.

Ward SJ (1990c) Reproduction in the western pygmy possum, *Cercartetus concinnus* (Marsupialia: Burramyidae), with notes on reproduction of some other small possum species. *Aust. J. Zool.* **38:** 423–438.

Watson DM, Croft DB & Crozier RH (1992) Paternity exclusion and dominance in captive red-necked wallabies, *Macropus rufogriseus* (Marsupialia: Macropodidae). *Aust. Mammal.* **15:** 31–36.

Wells RT (1973) Physiological and behavioural adaptations of the hairy-nosed wombat (*Lasiorhinus latifrons* (Owen)) to its arid environment. Ph.D. thesis, University of Adelaide, Australia.

Wells RT (1978) Field observations of the hairy-nosed wombat, *Lasiorhinus latifrons* (Owen). *Aust. Wildl. Res.* **5:** 299–303.

Wells RT (1989) Vombatidae. In *Fauna of Australia, Vol. 1b Mammalia*. DW Walton & BJ Richardson (eds), pp. 755–768. Australian Government Publishing Service, Canberra.

Wilson DE & Reeder DM (1993) *Mammal Species of the World*. Smithsonian IP, Washington, DC.

Winter J (1977) The behaviour and social organization of the brush-tail possum (*Trichosurus vulpecula* Kerr). Ph.D. thesis, University of Queensland, Brisbane, Australia.

Wittenberger JF (1979) The evolution of mating systems in birds and mammals. In *Handbook of Behavioural Neurobiology, Vol. 3. Social Behaviour and Communication*. P Marler and J Vandenberg (eds), pp. 271–349. Plenum Press, New York.

Woolley P (1966) Reproduction in *Antechinus* spp. and other dasyurid marsupials. *Symp. Zool. Soc. Lond.* **15:** 281–294.

Woolley P (1971a) Observations on the reproductive biology of the dibbler, *Antechinus apicalis* (Marsupialia: Dasyuridae). *J. Roy. Soc. West. Aust.* **54:** 99–102.

Woolley P (1971b) Maintenance and breeding of laboratory colonies of *Dasyuroides byrnei* and *Dasycercus cristicaudata*. *Int. Zoo Year.* **11:** 351–354.

Woolley P (1988) Reproduction in the *Ningbing Antechinus* (Marsupialia: Dasyuridae). Field and laboratoy observations. *Aust. Wildl. Res.* **15:** 149–156.

Woolley P (1990a) Reproduction in *Sminthopsis macroura* (Marsupialia: Dasyuridae). I The female. *Aust. J. Zool.* **38:** 187–205.

Woolley P (1990b) Reproduction in *Sminthopsis macroura* (Marsupialia: Dasyuridae). II. The male. *Aust. J. Zool.* **38:** 207–217.

Woolley P (1991) Reproductive pattern of captive Boullanger Island dibblers, *Parantechinus apicalia* (Marsupialia: Dasyuridae). *Wildl. Res.* **18:** 157–163.

Woolley P & Ahern LD (1983) Observations on the ecology and reproduction of *Sminthopsis leucopus* (Marsupialia: Dasyuridae). *Proc. Roy. Soc. Vict.* **95:** 169–180.

16 Sperm Competition in Mammals

Montserrat Gomendio[1]*, Alexander H. Harcourt*[2] *and Eduardo R. S. Roldán*[3]

[1] Departamento de Ecología Evolutiva, Museo Nacional de Ciencias Naturales (CSIC), José Gutierrez Abascal, G-28006 Madrid, Spain; [2] Department of Anthropology, University of California, Davis, CA 95616–8522, USA; [3] Instituto de Bioquímica (CSIC-UCM), Ciudad Universitaria, 28040 Madrid, Spain

'Usted replicará que la realidad no tiene la menor obligación de ser interesante. Yo le replicaré que la realidad puede prescindir de esa obligación, pero no las hipótesis'. J. L. BORGES, *La muerte y la brújula. 1942.*

I. HOW WIDESPREAD IS SPERM COMPETITION IN MAMMALS?

A. Definition

For nearly 100 years, Darwin's theory of selection in relation to sex, proposed two processes to explain males' ornaments and weapons that

Sperm Competition and Sexual Selection
ISBN 0-12-100543-7

we now term female choice, and male–male competition (Darwin 1871); either females choose particularly ornate males; or alternatively, relatively passive females accept the winner of fights among males. (Current knowledge of species in which the females are more brightly coloured or aggressive than males leads to a more general formulation of the principle of sexual selection in which instead of 'females', we write the sex with the lower potential reproductive rate' and for 'males', 'the sex with the higher potential reproductive rate' (Clutton-Brock and Parker 1992).) Ninety-nine years after Darwin distinguished sexual selection from natural selection, Parker (1970) suggested that, in addition to competing by being attractive and strong, males could compete by producing large numbers of sperm, arguing that the male that produced the most sperm would be the male most likely to fertilize the ova if more than one male mated with a fertile female.

In mammals, females are sexually receptive to males for limited periods of time known as 'oestrus' which vary in length from species to species. In this group, sperm competition takes place when receptive females copulate with several males and spermatozoa from rival males compete within the female reproductive tract to fertilize the available ova. For a long time, the expectation has been that sperm competition should be a rare phenomenon among eutherian mammals because most females do not possess proper sperm-storage organs and spermatozoa are short-lived (Parker 1984). However, recent studies have provided evidence that suggests that sperm competition is more common among eutherian mammals than previously thought.

In most mammalian groups there are species in which females commonly copulate with more than one male while in oestrus (rodents, Dewsbury 1984; Gomendio and Roldan 1991; Murie 1995; ungulates, Ginsberg and Rubenstein 1990; primates, Smuts *et al.* 1987; general, Ginsberg and Huck 1989; Møller and Birkhead 1989). In some mammals, copulations are difficult to observe in the wild, and in those cases in which males try to prevent sexual access to females by rival males, females tend to engage in copulations with males other than the guarding male in rather cryptic ways, suggesting that the information revealed by observational studies may be an underestimate of the number of species in which females actually copulate with multiple males.

B. Conditions that have to be met

To date, no observations of sperm competing within the female tract have been made. Obtaining this kind of evidence involves serious methodological problems, and physiological studies on mammalian reproduction have not yet focused on this topic. Thus, sperm competition is inferred from a number of factors. The most direct evidence of sperm competition is multiple paternity, or paternity by males other than the female's main sexual partner. As the use of molecular techniques to determine paternity becomes more widespread, this kind of evidence is accumulating slowly. For most of the species, however, evidence for the occurrence of sperm competition comes from observational studies carried out by behavioural ecologists, which tends to be indirect. To be certain that observational evidence is real proof of sperm competition, several criteria have to be satisfied.

For sperm competition to occur, the first obvious requisite is that females must copulate with more than one male. This type of observational evidence can be obtained in diurnal species where visibility is good, which is the case for many primate species. However, it is difficult to obtain reliable evidence for species with small body size, nocturnal species or species which live in habitats with poor visibility.

Copulations by rival males must also take place near the time of ovulation and within the life-span of spermatozoa. In mammals, all ova are produced simultaneously (irrespective of the number of ova) and remain fertile for about 24 h, while spermatozoa are apparently short-lived (see below). It follows that ejaculates will be able to participate in the competition to fertilize ova only if males copulate near the time of ovulation. If males copulate too early, their sperm will no longer be alive by the time the female ovulates. Alternatively, if males copulate too close to ovulation or after ovulation their spermatozoa may not be ready to fertilize during the short period in which ova are available. Thus, the optimal timing for males would be to copulate as close to ovulation as possible, allowing time for capacitation to occur (i.e. to undergo in the female tract an as yet poorly defined maturational process necessary for sperm to be able to fertilize). In several species of rodents (Dewsbury 1988a; Huck *et al.* 1989), rabbits (Chen *et al.* 1989), pigs (Dziuk 1970) and humans (Wilcox *et al.* 1995), there is evidence suggesting that the timing of copulations is crucial for the success of ejaculates. In species in which periods of oestrus are short in relation to sperm life-span (i.e. a few hours), most copulations will take place within the optimal time period. However, in species in which females are in oestrus for long periods of time in relation to sperm life-spans (i.e. periods over 1 week and up to the whole cycle), and in which females do not advertise reliably the timing of ovulation, it may be difficult for males to predict when copulations should take place to assure fertilization.

Spermatozoa from rival males must also coexist within the female tract for sperm competition to occur. Spermatozoa have limited energy reserves and have no functional cellular systems to repair degenerating or damaged cellular elements (Drobnis and Overstreet 1992). In addition, female mammals lack proper sperm-storage organs, which in other taxa such as insects and birds, extend sperm life-span for weeks, or even years (Parker 1970; Birkhead and Møller 1992, 1993). Thus, mammalian spermatozoa tend to live for short periods of time within the female tract, normally for a few days (Bishop 1969; Austin 1975; Hunter 1988). Therefore, if copulations by rival males are timed so that sperm from different males are not alive simultaneously in the female tract, sperm competition does not take place. Whether live sperm from rival males coexist in the female tract around the time of ovulation or not depends on sperm life-span, on the timing of copulations by rival males and on the duration of oestrus periods.

To determine sperm fertile life-span is not as straightforward as it may appear. Different parameters have been used and these need to be carefully evaluated. Some studies have relied on the mere presence of sperm within the female tract, and in many cases the methods used do not even allow discrimination between dead and live sperm. This type of evidence is clearly insufficient since sperm may remain in the female tract long after their death (Hunter 1987). Other studies have relied on evidence of sperm motility to determine sperm life-span. It is noteworthy that mammalian sperm remain motile for some time after they have lost their fertilizing capacity (Bishop 1969; Austin 1975; Drobnis and Overstreet 1992; Mortimer 1983). Gomendio and Roldan (1993b) proposed that, in the context of sperm competition, only sperm fertile life matters, since sperm which remain motile but are unable to fertilize are unlikely to participate in sperm competition and obviously cannot win the contest. However, Baker and Bellis (1995) have suggested that sperm which can no longer fertilize retain their motility to prevent rival sperm from fertilizing ova by other means (see Section III). These authors go on to suggest that the very fact that spermatozoa remain motile after losing their fertilizing capacity, suggests that they do indeed perform other roles in competition with rival ejaculates; however, direct information on the possible roles played by nonfertilizing motile sperm is lacking. This suggestion ignores the fact that although sperm may retain the ability to move when assessed *in vitro*, sperm remain quiescent in several parts of the female tract. In other words, the fact that sperm show motility under experimental conditions, does not necessarily mean that sperm continue to move inside the female tract. Any form of competition in which nonfertilizing motile sperm could participate, could be seen as a form of 'spite' since, after losing their fertilizing capacity, motile sperm would try to prevent rival sperm from fertilizing, but would themselves no longer be able to fertilize. There is, however, an alternative explanation to the mismatch between sperm fertile life and the period of time in which sperm remain motile. As sperm age, the likelihood that the genetic material carried by the sperm head will deteriorate increases and a point is reached in which fertilizing ability is lost to prevent the formation of nonviable embryos (Lanman 1968). It would be in the interest of both the male and the female to have a switch-off of this kind when embryo survival is at stake, and it makes sense that the loss of fertilizing ability should take place as soon as the risk of damage to the DNA appears, even if there is still energy to keep motility going for a while. The mismatch between loss of fertilizing capacity and loss of motility could result from stronger selection pressures acting on the former, or from the difficulty in determining the exact magnitude of the reserves needed to maintain motility, given that sperm go through phases of quiescence while in the female tract. Given the lack of evidence on the possible functions of sperm motility in the absence of fertilizing capacity, and the fact that sperm remain motile after losing fertilizing capacity in all the species studied and not only when sperm competition occurs, we will focus our discussion on sperm fertile life-spans.

Spermatozoa from most mammalian species remain fertile for about 48 h (Table 16.1). It is worth discussing in some detail other species which are widely considered as exceptions, since this may not be the case for some (for discussion on humans see Section VII). It has been suggested that in the hare, sperm may remain fertile throughout gestation (about 40 days) (Martinet and Raynaud 1975). The idea was developed as a way of explaining the phenomenon known as 'superfetation' (i.e. initiation and development of a new pregnancy in already pregnant females). In the wild, a few cases have shown that well-developed embryos coexist with fertilized ova at early developmental stages; the hare is an induced ovulator, and so new pregnancies are obviously the result of copulations during the late stages of the first pregnancy. Nearly all the evidence for the occurrence of superfetation in captivity has been based on the duration of inter-birth intervals, because it has been assumed that the occurrence of shorter than usual inter-birth intervals was a consequence of superfetation. However, artificial insemination experiments (Stavy and Terkel 1992) have shown that shorter inter-birth intervals after copulations during the late stages of pregnancy result largely from early deliveries and the prompt initiation of a new

Table 16.1. The duration of sperm fertile life-spans in several mammalian species.

Species	Sperm fertile life-span	References
Mus musculus	6 h	1, 2, 3
Rattus norvegicus	14 h	1, 2, 3
Cavia porcellus	22 h	1, 2, 3
Oryctolagus cuniculus	28–32 h	1, 2, 3
Sus scrofa	24–48 h	2
Mustela putorius	36–48 h	1
Ovis aries	24–48 h	1, 2, 3
Equus caballus	144 h	1, 2, 3
Bos taurus	24–50 h	1, 2, 3
Canis familiaris	84 h	4
Homo sapiens	33.6 h	5, 6
Myotis lucifugus	138 days	3
Eptesicus fuscus	156 days	3

[1] Bishop (1969); [2] Blandau (1969); [3] Parkes (1960); [4] Doak *et al.* (1967); [5] Weinberg and Wilcox (1995); [6] Wilcox *et al.* (1995).

pregnancy, rather than from superfetation. This suggests that superfetation can not be inferred solely from shorter than usual inter-birth intervals and that rates of superfetation in hares have been inflated. In their study on captive hares, Martinet and Raynaud (1975) suggested that superfetation may occur when sperm from the male who inseminated the female first survives in the female tract throughout pregnancy and fertilizes ova before the female gives birth. The hypothesis requires subsequent copulations towards the end of pregnancy because ovulation is induced in this species. This hypothesis is based on two cases in which females who had been allowed to copulate with a vasectomized male towards the end of pregnancy, were found to have embryos at early developmental stages a few days later. The authors suggested that, as a result of copulations with the vasectomized males, the pregnant females had ovulated, and that these ova had been fertilized by sperm from the male who initiated the ongoing pregnancy. Other alternative explanations may be more parsimonious.

First, is the obvious possibility that the males were not properly vasectomized which is a common occurrence. The authors do not specify if steps were taken to check whether the vasectomy had been successful (e.g. ejaculate examination or matings with nonpregnant females). Second, it is likely that the stimulus provided by copulation with the vasectomized male triggered parthenogenetic development of ova. In rabbits, oocytes can begin parthenogenetic activation and development without sperm participation (Kaufman 1983; Ozil 1990). As oocytes 'age' after ovulation they become more prone to start parthenogenetic development in response to a variety of stimuli if fertilization does not take place (Kaufman 1983; Kubiak 1989). These parthenogenetic embryos can implant and reach day 11 of gestation, after which normal development starts to fail (rabbit, Ozil 1990; mouse, Kaufman 1983). This would also explain why at least in one of the two cases described by Martinet and Raynaud (1975) ova were developing asynchronously. Thus, the evidence presented by Martinet and Raynaud (1975) does not demonstrate sperm life-spans in the hare of about 40 days, a result which was difficult to understand given that the hare needs to copulate in order to induce ovulation.

The domestic dog (*Canis familiaris*) has been widely presented as another extreme in sperm life-spans. In the dog, motile sperm have been found in the uterine glands up to 11 days after insemination and for an average duration of 7 days (Doak *et al.* 1967). However, motile spermatozoa were found in the oviducts up to only 3 days after mating, indicating that replenishment of oviductal reserves was not taking place, and thus strongly suggesting that sperm found in the uterine glands do not participate in fertilization. Some authors have used the value of 11 days as 'sperm-fertilizing life-span' but it is not known for how long they retain fertilizing capacity. Doak *et al.* (1967) suggested that if the dog follows the trend found in a few species (i.e. that fertile life-span in sperm tends to be about half the duration of sperm motility, Bishop 1969), it would follow that sperm fertile life in the dog would be about 5–6 days at the most, and 3.5 days on average. This value is not very different from the sperm fertile life-spans obtained for other mammalian species, and is in accordance with the peculiar reproductive biology of the female dog. In the bitch, ovulation takes place up to 60 h after the onset of oestrus, but the ova are ovulated as primary oocytes and are not capable of being fertilized until about 60 h after ovulation, when they undergo the first meiotic division and become secondary oocytes (Tsutsui 1989). Ova remain fertilizable from the time they undergo the first meiotic division, until about 48 h, which is unusually long for mammals. Thus, a male mating around the time of ovulation would have to produce sperm able to survive for at least 2.5 days and up to 4.5 days to have any chance of fertilizing ova. Thus, indirect evidence suggesting that the duration of sperm fertile life-spans may be slightly longer in the dog than in other mammalian species, is in accordance with the view that features of female reproductive biology influence to a great extent the duration of sperm life-spans (see Section II.B).

Bats appear to be a remarkable exception to the mammalian trend of short sperm life-spans, since in this group females may store sperm for months (Fenton 1984). It has been argued that female bats store sperm so that they can survive throughout hibernation. It is unclear, however, why females do not copulate after hibernation, or why they do not use other mechanisms more common in mammals, such as delayed implantation or embryonic diapause, to uncouple copulation and birth. These physiological alternatives would give females more control over copulations since they would cancel the opportunities for sneak copulations that some males achieve while females are in torpor (Fenton 1984). In any case, it seems unlikely that hibernation is a crucial factor linked to the evolution of long sperm life-spans in bats, since sperm may also survive for long periods in tropical species of bat. Thus, the selective forces favouring long sperm survival in bats remain to be elucidated.

Because there is so little reliable information on fertile life-spans of mammalian sperm, it is difficult to determine how close copulations by different males should be for sperm competition to occur. In some mammalian species, oestrus periods last for only a few hours (Gomendio and Møller in prep.). Assuming sperm life-spans of up to 48 h are common, copulations by several males within short oestrus periods appear likely to result in sperm competition. In several species of ground squirrels females are in oestrus for a few hours per year and still they commonly copulate with several males. In a study population of *Spermophilus columbianus* females were in oestrus for only 4 h each year, and all the females were observed to copulate with more than one male (an average of 4.4 males per female) (Murie 1995). Similarly, *Spermophilus beecheyi* females are in oestrus for only slightly longer periods of time (6.7 h) and females copulate with an average of 6–7 males (Boellstorf *et al.* 1994). Thus, short oestrus periods do not represent a constraint in terms of possibilities for multiple mating for females, provided that males can locate females while they are in oestrus. The ability to locate females will depend both on how dispersed individuals are and on the kinds of advertisement by receptive females. In other mammalian species, females are in oestrus for longer periods of time, and in this case it is important to determine both sperm life-spans and the timing of copulations in relation to ovulation. This is the case for many primate species. If oestrus periods are long, but females advertise ovulation reliably (such as occurs in many primates living in multimale groups, where females have sexual swellings), then males mating at short intervals around ovulation are more likely to enter sperm competition. There is scope, however, for males mating well before ovulation takes place to participate in sperm competition, if their sperm survive long enough. Finally, in species which have effectively long oestrus periods, because females do not confine oestrus behaviour closely to ovulation, and females do not advertise reliably the timing of ovulation (e.g. humans), it is crucial to determine, with more invasive methods, when females ovulate and for how long sperm survive in the female tract, in order to infer sperm competition.

C. Evidence of sperm competition in mammals

Because direct observations on female mating patterns (i.e. whether females are monandrous or polyandrous) are often difficult to obtain, indirect types of evidence have been used. The most widely used approach has been to infer likelihood of sperm competition from patterns of social organization. Thus, socially monogamous species and those with single-male groups have usually been assumed to experience no sperm competition, while species forming multimale groups have usually been assumed to experience high levels of sperm competition (Harcourt *et al.* 1981; Møller 1988a). This 'shortcut' has yielded very important results (such as differences in testes size and ejaculate features in primates, see Section IV.B) which suggest that, on the whole, the resulting categorization is meaningful. However, such categorization in terms of social organization may not reflect accurately mating systems in some species, and for species with unusual or poorly defined types of social organization (e.g. 'dispersed', 'solitary') it is difficult to predict which mating system is more likely. In primates, there is a wealth of information on social organization and mating systems, which allows more detailed scrutiny than in other groups. In monogamous species multiple mating by females seems less frequent than in other groups, but it has been observed, although rarely, in species such as gibbons (Reichard 1995). However, there are cases in which females forming monogamous social groups are highly promiscuous. This is the case in Mongolian gerbils which live in pairs or family groups, and where females incite other males to copulate, probably to avoid inbreeding (Ågren 1990). Similarly, Ethiopian wolves *Canis simensis* live in family packs, but 70%

of copulations take place between the dominant female and males from other packs (Sillero-Zubiri *et al.* 1996). Although most single-male species do appear to be essentially monandrous, in some species males cannot defend females successfully and polyandrous matings are common. The risk of polyandry increases with the number of females in oestrus at the same time, because harem holders are less likely to be able to defend several females simultaneously. For this same reason, few seasonal breeders have single male units, but those that do have them, such as patas monkeys, might have a high rate of polyandry (see discussion in Roldan and Gomendio 1995). This is also the case of greater spearnose bats in which males form harems which are difficult to defend because they contain numerous females that breed synchronously (McCraken and Bradbury 1981). Conversely, in some species forming multimale groups, such as baboons, females may be consorted around the time of ovulation by particular males, thus diminishing the chances that sperm competition will take place. Finally, solitary species such as orang-utans and black bears were assumed to be unlikely candidates for sperm competition, yet observational evidence has revealed that females tend to copulate with several males (Rodman and Mitani 1987; Schenk and Kovacs 1995). Thus, inferences about female mating patterns made solely on the basis of social organization may be misleading. The extent to which social organization and mating patterns can be unrelated has been dramatically demonstrated in birds, and mammalogists should be cautious.

Recently developed techniques to determine paternity provide the clear evidence of sperm competition (see Appendix I). In species with litter sizes greater than one, multiply fathered litters are unquestionable proof of sperm competition. In some cases, evidence of multiple paternity has been used to confirm that previous observations of multiple-male mating by females did indeed result in sperm competition. In several species of ground squirrels, multiple mating by females has been shown to result in a high percentage of litters sired by several males. In populations of *Spermophilus tridecemlineatus*, *Spermophilus beldingi* and *S. beecheyi*, 50%, 88% and 89% of litters, respectively, were found to be multiply sired (Hanken and Sherman 1981; Boellstorf *et al.* 1994; Murie 1995); in other species of ground squirrels in which females also copulate promiscuously, only a small percentage of litters are multiply sired. For example, in a population of *S. columbianus*, all receptive females were observed to copulate with more than one male during the 4-h oestrus period, yet only 16% litters were multiply sired (Murie 1995). Similarly, in black-tailed prairie dogs although 38% of females copulated with more than one male, only 3% of litters were multiply sired (Hoogland and Foltz 1982). It is still not known why in some cases high levels of promiscuity result in high levels of multiple paternity while in others it does not. One possible explanation is that in some species there are males which easily outcompete other males and thus, even if there has been sperm competition, these successful males father all the offspring in a high percentage of litters. Thus, while multiple paternity is definitely direct proof of sperm competition, females giving birth to litters sired by just one male may have still experienced sperm competition.

In other species, behavioural observations suggesting low or zero levels of sperm competition, have been proven wrong by paternity analyses. This is the case for grey seals, a species in which males defend harems against other males. Recent paternity analyses have shown that harem holders father fewer offspring than suggested by behavioural criteria because copulations by females with subordinate males had gone largely unnoticed to human observers, probably because they tend to occur in the sea or at night (Amos *et al.* 1993). Similarly, Gunninson's prairie dogs *Cynomys gunnisoni* are colonial ground squirrels, in which territories are occupied by monogamous pairs or polygynous groups. Based on knowledge of their social organization the assumption was that most copulations would occur within social groups. However, DNA fingerprinting analyses have revealed that 33% of litters were multiply sired and, even more surprising, that 61% offspring were sired by extraterritorial males (Travis *et al.* 1996). It is likely that extraterritorial copulations take place because females travel to neighbouring territories.

In species with one offspring, multiple paternity is obviously not a possibility, and sperm competition can be inferred if paternity analyses show that offspring are fathered by males other than the ones who are known to have copulated with the female around ovulation. This argument has been used to infer sperm competition in humans from instances in which the biological father was found to be different from the putative father (e.g. the husband). However, this information is insufficient because it needs to be established whether the putative father also copulated with the female around the time of ovulation and with an interval shorter than sperm fertile life-span (see Section VII.A).

In many cases, paternity analyses have revealed a surprising mismatch between expectations based on behavioural observations and patterns of genetic paternity. This mismatch is partly the consequence of the limitations of behavioural observations, but also of our ignorance about underlying physiological mechanisms.

II. THE FEMALE TRACT AS AN ARENA

Female reproductive physiology in general, and features of the female reproductive tract in particular, set the scenario in which rival ejaculates have to compete. The processes that ejaculates experience within the female tract determine which ejaculate features will make some males more successful than others. Thus, in species in which ejaculates have to survive for long periods of time in the female tract (e.g. insects), the ability to enter sperm-storage organs and to attain favourable positions within these may be crucial, while in the absence of storage organs the amount of energy reserves contained in each sperm cell may be the most important factor. Likewise, when sperm have actively to cover long distances, swimming speed and/or endurance may be crucial, while in species in which sperm have to negotiate thick ovum vestments, acrosomal enzymes may be one of the key factors. Female reproductive physiology and tract morphology show a striking degree of variation in the animal kingdom, as well as within mammals. Female reproductive physiology will thus define the context that will determine how selective forces will shape ejaculates.

Among mammals, the female tract represents a formidable barrier for spermatozoa. Out of the millions of spermatozoa contained in an ejaculate, most die or are phagocytozed before they reach the vicinity of the ova (Austin 1965; Hunter 1988). In most of the species which have been studied to date, millions of spermatozoa are ejaculated into the female tract, but only of the order of thousands reach the isthmus, and only 2–20 reach the ampulla, the site of fertilization (Suarez *et al.* 1990). Whether the drastic reduction in sperm numbers within the female tract represents some kind of selection on the part of females or whether it is a random process is still a controversial issue. Recently developed techniques of intracytoplasmic sperm injection allow fertilization by sperm chosen at random. As the use of these experimental procedures becomes more common, it is becoming increasingly clear that these randomly chosen sperm result in a higher incidence of chromosomal anomalies or even embryonic death (In't Veld *et al.* 1995). This strongly suggests that the reduction of sperm numbers in the female tract is a selective process on the part of the female that eliminates unfit sperm.

A. Interactions sperm/female tract

Much of the information presented in this section is based on the following reviews: Mortimer 1983, 1995; Hunter 1987, 1988; Bedford and Hoskins 1990; Overstreet and Katz 1990; Barrat and Cooke 1991; Bedford 1991; Drobnis and Overstreet 1992 and Yanagimachi 1994. More specific papers will only be cited where needed.

Mammalian males produce spermatozoa continuously (in seasonal breeders only throughout the breeding season) and more than 50% of sperm remain in the cauda epididymis (Robaire and Hermo 1988). It has been suggested that the cauda epididymis has evolved as a storage organ in the male to compensate for the low rate of sperm production in mammals compared with birds (Bedford 1979), but no detailed comparative study has been made. We have conducted a preliminary comparative analysis attempting to link epididymal transit time, or the time spent by spermatozoa in the cauda epididymis (data from Orgebin-Crist and Olson 1984), which would be an indication of storage capacity or need, with daily sperm production, or sperm production in relation to mass of parenchyma (data from Amman *et al.* 1976); no significant relationships were found.

Mature sperm available for ejaculation are maintained in the cauda epididymis and the vas deferens in an immotile state (Turner and Reich 1985). Sperm motility was thought to be initiated at ejaculation as a result of the mixture of sperm with secretions of the accessory male glands. However, recent evidence suggests that, at least in the rat and the hamster, sperm motility develops later (uterus and oviduct, respectively) and is triggered by fluids present in the female tract (Bedford and Yanagimachi 1992).

Ejaculates contain several fractions that differ in composition and sperm concentration, but the functional significance of this variation is unknown. All we know is that the majority of sperm are contained in the first portion, and that some seminal plasma constituents stimulate muscle contractions of the tract and thus facilitate sperm transport. Some seminal plasma coagulates may serve to retain sperm in the female tract, and plugs are critical for sperm transport in some species of rodents (Blandau 1945; Matthews and Adler 1978). A recent study has compared the relative importance of rat copulatory plugs (1) as a sperm reservoir, (2) as a block to prevent sperm leaking from the vagina, and (3) in sperm transport into the uterus, and concluded that only the third mechanism is important (Sofikitis *et al.* 1990).

Males can deposit their ejaculates either in the vagina, the cervix or the uterus, and this variation implies that spermatozoa from different species will have to swim different distances and face environments of different chemical and physical properties (Katz *et al.* 1989; Katz and Drobnis 1990). In addition, receptive females of some species develop sexual swellings which greatly increase the distance that ejaculates must cover (e.g. up to 50% of vaginal length in chimps, Dixson and Mundy 1994). In some species, large proportions of the ejaculate are expelled outside the female tract, up to 80% in pigs, sheep, cattle and rabbits (Overstreet 1983; Ginsberg and Huck 1989). The female responds to the appearance of sperm in the tract with a leucocytic invasion of the uterus and cervix whose prime function is to phagocytose sperm (Pandya and Cohen 1985; Barrat *et al.* 1990).

In species in which ejaculates are deposited in the vagina, such as humans, high acidity often creates a lethally hostile environment (Yanagimachi 1994). The acidity may have evolved to inhibit the growth of microorganisms in the section of the tract which is more vulnerable to the attack of external agents. Seminal plasma buffers this acidity to some extent, and sperm that penetrate into cervical mucus reach a section that is safer owing to its more optimal pH (Fordney-Settlage 1981).

Cervical mucus shows a cyclic receptivity to sperm: sperm are able to penetrate around the time of ovulation, but receptivity decreases rapidly thereafter. How successful sperm are in penetrating cervical mucus depends on a number of factors such as sperm concentration, sperm functional status (e.g. membrane destabilization), seminal enzymes and the presence of antisperm antibodies; however, the single most important factor is sperm motility (and sperm morphology linked to movement characteristics) which accounts for 85% of the variability in penetration (Barrat and Cooke 1991). It has been suggested that cervical crypts and/or uterine glands may act as sperm reservoirs in much the same way as sperm storage tubules in birds. There is evidence suggesting that sperm which enter cervical crypts or uterine glands remain motile and may be less susceptible to phagocytic attack. In some cases, such as dogs and humans, the number of sperm present in the uterine glands and cervical crypts, respectively, has been shown to decrease some time after copulation and some authors have assumed that these sperm have continued migrating up the female tract. However, in the dog, while sperm found in the uterine glands may retain motility for longer, no sperm were found in the oviduct after 3 days, suggesting that sperm in the uterine glands do not continue migrating up the female tract (Doak *et al.* 1967). Our knowledge on humans is restricted to one study (Insler *et al.* 1980) in which women were treated with oestrogen or gestagen, artificially inseminated and underwent hysterectomy. The decline in sperm numbers in the cervical crypts which took place 48 h after insemination, could have been due to migration but also to sperm loss or redistribution of sperm. Although human sperm collected from the cervical mucus seem to remain functional after 3 days, at least two recent reviews on the subject (Overstreet and Katz 1990; Mortimer 1995) conclude that there is no evidence whatsoever of sperm leaving the cervical crypts or the uterine glands and migrating up the female tract. We have been able to find no evidence in the literature in support of this claim either. Thus, it is likely that sperm inside the crypts or uterine glands are 'trapped' and will never participate in fertilization.

Sperm transport along the female tract has been found to occur in two phases, at least in some species. There is a first phase of rapid sperm transport, in which a few sperm are transported within minutes along the entire length of the female tract. This population of sperm tends to end up badly damaged and is not likely to participate in fertilization; most of them end up going into the peritoneal cavity. It is still unclear whether this initial phase of rapid transport is merely a by-product of contractions experienced by the female tract, or whether rapidly transported sperm do perform some function, such as delivering some kind of signal (e.g. about the presence of an ejaculate) to the upper sections of female tract.

A second phase of more gradual sperm transport takes place, which results in another population of sperm arriving at the oviduct a few hours after copulation. To a large extent, sperm are transported passively by muscular movements of the female tract and by the activity of the cilia that are lining the lumina along the first sections of the female tract. It has been suggested that even immotile particles can be transported in this way, but the biological meaning of these findings has been questioned (Mortimer 1983). The contractions of the female tract are stimulated by copulation and by some seminal constituents.

After being passively transported along the first sections, sperm must then actively swim to go through the uterotubal junction, which represents a physical barrier to spermatozoa owing to its narrow diameter and to the presence of highly infolded walls (Katz *et al.* 1989; Katz and Drobnis 1990; Suarez *et al.* 1990). Thus, vigorously motile sperm with normal morphology preferentially enter the oviduct; morphologically abnormal sperm are at a disadvantage because of their inefficient movement and immotile sperm do not enter the oviduct at all (see Fig. 16.1).

The lower isthmus of the oviduct appears to act as a 'reservoir' where spermatozoa become inactive and conserve sperm function (hamster, Smith and Yanagimachi 1990, 1991; guinea pig, Yanagimachi and Mahi 1976; rabbit, Overstreet and Cooper 1975; sheep, Hunter *et al.* 1982; pig, Hunter 1984; cow, Hunter and Wilmut 1984; Lefebvre *et al.* 1995; mice, Suarez 1987; human, Pacey *et al.* 1995). Sperm remain inactive in the lower isthmus of the oviduct until ovulation occurs and, thus, the length of time that spermatozoa spend here depends partly on the timing of mating in relation to ovulation (Smith *et al.* 1987). During this period of residence sperm attach to the mucosa of the oviductal wall via the acrosomal region of the head and those sperm that fail to attach to the wall die or lose their fertilizing capacity (Smith and Yanagimachi 1990; Pollard *et al.* 1991; Suarez *et al.* 1991a). Thus, there is also considerable loss of sperm in the oviduct. It is important to emphasize that this 'reservoir' is not equivalent to the 'sperm-storage organs' described for taxa such as birds and insects. Sperm remain in the oviductal reservoir for a few hours at the most and it serves only to maintain a functional population of sperm ready to fertilize when ovulation occurs.

In other taxonomic groups, sperm remain in highly specialized sperm storage organs for weeks, months and years: copulation and fertilization are therefore uncoupled (Birkhead and Møller 1993). Gomendio and Roldan (1993b) suggested that sequential ovulation of ova in birds may have favoured the evolution of storage organs in the female so that enough sperm are stored to ensure fertilization of several ova over a number of days and in which sperm are safe from being swept away as the egg about to be laid descends along the female tract. In contrast, female mammals ovulate all ova simultaneously and there is no need to ensure the survival of spermatozoa over a long period or to protect spermatozoa from being expelled by descending eggs. However, in mammals there is a narrow time-window in which fertilization must take place since ova fertile life-span is only a few hours long and sperm must become capacitated (a maturational process which can take several hours, see Table 16.2) before they are able to fertilize. Thus, female mammals, rather than evolving a way of ensuring long sperm survival, have evolved a way of improving synchronization between the sperm population most likely to participate in fertilization and ovulation. In this context, the isthmus of the oviduct plays the role of a sperm reservoir during the hours preceding ovulation. There is, however, an exception to the general eutherian trend: bats. In this group, sperm may remain

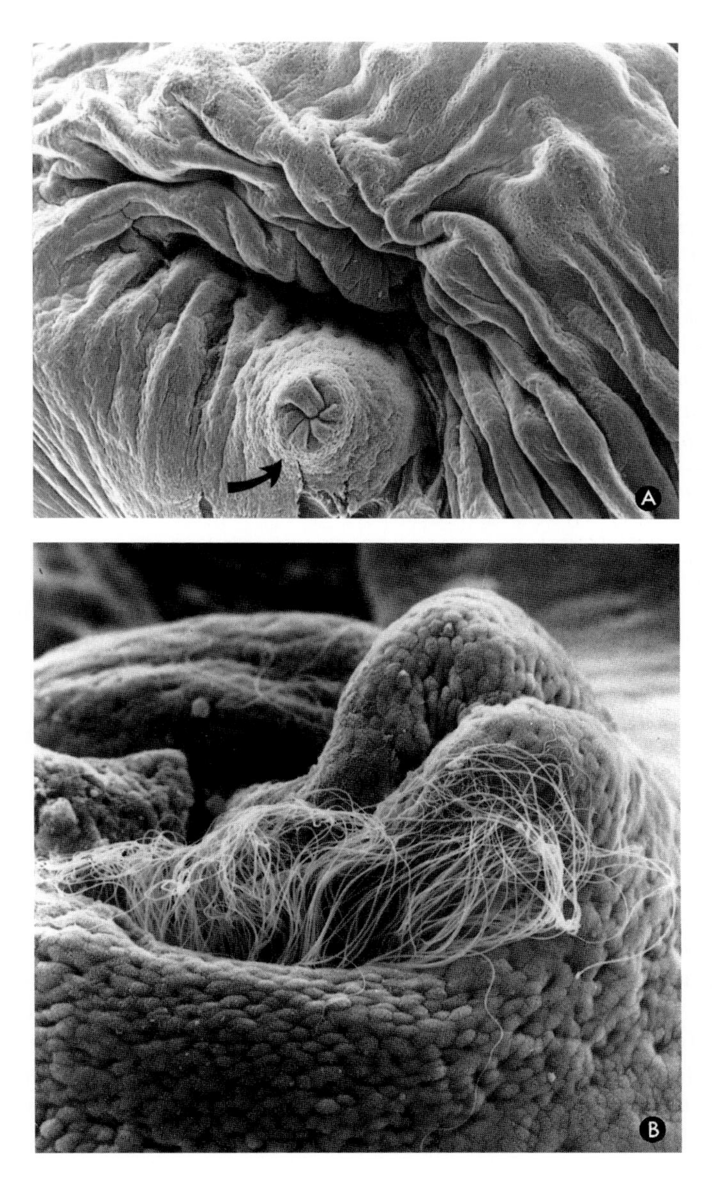

Fig. 16.1. *Scanning electron micrographs of the rat uterotubal junction viewed from its uterine aspect. (A) The junction, arrow, appears as a small papilla projecting into the uterine cavity (× 654). (B) The uterotubal junction, removed from a female, killed after mating, with spermatozoa, the heads of which are buried within the ostium (× 537). From P. Gaddum-Rosse (1981)* Am. J. Anat. **160,** *333–341, with permission from the author and John Wiley & Sons, Inc.*

Table 16.2. The duration of capacitation in several mammalian species (when more than one reference was used, average values were obtained).

Species	Capacitation time (h)	References
Sus scrofa	4.00	1
Bos taurus	6.75	2, 3
Mesocricetus auratus	2.50	4
Homo sapiens	6.50	1
Oryctolagus cuniculus	6.00	5
Ovis aries	5.25	3, 6
Rattus norvegicus	3.50	7
Mus musculus	3.25	8
Macaca mulatta	6.00	1

[1] Austin (1985); [2] Hunter (1988); [3] Fournier-Delpech and Thibault (1991); [4] Cherr and Drobnis (1991); [5] O'Rand and Nikolajczyk (1991); [6] Mattner (1963); [7] Shalgi (1991); [8] Storey and Kopf (1991).

viable in the female tract for months and, as in the isthmus of the oviduct, associations between sperm and the epithelial lining of the female tract (oviduct, uterotubal junction or uterus) appear to extend sperm survival (Fenton 1984).

Around the time of ovulation, and probably in response to an endocrine 'signal' released from the follicle (Hunter 1988), sperm present in the isthmus of the oviduct complete capacitation and detach from the wall (Smith and Yanagimachi 1991). Only a small fraction of sperm present in the isthmus continue their migration towards the ampulla. At this stage, spermatozoa acquire a peculiar type of motility, originally described as 'hyperactivation', which has now been shown to be an efficient method of forward propulsion within the viscous fluids encountered in this section of the female tract (Suarez *et al.* 1991b) and an efficient way of penetrating the cumulus matrix (Suarez and Dai 1992). The first spermatozoa to reach the vicinity of the ova appear to be those most likely to fertilize them (Cummins and Yanagimachi 1982). The ratio of sperm to ova in the ampulla, in those species in which *in vivo* studies have been carried out (hamster, rat, mouse, guinea pig and rabbit) is 1:1 (Cummins and Yanagimachi 1982; Smith *et al.* 1987; Shalgi and Phillips 1988; reviewed by Hunter 1993). The widespread idea of numerous sperm reaching the ovum and competing to penetrate the ovum vestments, is an artefact of *in vitro* fertilization experiments and is very misleading.

When the spermatozoon reaches the ovum, a tightly controlled sequence of events takes place. Before it can start to penetrate the cumulus oophorus, the spermatozoon must have become capacitated but should not have undergone the acrosome reaction (Suarez *et al.* 1984; Talbot 1985; Cummins and Yanagimachi 1986). Furthermore, the fertilizing spermatozoon should reach the surface of the zona pellucida with the acrosomal 'cap' still on; acrosome-reacted sperm cannot bind to the zona pellucida (Saling and Storey 1979; Florman and Storey 1982; Liu and Baker 1994). Thus, the acrosome reaction is completed on the surface of the zona pellucida, although it probably begins when the spermatozoon interacts with the progesterone secreted by the cumulus oophorus (Roldan *et al.* 1994): progesterone appears to prime the spermatozoon (initiating a series of molecular events) for the subsequent action of an additional zona pellucida agonist, the glycoprotein ZP3 (which triggers further cell signalling events ending in membrane fusion and the release of enzymes contained in the acrosome).

Since the spermatozoon penetrates the cumulus oophorus with the acrosome intact, the enzymes contained inside the acrosome do not appear to be required for cumulus penetration, although some surface enzymes, such as hyaluronidase, may facilitate sperm passage (Zao *et al.* 1985; Lin *et al.* 1994; Hunnicut *et al.* 1996). It is clear that the acrosomal enzymes aid sperm penetration through the zona pellucida, because spermatozoa lacking an acrosome cannot penetrate the zona (Liu and Baker 1994; Yanagimachi 1994) and specific blockade of certain acrosomal enzymes prevents sperm penetration of the zona pellucida (Liu and Baker 1994). However, the exact role played by acrosomal enzymes is still a matter of debate.

The acrosome contains a large array of powerful enzymes, some of which are located in the plasma membranes and may be involved in the generation of second messengers (Zaneveld and De Jonge 1991). In other taxa the acrosomal enzymes dissolve the ovum vestments producing a hole, through which the spermatozoon swims (Yanagimachi 1994): this is not the case in mammals. The hypotheses proposed to explain how a spermatozoon penetrates the zona pellucida differ in the emphasis put on enzymatic action, as opposed to mechanical force. Drobnis *et al.* (1988) suggested that the acrosomal contents released on the zona surface digest the cumulus matrix locally and that this enables the proximal region of the sperm tail to move more freely, allowing more efficient movement (Fig. 16.2). Conversely, Talbot (1985) speculated that part of the hyaluronidase released on the zona surface may diffuse into the perivitelline space and digest hyaluronic acid in the zona and perivitelline space. Another possibility is that acrosomal enzymes may be important only during sperm binding and initial entry into the zona. Once the sperm head is within the zona it may no longer require acrosomal enzymes and the mechanical force generated by the tail may be sufficient. Supporting evidence comes from the study by Jedlicki and Barros (1985) who noted that successful penetration of the zona pellucida was always preceded by successful sperm binding and the production of a hole by enzymatic activity in which the sperm head was inserted. The magnitude of such hole seems superficial in comparison with other taxa, but may nevertheless be sufficient for the sperm to begin successful penetration (Fig. 16.3). After this first step, penetration proceeds mainly by the movements of the sperm tail (Katz *et al.* 1989; Katz and Drobnis 1990).

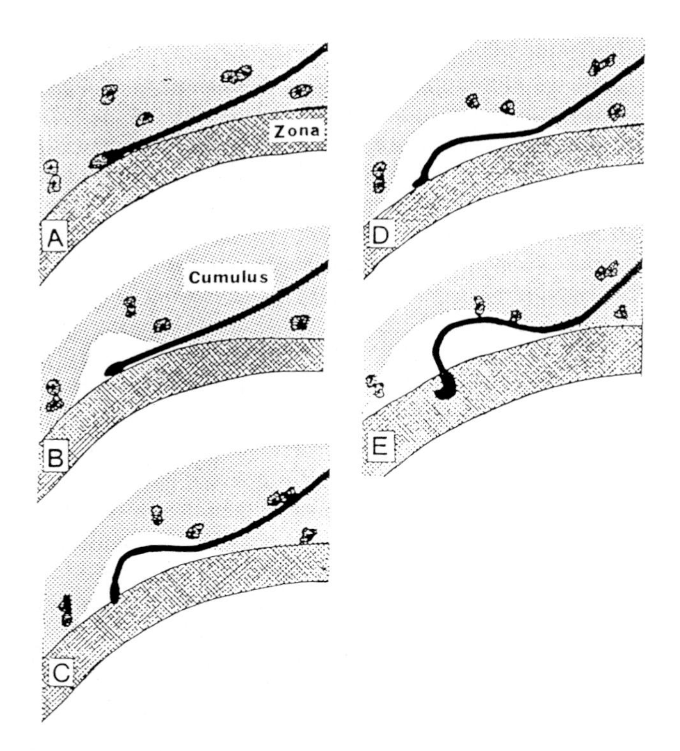

Fig. 16.2. *Diagrams showing that hyaluronidase released from the acrosome at the zona surface depolymerizes the cumulus matrix locally to enable the proximal region of the sperm tail to move more freely. From E. Z. Drobnis, A. I. Yudin, G. N. Cherr and D. F. Katz (1988)* Developmental Biology **130,** *311–323, with permission from the authors and Academic Press.*

1. What does the female tract select for?

(a) Sperm numbers

Short (1981) was the first to propose that because ejaculates tend to become diluted in larger female tracts, males should increase the number of sperm in each ejaculate as the size of the female tract increases (see also Dahl *et al.* 1993). The relationship between testes size and body weight may result, at least partly, from this effect. It has been found that males do indeed produce greater number of spermatozoa in relation to body size when females have long oviducts in relation to their body size (Fig. 16.4; Gomendio and Roldan 1993a). The oviduct was chosen as a variable because sperm need to swim actively along this section, and oviductal length is related to the distance between the site of sperm deposition and the site of fertilization.

Sperm numbers also appear to be related to the likelihood that spermatozoa will survive within the female tract. When sperm storage enhances sperm survival and/or sperm transport is particularly efficient, such as in

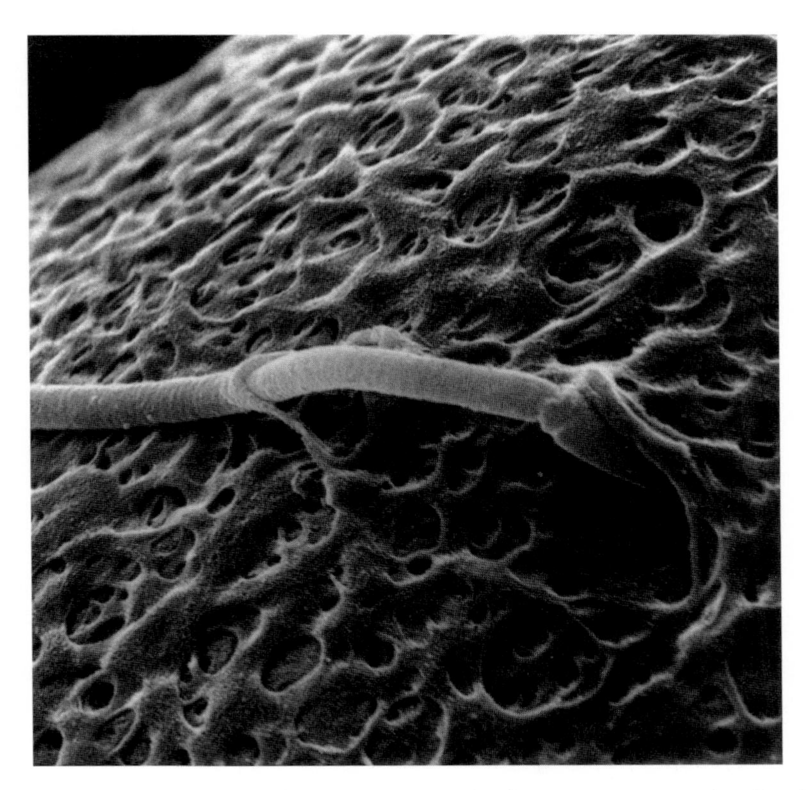

Fig. 16.3. *Scanning electron micrograph of a spermatozoon that has begun to penetrate the zona. Acrosomal ghosts are seen encircling the midpiece of the spermatozoon. From R. Yanagimachi and D. M. Phillips (1984)* Gamete Research **9,** *1–19, with permission from the authors and John Wiley & Sons, Inc.*

some marsupials, males produce ejaculates containing fewer sperm (Bedford *et al.* 1984; Taggart and Temple-Smith 1991, 1994).

(b) Sperm motility and sperm morphology

Good sperm motility in general is very important for active swimming along certain sections of the female tract and for penetration of physical barriers, such as the uterotubal junction and ova vestments. The percentage of motile sperm is one of the main factors influencing fertility rates (Drobnis and Overstreet 1992).

Sperm motility and sperm morphology are intricately linked because morphologically abnormal sperm swim more slowly or less efficiently (Katz *et al.* 1982) and are thus selected against at various levels: cervical mucus (Katz *et al.* 1989), uterotubal junction (Krzanowska 1974), oviduct and ova vestments (Krzanowska and Lorenc 1983; Meistrich *et al.* 1994). Thus, the percentage of morphologically normal sperm is also

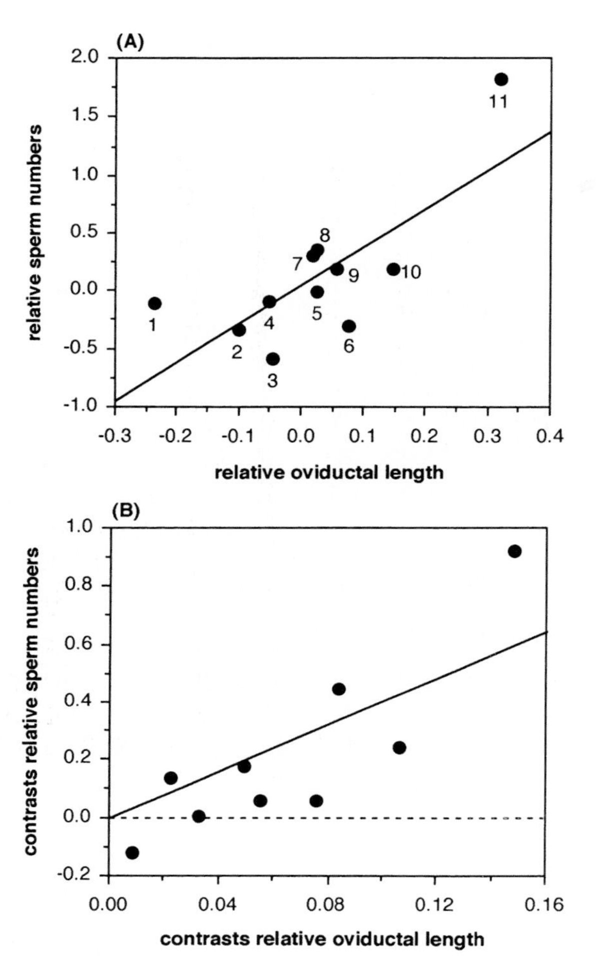

Fig. 16.4. *Relation between relative oviductal length and the relative number of spermatozoa per ejaculate. Oviductal length and number of spermatozoa were both corrected for body size. (A) Log values for each species: 1,* Papio cynocephalus*; 2,* Rattus norvegicus*; 3,* Homo sapiens*; 4,* Oryctolagus cuniculus*; 5,* Bos taurus*; 6,* Cavia porcellus*; 7,* Mus musculus*; 8,* Equus caballus*; 9,* Ovis aries*; 10,* Mesocricetus auratus*; 11,* Sus scrofa*. (B) Data points are contrasts for each variable. Contrasts are standardized linear differences in a trait between taxa at each node in the phylogeny. The use of contrasts removes phylogenetic effects in comparative studies. The regression line was forced through the origin. From M. Gomendio and E. R. S. Roldan (1993a)* Proceedings of the Royal Society of London, Series B, **252,** *7–12, with permission from The Royal Society.*

a good predictor of fertility rates in humans both *in vivo* and *in vitro* (Fig. 16.5; Liu and Baker 1994; Mortimer 1994).

Recent evidence suggests that even some morphologically normal spermatozoa may swim inefficiently and in these cases morphologically

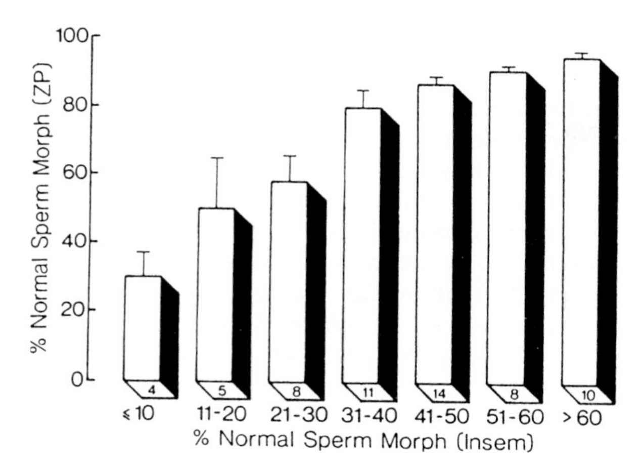

Fig. 16.5. *Correlation between percentage normal morphology of sperm in the insemination medium and bound to the zona pellucida (ZP). From D. Y. Liu and H. W. G. Baker (1994)* Male Factor in Human Infertility *(ed. J. Tesarik), pp. 169–185, Ares-Serono Symposia Publications, Rome, with permission.*

normal sperm may be unable to fertilize. Morales *et al.* (1988) found that human patients suffering from infertility not only had a lower proportion of normal sperm, but also their normal sperm were less likely to be motile and swam less efficiently than normal donor sperm.

(c) Ability to penetrate ova vestments

According to Drobnis and Overstreet (1992) the second most important factor determining fertility rates is acrosomal status, which is crucial for penetration of the ova vestments. Men with sperm without acrosomes are sterile and sperm without an acrosome are unable to fertilize either *in vivo* or *in vitro* (Liu and Baker 1994; Mortimer 1994). Thus, ejaculates in which sperm either undergo the acrosome reaction too early or fail to do so cannot fertilize ova. Because morphologically abnormal spermatozoa are less likely to undergo the acrosome reaction at the appropriate time, they are also selected against at this level (Fukuda *et al.* 1989).

Sperm also penetrate the ova vestments by thrusting movements of the flagella and by lateral oscillations of the head (Katz *et al.* 1989; Katz and Drobnis 1990; Bedford 1991). It has been suggested that sperm size might determine, to some extent, the ability of sperm to penetrate the ova vestments, because sperm with long flagella are able to generate greater forces. If this argument is correct, then sperm size should increase as the thickness of the zona pellucida increases. Contrary to this prediction, sperm length does not appear to be related to the thickness of the zona pellucida (Gomendio and Roldan 1993b). Thus, the forces generated by the flagellum when the spermatozoon penetrates the zona pellucida do not appear to have played an important role in selecting for sperm size. The lack of association between sperm size and the thickness of the zona pellucida suggests that either sperm penetration may be achieved primarily by enzymatic digestion of this ovum vestment, or that it involves synergy between enzyme activity and sperm forces. Moreover, it is likely that both the physical properties of the ova vestments and the enzyme contents of the acrosome vary between species.

Bedford (1983, 1991) has suggested that several traits of mammalian sperm have evolved in response to the thick layer of vestments surrounding the ova, a unique mammalian feature. These sperm traits would include: a large acrosome, the presence of hyperactivated motility, and enhanced stability of the sperm head, all of which would improve the physical thrust needed to penetrate the ova vestments.

Several experimental studies have shown that when males inseminate females in a competitive context, some males (or strains) are consistently more successful than others at fertilizing the female's ova, even when similar numbers of sperm are artificially inseminated ('differential fertilizing capacity') (Lanier *et al.* 1979; reviewed in Dewsbury 1984). We do not know what it is that makes particular males (or strains) more successful, but it is likely that their ejaculates will be of better quality in terms of one, or more, of the factors mentioned in this section.

B. Time constraints

The length of the oestrus cycle varies considerably between species: in some cases females are receptive for a few hours in each sexual cycle (e.g. mice, ground squirrels), whereas in others females are receptive for several days (e.g. macaques, gazelles), or throughout the whole cycle (e.g. vervet monkeys, humans). Parker (1984) originally suggested that, in order to maximize the chances of fertilizing the ova, sperm life-span should be positively related to the duration of oestrus. In support of this hypothesis, Parker found a positive correlation between sperm life-span and duration of oestrus when comparing nine mammalian species. This relationship, however, does not hold when phylogenetic effects are removed by using new statistical techniques but the analysis can be refined by taking into account that the maximum period of time that spermatozoa have to remain fertile is that between the onset of oestrus and ovulation. Although oestrus usually continues after ovulation takes place, the time period between ovulation and the end of oestrus is irrelevant in this context. Thus, spermatozoa should have been selected to remain fertile until ovulation takes place, rather than for the whole oestrus period. There is in fact a strong relationship between sperm fertile life-span and the interval between the onset of oestrus and ovulation, which remains highly significant after controlling for phylogenetic effects (Fig. 16.6) (Gomendio and Roldan 1993a).

Humans are a potential exception to a correlation between duration of

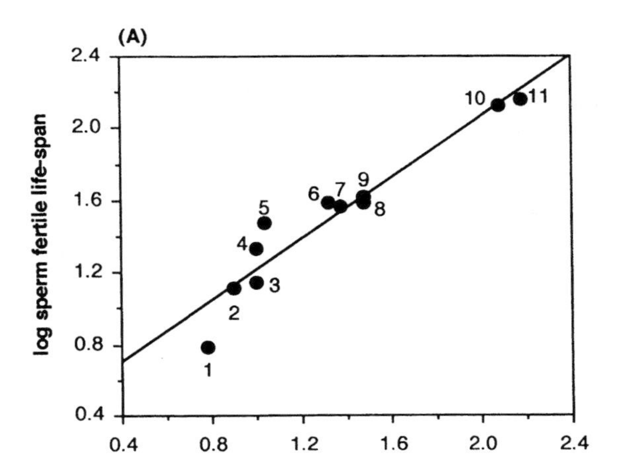

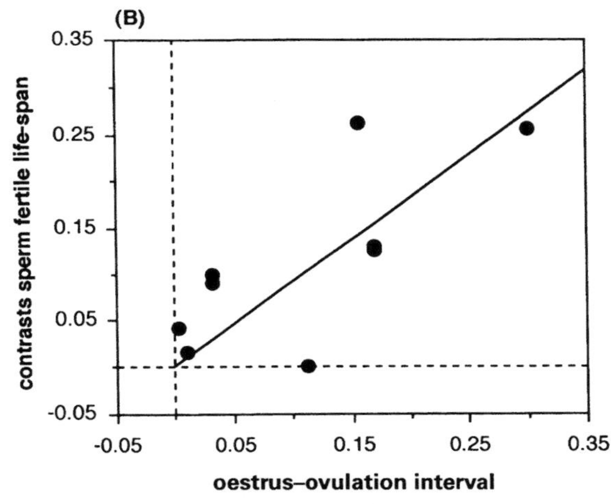

Fig. 16.6. *Relation between the interval between the onset of oestrus to ovulation and sperm fertile life span. (A) Log values for each species: 1,* Mus musculus; *2,* Mesocricetus auratus; *3,* Rattus norvegicus; *4,* Cavia porcellus; *5,* Oryctolagus cuniculus; *6,* Ovis aries; *7,* Sus scrofa; *8,* Bos taurus; *9,* Mustela putorius; *10,* Canis familiaris; *11,* Equus caballus. *(B) Data points are contrasts for each variable. Contrasts are standardized linear differences in a trait between taxa at each node in the phylogeny. The use of contrasts removes phylogenetic effects in comparative studies. The regression line was forced through the origin. From M. Gomendio and E. R. S. Roldan (1993a)* Proceedings of the Royal Society of London, Series B, **252,** *7–12, with permission from The Royal Society.*

oestrus and of sperm life-span, given that ovulation is extremely difficult to predict from behaviour, physiology, or morphology of females (Bancroft *et al.* 1983), and yet sperm are usually viable for only 2 or 3 days (Weinberg and Wilcox 1995; Wilcox *et al.* 1995). However, it seems that humans (by which is meant those involved in published large surveys, which are largely of those in Western populations) might copulate about once every 2–6 days, on average, although with considerable variation (Laumann *et al.* 1994, Table 3.4; Spira *et al.* 1994, Fig. 5.7; Wellings *et al.* 1994, Fig. 4.1), in which case sperm life-span is roughly equal to the average duration between mating and ovulation.

Eutherian spermatozoa are unique in that they need to become capacitated before they can fertilize the ova, and this process can take several hours (Yanagimachi 1988). Because sperm fertile life-span is so short, it may constrain the time available for capacitation. This appears to be the case: when sperm fertile life-span is long, capacitation takes longer than when it is short (Gomendio and Roldan 1993a). Interestingly, experiments in which spermatozoa have been placed in direct competition, and where there was no differential sperm transport, strongly suggest that spermatozoa which capacitate more quickly are more successful at fertilizing the ova, particularly when inseminated around or after ovulation (Krzanowska 1986).

Differences in sperm fertile life-span between different species may be related to differences in sperm size. Intergeneric comparisons have revealed a negative relationship between sperm size and sperm life-span: when females are receptive for long periods before ovulation takes place males produce short spermatozoa which may live for longer (Gomendio and Roldan 1993a). The obvious question to arise from these results is why should sperm size be related to sperm life-span. These authors have tentatively suggested an energetic argument based on Cummins and Woodall's (1985) finding that increases in total sperm length are achieved mainly through increases in the length of the principal piece of the tail, which are accompanied by relatively smaller increases in midpiece length. The midpiece contains the mitochondria which are the source of energy for the sperm cell. Gomendio and Roldan (1993a) have proposed that large spermatozoa may have relatively less energy to utilize and that this energy may be consumed more quickly owing to the greater requirements of larger cells and to their faster swimming speed. Alternatively, it is possible that sperm fertile life-span is determined through the interactions that take place between spermatozoa and the walls of the female tract.

C. Why have the female barriers evolved?

Given the magnitude of the barriers that spermatozoa encounter within the female tract, it is worth asking why these have evolved (Roldan *et al.* 1992; Birkhead *et al.* 1993; Gomendio and Roldan 1994). At least some of the barriers may have evolved to prevent infections, particularly those in the lower parts of the female tract which are exposed to the external environment and thus more easily colonized by external pathogens (see review by Profet 1993).

The barriers present in the female tract may also enable the females to avoid fertilizations which would result in nonviable embryos, such as those resulting from interspecific fertilizations or polyspermy, and to prevent abnormal sperm from reaching the ova. Several experimental studies have shown that the female tract acts as an effective barrier against interspecific fertilizations at different levels, from sperm transport to the ability to penetrate ova vestments (Maddock and Dawson 1974; Roldan *et al.* 1985; Smith *et al.* 1988; Roldan and Yanagimachi 1989; see review by Short 1976a). However, it is unclear whether this has been a strong selective pressure underlying the evolution of barriers in the female tract since there are widespread barriers at the behavioural level which prevent the occurrence of matings between individuals of different species. The prevention of polyspermy seems a more likely candidate (Polge *et al.* 1970; Hunter 1973), since this phenomenon is lethal in mammals (contrary to birds and reptiles, for example). Thus, avoidance of more than one spermatozoon penetrating the ova vestments is of crucial importance.

It is clear that the female tract selects against morphologically abnormal sperm (see above). This selection is based primarily on the fact that abnormal sperm tend to show deficiencies of flagellar activity and thus sperm movement, and as a consequence they swim more slowly and are less able to negotiate physical barriers. They show other significant deficiencies such as an inability to undergo the acrosome reaction in response to the appropriate signals and delayed capacitation (Fukuda *et al.* 1989; Long *et al.* 1996). The reason why morphologically abnormal sperm may suffer such strong selection on the part of the female tract is not obvious. It is possible that morphologically abnormal sperm carry poor or defective genetic material but there is little evidence which could be used to test this hypothesis. Sonta *et al.* (1991) showed that chromosomically abnormal sperm do not participate in fertilization. Experiments with aggregation chimaeras suggest that sperm head shape and the level of sperm abnormalities are determined largely by genes acting in the germ cells (Krzanowska 1991), and that genetic factors linked with the Y chromosome may influence the competitive ability of sperm (Krzanowska 1986). Another possibility is that the level of sperm abnormalities in an ejaculate is related to the quality of the male, and that by targeting abnormal sperm, females are ultimately selecting against a high proportion of sperm in ejaculates from low-quality males and thus diminishing their chances of fertilizing. There is evidence from a number of species that ejaculates containing a high proportion of abnormal sperm are less likely to fertilize (Oettle 1993; Liu and Baker 1994; Mortimer 1994). This effect would become exacerbated when sperm from males of different quality are in competition. A number of factors, such as inbreeding (Wildt *et al.* 1987a), diet and stress (Bronson 1989), related to male quality are known to influence the level of sperm abnormalities.

It is possible that females may select on a finer scale spermatozoa within a given ejaculate. One exciting but unexplored possibility is that some of the barriers present in the female tract may affect X- and Y-bearing sperm differently. One of the most obvious differences between X- and Y-bearing sperm, is the fact that the X-bearing sperm carry between 3 and 12.5% (depending on the species) more DNA than Y-bearing sperm (Johnson 1994). This difference has already been used successfully to separate X- and Y-bearing sperm by flow cytometric cell-sorting technology and could also be used by the female to sort sperm and perhaps favour one type. This would explain, at a proximate level, the biases in sex ratio at birth found in several mammalian species in which differential mortality during gestation does not appear to play a role (see review in Clutton-Brock 1991). It would be highly adaptive for females to be able to exercise such choice given the considerable reproductive advantages derived from producing offspring of the 'right' sex under some circumstances.

Finally, when spermatozoa from different males compete within the female tract, females may be able to select spermatozoa from particular males by forcing them to overcome a number of 'obstacles' (Birkhead *et al.* 1993; see Chapter 2). Even if direct selection on the part of the female was not possible, the female tract may provide a competitive arena where the 'abilities' of different ejaculates would be tested. Selection between ejaculates from different males would be highly adaptive if females could pass on the competitive traits of the successful ejaculates to their sons. On highly speculative grounds, it has also been suggested that females would benefit from choosing the best male genotypes on the basis of ejaculate features.

If the barriers present in the female tract allow females to test ejaculates from different males and to favour the most competitive ones, polyandrous females should possess barriers of a greater magnitude than monandrous females. It would make sense for example if polyandrous females had longer or more convoluted oviducts, so that sperm had to actively swim greater distances and thus selection against less vigorous sperm would be more intense. This question has not yet been addressed.

III. MECHANISMS OF SPERM COMPETITION IN MAMMALS

A. Scramble or contest competition?

Constraints to sperm production exist, despite mammalian sperm being extremely small compared with both the size of the male that produces them, and the ova (Dewsbury 1982). The evidence for costs of production include, among others, the effects of starvation, the existence of refractory periods, i.e. latencies to return to full production after a period of frequent copulations, and seasonal reduction in testes size and functioning (Dewsbury 1982; Bronson 1989; Harcourt 1991). The constraints on production mean that there will be constraints on mechanisms of sperm competition, and adaptations to it.

Theory, modelling, and evidence (below) all indicate that most mammalian sperm competition is scramble competition, as opposed to contest competition. In scramble competition, contestants do not actively prevent others from obtaining access to the resource, as in contest competition, rather they simply get to it first (or in other contexts use it up faster). Contest competition among sperm can be expected where: (1) the resource is defensible; (2) sperm come into close contact with one another; and (3) if contest competition involves death of sperm, an argument would have to be made that allowed considerable diploid control in order that altruistic sperm could exist. Scramble competition will occur if none of these conditions prevail.

In the case of sperm, the process of scramble competition has been likened to a raffle, or to bets in a race: either way, the greater the number of entries (tickets, horses, sperm), other things being equal, the greater is the chance of winning, i.e. fertilizing the ova (Parker 1970, 1982; Parker *et al.* 1972). With limits to production, larger numbers of sperm are achieved by producing smaller sperm (Parker 1982). Nevertheless, not all animals produce equivalently tiny sperm: considerable variation in sperm size occurs across taxa, within taxa, and within individuals (Seuanez *et al.* 1977; Cummins and Woodall 1985; Roldan *et al.* 1992). Some variation probably results from the variation within species and individuals and is a combination of normal and maladaptive variation (Cohen 1973; Manning and Chamberlain 1994); some might reflect interaction of haploid and diploid gene expression (Parker and Begon 1993), and some might indicate the possibility of adaptations to scramble competition that might more than compensate for a reduction in the number of sperm (Parker 1993; Parker and Begon 1993; Baker and Bellis 1995). Constraints on sperm number might mean that competitiveness can be enhanced only by an increase in sperm size (Gomendio and Roldan 1991; Parker 1993), or that not all sperm competition is by scramble (Baker and Bellis 1995).

Adaptive polymorphism of sperm is most evident in insects, some of which produce not just extremely different nucleate and nonnucleate sperm (pyrene and apyrene), but also (or instead) two obvious nucleate morphs, often a long and a short one (see Chapter 10). *Drosophila obscura* and associated species are an example, with short sperm averaging about 70 μm and long morphs about 200 μm (Bressac *et al.* 1991). Some long morphs are longer than the female's spermathecae – longer than either the male's or female's body in some species (Pitnick *et al.* 1995) – and the suggestion has been made that these long morphs might block subsequent sperm from entering the spermathecae. A correlation of sperm length with the length of females' sperm storage tubules in birds has also led to the suggestion that blocking might be a function of the sperm, in addition to fertilization (Briskie and Montgomerie 1992, 1993). In mammals, the ovum, let alone sperm-storage sites, or the female reproductive tract, are enormous in comparison with the size of the sperm, and thus contest competition is highly unlikely: the resource is indefensible, and sperm are unlikely to be in close contact with one another in large numbers for long periods. Some previous work indicated that high numbers of sperm reached the ovum. However, recent evidence indicates that very few sperm reach the ovum *in vivo* (Cummins and Yanagimachi 1982; Smith *et al.* 1987; Shalgi and Phillips 1988; Hunter 1993). Thus, sperm-storage sites remain the only likely places for contest competition. Only bats, American marsupials, and the Australian Dasyuridae appear to be able to store sperm for long periods (see Section I.B) (Bedford *et al.* 1984; Fenton 1984; see Chapter 15): only in these might contest competition among sperm be expected.

However, in birds, which have sperm-storage tubules where sperm live for weeks, and in which the mechanisms of sperm competition have been better studied (at least in the chicken) than in mammals, no contest competition has been demonstrated. Even passive blocking does not appear to occur. The last to copulate/first to fertilize (equivalent to a last-in/first-out) phenomenon, which was thought to reflect layering and hence potential blocking, now seems to be due simply to passive leaking of sperm reducing the numbers from previous ejaculates (Birkhead *et al.* 1995b; Colegreave *et al.* 1995). Whether sperm, as opposed to the total contents of a spermatheca, can be counted as blocking subsequent entry to a spermatheca depends on how much of a role the sperm, as opposed to the other spermathecal contents, play in the blocking. However, the paucity of apparent contest competition in birds implies that even in those mammals that store sperm for some time, we probably will not see much, if any, variety of adaptations to contest competition among sperm.

B. *Adaptive nonfertilizing sperm in mammals?*

In addition to blocking, sperm competition by contest could take the form of active displacement or chemical attack. All have been suggested for invertebrates (Sivinski 1984; Chapter 10). The most insistent argument that vertebrates, including mammals, produce sperm adapted for contest competition, in addition to ones adapted to scramble to reach the ovum, has been that of Baker and Bellis (1995). Most of their data in support of the hypothesis comes from humans, a species that, we will suggest later, are not often multiple-male maters. Argument exists about the reality and likelihood that mammals produce adaptive nonfertilizing sperm adapted for contest competition (Harcourt 1991). Our presentation of the argument will use Baker and Bellis's latest account of their 'kamikaze sperm' hypothesis (Baker and Bellis 1995).

Baker and Bellis's (1995) contest competition hypothesis is an alternative to the main hypothesis that explains variation in morphology of sperm. Cohen (1973) originally suggested that the abnormal sperm resulted from postsegregation mistakes in meiosis; for example, errors in chiasma formation. Baker and Bellis's initial proposal of their theory appeared to be stimulated by observations that far fewer sperm were needed during artificial insemination to achieve fertilization than were produced in any one ejaculate, and by observations of sperm in copulatory plugs. Sperm were adapted to, for example, improve the efficacy of blocking by plugs (Baker and Bellis 1988). Subsequently, they have argued, supported by more data than previous claims, that most sperm are adapted for contest competition. They suggest that, in humans, the main/only sperm adapted to fertilize are the micro and macro oval-headed sperm, a minority of all sperm; coiled sperm block subsequent sperm passage through the cervical mucus, as do other extreme morphs that hitherto were seen as pathologically abnormal. The majority of oval-headed sperm are not fertilizing sperm, but instead are adapted to seek, find and then destroy other sperm with their acrosomal lytic enzymes; tapering and pyriform sperm destroy the males' own sperm for control of offspring number (Baker and Bellis 1995; see Chapters 11 and 12). Four of their lines of evidence, almost all from humans, for a nonfertilizing, contest competitive role for sperm are that: (1) the proportion of coiled (abnormal) sperm in an ejaculate was inversely proportional to the percentage retention of sperm by the female of the next ejaculate, as if coiled sperm prevented entry of subsequent sperm (Baker and Bellis 1995, p. 260); (2) abnormal sperm increase with age of male, along with the probability of extra-pair copulations (Baker and Bellis 1995, p. 266); (3) sperm were more likely to agglutinate when mixed with sperm from other males than when mixed with fractions of sperm from the same original ejaculate (semen removed), as if they were adapted to destroy sperm from other males, using the powerful lytic enzymes in the acrosome (Baker and Bellis 1995, pp. 273, 297); and (4) because most normal sperm are nonfertilizing sperm, other findings about the relationship between sperm morphology and mating system (below) support rather than contradict their hypothesis (Baker and Bellis 1995, p. 289), an argument that will be made clear below. We suggest that except for one published set of correlations, for which the data have still to be presented, all of the evidence for nonfertilizing roles of mammalian sperm in contest competition is more easily explained by existing hypotheses. Any new hypothesis is usually an explanation of a phenomenon for which there are existing explanations.

Cohen's proposal that nonmodal sperm are meiotic mistakes is not an unsupported idea. For example, his analysis suggested a correlation between sperm redundancy and both mean chiasma frequency and haploid chromosome number, as would be expected if polymorphism in sperm indicated deviation from a main fertilizing morph, i.e. were meiotic mistakes, because numbers of chromosomes and frequency of chiasmata increase the chance for mistakes (Cohen 1969, 1973). With more data, especially on primates, and taking account of phylogeny, Manning and Chamberlain (1994) showed that Cohen's correlations still held, significantly so for the correlation of numbers of sperm with number of haploid chromosome. Their explanation for the correlations depended on the production of viable fertilizing sperm in conditions of both intra- and inter-ejaculate sperm competition, on the assumption that the more defective sperm there are the greater the number that will have to be produced to ensure at least one nondefective sperm reaches the ovum (Manning and Chamberlain 1994). Parker and Begon's (1993) alternative suggestion is that the variation could result from conflict between the male and the gametes over sperm size and number. The effect of the conflict varies with the risk of polyandry, but the models indicate that, other things being equal, polyandry should minimize variation; however, the effect has not been demonstrated (see below).

Countering Baker and Bellis's argument that more sperm were produced than were needed for fertilization, Harcourt (1991) suggested that sperm loss was often so great, and probability of fertilization low enough from one insemination, that far more sperm had to be produced than the one that could fertilize. Baker and Bellis have subsequently produced a lot of data demonstrating, in humans, how great the magnitude of sperm loss can be: 12% of ejaculates suffer 100% ejection from the female (Baker and Bellis 1995, p. 45) and, using their own argument, few sperm are fertilizing sperm. That nonfertilizing sperm are found in greater numbers near the entrance of the female reproductive tract, in plugs and in the cervical mucus cannot in itself be evidence that such sperm are adapted to block passage of subsequent sperm (Baker and Bellis 1995, p. 294), in the same way that finding contestants with broken legs nearer the start than the end of an obstacle course would not be evidence that they broke their legs to trip others. However, the possibility that the proportion of coiled (blocking) sperm in one ejaculate correlates inversely with retention of sperm from a subsequent ejaculate could be compelling, if the data were presented, as opposed to just the statistical values. In addition, because experimental procedure can itself induce coiling (Mortimer 1994), proper evaluation of the results also requires more details of methodology.

Baker and Bellis proposed that the increase in the proportion of abnormal sperm with age in humans matched the increase in probability of extra-pair copulations by males with age, and indicated an increase in nonfertilizing morphs adapted to contest competition (Baker and Bellis 1995, p. 261). First, Baker and Bellis's data showing an increase in probability of extra-pair copulations with age of male (Baker and Bellis 1995, p. 266) is not matched by all other studies. All indicate either no change, or a decrease in the frequency of extra-pair copulations with age (Laumann *et al.* 1994; Spira *et al.* 1994; Wellings *et al.* 1994). For example, Spira *et al.* (1994) state 'The older the men (among those who live as a couple), the more often their last intercourse was with the women with whom they lived' (Baker and Bellis 1995, p. 126). Thus, even Baker and Bellis premise might be wrong, let alone the conclusion from the premise. Second, proportions of abnormal sperm also increase with inbreeding and infection (Wildt *et al.* 1983), and the proportion of abnormal cells of all body parts increase with these three factors (and correlates with the infirmities of age, including reduced fertilizing ability).

For Baker and Bellis, one of their stronger lines of evidence for nonfertilizing sperm was their own observation that sperm were more likely to agglutinate when mixed with sperm from other males than when mixed with fractions of sperm from the same original ejaculate (Baker and Bellis 1995, pp. 272–276). They also speculated that sperm would destroy other sperm by undergoing an acrosome reaction and thus releasing lytic enzymes. Thus, they claim (Baker and Bellis 1995, p. 276) that in heterospermic mixes there is 'an increase ($P = 0.009$) in the number of diads, pairs of sperm joined at the head . . . of the order of 42% after 6 h (8.02 ± 3.1 in homospermic; 11.38 ± 3.8 in heterospermic)', or that 'sperm appear, in effect, simply to drop dead after having swum past another sperm' (Baker and Bellis 1995, p. 275). Leaving aside the issue of data handling and statistics, there are two main problems with their results and interpretations. Baker and Bellis supply little detail of the experimental conditions used (e.g. selection swim-up protocol for/against different sperm morphs, composition of the incubation medium, sperm concentrations, control of pH during incubations, measures taken to avoid sperm sticking to glass and/or plastic surfaces), although experimental conditions can strongly influence the outcome of such tests. Furthermore, it is also essential to know what type of sperm donors were employed, and how variables such as blood groups, rhesus factor, and absence/presence of antisperm antibodies were controlled. In the absence of these details, the information has to be considered as merely anecdotal.

The hypotheses they consider are not supported by current knowledge of mechanisms of fertilization and sperm function. The biochemistry and physiology of the acrosome reaction has been exhaustively studied both with regard to stimuli initiating it and the mechanisms underlying this process (Roldan 1990; Fénichel and Parinaud 1995). The acrosome reaction is initiated in response to the specific ovum-associated agonists progesterone and the zona pellucida glycoprotein ZP3 (Roldan *et al.* 1994). It is clear that the acrosome reaction is essential for penetration of the zona pellucida: sperm without an acrosome cannot fertilize (Liu and Baker 1994; Yanagimachi 1988, 1994). Furthermore, the acrosome reaction must be completed on the surface of the zona pellucida because only acrosome-reacted sperm can penetrate the zona pellucida but acrosome-reacted sperm cannot penetrate the cumulus oophorus or attach to the zona (Yanagimachi 1994). If, as argued by Baker and Bellis, the acrosome reaction is an adaptive suicidal response to proximity of sperm from other males, it is pertinent to ask how would the acrosome reaction be triggered in this context. If 'when sperm encounter sperm from other male, they release acrosomal enzymes into the surrounding medium' (Baker and Bellis 1995, p. 275) it is also necessary to consider how would sperm detect the proximity of other sperm and, moreover, distinguish that the approaching sperm belongs to a different male. Baker and Bellis's proposed function of the acrosome reaction as a 'sperm killer' is difficult to reconcile with the fact that the acrosome reaction takes place on the surface of the zona pellucida. If the main function of the acrosome reaction is to kill rival males' sperm, why has it been selected to wait until so late in the race, to wait for such a specific time and place in the race, and especially to wait until so few other sperm are present (sometimes none)? If as few sperm reach the vicinity of the ovum as current work indicates, suicide on the verge of success seems remarkably unadaptive.

Baker and Bellis (1995) have produced other evidence that they suggest supports a notion of nonfertilizing sperm adapted for contest competition, such as the fact that human females have crypts adapted for sperm storage. The evidence for this and other claims is so marginal, and so at variance with existing well-supported evidence for other explanations, that we will not discuss other problems with the biological mechanism of contest competition, but move on to consideration of predictions from the theory. Harcourt (1991) tested a range of what he suggested might be reasonable predictions from Baker and Bellis's contest competition hypothesis concerning the relation between sperm production and mating system. For example, the percentage of normal sperm and the coefficient of variation in sperm length should be lower and higher, respectively, in polyandrous taxa, if most sperm are nonfertilizing morphs adapted to contest competition. No correlation of mating system (or relative testes size) with proportion of normal sperm was found in small samples of primates (Møller 1988a; Harcourt 1991), and a far larger sample of mammals (Møller unpublished). In contrast, and potentially in contradiction to Baker and Bellis's hypothesis, the number of motile sperm correlated with polyandry in Møller's primate sample (with number of normal sperm, and percentage motility contributing independently to this result), a result confirmed by further analysis (Section IV.B.2). With respect to variability of sperm, while Harcourt's analysis indicated the possibility that polyandrous genera had less variable sperm lengths than did monandrous genera, he did not take sophisticated account of phylogenetic effects. We therefore performed the analysis using Purvis' (1995) primate phylogeny, and Purvis and Rambaut's (1995) comparative analysis by independent contrasts (CAIC) to search for an association between the coefficient of variation of sperm length with mating system, and with residual testes. We found no association. This result is sufficiently at variance with predictions to make it necessary to test it with a larger sample size, and on taxa other than primates.

Baker and Bellis's more general criticism of such tests of their hypothesis is that because their hypothesis now is that almost all sperm are nonfertilizing, including most of the normal sperm, tests that assume that only abnormal sperm are adaptively nonfertilizing are invalid (Baker and Bellis 1995, p. 289). Yet, at the same time, Baker and Bellis suggest that at least one measure of the proportion of normal sperm in an ejaculate could be an index of the proportion of fertilizing sperm (Baker and Bellis 1995, p. 304). That being the case, measures of normal sperm can be used to test their hypothesis, contrary to their suggestion.

Even if nonfertilizing sperm adapted for contest competition are unlikely in mammals (and we repeat that the evidence for them is negligible, and against them, considerable), the existence of a variety of fertilizing morphs is not impossible. Females differ from one another in that the female reproductive tract has widely varying conditions over time, and along its length and the timing of insemination in relation to ovulation varies from copulation to copulation. Perhaps the polymorphism is an adaptation to the variation and its unpredictability (Roldan and Gomendio 1992). Thus, Baker and Bellis (1995, p. 270) have suggested that while large oval-headed sperm of humans swim faster than small ones (beneficial if copulation occurs shortly before ovulation), the small oval-headed sperm of humans survive longer and are more likely to be retained by the female (beneficial under more adverse tract conditions than usual, or if insemination occurred some time before ovulation).

IV. WHAT HAS SPERM COMPETITION SELECTED FOR IN THE MALE MAMMAL?

Scramble competition will usually select for many highly viable sperm that reach the vicinity of the ovum quickly, and effectively activate it (Parker 1984; Chapter 1). Parker (1984) has suggested also that sperm should be selected to be viable for the time that it takes them to meet a fertilizable ovum, and hence that sperm life-span should equal the duration of oestrus (see above, Section II.B, for further discussion). A further prediction from Parker's hypothesis, and we think it is a novel one, is that the life-span of sperm should be longer in species where the males and females range separately than in those in which they range together, and among the former, longer in those without induced ovulation. This prediction is made on the assumption that when the sexes range separately they cannot monitor each other's reproductive readiness, and are not more able than the others to stimulate reproductive readiness (although see Section IV.D).

If we assume that the frequency and intensity of sperm competition is stronger in taxa in which several males copulate with a single potentially fertile female (polyandrous taxa), then adaptations to sperm competition should be more developed in such taxa than in monandrous taxa. A powerful way to test for the action of sexual selection is thus to compare characters of polyandrous and monandrous taxa (Short 1979; Harcourt *et al.* 1981). Monandrous taxa include both monogamous taxa (each male and female copulate with only one partner, usually their social partner), and one-male harem taxa (one male copulates with several females, who copulate only with that male). Another category of social system and mating system also exists. Dixson (1987) was the first to point out that primate species with a dispersed social system (male and female range separately) appear to exhibit anatomical correlates of multi-male mating. We know little about their mating system, because many of them are prosimians, and are small and nocturnal. Nevertheless, not only has polyandrous mating occasionally been seen in these species, but Dixson's original observations are being confirmed, both with more data, and with tighter phylogenetic analysis (Verrell 1992; Harcourt *et al.* 1995). Thus, dispersed primate taxa are now sometimes being lumped with polyandrous taxa and compared with the monandrous taxa.

In this section, we examine correlates of polyandrous mating (and hence sperm competition) in behaviour of males, numbers of sperm, nature of sperm, nature of semen constituents; and morphology of the penis. The female's role in sperm competition, whether she promotes or accepts it, and her reactions to it, are considered in the next section. Evidence suggests little difference among mammals in the nature of sperm competition, so we should expect concordance in adaptations to it, and therefore in, for example, ejaculate features (Møller 1991a). We suggest that in taxa with sperm competition males tend to initiate copulation, to mate-guard after copulation, to copulate frequently and rapidly, to have large sperm-producing organs with relatively large amounts of spermatogenic compared to interstitial tissue, to have large sperm reserve volume, and to have high numbers of motile, powerful sperm whose lifetime is proportional to the duration of oestrus. We also suggest that they probably have large amounts of seminal fluid in order to carry more sperm, to neutralize the females' chemical environment, and perhaps also to be used in plug formation, and have more obviously specialized penises.

A. Behaviour

1. Male initiation

Males who are most active in initiating copulations should achieve more inseminations than others and a comparison of monandrous and polyandrous taxa indicates that males of polyandrous taxa are more active initiators. Thus, among rodents, males of polyandrous taxa waited less to copulate after being placed with a female than did males of monandrous taxa (seven polyandrous species (five genera); three monandrous genera and species) (Dewsbury 1981). Mating systems were from Gomendio and Roldan (1991). While Dewsbury's explanation was not in terms of sperm competition, but rather that choice of the correct mate might make more difference to the reproductive success of monogamous males than of others, the data can obviously be interpreted in terms of sperm competition. The same contrast is seen among apes and baboons (Harcourt 1981). Thus, chimpanzee males actively initiate copulations, sometimes with a specific copulatory display, while gorilla females are more active at soliciting copulations, even if males tend to be more aggressive to females on days that females are in oestrus. The relationship between the monandrous gelada *Theropithecus gelada* males and females barely changes on days that the females are in oestrus, while polyandrous *Papio* males significantly increase their grooming of females, and become more responsible for maintaining proximity to them. While active courtship displays are rare in mammals, an obvious prediction is that in all taxa, the displays would be more obvious in the polyandrous species, as they appear to be among the African apes. This prediction has still to be tested.

2. Multiple ejaculation

Parker (1984) argued that a single ejaculation is optimal when males cannot detect ovulation, when most (>90%) sperm live for the duration of a female's oestrus, and when males are biologically capable of inseminating as many sperm in a single ejaculate as they could in multiple ejaculates. However, when there are limits to production and storage, and hence limits to the size of a single ejaculate, as there probably are (Section IV.B), then multiple ejaculation is an obvious way of increasing the number of sperm inseminated. Second, if sperm do not last for the duration of oestrus, as might be the case (Section II.B), multiple ejaculations give a greater chance of optimally timing insemination in relation to ovulation. Third, in multi-male systems, where several males might be present around a fertile female, it might pay to inseminate a large amount of sperm by several brief completed ejaculations rather than attempt it by a (presumably) long single mount that could well be interrupted before completion, especially as subordinate males and even juveniles can interrupt copulations (Smuts *et al.* 1987).

The majority of mammals are capable of multiple ejaculation; species counts indicate over 80% (Dewsbury 1972). Experimental results on the reproductive consequences of multiple ejaculation consistently indicates that where multiple ejaculation makes a difference, it is associated with increased number offspring sired by the multiply ejaculating male, even when only one male copulates, but especially when more than one copulates (Gibson and Jewell 1982; Bedford *et al.* 1984; Dewsbury 1984; Robl and Dziuk 1988; Ågren 1990). Most studies indicate that sperm competition promotes such multiple ejaculation. Thus, Dewsbury's (1981) data on number of ejaculations before satiety in muroid rodents indicate that the seven polyandrous species ejaculated more often in a bout than did the three known monandrous species. Dewsbury (1981) also indicated that all nonmonogamous species displayed a stronger Coolidge effect (ability to remate in the presence of a strange female) than did all the monogamous species (n = 3 polyandrous species and genera, n = 3 monandrous species and genera). Most analyses for primates show the same effect (with Dewsbury and Pierce's (1989) being an exception): polyandrous primates ejaculate more often than monandrous ones (Short 1979; Harcourt 1981; Dixson 1995a). Thus, Dixson found that polyandrous primate species ejaculate over five times as often as do monandrous ones (once every 1.1 h compared with once every 7.6 h; Dixson 1995a). Controlling for phylogenetic independence, using Ridley (1983) and CAIC (Purvis and Rambaut 1995), Dixson's sample is reduced to only six independent taxa, and four independent contrasts, but the result is still the same. Polyandrous and monandrous taxa do not overlap in mean ejaculatory frequency, and within all four independent pairs of related taxa, the polyandrous taxon ejaculated more frequently than did the monandrous one. The average frequencies themselves also remained very similar: the median polyandrous taxon ejaculated nearly once per hour (0.9 times), compared with once every 6 h per monandrous taxon (0.15 times). Furthermore, Dixson's data indicate that polyandrous taxa have consistently higher ejaculatory rates than polygynous ones (three contrasts, mean of 0.9 ejaculations per hour compared with 0.2), again indicating that sperm competition rather than sperm depletion might be the better explanation for differences between taxa in sperm output. Nevertheless, some effect of sperm depletion probably exists because the two polygynous taxa ejaculated at four times the frequency of the three monogynous ones (two contrasts). In mammals as a whole, a crude comparison of relative frequency of copulations showed 80% of 23 multimale species with frequently copulating males, compared with only 15% of 13 monandrous species (Møller and Birkhead 1989). A similar contrast is seen within species (below).

As expected, if sperm competition makes brief intromissions advantageous, the polyandrous chimpanzee has shorter mounts (7.5 s) than the monandrous apes (e.g. the gorillas 1.5 min) (Short 1979; Harcourt *et al.* 1980). Perhaps where prolonged intromissions are associated with polyandry, and the prolongation occurs after ejaculation, as it does in at least some of the species (Dixson 1987), then the prolonged intromission can be seen as a form of postcopulatory mate-guarding (Section IV.A.4).

A number of reports from various species of mammal indicate that males are particularly likely to copulate on first meeting a female, or on re-meeting a known female after an absence (Møller and Birkhead 1989; Baker and Bellis 1995). Whatever the proximate mechanism, an obvious ultimate explanation is sperm competition. In the absence of knowledge of previous nonmating, the male's reaction to possible mating should be to copulate. At the same time, one could expect such a behaviour in the absence of sperm competition: with separation comes lack of knowledge of female reproductive state, and therefore an advantage of mating as soon as the female is met in case she is ovulating. Better evidence that males react behaviourally to the probability of sperm competition are the many observations of males being more likely to copulate, and to copulate repeatedly, in the presence of other males (Hrdy and Whitten 1987; Møller and Birkhead 1989). For example, male pigtail macaques *Macaca nemestrina* can reduce their normal 43 min interejaculatory interval to 17.5 min if they see another male copulate with the female with whom they had been mating (Busse and Estep 1984); dominant bighorn sheep

Ovis canadensis copulate at six times their mean rate immediately following their takeover of a female from another male (Hogg 1988); and Grevy zebra *Equus grevyi* males mating polyandrously copulated nearly five times more frequently than monandrous males, at 1.13 times per hour compared with 0.23 times (Ginsberg and Rubenstein 1990).

3. Multiple mounting

Multiple-mounting and multiple ejaculation need to be distinguished in the context of sperm competition. The former refers to the apparent physiological requirement for several mounts before ejaculation and the latter is merely more than one insemination per female per fertile period, or some measure of more or less frequent ejaculation. Of the suggestions for the function of multiple-mounting, none seem particularly advantageous under conditions of polyandry. Multiple mounting might stimulate reproductive readiness in the female, but not obviously more so than thrusting during a single episode of mounting. Indeed, multiple mounting might be particularly disadvantageous in polyandrous mating systems if it allows supplantation before ejaculation (Rood 1972; Harcourt 1981; Hrdy and Whitten 1987).

Dewsbury and Pierce's (1989) data for primates indicate little or no relation between mating system and various measures of copulatory behaviour, including multiple mounting. In their conclusions, they did not correct for phylogenies, but the data themselves show so little relation that correction would probably not change the conclusion. Dixson (1991) differed; in his analysis of a sample of 12 species who multiply mounted, 11 were polyandrous, or had a very recent evolutionary history of being polyandrous (*Papio hamadryas*). While multiple mounting might be concentrated among polyandrous primate species and be rare among monandrous primate species, many polyandrous species are also single-mounters. Dixson's functional explanation for multiple mounts was that they dislodged plugs and evidence exists from rodents to show that they could so function (Hartung and Dewsbury 1978). However, vigorous thrusting during a single mount would have a similar effect, and the fact that many polyandrous taxa are single mounters, and that almost all mammals thrust, makes this hypothesis weak.

Shively *et al.* (1982) made a more sophisticated comparison of single-mount or multiple-mount inseminators than simply whether the species were polyandrous or monandrous. They suggested that the distinction was based not so much on the number of males mating, but more on the degree of competition between males for access to females. Where dominant males tended to obtain a disproportionate share of matings, multiple mounting was prevalent; where males were more tolerant, and presumably therefore where there was more sperm competition, single mounting was the prevalent ejaculatory behaviour. If Shiveley *et al.*'s (1982) distinction holds (different authors classify species differently), their comparison contradicts the suggestion that competition among males for access to females in a multi-male group should make rapid ejaculation advantageous. An alternative possibility is that single-mount ejaculation allows frequent ejaculations (see previous section), but data from Dixson's analysis of ejaculatory frequencies in primates (Dixson 1995a) suggests the opposite: the one multiple-mount species ejaculated at three times the median frequency of the three single-mount species.

The selective pressures on multiple mounting remain unclear. We clearly need biological data on, for example, whether multiple mounting is more efficacious than thrusting for stimulating the female, on whether multiple mounting is more likely to dislodge previous semen coagula, and if so whether it is more likely to result in fertilization, whether long mounts are more likely to be disrupted; and so on.

4. Postcopulatory mate guarding

Postcopulatory mate guarding can be interpreted as a reaction to potential sperm competition. If the first male to copulate were always the one to fertilize the female, there would be no need for the first male to prevent subsequent copulations. However, as we have stated (Section I.C), copulations within the overlapping lifetimes of sperm and ovum are frequent in mammals, and we can expect certain regularities in the frequency and intensity of mate guarding in relation to whether ovulation is induced or not, and the length of oestrus (Section VI). Postcopulatory mate guarding can take several forms, from fighting other males to prevent access to the female, to prolonged mounting, to a copulatory lock, to deposition of a copulatory plug, formed from semen constituents, in the female's tract.

In the case of prolonged mounting in primates (>3 min), it appears that at least among the prosimians, most of the prolongation occurs after ejaculation, as if the animals are postcopulatory mate guarding (Dewsbury and Pierce 1989). Dixson (1991) suggested that relative freedom from predation (the primates with prolonged copulations are almost all arboreal, and the prosimians are nocturnal), and postural stability on flimsy branches (the prosimians are small, and the atelines have prehensile tails) correlated with prolonged copulations being a means of sperm competition, implying that predation and instability prevented their use. However, *Lemur catta*, a polyandrous species, which is terrestrial and hence posturally stable, but lives on Madagascar and is therefore relatively free of predation, nevertheless does not have prolonged intromission (Dixson 1991). We cannot yet tell whether prolonged intromission is a form of postcopulatory mate guarding, or is simply acting as a plug to prevent leakage, or as a means of inducing peristalsis in the female to carry the sperm out of the vagina (and therefore also prevent leakage) and toward the ovum. All of these effects would be more advantageous under conditions of polyandry, but are not strictly mate-guarding.

Whether postcopulatory mate guarding is a beneficial tactic will also depend on the ease of finding other females. If they are difficult to find, fertilization of the current one should be assured by guarding, as Sherman (1989) showed when he compared two species of ground squirrels.

A very specific behavioural (and sometimes anatomical – see Section IV.D) form of postcopulatory mate-guarding is the copulatory lock. Dogs are perhaps the best known example, but other taxa also lock, for example, several rodents and also some marsupials, bats, and other insectivores (Dewsbury 1972; Hartung and Dewsbury 1978; Langtimm and Dewsbury 1991). However, is the lock an adaptation to sperm competition, or is it simply a means of preventing leakage of semen, and therefore sperm, by acting as a plug, or stimulating the female tract to transport the ejaculate? If locks are an adaptation to sperm competition, they would presumably work best if they lasted a significant proportion of the time that the female is potentially fertile, and/or that the sperm live. In the case of the dog, they do not: the lock does not last the 9 days of oestrus or even the 3 days of a sperm's minimum life-span (see Section I.B). Some lock might be better than no lock, however, especially if the function was stimulation, when they need last only long enough for the bulk of the sperm to be carried further than the deposition site of a subsequent insemination. Are locks more prevalent in multi-male than single-male taxa? We have no data. At the same time, if locks prevent leakage, a difference between the two mating systems might not be expected, especially if in the single-male taxa the male copulates only once with a female, and especially if females are widely distributed, because search time means that the male can then less afford loss of sperm. Whatever their function, locks might have costs, as Dixson (1991) suggested for prolonged copulations in primates. One piece of evidence for costs is that in New World neotomine-peromyscine rodents (Sigmodontinae, Muridae), the later evolved taxa produce copulatory plugs instead of locking (Langtimm and Dewsbury 1991). That at least rodent species appear to either lock or form plugs implies similarity in function (Voss 1979). However, the function of plugs has not been established (Section IV.C), and while locks cannot aid storage of sperm, they could function in all the other ways suggested for plugs.

5. Order effects

For some time, arguments continued about whether in mammals the first or subsequent male to copulate was the one most likely to fertilize the female. Data supported both camps (Dewsbury 1984). The argument is now largely resolved, with the realization that when both sperm and ovum are fertilizable only for a relatively brief time in relation to the period that the female is receptive, the male that copulates closest to the interval before ovulation that allows the sperm to become capacitated at the optimum time for fertilization is the male most likely to fertilize the female (see Section I.B) (Huck *et al.* 1989). Observations on a number of species indicate that males compete most to mate in the day or two before ovulation, as, for example, Bercovitch (1989) and Hogg (1984) have shown for a baboon *Papio cynocephalus* and a bighorn sheep *Ovis canadensis* population, respectively.

B. Physiology (sperm)

1. Testes size

Other things being equal, insemination of a large number of sperm (which is the major adaptation to sperm competition; see above) will be helped by a large amount of sperm producing tissue, and hence large sperm-producing organs, i.e. testes. All comparisons available indicate that polyandrous taxa have larger testes in relation to their body size than monandrous taxa. If testes size is plotted against body size, taxa above the overall regression line tend to be polyandrous; taxa on or below it tend to be monandrous (see Fig. 16.7 for primates). The relationship holds across the mammalian class, within orders (e.g. primates, ungulates), within families (e.g. cervids, equids, pongids), within subfamilies (e.g. cercopithecines), and even within species (Popp and DeVore 1979; Short 1979; Harcourt *et al.* 1981, 1995; Clutton-Brock *et al.* 1982; Kenagy and Trombulak 1986; Dixson 1987; Møller 1988a, 1989; Ginsberg and Rubenstein 1990; Dahl *et al.* 1993; Stockley and Purvis 1993). Kappeler (Kappeler 1993) suggested that the relationship did not hold for lemurs. However, if solitary species (which are equivalent to Dixson's dispersed category, and therefore perhaps polyandrous, but in fact have an unknown mating system) are excluded, all four group-living (presumably polyandrous) species had a positive relative testes size (volume), and three of the pair-living four had a negative one. The exception was *Varecia variegata*, which might not be the monandrous species that earlier reports indicated (Morland 1993). The relationship between relative testes size and mating system is so strong in mammals that relative testes size can be used as a good indicator of mating system. In the case of the prosimians analysed by Kappeler, for example, we can predict that of the three dispersed species, the one with the largest relative testes size, *Mirza coquereli*, will be polyandrous. Among baleen whales, the right whale *Eubalaena glacialis* must be polyandrous, whereas the blue whale *Balaenoptera musculus* is probably monandrous (Brownwell and Ralls 1986).

The relationship also holds within and between families of many other taxa, such as birds and butterflies (Møller 1988b, 1991b; Svärd and Wiklund 1989; Birkhead and Møller 1992; Gage 1994; Møller and Briskie 1995). The fact that exactly the same relationship is found in such extremely different taxa – taxa with different lifestyle, physiology, female anatomy, sperm storage, and so on – suggests that a main factor,

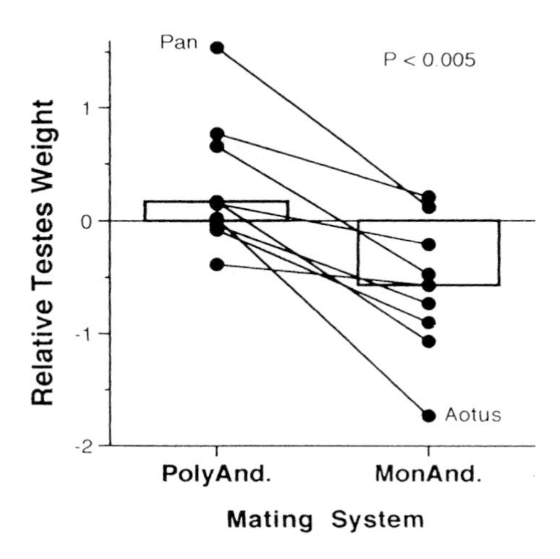

Fig. 16.7. *Relative testes weight (i.e. residuals from body weight × testes weight regression) in relation to mating system, comparing pairs of polyandrous and monandrous primate taxa identified as related by CAIC (Purvis 1995) from a sample of 58 primate species, using data from Harcourt* et al. *(1995, Table 2). Extreme taxon of each mating system is identified. (Wilcoxon signed-ranks test,* T = *45,* P = *0.004.)*

possibly the main factor, in the relationship is the single factor common to all the taxa, namely intensity of sperm competition. At the same time, polyandrous females in all the taxa have perhaps especially hostile reproductive tracts as a means of selecting for especially vigorous sperm.

One way to inseminate a large number of sperm is to store sperm over a long period, and then release the sperm all at once. If this occurred, testes would not necessarily have to be large. However, across nine species of mammal, relative numbers of sperm stored (and presumably epididymal volume) correlated with relative testes size in mammals (Møller 1989); in a highly polyandrous species, the primate rhesus macaque *Macaca mulatta*, Bercovitch and Rodriguez (1993) found a positive ($P < 0.05$) correlation between weight of epididymis and weight of one testis in a sample of 10 sexually mature animals ($r^2 = 0.44$). The implication is that storage correlates with production, and that there is no trade-off of storage with production.

All the studies that found correlations between probabilities of polyandry and relative testes size were, in effect, generalizations from Short's original statement of the relationship in great apes (pongids) (Short 1979), and are thus perhaps not strong tests of the explanation for the relationship. However, Parker's theoretical analyses suggested a novel prediction of sperm competition theory (Parker 1990a,b), which Stockley and Purvis (1993) used in a novel test of it. In species in which females are monopolizable, small males might be expected to compete more by sperm competition than by aggressive competition (Parker 1990b), and dominant males to compete less by sperm competition, especially if sneak matings are rare (Parker 1990a); in contrast, when females are not monopolizable, all males might use both forms of competition relatively equally. Females should be more monopolizable in nonseasonally breeding species than in seasonally breeding ones (Trivers 1972; Emlen and Oring 1977). Therefore, small males of nonseasonally breeding species should have relatively larger testes compared with large males than would be the case for seasonally breeding species (accepting, of course, that seasonality and body size are only rough measures of monopolizability and competitive ability). In other words, the slope of the regression line of testes weight on body weight should be shallower in the nonseasonal species (Stockley and Purvis 1993). Using a comparative analysis by independent contrasts, which compares members of independent pairs of related species that differ in the character of interest (Pagel 1992; Purvis and Rambaut 1995), Stockley and Purvis (1993) found that in all six taxonomically independent pairs of mammalian species (two of them primates), in which one member was nonseasonally breeding and the other member was seasonally breeding, the regression slope of the former was shallower or more negative (and the correlation coefficient lower) than of the latter, i.e. the small males of the nonseasonal species had relatively larger testes compared with the large males than was the case in the related seasonally breeding species. (The difference emerged despite the fact that seasonality is clearly an extremely crude index of the nature of competition.) For example, in a nonseasonal population of a possum *Trichosurus vulpecula*, testes weight correlated with body weight with a negative slope of 0.06, and a nonsignificant coefficient of –0.06, whereas in a seasonal population, the slope was 1.16 and the coefficient was a significant 0.58; and the nonseasonal primate *Alouatta palliata* had a slope of 0.37 and a nonsignificant correlation of 0.12, whereas the seasonal *Macaca fascicularis* had a slope of 1.87 and a significant correlation of 0.80.

Not only were the slopes of the nonseasonal species all less than the slopes of the seasonal species, but the former were all less than 1.0, and the latter all greater than 1.0. A possible implication is not simply that there was stronger selection on small males of the nonseasonal species to have large testes, but that in the seasonal species, where selection on sperm-producing capabilities was equal on all males, the larger males were able to devote a greater proportion of body resources to testes than were the smaller males. If so, the relation could be seen as providing yet more evidence of constraints on sperm production (see Section III). However, this very result and its interpretation in terms of constraints suggests an alternative interpretation than Stockley and Purvis's (1993). Competing by sperm production might be cheaper than competing by fighting, especially where sperm production is only for a limited period. In that case, the positive slopes of the seasonal species reflects merely the greater energetic resources of bigger males that can be devoted to sperm production, not that small males devote more to sperm production in nonseasonal species. A test of this explanation would be the relative testes size of the large males, which should be greater in the seasonal than nonseasonal species. If Stockley and Purvis's (1993) results and the explanation for them are confirmed, we then have to ask why we do not see an 'arms race' between the large and small males in the nonseasonal species, with the large males responding to the small males' greater sperm output by increasing their own (cf. Rice 1996), especially if the larger males are better able to bear the cost of sperm production. Perhaps the fact that we do not is some confirmation of their explanation?

Short (1977) suggested that one selective pressure on the evolution of larger testes was frequency of mating. If it is the case that seasonally breeding species copulate more frequently than nonseasonally breeding ones, because oestrus females are concentrated in time then, independently of mating system, an effect of breeding season ought to be seen on sperm production and testes size. Such an effect could both exaggerate and confound the effect of mating system. It could exaggerate it if males of seasonal species copulated more frequently, or if females of seasonal species copulated more often with more than one male, because single males would find it difficult to defend several simultaneously oestrus females (Trivers 1972; Emlen and Oring 1977). While the seasonally breeding polyandrous macaques do indeed have larger testes than the nonseasonal polyandrous baboons (Dunbar and Cowlishaw 1992), across primates as a whole, seasonality has no influence on either mating system, or relative testes size, independent of mating system (Harcourt *et al.* 1995). However, as Harcourt *et al.* (1995) pointed out, very few primates species have breeding seasons as short as those of a number of mammals: months, rather than weeks, is the normal duration of the season for primates (Smuts *et al.* 1987). Nevertheless, the lemur populations that apparently have seasons of only a few days (*Lemur catta*, *Propithecus verreauxi*) do not have obviously large testes.

2. Sperm production and numbers

In the absence of confirmatory correlations between size and activity the size of an organ is only a crude index at best of its potential activity. In the case of the testes, much testicular tissue is interstitial tissue, and has little to do directly with sperm production. Nevertheless, there is now considerable evidence that relative testes size is a good measure of spermatogenic activity, evidence that comes not just from mammals, as will be described below, but also from birds (Birkhead and Møller 1992) and insects (Svärd and Wiklund 1989).

The explanation for the relation between mating system and relative testes size demands that various measures of sperm output correlate with relative testes size and mating system. A way of increasing sperm production without increasing testes size would be to increase production rate per amount of spermatogenic tissue. However, no relation appears to exist between such efficiency and relative testes size in a taxonomically wide, if small, sample of mammals (nine genera, four orders) (Møller 1989). Instead, more sperm are produced by more spermatogenic tissue. Thus, Schultz's (1938) sample of three primate genera with large relative testes size (which are all polyandrous) had ratios of spermatogenic to interstitial tissue of 2.2–2.8, while the three with small relative testes size (all monandrous) had ratios of 0.9–1.3. We know of no other such comparisons.

With respect to sperm numbers, Møller (1988) reported that in about 20 species of primates (the number varied with the precise measure), residual testes size (body weight removed) varied significantly with the residuals of ejaculate volume, number of sperm per ejaculate, sperm motility (percentage motile) and number of motile sperm per ejaculate (see Fig. 16.8). Sperm concentration did not vary in this way but it would not necessarily be expected to do so under sperm competition theory, where total number of sperm, not their degree of packing in delivery, is what is presumably selected for. Mating system, as opposed to relative testes size, varied significantly with the residuals of motility and number of motile sperm per ejaculate, both being higher in polyandrous species. In Møller's analysis, phylogenetic effects were tested by comparing slopes and intercepts, using species, genera and families as data points. No significant differences appeared between the taxonomic levels and he therefore used species for analysis. To account more precisely for phylogenetic effects, we ran a comparative analysis by independent contrasts (CAIC; Purvis and Rambaut 1995) using Purvis's (1995) primate phylogeny, and data from Harcourt *et al.* (1995) on mating systems and relative testes size, and from Harcourt (1991) on sperm numbers and percentage of motile sperm. When mating system was compared with sperm numbers, or numbers of motile sperm, all contrasts were in the predicted direction (more sperm in multi-male taxa), but each comparison gave only three independent contrasts. However, we found no correlation between residual testes weight and either residual sperm numbers, or residual number of motile sperm (Wilcoxon signed-ranks test, n = 17 contrasts, n.s., two-tailed. These results clearly contradict any interpretations of taxonomic differences in testes weight that depend on sperm competition.

In a comparison for mammals as a whole (nine genera, four orders), residuals of daily sperm production rate, and of numbers of sperm in reserves correlated with relative testes size and with number of sperm per ejaculate (Møller 1989). For example, the monandrous vole *Microtus pinetorum* has smaller testes and fewer sperm per ejaculate than the polyandrous *Microtus pennsylvanicus* and *Peromyscus maniculatus* (Pierce *et al.* 1990). Prairie voles *Microtus ochrogaster* might need more investigation because although they have been classed as monandrous on the basis of behavioural observations of mating patterns (Gomendio and Roldan 1991), they have relatively large testes and a relatively large number of sperm per first ejaculate (Pierce *et al.* 1990). These reproductive measures support the observation of multiply fathered litters in this species (Carter and Getz 1993).

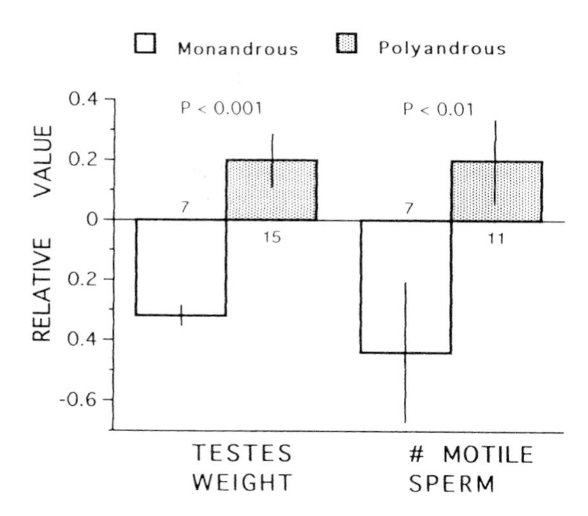

Fig. 16.8. *Comparison of relative testes weight (i.e. residuals from body weight × testes weight regression) and relative number of motile sperm per ejaculate of monandrous and polyandrous primates species, using data and analysis from Møller (1988a). (A CAIC on the same data gave too few comparisons for presentation – see text – and thus the relative testes weight comparison is shown to demonstrate the same result as in Fig. 16.7, with an alternative form of control of taxonomic independence). Numbers on zero x-axis are numbers of species (with species as unit of analysis justified because neither intercept nor slope of regression differed when genera or families were used as level of analysis). (Student's t- tests, t = 6.86, 3.71).*

Ejaculate parameters also appear to correlate with one another: an analysis of 23 phylogenetically independent contrasts in mammals showed that ejaculate volume, number of sperm per ejaculate, and proportions of motile and of normal sperm all correlated with each other (Møller 1991a). Such correlations across species appear to be matched by the same correlations within species. Individuals with few sperm, also appear to have abnormal and immotile sperm (Freund 1962; Gibson and Jewell 1982; Harrison and Wolf 1985).

The correlations differ in some ways from those found in birds for reasons not yet understood. For example, sperm concentration correlates with relative testes size in birds but not in mammals (Møller 1988b). However, in birds, as in mammals, sperm number also correlate with relative testes size, which is the main requirement of sperm competition theory.

Some evidence exists to show that polyandrous species can maintain higher outputs of sperm for longer than monandrous species can, which is in agreement with the data indicating that they can ejaculate more often (above; see also Svärd and Wiklund (1989) for Pierid butterflies). Thus, Pierce *et al.* (1990) found that sperm counts with repeated ejaculations dropped faster in the one monandrous species of vole in his sample than in two polyandrous species. Among the hominoids, the polyandrous chimpanzee appears to be able to maintain greater output than the monandrous human. Thus, two chimpanzees were not only able to ejaculate once an hour for 5 h, but experienced only a halving of sperm numbers by the end (Marson *et al.* 1989), whereas human males (n = 5–10 varying with the period since last ejaculation) experienced a more than halving of sperm output with ejaculations at intervals of 3 days or fewer compared with greater than 3-day intervals (Baker and Bellis 1995, p. 206).

As discussed in the section on testes size, Parker (1990a,b) has suggested that there should be differential investment in ejaculates in two contexts (Parker 1990a,b): (1) if males are assigned favoured and disfavoured roles in a nonrandom fashion, the male in the disfavoured role should invest more in ejaculates; or (2) if there are sneakers and guarders, and sneak matings are rare, sneakers should invest more in ejaculates. A recent study on common shrews *Sorex araneus* has addressed these questions (Stockley *et al.* 1994). In the study population, females tended to copulate multiply and most litters were multiply sired. Males followed two alternative mating tactics. Type A males occupied small ranges and made long distance movements during the breeding season to visit female ranges. Type B males established large overlapping ranges in areas of high female density and remained in them during the breeding season. Type B males inseminated more females and produced more offspring than type A males. Both types of males did not differ in body weight, testes weight or seminal vesicle weights. However, type A males had greater epididymal sperm numbers than type B males. The authors argue that type A and B males could fit into one of Parker's models, i.e. type A males may tend to copulate more often in a disfavoured role, and that the results support the prediction that type A males may produce more sperm. We suggest an alternative explanation. Because type B males inseminate more females than type A males during the breeding season, they may empty their sperm reserves to a greater extent. Thus, the difference may not lie in greater sperm production by type A males (which seems unlikely given that there are no differences in testes weight), but rather on greater sperm reserves among type A males due to their lower copulation frequency during the breeding season. Although males were kept in captivity for 5 days before being killed, sperm production during 5 days at the end of the breeding season may not compensate for major differences in mating activity during the whole breeding season. This possibility provides a very simple mechanism that would explain differences in ejaculate quality by males following alternative tactics. Thus, in Stockley's shrews type A males may copulate less frequently and will therefore have greater sperm reserves than type B males; this means, that type A males will have more sperm available for ejaculation, despite having the similar levels of sperm production. The same mechanism may be at work when sneak matings are rare, since sneakers will have more sperm ready for ejaculation when the opportunity to copulate arises, and when small/juvenile males have few mating opportunities. In conclusion, Parker's predictions could be met without differences in sperm production by males.

A variation on Parker's hypothesis is that males should adjust the ejaculate quality to the risk of sperm competition. In support of this hypothesis, Bellis *et al.* (1990) have provided evidence that male rats deliver fewer sperm when they are given the opportunity to guard females than when they are not. This result is not easily explained by Parker's models since female rats are highly promiscuous and males should always deliver high-quality ejaculates under the assumption that females are likely to have copulated/or to copulate in the future with other males. In other words, in a highly promiscuous species, such as the rat, the risk of sperm competition is so high that no variation should be expected.

3. Sperm length

An extension of sperm competition theory concerns the size of sperm. Polyandrous males should face greater selection to produce not just many sperm, but sperm adapted to reach the ovum first and fertilize it first. If power of swimming is correlated with length of sperm (Section II.A.1) (Katz and Drobnis 1990; Gomendio and Roldan 1991), polyandrous taxa should have longer sperm than monandrous taxa. In examining sperm length in relation to mating system, Gomendio and Roldan (1991) took more account of phylogenetic effects than Harcourt (1991), who used genera as the data points. Gomendio and Roldan (1991) found that polyandrous primates and rodents had significantly longer sperm than monandrous genera; CAIC was used for primates but, owing to the lack of good phylogenies, CAIC could not be used for muroid rodents. We have repeated the analysis using the latest version of CAIC (Purvis and Rambaut 1995), which apparently is better at dealing with categorical data than older versions, and found that out of four contrasts three are in the predicted direction, i.e. longer sperm among polyandrous taxa. The exception is *Theropithecus gelada*, as already noted by Gomendio and Roldan (1991). The fact that *T. gelada* has long sperm despite having single-male units, may be a feature retained from its multimale *Papio* ancestors (Dunbar 1988).

Dixson (1993) later repeated the analysis for primates, but compared sperm length with relative testes size, arguing that relative testes size was a more reliable measure of polyandry than the observed female mating patterns used by Gomendio and Roldan (1991) (see also Chapter 14). While species as data points showed a highly significant correlation – taxa with relatively large testes had long sperm – the problem was that many of the taxa were closely related. We conducted a CAIC (Purvis and Rambaut 1995) using Harcourt *et al.*'s (1995) residual testes weight data for primates and Dixson's (1993) data on primate sperm length, and found no relation when relative testes weight was used as a continuous variable instead of the categorical one that Dixson used ($n = 20$ contrasts,

Wilcoxon signed-ranks, n.s.). Roldan and Gomendio (1995) argued that, at least among the intensively observed primates, female mating patterns may be a more reliable measure of the intensity of sperm competition than testes size, because the latter is also related to aspects of female reproductive biology (see Gomendio and Roldan 1993a). Until we have DNA-fingerprinting evidence on paternity for the species included in these analyses, these conflicting interpretations will not be resolved. Gomendio and Roldan's (1991) analysis indicated that the contrasts in sperm length according to female mating patterns were greater in rodents than in primates, but we still await a good phylogeny for the rodents before analysis with CAIC can be done. The fact that exactly the same relationship – longer sperm in more polyandrous taxa – is found in birds (Briskie *et al.* 1997) and several families of butterflies (Gage 1994) – taxa with extremely different lifestyle, physiology, female anatomy, sperm storage, etc. – suggests, as argued before for testes size, that a main factor in the relationship is the single factor common to all the taxa, namely intensity of sperm competition.

So, what are the advantages of producing longer sperm under sperm competition for mammals? Gomendio and Roldan (1991) showed that sperm length is closely and positively correlated with swimming speed. While swimming speed or power may be irrelevant in the sections of the female tract where sperm are passively transported, the available evidence suggests that passing through the uterotubal junction and being the first ones to reach the lower isthmus of the oviduct may be crucial (Katz and Drobnis 1990) because, once here, sperm which attach to the epithelium survive until ovulation takes place and are not displaced by further sperm reaching this site (Hunter 1987). In addition, once ovulation takes place the spermatozoon which reaches the ovum first is the one most likely to fertilize it, implying that swimming speed may be crucial at this very last stage. In other taxa with storage organs, sperm length may determine the ability to reach the storage organs first or to race to the ovum from the storage organs once ovulation takes place.

It appears that the benefit of fast sperm outweighs the cost of their having a shorter life (see Section II.B). Thus, polyandrous mammals appear to be following the strategy of multiple ejaculations of fast, short-lived sperm, rather than infrequent ejaculations of slow, long-lived sperm. In other words, sperm competition is a race.

4. Sperm life-span

While the sperm of most mammalian taxa live for only a few days at most (Section I.B), there is nevertheless variation, and it appears that sperm life-span might match the duration of the interval between the start of oestrus and ovulation (Gomendio and Roldan 1993a). If so, and if oestrus tends to be longer in polyandrous taxa (Gomendio and Møller unpublished results), then polyandrous taxa should have sperm that live longer than those of monandrous taxa. However, Carter and Getz (1993) indicate that the apparently extremely monogamous prairie vole *Microtus ochrogaster* has a longer oestrus than several other more obviously polyandrous species, although there is evidence that some prairie vole litters are multiply sired.

C. Physiology (non-sperm)

If males facing sperm competition benefit by the production and insemination of greater numbers of sperm than males not facing sperm competition, then, presumably, the former also benefit from the production of greater amounts of fluid to carry the sperm and dilute female influence on the sperm (Section II.A). Is there evidence that polyandrous species have larger accessory reproductive glands than monandrous species? We know of only one comparison, that of the great apes. Short (1979) reported that the seminal vesicles of the polyandrous common chimpanzee are far larger than those of the monandrous orang-utan, human and gorilla. Secretions from the accessory glands almost certainly have many functions other than mere carriage of sperm (Mann and Lutwak-Mann 1981; Smith 1984a; Eberhard and Cordero 1995). Most of these functions have been demonstrated in invertebrates. However, semen contains many hormones that could stimulate the female tract (Smith 1984a; Eberhard and Cordero 1995). Human semen, for instance, carries higher concentrations of prostaglandin E, which is known to stimulate smooth muscle, than are found in any other human reproductive tissue (Smith 1984a). This finding, along with the observation that infertile men can have abnormally low concentrations of prostaglandins in their semen, led Smith (1984a) to suggest that the semen constituents might be specifically adapted to stimulate peristalsis in the female's reproductive tract. Other hormones in semen have been shown in other contexts to influence ovulation (Eberhard and Cordero 1995). We also know that semen constituents are responsible for the formation of copulatory plugs (van Wagenen 1936; Mann and Lutwak-Mann 1981).

As is the case for carriage of larger numbers of sperm, all of these processes are likely to be of greater benefit under conditions of sperm competition, and therefore to have evolved to a greater extent in polyandrous taxa. However, since both monandrous and polyandrous species could benefit from a number of the actions of semen (buffering, carriage of sperm, plugging, etc.) mere recording of the presence or absence of these actions in different taxa is insufficient to test ideas about the influence of sperm competition on their evolution or function, unless there is absolute separation between the mating systems. Thus, the fact that human semen contains high concentrations of prostaglandins does not imply sperm competition in humans. If prostaglandins stimulate female peristalsis, and so prevent otherwise large volumes of flowback (Baker and

Bellis 1995), even monandrous males could benefit. Instead of records of presence/absence, we need quantitative data on occurrence or production in relation to mating system.

Studying direct competitive interactions between males via their seminal products, Eberhard and Cordero wrote, 'The data presently available on seminal product functions firmly place the female between competing males: the males interact only via their abilities to elicit favourable responses from females' (Eberhard and Cordero 1995, p. 495). An exception to this generalization might be copulatory plugs. Many mammals produce them (Fenton 1984) and many functions have been suggested (Voss 1979; Overstreet 1983), including chastity enforcement, storage of sperm, prevention of sperm leakage, inducement of pseudopregnancy, and stimulation of sperm transport. Most would become more beneficial under conditions of sperm competition. If postcopulatory mate-guarding is a reaction to potential sperm competition, the fact that rodent species without copulatory locks have plugs implies that plugs are related to sperm competition (Voss 1979). In addition, accessory glands are apparently relatively small in species that lock and, correspondingly, do not produce plugs (Hartung and Dewsbury 1978). Sperm competition is more likely, the longer sperm remain viable in the female's tract. The fact that copulatory plugs are apparently absent in bats that do not store sperm (Fenton 1984) is further indication therefore that plugs are related to sperm competition. In his review of the evolution of copulatory plugs in rodents, Voss (1979) concluded that chastity enforcement was their most likely function. However, in rats *Rattus rattus* most of the evidence points to their function being carriage of the sperm into the uterus (Sofikitis *et al.* 1990; Cukierski *et al.* 1991). In contrast, in a bat that forms a plug in the usual way, by coagulation of accessory gland fluids, so few sperm are found in the plug that it almost certainly does not act as a sperm storage device (Fenton 1984). While few data exist on whether plugs prevent leakage, it seems highly likely that they could do so in many species, given their (sometimes) close fit to the female reproductive tract, including projections into the cervix (Hartung and Dewsbury 1978). With respect to preventing further inseminations, it appears that sometimes they work, but sometimes they clearly do not (Martan and Shepherd 1976; Hartung and Dewsbury 1978; Milligan 1979; Dewsbury 1988b). At present, therefore, despite Voss's (1979) conclusion, the safest conclusion might be that we have still too little clear experimental evidence to be able to negate any of the functions, and certainly to negate them in all species.

D. Penile anatomy

Mammalian (and much invertebrate) penile anatomy is too complex for the genital organ to function only for insemination (Eberhard 1985,

1990). Mammalian penises have collars, labia, fingers, corrugations, spades, hard spines, and some are shaped like corkscrews. The variation has been known about and used in taxonomic classification for decades, and there has been much speculation about the function of elaborate penises (Vinogradov 1925; Hill 1953; Fooden 1980; Eberhard 1985). Proposed functions include lock and key species separation. However, Eberhard (1985) argued persuasively that no function but the operation of sexual selection by female choice could explain the variation. Females resist males not just behaviourally but also mechanically, thereby increasing their chances of being mated, inseminated and fertilized by better males; males respond to the anatomical and behavioural challenges of the females, and a mating arms race results. Most of Eberhard's argument was based on analysis of invertebrate behaviour and anatomy, but most of it applies equally to mammals, and he made no distinction.

Dixson (1987) was the first to bring quantitative analysis to bear on Eberhard's ideas. He asked whether the penile anatomy of primates differed according to mating system – in other words to intensity of sperm competition – and found that it did. The penis of polyandrous species (including species with a dispersed social system) had a longer baculum and pars libera, and showed greater complexity of the distal penis. Dixson (1987) did not control for phylogeny, but his contrasts were later substantiated by Verrell (1992), who used Dixson's data, but added far tighter phylogenetic control. For example, Verrell (1992) showed within the New World Cebidae (capuchin and squirrel monkeys) and within the Old World Cercopithecidae (baboons, guenons, colobus, macaques, etc.) the same contrasts that Dixson reported. Nevertheless, the relationships are not straightforward. For example, the correlation between mating system and length of bacula appears to be driven more by short bacula in monandrous taxa, rather than long bacula in polyandrous ones. Thus: the polyandrous atelines (spider and howler monkeys) and *Saguinus* (tamarin) have relatively short bacula; all papionines (baboons) have relatively short bacula, whatever their mating system; and *Macaca sylvanus*, a species in which males are particularly tolerant of one another, and therefore in which sperm competition is probably intense, have (relatively) the shortest baculum of all the non-pongid primates (Dixson 1987). Furthermore, Harcourt and Gardiner (1994) showed that any association between mating system and penile spinosity was merely due to prosimians having highly spinous penises, and tending to have dispersed (and paired) mating systems.

The selective advantage of the various aspects of mammalian penile morphology discussed so far has not been established. If deposition of sperm near the cervix is advantageous, and if it is particularly advantageous when there is sperm competition, then a correlation between polyandry and length of penis is expected (Smith 1984b; Baker and Bellis 1995). However, the females of a number of Old World polyandrous primate taxa have sexual swellings (Clutton-Brock and Harvey 1976; Sillén-Tullberg and Møller 1993) and, therefore, particularly long penises (and bacula) might be expected in such species. In at least one such species, *Pan troglodytes*, the length of the penis (*c.* 14 cm) closely matches the length of the vagina (*c.* 17 cm) (Dixson and Mundy 1994). An explanation for the short baculum of the atelines, despite their being polyandrous, could be that New World monkey females generally lack sexual swellings; the situation is similar in *Cercopithecus aethiops*, a polyandrous species with no sexual swelling. That being the case, the long penis of polyandrous species is not necessarily explained by extra selection pressure to deposit ejaculate close to the cervix, but by the same selection pressure as in all species (presumably as a means of preventing flowback), with the long penis of a number of polyandrous taxa explained merely by the sexual swellings. Clearly, a mating system × sexual swelling × penile length analysis is required.

Morphology appears to correlate with copulatory behaviour and with mating system. Primate species with long bacula appear to have prolonged intromissions (Dixson 1987), as do carnivores and pinnipeds (Dixson 1995b). However, taxonomy severely confounds interpretation in all the orders. Prosimians as a whole have dispersed social systems (and are probably polyandrous), and have prolonged intromissions and long and complex penises. Similarly, all the carnivores and pinnipeds measured by Dixson have long intromissions and long bacula, except the felids, which have short bacula and short intromissions. Prolonged intromissions are not necessarily an adaptation to sperm competition, but whether they are or not, the functional relationship between them and penile morphology, if any, is debatable. For example, *Macaca arctoides* supports Dixson's suggestions of associations between long bacula and mating system and copulatory behaviour because it has a particularly long and complex penis, including a long baculum, and is not only polyandrous, as are all the other macaques, but also has particularly prolonged copulations for a macaque. However, the polyandrous Atelinae, which also have prolonged copulations, have very short bacula and pars libera, and many taxa with short intromission times have long bacula. The suggestion has been made for some insects that the form of the distal end of the male genitalia is adapted for removing a previous male's sperm (Waage 1979). Such a function cannot be generalized in invertebrates, because the penis does not usually reach to the sperm-storage sites (Eberhard 1985); in mammals, where peristalsis of the female tract occurs, sperm are transported further along the female tract so quickly that consecutive copulations might rarely be close enough together in time for any removal of previous sperm (Ginsberg and Huck 1989). Plug removal remains a remote possibility, although the spines would not be useful and, in a number of species, including all primates, the efficacy of the plugs as an obstruction is debatable at best (Harcourt and Gardiner 1994).

Finally, courtship and copulation stimulate reproductive readiness, synchrony, and peristalsis (Dixson 1986; Eberhard 1990). Such stimulation might be more beneficial to achieve, or to achieve more quickly, in polyandrous species. That is not to say that they are not beneficial in monandrous species. A monandrous male that produced only small numbers of sperm slowly would benefit from stimulating reproductive readiness in a female at the time that he copulates. If stimulation were a main benefit, then it seems likely that it might be provided most readily by spines (Dixson 1986). Nevertheless, despite suggestions to the contrary (Dixson 1987; Eberhard 1990; Verrell 1992), spinosity is not greatest, at least in primates, among polyandrous species (Harcourt and Gardiner 1994). Either it is not more advantageous to stimulate reproductive readiness rapidly in polyandrous taxa, or stimulation of reproductive readiness is not an important function of long, complex or spinous penises. Harcourt and Gardiner (1994) have tentatively suggested that dispersed species, in which the sexes cannot readily keep track of another's reproductive state, might benefit more than group-living species from stimulating reproductive readiness (Harcourt and Gardiner 1994).

Many suggestions for the functions of various aspects of penile anatomy exist, including their relation to sperm competition. We now are in need of predictions that separate hypotheses, and experiments that might tell us whether a proposed function will work at all. For example, does penile morphology influence the efficacy of stimulation of the female tract, of plug or ejaculate removal, or of sperm deposition? X-ray photography and ultra-sound scanning of copulating large mammals is possible; Dixson has artificially inseminated mock radio-opaque semen into chimpanzees to investigate transport through the os cervix (Dixson and Mundy 1994). Such studies should tell us more about how copulation behaviour correlates with the potential functions of anatomical variation. Far more work has been done by invertebratologists in correlating female anatomy with male anatomy, and usually finding, for example, greater variation among males than among females across taxa (Eberhard 1985), as is the case for variation in testicular and ovary size in primates (Short 1979), and as would be expected were male–male competition driving the variation (Short 1979). Clearly, more work needs to be done in vertebrates, including mammals. Are there obvious specializations in the female tract of species in which the male has massive penile saws or spines, such as in *Lemur fulvus* (Petter-Rousseaux 1964), or some rodents (Breed 1986)? It appears that it is in rodents that mammalian penile morphology takes on its most bizarre forms, and therefore this taxon is perhaps where lies the greatest possibility of finding correlates between male and female anatomy and their behaviour and mating systems.

E. Conclusion

Almost all parameters related to sperm production match predictions from sperm competition theory: polyandrous taxa have large testes, with high proportions of sperm-producing tissue; hence, they produce sperm at a high rate, with the result that their ejaculates and their sperm reserves contain large numbers of sperm, including motile sperm and longer sperm, compared with monandrous taxa. Within species, males who compete more by sperm competition than by monopolization of females appear to have relatively larger testes. Some of the results match new predictions from theory, and are not mere extensions of the original finding of a contrast between mating systems in primates of testes size. They are thus particularly strong support of the theory. In many areas we still know far too little: we know almost nothing about the role of semen constituents in sperm competition and therefore it is difficult to predict how accessory glands should vary with mating system. Reasons for differences between males in fertilizing ability after copulation are almost unknown, even if the considerable individual variation in semen and sperm quality indicates where the answer might lie. Indeed, the search for specific physiological and anatomical differences that correlate with differential fertilizing ability can itself be a good means of testing ideas about the characters that might influence fertilizing ability and sperm competition (Birkhead *et al.* 1995a).

V. DO FEMALES PROMOTE OR ACCEPT SPERM COMPETITION?

The issue of sperm competition has been studied mostly from the male point of view. However, for sperm competition to occur females must copulate with more than one male, and only recently have researchers begun to ask why they should do so.

Mating involves a number of costs for females such as the time and energy devoted to courtship and copulation, increased predation risk, decreased foraging efficiency, increased risk of transmission of parasites, increased risk of contracting venereal diseases, and increased sexual harassment from males and the consequent risk of injury.

The issue here is whether any of these costs would increase as a consequence of mating with several males. It seems reasonable to assume that time and energy devoted to mating, increased predation risk, decreased foraging efficiency associated with mating, and perhaps increased harassment from males, are costs which are related to the number of copulations achieved by each female, irrespective of whether such copulations involve one male or several males. However, increased risk of transmission of parasites and of contracting venereal diseases probably increase with the number of mates rather than with the number of copulations. Mating with several males probably has other additional costs such as the risk of being punished by the guarding male (provided that there is one) (Clutton-Brock and Parker 1995) and the cost associated with increased sibling competition in multiply sired litters. In some species mating with multiple males results in 'pregnancy blockage' and implies high costs for females (Dewsbury 1985; Wynne-Edwards and Lisk 1984).

Given that there seem to be costs associated with mating with more than one male, there are two scenarios which could explain the evolution of mating with multiple males:

1. Females do not benefit from mating with multiple males, but incur high costs if they refuse to do so. These costs would be mainly the result of physical injury by larger males, particularly in sexually dimorphic species (Smuts and Smuts 1993; Clutton-Brock and Parker 1995). If this is the case, females would merely be 'forced to accept' sperm competition as part of male/male competition.
2. Females do benefit from mating with multiple males and the benefits outweigh the costs. Possible benefits of multiple-male mating include the following (reviewed by Halliday and Arnold 1987):
   - 2.1. Obtaining extra material benefits. In mammals the ejaculate does not appear to provide nutrients to females or ova. Thus, this hypothesis is not applicable to mammals.
   - 2.2. Assurance of fertilization. Females copulate with more than one male to avoid the possibility of not being fertilized if they copulate with sterile males. This hypothesis is not supported by the available evidence, since male sterility appears to be infrequent. However, more precise information on the prevalence of male sterility is necessary to test this hypothesis rigorously.
   - 2.3. In unpredictable environments, females may benefit from producing offspring as genetically heterogeneous as possible.
   - 2.4. By promoting sperm competition females increase the number of potential fathers and, thus, may increase the number of males predisposed to provide parental care and/or diminish the number of potential infanticidal males (Hrdy 1979, 1981; O'Connell and Cowlishaw 1994; Bercovitch 1995). In clear contradiction with this hypothesis it has also been suggested that when females copulate with multiple males, all partners will be uncertain of their paternity and thus unlikely to provide care. However, in the promiscuous baboons some males form consortships during the period in which ovulation is more likely, and it is precisely these males that are more likely to form affiliative bonds with the females after they become pregnant and to contribute, albeit in indirect ways, to infant care (Bercovitch 1995). Presumably, in this case paternity certainty is related to the level of care which the male is prepared to provide. This factor would only be important in the few mammalian species in which males provide par-

ental care. Diminishing the risk of infanticide may be more common.

2.5. By promoting sperm competition females enhance the genetic quality of offspring. This hypothesis assumes that genetically superior males produce sperm that perform better under sperm competition, and also that the offspring inherit the qualities which made their father successful. Given the costs associated with multiple mating, female choice of genetically superior males via sperm competition would be more likely to occur in species in which females cannot detect male quality by assessing male phenotype, or species in which females cannot prevent copulations by several males. If females can detect male genetic quality on the basis of male phenotype, and females may choose with whom they copulate, it would appear less costly for females to copulate exclusively with the superior male. An added benefit is that the process is certain, whereas multiple mating always carries the risk of fertilization by inferior males.

It has also been suggested that females would benefit from promoting sperm competition if, by being fertilized by males producing the most competitive ejaculate they pass on such traits to their sons (Harvey and Bennett 1985; Harvey and May 1989). This hypothesis may explain how could females benefit from being fertilized by males with competitive ejaculates in a sperm competition context, but not why multiple male mating evolved. Curtsinger (1991) elaborated a genetic model and concluded that this scenario is unlikely. However, more recently Keller and Reeve (1995) elaborated a model based on different assumptions which led to the conclusion that multiple mating can spread rapidly provided that there is a small fraction of females that originally copulate multiply (either because there are advantages to females or because the copulations are forced by males) and there are heritable differences among males in one or several of the following: (1) number of sperm, (2) success of sperm in fertilization, (3) ability to displace sperm from previous copulations, and (4) ability to prevent further copulations. As discussed in the section on mechanisms, (1) and (2) are more likely in mammals. In this situation, sons will inherit high fertilization efficiency and the trait will spread.

Halliday and Arnold (1987) proposed that multiple mating by females may merely be a by-product of selection on males to copulate with as many females as possible. This hypothesis assumes a correlated effect of selection acting on one sex or the other, and thus requires no benefits for females. Sherman and Westneat (1988) and Cheng and Siegel (1990) argued that a correlated effect would not be expected if multiple mating implies costs to females; breeding evidence from mammals, birds and insects demonstrates no correlated response.

The observational evidence which is beginning to accumulate shows that, in many cases, females are not just accepting copulations by eager males, but rather that it is precisely females who solicit copulations by several males. This evidence strongly suggests that females promote sperm competition because there are benefits associated with it. Columbian squirrel females seek additional mates despite the apparent reluctance on the part of males (Murie 1995). Among chacma baboons, females utter copulation calls which may elicit sperm competition in order to prevent infanticide and/or obtain good genes for sons (O'Connell and Cowlishaw 1994). It has been suggested that the prominent sexual swellings that some primate females develop around the time of ovulation, have evolved to advertise oestrus and hence incite competition among males or promote sperm competition (Hrdy and Whitten 1987). Sauther (1991) has suggested that *Lemur catta* females seek multiple mating (particularly with extra-troop males) to avoid inbreeding. Socially monogamous Mongolian gerbil females incite other males in the vicinity to copulate, and inbreeding avoidance has also been suggested as the main benefit for females (Ågren 1990). Ethiopian wolves *Canis simensis* live in family packs, but females facilitate copulations with extra-pack males, again to avoid inbreeding (Sillero-Zubiri *et al.* 1996). When females live in close family groups, mating with outsiders is probably the best way to reduce inbreeding in offspring, while avoidance of mating with familiar males is likely to be costly. Thus, females may end up copulating with both. Conversely, in populations with high levels of inbreeding the benefits of sperm competition to females would be quite straightforward. High levels of inbreeding in males are related to poor quality ejaculates (Wildt *et al.* 1987a,b; Pusey and Wolf 1996). Thus, when females copulate with several males they are likely to be fertilized by the less inbred male. This is one of the cases in which selection through sperm competition may be more efficient than selection at the phenotypic level, which may not show as clearly the effects of inbreeding. Hence, promoting sperm competition would be a hitherto ignored way of avoiding inbreeding by females. In this context, offspring would benefit from low levels of inbreeding and sons in particular would have more competitive ejaculates owing to their low levels of inbreeding, even if they do not inherit ejaculate features from their fathers.

VI. IMPLICATIONS FOR MATING SYSTEMS

Mating systems are the outcome of the reproductive strategies of individuals. Because each individual will pursue its own reproductive interests there will be conflict of interests between males and females. Traditionally, mating systems were seen as being the result of mate acquisition tactics by males which were aimed at copulating with as many females as possible. In this framework, male reproductive success was assumed to correlate with the number of females mated. Studies on sperm competition have revealed that this is only half of the story: in polyandrous species sperm from rival males will compete within the female tract and females might also exercise some choice at this level. This new facet has a number of implications. First, male reproductive success cannot be equated to male mating success. Second, male and female reproductive strategies will also be moulded, to some extent, by sperm competition mechanisms which will ultimately determine paternity. In monandrous species, some aspects of male mating strategies can best be understood as being aimed at avoiding sperm competition during the period in which ejaculates are most vulnerable.

In this context, male mating strategies may be seen as a mixture of mate acquisition tactics and of mate-guarding tactics. Thus, there will be a trade-off between the number of females mated and how much effort is put into protecting ejaculates under the risk of sperm competition. This implies that, in addition to the traditional factors considered to influence mating patterns (i.e. females spatial and temporal distribution; Clutton-Brock 1989), other factors related to the way in which sperm competition mechanisms work will also be important. Gomendio and Roldan (1993b) have suggested that differences in sperm competition mechanisms between birds and mammals may have influenced to a great extent the evolution of disparate mating systems. In birds, females lay their eggs in chronological sequence over a number of days (roughly at the rate of one egg per day) and may store sperm for weeks in the 'sperm-storage tubules'. Sperm competition mechanisms operate in such a way that it is the last male to copulate with the female that is the one most likely to fertilize the next ovum. Thus, males tend to copulate most frequently before egg laying starts, and fill the sperm-storage tubules with their own sperm; however, they need to mate guard the female over the whole egg laying sequence (i.e. days) to ensure paternity (Chapter 14). The benefits of prolonged mate guarding are likely to have contributed to the evolution of the pair bond which is so common in birds. The last male advantage provides a relatively easy way of obtaining extra-pair offspring, since males need to copulate only once at the right time to outcompete the female's partner; it is also a relatively easy way for females to manipulate the paternity of their offspring to their own advantage. In addition, by putting some extra effort into the ejaculates produced during extra-pair copulations males may gain disproportionate benefits in terms of offspring (Birkhead *et al.* 1995a). This fact may have contributed to the widespread occurrence of extra-pair copulations in birds (see Fig. 16.9).

The mammalian scenario is radically different. Females are fertile for very limited periods of time because all ova are ovulated at the same time, and remain viable for about 24 h, and females do not possess sperm-storage organs. As a consequence, mammalian spermatozoa are short-lived. Sperm competition mechanisms are such that there are no order effects. Instead, sperm numbers (either in one or several ejaculates),

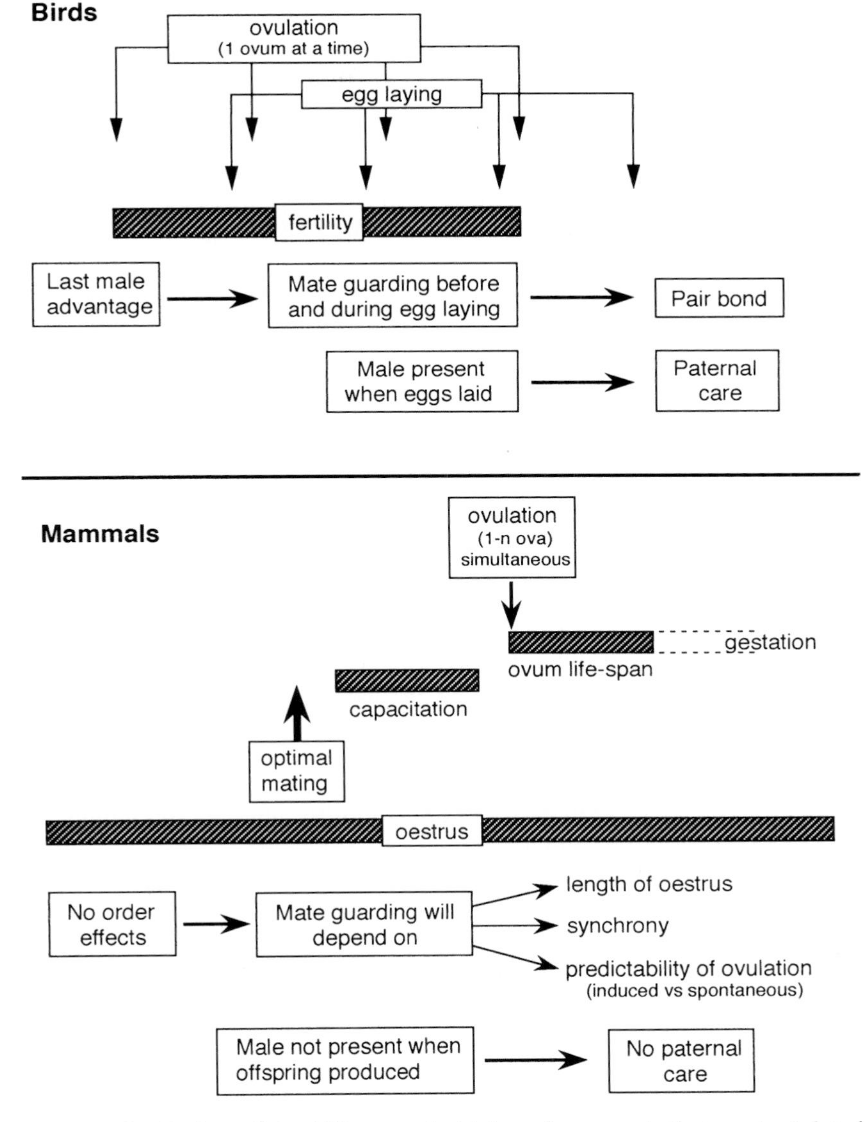

Fig. 16.9. *Outline of the differences in birds and mammals that are postulated to have influenced the evolution of sexual behaviour and mating systems. From M. Gomendio and E. R. S. Roldan (1993b)* Trends in Ecology and Evolution **8,** *95–100, with permission from Elsevier Science Ltd.*

timing of mating in relation to ovulation, and perhaps ejaculate quality, determine paternity to a great extent. The best strategy for males is thus to copulate close to ovulation, but allowing time for capacitation to take place. There is far less need than in birds for males to defend females after ovulation, and since females undergo gestation and lactation, males gain little from staying with a female. Because it is female mammals that care for the offspring, there is little room for mammalian males to improve offspring survival and much to be gained from mating with other females. Taking these factors into consideration it is understandable that polygyny is by far the most common mating system in mammals.

Much variation in male mating behaviour has been attributed to four characteristics of females (Clutton-Brock 1989): (1) the extent to which female reproductive rate may be increased by male assistance in rearing offspring; (2) the size of female ranges and core areas; (3) the size and stability of female groups; and (4) the density and distribution of females in space. We suggest that some aspects of female reproductive physiology in conjunction with sperm competition mechanisms will determine to an important extent mating strategies. These factors include the length of the oestrus cycle, whether ovulation is spontaneous or induced, whether ovulation is predictable or not, and the degree of female synchrony in sexual receptivity.

A. Induced ovulators

The first male to copulate triggers ovulation. It seems likely that males will have evolved sperm fertile life-spans (normally ovulation takes place a few hours after copulation) and capacitation times that synchronize these events so that the sperm from the first male will be ready to fertilize at the time ovulation occurs. If this is the case, then the first male may have an advantage over males mating later in the female cycle. If synchrony is so tight that males mating later have a very slim chance of fertilizing, then the first male to copulate should not guard the female. However, if males mating after the first male do have a chance of fertilizing, then the first male should mate guard until ovulation occurs and fertilization by other males is no longer possible. Because ovulation tends to occur a few hours after copulation, mate guarding should in any case be short. *Spermophilus tridecemlineatus* is an induced ovulator in which ovulation takes place 8–36 h after copulation; the first male to copulate fathers most of the offspring and males copulate several times with a female but do not mate guard (Foltz and Schwagmeyer 1989), suggesting that it fits into the first model proposed. In this species males leave their mates after they have achieved a copulation longer than 9 min (Schwagmeyer and Parker 1994). It is possible that a copulation this long is necessary to induce ovulation. Laboratory studies have shown that, as expected, delays between copulations by two successive males favour the first one (Schwagmeyer and Foltz 1990). In accordance with these findings, field data show that males that do not find an oestrus female until at least 4 h after her first copulation, tend not to copulate with her (Schwagmeyer and Parker 1990). A first-male advantage has also been reported in the rabbit (Dziuk 1965), which is also an induced ovulator.

When ovulation is induced, and there is an advantage associated with mating first, there will be a strong selection on males to be the first to locate oestrus females (a form of precopulatory guarding). In *S. tridecemlineatus*, males are extremely mobile during the mating season (Schwagmeyer 1988) and seem capable of anticipating the onset of female receptivity from one day to the next (Schwagmeyer 1995). Thus, males appear to use current information on female reproductive state to decide how much effort they put into searching for particular females on the following day. It has been argued that this process may require complex cognitive abilities.

B. Spontaneous ovulators

1. Short oestrus

Most males mating within an oestrus period will have a chance of fertilizing the ova.

1. If there is a high degree of synchrony in sexual receptivity between females, and these are spatially close, males have plenty of opportunities to copulate with other females and may be unable to defend a large group of females. No mate guarding should be expected. In *Spermophilus beldingi* receptive females are clustered, and males do not guard females, presumably because females can be easily located and males can gain more from resuming searching of other oestrus females after copulation (Sherman 1989).
2. If females are asynchronous, or spatially scattered, males may be able to defend one oestrus female at a time, and are unlikely to find other females in oestrus. Thus, short periods of mate guarding should be observed since there would not be any loss of mating opportunities. This seems to be the case in *Spermophilus brunneus*. Females are in oestrus for one afternoon each year and males guard for about 3 h after copulation to improve their chances of fathering offspring (Sherman 1989).

2. Long oestrus

Only males mating around the time of ovulation will have chances of fertilizing the ova.

1. Females advertise ovulation: males should attempt to copulate with females around ovulation. In *Papio cynocephalus*, males compete to copulate when sexual swellings reach their maximum size (which pinpoints to a day or two the timing of

ovulation; Bercovitch 1989). Outside this period, dominant males tend to ignore females, and subordinate or juvenile males take advantage of the opportunity to copulate with females. Whether there is mate guarding or not will depend on whether there is loss of opportunity to copulate with other females and thus on the degree of female synchrony. For this reason, consortships are more prevalent in baboons (nonseasonal breeders) than in macaques (seasonal breeders).

2. No advertisement of ovulation: males will be unable to predict when ovulation will take place, and thus will be unable to assess the value of copulations at different times throughout the cycle. In this case, females will always be synchronous to some extent because they will be receptive most of the time and males will not be able to pinpoint ovulation time. This situation will favour copulations at short intervals over long periods of time (to ensure fertilization) and possibly extended mate guarding. This appears to be the case in humans.

VII. HUMANS

A. Likelihood of sperm competition

If there is multi-male insemination of live sperm within the overlapping lifetime of sperm and ovum, then there is sperm competition (see Section I). To judge the likelihood of sperm competition, two measures are needed: frequency of multiple-mating and the duration of the overlap period. Smith (1984a) suggested that rape, maybe especially in war, might be one of the more common contexts of multi-male mating in humans, and suggested that it might have been more frequent in the past than now. With respect to consensual copulations, the marriage system of polyandry is extremely rare. It apparently exists only in quite tightly defined environmental circumstances, and breaks down very rapidly when those environmental constraints are removed (Crook 1980). That being the case, it might not be worthwhile looking for biological differences between the few societies that show regular marriage polyandry and other societies, especially as we have little to no idea of whether polyandry and its associated multiple matings has existed for long enough for biological adaptations to have evolved. In other societies, the existence of extra-pair copulation, perhaps often initiated by the female (Hrdy 1981), is probably the most frequent context of multiple mating. Its frequency, and paternity by males other than the main current mate, have been taken as indications of sperm competition (Smith 1984b; Baker and Bellis 1995), with values of 30% non-main partner paternity in some studies being used to imply high degrees of sperm competition. However, ovulation is so concealed in humans, and length of gestation so variable, that copulations that occur many days, even weeks, distant from the next nearest mating with the principal mate could result in confused paternity. Without information on the interval between copulations by the main partner and copulations by extra-pair partners, and information on how close copulations were to ovulation, it is impossible to determine whether cases of extra-pair paternity in humans have involved sperm competition or not (see below for discussion on human sperm life-span).

High degrees of multiple mating by young females is reported for some noncontracepting traditional societies (see Shostak 1981, Chapter 12). Before sperm competition is invoked as a possible consequence, it would need to be determined whether in fact the females were ovulating, because women in traditional societies without Western nutrition can mature later than well-fed Western women, and have a longish period of adolescent sterility after menarche (Marshall 1970; Short 1976b). With respect to Western societies, three recent surveys with larger sample sizes than any before, in America, Britain and France, especially those from Britain and France, all produce fairly similar figures for probability of multiple mating within the lifetime of sperm and ovum and therefore of sperm competition (Spira *et al.* 1994; Laumann *et al.* 1994; Wellings *et al.* 1994; see also Johnson *et al.* 1992; Spira *et al.* 1992). Unfortunately few figures are exactly comparable among the studies. We present values for young females because: (1) the young are the most sexually active section of the population, and it is therefore among them that we might expect to see the strongest adaptations to polyandry; (2) if the young reproduce as a consequence of their activities, then in an increasing population (as the human population is), their adaptations will evolve most rapidly; (3) the sample in Baker and Bellis's (1995) study (which makes the strongest claims so far for polyandry and sperm competition in humans) was young (mean = 21 years); and (4) given that much of the information comes from questionnaires with, to some extent, fairly open-ended answers possible, the more confined the sample population the more comparable the answers should be (self reporting on frequency of intercourse with 'concurrent' partners might be more comparable across 21-year-olds from different countries than between 21- and 51-year-olds in the same country).

Among young (18–24 years) French females, the mean number of sexual partners in the past year was less than 1.0, and was never more than one at any age (Spira *et al.* 1994, Fig. 5.4). Only 10% of young French females reported having multiple partners in the previous year (Spira *et al.* 1994, Table 5.20); the comparable figure for Britain was 15% (Wellings *et al.* 1994, Table 3.1); and for the US, about 20% (Laumann *et al.* 1994, Table 5.4). More than one partner is not the same as simultaneous partners. Among young British females who had at least one partner in the past 5 years (roughly 60%; Wellings *et al.* 1994, p. 96), 15% reported concurrent partners of either sex (Wellings *et al.* 1994, p. 115). In France, 5% of the 10% of young females with more than one partner had at least two sexual partners at the same time (Spira *et al.* 1992). The US survey reported on concurrent partners, but did not appear separate by sex, although the British and French surveys reported roughly twice the frequency of concurrent partners in men as in women. All reports indicate that serial partners were more common than concurrent ones. For example, the US survey found that people with three partners averaged 9 months with only one partner, even though 60% of those with three partners reported any simultaneity of partners (Laumann *et al.* 1994, Table 5.2). All these figures are much lower than those from Baker and Bellis's (1995) sample of female readers of Company magazine. Their sample had a median age of 21 years, putting its values mainly among the sexually active section of what was probably a more sexually active section of the population anyway, judging from the nature of Company magazine. Among Baker and Bellis's respondents who had had some sexual experience, 87% had had more than one partner after 201–500 copulations (or roughly 4–10 years of sexual activity). This compares with about 50% in the other samples. The UK figure of 15% of those with more than one partner reporting concurrent partners at some time in the last 5 years gives a value of about 2% with concurrent partners in the last year. The French figures indicate that only 0.5% of young women had concurrent partners. Alternatively, if we assume that only 3 months per year are spent with more than one partner, then about 2–5% of females have concurrent partners in a year.

Of course, even this figure exaggerates the chances of multiple-mating, because not all of that 2–5% is perpetually at risk of multiple insemination within the window of fertilization. How long is the window? Both Smith (1984a) and Baker and Bellis (1995) suggested upwards of a week, using data that indicated that viable sperm could still be found in a female's upper reproductive tract. However, there will not be selection for adaptations to sperm competition unless the sperm fertilize, and thus probability of overlap of lifetime of viable sperm and ovum, rather than simply lifetime of sperm in the female, might be a closer measure of opportunity for functional sperm competition. First, while some sperm might be viable for 1 week, other data indicate a far shorter period for the vast majority of sperm. A recent detailed quantitative analysis of lifetime of human sperm and ovum in relation to probability of conception concluded that the mean lifetime of sperm was 1.4 days, although some sperm (less than 1%) can survive 6 days; the average lifetime of the ovum was judged to be less than 1 day (Weinberg and Wilcox 1995). In a study of 221 healthy women (625 menstrual cycles, 192 pregnancies, 129 live births), conception occurred only when intercourse took place during the 6 days before the estimated day of ovulation. Probability of conception per intercourse changed from 10% 5 days before ovulation to 33% on the day of ovulation itself, but only 6% of the pregnancies could be firmly attributed to sperm that were three or more days old (Wilcox *et al.* 1995). Over 60% of conceptions resulted from copulations on the day of ovulation and the previous day, with less than 20% resulting from copulations more than 2 days before ovulation (Weinberg and Wilcox 1995). Double matings that might lead to a reasonable chance of selection for adaptations to sperm competition would therefore have to occur within 2 or 3 days of one another, with the first being within 3 days before ovulation. In other words, polyandrous females would have to copulate, on average, with a different male each 1–3 days for there to be a high chance of sperm competition. Even in Baker and Bellis's (1995) sample of probably fairly sexually active females, less than 20% of those with 500 copulations (presumably the younger and most sexually active section of the sample) said that they had ever, let alone regularly, copulated with two males within a day of each other.

Thus, polyandry in humans that might lead to adaptations to sperm competition is probably a lot less common than indicated by Smith (1984a) and by Baker and Bellis (1995). Nevertheless, it must sometimes occur. The question, therefore, is whether the frequency of such polyandry is sufficient for adaptations to sperm competition; we do not know the answer because, at present, the calculation depends on knowing the total energy devoted to mating (Parker 1984), which is next to impossible to determine. However, we can ask whether we see any anatomical, behavioural or physiological adaptations to sperm competition in humans.

B. *Adaptations to sperm competition*

1. Behaviour

Female and male humans sometimes copulate polyandrously. Is there evidence that the behaviour of either is adapted to functional polyandry, i.e. encouraging or reacting to polyandry that might result in sperm competition? Baker and Bellis have suggested that human British females are particularly likely to copulate with males other than their main partner in the highly fertile period just before ovulation rather than after it (Baker and Bellis 1995, p. 161). However, the fact that females were most likely of all to be unfaithful during menstruation, when clearly no fertilizations can result, must call into question arguments for the adaptiveness of the timing of extra-pair copulations by human females.

2. Anatomy and physiology

We consider the female first. It has been suggested that hidden ovulation and sexual swellings are adaptations to multiple matings (see Sillén-Tullberg and Møller (1993) for one of the latest of several reviews). Humans do both, in effect having a permanent display of fertility after adolescence, namely enlarged breasts and buttocks, with both appearing in traditional societies, which are not overfed, years before fertile ovulation

(Short 1976b). Hiding ovulation makes mate-guarding difficult; attractive sexual swellings makes it especially difficult. The alternative strategy of signalling obvious ovulation, and hence making it impossible for a single male to defend a consequently attractive female, might not be advantageous for a species in which females might benefit from partnership with a male, if the partner's reaction is to abandon the female. Indeed, analysis of the timing of the appearance of signals of ovulation and of monogamy in the primate evolutionary record indicate that lack of ovulatory signals promotes monogamy (Sillén-Tullberg and Møller 1993). However, we suggest that it is still difficult to differentiate promotion of polyandry as the function of hidden ovulation from prevention of the harassment that would occur around ovulation if the event were obvious in a multi-male society (Smuts and Smuts 1993; Clutton-Brock and Parker 1995).

Smith (1984a) suggested that because human males had testes that were relatively large compared with gorillas and orang-utans, humans were polyandrous, even though he also pointed out that humans testes were far smaller than those of the known polyandrous chimpanzee. Indeed, human testes as a proportion of body weight are far closer to the monandrous apes than they are to the chimpanzee, and if female body weight (as opposed to the mean of female and male, or male body weight) is used to take account more closely of the possible dilution effect of the female reproductive tract, then the human measure is the same as gibbons and orang-utans, and larger than only gorillas (Dahl *et al.* 1993). In comparison with all primates, humans have relatively small testes (testes size is below the regression line for testes size × body size). Human sperm numbers are also low, at 48–134 million per ejaculate (Fisch and Goluboff 1996) compared with over 650 million for the chimpanzee (Short 1979; Møller 1988a). Variability in human sperm length is relatively high (Seuanez *et al.* 1977; Harcourt 1991) but the sperm length is the shortest of all the hominoids (Gomendio and Roldan 1991). All the data indicate that, compared with other primates, including apes, humans are monandrous.

However, a split into monandry and polyandry is clearly crude. In birds, for example, taxa that are very monogamous have lower relative testes size than those that have some degree of extra-pair copulation, which in turn have lower relative testes size than colonial birds (Birkhead and Møller 1992). Indeed, a good correlation between degree of extra-pair paternity and relative testes mass has been demonstrated in birds (Møller and Briskie 1995). In mammals and primates, the variation in relative testes size within the polyandrous and monandrous categories could reflect variation in intensity of sperm competition. For example, *Macaca* species have larger relative testes size than *Papio* species, possibly reflecting a greater intensity of sperm competition in seasonal species (Dunbar and Cowlishaw 1992; but see Stockley and Purvis 1993; Harcourt *et al.* 1995). Since humans have a large relative testes size for a monandrous primate, perhaps there is some sperm competition. Baker and Bellis have argued that their observation that more sperm are produced the longer the male has been away from his main partner, with hours since last ejaculation taken into account (Baker and Bellis 1995, p. 206), indicates adaptive reaction to the possibility of multiple mating. An alternative interpretation is that the male is using past absence as an indication of possible future absence (perhaps in the same way as birds apparently use previous time travelling to the current food patch as an indication of its worth (Cuthill *et al.* 1990), in which case the increased numbers of sperm inseminated could be explained by reduced future probability of being able to copulate on the day of ovulation, whether there was sperm competition or not.

Humans have a longer penis than any of the other apes (Short 1979). That statement has been taken as an indication that humans have an unusually long penis (Short 1979; Smith 1984a; Baker and Bellis 1995) and the argument has been made that sperm competition is involved, the benefit being placement of the ejaculate near the cervix (Smith 1984a; Baker and Bellis 1995). However, without an allometric analysis we cannot know how unusual the length of the human's penis is. The same applies to the apparent unusual width of the human penis. (Dixson (1987) did not include humans in his analysis of penile morphology in relation to mating system.) Within the apes, a very simple comparison of penile length to body weight indicates that the human penis is, relatively, not longer than the chimpanzee's. Also, all species, whatever their mating system, would benefit from placement of the ejaculate close to the cervix. Without a comparison of length of vagina with length of penis across the apes and primates, we cannot know whether there has been extra selection on polyandrous species to place the ejaculate closer to the cervix.

Whether the human penis (and thrusting during copulation) acts to suck out previous inseminations (Baker and Bellis 1995), or to stimulate peristalsis in the female to suck the sperm in (Eberhard 1990), its apparent unusual width could be explained by the unusual width of the vagina of a species that produces unusually large young (Baker and Bellis 1995). Here, a test would be to compare humans with other species that produce unusually large young, such as squirrel monkeys (*Saimiri*; Harvey *et al.* 1987). The full test would be to compare polyandrous and monandrous taxa with large and small newborns. The size of the newborn, stimulation of the female, and sperm competition might then be distinguishable explanations of penile morphology.

3. Conclusion

Compared with other mammals, including primates, humans are behaviourally, anatomically, and physiologically monandrous. Nevertheless, multi-male matings within the window of life-span of sperm and ovum undoubtedly sometimes occur. Currently, we have little convincing evidence that humans show adaptations to the possibility of sperm competition, because most suggested adaptations can be explained otherwise. The issue of sperm competition in humans still needs predictions that can separate sperm competition from the alternative hypotheses.

VIII. CONCLUSION

Accumulating evidence suggests that sperm competition has been an important selective pressure in the evolution of mammalian reproductive biology and sexual behaviour. As the use of molecular techniques to determine paternity becomes more widespread two facts are becoming clear: that sperm competition is common among mammals, and that the use of patterns of social organization as indexes of the intensity of sperm competition may be misleading. These conclusions call into question some of the analyses that have been carried out in the past, and show that there is an urgent need for paternity analyses among mammals. In this respect, the study of sperm competition in mammals lags far behind the study of, for example, birds.

Mammalogists have an important advantage, however. Basic mammalian reproductive physiology is reasonably well understood (albeit in a few laboratory and domestic species), including aspects of reproductive cycles, interactions between sperm and the female tract, sperm function and fertilization. These physiological studies have not addressed the issue of sperm competition as such, but their findings are of great value in understanding the mechanisms who may underlie sperm competition in mammals. However, to integrate this information requires some effort on the part of behavioural ecologists, which tend not to be familiar with the methodology and jargon of reproductive physiology. To ignore this vast source of information can only lead to the elaboration of unsubstantiated hypotheses. On the basis of this information we can conclude that in eutherian mammals sperm fertile life-span tends to be short (except in bats), that there are no sperm-storage organs, that the interactions between sperm and the female tract are complex, that competition between ejaculates is in the nature of scramble competition not contest competition, that sperm face major barriers within the female tract (which select for large sperm numbers, a large proportion of morphologically normal sperm, and for sperm that swim efficiently along certain sections of the female tract and which are able to undergo the acrosome reaction at the right time and place in order to penetrate the ova vestments), and that the timing of mating in relation to ovulation determines to a great extent male success at fertilization. Thus, mechanisms of sperm competition in mammals appear to differ to great extent from other taxa, in which sperm are stored within the female tract for long periods of time.

Over a decade of work on sperm competition in mammals suggests that it exaggerates the characters that the female tract already selects for, and thus selects for large testes because these produce more sperm, and also selects for a large proportion of morphologically normal and motile sperm. Sperm competition also seems to select for longer sperm because these are able to swim more powerfully and may outcompete other sperm in the race to fertilize the ova. In primates, sperm competition is associated with longer penile pars libera and a more complex penile distal end, but no equivalent analysis has been made for any other taxon. Fewer, and less statistically verified data indicate that sperm competition also selects for ardent mammalian males who copulate often. However, other aspects of copulatory behaviour remain to be firmly related to sperm competition, such as obligatory multiple mounting before ejaculation, prolonged intromissions, and copulatory locks; many aspects of penile morphology have yet to be explained. Indeed, if we try to summarize the main advances in the study of sperm competition in mammals of the last decade, the conclusion is gloomy. Fifteen years ago, Short (1979) and Harcourt *et al.* (1981) showed that sperm competition favours increases in testes size in primates. Today, we cannot firmly conclude much more because of the poor quantity and quality of the data, particularly after the drastic reductions in sample size that can take place when the new methods to account for phylogenetic effects are used. Most analyses have made extensive use of data extracted from the literature, which in most cases involve minuscule sample sizes, whose quality is sometimes difficult to assess. This state of affairs reflects the urgent need for detailed information on a greater number of species, and for more data collected with modern methods specifically to test existing hypotheses.

Only recently has attention been turned to the female perspective. What little evidence we have suggests that, in some cases, females are promoting sperm competition for their own interests, which may include manipulating certainty of paternity to diminish the risk of infanticide, ensuring that their sons inherit competitive ejaculates and/or inbreeding avoidance. Most of this evidence is circumstantial and, again, more data are required to test the hypotheses. In addition, we suggest that female reproductive physiology may be an important determinant, not only of the morphology and function of the male reproductive organs and of sperm, but also of male mating strategies. However, we still lack data on the nature of morphological and physiological interactions between the female and the male as they relate to sperm competition.

Finally, theoretical modelling suggests that intraspecific variation in male adaptations to sperm competition should also be considerable, both at the behavioural and the physiological level. This remains largely unexplored.

While this review has concentrated on what we know about sperm competition, it should be evident that many areas remain unexplored, and that sometimes what we claim to know is based on slender evidence. We hope that this review will stimulate and orientate researchers towards the most exciting and promising areas within the study of sperm competition in mammals, and to highlight just how much more data we need.

ACKNOWLEDGEMENTS

We are grateful to Roger Short for extremely useful comments on an earlier draft, and to Alfredo Vitullo for discussion of some issues. M.G. received financial support from the Ministerio de Educación y Ciencia – DGICYT (grant PB93–0186).

REFERENCES

Ågren G (1990) Sperm competition, pregnancy initiation and litter size: influence of the amount of copulatory behaviour in Mongolian gerbils, *Meriones unguiculatus*. *Anim. Behav.* **40:** 417–427.

Amman RP, Johnson L, Thompson DL & Pickett BW (1976) Daily spermatozoal production, epididymal spermatozoal reserves and transit time of spermatozoa through the epididymis of the rhesus monkey. *Biol. Reprod.* **15:** 586–592.

Amos W, Twiss S, Pomeroy PP & Anderson SS (1993) Male mating success and paternity in the grey seal, *Halichoerus grypus*: a study using DNA fingerprinting. *Proc. Roy. Soc. Lond. Ser. B* **252:** 199–207.

Austin CR (1965) *Fertilization*. Prentice-Hall, Austin, New Jersey.

Austin CR (1975) Sperm fertility, viability and persistence in the female tract. *J. Reprod. Fertil. (Suppl.)* **22:** 75–89.

Austin CR (1985) Sperm maturation in the male and female genital tracts. In *Biology of Fertilization*, Vol. 2. CB Metz & A Monroy (eds), pp. 121–155. Academic Press, Orlando.

Baker RR & Bellis MA (1988) 'Kamikaze' sperm in mammals? *Anim. Behav.* **36:** 936–939.

Baker RR & Bellis MA (1995) *Human Sperm Competition*. Chapman and Hall, London.

Bancroft J, Sanders D, Davidson DW & Warner P (1983) Mood, sexuality, hormones and the menstrual cycle. III. Sexuality and the role of androgens. *Psychos. Med.* **45:** 509–516.

Barrat CLR, Bolton AE & Cooke ID (1990) Functional significance of white blood cells in the male and female reproductive tract. *Hum. Reprod.* **5:** 639–648.

Barrat CLR & Cooke ID (1991) Sperm transport in the human female reproductive tract – a dynamic interaction. *Int. J. Androl.* **14:** 394–411.

Bedford JM (1979) Evolution of the sperm maturation and sperm storage functions of the epididymis. In *The Spermatozoon*. DW Fawcett & JM Bedford (eds), pp. 7–21. Urban and Schwarzenberg, Baltimore-Munich.

Bedford JM (1983) Form and function of eutherian spematozoa in relation to the nature of egg vestments. In *Fertilization of the Human Egg In Vitro*. HM Beier & HR Lindner (eds), pp. 133–146. Springer-Verlag, Berlin.

Bedford JM (1991) The co-evolution of mammalian gametes. In *A Comparative Overview of Mammalian Fertilization*. BS Dunbar & MG O'Rand (eds), pp. 3–35. Plenum Press, New York.

Bedford JM & Hoskins DD (1990) The mammalian spermatozoon: morphology, biochemistry and physiology. In *Marshalls Physiology of Reproduction*, Vol. 2. GE Lamming (ed.), pp. 379–568. Churchill Livingstone, London.

Bedford JM & Yanagimachi R (1992) Initiation of sperm motility after mating in the rat and hamster. *J. Androl.* **13:** 444–449.

Bedford JM, Rodger JC & Breed WG (1984) Why so many mammalian spermatozoa – a clue from marsupials? *Proc. Roy. Soc. Lond. Ser. B* **221:** 221–233.

Bellis RR, Baker MA & Gage MJG (1990) Variation in rat ejaculates consistent with the kamikaze sperm hypothesis. *J. Mammal.* **71:** 479–480.

Berard JD, Nürnberg P, Epplen JT & Schmidtke J (1993) Male rank, reproductive behaviour and reproductive success in free-ranging rhesus macaques. *Primates* **34:** 481–489.

Bercovitch FB (1989) Body size, sperm competition, and determinants of reproductive success in male savanna baboons. *Evolution* **43:** 1507–1521.

Bercovitch FB (1995) Female cooperation, consortship maintenance, and male mating success in savanna baboons. *Anim. Behav.* **50:** 137–149.

Bercovitch FB & Nürnberg P (1996) Socioendocrine and morphological correlates of paternity in rhesus macaques (*Macaca mulatta*). *J. Reprod. Fertil.* **107:** 59–68.

Bercovitch FB & Rodriguez JF (1993) Testis size, epididymis weight, and sperm competition in rhesus macaques. *Am. J. Primatol.* **30:** 163–168.

Biquand S, Boug A, Biquand-Guyot V & Gautier JP (1994) Management of commensal baboons in Saudi Arabia. In *Proceedings of the Symposium 'Commensal Primates'*. S Biquand & JP Gautier (eds). *Rev. Ecol.* **49:** 213–222.

Birdsall DA & Nash D (1973) Occurrence of successful multiple insemination of females in natural populations of deer mice (*Peromyscus maniculatus*). *Evolution* **27:** 106–110.

Birkhead TR & Møller AP (1992) *Sperm Competition in Birds*. Academic Press, London.

Birkhead TR & Møller AP (1993) Sexual selection and the temporal separation of reproductive events: sperm storage data from reptiles, birds and mammals. *Biol. J. Linn. Soc.* **50:** 295–311.

Birkhead TR, Møller AP & Sutherland WJ (1993) Why do females make it so difficult for males to fertilize their eggs? *J. Theor. Biol.* **161:** 51–60.

Birkhead TR, Fletcher F, Pellatt EJ & Staples A (1995a) Ejaculate quality and the success of extra-pair copulations in the zebra finch. *Nature* **377:** 422–423.

Birkhead TR, Wishart GJ & Biggins JD (1995b) Sperm precedence in the domestic fowl. *Proc. Roy. Soc. Lond. Ser. B* **261:** 285–292.

Bishop DW (1969) Sperm physiology in relation to the oviduct. In *The Mammalian Oviduct*. ESE Hafez & RJ Blandau (eds), pp. 231–250. The University of Chicago Press, Chicago.

Blandau RJ (1945) On factors involved in sperm transport through the cervix uteri of the albino rat. *Am. J. Anat.* **77:** 253–272.

Blandau RJ (1969) Gamete transport: comparative aspects. In *The Mammalian Oviduct*. ESE Hafez & RJ Blandau (eds), pp. 129–162. The University of Chicago Press, Chicago.

Boellstorff DE, Owings DH, Penedo MCT & Hersek MJ (1994) Reproductive behaviour and multiple paternity of California ground squirrels. *Anim. Behav.* **47:** 1057–1064.

Boonstra R, Xia X & Pavone L (1993) Mating system of the meadow vole, *Microtus pennsylvanicus*. *Behav. Ecol.* **4:** 83–89.

Breed WG (1986) Comparative morphology and evolution of the male reproductive tract in the Australian hydromyine rodents (Muridae). *J. Zool.* **209:** 607–619.

Bressac C, Joly D, Devaux J, Serres C, Feneux D & Lachaise D (1991) Comparative kinetics of short and long sperm in sperm dimorphic Drosophila species. *Cell Motil. Cytoskel.* **19:** 269–274.

Briskie JV & Montgomerie R (1992) Sperm size and sperm competition in birds. *Proc. Roy. Soc. Lond. Ser. B* **247:** 89–95.

Briskie JV & Montgomerie R (1993) Patterns of sperm storage in relation to sperm competition in passerine birds. *Condor* **95:** 442–454.

Briskie JV, Montgomerie R & Birkhead TR (1997) The evolution of sperm size in birds. *Evolution* **51:** 937–945.

Bronson FH (1989) *Mammalian Reproductive Biology.* The University of Chicago Press, Chicago.

Brotherton PNM, Pemberton JM, Komers PE & Malarky G (1997) Genetic and behavioural evidence of monogamy in a mammal, Kirk's dik-dik (*Maquoda kirkii*). *Proc. Roy. Soc. Lond. Ser. B* **264:** 675–681.

Brownwell RL & Ralls K (1986) *Potential for Sperm Competition in Baleen Whales No. 8*. International Whaling Commission.

Busse CD & Estep DQ (1984) Sexual arousal in male pigtailed monkeys (*Macaca nemestrina*): effects of serial matings by two males. *J. Comp. Psychol.* **98:** 227–231.

Carter CS & Getz LL (1993) Monogamy and the prairie vole. *Sci. Am.* **1993:** 70–76.

Chen Y, Li J, Simkin ME, Yang X & Foote RH (1989) Fertility of fresh and frozen rabbit semen inseminated at different times is indicative of male differences in capacitation time. *Biol. Reprod.* **41:** 848–853.

Cherr GN & Drobnis EZ (1991) Fertilization in the golden hamster. In *A Comparative Overview of Mammalian Fertilization.* BS Dunbar & MG O'Rand (eds), pp. 217–243. Plenum Press, New York.

Clutton-Brock TH (1989) Mammalian mating systems. *Proc. Roy. Soc. Lond. Ser. B* **236:** 339–372.

Clutton-Brock TH (1991) *The Evolution of Parental Care.* Princeton University Press, Princeton.

Clutton-Brock TH & Harvey PH (1976) Evolutionary rules and primate societies. In *Growing Points in Ethology.* PPG Bateson & RA Hinde (eds), pp. 195–237. Cambridge University Press, Cambridge.

Clutton-Brock TH & Parker GA (1992) Potential reproductive rates and the operation of sexual selection. *Quart. Rev. Biol.* **67:** 437–456.

Clutton-Brock TH & Parker GA (1995) Sexual coercion in animal societies. *Anim. Behav.* **49:** 1345–1365.

Clutton-Brock TH, Guinness FE & Albon SD (1982) *Red Deer. Behavior and Ecology of Two Sexes.* Edinburgh University Press, Edinburgh.

Cohen J (1969) Why so many sperms? An essay on the arithmetic of reproduction. *Sci. Progr.* **57:** 23–41.

Cohen J (1973) Crossovers, sperm redundancy, and their close association. *Heredity* **31:** 408–413.

Colegreave N, Birkhead TR & Lessels CM (1995) Sperm precedence in zebra finches does not require special mechanisms of sperm competition. *Proc. Roy. Soc. Lond. Ser. B* **259:** 223–228.

Crook JH (1980) Social change in Indian Tibet. *Soc. Sci. Inf.* **19:** 139–166.

Cukierski MA, Sina JL, Prahalada S, Wise LD, Antonello JM, MacDonald JS & Robertson RT (1991) Decreased fertility in male rats administered the 5 alpha-reductase inhibitor, finasteride, is due to deficits in copulatory plug formation. *Reprod. Toxicol.* **5:** 353–362.

Cummins JM & Woodall PF (1985) On mammalian sperm dimensions. *J. Reprod. Fertil.* **75:** 153–175.

Cummins JM & Yanagimachi R (1982) Sperm–egg ratios and the site of the acrosome reaction during *in vivo* fertilization in the hamster. *Gam. Res.* **5:** 239–256.

Cummins JM & Yanagimachi R (1986) Development of the ability to penetrate the cumulus oophorus by hamster spermatozoa capacitated *in vitro*, in relation to the timing of the acrosome reaction. *Gam. Res.* **15:** 187–212.

Curie-Cohen M, Yoshihara D, Blystad C, Luttrell L, Benforado K & Stone WH (1981) Paternity and mating behaviour in a captive troop of rhesus monkeys. *Am. J. Primatol.* **1:** 335.

Curtsinger JW (1991) Sperm competition and the evolution of multiple mating. *Am. Nat.* **138:** 93–102.

Cuthill IC, Kacelnik A, Krebs JR, Haccou P & Isawa Y (1990) Patch use by starlings: the effect of recent experience on foraging decisions. *Anim. Behav.* **40:** 625–640.

Dahl JF, Gould KG & Nadler RD (1993) Testicle size of orang-utans in relation to body size. *Am. J. Phys. Anthropol.* **90:** 229–236.

Darwin C (1871) *The Descent of Man, and Selection in Relation to Sex*. John Murray, London.

Dewsbury DA (1972) Patterns of copulatory behaviour in male mammals. *Quart. Rev. Biol.* **47:** 1–33.

Dewsbury DA (1981) An exercise in the prediction of monogamy in the field from laboratory data on 42 species of muroid rodents. *Biologist* **63:** 138–162.

Dewsbury DA (1982) Ejaculate cost and female choice. *Am. Nat.* **119:** 601–610.

Dewsbury DA (1984) Sperm competition in muroid rodents. In *Sperm Competition and the Evolution of Animal Mating Systems*. RL Smith (ed.), pp. 547–571. Academic Press, Orlando.

Dewsbury DA (1985) Interactions between males and their sperm during multimale copulatory episodes of deer mice (*Peromyscus maniculatus*). *Anim. Behav.* **33:** 1266–1274.

Dewsbury DA (1988a) Sperm competition in deer mice (*Peromyscus maniculatus bairdi*). *Behav. Ecol. Sociobiol.* **22:** 251–256.

Dewsbury DA (1988b) A test of the role of copulatory plugs in sperm competition in deer mice (*Peromyscus maniculatus*). *J. Mammal.* **69:** 854–857.

Dewsbury DA & Pierce JD (1989) Copulatory patterns of primates as viewed in broad mammalian perspective. *Am. J. Primatol.* **17:** 51–72.

Dixson AF (1986) Genital sensory feedback and sexual behavior in male and female marmosets (*Callithrix jacchus*). *Physiol. Behav.* **37:** 447–450.

Dixson AF (1987) Observations on the evolution of the genitalia and copulatory behaviour in male primates. *J. Zool.* **213:** 423–443.

Dixson AF (1991) Sexual selection, natural selection and copulatory patterns in male primates. *Folia Primatol.* **57:** 96–101.

Dixson AF (1993) Sexual selection, sperm competition and the evolution of sperm length. *Folia Primatol.* **61:** 221–227.

Dixson AF (1995a) Sexual selection and ejaculatory frequencies in primates. *Folia Primatol.* **64:** 146–152.

Dixson AF (1995b) Baculum length and copulatory behaviour in carnivores and pinnipeds (Grand Order Ferae). *J. Zool.* **235:** 67–76.

Dixson AF & Mundy NI (1994) Sexual behavior, sexual swelling, and penile evolution in chimpanzees (*Pan troglodytes*). *Arch. Sexual Behav.* **23:** 267–280.

Doak RL, Hall A & Dale HE (1967) Longevity of spermatozoa in the reproductive tract of the bitch. *J. Reprod. Fertil.* **13:** 51–58.

Drobnis EZ & Overstreet JW (1992) Natural history of mammalian spermatozoa in the female reproductive tract. *Oxf. Rev. Reprod. Biol.* **14:** 1–45.

Drobnis EZ, Yudin AI, Cherr GN & Katz DF (1988) Kinematics of hamster sperm during penetration of the cumulus cell matrix. *Gam. Res.* **21:** 367–383.

Dunbar RIM (1988) *Primate Social Systems.* Croom Helm, London.

Dunbar RIM & Cowlishaw G (1992) Mating success in male primates: dominance rank, sperm competition and alternative strategies. *Anim. Behav.* **44:** 1171–1173.

Duvall SW, Bernstein IS & Gordon TP (1976) Paternity and status in a rhesus monkey group. *J. Reprod. Fertil.* **47:** 25–31.

Dziuk PJ (1965) Double mating of rabbits to determine capacitation time. *J. Reprod. Fertil.* **10:** 389–395.

Dziuk P (1970) Estimation of optimum time for insemination of gilts and ewes by double-mating at certain times relative to ovulation. *J. Reprod. Fertil.* **22:** 277–282.

Eberhard WG (1985) *Sexual Selection and Animal Genitalia.* Harvard University Press, Cambridge, MA.

Eberhard WG (1990) Animal genitalia and female choice. *Am. Sci.* **78:** 134–141.

Eberhard WG & Cordero C (1995) Sexual selection by cryptic female choice on male seminal products – a new bridge between sexual selection and reproductive physiology. *Trends Ecol. Evol.* **10:** 493–496.

Emlen ST & Oring LW (1977) Ecology, sexual selection, and the evolution of mating systems. *Science* **197:** 215–223.

Fénichel P & Parinaud J (eds) (1995) *The Human Sperm Acrosome Reaction.* John Libbey Eurotext, Montrouge.

Fenton MB (1984) Sperm competition? The case of vespertilionid and rhinolophid bats. In *Sperm Competition and the Evolution of Animal Mating Systems.* RL Smith (ed.), pp. 573–587. Academic Press, London.

Fisch H & Goluboff ET (1996) Geographical variations in sperm counts: a potential cause of bias in studies of semen quality. *Fertil. Steril.* **65:** 1044–1046.

Florman HM & Storey BT (1982) Mouse gamete interactions: the zona pellucida is the site of the acrosome reaction leading to fertilization *in vitro*. *Dev. Biol.* **91:** 121–130.

Foltz DW (1981) Genetic evidence for long-term monogamy in a small rodent, *Peromyscus polionotus*. *Am. Nat.* **117:** 665–675.

Foltz DW & Schwagmeyer PL (1989) Sperm competition in the 13-lined ground squirrel: differential fertilization success under field conditions. *Am. Nat.* **133:** 257–265.

Fooden J (1980) Classification and distribution of living macaques (Macaca Lacepede, 1799). In *The Macaques: Studies in Ecology, Behavior and Evolution.* DG Lindberg (ed.), pp. 1–9. Van Nostrand Rheinhold, New York.

Fordney-Settlage D (1981) A review of cervical mucus and sperm interactions in humans. *Int. J. Fertility* **26:** 161–169.

Fournier-Delpech S & Thibault C (1991) Acquisition de la fécondance du spermatozoide. In *La Reproduction chez les Mammiléres et l'Homme.* C Thibault & M-C Levasseur (eds). Ellipses, Paris.

Freund M (1962) Interrelationships among the characteristics of human semen and factors affecting semen-specimen quality. *J. Reprod. Fertil.* **4:** 143–159.

Fukuda M, Morales P & Overstreet JW (1989) Acrosomal function of human spermatozoa with normal and abnormal head morphology. *Gam. Res.* **24:** 59–65.

Gage M (1994) Associations between body size, mating pattern, testis size and sperm lengths across butterflies. *Proc. Roy. Soc. Lond. Ser. B* **258:** 247–254.

Gagneux P, Woodruff DS & Boesch C (1997) Furtive mating in female chimpanzees. *Nature* **387:** 358–359.

Gibson RM & Jewell PA (1982) Semen quality, female choice, and multiple mating in domestic sheep: a test of Trivers' sexual competence hypothesis. *Behaviour* **80:** 9–31.

Gilbert DA, Packer C, Pusey AE, Stephens JC & O'Brien SJ (1991) Analytical DNA fingerprinting in lions: parentage, genetic diversity and kinship. *J. Hered.* **82:** 378–386.

Ginsberg JR & Huck UW (1989) Sperm competition in mammals. *Trends Ecol. Evol.* **4:** 74–79.

Ginsberg JR & Rubenstein DI (1990) Sperm competition and variation in zebra mating behavior. *Behav. Ecol. Sociobiol.* **26:** 427–434.

Gomendio M & Roldan ERS (1991) Sperm competition influences sperm size in mammals. *Proc. Roy. Soc. Lond. Ser. B* **243:** 181–185.

Gomendio M & Roldan ERS (1993a) Co-evolution between male ejaculates and female reproductive biology in eutherian mammals. *Proc. Roy. Soc. Lond. Ser. B* **252:** 7–12.

Gomendio M & Roldan ERS (1993b) Mechanisms of sperm competition: linking physiology and behavioural ecology. *Trends Ecol. Evol.* **8:** 95–100.

Gomendio M & Roldan ERS (1994) The evolution of gametes. In *Principles of Medical Biology*, Vol. 1B. EE Bittar & N Bittar (eds), pp. 115–151. JAI Press, Greenwich, Connecticut.

Halliday T & Arnold SJ (1987) Multiple mating by females: a perspective from quatitative genetics. *Anim. Behav.* **35:** 939–941.

Hanken J & Sherman PW (1981) Multiple paternity in Belding's ground squirrels. *Science* **212:** 351–353.

Harcourt AH (1981) Intermale competition and the reproductive behavior of the great apes. In *Reproductive Biology of the Great Apes*. CE Graham (ed.), pp. 301–318. Academic Press, New York.

Harcourt AH (1991) Sperm competition and the evolution of nonfertilizing sperm in mammals. *Evolution* **45:** 314–328.

Harcourt AH & Gardiner J (1994) Sexual selection and genital anatomy of male primates. *Proc. Roy. Soc. Lond. Ser. B* **255:** 47–53.

Harcourt AH, Fossey D, Stewart KJ & Watts DP (1980) Reproduction in wild gorillas and some comparisons with chimpanzees. *J. Reprod. Fertil. Suppl.* **28:** 59–70.

Harcourt AH, Harvey PH, Larson SG & Short RV (1981) Testis weight, body weight and breeding system in primates. *Nature* **293:** 55–57.

Harcourt AH, Purvis A & Liles L (1995) Sperm competition: mating system, not breeding season, affects testes size of primates. *Funct. Ecol.* **9:** 468–476.

Harrison RM & Wolf RH (1985) Sperm parameters and testicular volumes in *Saguinus mystax*. *J. Med. Primatol.* **14:** 281–284.

Hartung TG & Dewsbury DA (1978) A comparative analysis of copulatory plugs in muroid rodents and their relationship to copulatory behavior. *J. Mammal.* **59:** 717–723.

Harvey PH & Bennett PM (1985) Sexual dimorphism and reproductive strategies.

In *Human Sexual Dimorphism*. J Ghesquire, RD Martin & F Newcombe (eds), pp. 43–59. Taylor and Francis, London.

Harvey PH & May RM (1989) Copulation dynamics: out for the sperm count. *Nature* **337:** 508–509.

Harvey PH, Martin RD & Clutton-Brock TH (1987) Life histories in comparative perspective. In *Primate Societies*. BB Smuts, DL Cheney, RM Seyfarth, RW Wrangham & TT Struhsaker (eds), pp. 181–196. University of Chicago Press, Chicago.

Hill WCO (1953) *Primates: Comparative Anatomy and Taxonomy. 1. Strepsirhini*. University Press, Edinburgh.

Hogg JT (1984) Mating in bighorn sheep: multiple creative male strategies. *Science* **225:** 526–529.

Hogg JT (1988) Copulatory tactics in relation to sperm competition in Rocky Mountain bighorn sheep. *Behav. Ecol. Sociobiol.* **22:** 49–59.

Hoogland JL (1995) *The Black-tailed Prairie Dog. Social Life of a Burrowing Mammal.* University of Chicago Press, Chicago.

Hoogland JL & Foltz DW (1982) Variance in male and female reproductive success in a harem-polygynous mammal, the black-tailed prairie dog (Sciuridae: *Cynomys ludovicianus*). *Behav. Ecol. Sociobiol.* **11:** 155–163.

Hrdy SB (1979) Infanticide among animals: a review, classification, and examination of implications for the reproductive strategies of females. *Ethol. Sociobiol.* **1:** 13–40.

Hrdy SB (1981) *The Woman That Never Evolved*. Harvard University Press, Cambridge, MA.

Hrdy SB & Whitten PL (1987) Patterning of sexual activity. In *Primate Societies*. BB Smuts, DL Cheney, RM Seyfarth, RW Wrangham & TT Struhsaker (eds), pp. 370–384. University of Chicago Press, Chicago.

Huck UW, Tonias BA & Lisk RD (1989) The effectiveness of competitive male inseminations in golden hamsters, *Mesocricetus auratus*, depends on an interaction of mating order, time delay between males, and the time of mating relative to ovulation. *Anim. Behav.* **37:** 674–680.

Hunnicutt GR, Mahan K, Lathrop WF, Ramarao CS, Myles DG & Primakoff P (1996) Structural relationship of sperm soluble hyaluronidase to the sperm membrane protein PH-20. *Biol. Reprod.* **54:** 1343–1349.

Hunter RHF (1973) Polyspermic fertilization in pigs after tubal deposition of excessive numbers of spermatozoa. *J. Exp. Zool.* **183:** 57–64.

Hunter RHF (1984) Pre-ovulatory arrests and peri-ovulatory redistribution of competent spermatozoa in the isthmus of the pig oviduct. *J. Reprod. Fertil.* **72:** 203–211.

Hunter RHF (1987) Human fertilization *in vivo*, with special reference to progression, storage and release of competent spermatozoa. *Hum. Reprod.* **2:** 329–332.

Hunter RHF (1988) *The Fallopian Tubes. Their Role in Fertility and Infertility*. Springer-Verlag, Berlin.

Hunter RHF (1993) Sperm: egg ratios and putative molecular signals to modulate gamete inetractions in polytocous mammals. *Molec. Reprod. Devel.* **35:** 324–327.

Hunter RHF & Wilmut I (1984) Sperm transport in the cow: peri-ovulatory redistribution of viable cells within the oviduct. *Reprod. Nutr. Devel.* **24:** 597–608.

Hunter RHF, Barwise L & King R (1982) Sperm transport, storage and release in the sheep oviduct in relation to the time of ovulation. *Br. Vet. J.* **138:** 225–232.

Inoue M, Mitsunaga F, Ohsawa H, Takenaka A, Sugiyama Y, Gaspard SA &

Takenaka O (1991) Male mating behaviour and paternity discrimination by DNA fingerprinting in a Japanese macaque group. *Folia Primatol.* **56:** 202–210.

Insler V, Glezerman M, Zeidel L, Bernstein D & Misgav N (1980) Sperm storage in the human cervix: a quantitative study. *Fertil. Steril.* **33:** 288–293.

In't Veld P, Brandenburg H, Verhoeff A, Dhort M & Los F (1995) Sex chromosomal abnormalities and intra-cytoplasmic sperm injection. *Lancet* **346:** 773.

Jedlicki A & Barros C (1985) Scanning electron microsocope study of *in vitro* prepenetration gamete interactions. *Gam. Res.* **11:** 121–131.

Johnson AM, Wadsworth J, Wellings K, Bradshaw S & Field J (1992) Sexual lifestyles and HIV risk. *Nature* **360:** 410–412.

Johnson LA (1994) Isolation of X- and Y-bearing sperm for sex preselection. *Oxf. Rev. Reprod. Biol.* **16:** 303–326.

Kappeler PM (1993) Sexual selection and lemur social systems. In *Lemur Social Systems and their Ecological Basis.* PM Kappeler & JU Ganzhorn (eds), pp. 225–242. Plenum Press, New York.

Katz DF & Drobnis EZ (1990) Analysis and interpretation of the forces generated by spermatozoa. In *Fertilization in Mammals.* BD Bavister, J Cummins & ERS Roldan (eds), pp. 125–137. Serono Symposia, Norwell.

Katz DF, Diel L & Overstreet JW (1982) Differences in the movement of morphologically normal and abnormal human seminal spermatozoa. *Biol. Reprod.* **26:** 566–570.

Katz DF, Drobnis EZ & Overstreet JW (1989) Factors regulating mammalian sperm migration through the female reproductive tract and oocyte vestments. *Gam. Res.* **22:** 443–469.

Kaufman MH (1983) *Early Mammalian Development: Parthenogenetic Studies. Developmental and Cell Biology Series.* Cambridge University Press, Cambridge.

Kawata M (1985) Mating system and reproductive success in a spring population of the red-backed vole, *Clethrionomys rufocanus bedfordiae. Oikos* **45:** 181–190.

Kawata M (1988) Mating success, spatial organisation and male characteristics in experimental field populations of the red-backed vole, *Clethrionomys rufocanus bedfordiae. J. Anim. Ecol.* **57:** 217–235.

Keane B, Waser PM, Creel SR, Creel NM, Elliott LF & Minchella DJ (1994) Subordinate reproduction in dwarf mongooses. *Anim. Behav.* **47:** 65–75.

Keller L and Reeve HK (1995) Why do females mate with multiple males? The sexually selected sperm hypothesis. *Adv. Stud. Behav.* **24:** 291–315.

Kenagy GJ & Trombulak SC (1986) Size and function of mammalian testes in relation to body size. *J. Mammal.* **67:** 1–22.

Krzanowska H (1974) The passage of abnormal spermatozoa through the uterotubal junction of the mouse. *J. Reprod. Fertil.* **38:** 81–90.

Krzanowska H (1986) Interstrain competition among mouse spermatozoa inseminated in various proportions, as affected by the genotype of the Y chromosome. *J. Reprod. Fertil.* **77:** 265–270.

Krzanowska H (1991) Phenotype and fertilizing capacity of spermatozoa of chimaeric mice produced from two strains that differ in sperm quality. *J. Reprod. Fertil.* **91:** 667–676.

Krzanowska H & Lorenc E (1983) Influence of egg investments on *in-vitro* penetration of mouse eggs by misshapen spermatozoa. *J. Reprod. Fertil.* **68:** 57–62.

Kubiak J (1989) Mouse oocytes gradually develop the capacity for activation during the metaphase II arrest. *Dev. Biol.* **136:** 537–545.

Langtimm CA & Dewsbury DA (1991) Phylogeny and evolution of rodent copulatory behaviour. *Anim. Behav.* **41:** 217–225.

Lanier DL, Estep DG & Dewsbury DA (1979) The role of prolonged copulatory behaviour in facilitating reproductive success in a competitive mating situation in laboratory rats. *J. Comp. Physiol. Psychol.* **93:** 781–792.

Lanman JT (1968) Delays during reproduction and their effects on the embryo and fetus. 1. Aging of sperm. *N. Engl. J. Med.* **278:** 993–999.

Laumann EO, Gagnon JH, Michael RT & Michaels S (1994) *The Social Organization of Sexuality. Sexual Practices in the United States.* University of Chicago Press, Chicago.

Lefebvre R, Chenoweth PJ, Drost M, Le Clear CT, MacCubbin M, Dutton JT & Suarez SS (1995) Characterization of the oviductal sperm reservoir in cattle. *Biol. Reprod.* **53:** 1066–1074.

Lin Y, Mahan K, Lathrop WF, Myles DG & Primakoff P (1994) A hyaluronidase activity of the sperm plasma membrane protein PH-20 enables sperm to penetrate the cumulus cell layer surrounding the egg. *J. Cell Biol.* **125:** 1157–1163.

Liu DY & Baker HWG (1994) Tests for human sperm-zona pellucida binding and penetration. In *Male Factor in Human Infertility.* J Tesarik (ed.), pp. 169–185. Ares-Serono Symposia Publications, Rome.

Long JA, Wildt DE, Wolfe BA, Critser JK, DeRossi RV & Howard J (1996) Sperm capacitation and the acrosome reaction are compromised in teratospermic domestic cats. *Biol. Reprod.* **54:** 638–646.

MacIntyre S & Sooman A (1991) Non-paternity and prenatal genetic screening. *Lancet* **338:** 869–871.

Maddock MB & Dawson WD (1974) Artificial insemination of deer-mice (*Peromyscus maniculatus*) with sperm from other rodent species. *J. Embryol. Exp. Morphol.* **31:** 621–634.

Mann T & Lutwak-Mann C (1981) *Male Reproductive Function and Semen.* Springer-Verlag, Berlin.

Manning JT & Chamberlain AT (1994) Sib competition and sperm competitiveness: an answer to 'Why so many sperms?' and the recombination/sperm number correlation. *Proc. Roy. Soc. Lond. Ser. B* **256:** 177–182.

Marshall WA (1970) Sex differences at puberty. *J. Biosoc. Sci. Suppl.* **2:** 31–41.

Marson J, Gervais D, Meuris S, Cooper RW & Jouannet P (1989) Influence of ejaculation frequency on semen characteristics in chimpanzees. *J. Reprod. Fertil.* **85:** 43–50.

Martan J & Shepherd BA (1976) The role of copulatory plugs in reproduction in guinea pigs. *J. Exp. Zool.* **196:** 79–83.

Martinet L & Raynaud F (1975) Prolonged spermatozoan survival in the female hare uterus: explanation of superfetation. In *The Biology of Spermatozoa.* ESE Hafez & CG Thibault (eds), pp. 134–144. Karger, Basel.

Matthews MK, Jr & Adler NT (1978) Systematic interrelationship of mating, vaginal plug position, and sperm transport in the rat. *Physiol. Behav.* **20:** 303–309.

Mattner PE (1963) Capacitation of ram spermatozoa and penetration of the ovine egg. *Nature* **199:** 772–773.

McCracken GF & Bradbury JW (1977) Paternity and genetic heterogeneity in the polygynous bat, *Phyllostomus hastatus. Science* **198:** 303–306.

McCracken GF & Bradbury JW (1981) Social organization and kinship in the polygynous bat, *Phyllostomus discolor. Behav. Ecol. Sociobiol.* **8:** 11–34.

Meistrich ML, Kasai K, Olds-Clarke P, MacGregor GR, Berkowitz AD & Tung KSK

(1994) Deficiency in fertilization by morphologically abnormal sperm produced by *azh* mutant mice. *Molec. Reprod. Devel.* **37:** 69–77.

Milligan SR (1979) The copulatory pattern of the bank vole (*Clethrionomys glareolus*) and speculation on the role of penile spines. *J. Zool.* **188:** 279–283.

Møller AP (1988a) Ejaculate quality, testes size and sperm competition in primates. *J. Hum. Evol.* **17:** 479–488.

Møller AP (1988b) Testes size, ejaculate quality and sperm competition in birds. *Biol. J. Linn. Soc.* **33:** 273–283.

Møller AP (1989) Ejaculate quality, testes size and sperm production in mammals. *Funct. Ecol.* **3:** 91–96.

Møller AP (1991a) Concordance of mammalian ejaculate features. *Proc. Roy. Soc. Lond. Ser. B* **246:** 237–241.

Møller AP (1991b) Sperm competition, sperm depletion, paternal care, and relative testis size in birds. *Am. Nat.* **137:** 882–906.

Møller AP & Birkhead TR (1989) Copulation behaviour in mammals: evidence that sperm competition is widespread. *Biol. J. Linn. Soc.* **38:** 119–131.

Møller AP & Briskie JV (1995) Extra-pair paternity, sperm competition and the evolution of testis size in birds. *Behav. Ecol. Sociobiol.* **36:** 357–365.

Morales P, Katz DF, Overstreet JW, Samuels SJ & Chang RJ (1988) The relationship between the motility and morphology of spermatozoa in human semen. *J. Androl.* **9:** 241–247.

Morland HS (1993) Reproductive activity of ruffed lemurs (*Varecia variegata variegata*) in a Madagascar rain forest. *Am. J. Phys. Anthropol.* **91:** 71–82.

Mortimer D (1983) Sperm transport in the human female reproductive tract. *Oxf. Rev. Reprod. Biol.* **5:** 30–61.

Mortimer D (1994) *Practical Laboratory Andrology*. Oxford University Press, Oxford.

Mortimer D (1995) Sperm transport in the female genital tract. In *Gametes–The Spermatozoon*. JG Grudzinkas & JL Yovich (eds), pp. 157–174. Cambridge University Press, Cambridge.

Murie JO (1995) Mating behavior of Columbian ground squirrels. I. Multiple mating by females and multiple paternity. *Can. J. Zool.* **73:** 1819–1826.

O'Connell SM & Cowlishaw G (1994) Infanticide avoidance, sperm competition and female mate choice: the function of copulation calls in female baboons. *Anim. Behav.* **48:** 687–694.

Oettle EE (1993) Sperm morphology and fertility in the dog. *J. Reprod. Fertil. Suppl.* **47:** 257–260.

O'Rand MG & Nikolajczyk BS (1991) Fertilization in the rabbit. In *A Comparative Overview of Mammalian Fertilization*. BS Dunbar & MG O'Rand (eds), pp. 271–279. Plenum Press, New York.

Orgebin-Crist MC & Olson GE (1984) Epididymal sperm maturation. In *The Male in Farm Animal Reproduction*. M Courot (ed.), pp. 80–102. Martinus Nijhoff Publishers, Boston.

Overstreet JW (1983) Transport of gametes in the reproductive tract of the female mammal. In *Mechanism and Control of Animal Fertilization*. JF Hartmann (ed.), pp. 499–543. Academic Press, London.

Overstreet JW & Cooper GW (1975) Reduced sperm motility in the isthmus of the rabbit oviduct. *Nature* **258:** 718–719.

Overstreet JW & Katz DF (1990) Interaction between the female reproductive tract and spermatozoa. In *Controls of Sperm Motility: Biological and Clinical Aspects*. C Gagnon (ed.), pp. 63–75. CRC Press, Boca Raton.

Ozil JP (1990) The parthenogenetic development of rabbit oocytes after repetitive pulsatile electrical stimulation. *Development* **109:** 117–127.

Pacey AA, Hill CJ, Scudamore IW, Warren MA, Barratt CLR & Cooke ID (1995) The interaction *in-vitro* of human spermatozoa with epithelial cells from the human uterine (fallopian) tube. *Hum. Reprod.* **10:** 360–366.

Pagel MD (1992) A method for the analysis of comparative data. *J. Theor. Biol.* **156:** 431–442.

Pandya IJ & Cohen J (1985) The leukocytic reaction of the human uterine cervix to spermatozoa. *Fertil. Steril.* **43:** 417–421.

Parker GA (1970) Sperm competition and its evolutionary consequences in insects. *Biol. Rev.* **45:** 525–567.

Parker GA (1982) Why so many tiny sperm? The maintenance of two sexes with internal fertilization. *J. Theor. Biol.* **96:** 281–294.

Parker GA (1984) Sperm competition and the evolution of animal mating strategies. In *Sperm Competition and the Evolution of Animal Mating Systems.* RL Smith (ed.), pp. 1–60. Academic Press, Orlando.

Parker GA (1990a) Sperm competition games: guards and extra-pair copulations. *Proc. Roy. Soc. Lond. Ser. B* **242:** 127–133.

Parker GA (1990b) Sperm competition games: raffles and roles. *Proc. Roy. Soc. Lond. Ser. B* **242:** 120–126.

Parker GA (1993) Sperm competition games: sperm size and sperm number under adult control. *Proc. Roy. Soc. Lond. Ser. B* **253:** 245–254.

Parker GA & Begon ME (1993) Sperm competition games: sperm size and number under gametic control. *Proc. Roy. Soc. Lond. Ser. B* **253:** 255–262.

Parker GA, Baker RR & Smith VGF (1972) The origin and evolution of gamete dimorphism and the male-female phenomenon. *J. Theor. Biol.* **36:** 529–553.

Parkes AS (ed.) (1960) *Marshall's Physiology of Reproduction,* Vol. I, part 2. Longmans, London.

Pemberton JM, Albon SD, Guinness FE, Clutton-Brock TH & Dover GA (1992) Behavioural estimates of male mating success tested by DNA fingerprinting in a polygynous mammal. *Behav. Ecol.* **3:** 66–75.

Petter-Rousseaux A (1964) Reproductive physiology and behavior of the Lemuroidea. In *Evolutionary and Genetic Biology of Primates.* J Buettner-Janusch (ed.), pp. 91–132. Academic Press, New York and London.

Pierce JD, Ferguson B, Salo AL, Sawrey DK, Shapiro LE, Taylor SA & Dewsbury DA (1990) Patterns of sperm allocation across successive ejaculates in four species of voles (*Microtus*). *J. Reprod. Fertil.* **88:** 141–149.

Pitnick S, Spicer GS & Markow TA (1995) How long is a giant sperm? *Nature* **375:** 109.

Polge C, Salamon S & Wilmut I (1970) Fertilizing capacity of frozen boar sperm following surgical insemination. *Vet. Rec.* **87:** 424–428.

Pollard JW, Plante C, King WA, Hansen PJ, Betteridge KJ & Suarez SS (1991) Fertilizing capacity of bovine sperm may be maintained by binding to oviductal epithelial cells. *Biol. Reprod.* **44:** 102–107.

Pope TR (1990) The reproductive consequences of male cooperation in the red howler monkey: paternity exclusion in multi-male and single-male troops using genetic markers. *Behav. Ecol. Sociobiol.* **27:** 439–446.

Popp JL & DeVore I (1979) Aggressive competition and social dominance theory: synopsis. In *The Great Apes.* DA Hamburg & ER McCown (eds), pp. 317–338. The Benjamin/Cummings Publ. Co, Menlo Park.

Potts WK, Manning CJ & Wakeland EK (1991) Mating patterns in seminatural populations of mice influenced by MHC genotype. *Nature* **352:** 619–621.

Profet M (1993) Menstruation as a defense against pathogens transported by sperm. *Quart. Rev. Biol.* **68:** 335–381.

Purvis A (1995) A composite estimate of primate phylogeny. *Phil. Trans. Roy. Soc. Lond. B* **348:** 405–421.

Purvis A & Rambaut A (1995) Comparative analysis by independent contrasts (CAIC) – an Apple Macintosh application for analysing comparative data. *Comput. Appl. Biosci.* **11:** 247–251.

Pusey A & Wolf M (1996) Inbreeding avoidance in animals. *Trends Ecol. Evol.* **11:** 201–206.

Reichard U (1995) Extra-pair copulations in a monogamous gibbon (*Hylobates lar*). *Ethology* **100:** 99–112.

Ribble DO (1991) The monogamous mating system of *Peromyscus californicus* as revealed by DNA fingerprinting. *Behav. Ecol. Sociobiol.* **29:** 161–166.

Rice WR (1996) Sexually anatagonistic male adaptation triggered by experimental arrest of female evolution. *Nature* **381:** 229–232.

Ridley M (1983) *The Explanation of Organic Diversity. The Comparative Method and Adaptations for Mating.* Clarendon Press, Oxford.

Robaire B & Hermo L (1988) Efferent ducts, epididymis and vas deferens: structure, functions and their regulation. In *The Physiology of Reproduction*, Vol. 2. E Knobil & JD Neill (eds), pp. 999–1080. Raven Press, New York.

Robl JM & Dziuk PJ (1988) Comparison of heterospermic and homospermic inseminations as measures of male fertility. *J. Exp. Zool.* **245:** 97–101.

Rodman PS & Mitani JC (1987) Orangutans: sexual dimorphism in a solitary species. In *Primate Societies*. BB Smuts, DL Cheney, RM Seyfarth, RW Wrangham & TT Struhsaker (eds), pp. 146–154. The University of Chicago Press, Chicago.

Roldan ERS (1990) Physiological stimulators of the acrosome reaction. In *Fertilization in Mammals*. BD Bavister, J Cummins & ERS Roldan (eds), pp. 197–205. Serono Symposia, Norwell.

Roldan ERS & Gomendio M (1992) Morphological, functional and biochemical changes underlying the preparation and selection of fertilising spermatozoa '*in vivo*'. *Anim. Reprod. Sci.* **28:** 69–78.

Roldan ERS & Gomendio M (1995) Sperm length and sperm competition in primates: a rebuttal of criticism. *Folia Primatol.* **64:** 225–230.

Roldan ERS & Yanagimachi R (1989) Cross fertilization between Syrian and Chinese hamsters. *J. Exp. Zool.* **250:** 321–328.

Roldan ERS, Vitullo AD, Merani MS & von Lawzewitsch I (1985) Cross fertilization *in vivo* and *in vitro* in species of vesper mice, *Calomys* (Rodentia, Cricetidae). *J. Exp. Zool.* **233:** 433–442.

Roldan ERS, Gomendio M & Vitullo AD (1992) The evolution of eutherian spermatozoa and underlying selective forces: female selection and sperm competition. *Biol. Rev.* **67:** 551–593.

Roldan ERS, Murase T & Shi Q-X (1994) Exocytosis in spermatozoa in response to progesterone and zona pellucida. *Science* **266:** 1578–1581.

Rood JP (1972) Ecological and behavioral comparisons of three genera of Argentine cavies. *Anim. Behav. Monogr.* **5:** 1–83.

de Ruiter JR & van Hooff JARAM (1993) Male dominance rank and reproductive success in primate groups. *Primates* **34:** 513–523.

Saling PM & Storey BT (1979) Mouse gamete interactions during fertilization *in vitro.* Chlortetracycline as a fluorescent probe for the mouse sperm acrosome reaction. *J. Cell Biol.* **83:** 544–555.

Sauther ML (1991) Reproducive behavior of free-ranging *Lemur catta* at Beza Mahafaly special reserve, Madagascar. *Am. J. Phys. Anthropol.* **84:** 463–477.

Schenk A & Kovacs KM (1995) Multiple mating between black bears revealed by DNA fingerprinting. *Anim. Behav.* **50:** 1483–1490.

Schultz AH (1938) The relative weight of the testes in primates. *Anat. Rec.* **72:** 387–394.

Schwagmeyer PL (1988) Scramble-competition polygyny in an asocial mammal: male mobility and mating success. *Am. Nat.* **131:** 885–892.

Schwagmeyer PL (1995) Searching today for tomorrow's mates. *Anim. Behav.* **50:** 759–767.

Schwagmeyer PL & Foltz DW (1990) Factors affecting the outcome of sperm competition in 13-lined ground squirrels. *Anim. Behav.* **39:** 156–162.

Schwagmeyer PL & Parker GA (1990) Male mate choice as predicted by sperm competition in 13-lined ground squirrels. *Nature* **348:** 62–64.

Schwagmeyer PL & Parker GA (1994) Mate-quitting rules for male 13-lined ground squirrels. *Behav. Ecol.* **5:** 142–150.

Seuanez HN, Carothers AD, Martin DE & Short RV (1977) Morphological abnormalities in spermatozoa of man and great apes. *Nature* **270:** 345–347.

Shalgi R (1991) Fertilization in the rat. In *A Comparative Overview of Mammalian Fertilization.* BS Dunbar & MG O'Rand (eds), pp. 245–255. Plenum Press, New York.

Shalgi R & Phillips DM (1988) Motility of rat spermatozoa at the site of fertilization. *Biol. Reprod.* **39:** 1207–1213.

Sherman PW (1989) Mate guarding as paternity insurance in Idaho ground squirrels. *Nature* **338:** 418–420.

Sherman PW & Westneat DF (1988) Multiple mating and quantitative genetics. *Anim. Behav.* **36:** 1545–1547.

Shively C, Clarke S, King N, Schapiro S & Mitchell G (1982) Patterns of sexual behavior in male macaques. *Am. J. Primatol.* **2:** 373–384.

Short RV (1976a) The origin of species. In *Reproduction in Mammals. 6. The Evolution of Reproduction.* CR Austin & RV Short (eds), pp. 110–131. Cambridge University Press, Cambridge.

Short RV (1976b) The evolution of human reproduction. *Proc. Roy. Soc. Lond. Ser. B* **195:** 3–24.

Short RV (1977) Sexual selection and the descent of man. In *Reproduction and Evolution.* JH Calaby & CH Tyndale-Biscoe (eds), pp. 3–19. Australian Academy of Science, Canberra.

Short RV (1979) Sexual selection and its component parts, somatic and genital selection, as illustrated by man and the great apes. *Adv. Stud. Behav.* **9:** 131–158.

Short RV (1981) Sexual selection in man and the great apes. In *Reproductive Biology of the Great Apes.* CE Graham (ed.), pp. 319–341. Academic Press, New York.

Shostak M (1981) *Nisa. The Life and Words of a !Kung Woman.* Harvard University Press, New York.

Siegel S (1956) *Nonparametric Statistics for the Behavioural Sciences.* McGraw-Hill Kogakusha Ltd., Tokyo.

Sillén-Tullberg B & Møller AP (1993) The relationship between concealed ovulation and mating systems in anthropoid primates – a phylogenetic analysis. *Am. Nat.* **141:** 1–25.

Sillero-Zubiri C, Gotelli D & Macdonald DW (1996) Male philopatry, extra-pack copulations and inbreeding avoidance in Ethiopian wolves (*Canis simensis*). *Behav. Ecol. Sociobiol.* **38:** 331–340.

Sivinski J (1984) Sperm in competition. In *Sperm Competition and the Evolution of Animal Mating Systems.* RL Smith (ed.), pp. 86–115. Academic Press, San Diego.

Small MF & Smith DG (1982) The relationship between maternal and paternal rank in rhesus macaques (*Macaca mulatta*). *Anim. Behav.* **30:** 626–633.

Smith RL (1984a) Human sperm competition. In *Sperm Competition and the Evolution of Animal Mating Systems.* RL Smith (ed.), pp. 601–659. Academic Press, San Diego.

Smith, RL (ed.) (1984b) *Sperm Competition and the Evolution of Animal Mating Systems.* Academic Press, San Diego.

Smith TT & Yanagimachi R (1990) The viability of hamster spermatozoa stored in the isthmus of the oviduct: the importance of sperm–epithelium contact for sperm survival. *Biol. Reprod.* **42:** 450–457.

Smith TT & Yanagimachi R (1991) Attachment and release of spermatozoa from the caudal isthmus of the hamster oviduct. *J. Reprod. Fertil.* **91:** 567–573.

Smith TT, Koyanagi F & Yanagimachi R (1987) Distribution and number of spermatozoa in the oviduct of the golden hamster after natural mating and artificial insemination. *Biol. Reprod.* **37:** 225–234.

Smith TT, Koyanagi F & Yanagimachi R (1988) Quantitative comparison of the passage of homologous and heterologous spermatozoa through the uterotubal junction of the golden hamster. *Gam. Res.* **19:** 227–234.

Smuts BB (1987) Sexual competition and mate choice. In *Primate Societies.* BB Smuts, DL Cheney, RM Seyfarth, RW Wrangham & TT Struhsaker (eds), pp. 385–399. University of Chicago Press, Chicago.

Smuts BB & Smuts RW (1993) Male aggression and sexual coercion of females in nonhuman primates and other mammals: evidence and theoretical implications. *Adv. Stud. Behav.* **22:** 1–63.

Smuts BB, Cheney DL, Seyfarth RM, Wrangham RW & Struhsaker TT (eds) (1987) *Primate Societies.* The University of Chicago Press, Chicago.

Sofikitis N, Takahaski C, Nakamura I, Kadowaki H, Okazaki T, Shimamoto T & Miyagawa I (1990) The role of rat copulatory plug for fertilization. *Acta Eur. Fertil.* **21:** 155–158.

Sonta S, Yamada M & Tsukasaki M (1991) Failure of chromosomally abnormal sperm to participate in fertilization in the Chinese hamster. *Cytogen. Cell Genet.* **57:** 200–203.

Spira A, Bajos N, Bejin A, Beltzer N, Bozon M, Ducot B, Durandeu A, Ferrand A, Giami A, Gilloire A, Giraud M, Leridon H, Messiah A, Ludwig D, Moatti JP, Mounnier L, Olomucki H, Poplavsky J, Riandey B, Spencer B, Sztalryd JM & Touzard H (1992) AIDS and sexual behaviour in France. *Nature* **360:** 407–409.

Spira A, Bajos N, Bejin A, Beltzer N, Bozon M, Ducot B, Durandeu A, Ferrand A, Giami A, Gilloire A, Giraud M, Leridon H, Messiah A, Ludwig D, Moatti JP, Mounnier L, Olomucki H, Poplavsky J, Riandey B, Spencer B, Sztalryd JM & Touzard H (1994) *Sexual Behaviour and AIDS.* Ashgate Publ Co, Brookfield.

Stavy M and Terkel J (1992) Interbirth interval and the duration of pregnancy in hares. *J. Reprod. Fertil.* **95:** 609–615.

Stockley P & Purvis A (1993) Sperm competition in mammals: a comparative study of male roles and relative investment in sperm production. *Funct. Ecol.* **7:** 56–570.

Stockley P, Searle JB, Macdonald DW & Jones CS (1993) Female multiple mating behaviour in the common shrew as a mechanism to reduce inbreeding. *Proc. Roy. Soc. Lond. Ser. B* **254:** 173–179.

Stockley P, Searle JB, MacDonald DW & Jones CS (1994) Alternative reproductive tactics in male common shrews: relationships between mate-searching behaviour, sperm production, and reproductive success as revealed by DNA fingerprinting. *Behav. Ecol. Sociobiol.* **34:** 71–78.

Storey BT & Kopf GS (1991) Fertilization in the mouse: II. Spermatozoa. In *A Comparative Overview of Mammalian Fertilization.* BS Dunbar & MG O'Rand (eds), pp. 167–216. Plenum Press, New York.

Suarez S (1987) Sperm transport and motility in the mouse oviduct: observations *in situ. Biol. Reprod.* **36:** 203–210.

Suarez SS & Dai X (1992) Hyperactivation enhances mouse sperm capacity for penetrating viscoelastic media. *Biol. Reprod.* **46:** 686–691.

Suarez S, Drost M, Redfern K & Gottlieb W (1990) Sperm motility in the oviduct. In *Fertilization in Mammals.* BD Bavister, J Cummins & ERS Roldan (eds), pp. 111–124. Serono Symposia, Norwell.

Suarez SS, Katz DF & Meizel S (1984) Changes in motility that accompany the acrosome reaction in hyperactivated hamster spermatozoa. *Gam. Res.* **10:** 253–265.

Suarez S, Redfern K, Raynor P, Martin F & Phillips, DM (1991a) Attachment of boar sperm to mucosal explants of oviduct *in vitro*: possible role in formation of a sperm reservoir. *Biol. Reprod.* **44:** 998–1004.

Suarez SS, Katz DF, Owen DH, Andrew JB & Powell RL (1991b) Evidence for the function of hyperactivated motiliy in sperm. *Biol. Reprod.* **44:** 375–381.

Svärd L & Wiklund C (1989) Mass and production rate of ejaculates in relation to monandry/polyandry in butterflies. *Behav. Ecol. Sociobiol.* **24:** 395–402.

Taggart DA & Temple-Smith PD (1991) Transport and storage of spermatozoa in the female reproductive tract of the brown marsupial mouse, *Antechinus stuartii* (Dasyuridae). *J. Reprod. Fertil.* **93:** 97–110.

Taggart DA & Temple-Smith PD (1994) Comparative studies of epididymal morphology and sperm distribution in dasyurid marsupials during the breeding season. *J. Zool.* **232:** 365–381.

Talbot P (1985) Sperm penetration through oocyte investments in mammals. *Am. J. Anat.* **174:** 331–346.

Tsutsui T (1989) Gamete physiology and timing of ovulation and fertilization in dogs. *J. Reprod. Fertil. Suppl.* **39:** 269–275.

Travis SE, Slobodchikoff CN & Keim P (1996) Social assemblages and mating relationships in prairie dogs: a DNA fingerprint analysis. *Behav. Ecol.* **7:** 95–100.

Trivers R (1972) Parental investment and sexual selection. In *Sexual Selection and the Descent of Man.* B Campbell (ed.), pp. 136–179. Heinemann, London.

Turner TT & Reich GW (1985) Cauda epididymal sperm motility: a comparison among five species. *Biol. Reprod.* **32:** 120–128.

Tutin CEG (1980) Reproductive behaviour of wild chimpanzees in the Gombe National Park, Tanzania. *J. Reprod. Fertil. Suppl.* **28:** 43–57.

van Staaden MJ, Chesser RK & Michener GR (1994) Genetic correlations and matrilineal structure in a population of *Spermophilus richardsonii*. *J. Mammal.* **75:** 573–582.

van Wagenen G (1936) The coagulating function of the cranial lobe of the prostate gland in the monkey. *Anat. Rec.* **66:** 411–421.

Verrell PA (1992) Primate penile morphologies and social systems: further evidence for an association. *Folia Primatol.* **59:** 114–120.

Vinogradov BS (1925) On the structure of the external genitalia in Dipodidae and Zapodidae (Rodentia) as a classificatory character. *Proc. Zool. Soc. Lond.* **1925:** 577–585.

Voss R (1979) Male accessory glands and the evolution of copulatory plugs in rodents. *Occ. Pap. Mus. Zool. Univ. Mich.* **689:** 1–27.

Waage JK (1979) Dual function of the damselfly penis: sperm removal and transfer. *Science* **203:** 916–918.

Weinberg CR & Wilcox AJ (1995) A model for estimating the potency and survival of human gametes *in vivo*. *Biometrics* **51:** 405–412.

Wellings K, Field J, Johnson AM & Wadsworth J (1994) *Sexual Behaviour in Britain*. Penguin Books Ltd, Harmondsworth.

Wilcox AJ, Weinberg CR & Baird DD (1995) Timing of sexual intercourse in relation to ovulation – effects on the probability of conception, survival of the pregnancy, and sex of the baby. *N. Engl. J. Med.* **333:** 1517–1521.

Wildt DE, Bush M, Howard JG, O'Brien SJ, Meltzer D, van Dyk A, Ebedes H & Brand DJ (1983) Unique seminal quality in the South African cheetah and a comparative evaluation in the domestic cat. *Biol. Reprod.* **29:** 1019–1025.

Wildt DE, Bush M, Goodrowe KL, Packer C, Pusey AE, Brown JL, Joslin P & O'Brien SJ (1987a) Reproductive and genetic consequences of founding isolated lion populations. *Nature* **329:** 328–331.

Wildt DE, O'Brien SJ, Howard JG, Caro TM, Roelke ME, Brown JL and Bush M (1987b) Similarity in ejaculate–endocrine characteristics in captive vs. free-ranging cheetahs of two subspecies. *Biol. Reprod.* **36:** 351–360.

Wynne-Edwards KE & Lisk RD (1984) Djungarian hamsters fail to conceive in the presence of multiple males. *Anim. Behav.* **32:** 626–628.

Xia X & Millar JS (1991) Genetic evidence of promiscuity in *Peromyscus leucopus*. *Behav. Ecol. Sociobiol.* **28:** 171–178.

Yanagimachi R (1988) Mammalian fertilization In *The Physiology of Reproduction*. E Knobil & J Neill (eds), pp. 135–185. Raven Press, New York.

Yanagimachi R (1994) Mammalian fertilization. In *The Physiology of Reproduction*, 2nd edn. E Knobil & JD Neill (eds), pp. 189–317. Raven Press, New York.

Yanagimachi R & Mahi CA (1976) The sperm acrosome reaction and fertilization in the guinea-pig: a study *in vivo*. *J. Reprod. Fertil.* **46:** 49–54.

Zaneveld LJD & De Jonge CJ (1991) Mammalian sperm acrosomal enzymes and the acrosome reaction. In *A Comparative Overview of Mammalian Fertilization*. BS Dunbar & MG O'Rand (eds), pp. 63–79. Plenum Press, New York.

Zao PZR, Meizel S & Talbot P (1985) Release of hyaluronidase and beta-acetylhexosaminidase during *in vitro* incubation of hamster sperm. *J. Exp. Zool.* **134:** 63–71.

APPENDIX 1

Multiple paternity in mammals (compiled by T. R. Birkhead and A. C. Appleton)

The table is in two parts. The upper part deals with species with litters of more than one offspring and records the percentage of litters with multiple paternity. The lower part deals with litters of one and records either (a) the number or percentage of instances when the male assumed to be the father was not, and/or (b) the percentage offspring fathered by males other than alpha, dominant or consorting male.

Family	Species	Percentage (n) litters with multiple paternity	Reference
Litter size > 1			
Felidae	Lion *Panthera leo*	4.2 (1/24)	1
Canidae	Ethiopian wolf *Canis simensis*	5.4 (2/37)	2
Ursidae	Black bear *Ursus americanus*	50.0 (1/2)	3
Viveridae	Dwarf mongoose *Helogale parvula*	19.1 (4/21)	4
Scuridae[a]	California ground squirrel *Spermophilus beecheyi*	89.0 (8/9)	5
	Columbian ground squirrel *S. columbianus*	15.8 (26/165)	6
	Belding's ground squirrel *S. beldingi*	77.8 (21/27)	7
	Richardson's ground squirrel *S. richardsonii*	28.6 (2/7)	8

	Idaho ground squirrel	71.4 (5/7)	9
	S. brunneus		
	Gunnison's prairie dog	33.3 (7/21)	10
	Cynomys gunnisoni		
	Black-tailed prairie dog	2.9 (3/102)	11
	C. ludovicianus	5.0 (13/259)	12
Muridae	Deer mouse	10.3 (11/107)	13
	Peromyscus maniculatus		
	White-footed mouse	11.5 (7/61)	14
	P. leucopus		
	California mouse	0 (0/22)[b]	15
	P. californicus		
	Oldfield mouse	0 (0/220)	16
	P. polionotus		
	House mouse	51.7 (30/58)	
	Mus musculus domesticus	involved extra-territorial copulations[c]	17
	Red-backed vole	20.7 (6/29)	18
	Clethrionomys rufocanus bedfordiae	0 (0/35)	19
	Meadow vole	14.6 (11/75)	20
	Microtus pennsylvanicus		
Soricidae	Common shrew	88.9 (8/9)	21
	Sorex araneus		

Continued.

Family	Species	Mismatch between behavioural observations and paternity analysis	Percentage offspring fathered by males other than alpha/dominant/ consorting male	Reference
Litter size 1				
Cervidae	Red deer *Cervus elaphus*	11% (1/9) offspring: copulating male was not the father	32.0–35.0% (26–28/80)	22
Bovidae	Kirk's dik-dik *Modoqua kirkii*		0 (0/12)	23
Phocidae	Grey seal *Halichoerus grypus*	36% cases: candidate male based on behavioural data did not father the offspring		24
Cercopithecidae	Red howler monkey *Alouatta seniculus*		In 1/5 single male troops: 6.3% (1/16) fathered by intruding extra group male	25
			Multimale (2–3 males) troops: 66.7 (2/3) to 100% (3/3) offspring fathered by only one male of the troop	25
	Rhesus macaque *Macaca mulatta*		75.9 (22/29)	26
			46.1 (41/89)	27
			36.4 (4/11) sired by extra-group males	28
			All data included: low-ranking males sired 73% (11/15) offspring	28
			73% offspring	29
			1973–76: 54–83% (21–40/46) (alpha male had 80% copulations)	30
			1977–80: 70–93% (32–43/46) (alpha male had 60% copulations)[d]	
	Japanese macaque *M. fuscata*		87.5% (7/8)	31

	Long-tailed macaque *M. fascicularis*	Group 1: 38% (8/21) Group 2: 27% (3/11) Group 3: 8% (1/13)	32
	Hamadryas baboon *Papio cynocephalus hamadryas*	0[e]	33
Phyllastomatidae	Greater spear-nosed bat *Phyllostomus hastatus*	6–72% offspring fathered by males other than harem male	34
		10–40% offspring fathered by males other than harem male	35
Pongidae	Chimpanzee *Pan troglodytes*	54% (7/13) extra-group paternity	36
Hominidae		67% (4/6) fathered by non-consort male	36
	Human *Homo sapiens*	said to be about 9%	37, 38, 39

[a] Care is needed in using data from ground squirrel studies in comparative studies since data are not always comparable: some authors include paternity data only for those females known to copulate with >1 male.
[b] Eighty-two offspring from 22 complete families.
[c] In captive, seminatural conditions.
[d] Captive troop.
[e] 4/5 Harem males vasectomized: 0/6 females of vasectomized males reproduced after nearly 4 years (two females of intact male gave birth to six infants in same time period).
[1] Gilbert *et al.* (1991); [2] Sillero-Zubiri *et al.* (1996); [3] Schenk and Kovacs (1995); [4] Keane *et al.* (1994); [5] Boellstorf *et al.* (1994); [6] Murie (1995); [7] Hanken and Sherman (1981); [8] van Staaden *et al.* (1994); [9] Sherman (1989); [10] Travis *et al.* (1996); [11] Hoogland and Foltz (1982); [12] Hoogland (1995); [13] Birdsall and Nash (1973); [14] Xia and Millar (1991); [15] Ribble (1991); [16] Foltz (1981); [17] Potts *et al.* (1991); [18] Kawata (1988); [19] Kawata (1985); [20] Boonstra *et al.* (1993); [21] Stockley *et al.* (1993); [22] Pemberton *et al.* (1992); [23] Brotherton *et al.* (1997); [24] Amos *et al.* (1993); [25] Pope (1990); [26] Duvall *et al.* (1976); [27] Small and Smith (1982); [28] Berard *et al.* (1993); [29] Bercovitch and Nürnberg (1996); [30] Curie-Cohen *et al.* (1981); [31] Inoue *et al.* (1991); [32] de Ruiter and van Hooff (1993); [33] Biquand *et al.* (1994); [34] McCracken and Bradbury (1977); [35] McCracken and Bradbury (1981); [36] Gagneux *et al.* (1997); [37] Baker and Bellis (1995); [38] MacIntyre and Sooman (1991); [39] Smith (1984a).

17 Sperm Competition, Sexual Selection and Different Routes to Fitness

T. R. Birkhead[1] *and A. P. Møller*[2]

[1] Department of Animal and Plant Sciences, The University, Sheffield S10 2TN, UK; [2] Laboratoire d'Ecologie, CNRS URA 258, Université Pierre et Marie Curie, Bât. A, 7ème étage, 7 quai St Bernard, Case 237, F-75252 Paris Cedex 05, France

I. INTRODUCTION

The role of sperm competition in sexual selection has been largely ignored (Darwin 1871) or underestimated (Andersson 1994). This has occurred for two reasons. First, because it has been convenient to assume that females are sexually monogamous and, second, because the frequency with which females copulate with more than one male during a single reproductive cycle has been difficult to measure. As this volume testifies, the evidence that multiple mating by females is routine and that sexual monogamy is the exception rather than the rule, is now overwhelming. Sperm competition in its broadest sense (comprising both the differential competitiveness of ejaculates and the differential utilization of sperm by females) is thus a central part of sexual selection. Moreover, there is good evidence that its effects can extend far beyond the processes that take place within the female reproductive tract.

Sexual monogamy is rare and can occur for at least three reasons: (1) lack of opportunity for multiple mating by females, (2) the opportunity

Sperm Competition and Sexual Selection
ISBN 0-12-100543-7

for multiple mating by females exists but is not exploited, and (3) the opportunity for multiple mating by females is exploited but does not result in mixed paternity. The last case does not strictly comprise sexual monogamy but does comprise genetic monogamy.

One of the best examples of reduced opportunity for multiple mating by females is the platyhelminth *Diplozoon gracile*, a parasite of freshwater fish, in which immature individuals fuse together for life and once mature exchange gametes only between themselves (Lambert *et al.* 1987). Among vertebrates, sexual monogamy is likely among the Syngnathidae (seahorses and those pipefish in which females transfer their eggs into the male's enclosed brood pouch where they are fertilized; Greenwood 1975). Jones and Avise (1997) have recently confirmed by the absence of cuckoldry in one species, the dusky pipefish *Syngnathus floridae*. Another possible example involves the catfish *Corydoras aeneus* (Kohda *et al.* 1995). The female attaches herself to a male's genital region and swallows his sperm which, in less than 10 seconds, passes through her gut and is released onto her recently shed ova, which are held in a 'pouch' formed by her pelvic fins. The reason for this bizarre fertilization mechanism, which may ensure a high probability of monogamy, is associated with life in fast-flowing streams in which sperm would be washed away if a more conventional fertilization process was used. In both *Corydoras* and the Syngnathidae relative testes size is minute, as might be predicted from a lack of sperm competition (M. Kohda, pers. comm., see also Stockley *et al.* 1997).

The most striking example of monogamy involves females passing up an opportunity to copulate with multiple males when it would clearly be to their advantage to do so. In the Hamadryas baboons *Papio cynocephalus hamadryas* four out of five harem-leading males in an isolated group were vasectomized. After almost 4 years both females of the intact male had given birth whereas none of the six females paired to the vasectomized males had done so (Biquand *et al.* 1994). Just as in the monogamous fish above, Hamadryas baboons have relatively small testes (1.3% of body mass) compared with four other more promiscuous *Papio* species (2.1–3.5%) (Kummer 1995, p. 112).

The final category of sexual monogamy includes those socially monogamous birds in which females are either never seen to engage in extra-pair copulations (e.g. Capricorn silvereye *Zosterops lateralis chlorocephala*; Robertson 1996), or where (albeit infrequent) extra-pair copulations fail to result in extra-pair paternity; as in many long-lived, colonial seabirds and some raptors (Birkhead and Møller 1996).

When females are not sexually monogamous, the opportunities for sexual selection are increased. As Møller (see Chapter 2: Fig. 17.1) has suggested, sexual selection can affect fitness through two different routes. The mating success route involves the relationship between secondary sexual characters and the numbers of sexual partners an individual has and, hence, includes sperm competition. The second route relates to the way secondary sexual characters and individual female

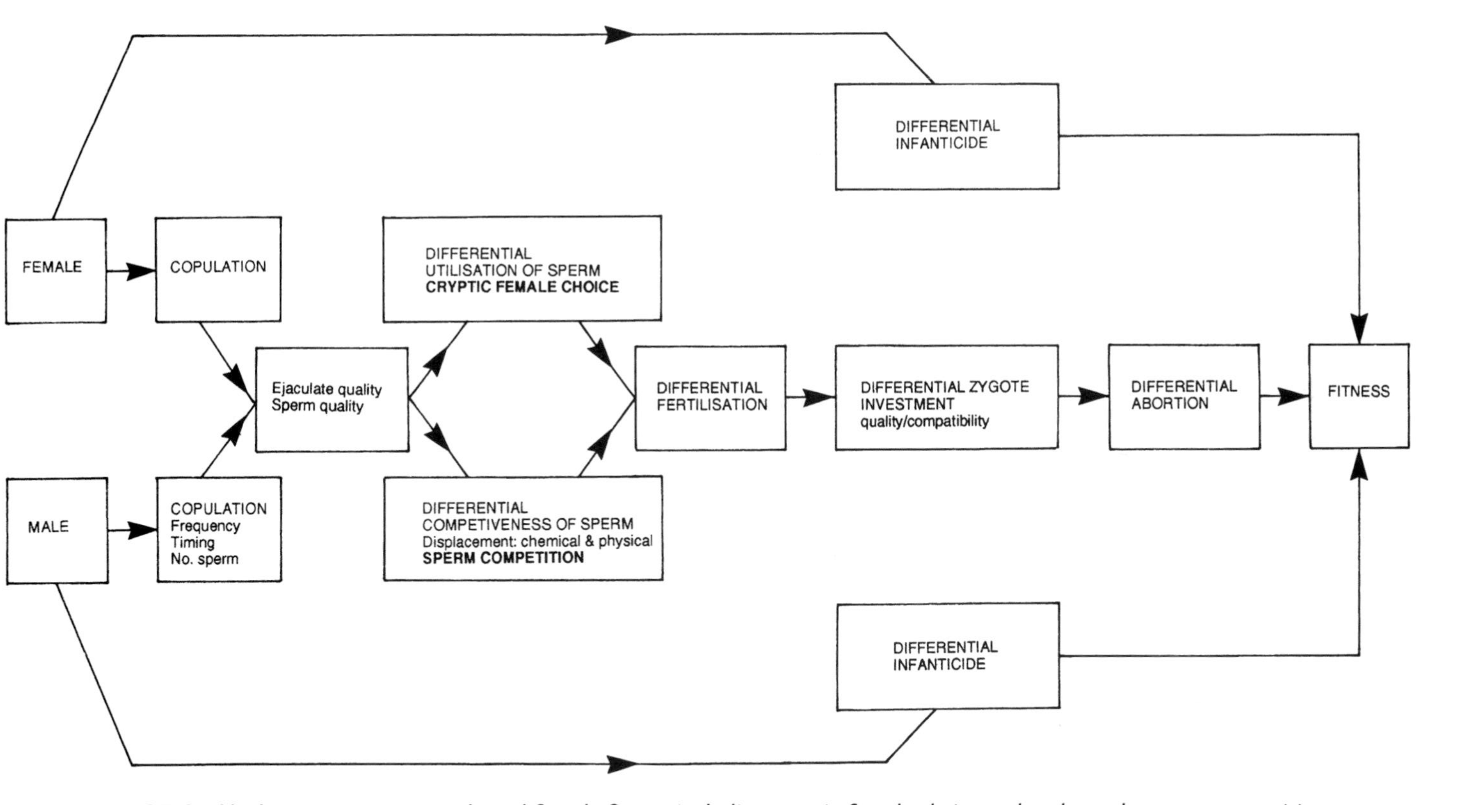

Fig. 17.1. *Mechanistic routes to male and female fitness including cryptic female choice and male–male sperm competition.*

fecundity interact and includes factors such as courtship feeding, differential parental investment and differential abortion. This route is less linked with sperm competition than the mating success route (however, see below for possible exceptions), although sperm quality may affect a female's fecundity.

There are two key questions relating to sperm competition and sexual selection. The first is how copulation with more than one partner enhances fitness. For males the benefit of copulating with and fertilizing several females has been intuitively obvious since Bateman's (1948) results were discussed by Trivers (1972), but for females the benefits of copulating with more than one male are far less obvious and continue to be controversial (see Chapter 2). The second question concerns the underlying processes that determine patterns of paternity when females are inseminated by more than one male. This question encompasses the competition between the sperm from different males and the differential utilization of sperm by females. Although these questions deal with different levels of analysis (see Sherman 1988; Alcock and Sherman 1994), they are closely inter-related. Indeed, as Simmons and Siva-Jothy (Chapter 10) have emphasized, we cannot hope to answer the question of function without considering the question of mechanisms and vice versa. Here, we discuss each question separately initially, and then consider the interaction between them.

II. MECHANISMS

A. General

To identify the processes that determine reproductive success and hence where sexual selection operates, we need to understand the physiological processes that determine which sperm fertilize particular ova and produce viable zygotes. Although we have made considerable progress, understanding the mechanisms of sperm competition has not been straightforward. Most of those interested in this field are (or were) behavioural ecologists rather than reproductive physiologists. Most reproductive physiologists are interested in the process of fertilization *per se*, but (presumably because they assume females to be sexually monogamous) rarely consider the factors that determine which of several copulating males fertilize a female's ova. The only studies by reproductive physiologists which approach this are those that use heterospermic inseminations (mixed inseminations of sperm from different males) in domestic mammals and birds to identify males of superior fertilizing ability (Dziuk 1996). In turn, this has identified differential fertilizing capacity (DFC) as a widespread phenomenon, but there has been little attempt to understand its underlying causes. In particular, it needs to be established to what extent DFC represents intrinsic differences in sperm quality, female control, or compatibility of genomes or gametes (see below).

When comparing the chapters in this volume with those of its predecessor (Smith 1984), a change in emphasis is apparent. Behavioural ecologists have started to interact much more with reproductive physiologists and use their techniques to understand the mechanisms of sperm competition. Indeed, this has been responsible in part for the general encompassing of physiological mechanisms by the study of behavioural ecology (Krebs and Davies 1997). However, studying gametes in the female reproductive tract is difficult and a far cry from watching whole organisms, because most events take place at a microscopic level and are extremely difficult to see (but see Suarez 1987). As an additional tool to help overcome this difficulty, investigators use mathematical models to identify plausible physiological processes (see Chapter 1).

In some species and some higher taxa we now have a reasonable, if far from complete, understanding of the basic mechanisms of sperm competition (e.g. certain insects and birds; Birkhead and Parker 1997; see also Chapters 10 and 14). The reproductive processes and anatomy in insects are remarkably diverse; given the abundance and diversity of insect species this is hardly surprising, but it means that it is difficult to make any generalizations about sperm competition mechanisms. The components that have been identified show a remarkable range. The factors known or thought to affect the outcome of sperm competition across all the taxa included in this book are summarized in Table 17.1.

B. *Sexual conflict*

When the attempts of one sex to maximize its fertilization success result in a reduction in the fitness of the other, the result is sexual conflict (Parker 1979, 1984; Stockley 1997). Although we typically think of sexual conflict in terms of male adaptations reducing female fitness, the reverse may also be true. Sexual conflict can occur in both external and internal fertilizers. Among the former, Warner *et al.* (1995) have shown that among blue headed wrasse *Thalassoma bifasciatum* large territorial males do best by defending a territory and spawning singly with a large number of females in succession, but releasing fewer sperm per spawning. This results in reduced fertility per female, indicating that, in this instance, sexual conflict is resolved in the male's favour. The opportunities for sexual conflict are greater and potentially more complex in internal fertilizers. The processes that take place between copulation, fertilization and beyond, which determine the fitness of either a male or a female, hinge on a series of steps over which each partner has different degrees of control. Sexual selection can operate at each of these stages

Table 17.1. Factors known to affect the outcome of sperm competition in various taxa.[a]

Taxon	Marine inverte- brates	Her- maphro- dites	Mol- luscs	Insects	Arach- nids	Fish (internal)[b]	Fish (external)	Am- phibian (internal)	Am- phibian (external)	Reptiles	Birds	Mar- supials	Mammals
Ejaculate													
Number of sperm: general	+	+	+	+		+	+		+	+	+		+
strategic changes				+			+	+		+	−		?
Frequency			+										
Quality											+		
Sperm: proportion live											+		+
proportion motile											+		+
velocity											?		+
survival			+										+
size					+						+		+
morphs			?	+							−		
capacitation time				−						?	−		+
Copulation													
Frequency	N/A			+			N/A		N/A	?	+		+
Timing relative to female cycle	N/A						N/A	+	N/A	?	+		+
Duration	N/A		?	+			N/A		N/A	?			
Sperm displacement													
physical	N/A			+			N/A		N/A				
removal	N/A		+/?	+			N/A		N/A				
repositioning	N/A			+			N/A		N/A				
chemical	N/A			+			N/A		N/A				
Female differential retention	N/A			+			N/A		N/A		+		?
storage	N/A		+	+			N/A		N/A	?			
use at fertilization	N/A		+		+		N/A		N/A				
Sperm loss from female tract	N/A				+		N/A		N/A				
Mortality	N/A	+	+	+			N/A	+	N/A				
Not mortality	N/A			+			N/A		N/A		+		

[a] + = Evidence for its occurrence; − = evidence for it not being important; ? = possibly some evidence; blank = not known. The information in this table has been obtained directly from the authors of the respective chapters (even where the entire column is blank).
[b] Internal and external refer to mode of fertilization.

(Fig. 17.1). Even when a female copulates with just one male, the fact that in most species only a tiny proportion of sperm reaches the site of fertilization, suggests that females select sperm from within ejaculates. This in itself indicates intersexual conflict, and that male and female reproductive features have co-evolved (e.g. Bedford 1991; Stockley 1997). When females are inseminated with the sperm from more than one male, the situation is even more complex because intrasexual conflict may result in extreme adaptations which, in turn, intensify intersexual conflict.

This appears to be the case in *Drosophila melanogaster*: sperm competition has resulted in strong selection on males to produce substances in their seminal fluid that apparently disables the stored sperm of rival males, but which also reduces female survival (Chapman *et al.* 1995). Moreover, Rice (1996) has showed experimentally that male *Drosophila* rapidly adapt to an evolutionary 'static' female phenotype and evolve increased net fitness when in competition with control males. The adaptation by males resulted in increased female mortality through: (1) more frequent copulation by adapted males, and (2) higher mortality per insemination. Rice's (1996) results clearly show that male and female reproductive features coevolve, and Chapman *et al.*'s (1995) study show that males are ahead in this particular arms race.

In both *Drosophila* and the blue headed wrasse, sexual conflict appears to have been resolved in favour of the male. In fact there appear to be no cases where the reverse is true (e.g. see Stockley 1997). Several authors have proposed that females should retain some control over which males fertilize their eggs, i.e. cryptic female choice (see Chapter 3) and in one sense this appears to be true. It is true in that female traits have co-evolved with male traits. However, because selection operates more intensively on males (Parker 1984), females are usually behind in the arms race. Sexual conflict is thus equivalent to Dawkins and Krebs' (1979) life-dinner principle: the outcome of sperm competition determines the number offspring a male produces, whereas for a female it determines only their quality (Parker 1984). Eberhard's (1996) view is different: he has argued that many of the traits we think of as having evolved through the effects of sexual selection on males, such as ejaculate size, may in fact have co-evolved in response to sexual selection on females. For example, females may have evolved sperm stores large enough to allow sperm to mix within them, and in response to this males evolved ejaculates large enough to fill the sperm store.

The idea of cryptic female choice (Thornhill 1983), particularly in terms of physiological processes such as the manipulation of sperm and the differential utilization of sperm by females, remains controversial. Although there is a considerable body of evidence consistent with the occurrence of cryptic female choice for sperm (Eberhard 1996), in practice it can be extremely difficult to establish the relative importance of female and male factors in the utilization of sperm (see Simmons *et al.* 1996). In order to demonstrate unequivocally the occurrence of cryptic female choice for sperm we suggest (see also Eberhard 1996) that three basic criteria must be met.

1. There must be variance in the proportion offspring fathered by the last male (P_2) across females.
2. Some of this variance must be attributable to females, and experiments must be performed to partition the variance between males and females.
3. Some of the variation in P_2 must be linked to one or more male characteristics, otherwise females have nothing to base their choice on.

So far, in our opinion, very few studies have convincingly met all three of these criteria.

C. Sexual selection

The interaction between: (1) the ejaculates (sperm and seminal substances) of different males, (2) ejaculates and the female reproductive tract, and eventually (3) male and female gametes, takes place in a series of stages. At each stage, males and females have different degrees of control over the likelihood of sperm reaching the next stage. A general model of the successive stages for an internal fertilizer is shown in Fig. 17.2: it comprises copulations, with or without insemination, followed by transport to storage sites, followed by further transport to the site of fertilization, where the gametes interact directly. As Fig. 17.1 shows,

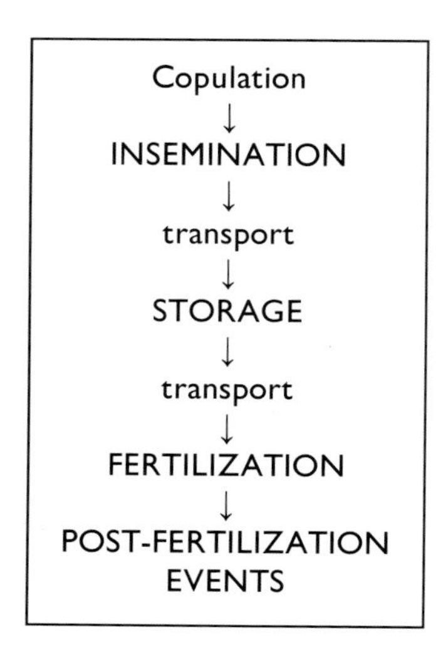

Fig. 17.2. *General model of the main stages between copulation, fertilization and post-fertilization events.*

even after fertilization has occurred, male and female fitness may be affected by whether the zygote is aborted or maintained and, in the latter case, by the degree of parental investment. Finally, males may still retain some control over both their own and female fitness via differential infanticide, as occurs in a number of taxa.

I. Copulation, insemination and sperm numbers

One of the most consistent patterns to emerge in this book is the fact that across a range of taxa, those species that experience the most intense sperm competition have the largest testes relative to their body size. Relative testes size determines the rate of sperm production and ultimately the number of sperm a male can inseminate into females. The other correlate of sperm competition, which less often examined, is the relative size of the male's sperm store. In reptiles and mammals this is the epididymis (see Chapters 13 and 16), and in passerine birds, the seminal glomera (see Chapter 14). In one sense the virtually ubiquitous nature of the testes and sperm store size and sperm competition relationship is not surprising because, in many situations where females copulate with more than one male, the relative numbers of sperm from a particular male at the site of fertilization has a marked effect on his probability of paternity. However, this obviously results in males producing and inseminating numbers of sperm far in excess of what is needed for fertilization. Parker (1984) has attributed this sperm redundancy to sperm competition, but several different hypotheses have also been proposed (Cohen 1967; Hurst 1990; Manning and Chamberlain 1994; Møller 1997b).

Given that sperm numbers are so important in determining male reproductive success and that sperm competition is so widespread, we might ask why all males do not hedge their bets and produce similarly high numbers of sperm. This, in turn, raises the question of what sets a limit on sperm numbers? The answer must be a combination of sperm production tissue and space for sperm storage in the male tract. The fact that not all males, either within or between species, invest similarly in these two types of tissue suggests that sperm production and storage carry costs. These costs are largely unknown, but must include the metabolic costs of producing and storing sperm and the energetic costs of carrying additional tissue and the possible interactions between androgen-based sperm production and immune function (Folstad and Skarstein 1997).

The number of sperm a male inseminates into a particular female will be a function of the number and size of ejaculates. Mean ejaculate size for any particular species may have evolved in relation to the intensity of sperm competition (Møller 1988), although in some species at least, males may be able to make strategic changes to ejaculate size (e.g. fish, see Chapter 11; some insects, see Chapter 10). In most cases males may determine the number of sperm they inseminate, but when males deliver sperm in spermatophores females have the opportunity to control the numbers of sperm they receive (Sakaluk and Eggert 1996). The number and timing of ejaculates may also be important. In some species males inseminate each female only once, but in others females may receive numerous ejaculates from the same male: the evolution of multiple vs. single ejaculates are discussed by Parker in Chapter 1. Although females of many species store sperm prior to using it to fertilize their ova, the timing of inseminations can also have an important effect on the likelihood of fertilization simply because timing determines the numbers of sperm either inseminated or retained by the female tract. Under these circumstances both male and female factors can be important: males must be ready to inseminate and females prepared to copulate at the appropriate time. In some taxa (e.g. birds, see Chapter 14), the timing of insemination in relation to various aspects of the female cycle has an important effect on the retention of sperm by females, which then affects the probability of fertilization. In mammals, where sperm storage is minimal, the timing of inseminations – by both male and female – in relation to ovulation and the time required for sperm to capacitate and reach the ova, is critical to fertilization success (see Chapter 16).

2. Retention of inseminated sperm by the female

Once sperm are introduced into the first part of the female tract several factors are known to affect their likelihood of being transported to the next stage (usually a storage site: Fig. 17.2). Again, sperm numbers are important: in birds, for example, a significant positive correlation exists between the numbers of sperm inseminated and the numbers subsequently found on the perivitelline layers of the ovum (Brillard 1993). In addition to this, various ejaculate features (e.g. the proportion of live sperm, sperm velocity, sperm size) are important. The proportion of sperm retained in the female tract has traditionally been regarded as an entirely male attribute, and the fact that in a number of taxa different individual males or the males of different genetic strains exhibit differential fertilizing capacity (Lanier *et al.* 1979; Lewis and Austad 1990; Dziuk 1996) is consistent with this. The physiological basis for DFC remains unknown, but in a number of studies of birds and mammals DFC correlates well with standard measures of semen quality when sperm are not in competition (Dziuk 1996). There is also some evidence that DFC has a genetic basis (Dziuk 1996). Although these studies suggest a strong male effect, we cannot exclude the possibility that DFC also has a female component. There are two possibilities: (1) females may prefer the sperm of certain males and retain and transport a higher proportion to the storage sites than those of some other male; and (2) the sperm of certain males may be more compatible with the reproductive tract or ova of certain females (Trivers 1972; Zeh and Zeh 1997; Møller 1997b). Recently,

Wilson *et al.* (1997) have shown that the female can have a pronounced effect on the proportion of ova that males fertilize in a sperm competition situation. In the numerous studies of DFC in domestic birds and mammals the role of the female has rarely been considered, and the experimental design does not permit us to partition the variance into male and female effects. However, in a study of mallard ducks *Anas platyrhynchos* (Cheng *et al.* 1983) the same females were inseminated with sperm from the same two males in three successive trials but there was no evidence for any female effect (Cunningham 1998).

This indicates that in some species the effect of male/semen quality may be most important (either because the sperm of certain males overcome any defences a female might have, or because females prefer the sperm of these males), whereas in other situations, male–female compatibility will be important. Either way, this raises some interesting issues regarding the ways in which sperm are 'recognized' as being either high quality or compatible by the female tract (see below).

3. Stored sperm and sperm–sperm (and female–sperm) interactions

In the majority of species the ejaculates from different males are most likely to interact once they have been accepted beyond the first stages of the female tract (Fig. 17.2). However, if two males inseminate a female in rapid succession or if ejaculates reside in the first part of the female tract for a long time, they may be able to interact directly. Various types of ejaculate interactions are possible: Baker and Bellis (1988) suggested that the sperm of one male may kill or otherwise disable the sperm from another: they referred to this as the kamikaze sperm hypothesis. In their subsequent accounts of their work (Baker and Bellis 1995; Baker 1996) this idea evolved from hypothesis to fact (Birkhead *et al.* 1997), but with virtually no supporting evidence. Indeed, subsequent tests of the hypothesis in humans and birds provided no evidence for this mechanism (Moore *et al.* in prep.; Fletcher and Birkhead, unpublished data).

However, there is no evidence for direct interactions between the sperm of different males within the female tract. Rather, interactions appear to be between ejaculates and fall into at least four categories: (1) sealing-in, as in ghost crab *Inachus phalangium*, which uses a glue-like substance in the seminal fluid to seal-in any previously stored sperm and prevent it from fertilizing (Diesel 1990); (2) repositioning, as in some odonates (Siva-Jothy 1988); (3) displacement, as in *Scatophaga* where a male uses his own ejaculate to physically displace previously stored sperm (see Chapter 10); and (4) chemical, as in *Drosophila* (above).

Some mechanisms appear to comprise only male–male competition but, as Eberhard (1996) has pointed out, such simple mechanisms potentially belie much more complex male–female interactions. For example, the (apparently) straightforward process of sealing in previously stored sperm in the ghost crab does not mean that females are entirely without control. A female may be able to use behavioural processes to influence which male inseminates her. If females can gain sufficient control simply through the use of behavioural tactics, there may have been little selection pressure to develop physiological control mechanisms. At the other extreme, where females have relatively little behavioural control, as in those cases where females can be forcibly inseminated, as in *Scatophaga*, there may have been strong selection for them to develop physiological mechanisms. Forced copulations occur in a range of taxa, including insects (see Chapter 10), fish (see Chapter 11), amphibians (see Chapter 12), reptiles (see Chapter 13), birds (see Chapter 14), and mammals (see Chapter 16) and while some authors have suggested that female resistance to such 'rape' attempts might be a ploy to assess male quality (e.g. in birds; Westneat *et al.* 1990), the costs of such male behaviour to females appear so high that this seems unlikely. When females are forcibly copulated by males, if males differ in quality and females gain some benefit from being fertilized by a particular male, then selection will have favoured females with the ability to control paternity by physiological means. However, as we have stated, the evidence for cryptic female choice is limited.

4. Sperm–egg interactions and zygote fitness

For externally fertilizing species sperm–egg interaction is the most important form of discrimination, and there is evidence from a number of taxa for nonrandom fertilization (Carré and Sardet 1984; Grosberg 1988; Carré *et al.* 1991; Bishop *et al.* 1996). For internally fertilizing species the penetration of the ovum by one or more sperm is almost the final stage at which sperm might compete or females discriminate between sperm. Given that fertilization commits an ovum, and that ova are often well provisioned and hence costly to produce, we might expect most discrimination to have occurred at an earlier stage. Nevertheless, there is good evidence for differential fusion between male and female gametes (Wedekind *et al.* 1995).

5. Postfertilization control

Even after fertilization has occurred, females of internally fertilizing species may still retain some control. Females may invest differentially in ova fertilized by poor quality or incompatible sperm, or poor quality zygotes or embryos may be resorbed or aborted. In taxa where ova are not especially costly to produce, resorption or abortion of particular zygotes might be a viable option (Wasser and Barash 1982; Eberhard 1996, p. 162; Møller 1997b). Indeed, there is evidence for such effects: for example, female mice eliminate up to one-third of fertilized eggs without affecting the total litter size (Hull 1964). In contrast, in those taxa, such as reptiles and birds, which produce large, yolky ova differential abortion may be too costly to be a viable option.

In cases where offspring are fathered by a competitor's sperm, males may increase their fitness, at the female's expense through infanticide. Well-documented examples include those in which males take over females that are already inseminated by previous males (lions, langurs, rodents; see Wittenberger 1981). It is also possible that maternal infanticide, for example, in rodents (Bronson 1989; Eberhard 1996, p. 165) and some birds (St Clair *et al.* 1995) may provide females with the ultimate means of discriminating between males.

III. FUNCTIONAL SIGNIFICANCE

A. General

The functional aspects of sperm competition have received considerable attention during the last two decades. While males in most systems potentially benefit from copulations with additional females in terms of increased reproductive success, the benefits for females are often much less obvious (see Chapter 2). Once a female has obtained sufficient sperm to fertilize her eggs, it is not evident why she should copulate with additional males. Bateman's (1948) study of *Drosophila melanogaster* demonstrated that while males continue to increase their fitness by copulating with additional females, resulting in an increase in the variance of male reproductive success, this was not the case for females whose fitness after copulations with a couple of males rapidly levelled off. A number of potential benefits for females of copulations with multiple males, both direct and indirect ones, are listed in Table 17.2. Fitness associated with sperm competition can be considered to be achieved in two different ways: (1) variance in copulation success at the level of fertilization, and (2) variance in the number offspring per partner (see Fig. 2.1 in Chapter 2).

There is accumulating evidence that sperm competition leads to an increase in the variance of male reproductive success in several species of birds, since males that have a high probability of fathering offspring in their own nests also tend to father offspring in other nests (see Chapter 2). Potentially, sperm competition in other taxa may also result in an increase in the variance of male reproductive success. This indicates that sperm competition is an important component of variance of reproductive success in birds and, because male success in obtaining copulation partners is often directly related to the expression of secondary sexual characteristic, this leads to sexual selection. Species with intense sperm

Table 17.2. A summary of female benefits of multiple mating.

Benefit	Reference
Direct	
Fertility insurance	Walker (1980)
Male parental care	Davies (1992)
Courtship feeding	Thornhill and Alcock (1983)
Ejaculatory nutrients	Friedel and Gillot (1977)
Avoidance of infanticide	Hrdy (1977)
Indirect	
Genetic diversity	Williams (1975), Brown (1997)
Genetic complementarity	Trivers (1972), Zeh and Zeh (1996, 1997)
Viability genes	Møller (1994)
Attractiveness genes	Fisher (1930)

competition therefore have a greater potential for sexual selection than species in which sperm competition is limited or absent.

Superficially, it may appear as if only variance in male mating success (number of sexual partners), but not variance in reproductive success per partner, should be affected by sperm competition. However, this is not necessarily true. Variance in the number of offspring per partner may be affected by parental investment, differential parental investment, abortion and infanticide (Fig. 2.1 in Chapter 2). There is little evidence for any of these components of variance in reproductive success generally being affected by sperm competition. There is some evidence that parental investment at the level of the entire brood depends on sperm competition. Males of bird species with intense sperm competition provide relatively less parental care than males that have not experienced sperm competition (reviewed in Chapters 2 and 4). Comparative analyses have demonstrated a similar effect among species, with males providing less parental care in species that have a high frequency of extra-pair paternity (Møller and Birkhead 1993). If nestlings differ in the ability to compete for food, and if this competitive ability is related to male attractiveness, multiple paternity may result in intensified competition among half-sibs (Briskie *et al.* 1994). The variance in the number offspring per brood may therefore increase further under intense sperm competition. Such an effect has been demonstrated in partial cross-fostering experiments in the barn swallow *Hirundo rustica* (J. Shykoff and A. P. Møller unpublished data). Thus, differences in male parental care related to sperm competition may affect the quality of offspring and hence variance in the number offspring per partner. There is little evidence of abortion or infanticide being related to sperm competition, with the exception of the Bruce effect (Bruce 1959) and infanticide by replacement males (both of which occur in the absence of sperm competition; see Chapter 16).

Sperm competition may affect the two main routes to fitness (number of females per male and number offspring per partner), as outlined above, but there may also be covariation between these two components. The most likely outcome is positive covariance. For example, if attractive males acquire more copulation partners than less attractive males, and their partners are less likely to engage in copulations with additional males, this will result in an increase in the variance in male mating success. Females mated to attractive males may also invest differentially in their offspring and refrain from abortion and infanticide, thereby causing positive covariance between the two components of fitness. This is only hypothetical at present, but recent developments of analytical techniques provide a way of identifying such effects in future studies (Webster *et al.* 1995).

What is the relative importance of the two routes to fitness, as affected by sperm competition in different taxa? The component due to variance in the number of mates may be more important in taxa with greater potential for fertilization of a large number of females, such as in group-spawning fish and invertebrates with external fertilization. Variation in reproductive success per mate may be predicted to be more important among taxa with parental care (some invertebrates, fish, amphibians and reptiles, and most birds and mammals). For species with biparental care, the effect of care on offspring quality will depend on the level of care provided by the two sexes (Clutton-Brock 1991).

B. Sperm competition and direct vs. indirect fitness benefits

Females may engage in copulations with multiple males to obtain direct or indirect fitness benefits. The former type affect the quality or the quantity offspring produced in this generation and appear to play an important role in some species (see Chapter 2). For example, several studies have demonstrated that females experience enhanced reproductive success from multiple copulations because of higher fertilization rates (Ridley 1989). Although this appears superficially to be a situation where females obtain direct fitness benefits from copulations with multiple males, this is not necessarily the case. Why do males not provide females with sufficient sperm to fertilize all eggs? Unless males are unable to increase their sperm production or ejaculate size for physiological or other reasons, any male that increases the number of sperm allocated to a female would be favoured by selection. Selection experiments and comparative studies indicate that sperm production and ejaculate size are traits that readily respond to selection and evolve rapidly (see Chapter 2). Hence, the question asked may not be the right one. Instead, we should ask why do females not store sufficient sperm to fertilize all their eggs? There are two possible answers: it is due to either inadequate sperm stores, or 'too many' eggs? This may be the case at the proximate, but probably not at the functional level. These puzzling questions may more readily be understood from a sexual conflict perspective (Stockley 1997). If female reproductive behaviour, physiology and anatomy have evolved in response to facilitate female reproductive decisions, given previous adaptations of males, features that at face value appear to give rise to direct fitness benefits may actually result in indirect fitness benefits. The increase in the number of eggs, or the decrease in the size of sperm storage organs, are potentially secondary adaptations to facilitate female manipulation of paternity.

The association between what appears to be direct fitness benefits for females copulating with multiple males to increase their fertilization success can also be interpreted from a life history perspective. Although the relationship between sperm competition and life history theory has been discussed only in the most general terms in the past (Birkhead and Møller 1992, pp. 242–244), this may underestimate the potential importance of the subject. If fecundity per mate is directly related to sperm competition, the maintenance and the evolution of sperm competition can be fully understood only from a life history perspective. Previous intraspecific and comparative studies have indicated that male parental care, and hence female parental effort, is directly associated with sperm competition in birds and, potentially, in other groups of organisms. This should have direct consequences for optimal life history under these conditions. Finally, direct fitness benefits to females from multiple copulations may give rise to changes in fecundity and hence life history features (Birkhead and Møller 1992).

A second type of direct fitness benefit for multiply copulating females is the acquisition of resources that are directly channelled into the production offspring. A similar question about the nature of the fitness benefits may be asked for this type of apparent direct fitness benefit. However, it is less evident how provisioning during copulation may serve as a mechanism that gives rise to indirect fitness benefits.

The taxonomic distribution of direct and indirect fitness benefits associated with sperm competition is currently unknown. A reasonable prediction seems to be that indirect fitness benefits may prove to be more important in taxa with intense sexual selection and extravagant secondary sexual characters.

C. *Sperm competition and offspring quality*

In this section we briefly discuss ways in which sperm competition may affect the quality of offspring. This may occur in terms of (1) offspring diversity, (2) offspring attractiveness, (3) offspring viability, or (4) direct fitness effects on offspring quality. Although these mechanisms are not necessarily mutually exclusive, we briefly discuss each in turn.

1. Offspring diversity

Sperm competition may result in the production of genetically more diverse offspring that, under certain environmental conditions, may have an enhanced probability of survival and reproduction (Williams 1975; Loman *et al.* 1988; Ridley 1993; Brown 1997). Recombination in sexually reproducing organisms by itself results in the generation of immense genetic diversity, even in the absence of any sperm competition (Williams 1975). Hence, sperm competition is certainly not the only way to produce genetically diverse offspring. We know of only two species where this mechanism may be important: both the tree swallow *Tachycineta bicolour* and the aquatic warbler *Acrocephalus paludicola* produce broods of a maximum of six offspring sired by up to five different males (Dunn and Robertson 1993; Lifjeld *et al.* 1993; Schulze-Hagen *et al.* 1993). It is difficult to imagine how any explanation other than genetic diversity could account for such extreme variance in paternity. However, based on theoretical arguments and current empirical findings we believe that this functional explanation is of little general importance.

2. Offspring attractiveness

Fisher (1930) proposed that sexual selection may act through pure arbitrary attractiveness if both male trait and female mate preference has a genetic basis. 'Choosy' females will benefit from their mate choice through the attractiveness of sons and daughters if they copulate with the most attractive males. Although Fisher envisioned this process at the precopulation level, there is no a priori reason why similar mechanisms could not work at the postcopulation level. Indeed, Birkhead *et al.* (1993) and Keller and Reeve (1995) suggested that sexually attractive sperm may interact with postcopulation female mate choice mechanisms to facilitate reproductive competition. Although this mechanism potentially would result in fitness advantages to certain individuals, there is no observational, experimental or comparative evidence to support the existence of such a mechanism. One might predict that such pure attractiveness would be more important in taxa with a potentially large variance in male reproductive success, such as group spawners and externally fertilizing organisms. However, in practice, it would be extremely difficult to discriminate between this mechanism and the following one.

3. Offspring viability

Females may benefit from sperm competition by the production of offspring with a genetically determined greater viability as signalled by the expression of male secondary sexual characters. Again, this mechanism could work both at the pre- and post-copulation levels. Sexual selection theory at the precopulation level suggests that such benefits may give rise to general viability effects or more specific effects in terms of, for example, parasite resistance (reviewed by Andersson 1994). However, other mechanisms are also possible, mediated by growth and developmental performance of gametes, embryos and offspring. We will briefly discuss these possibilities.

(a) Developmental selection

Developmental selection arises from differential growth and performance being used by parents to discriminate between gametes and zygotes of a single or multiple parental origins (reviewed by Møller 1997b). Developmental stability reflects the ability of individuals to express a stable phenotype under given environmental conditions. Numerous studies have shown that sexual selection often acts against potential mates with a developmentally unstable phenotype in terms of asymmetry or presence of phenodeviants (Møller and Swaddle 1997). The level of developmental stability of individuals is determined by a diverse array of genetic and environmental factors (Møller and Swaddle 1997). Offspring resemble their parents with respect to asymmetry and the frequency of phenodeviants (Møller and Thornhill 1997); there also appears to be a general mechanism of developmental stability expressed during developmental processes at different stages of the life cycle (Møller 1997b). For example, pollen from flowers with asymmetric petals in *Epilobium angustifolium* and *Lychnis viscaria* give rise to more embryo abortion due to developmental errors during early embryogenesis than pollen from symmetric flowers (Møller 1997b; Eriksson 1996). These results imply that developmental problems during embryogenesis and flower production are positively associated and potentially reflect overall developmental instability. If growth rate and developmental performance measured as developmental stability has an additive genetic component (as suggested above), choosy individuals either at the within-ejaculate or between-ejaculate level will experience a selective advantage in terms of production offspring with superior growth or developmental performance. This mechanism could work at the level of gametes, embryos and growing offspring, since often only a small fraction of all gametes and zygotes give rise to mature offspring. In other words, growth rate and stability of development may act as mechanisms giving rise to sexual selection for genetically based viability. Inbreeding can be viewed as a special case of this phenomenon because it often results in poor growth performance or development of offspring with abnormal phenotypes (Møller 1997b). For example, sperm competition in certain geographically isolated populations of reptiles occurs in the presence of frequent abnormalities, and females may avoid complete lack of reproductive success by copulating with more than a single male (Madsen *et al.* 1992; Olsson *et al.* 1996). Obviously, similar mechanisms may act in populations with less inbreeding, or in populations that have been subject to intense inbreeding, and hence purging of deleterious recessive alleles, for extended periods of time

(Brooker *et al.* 1990; Stockley and Purvis 1993; Bensch *et al.* 1994; Zeh and Zeh 1996).

(b) Incompatibility

Genetic compatibility between mates may affect the outcome of sperm competition because only compatible combinations of gametes will give rise to viable offspring. Superficially, it may appear incredibly wasteful to engage in copulation and fertilization and subsequently discard offspring of incompatible genotypes. This is only partly the case. For example, the major histocompatibility complex (MHC), which is associated with disease resistance in vertebrates, also plays an important role in mate choice in both birds and mammals, including humans (Potts *et al.* 1991; Wedekind *et al.* 1995; von Schantz *et al.* 1996). Individuals with complementary MHC haplotypes produce viable offspring, whereas zygotes of incompatible genotypes are aborted. Interestingly, there is a direct link between sperm competition and MHC in mice *Mus musculus* living under semi-natural conditions (Potts *et al.* 1992). Female mice that live in the territories of males with incompatible MHC haplotypes engage in extra-territorial copulations with males of different haplotypes as a way of increasing the genetic quality of their offspring (Potts *et al.* 1991). Although this finding has been disputed (Pomiankowski and Pagel 1992; Potts *et al.* 1992), the result has been upheld in subsequent studies of the same population. Plant incompatibility systems may be involved in similar types of subtle mate choice at the post-fertilization level (Willson and Burley 1983).

(c) General viability

Indirect fitness benefits arising from sexual selection include general viability, but also specific types of viability such as those resulting from resistance to parasites (Hamilton and Zuk 1982; Folstad and Karter 1992; Andersson 1994). There is some evidence consistent with the parasite-resistance idea. Intraspecific studies of the barn swallow indicate that males preferred by females as extra-pair copulation partners have better immune responsiveness than less preferred males (Saino and Møller 1995; Saino *et al.* 1997). Similarly, comparative analyses of the association between extra-pair paternity (a component of sexual selection) and sexual dichromatism (a signal that has evolved owing to sexual selection) and immune function (measured as the relative size of the spleen), respectively, indicate that secondary sexual characters signal immune responsiveness (Møller 1997a). When relative spleen size is used to predict extra-pair paternity and sexual dichromatism, only the former variable is related to immune function. This finding implies that once immune function is accounted for by current sexual selection, as measured by extra-pair paternity, there is no further information in the secondary sexual character (Møller 1997a).

4. Direct fitness effects on offspring quality

Direct fitness benefits from multiple copulations (including those with multiple males), such as those arising from acquisition of more sperm or other resources during copulation, may potentially alter the quality of offspring. Females usually produce a greater number of eggs than that giving rise to mature offspring, and the quality of such eggs may determine whether they eventually end up as offspring (reviewed by Møller 1997b). A greater fertilization frequency may reduce the number of eggs discarded and hence reduce the average quality of eggs being fertilized. A similar argument may be raised for the quality of sperm (Møller 1997b). Females that are limited by availability of sperm potentially run the risk of using sperm of inferior quality for fertilization (Lodge *et al.* 1971). Hence, as a larger fraction of stored sperm is used for fertilization, the quality of fertilizing sperm and hence the quality offspring may decrease as a consequence. Therefore, increased developmental selection, as described above, may be a consequence of egg and sperm utilization patterns. Again, what appears to be a clear case of direct fitness benefits acquired by females may, upon scrutiny, prove to be indirect fitness benefits.

Females of many insects obtain nuptial gifts during copulation, and a direct fitness benefit from multiple copulations is resources that can be used for maintenance or production of additional eggs (see Chapters 2 and 10). Surprisingly, it remains unknown whether greater amounts of such resources from multiple copulations by a single female enhances offspring quality.

In conclusion, there is little general agreement about the functional explanation for female copulations with multiple males. The relative importance of different types of benefits in different groups of organisms still remain elusive, but future studies may rectify this situation.

IV. CONCLUSIONS

The link between the adaptive significance of why females copulate with more than one male and the mechanistic processes that determine the outcome of multiple mating may be particularly close. For several of the hypotheses listed in Table 17.2 the mechanisms which result in the differential success or differential utilization of sperm depend on the recognition of gametes – processes that in many cases are likely to be immunological (Wedekind *et al.* 1995; see references in Zeh and Zeh 1997). Therefore, integration of the mechanisms which result in differential fertilization success and the functional significance of multiple mating by females require us to broaden our horizons and interact with those in other disciplines. As the chapters in this book demonstrate, such interdisciplinary collaboration has already started, but as the field we know as sperm competition continues to expand we will need to do this more and more. As a result of this integration we should, in the next decade or so, be able to establish the functional significance of female multiple mating and achieve a better understanding of the mechanisms of reproduction. We look forward to this volume's successor.

REFERENCES

Alcock J & Sherman PW (1994) The utility of the proximate-ultimate dichotomy in ethology. *Ethology* **96:** 58–62.

Andersson M (1994) *Sexual Selection*. Princeton University Press, Princeton.

Baker RR (1996) *Sperm Wars*. Fourth Estate, London.

Baker RR & Bellis MA (1988) 'Kamikaze' sperm in mammals? *Anim. Behav.* **36:** 936–939.

Baker RR & Bellis M (1995) *Human Sperm Competition*. Chapman and Hall, London.

Bateman AJ (1948) Intra-sexual selection in *Drosophila*. *Heredity* **2:** 349–368.

Bedford JM (1991) The coevolution of mammalian gametes. In *A Comparative Overview of Mammalian Fertilization*. BS Dunbar & MG O'Rand (eds), pp. 3–35. Plenum Press, New York.

Bensch S, Hasselquist D & von Schantz T (1994) Genetic similarity between parents predicts hatching failure: non-incestuous inbreeding in the great reed warbler? *Evolution* **48:** 317–326.

Biquand S, Boug A, Biquand-Guyot V & Gautier JP (1994) Management of commensal baboons in Saudi Arabia. Proc. Symp. 'Commensal Primates'. *Rev. Ecol.* **49:** 213–222.

Birkhead TR & Møller AP (1992) *Sperm Competition in Birds: Evolutionary Causes and Consequences*. Academic Press, London.

Birkhead TR & Møller AP (1996) Monogamy and sperm competition in birds. In *Partnerships in Birds: the Ecology of Monogamy*. JM Black (ed.), pp. 323–343. Oxford University Press, Oxford.

Birkhead TR & Parker GA (1997) Sperm competition and mating systems. In *Behavioural Ecology: An Evolutionary Approach*, 4th edn. JR Krebs & NB Davies (eds), pp. 121–145. Blackwell, Oxford.

Birkhead TR, Møller AP & Sutherland WJ (1993) Why do females make it so difficult for males to fertilize their eggs? *J. Theor. Biol.* **161:** 51–60.

Birkhead TR, Moore HDM & Bedford JM (1997) Sex, science and sensationalism. *Trends Ecol. Evol.* **12:** 121–122.

Bishop JDD, Jones CS & Noble LR (1996) Female control of paternity in the internally fertilizing compound ascidian *Diplosoma listerianum*. II. Investigation of male mating success using RAPD markers. *Proc. Roy. Soc. Lond. Ser. B* **263:** 401–407.

Brillard JP (1993) Sperm storage and transport following natural mating and artificial insemination. *Poultry Sci.* **72:** 923–928.

Briskie JV, Naugler CT & Leech SM (1994) Begging intensity of nestling birds varies with sibling relatedness. *Proc. Roy. Soc. Lond. Ser. B* **258:** 73–78.

Bronson FH (1989) *Mammalian Reproductive Biology*. University of Chicago Press, Chicago.

Brooker MG, Rowley I, Adams M & Baverstock PR (1990) Promiscuity: an inbreeding avoidance mechanism in a socially monogamous species? *Behav. Ecol. Sociobiol.* **26:** 191–199.

Brown JL (1997) A theory of mate choice based on heterozygosity. *Behav. Ecol.* **8:** 60–65.

Bruce HM (1959) An exteroceptive block to pregnancy in the mouse. *Nature* **184:** 105.

Carré D & Sardet C (1984) Fertilization and early development in *Beroe ovata*. *Develop. Biol.* **105:** 188–195.

Carré D, Rouviere C & Sardet C (1991) *In vitro* fertilisation in ctenophores: sperm entry, mitosis, and the establishment of bilateral symmetry in *Beroe ovata*. *Develop. Biol.* **147:** 381–391.

Chapman T, Liddle LF, Kalb JM, Wolfner MF & Partridge L (1995) Cost of mating in *Drosophila melanogaster* females is mediated by male accessory gland products. *Nature* **373:** 241–244.

Cheng KM, Burns JT & McKinney F (1983) Forced copulation in captive mallards. III. Sperm competition. *Auk* **100:** 302–310.

Clutton-Brock TH (1991) *The Evolution of Parental Care*. Princeton University Press, Princeton.

Cohen J (1967) Correlation between chiasma frequency and sperm redundancy. *Nature* **215:** 862–863.

Cunningham EJA (1998) Forced copulation and sperm competition in mallard. Unpublished Ph.D. thesis, University of Sheffield, Sheffield.

Darwin C (1871) *The Descent of Man, and Selection in Relation to Sex*. John Murray, London.

Davies NB (1992) *Dunnock Behaviour and Social Evolution*. Oxford University Press, Oxford.

Dawkins R & Krebs JR (1979) Arms races between and within species. *Proc. R. Soc. Lond. Ser. B* **205:** 489–511.

Diesel R (1990) Sperm competition and reproductive success in the decapod *Inachus phalangium* (Majidae): a male ghost spider crab that seals off rivals' sperm. *J. Zool.* **220:** 213–223.

Dunn PO & Robertson RJ (1993) Extra-pair paternity in polygynous tree swallows. *Anim. Behav.* **45:** 231–239.

Dziuk PJ (1996) Factors that influence the proportion offspring sired by a male following heterospermic insemination. *Anim. Reprod. Sci.* **43:** 65–88.

Eberhard WG (1996) *Female Control: Sexual Selection by Cryptic Female Choice*. Princeton University Press, Princeton.

Eriksson M (1996) Consequences for plant reproduction of pollinator preference for symmetric flowers. Ph.D. thesis, Department of Zoology, Uppsala University, Sweden.

Fisher RA (1930) *The Genetical Theory of Natural Selection*. Clarendon Press, Oxford.

Folstad I & Karter AJ (1992) Parasites, bright males, and the immunocompetence handicap. *Am. Nat.* **139:** 603–622.

Folstad I & Skarstein F (1997) Is male germline control creating avenues for female choice? *Behav. Ecol.* **8:** 109–112.

Friedel T & Gillott C (1977) Contribution of male-produced proteins to vitellogenesis in *Melanoplus sanguipes*. *J. Insect Physiol.* **23:** 145–151.

Greenwood PH (1975) *A History of Fishes*. Wiley and Sons, New York.
Grosberg K (1988) The evolution of allorecognition specificity in clonal invertebrates. *Quart. Rev. Biol.* **63:** 377–412.
Hamilton WD & Zuk M (1982) Heritable true fitness and bright birds: a role for parasites? *Science* **218:** 384–387.
Hrdy SB (1977) *The Langurs of Abu*. Harvard University Press, Cambridge, MA.
Hull P (1964) Partial incompatibility not affecting total litter size in the mouse. *Genetics* **50:** 563–570.
Hurst LD (1990) Parasite diversity and the evolution of diploidy, multicellularity and anisogamy. *J. Theor. Biol.* **144:** 429–433.
Jones AG & Avise JC (1997) Polygynandry in the dusky pipefish *Syngnathus floridae* revealed by microsatellite DNA marking. *Evolution* **51:** 1611–1622.
Keller L & Reeve HK (1995) Why do females mate with multiple males? The sexually selected sperm hypothesis. *Adv. Stud. Behav.* **24:** 291–315.
Kohda M, Tanimura M, Kikue-Nakamura M & Yamagishi S (1995) Sperm drinking by female catfishes: a novel mode of insemination. *Environ. Biol. Fish.* **42:** 1–6.
Krebs JR & Davies NB (ed) (1997) *Behavioural Ecology: An Evolutionary Approach*. Blackwell, Oxford.
Kummer H (1995) *In Quest of the Sacred Baboon*. Princeton University Press, Princeton.
Lambert A, Le Brun N & Renaud F (1987) L'étrange reproduction d'un ver parasite. *Recherche* **18:** 1548–1551.
Lanier DL, Estep DG & Dewsbury DA (1979) Role of prolonged copulatory behaviour in facilitating reproductive success in a competitive mating situation in laboratory rats. *J. Comp. Physiol. Psychol.* **93:** 781–792.
Lewis SM & Austad SN (1990) Sources of intraspecific variation in sperm precedence in red flour beetles. *Am. Nat.* **135:** 351–359.
Lifjeld J, Dunn PO, Robertson RJ & Boag PT (1993) Extra-pair paternity in monogamous tree swallows. *Anim. Behav.* **45:** 213–229.
Lodge JR, Fechheimer NS & Jaap RG (1971) The relationship of *in vivo* sperm storage interval to fertility and embryonic survival in the chicken. *Biol. Reprod.* **5:** 252–257.
Loman J, Madsen T & Hakansson T (1988) Increased fitness from multiple matings and genetic heterogeneity: a model of a possible mechanism. *Oikos* **52:** 69–72.
Madsen T, Shine R, Loman J & Håkansson T (1992) Why do female adders copulate so frequently? *Nature* **355:** 440–441.
Manning JT & Chamberlain AT (1994) Sib-competition and sperm competitiveness: an answer to 'Why so many sperms?' and the recombination/sperm number correlation. *Proc. Roy. Soc. Lond. Ser. B* **256:** 177–182.
Møller AP (1988) Testis size, ejaculate quality, and sperm competition in birds. *Biol. J. Linn. Soc.* **33:** 273–383.
Møller AP (1994) *Sexual Selection and the Barn Swallow*. Oxford University Press, Oxford.
Møller AP (1997a) Immune defence, extra-pair paternity, and sexual selection in birds. *Proc. Roy. Soc. Lond. Ser. B* **264:** 561–566.
Møller AP (1997b) Developmental selection against developmentally unstable offspring and sexual selection. *J. Theor. Biol.* **185:** 415–422.
Møller AP & Birkhead TR (1993) Certainty of paternity covaries with paternal care in birds. *Behav. Ecol. Sociobiol.* **33:** 261–268.

Møller AP & Swaddle JP (1997) *Asymmetry, Developmental Stability and Evolution Biology*. Oxford University Press, Oxford.

Møller AP & Thornhill R (1997) A meta-analysis of the heritability of developmental stability. *J. Evol. Biol.* **10:** 1–16.

Olsson M, Shine R, Madsen T, Gullberg A & Tegelström H (1996) Sperm selection by females. *Nature* **383:** 585.

Parker GA (1979) Sexual selection and sexual conflict. In *Sexual Selection and Reproductive Competition in Insects*. MS Blum & NA Blum (eds), pp. 123–166. Academic Press, New York.

Parker GA (1984) Sperm competition and the evolution of animal mating strategies. In *Sperm Competition and the Evolution of Animal Mating Systems*. RL Smith (ed.), pp. 1–60. Academic Press, Orlando.

Partridge L (1994) Genetic and nongenetic approaches to questions about sexual selection. In *Quantitative Genetic Studies of Behavioural Evolution*. RB Boake (ed.), pp. 126–141. University of Chicago Press, Chicago.

Pomiankowski A & Pagel M (1992) Sexual selection and MHC genes. *Nature* **356:** 293–294.

Potts WK, Manning CJ & Wakeland EK (1991) Mating patterns in semi-natural populations of mice influenced by MHC genotype. *Nature* **352:** 619–621.

Potts WK, Manning CJ & Wakeland EK (1992) Reply. *Nature* **356:** 294.

Rice WR (1996) Sexually antagonistic male adaptation triggered by experimental arrest of female evolution. *Nature* **381:** 232–243.

Ridley M (1989) The incidence of sperm displacement in insects: four conjectures, one corroboration. *Biol. J. Linn. Soc.* **38:** 349–367.

Ridley M (1993) *Evolution*. Blackwell, Oxford.

Robertson BC (1996) The mating system of the capricorn silvereye, *Zosterops lateralis chlorocephala*: a genetic and behavioural assessment. Ph.D. thesis, University of Queensland, Australia.

Saino N & Møller AP (1995) Sexual ornamentation and immunocompetence in the barn swallow. *Behav. Ecol.* **7:** 227–232.

Saino N, Primmer CR, Ellegren H & Møller AP (1997) An experimental study of paternity and tail ornamentation in the barn swallow *Hirundo rustica*. *Evolution* **51:** 562–570.

Sakaluk SK & Eggert A-K (1996) Female control of sperm transfer and intraspecific variation in sperm precedence: antecedents to the evolution of a courtship food gift. *Evolution* **50:** 694–703.

Schulze-Hagen K, Swatschek I, Dyrcz A & Wink M (1993) Multiple Vaterschaften in Bruten des Seggenrohrsängers *Acrocephalus paludicola*: erste Ergebnisse des DNA-Fingerprintings. *J. Ornithol.* **134:** 145–154.

Sherman PW (1988) The levels of analysis. *Anim. Behav.* **36:** 616–619.

Simmons LW, Stockley P, Jackson RL & Parker GA (1996) Sperm competition or sperm selection: no evidence for female influence over paternity in yellow dung flies *Scatophaga stercoraria*. *Behav. Ecol. Sociobiol.* **38:** 199–206.

Siva-Jothy MT (1988) Sperm 'repositioning' in *Crocothemis erythraea*, a libellulid dragonfly with a brief copulation. *J. Insect Behav.* **1:** 235–245.

Smith RL (1984) *Sperm Competition and the Evolution of Animal Mating Systems*. Academic Press, Orlando.

St Clair CC, Waas JR, St Clair RC & Boag PT (1995) Unfit mothers? Maternal infanticide in royal penguins. *Anim. Behav.* **50:** 1177–1185.

Stockley P (1997) Sexual conflict resulting from adaptations to sperm competition. *Trends Ecol. Evol.* **12:** 154–159.

Stockley P, Gage MJG, Parker GA & Møller AP (1997) Sperm competition in fishes: the evolution of testis size and ejaculate characteristics. *Am. Nat.* **149:** 933–954.

Stockley P & Purvis A (1993) Sperm competition in mammals: a comparative study of male roles and relative investment in sperm production. *Trends Ecol. Evol.* **7:** 560–570.

Suarez SS (1987) Sperm transport and motility in the mouse oviduct: observation *in situ. Biol. Reprod.* **36:** 203–210.

Thornhill R (1983) Cryptic female choice and its implications in the scorpionfly *Harpobittacus nigriceps. Am. Nat.* **122:** 765–788.

Thornhill R & Alcock J (1983) *The Evolution of Insect Mating Systems.* Harvard University Press, Cambridge, MA, and London.

Trivers RL (1972) Parental investment and sexual selection. In *Sexual Selection and the Descent of Man, 1871–1971.* B Campbell (ed.), pp. 136–179. Aldine-Atherton, Chicago.

von Schantz T, Wittzell H, Goransson G, Grahn M & Persson K (1996) MHC genotype and male ornamentation: genetic evidence for the Hamilton-Zuk model. *Proc. Roy. Soc. Lond. Ser. B* **263:** 265–271.

Walker WF (1980) Sperm utilization strategies in nonsocial insects. *Am. Nat.* **115:** 780–799.

Warner RR, Shapiro DY, Marcanato A & Petersen CW (1995) Sexual conflict: males with highest mating success convey the lowest fertilization benefits to females. *Proc. Roy. Soc. Lond. Ser. B* **262:** 135–139.

Wasser SK & Barash DP (1982) Reproductive suppression among female mammals: implications for biomedicine and sexual selection theory. *Quart. Rev. Biol.* **58:** 513–538.

Webster MS (1995) Effects of female choice and copulations away from colony on fertilization success of male Montezuma oropendulas (*Psarocolius montezuma*). *Auk* **112:** 659–671.

Webster MS, Pruett-Jones S, Westneat DF & Arnold SJ (1995) Measuring the effects of pairing success, extra-pair copulations and mate quality on the opportunity for sexual selection. *Evolution* **49:** 1147–1157.

Wedekind C, Chapuisat M, Macas E & Rulicke T (1995) Non-random fertilization in mice correlates with MHC and something else. *Heredity* **77:** 400–409.

Westneat DF, Sherman PW & Morton ML (1990) The ecology and evolution of extra-pair copulations in birds. *Curr. Ornithol.* **7:** 331–369.

Williams GC (1975) *Sex and Evolution.* Princeton University Press, Princeton.

Willson MF & Burley N (1983) *Mate Choice in Plants.* Princeton University Press, Princeton.

Wilson N, Tubman S & Eady P (1997) Female genotype affects male success in sperm competition. *Proc. Roy. Soc. Lond. Ser. B* **264:** 1491–1495.

Wittenberger JL (1981) *Animal Social Behaviour.* Duxbury, Boston.

Zeh JA & Zeh DW (1996) The evolution of polyandry I: intragenomic conflict and genetic incompatability. *Proc. Roy. Soc. Lond. Ser. B* **263:** 1711–1717.

Zeh JA & Zeh DW (1997) The evolution of polyandry II: post-copulatory defences against genetic incompatibility. *Proc. Roy. Soc. Lond. Ser. B* **264**, 69–75.

Index